P9-EMP-247

SPIN LABELING

THEORY AND APPLICATIONS

Edited by LAWRENCE J. BERLINER

Department of Chemistry
The Ohio State University
Columbus, Ohio

ACADEMIC PRESS New York San Francisco London 1976

A Subsidiary of Harcourt Brace Jovanovich, Publishers

ACADEMIC PRESS, INC.
111 Fifth Avenue, New York, New York 10003

United Kingdom Edition published by
ACADEMIC PRESS, INC. (LONDON) LTD.
24/28 Oval Road, London NW1

Library of Congress Cataloging in Publication Data
Main entry under title:

Spin labeling.

(Molecular biology, an international series of monographs and textbooks)
Includes bibliographies and index.
1. Electron paramagnetic resonance. I. Berliner, Lawrence J. II. Series.
QH324.9.E36S64 574.1'92'028 75-3587
ISBN 0-12-092350-5

PRINTED IN THE UNITED STATES OF AMERICA

SPIN LABELING

Theory and Applications

MOLECULAR BIOLOGY

An International Series of Monographs and Textbooks

Editors: BERNARD HORECKER, NATHAN O. KAPLAN, JULIUS MARMUR, AND HAROLD A. SCHERAGA

A complete list of titles in this series appears at the end of this volume.

To Marian and my parents

Contents

List of Contributors

Numbers in parentheses indicate the pages on which the authors' contributions begin.

Lawrence J. Berliner (1), Department of Chemistry, The Ohio State University, Columbus, Ohio

Keith W. Butler (411), Division of Biological Sciences, National Research Council of Canada, Ottawa, Canada

Jack H. Freed (53), Department of Chemistry, Cornell University, Ithaca, New York

Betty Jean Gaffney (183, 567), Department of Chemistry, The Johns Hopkins University, Baltimore, Maryland

O. Hayes Griffith (251, 453), Institute of Molecular Biology and Department of Chemistry, University of Oregon, Eugene, Oregon

Patricia Jost (251, 453), Institute of Molecular Biology and Department of Chemistry, University of Oregon, Eugene, Oregon

Thomas R. Krugh (339), Department of Chemistry, University of Rochester, Rochester, New York

J. Lajzerowicz-Bonneteau (239), Laboratory of Physical Spectrometry, University of Science and Medicine of Grenoble, Grenoble, France

Geoffrey R. Luckhurst (133), Department of Chemistry, The University, Southampton, England

Harden M. McConnell (525), Department of Chemistry, Stanford University, Stanford, California

Joel D. Morrisett (273), Department of Medicine, Baylor College of Medicine, Houston, Texas

Pier Luigi Nordio (5), Institute of Chemical Physics, University of Padua, Padua, Italy

Joachim Seelig (373), Department of Biophysical Chemistry, Biocenter of the University of Basel, Basel, Switzerland

Ian C. P. Smith (411), Division of Biological Sciences, National Research Council of Canada, Ottawa, Canada

Preface

The spin label technique has received recognition to date in several specialized review journals.† In each case, either a broad review of the state of the art or a detailed account of a specific application was presented. However, nowhere has it been possible to treat comprehensively theory, techniques, and applications of the whole field of biomedical research, due either to the physical limitations of space or to the limits defined by the individual author.

We have attempted here to compile in one text most of the necessary background, theory, and applications of spin labeling. Our main intent has been to aim this book toward the level of the graduate student, that is, a level of understanding which can include readers from the medically oriented through the biochemically to the physically oriented backgrounds. Our aim is to elucidate for the reader the essential principles of spin labeling and to communicate not only its promises but also the limitations and pitfalls which occasionally arise. Topics are generally covered not necessarily as reviews of the current literature, but from a pedagogical point of view.

Much appreciation is due to the authors, all recognized experts in their fields, for their total and generous cooperation. I am also indebted to Ms. Barbara Cassity for her excellent secretarial and editorial assistance, and to the editorial staff of Academic Press for their continued help and cooperation.

† A very recent monograph in Russian has been published [Likhtenstein, G.T. (1974), "Spin Labeling," Nauka, Moscow, USSR].

1

Introduction

LAWRENCE J. BERLINER

DEPARTMENT OF CHEMISTRY
THE OHIO STATE UNIVERSITY
COLUMBUS, OHIO

I. ELECTRON SPIN RESONANCE IN BIOLOGY

Electron spin resonance (ESR) or electron paramagnetic resonance (EPR) spectroscopy has constantly found applications to biochemical and biomedical problems as a highly sensitive tool for the detection of free radical species. This applies to the detection (and identification) of free radical intermediates in metabolic reactions, to the observation of stable, naturally occurring paramagnetic species (e.g., transition metal ions), to the detection of free radicals produced by external radiation, and to the analysis of paramagnetic probes introduced into specific biological systems (spin labeling). A recent text on the general subject of ESR in biology is Swartz *et al.* (1972).

While there is great potential for the direct observation of paramagnetic free radical species associated with specific biochemical phenomena, the natural occurrence of paramagnetism in biological systems is relatively low. A great advantage is to be gained by the introduction of some versatile probe containing a free radical center.

II. SPIN LABELING—A REPORTER GROUP TECHNIQUE

The concept of labeling a biological system with some external probe molecule was termed the "reporter group" technique by Burr and Koshland (1964). The requirements of such a group are:

(1) It must be some environmentally sensitive moiety that can be introduced into specific centers of the system of interest and it must subsequently "report" changes in its environment to an appropriate detector.

(2) The physical properties of the reporter group that are detected must be either unique or distinct from the properties of the system under investigation. A pictorial representation of a reporter group in an enzyme is given in Fig. 1.

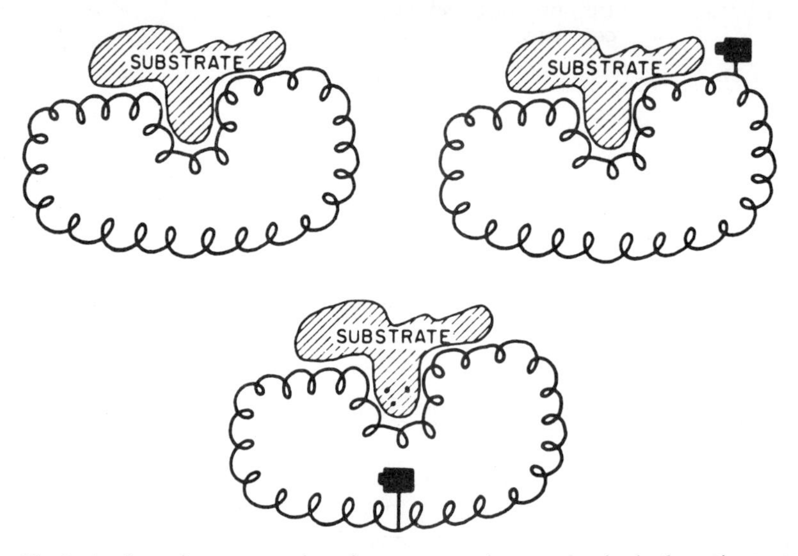

Fig. 1. A schematic representation of an enzyme–substrate complex in the native protein, the protein containing a reporter group (solid black area) adjacent to the substrate binding area, and the protein containing a reporter group distant from the substrate binding area. (Burr and Koshland, 1964.)

(3) Last, it is obvious that an additional requirement of any reporter group must be that the biological system suffer little or no perturbation(s) in its structure and function as a result of the incorporation of the probe. ("One must ensure that the reporter group is *reporting* the news, not *making* the news.")

By virtue of the relatively low content of stable paramagnetic species in biological materials, an ESR-sensitive reporter group will usually be physically distinct from the rest of the system. An ideal spin label is a stable (organic) free radical of a structure and/or reactivity that facilitates its introduction to a specific target site in a biological system or macromolecule. While this objective might at first seem easily obtainable, most spin labels used at present have been protected nitroxide radicals whose syntheses were

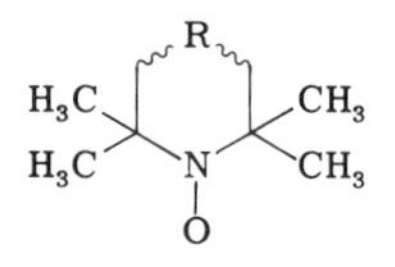

achieved only recently in the chronological development of organic chemistry (Hoffman and Henderson, 1961; Rozantsev and Neiman, 1964; Rozantsev and Krinitzkaya, 1965; Briere *et al.*, 1965).

Although spin labeling generally refers to the use of nitroxide radicals, other examples, such as nitric oxide, Mn^{2+}, other paramagnetic transition metal ions, lanthanide ions, and a few other organic radicals, have been used as probes of specific environments. For example, the chlorpromazine radical cation was demonstrated to be an accurate reporter of DNA intercalation geometry (Ohnishi and McConnell, 1965). However, due to the great versatility, sensitivity, and wealth of information derived from nitroxide free radicals, this class of molecules has found the greatest use in spin label applications. A properly designed spin label may be sensitive to several aspects of its physicochemical environment: molecular motion, orientation, and local electric and magnetic fields. These features are developed and discussed throughout the text. The chapters to follow will illustrate the extreme power, potential, and future promise of this valuable technique.

III. PRELUDE

The text is arranged so that a cover to cover reading should smoothly develop the necessary theory and background preceding the various applications to follow. The next three chapters are primarily theoretical. They begin with the theory of magnetic resonance and then lead into detailed theoretical discussions of molecular motion and spectral simulations. Then the theoretical aspects of spin–spin interactions are covered in a separate chapter on nitroxide biradicals. These opening chapters are followed by discussions of the chemistry and molecular structure of nitroxide spin labels. Several syntheses are included, many of which either have never been published or have been modified since their original publication. Following these two chapters, a special chapter is included on instrumental methods, covering the operation of ESR spectrometers, the applications of computers, and other aspects of spin label methodology.

The text then divides its coverage into the two principal areas of biomedical research where spin labeling finds most application: proteins (and enzymes) and lipids (and membranes). Chapter 8 comprehensively covers techniques and applications to enzyme problems and Chapter 9 develops the theoretical and experimental background for the powerful complementary approach: nuclear magnetic resonance relaxation methods.

The last four chapters are devoted to an integrated coverage of spin labeling as applied to lipid and membrane systems: first, theory and applications to liquid crystals as model systems; second, applications to artificial lipid bilayers; and last, discussions of molecular motion, polarity, and fluidity in membranes, and of applications to phenomena at the level of membrane suprastructure.

Since the spin label literature contains a number of structural names for various nitroxide spin labels, we have attempted to interrelate the different structural nomenclatures. Each author has tended to use the nomenclature familiar to his own laboratory, while at the same time including several alternate structural descriptions that the reader may expect to find both in the literature and elsewhere within this text. Appendix V (p. 572) contains a list of symbols and abbreviations used in the text which the reader is strongly urged to make use of.

References

Briere, R., Lemaire, H., and Rassat, A. (1965). Nitroxides XV: Synthèse et étude de radicaux libres stables pipéridiniques et pyrrolidinique, *Bull Soc. Chim. Fr.* **1965**, 3273–3283.

Burr, M., and Koshland, D. E. Jr. (1964). Use of "reporter groups" in structure-function studies of proteins, *Proc. Nat. Acad. Sci. U.S.* **52**, 1017–1024.

Hoffmann, A. K., and Henderson, A. T. (1961). A new stable free radical: Di-*t*-Butyl nitroxide, *J. Amer. Chem. Soc.* **83**, 4671.

Ohnishi, S., and McConnell, H. M. (1965). Interaction of the radical ion of chlorpromazine with deoxyribonucleic acid, *J. Amer. Chem. Soc.* **87**, 2293.

Rozantsev, E. G. and Krinitzkaya, L. A. (1965). Free iminoxyl radicals in the hydrogenated pyrrole series, *Tetrahedron* **21**, 491–499.

Rozantsev, E. G., and Neiman, M. B. (1964). Organic radical reactions involving no free valence, *Tetrahedron* **20**, 131–137.

Swartz, H., Bolton, J. R., and Borg, D. (eds.) (1972). "Biological Applications of Electron Spin Resonance Spectroscopy." Wiley (Interscience), New York.

2

General Magnetic Resonance Theory

PIER LUIGI NORDIO

INSTITUTE OF CHEMICAL PHYSICS
UNIVERSITY OF PADUA
PADUA, ITALY

I. INTRODUCTION

Electron spin resonance spectroscopy has developed at an outstanding pace since the late fifties, and now it can be safely stated that the technique is very well understood in its many aspects.

Its contribution has been determinant in clarifying the structures of free radicals, paramagnetic metal complexes, and molecules in excited triplet states. In a continuous search for new applications, it has proven to be a powerful tool in investigations covering numerous fields of chemistry and physics, such as the kinetics of radical reactions, polymerization mechanisms, energy transfer in molecular crystals, and liquid crystal structures. Its wide utilization in biological studies is well illustrated in this book.

The ease of analyzing the spectra in terms of magnetic parameters and of interpreting these parameters on the basis of molecular structures is the main reason for the vast versatility of this branch of spectroscopy. Another important factor is the sensitivity of the spectral features (resonance frequencies, multiplet splittings, and line shapes) to the electronic distribution, molecular orientations, nature of the environment, and molecular motions.

The analysis of spectral profiles that provide information of molecular motions is very often, in biological applications, the only available approach to the study of conformational changes or biological functions. As is common when dealing with time-dependent phenomena, one needs in these cases rather sophisticated theoretical techniques to extract valuable information from the experimental data. This is the reason why in this necessarily concise account of the basic principles of the magnetic resonance, particular attention will be directed toward the factors influencing the spectral line shapes.

This chapter is divided into two distinct parts. In the first part the general aspects of the ESR spectroscopy and the information made available from the interpretation of the spectra will be discussed. The presentation takes into account the fact that some readers may lack an adequate background in quantum mechanics, and therefore the theoretical derivations and formulas have been kept at a minimum.

In the second part an outline of the quantum mechanical description of the line shape problem will be found, and it is addressed to those who are interested in gaining a critical understanding of the underlying theoretical treatment.

II. PRINCIPLES OF ELECTRON SPIN RESONANCE

Electron paramagnetic resonance is a spectroscopic technique that deals with the transitions induced between the Zeeman levels of a paramagnetic system situated in a static magnetic field. It is well known that an atom or molecule will possess a permanent magnetic dipole moment only if it has nonzero electronic angular momentum **J**, and contributions to this may come from the spin angular momentum **S** as well as the orbital angular momentum **L**. According to quantum theory, the magnetic moment of an atom in an electronic state specified by the quantum numbers L, S, and J is

$$\boldsymbol{\mu} = -g\beta_e \mathbf{J} \tag{1}$$

where g is the spectroscopic splitting factor

$$g = 1 + \frac{J(J+1) + S(S+1) - L(L+1)}{2J(J+1)} \tag{2}$$

and β_e is the Bohr magneton, $|e|\hbar/2mc$. The interaction energy (Zeeman energy) of the magnetic moment with an external field **H** of uniform intensity H_0 is calculated by the quantum mechanical operator

$$\mathscr{H} = -\boldsymbol{\mu} \cdot \mathbf{H} = g\beta_e \mathbf{H} \cdot \mathbf{J} = g\beta_e H_0 J_z \tag{3}$$

where J_z is the operator corresponding to the projection of the angular momentum along the field direction. The expectation values of J_z, which shall be denoted by M_J, range between $-J$ and $+J$ by integer steps, and therefore the effect of the magnetic field is to produce $2J + 1$ levels, each of which has an energy $W_{M_J} = g\beta_e H_0 M_J$ and a population given by the Boltzmann distribution law

$$P_{M_J} = \exp(-W_{M_J}/kT) \Big/ \sum_{M_J} \exp(-W_{M_J}/kT) \tag{4}$$

Transitions can occur between the levels if the sample is irradiated with an electromagnetic field of proper frequency to match the energy difference. In the normal ESR experiments, this field is polarized in a plane perpendicular to the static field direction. The transitions are induced only between adjacent levels characterized by the quantum numbers M_J and $M_J \pm 1$, hence the resonance condition leading to energy absorption by the sample from the field is met when

$$h\nu = \Delta W = g\beta_e H_0 \tag{5}$$

In the following, we shall treat only molecular systems in which the magnetic properties are principally due only to the spin angular momentum **S**. Actually, it can be shown that a magnetic moment caused by the orbital motion can result only in the cases of atoms, linear molecules, or systems with orbitally degenerate electronic states. Even in these cases, intermolecular interactions in the condensed phases or molecular distortions lifting the degeneracy may quench the orbital angular momentum. In any other nonlinear, nondegenerate molecule an effective quenching of the orbital angular momentum occurs, although small residual contributions still remain, due to a particular interaction between the spin and the orbital motion called spin–orbit coupling. As a result, the measured g value deviates from the value of two expected for free-spin systems, and the amount of this deviation can be a characterization of the molecule under investigation, as will be seen in the following section.

Electron spin resonance experiments are usually carried out at a fixed frequency, and the resonance condition is found by varying the intensity of the static magnetic field. Most ESR spectrometers operate with microwave fields at 9.5 GHz (X band) or 35 GHz (Q band), the resonant fields in these two cases being about 3,400 and 12,500 G, respectively. The magnetic energy is 0.3–1 cm^{-1}, some hundred times smaller than kT at ordinary temperatures, so that the difference in population of the magnetic levels, as given in Eq. (4), is very minute. However, this population difference is responsible for the detection of the ESR signal. In fact, when the sample is irradiated with the microwave field, it absorbs energy from the field and it is excited to higher energy levels. At the same time, the inverse transition occurs by stimulated emission, but since the rates of both processes are proportional to the population of the level from which the transition starts, a net energy absorption results. These circumstances, nevertheless, do not suffice to permit continuous detection of the signal, since the prevailing absorption process would eventually equalize the population of the magnetic levels, causing the signal to disappear (saturation). Actually, there are relaxation mechanisms, which bring the system back to the Boltzmann equilibrium populations after it has been disturbed by the absorption of radiation. The equilibrium situation is restored by means of nonradiative transitions from the higher to the lower energy states and the consequent transfer to the environment of the magnetic energy, which is dissipated as thermal energy. The rate at which thermal equilibrium is restored is defined by a characteristic time called the *spin–lattice* or *longitudinal relaxation time* T_1, assuming the process to be represented by a single exponential decay:

$$d\mathcal{M}_z/dt = -(\mathcal{M}_z - \mathcal{M}_{eq})/T_1 \tag{6}$$

where $\mathcal{M}_z$ is the macroscopic magnetization of the sample and $\mathcal{M}_{eq}$ is given by the *Curie law*

$$\mathcal{M}_{eq} = N g_e^2 \beta_e^2 S(S+1) H_0/3kT \tag{7}$$

where N is the number of spins, g_e is the free-spin g value, and T is the absolute temperature.

The spin–lattice relaxation process shortens the lifetime of the magnetic levels, and therefore broadening of the spectral lines may occur because of the Heisenberg uncertainty principle. According to this principle, if a system maintains a particular state not longer than a time Δt, the uncertainty in the energy of the state cannot be less than $\Delta W \simeq h/\Delta t$. This means that the spectral linewidth, in frequency units, must be at least of the order of $1/T_1$.

However, under conditions of low microwave power, so as to avoid saturation effects, the linewidths are usually caused by other relaxation mechanisms, which produce modulation of the magnetic levels without causing transitions between them. These processes, which keep the total Zeeman energy constant in contrast with the spin–lattice relaxation mechanisms previously discussed, are characterized by a relaxation time T_2 called the *transverse relaxation time.*

Due to these processes, the perpendicular component of the magnetization, which in absence of relaxation would follow the oscillating microwave field with the same angular velocity $\omega = 2\pi\nu$, decays toward zero with the characteristic time T_2 according to the rate equation

$$d\mathscr{M}_x/dt = -\mathscr{M}_x/T_2 \tag{8}$$

Equations (6) and (8) are known as the *phenomenological Bloch equations* for the macroscopic magnetization. As will be seen in some detail in Section V, the transverse relaxation processes produce an absorption curve that is described by a *Lorentzian function.* On an angular frequency scale, the normalized shape function for a resonance line centered at ω_0 is in this case

$$f(\omega) = \frac{T_2}{\pi} \frac{1}{1 + T_2^2(\omega - \omega_0)^2} \tag{9}$$

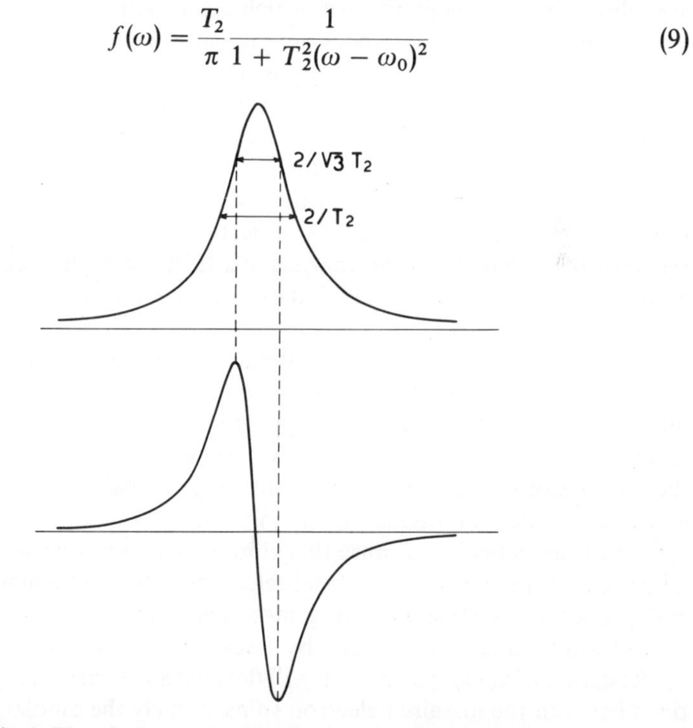

Fig. 1. Plot of a Lorentzian function and of its first derivative.

Experimentally, T_2 is obtained from the width of the curve at half-height, which is $2/T_2$. In the common ESR spectrometer, however, a 100-kHz field modulation is used in order to obtain more effective signal amplification, and this causes the displayed curve to be the derivative of the absorption curve. If the peak-to-peak distance of the derivative curve, corresponding to the width of the absorption curve at the points of maximum slope, is denoted by δ, we find that

$$1/T_2 = (\sqrt{3}/2)\delta \qquad (10)$$

A Lorentzian curve and its first derivative are shown in Fig. 1.

III. THE MAGNETIC INTERACTIONS

There are several interactions occurring in a paramagnetic system, which have to be considered. These interactions can manifest themselves depending on the nature of the system or of the medium in which it is embedded. The magnetic interactions can be intra- or intermolecular; however, the latter can be avoided by using samples in which the paramagnetic molecules are diluted in a diamagnetic host, which can be either a solvent or a crystal lattice. The need for dilution does not pose any particular problem for signal detection, given the high sensitivity of the ESR spectrometers, and use of concentrations in the range of 10^{-3}–10^{-5} mole/liter of paramagnetic species represents optimum conditions for routine experiments.

The inherent magnetic interactions of the isolated molecules are responsible for the complex structure of the spectra. If their magnitude is known, the ESR spectrum of a particular molecular species can be reconstructed theoretically. Conversely, the analysis of a ESR spectrum makes available a number of magnetic parameters that can be related to a particular molecular structure.

In addition to the Zeeman interaction, which has been mentioned previously, we will discuss in detail the interactions between electron and nuclear magnetic moments, which give rise to the multiplet structures of the spectra, known as hyperfine structures. These are of utmost importance in the analysis of the spectra and in identifying the paramagnetic species. We will concentrate our discussion mainly on molecules with one unpaired electron (free radicals), because this kind of system is normally used in spin labeling techniques. However, brief reference will also be made to the general properties of systems having more than one unpaired electron. This subject will be treated more fully by Luckhurst in Chapter 4 on biradicals.

Referring to these systems, we will deal with the various types of interactions between the unpaired electron spins, namely the dipolar coupling and the so-called exchange interactions.

A. Zeeman Interaction

From the discussion of Section II, it follows that the Zeeman interaction energy of the magnetic moment associated to the electronic spin **S** with an external field is expressed by the Hamiltonian

$$\mathscr{H}_Z = g_e \beta_e \mathbf{H} \cdot \mathbf{S} = g_e \beta_e H_0 S_z \tag{11}$$

If the values of the components of the spin angular momentum **S** along the magnetic field direction are denoted by m_s, then it follows that the energy of the permissible levels of the system (or, in quantum mechanical terms, the eigenstates of the operator $\mathscr{H}_Z$) are

$$W_{m_s} = g_e \beta_e H_0 m_s \tag{12}$$

For a system with only one unpaired electron, $S = \frac{1}{2}$ and m_s can assume only the values $\pm\frac{1}{2}$. The difference in energy between the two levels produced by the magnetic field is therefore (Fig. 2)

$$\Delta W = g_e \beta_e H_0 \tag{13}$$

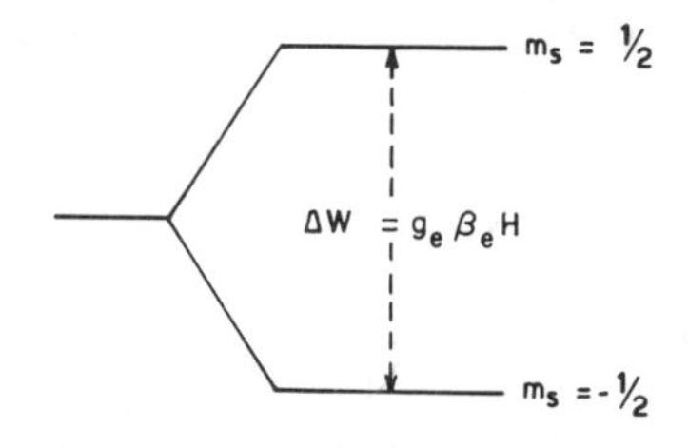

Fig. 2. Zeeman splitting of an $S = \frac{1}{2}$ state.

If the sample is irradiated with an electromagnetic radiation of frequency ν_0, energy absorption will occur at a field value satisfying the resonance condition

$$H = h\nu_0 / g_e \beta_e \tag{14}$$

If the orbital angular momentum is completely quenched, $L = 0$, and so we expect from Eq. (2) the value of g_e to be equal to 2, or more exactly to 2.00232 if quantum electrodynamic corrections are taken into account.

The reason for the quenching of the orbital angular momentum can be understood qualitatively from the following example.

Consider an electron in the atomic p orbital. If the atom is isolated, three degenerate states, i.e., three states with the same energy, are possible. They are represented by the three solutions p_1, p_0, and p_{-1} of the Schrödinger equation for the atomic system, and they have the complex form

$$[(x + iy)/\sqrt{2}]f(r), \qquad zf(r), \qquad [(x - iy)/\sqrt{2}]f(r)$$

where $f(r)$ describes the radial part of the orbital and it is a spherically symmetric function. In the presence of an external field applied in the z direction, the three functions correspond to states in which the component of the angular momentum along this direction has a value of 1, 0, and -1 (in $h/2\pi$ units), respectively. In other words, they are eigenfunctions of the projection operator

$$L_z = -i\hbar[x(\partial/\partial y) - y(\partial/\partial x)] \tag{15}$$

and 1, 0, and -1 are the corresponding eigenvalues.

The system therefore possesses an orbital angular momentum, and hence an associated magnetic moment. This conclusion, however, is valid only for the isolated atom. If it is introduced into a molecule, the p electron is then subjected to the electrostatic interactions arising from other electrons and nuclei composing the molecule. If we fix a set of Cartesian axes (x, y, z) to the atom under consideration, the interactions in the directions of these axes will be generally different, as a consequence of the fact that the local symmetry of the atom in the molecule is lowered with respect to the spherical symmetry of the isolated atom. As a result, the primitive degeneration of the three p states is lifted. The atomic functions that represent the new situation are the "symmetry-adapted" functions p_x, p_y, and p_z:

$$p_x = xf(r), \qquad p_y = yf(r), \qquad p_z = zf(r)$$

These functions are real, and for all three the "expectation value" of the z component of the angular momentum, defined as

$$\langle p_i \,|\, L_z \,|\, p_i \rangle = \int p_i L_z p_i \; dx\, dy\, dz, \qquad i = x, y, z \tag{16}$$

is zero, as found by direct integration on the angular part of the p functions. The same is true for the other angular momentum components, L_x and L_y being obtained from L_z of Eq. (15) by cyclic permutation of the variables (Slichter, 1963). Under these circumstances, it is said that the orbital angular momentum is "quenched."

When the molecule is subjected to the action of a magnetic field, the combined effect of the field and of the spin–orbit coupling reinstates small contributions of orbital angular momentum into the ground state of the system. In this way the magnetic properties of the system are not due only to the spin angular momentum; nevertheless, it is convenient that we continue to regard the resulting magnetic moment as produced by the pure spin, defining consequently an effective g factor for the system. Deviations of the g factor from the free-spin value are small (~ 1‰) in the case of aromatic free radicals, but they can be relatively large in molecules containing transition metal ions. Furthermore, the magnitude of such corrections depends on the orientation of the magnetic field with respect to the molecular axis system. This implies that the g factor of a molecule cannot be represented by a scalar, but must be represented by a second-rank tensor.

The detailed theoretical treatment leads to the following expression for the g-tensor components (Slichter, 1963):

$$g_{ij} = g_e \delta_{ij} - 2\lambda \sum_l \frac{\langle \psi_0 | L_i | \psi_l \rangle \langle \psi_l | L_j | \psi_0 \rangle}{E_l - E_0} \tag{17}$$

where δ_{ij} is the Kronecker symbol, which is equal to one if $i = j$ and zero otherwise; λ is the spin–orbit coupling; and ψ_0 and ψ_l are the wave functions that represent the ground and the excited states of the system, respectively, E_0 and E_l being the corresponding energies.

For those who are unfamiliar with tensor algebra, we note that a three-dimensional tensor of rank p is represented by 3^p components in the Cartesian space. Thus a scalar, which can be thought of as a zeroth-rank tensor, is defined by only one number, or component; a vector, or first-rank tensor, is defined by three components; and a second-rank tensor is defined by nine components.

The latter can be arranged for the sake of convenience to form a three-by-three matrix. Hence a general second-rank tensor $\mathbf{T}$ is represented by the matrix

$$\begin{pmatrix} T_{xx} & T_{xy} & T_{xz} \\ T_{yx} & T_{yy} & T_{yz} \\ T_{zx} & T_{zy} & T_{zz} \end{pmatrix} \tag{18}$$

If the tensor is symmetric (and we shall always deal with cases of this kind), $T_{xy} = T_{yx}$, $T_{xz} = T_{zx}$, and $T_{yz} = T_{zy}$, and then the number of independent components of the tensor is reduced to six.

The specification of the components of a tensor presupposes a definite choice of the coordinate axis system. If the axis system (x, y, z) is rotated into the new set (x', y', z'), then the components of a vector (first-rank tensor) in the primed system are related to the components in the original system by the linear transformation

$$T'_i = \sum_j R_{ij} T_j, \qquad i, j = x, y, z \tag{19}$$

The nine coefficients R_{ij} form a three-by-three matrix, known as the Euler matrix, which defines the relationship between the two coordinate sets. The rows of the matrix give the direction cosines of the new axes with respect to the original reference system, while the columns of the matrix are formed by the direction cosines of the unprimed axes in the rotated system. Analogous to Eq. (19), the transformation property of a second-rank tensor under rotation of the axis system is expressed as follows:

$$T'_{ij} = \sum_{kl} R_{ik} R_{jl} T_{kl} \tag{20}$$

The result follows immediately from Eq. (19) if the nine components T_{kl} of the second-rank tensor are thought to be constructed from the products $a_k b_l$ between the components of two vectors **a** and **b** (Brink and Satchler, 1968).

An important property, derived immediately from Eq. (20) by taking into account the orthogonality properties of the direction cosines, is the following: Any transformation of the type (20) will not affect the sum of the diagonal elements of a second-rank tensor. This sum is called the "trace" of the tensor, and it is indicated by the symbol Tr:

$$\mathrm{Tr}\,\mathbf{T} = \sum_{i} T_{ii} \tag{21}$$

Furthermore, there exists a unique choice of the axis system in which a second-rank tensor takes on a particularly simple form, all the off-diagonal elements vanishing and only the diagonal elements being different from zero. The axis system that puts the tensor in diagonal form is called the principal axis system of the tensor, and the diagonal elements are known as its principal values. In the case of the g tensor, the principal axes always coincide with the axes of molecular symmetry, if these exist.

We conclude this digression on the tensor algebra by recalling some tensor multiplication rules. If $\mathbf{T}$ is a second-rank tensor and $\mathbf{v}$ is a vector, the product $\mathbf{T} \cdot \mathbf{v}$ is still a vector, whose components are given by the matrix multiplication law:

$$\begin{pmatrix} T_{xx} & T_{xy} & T_{xz} \\ T_{yx} & T_{yy} & T_{yz} \\ T_{zx} & T_{zy} & T_{zz} \end{pmatrix} \begin{pmatrix} v_x \\ v_y \\ v_z \end{pmatrix} = \begin{pmatrix} T_{xx} v_x + T_{xy} v_y + T_{xz} v_z \\ T_{yx} v_x + T_{yy} v_y + T_{yz} v_z \\ T_{zx} v_x + T_{zy} v_y + T_{zz} v_z \end{pmatrix} \tag{22}$$

Inner multiplication of the resulting vector by another vector **u** gives a scalar, or zeroth-rank tensor. Thus the product

$$\mathbf{u} \cdot \mathbf{T} \cdot \mathbf{v} \tag{23}$$

is a scalar, and it is invariant under rotation of the coordinate system.

Following from the above considerations, we can write the Zeeman Hamiltonian for a paramagnetic molecule in the form

$$\mathcal{H}_Z = \beta_e \mathbf{H} \cdot \mathbf{g} \cdot \mathbf{S} \tag{24}$$

The physical observable corresponding to the Hamiltonian operator is the energy, and obviously it cannot depend on the orientation of the reference frame. It follows that the choice of the reference system is a completely arbitrary one. However, there are two particularly convenient possibilities to be considered:

(a) The laboratory axis system (x, y, z) with the z axis coinciding with the direction of the magnetic field. In this system $\mathbf{H} = \mathbf{k}H_0$, and the spin operators are well defined, given that the direction of $\mathbf{H}$ corresponds to the quantization axis. When expressed in the laboratory axis system, the Zeeman Hamiltonian becomes

$$\mathcal{H}_Z = \beta_e H_0(g_{zx} S_x + g_{zy} S_y + g_{zz} S_z) \tag{25}$$

(b) The molecular axis system (p, q, r) where the g tensor is diagonal. In this case we obtain

$$\mathcal{H}_Z = \beta_e(g_{pp} H_p S_p + g_{qq} H_q S_q + g_{rr} H_r S_r) \tag{26}$$

where H_p, H_q, are H_r are the components of $\mathbf{H}$ in the molecular axis system.

It is obviously possible to pass from one formulation to the other by making use of the relationships (19) and (20). Let us now consider as an example a case in which the molecule is oriented with one of the principal axes parallel to the field direction. The resonance condition will be met at field values

$$h\nu_0/g_{pp}\beta_e, \qquad h\nu_0/g_{qq}\beta_e, \qquad h\nu_0/g_{rr}\beta_e$$

depending on which one of the principal axes is directed along the field. For any arbitrary orientation of the molecule, the resonant field is still given by the relation

$$H = h\nu_0/g_{\text{eff}}\beta_e \tag{27}$$

where now the *effective* g factor is

$$g_{\text{eff}} = [g_{pp}^2 l^2 + g_{qq}^2 m^2 + g_{rr}^2 n^2]^{1/2} \tag{28}$$

(l, m, n) being the direction cosines of $\mathbf{H}$ in the molecular axis system.

Let (α, β, γ) be the Euler angles that carry the laboratory axis system (x, y, z) into coincidence with the molecular system (p, q, r). These angles are defined by the three successive rotations (Rose, 1957; Edmonds, 1957): (i) rotation through α about the original z axis; (ii) rotation through β about the y' direction obtained after the first step; (iii) rotation through γ about the final r direction. According to this convention, one finds

$$g_{\text{eff}}^2 = g_{pp}^2 \sin^2 \beta \cos^2 \gamma + g_{qq}^2 \sin^2 \beta \sin^2 \gamma + g_{rr}^2 \cos^2 \beta \qquad (29)$$

This equation is further simplified if the tensor has an axial symmetry, i.e. if $g_{pp} = g_{qq} \neq g_{rr}$. By denoting g_{pp} and g_{qq} by $g_\perp$, and g_{rr} with $g_\parallel$, we obtain for this case

$$g_{\text{eff}}^2 = g_\parallel^2 \cos^2 \beta + g_\perp^2 \sin^2 \beta \qquad (30)$$

The analysis of the angular dependence of the g factor for radical molecules trapped in a crystal lattice leads to the determination of the principal values of the g tensor and of the spatial disposition of the molecule relative to the crystal axes (Carrington and McLachlan, 1967).

B. Interaction with the Microwave Field

In addition to the effect of the static magnetic field, which determines the separation of the Zeeman levels, we have to consider the presence of the microwave field, which is responsible for the ESR transitions. The magnetic field associated with the radiation oscillates along the x axis perpendicular to the main field direction, and thus it has components

$$H_{1x} = H_1 \cos \omega t, \qquad H_{1y} = H_{1z} = 0 \qquad (31)$$

The corresponding Hamiltonian for the interaction with the spin magnetic moment, neglecting for simplicity the g-factor anisotropy, is

$$\mathscr{H}' = g_e \beta_e H_1 S_x \cos \omega t \qquad (32)$$

Since H_1 is much smaller than the main field H_0 (it is normally of the order of few milligauss to avoid saturation phenomena), $\mathscr{H}'$ can be treated as a time-dependent perturbation on the eigenstates of $\mathscr{H}_Z$. From standard perturbation theory, it can be easily verified that the effect of $\mathscr{H}'$ is to induce transitions between the states $|\alpha\rangle$ and $|\alpha'\rangle$ of $\mathscr{H}_Z$ of energies E_α and $E_{\alpha'}$ at the rate given by the transition probability (Hameka, 1965; Slichter, 1963)

$$w_{\alpha\alpha'} = (2\pi/\hbar) g_e^2 \beta_e^2 H_1^2 \, |\langle \alpha | S_x | \alpha' \rangle|^2 \, \delta(E_\alpha - E_{\alpha'} - \hbar\omega) \qquad (33)$$

Here δ is the Dirac delta function, which imposes the condition $E_\alpha - E_{\alpha'} = \hbar\omega$. The condition for nonvanishing values of $\langle \alpha | S_x | \alpha' \rangle$ is found by rewriting the spin operator S_x in terms of the "shift" operators S_+ and S_-:

$$S_\pm = S_x \pm iS_y, \qquad S_x = \tfrac{1}{2}(S_+ + S_-)$$

The effect of these operators on the basis spin functions is

$$S_{\pm} \mid S, m_s\rangle = [S(S+1) - m_s(m_s \pm 1)]^{1/2} \mid S, m_s \pm 1\rangle \tag{34}$$

Therefore transitions will occur only between levels with energy separation $\Delta E = \hbar\omega$, satisfying the selection rule $\Delta m_s = \pm 1$.

C. Hyperfine Couplings

From the results of Section III.A one might think that the g values could be used to characterize individual radical species, but this is not always the case, because of the smallness of the g deviations from the free-spin value, and the difficulty of an accurate theoretical interpretation in terms of molecular structure. By far more important for the identification of the paramagnetic molecules are the conspicuous multiplet structures, called *hyperfine structures*, that are very often present in the ESR spectra. The hyperfine structure is caused by the interaction of the electron spin magnetic moment with the magnetic moments of nuclei possessing nonvanishing nuclear spin angular momentum, such as hydrogen ($I_{\mathrm{H}} = \frac{1}{2}$), nitrogen ($I_{\mathrm{N}} = 1$), and fluorine ($I_{\mathrm{F}} = \frac{1}{2}$). Carbon and oxygen in their normal isotopic states, ^{12}C and ^{16}O, have zero nuclear spin, but the rare isotopes ^{13}C (natural abundance 1%) and ^{17}O (natural abundance 0.04%) have nuclear spins of $\frac{1}{2}$ and $\frac{5}{2}$, respectively.

The multiplet structure of the ESR spectra arises from the fact that the electron spin magnetic moment interacting with the nucleus "feels" a different total field according to which of the $2I + 1$ allowable orientations is assumed by the nuclear spin in the static magnetic field.

The magnetic interactions between electron and nuclear spins are represented by the Hamiltonian

$$\mathscr{H} = -g_e \beta_e g_N \beta_N \left\{ \frac{(\mathbf{I} \cdot \mathbf{S})r^2 - 3(\mathbf{I} \cdot \mathbf{r})(\mathbf{S} \cdot \mathbf{r})}{r^5} - \frac{8\pi}{3} (\mathbf{I} \cdot \mathbf{S})\, \delta(\mathbf{r}) \right\} \tag{35}$$

where $\mathbf{r}$ is the electron–nucleus distance vector and $\delta(\mathbf{r})$ is the Dirac delta function. The first term appearing in Eq. (35) describes the electron–nucleus *dipolar* interaction, and it is readily derived by classical arguments (Hameka, 1965). The second term, which gives rise to the so-called *Fermi contact coupling*, can also be obtained by a simple classical treatment, although the derivation based on the relativistic Dirac equation is more advisable in this case (Slichter, 1963).

From the Hamiltonian (35) we can obtain a *spin-Hamiltonian* $\mathscr{H}_{\mathrm{hf}}$, acting on spin variables only, by integration over the electron spatial coordinates, i.e., by averaging over the electron probability distribution $\psi^2(r)$ corresponding to the ground electronic state.

Let us examine separately the results obtained by the two terms of the Hamiltonian given in Eq. (35). The dipolar interaction takes the following form (Carrington and McLachlan, 1967):

$$\mathscr{H}_1 = \mathbf{I} \cdot \mathbf{A}' \cdot \mathbf{S} \tag{36}$$

where $\mathbf{A}'$ is again a second-rank tensor, with elements given by the relationship

$$A'_{ij} = -g_e \beta_e g_N \beta_N \langle (r^2 \delta_{ij} - 3x_i x_j) r^{-5} \rangle \tag{37}$$

and the brackets denote the integration over the electron distribution.

The dipolar term is clearly symmetric, and its trace vanishes because $\Sigma x_i^2 = r^2$. Recalling what was previously said regarding the g tensor, we can see that the magnitude of the dipolar interaction depends on the orientation of the molecule relative to the field direction.

For the contact interaction one obtains

$$\mathscr{H}_2 = a\mathbf{I} \cdot \mathbf{S} \tag{38}$$

where the isotropic coupling constant a is a scalar defined by

$$a = (8\pi/3) g_e \beta_e g_N \beta_N \psi^2(0) \tag{39}$$

and $\psi(0)$ is the value assumed by the unpaired electron wave function at the nuclear position. Finally, the complete hyperfine Hamiltonian $\mathscr{H}_{\text{hf}}$, obtained by summing $\mathscr{H}_1$ and $\mathscr{H}_2$, can be written in the compact form

$$\mathscr{H}_{\text{hf}} = \mathbf{I} \cdot \mathbf{A} \cdot \mathbf{S} \tag{40}$$

by defining a hyperfine tensor $\mathbf{A}$ with components

$$A_{ij} = A'_{ij} + \delta_{ij} a \tag{41}$$

In the principal axis system of the hyperfine tensor, the Hamiltonian assumes the simpler form

$$\mathscr{H}_{\text{hf}} = A_{pp} I_p S_p + A_{qq} I_q S_q + A_{rr} I_r S_r \tag{42}$$

It must be noted that if the electron distribution has a spherical symmetry, i.e., it is represented by an s-type function, then the integration of the dipole operator over the electron coordinates gives

$$\langle x^2/r^5 \rangle = \langle y^2/r^5 \rangle = \langle z^2/r^5 \rangle = \tfrac{1}{3}\langle 1/r^3 \rangle \tag{43}$$

and so the dipolar terms vanish. Thus, only if the unpaired electron is in a p, d, f, ... state will the dipolar interaction manifests itself. On the contrary, the contact term requires a finite electron density at the nuclear position, and only the s-type atomic wave functions have nonzero values at the origin. If the electron is described by a wave function that is a mixture of s and p states, both the dipolar and contact interactions are present simultaneously.

In most organic free radicals, the unpaired spin distribution is described by p-type orbitals. Among the very many examples are the planar $\cdot CH_3$ and NH_3^+ radicals, where the electron is concentrated on a 2p atomic orbital centered on the carbon or nitrogen atom, respectively; or the positive and negative ions of the aromatic hydrocarbons, where the electron is delocalized on π molecular orbitals, which still can be constructed as linear combinations of 2p atomic orbitals. It is well known that the solution spectra of all of these systems exhibit a rich hyperfine structure due to contact interactions with the nuclei lying in the nodal plane of the π-electron distribution. This situation is not surprising, since it is well understood that the simple notion of electrons described by definite molecular shells, implied by the molecular orbital treatments, can only give an approximate picture of the electronic structures.

In fact, due to the presence of the unpaired electron, which interacts differently with electrons having opposite spin, a polarization of the doubly occupied σ orbitals may occur, inducing a net spin density at the positions of the nuclei. This effect can be accounted for quantitatively in accurate calculations by refining the molecular functions with a technique known as configuration interaction. It consists in mixing to the first-approximation ground-state wave function, by a variational procedure, contributions from excited wave functions obtained by promotion of σ electrons into the π system. The adjustment of the electronic distribution produced in this way can be insignificant as far as the total electron density is concerned, although it is totally responsible for the observed isotropic couplings in π-electron systems. In fact it can be calculated that one unpaired electron in a hydrogen 1s atomic orbital or a nitrogen 2s orbital would produce a hyperfine interaction of 1420 or 1540 MHz, respectively. The observed coupling constants for the NH_3^+ radical ($a_H = -77$ MHz, $a_N = 55$ MHz) show that the induced unpaired electron density on each nucleus amounts to a small fraction of an electron. The negative sign of the hydrogen coupling constant means that on this nucleus there is an excess of electron with spin opposite to that of the unpaired electron on the $2p_N$ orbital. The absolute sign of the coupling constants cannot be obtained directly from the observed spectra, but it is inferred from theoretical arguments. Experimental verification of the signs can be available only from the line shifts measured in the nuclear magnetic resonance spectra of paramagnetic molecules.

In contrast to the contact term, the dipolar interaction is a first-order effect that can be calculated immediately if an approximate π-electron distribution is known. Due to the r^{-3} dependence, it is sensitive only to the local p-electron population.

Table I lists the principal values of the hyperfine dipolar tensor, together with the isotropic coupling constants, for the malonic acid radical $^{13}CH(COOH)_2$ and NH_3^+ radical. These values are rather typical for any C–H or N–H fragment. The *p* and *r* molecular axes are parallel to the X–H bond direction and to the 2p orbital axis, respectively. The reader is referred to the book by Carrington and McLachlan (1967) for a discussion of the signs of the dipolar couplings.

TABLE I
CONTRIBUTIONS TO HYPERFINE SPLITTINGS[a]

	^{13}C–H		^{14}N–H	
	H	^{13}C	H	^{14}N
A'_{pp}	30	−50	61	−35
A'_{qq}	−30	−70	−55	−35
A'_{rr}	0	120	−6	70
a	−60	93	−77	55

[a] Typical values (in MHz) of the anisotropic and isotropic contributions to the hyperfine splittings due to the magnetic nuclei present in the $^{13}\dot{C}$–H and $^{14}\dot{N}$–H molecular fragments.

If a C–H or N–H molecular fragment is inserted in a conjugated molecule where the unpaired electron is delocalized on several atoms, the isotropic coupling constant of a ^{13}C or ^{14}N nucleus is only roughly proportional to the unpaired π-electron density on that particular atom. On the contrary, the isotropic hyperfine coupling constants of hydrogen nuclei are strictly correlated to the unpaired π-electron density ρ^π on the conjugated atoms to which they are bound, according to the well-known McConnell relationship (Carrington and McLachlan, 1967)

$$a_H = Q_{XH}^H \rho_X^\pi \tag{44}$$

where Q_{XH}^H is a constant that depends exclusively on the nature of the X atom to which the hydrogen is attached. The value of the constant can be obtained from the experimental value of a_H in molecules where the electron density on the atoms carrying the unpaired electron is known from symmetry considerations. From the examples given in Table I, where the unpaired electron density on the carbon and nitrogen atoms is equal to unity, it therefore follows that

$$Q_{CH}^H = -60 \text{ MHz}, \qquad Q_{NH}^H = -77 \text{ MHz}$$

D. Interactions between Electron Spins

When a molecule possesses more than one unpaired electron, we need to consider the interactions between the electron spins in addition to the interaction of the total spin **S** with the magnetic field and the hyperfine coupling with the nuclei. As examples of this type of system we refer to molecules in a triplet state, which have two unpaired electrons. In this case $S = 1$, and in the presence of an external field the allowed orientations of the magnetic moment correspond to the three values $+1, 0$, and -1 of the spin projection quantum number.

The triplet state can be either the ground state of the molecule, as in the case of normal molecular oxygen, or an energetically excited state. Triplet ground states result whenever the unpaired electrons can occupy distinct atomic or molecular orbitals of the same energy. The interelectronic repulsions lead in this case to the stabilization of the state with the highest spin multiplicity, in agreement with the Hund rule for atoms. Well-known examples of excited triplet states are provided by the phosphorescent states resulting from photoexcitation of aromatic molecules. The decay to the ground singlet state ($S = 0$) is spin-forbidden, and this is the reason for the relatively long lifetime of the paramagnetic state.

An important interaction between the electron spins is the dipolar interaction, analogous to the electron spin–nuclear spin dipolar coupling. The dipolar Hamiltonian is in this case, neglecting g-factor anisotropy,

$$\mathscr{H} = g_e^2\beta_e^2[(\mathbf{S}_1 \cdot \mathbf{S}_2)r^2 - 3(\mathbf{S}_1 \cdot \mathbf{r})(\mathbf{S}_2 \cdot \mathbf{r})]/r^5 \tag{45}$$

where **r** is the interelectronic distance. From this expression it is possible to derive a dipolar spin Hamiltonian $\mathscr{H}_d$, after averaging over the electron coordinates by a procedure similar to that used for the dipolar hyperfine coupling:

$$\mathscr{H}_d = \mathbf{S} \cdot \mathbf{D} \cdot \mathbf{S} \tag{46}$$

where $\mathbf{S} = \mathbf{S}_1 + \mathbf{S}_2$, and

$$D_{ij} = \tfrac{1}{2}g_e^2\beta_e^2\langle(r^2\,\delta_{ij} - 3x_ix_j)/r^5\rangle \tag{47}$$

the brackets indicating the average over the electronic wave function.

The electron dipolar interaction lifts the degeneracy of a triplet state even in the absence of the field, giving rise to the so-called "zero-field splitting" (Fig. 3). As a result, in the presence of an applied field the two allowed transitions $-1 \rightarrow 0$ and $0 \rightarrow 1$ do not have the same energy at a fixed field value. The spectrum will always consist of two lines, whose separation at sufficiently high field is independent of the field strength and dependent on

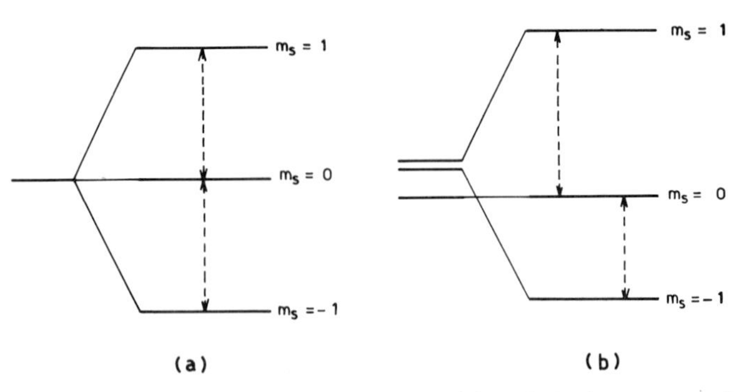

Fig. 3. (a) Zeeman levels of an $S = 1$ state, and (b) levels in the presence of zero-field splitting.

the principal values of $\mathbf{D}$ and the orientation of the field direction with respect to the molecular axes. By introducing the new zero-field parameters

$$D = D_{rr} - \tfrac{1}{2}(D_{pp} + D_{qq}), \qquad E = \tfrac{1}{2}(D_{pp} - D_{qq}) \tag{48}$$

we find that the separation between the line pair, called "fine splitting," is $D - 3E$, $D + 3E$, and $2D$ according to whether the field is directed along the p, q, or r molecular axis, respectively.

In addition, a "half-field" transition of weaker intensity is usually observed. It is called a $\Delta m_s = \pm 2$ transition, but this definition is only conventional because the states involved are not pure states with $m_s = \pm 1$, due to the mixing produced by the zero-field interaction. When the magnetic field is directed along the principal axes of the D tensor, the half-field transition is forbidden if the conventional equipment with the rf field polarized perpendicularly to the static field is adopted in the ESR spectrometer, but it can be observed with a parallel arrangement (van der Waals and De Groot, 1959). For any other orientation of the magnetic field relative to the molecular axes, the transition is weakly allowed for both the parallel and perpendicular dispositions of the microwave field. Figure 4 shows the effect on the energy levels of the magnetic field when directed along the molecular r axis. The heavy arrows indicate the "allowed" transitions and the dashed arrow the half-field transition.

In particular cases, especially when there is a pair of electrons on nearby sites with appreciably overlapping distributions, an isotropic coupling between the spins S_1 and S_2 must be introduced. This originates from the electrostatic interactions, which tend to couple the spins into a singlet and a triplet state. Following the Heitler–London description of the bonding, this interaction is termed *exchange interaction*, and it has the form

$$\mathcal{H}_e = J\,\mathbf{S}_1 \cdot \mathbf{S}_2 = \tfrac{1}{2}J(S^2 - S_1^2 - S_2^2) = \tfrac{1}{2}J(S^2 - \tfrac{3}{2})$$
$$\text{with} \quad S^2 = (\mathbf{S}_1 + \mathbf{S}_2)^2 \tag{49}$$

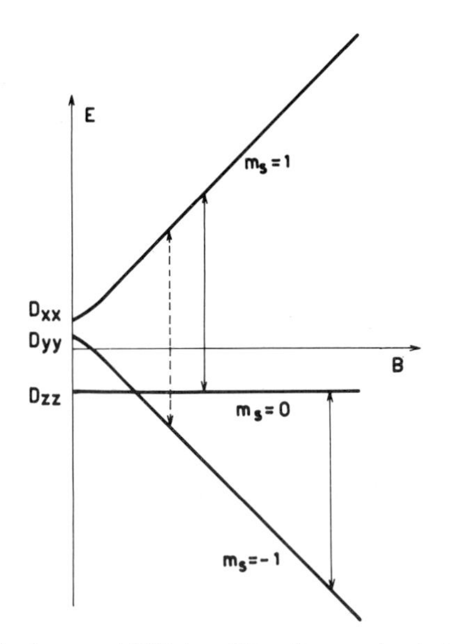

Fig. 4. Energy level scheme and ESR transitions for a molecule in the triplet state (static field applied along the molecular r axis).

The eigenfunctions of $\mathscr{H}_e$ are therefore the familiar singlet and triplet spin functions, which are eigenfunctions of S^2 with eigenvalues $S(S + 1)$ equal to 0 and 2, respectively. The exchange energy J corresponds to the singlet–triplet energy separation. Positive values of J lead to a state of lowest energy with antiparallel disposition of the spins. Complex situations arise when the magnitude of J is of the order of the hyperfine coupling a. The effects deriving from the presence of the exchange term is discussed in Chapter 4 on biradicals.

IV. ANALYSIS OF THE ESR SPECTRA

In the following sections we outline the methods of analysis of the ESR spectra for systems in the liquid phase, in single crystals, and in randomly oriented solids. In each of these cases the paramagnetic molecule is diluted in a diamagnetic host (the solvent, or a suitable crystal lattice) to avoid interactions between nearby spins, which would otherwise produce line broadening due to magnetic dipole interactions, or spin correlations due to exchange couplings.

We shall only treat systems with a single unpaired electron spin. The reader is referred to Chapter 4 for the discussion of the spectral features of systems with two or more interacting spins.

A. Solution Spectra

For a free radical in which the electron spin interacts via hyperfine coupling with N nuclei of individual nuclear spin quantum number I_k, the magnetic Hamiltonian is

$$\mathcal{H} = \beta_e \mathbf{H} \cdot \mathbf{g} \cdot \mathbf{S} + \sum_{k=1}^{N} \mathbf{S} \cdot \mathbf{A}^{(k)} \cdot \mathbf{I} \tag{50}$$

If the paramagnetic species is dissolved in a low-viscosity solvent, then the very fast molecular reorientations due to thermal motions average out the anisotropic terms of the spin Hamiltonian. The position of the spectrum in the field and the magnitude of the hyperfine splittings are thus determined by the average values of the diagonal elements of the $\mathbf{g}$ and $\mathbf{A}$ tensors. The energy levels of the system can then be obtained in terms of the *isotropic* Hamiltonian

$$\begin{aligned}\mathcal{H}_0 &= g\beta_e \mathbf{H} \cdot \mathbf{S} + \sum_{k=1}^{N} a_k \mathbf{S} \cdot \mathbf{I}_k \\ &= g\beta_e H_0 S_z + \sum_{k=1}^{N} a_k (S_x I_{kx} + S_y I_{ky} + S_z I_{kz})\end{aligned} \tag{51}$$

where

$$g = \tfrac{1}{3} \operatorname{Tr} \mathbf{g} \qquad \text{and} \qquad a_k = \tfrac{1}{3} \operatorname{Tr} \mathbf{A}^{(k)} \tag{52}$$

Since the hyperfine interactions are always much smaller than the Zeeman interaction energy, a perturbation procedure can be used to compute the energy levels, by treating the hyperfine term as a perturbation on the eigenstates of the Zeeman term. The eigenfunctions of the Zeeman Hamiltonian can be written as simple products of the usual electron spin and nuclear spin eigenfunctions of the spin operators S_z and I_{kz}. These product functions are also eigenfunctions of the simplified Hamiltonian

$$\mathcal{H}'_0 = g\beta_e H_0 S_z + \sum_k a_k S_z I_{kz} \tag{53}$$

and the required first-order energies are the corresponding eigenvalues

$$E = g\beta_e H_0 m_s + \sum_k a_k m_s M_k \tag{54}$$

where $m_s = \pm\frac{1}{2}$ and $M_k = -I_k, -I_k + 1, \ldots, I_k$.

Due to the effect of the static field and of the hyperfine couplings, a number of sublevels are generated, with energy given by Eq. (54). Some of these levels will have the same energy, i.e., they will be degenerate, if the hyperfine interaction is the same for two or more nuclei. In this case the nuclei are said to be magnetically equivalent, and this property is immediately recognizable from the molecular geometry.

The rf field applied to the system in a normal ESR experiment induces transitions among the sublevels. However, not all transitions are allowed. First of all, the states involved in the transitions must satisfy the condition $\Delta m_s = \pm 1$, as already seen from Eq. (33). Moreover, the operator S_x acts on electron spin variables only, and so the disposition of the nuclear spins of the two states between which the transition takes place has to be the same. If this was not the case, the matrix element appearing in Eq. (33) would vanish, given the orthogonality of the nuclear spin functions.

In conclusion, the allowed ESR transition must satisfy the selection rules $\Delta m_s = \pm 1$, $\Delta M_k = 0$ for all nuclei.

Let us suppose for the sake of simplicity that all the N nuclei are equivalent, as in the methyl radical $\cdot CH_3$ or in the benzene radical ions $(C_6H_6)^+$ and $(C_6H_6)^-$. The energy of the hyperfine levels is then

$$E(m_s, M) = g\beta_e H_0 m_s + a m_s M \tag{55}$$

where M is the vectorial sum of the individual quantum numbers M_k. The energy difference of the levels involved in the allowed transitions is

$$\Delta E(M) = g\beta_e H_0 + aM \tag{56}$$

The spectrum will consist of a number of lines equal to the $2NI_k + 1$ permissible values of M, with intensity proportional to the degeneracy of each nuclear M state, i.e., to the number of ways in which the individual spin components M_k can be combined to give the resultant M. If $I_k = \frac{1}{2}$, so that M_k can assume only the two values $\pm\frac{1}{2}$, the intensities are proportional to the coefficients of the binomial expansion. The spectral lines are equally spaced, and the separation between adjacent lines is a/h in frequency units. In the ESR experiments, where the field is scanned at fixed frequency ν_0 to meet the resonance conditions, the transitions will be symmetrically disposed about the center $H_0 = h\nu_0/g\beta_e$, and will occur at field values spaced by an amount $a/g\beta_e$. For this reason, it is customary to give the coupling constants in frequency units or in equivalent gauss, instead of in energy units. From relation (5) we find that 1 G corresponds to 2.8 MHz for $g = 2$.

Figure 5 shows the energy diagrams for the case of an electron spin interacting with a nucleus of spin $I = \frac{1}{2}$, with two equivalent nuclei of spin $\frac{1}{2}$, and with a single nucleus of spin 1. The dependence on the magnetic field strength of the energy levels of the last two situations, and the corresponding simulated spectra, are displayed in Fig. 6.

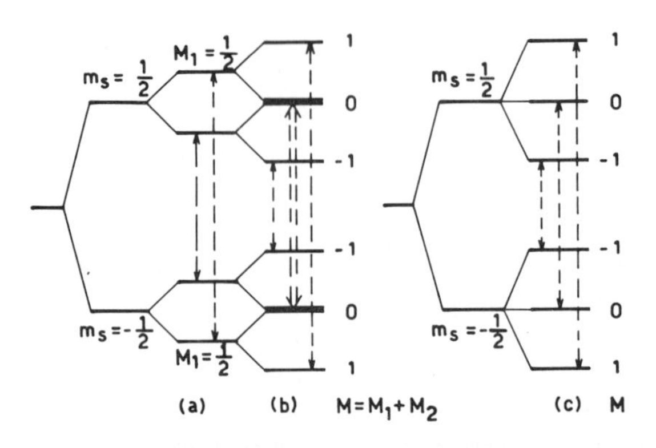

Fig. 5. Energy level scheme for an electron spin in a static magnetic field, interacting: (a) with a single nucleus of spin 1/2; (b) with two equivalent nuclear spins 1/2; (c) with a nucleus of spin 1. The coupling constant is assumed to be positive. Arrows indicate the allowed transitions.

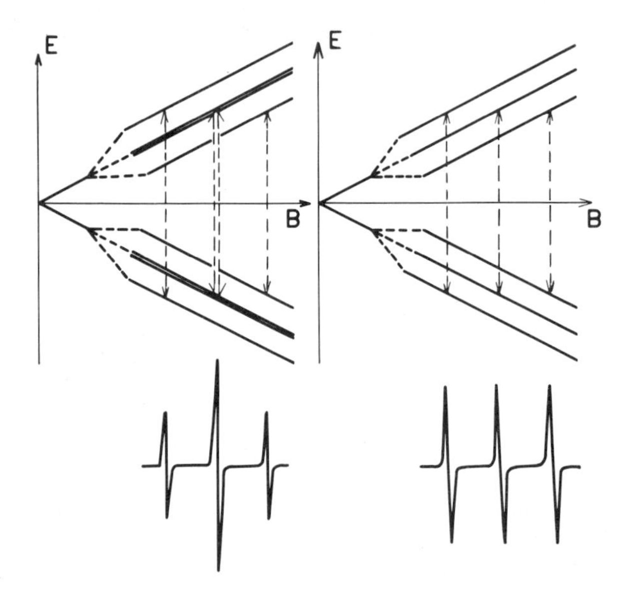

Fig. 6. Dependence of the energy levels on the field strength for the cases (b) and (c) of Fig. 5, and the corresponding simulated spectra.

B. Spectra in Oriented Crystals

In order to find the angular dependence of the spectra for free radicals oriented in single crystals, we have to compute the energy levels of the complete Hamiltonian given in Eq. (50). Several procedures can be used to obtain the eigenstates of the Hamiltonian, with the degree of accuracy desired, for the complex case of many interacting nuclei (Poole and Farach,

1972). However, given that usually $|A_{ij}^{(k)}| \ll g\beta_e H_0$, a first-order perturbation approach can again be adopted. We shall limit ourselves to the discussion of the results for the simple case of hyperfine interaction with a single nucleus of spin I, with a further condition of coincidence for the principal axis systems of the **g** and hyperfine tensors. Under these circumstances the transition energies for any fixed orientation of the magnetic field relative to the molecular axes, specified by the Euler angles (β, γ), are calculated by a relation analogous to Eq. (56):

$$\Delta E(M) = g_{\text{eff}}\beta_e H_0 + A_{\text{eff}} M \tag{57}$$

where g_{eff} has been already given in Eq. (29) and A_{eff} is similarly defined as

$$\begin{aligned} A_{\text{eff}}^2 &= A_{pp}^2 l^2 + A_{qq}^2 m^2 + A_{rr}^2 n^2 \\ &= A_{pp}^2 \sin^2\beta\cos^2\gamma + A_{qq}^2 \sin^2\beta\sin^2\gamma + A_{rr}^2\cos^2\beta \end{aligned} \tag{58}$$

When working at a fixed frequency, absorption will occur at field values determined by the resonance condition

$$h\nu_0 = g_{\text{eff}}(\beta, \gamma)\beta_e H + A_{\text{eff}}(\beta, \gamma) M \tag{59}$$

and thus the spectrum consists of the usual $2I + 1$ lines, equally spaced with a separation in field units of $A_{\text{eff}}/g_{\text{eff}}\beta_e$, and symmetrically disposed about a center determined by the value of g_{eff} according to Eq. (27).

If the hyperfine interaction tensor has axial symmetry, then the angular dependence of the splitting follows the relation

$$A_{\text{eff}}^2 = A_{\parallel}^2\cos^2\beta + A_{\perp}^2\sin^2\beta \tag{60}$$

β being the angle between the field direction and the symmetry axis of the magnetic tensor.

It is important to note that the experimental measurement of ΔE in Eq. (57) can determine only the absolute values of the principal components of the hyperfine tensor. However, these are related to the isotropic coupling constant a by the relation (52):

$$a = \tfrac{1}{3}(A_{pp} + A_{qq} + A_{rr})$$

The comparison between the value of a measured in solution and the values of the principal components of the tensor enables us to determine the *relative* sign of the isotropic and the purely anisotropic (or dipolar) part of the hyperfine interaction. On the other hand, the estimate of the absolute sign of the dipolar coupling is firmly established on theoretical grounds (Carrington and McLachlan, 1967; Atherton, 1973), so knowledge of the signs of the various terms of the hyperfine interaction can be safely assumed in most cases.

C. Spectra of Nonoriented Systems

In polycrystalline or vitreous samples, the principal axes of the paramagnetic molecules assume all possible angles relative to the magnetic field direction. Under these conditions, the ESR spectrum is expected to be the superposition of the spectra corresponding to all the possible orientations, and to be spread over the entire field range determined by the g-factor and hyperfine anisotropy. The distribution of the absorption intensity, however, is not uniform in this range. This fact can be visualized if we consider a system with an axially symmetric g factor, the symmetry axis being the r axis. As a consequence of the random orientation, there will be many more molecules with the r axis nearly perpendicular to the direction of the magnetic field than there are with the axis parallel to the field. Therefore there will be more paramagnets absorbing at the resonance field determined by $g_\perp$ than those absorbing in the field regions determined by $g_\parallel$. At these two field values, which correspond to the extrema of the range in which transitions can occur, turning points with different intensity result in the absorption spectrum, and therefore maxima or minima appear in the first derivative presentation (Wertz and Bolton, 1972).

The shape of the spectral envelope arising from both g-factor and hyperfine anisotropy can be computed with reasonable accuracy if each component line is assumed to be described by a shape function $\mathscr{L}$ independent of orientation. The amount of energy absorbed at a field H_r is given by

$$A(H_r) = \int_{H_{\min}}^{H_{\max}} P(H)\mathscr{L}(H_r - H)\, dH \tag{61}$$

where $P(H)\, dH$ represents the probability of finding molecules absorbing at the resonant field H, i.e., with orientation specified by the angles (β, γ) satisfying the resonance condition according to Eq. (59). For randomly dispersed samples, this probability is proportional to the elementary solid angle $\sin\beta\, d\beta\, d\gamma$. After the change to angular variables, numerical computation of the integral (61) is performed to obtain the overall absorption pattern. If the g-factor anisotropy is small and if the hyperfine component A_{rr} is larger than the other components, the outermost lines of the derivative spectrum have the appearance of absorption lines, centered at field values

$$(h\nu_0 \pm A_{rr}I)/g_{rr}\beta_e$$

A spectrum of this sort is shown in Fig. 7e.

Even in more complex situations the spectral parameters are obtainable from the analysis of the powder pattern, provided the spectrum is not too distorted by overlapping components (Weil and Hecht, 1963).

D. Linewidths in Solution Spectra

We conclude this section concerning the analysis of the spectral features with some considerations on the linewidth effects very often observed in solution spectra.

Following from the results attained in the preceding discussion, if an unpaired electron of a radical molecule interacts with a nucleus of spin I, the observed spectrum will consist of $2I + 1$ hyperfine components, each of which is identified by the nuclear quantum number M characterizing the sublevels involved in the transitions. These components are expected to have the same intensity, given that the energy separation between the magnetic levels is very small compared to the thermal energy kT, and so the populations of the levels, as seen from Eq. (4), are practically equal. In actual fact, we find that in most cases the heights of the various lines do not appear to be the same. Examination of the line shape reveals that this is due to the fact that the lines are not of the same width. For a Lorentzian curve as given in Eq. (9), the peak-to-peak height of the first derivative varies with the inverse square of the width, and therefore the relative height of two lines having the same intensity but different widths is given by

$$I_1/I_2 = [T_2(1)/T_2(2)]^2 \tag{62}$$

The linewidth variations have to be ascribed in most cases to the fluctuations of the anisotropic terms of the magnetic Hamiltonian caused by the molecular motions (Freed and Fraenkel, 1963; Hudson and Luckhurst, 1969). Even if the anisotropic terms do not contribute to the magnetic parameters measured in solution, they do constitute an important source of line broadening. The Hamiltonian (50) can in fact be considered as the sum of two contributions: an isotropic, orientationally invariant part $\mathscr{H}_0$, which has been given in Eq. (51), and a purely anisotropic, angular dependent part $\mathscr{H}_1$, which can be written as

$$\mathscr{H}_1 = \beta_e \mathbf{H} \cdot \mathbf{g}' \cdot \mathbf{S} + \sum_{k=1}^{N} \mathbf{S} \cdot \mathbf{A}'^{(k)} \cdot \mathbf{I} \tag{63}$$

where $\mathbf{g}'$ and $\mathbf{A}'^{(k)}$ are traceless tensors.

In liquid phase, the molecular tumbling makes $\mathscr{H}_1$ a random function of time. As a consequence, there is a random modulation of the energy levels and of the transition frequencies. In spite of the vanishing average value of $\mathscr{H}_1$, broadening of the absorption lines is expected to occur.

The frequency fluctuations can be characterized through their amplitude and coherence. The amplitude Δ is defined by the mean square value of the anisotropic interactions (in angular frequency units), and the coherence is given by the *correlation time* τ_c of the random motion. In the cases discussed so far of anisotropic interactions modulated by Brownian motions, the correlation time is a measure of the length of time over which the molecules persist in a given orientation. In other cases, in which modulation of the magnetic parameters of the system results from solvent interactions or conformational changes, the correlation time will be related to the mean time of existence of any particular molecular configuration.

A random process is considered to be *fast* if $\Delta\tau_c < 1$. Under this condition the anisotropic terms are averaged to zero and the spectrum is composed of sharp Lorentzian lines, characterized by the isotropic values of the magnetic parameters. The effect of the time-dependent part of the Hamiltonian manifests itself through the appearance of more or less pronounced differences in linewidth among the hyperfine components of the spectrum. This fact causes the relative height of the spectral lines to deviate from the expected value.

For $\Delta\tau_c > 1$, the motions are slow and the line shape reflects directly the random frequency distribution, approaching the limiting case of polycrystalline spectra.

In the following, we shall assume that the fast motional condition always holds.

For the case discussed above in which the spectral lines result from hyperfine interaction with a single nucleus, one finds that the dependence of the linewidth $1/T_2$ upon the nuclear quantum number M specifying the transition can be fitted by the equation

$$[T_2(M)]^{-1} = A + BM + CM^2 \tag{64}$$

The linewidth parameters A, B, and C depend on the magnitude of the magnetic anisotropies and on the rate of the molecular reorientations in the liquid. If these are sufficiently fast, the anisotropies are efficiently averaged out and linewidth effects are no longer observable. According to the Debye diffusion model (Debye, 1945), the correlation time for rotational diffusion can be calculated in terms of molecular dimensions, temperature, and viscosity of the medium if the molecules can be approximated to spheres of radius R:

$$\tau_c = (6D)^{-1}, \qquad D = kT/8\pi\eta R^3 \tag{65}$$

where D is the rotational diffusion constant and η is the viscosity. Typical values of τ_c for normal molecules in low-viscosity solvents are in the range 10^{-10}–10^{-11} sec.

E. An Instructional Example: The Nitroxide Free Radicals

The nitroxide free radicals are molecules containing the paramagnetic moiety

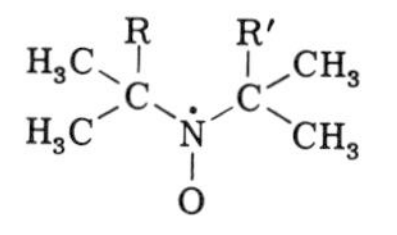

These species are remarkably stable and inert because of the protective effect exerted by the four methyl groups. The ESR spectra of these compounds comprise three sharp, well-resolved hyperfine lines resulting from the coupling of the unpaired electron spin with the nitrogen nuclear spin. Typical values of the magnetic parameters, obtained from single-crystal studies with the free radicals incorporated at low concentration into a diamagnetic host, are displayed in Table II (Griffith *et al.*, 1965), the molecular axis system† being chosen as shown:

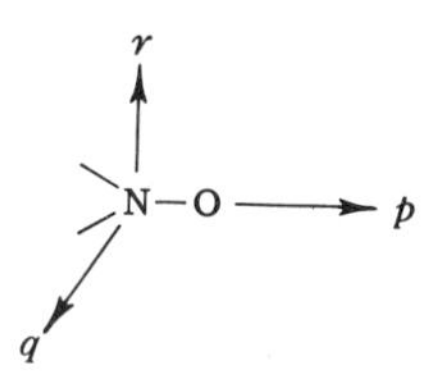

The value of the isotropic coupling constant (38 MHz compared with 55 MHz for the NH_3^+ radical) and the angular dependence of the dipolar interaction confirm that the unpaired electron is largely localized on a $2p\pi$ orbital of the nitrogen atom.

TABLE II

MAGNETIC PARAMETERS FOR NITROXIDE FREE RADICALS[a]

g_{pp}	2.0089	$A_{pp}(A_\perp)$	14 MHz
g_{qq}	2.0061	$A_{qq}(A_\perp)$	14 MHz
g_{rr}	2.0027	$A_{rr}(A_\parallel)$	87 MHz
g	2.0059	a	38 MHz

[a] Editor's note: A complete table of all nitroxide **g** and hyperfine tensors reported to date is compiled in Appendix II (p. 564).

† Editor's note: This axis system is, of course, equivalent to the common Cartesian axis system for nitroxides, where $p \equiv x$, $q \equiv y$, $r \equiv z$.

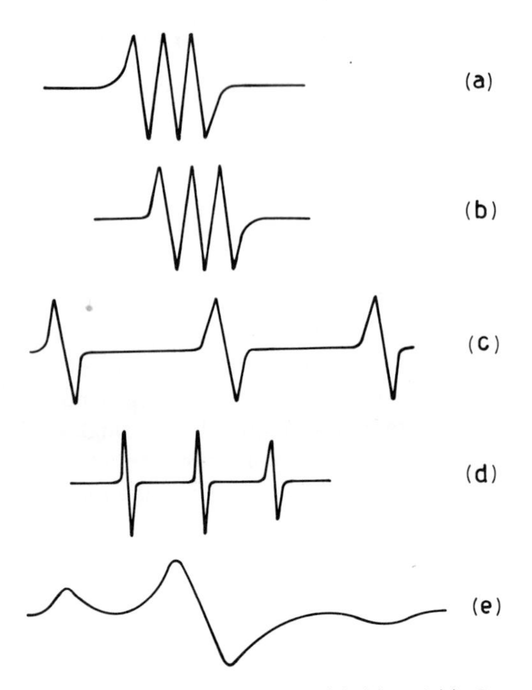

Fig. 7. X band spectra of nitroxide free radicals. (a), (b), and (c): Crystal spectra with the field along the principal axes of the g and A tensors. (d) Solution spectrum in aqueous glycerol, 25°C. (e) Spectrum of glassy solution of free radicals in glycerol at -80°C. The field increases to the right.

Figure 7 shows a series of spectra obtained in different conditions, namely in crystals, in solution, and in a glassy matrix. Spectra (a), (b), and (c) refer to molecules oriented in a crystal lattice with the magnetic field parallel to each of the three principal directions of the magnetic tensors. The hyperfine splittings are those listed in Table II, which correspond to field separations of 5, 5, and 31 G, respectively. Due to the g-factor anisotropy, the centers of the spectra fall at different field values, depending on the molecular orientation. If the resonance is occurring at about $H_0 = 3{,}400$ G for molecules oriented with their r axis along the field, then the positions of the centers for the spectra (a) and (b), when operating at constant frequency, are calculated from the relations

$$g_{pp}(H_0 + \delta H_p) = g_{qq}(H_0 + \delta H_q) = g_{rr} H_0$$

In this way one finds that the spectra (a) and (b) are shifted 10.2 and 5.6 G downfield with respect to the (c) spectrum, respectively.

Spectra (d) and (e) refer to free radicals dissolved in aqueous glycerol at 25°C and in a glassy matrix obtained by freezing a glycerol solution at −80°C.

Figure 8 shows the satellite lines revealed under high amplification due to the natural abundance of ^{13}C and ^{15}N nuclei. The three doublets come from the further subdivision of the nitrogen hyperfine levels in those molecules containing one ^{13}C nucleus, the probability of finding two ^{13}C on the same molecule being negligibly small. The two weakest lines come from the 0.365% of the molecules containing a ^{15}N nucleus, which has spin 1/2. Note that the gyromagnetic ratio $g_{^{15}N}\beta_{^{15}N}/g_{^{14}N}\beta_{^{14}N}$ is 1.40, and so the coupling constant for ^{15}N is correspondingly 1.40 times the coupling constant for the normal isotope.

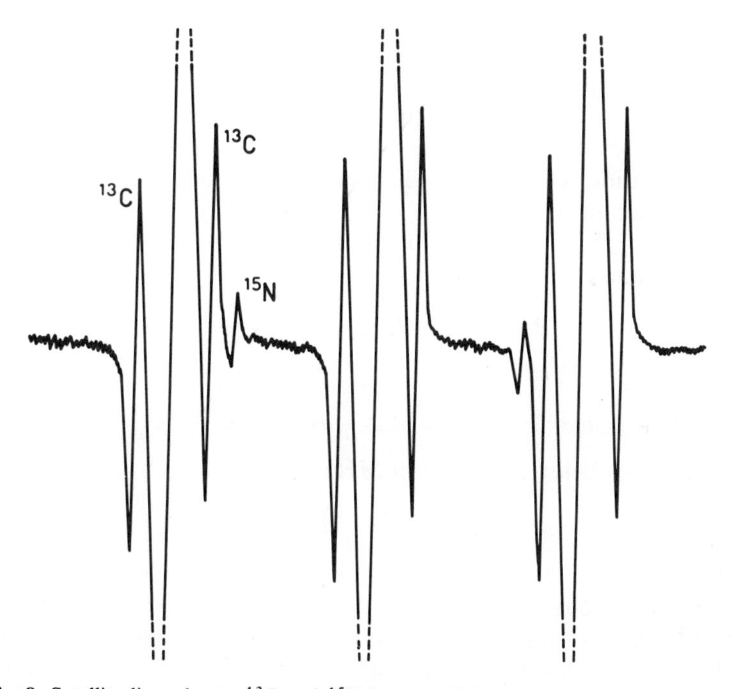

Fig. 8. Satellite lines due to ^{13}C and ^{15}N in natural abundance in the ESR spectrum of nitroxide free radicals.

As far as the width of the three lines of the spectra in solution is concerned, we observe that the high-field component of the nitrogen hyperfine structure is always broader than the two other lines, which have approximately equal width. This fact can be rationalized if we refer to Eq. (64) and calculate the explicit expressions for the linewidth parameters A, B, and C in terms of the magnetic anisotropies. The underlying theory will be developed in a following section, but here we shall anticipate some results.

Term A includes broadening contributions resulting from mechanisms that are different from the modulation of the anisotropic interactions considered so far. These broadening effects arise, for example, from the presence of oxygen in the solution or from instrumental factors, and they are all expected to be independent of line index M. For the above reason, and also because it is easier to measure the relative heights of the lines instead of the breadth of any single line, it is convenient to rewrite expression (64) in the following way:

$$[T_2(0)]^{-1}[(T_2(0)/T_2(\pm 1)) - 1] = C \pm B \tag{64'}$$

where $[T_2(0)]^{-1}$ is the breadth of the central line, which corresponds to $M = 0$, as seen from Fig. 5c. Experimental values of $T_2(0)/T_2(\pm 1)$ are obtained from the square root of the ratios of the experimental derivative curve peak heights, according to Eq. (62). Theoretical considerations lead to a positive value of the nitrogen isotropic coupling constant, and this permits the identification of the $M = 1$ nuclear spin state with the low-field hyperfine component.

In terms of the g- and hyperfine-tensor anisotropies, one obtains for the B and C parameters

$$B = \tfrac{4}{15} b(\Delta\gamma) H_0 \tau_c, \qquad C = \tfrac{1}{8} b^2 \tau_c \tag{66}$$

where

$$b = 2\pi A'_{rr} = 2\pi(A_{rr} - a) = \tfrac{4}{3}\pi(A_{\parallel} - A_{\perp}) \tag{67}$$

all the A values being expressed in MHz units, and

$$\Delta\gamma = \beta_e \hbar^{-1}[g_{rr} - \tfrac{1}{2}(g_{pp} + g_{qq})] \tag{68}$$

By introducing the values listed in Table II, one finds that $C \simeq -B$ for field values of about 3,400 G. In conclusion, one can see from Eq. (64′) that, due to the negative sign of B, only the high-field line (having $M = -1$) will appear to be broader than the central one in the X-band spectra of nitroxide free radicals. In the case where field intensity was increased to 12,500 G (Q band), the B term would become accordingly more negative, and the spectrum of Fig. 9b would be the result.

The very simple expressions for B and C given above are correct if (i) the molecular motion is isotropic, and (ii) the correlation time falls in a range satisfying the conditions $\omega_0^2 \tau_c^2 > 1$ and $\Delta\tau_c < 1$, ω_0 being the resonance frequency and Δ being practically the largest of the magnetic anisotropies. The first of the two conditions is discussed in Section V, where general

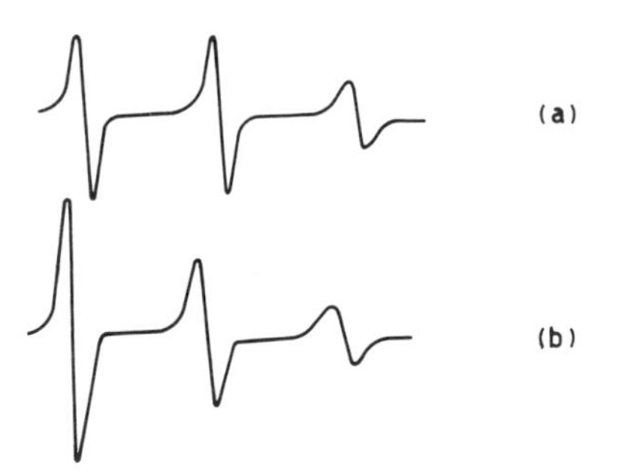

Fig. 9. (a) X-band and (b) Q-band spectra of nitroxide free radicals in solution. The field increases to the right.

expressions for the linewidth parameters, which are valid also for the case of anisotropic motion, are presented.

In an X-band experiment the resonance frequency is 9.5 GHz, which is equivalent to 6×10^{10} sec^{-1} in angular frequency units. On the other hand, $\Delta \simeq b$ is calculated from Eq. (67) to be 3×10^8 sec^{-1}. Thus the above treatment is applicable to X-band spectra of nitroxide free radicals if the correlation times fall in the range 5×10^{-11}–10^{-9} sec.

The paramagnetic resonance spectra of the nitroxide free radicals are sensitive not only to molecular motions, but also to the nature of the medium in which they are dissolved. This arises from the fact that the magnetic parameters of any radical species are very sensitive functions of the electronic distribution in the molecule, and therefore are influenced by perturbations due to the environment. Thus the isotropic coupling constant of the nitrogen atom in the nitroxide group increases by more than 10% when we go from hydrophobic to hydrophilic solvents. A noticeable decrease of the isotropic g factor is correspondingly observed. These effects can be understood qualitatively in terms of specific interactions of the polar solvents with the lone pairs of electrons on the oxygen atom. These interactions are expected to lower the energy of the nonbonded electrons and increase the electron affinity of the oxygen atom. In this way the three-π-electron distribution of the N–O group is shifted toward a situation which is well represented by the limiting structure corresponding to a maximum unpaired electron density on the nitrogen atom:

$$\begin{array}{c} \diagdown\ \dot{\text{N}}^{(+)}\ \diagup \\ | \\ |\text{O}|^{(-)} \end{array}$$

On the other hand, a rather crude estimate of the g factor based on Eq. (17) of Section III.A shows that the deviation from the free-spin value arises essentially from the term

$$\lambda_{\text{O}} \rho_{\text{O}}^{\pi} / \Delta E_{n \rightarrow \pi*} \tag{69}$$

where λ_O is the spin–orbit coupling of the oxygen atom, ρ_O^π is the spin density on the oxygen, and $\Delta E_{n \to \pi*}$ is the $n \to \pi$ excitation energy (Brière *et al.*, 1965). It is evident that the factors that cause the increase of a_N lead to a parallel reduction of the observed g value. The relation between the g value and the n–π energy difference is well demonstrated by the experimentally observed blue shift of the n–π transition in the UV spectra on passing from aprotic to polar solvents.†

V. LINE SHAPE THEORY

As already mentioned in the previous sections, line shape analysis is very important in biological applications of ESR spectroscopy because it represents the only way to obtain information on molecular motions. Unfortunately, the theoretical approach to this matter remains undoubtedly the most difficult problem of magnetic resonance theory. For this reason, we find it useful to give an outline of the quantum mechanical description of the problem, with the aim of providing a guide to readers who desire to go deeper into this area. They will then find exhaustive answers to the many questions that this concise account will certainly raise in the textbooks by Abragam (1961), Slichter (1963), and Atherton (1973), in the basic papers of Freed (1964), Freed and Fraenkel (1963), and Kivelson (1972), and in an excellent review by Hudson and Luckhurst (1969).

In the following we shall briefly introduce the density matrix formalism that plays an important role in relaxation theory. In addition, the spherical coordinate representation of the magnetic interaction tensors shall be discussed. Tensor components expressed in this way correspond to definite projection quantum numbers, and this makes their use particularly convenient when the averages of the anisotropic terms of the Hamiltonian are calculated. The basic definitions of the statistical distribution functions that are needed to characterize the random motion in liquids shall also be given.

The solution of the line shape problem derived here is valid only in the fast motional region, with an upper limit for correlation times of about 10^{-8} sec. Beyond this limit, the methods presented in Chapter 3 have to be adopted.

A. Density Matrix

As known from the basic principles of quantum theory, any physical observable is associated with a quantum mechanical operator. Thus, if a

† Editor's note: A compilation of isotropic hyperfine and **g** values for a nitroxide in several solvents of varying polarity is found in Chapter 8, Table II (p. 307).

system is described by a wave function ψ_k, the expectation value of the physical property $\mathscr{A}$ will be

$$A_k = \int \psi_k^* A \psi_k \, dv \tag{70}$$

If ψ_k is expanded in a complete set of orthonormal time-independent functions u_n, then we have

$$\psi_k = \sum_n c_n^k u_n$$

$$A_k = \sum_{n,m} c_n^{k*} c_m^k \int u_n^* A u_m \, dv = \sum_{n,m} c_n^{k*} c_m^k \langle n | A | m \rangle \tag{71}$$

This is the case in which maximal information on the property $\mathscr{A}$ can be achieved. In many instances, however, our knowledge of the system is not so complete. Consider for example, a system composed of a large number of spins, 10^{23} say, interacting among themselves or with the environment in which they are embedded, in the presence of a static magnetic field. The angular momentum and spin component of an individual spin are no longer defined, and if we wish to compute the value of the macroscopic magnetic moment induced by the field, we must take into account all possible spin orientations and add up the expectation values of S_z for each of these situations weighted with some probability factors.

Going back to Eq. (71), we see that in a composite system the products $c_n^{k*} c_m^k$ can be different for the various components of the system because they are represented by different functions ψ, but the matrix elements of the operator A on the basis set u_n are always the same.

We define therefore the ensemble average of $\mathscr{A}$ by the relation

$$\langle A \rangle = \sum_k p_k A_k = \sum_k \sum_{nm} p_k c_n^{k*} c_m^k \langle n | A | m \rangle \tag{72}$$

where p_k is the statistical weight of ψ_k in the specification of the system. It is convenient to define a density operator ρ through its matrix elements:

$$\langle m | \rho | n \rangle = \sum_k \sum_{nm} p_k c_n^{k*} c_m^k \tag{73}$$

so that a more compact form for $\langle A \rangle$ results:

$$\langle A \rangle = \sum_{nm} \langle m | \rho | n \rangle \langle n | A | m \rangle = \sum_m \langle m | \rho A | m \rangle = \mathrm{Tr}\, \rho A \tag{74}$$

Here Tr stands for the trace, and its value can be shown to be independent of the choice of the basis set functions, as expected on physical grounds.

If the system is at equilibrium, the explicit form for the density matrix operator ρ is

$$\rho_{eq} = e^{-\mathscr{H}/kT}/\mathrm{Tr}\; e^{-\mathscr{H}/kT} \tag{75}$$

its diagonal matrix elements being the Boltzmann factors given in Eq. (4).

If the state of the system develops in time, the wave functions ψ_k describing the system will then be time dependent. Since the basis functions u_n are time independent, the coefficients c_n must carry the time dependence. As a consequence, the density matrix will also vary with time, and its time evolution is expressed in terms of the Hamiltonian $\mathscr{H}$ of the system by the equation of motion (Slichter, 1963)

$$d\rho(t)/dt = (i/\hbar)[\rho(t), \mathscr{H}] = (i/\hbar)[\rho(t)\mathscr{H} - \mathscr{H}\rho(t)] \tag{76}$$

Accordingly, the time variation of the macroscopic observable $\mathscr{A}$ is determined by:

$$\langle A(t)\rangle = \mathrm{Tr}\; \rho(t)A \tag{77}$$

In the development of the theory of spin relaxation, we are interested in the calculation of the time dependence of the macroscopic magnetization $\mathscr{M}$. According to Eqs. (1) and (77), the components of the macroscopic magnetization are related to the quantum mechanical spin operators S_x, S_y, and S_z by the expression

$$\mathscr{M}_i(t) = -Ng\beta_e\langle S_i\rangle = -Ng\beta_e\; \mathrm{Tr}\; \rho(t)S_i, \qquad i = x, y, z \tag{78}$$

where N is the number of spins.

We have seen that the Hamiltonian for paramagnetic species in liquids always consists of a time-independent part $\mathscr{H}_0$ and a part $\mathscr{H}_1(t)$ that is a random function of time. Therefore the equation for the density matrix of the spin system becomes

$$d\rho(t)/dt = (i/\hbar)[\rho(t), \mathscr{H}_0 + \mathscr{H}_1(t)] \tag{79}$$

It is not possible to obtain exact solutions to this equation, and thus we must make use of time-dependent perturbation expansions. According to a method proposed by Redfield, the procedure to solve this equation consists in an iterative integration by successive approximations, followed by the statistical averaging over the ensemble defined by the probability distribution characteristic of $\mathscr{H}_1(t)$. To second order of approximation, we end up with the following system of linear differential equations (Slichter, 1963; Redfield, 1966):

$$\begin{aligned} d\rho_{\alpha\alpha'}/dt &= (i/\hbar)[\rho, \mathscr{H}_0]_{\alpha\alpha'} + \sum_{\beta\beta'} R_{\alpha\alpha'\beta\beta'}\rho_{\beta\beta'} \\ &= (i/\hbar)(E_{\alpha'} - E_\alpha)\rho_{\alpha\alpha'} + \sum_{\beta\beta'} R_{\alpha\alpha'\beta\beta'}\rho_{\beta\beta'} \end{aligned} \tag{80}$$

The matrix elements of the spin density operator ρ are calculated on the basis of the eigenfunctions of $\mathscr{H}_0$, and the $R_{\alpha\alpha'\beta\beta'}$ are constant coefficients. The summation is restricted to states that satisfy the condition among the energies $E_\alpha - E_{\alpha'} = E_\beta - E_{\beta'}$. The coefficients $R_{\alpha\alpha'\beta\beta'}$ form a matrix which is known as *relaxation matrix*. The derivation of Eq. (80) is rather lengthy and the reader is referred to the texts of Slichter and Atherton for the discussion of the various approximations employed. Furthermore, it has to be understood that Eq. (80) only gives an approximate time dependence for $\rho(t)$ and therefore cannot be used if the perturbation expressed by $\mathscr{H}_1(t)$ is too strong, or if it acts for too long a time. In fact, the applicability of Eq. (80) is restricted to the region of fast frequency modulation.

Equation (80) does not consider the effect of the applied alternating field, which induces the transitions between the levels. In order to include this effect, we must add to $\mathscr{H}_0 + \mathscr{H}_1(t)$ of Eq. (79) an extra term $\mathscr{H}_2(t)$ in the form given by Eq. (32) of Section III.B. To first order, the differential equation for the density matrix is modified in the following way (Slichter, 1963, p. 156):

$$\begin{aligned} d\rho_{\alpha\alpha'}/dt &= (i/\hbar)[\rho, \mathscr{H}_0 + \mathscr{H}_2(t)]_{\alpha\alpha'} + \sum_{\beta\beta'} R_{\alpha\alpha'\beta\beta'}\rho_{\beta\beta'} \\ &= (i/\hbar)(E_{\alpha'} - E_\alpha)\rho_{\alpha\alpha'} + (i/\hbar)\sum_{\alpha''}[\rho_{\alpha\alpha''}\langle\alpha''|\mathscr{H}_2(t)|\alpha'\rangle \\ &\quad - \langle\alpha|\mathscr{H}_2(t)|\alpha''\rangle\rho_{\alpha''\alpha'}] + \sum_{\beta\beta'} R_{\alpha\alpha'\beta\beta'}\rho_{\beta\beta'} \end{aligned} \tag{81}$$

Explicit expressions for the coefficients $R_{\alpha\alpha'\beta\beta'}$ will be given in the next section. We will limit ourselves for the moment to examine, with a specific example, the effect of the relaxation terms in the density matrix equation.

Let us consider a system of spin $S = \frac{1}{2}$, for which there are only two levels α and β, corresponding to the eigenvalues $+\frac{1}{2}$ and $-\frac{1}{2}$ for S_z, respectively. In this case, complete knowledge of the system is provided by the four elements $\rho_{\alpha\alpha}$, $\rho_{\beta\beta}$, $\rho_{\alpha\beta}$, and $\rho_{\beta\alpha}$ of the density matrix.

We want now to calculate the value of the $\mathscr{M}_x$ component of the magnetization with the aid of Eq. (78). Keeping in mind that the only nonvanishing matrix elements of S_x are $\langle\alpha|S_x|\beta\rangle$ and $\langle\beta|S_x|\alpha\rangle$, both having a value of $\frac{1}{2}$, and that ρ is Hermitian, so that $\rho_{\alpha\beta} = \rho^*_{\beta\alpha}$, we obtain

$$\langle S_x\rangle = \mathrm{Tr}\,\rho S_x = \rho_{\alpha\beta}\langle\beta|S_x|\alpha\rangle + \rho_{\beta\alpha}\langle\alpha|S_x|\beta\rangle = \tfrac{1}{2}(\rho_{\alpha\beta} + \rho_{\beta\alpha}) = \mathrm{Re}\,\rho_{\beta\alpha} \tag{82}$$

where Re $\rho_{\beta\alpha}$ means the real part of $\rho_{\beta\alpha}$, which in general will be a complex function.

The magnetic moment $\mathscr{M}_x$ is induced by the oscillating field applied in the direction perpendicular to the static field:

$$H_x(t) = H_1 \cos \omega t$$

If this field is thought of as being removed at time $t = 0$, the time development of the transverse magnetization is obtained by solving the equation for $\rho_{\beta\alpha}$ in the absence of the rf field, i.e., in the form given by Eq. (80) (Kivelson, 1972):

$$d\rho_{\beta\alpha}/dt = (i\omega_0 + R_{\beta\alpha\beta\alpha})\rho_{\beta\alpha} \tag{83}$$

where $\omega_0 = (E_\alpha - E_\beta)/\hbar$.

This equation is solved immediately, and by setting

$$-R_{\beta\alpha\beta\alpha} = 1/T_2$$

we find for $\mathcal{M}_x(t)$

$$\mathcal{M}_x(t) \propto \cos \omega_0 t \exp(-t/T_2) \tag{84}$$

We see in this way that the macroscopic magnetization component $\mathcal{M}_x(t)$, after the rf field is turned off, executes a damped precession and relaxes finally to zero with a decay time constant T_2. Equation (84) gives theoretical support to the physical intuitions leading to the phenomenological equation (8), and permits us to relate the experimentally observable relaxation time T_2 to microscopic properties of the system through the calculation of the relaxation matrix element $R_{\beta\alpha\beta\alpha}$.

B. Relaxation and Line Shapes

In a magnetic resonance experiment the shape of the resonance absorption curve is determined by the profile of the intensity of the energy absorbed as a function of the frequency ω of the exploring rf field (the experiment is thought of as being performed by keeping the main magnetic field at a fixed value H_0 and by varying the frequency of the oscillating field). As previously noted, the oscillating field $H_x(t)$ induces a magnetic moment $\mathcal{M}_x(t)$ in the macroscopic sample, and the power absorbed by the sample from the field is

$$P = -\mathcal{M} \cdot d\mathbf{H}/dt = -\mathcal{M}_x(dH_x/dt) \tag{85}$$

The actual spectrum is determined by the average power absorbed per cycle (Abragam, 1961; Pake, 1962)

$$\overline{P(\omega)} = -(\omega/2\pi) \int_0^{2\pi/\omega} \mathcal{M}_x(dH_x/dt)\, dt \tag{86}$$

$\mathcal{M}_x(t)$ can still be obtained by solving the differential equation for the density matrix, but this time the equation must be written in the complete form given in (81), to account for the presence of the rf field. The details of the calculation are given by Atherton (1973) and Slichter (1963); we shall only give here the final result. In the limit of small H_1 (that is, in the absence of saturation), $\mathscr{M}_x(t)$ is found to be proportional to H_1 and to be composed of a part in phase with the oscillating field and another part out of phase with it:

$$\mathscr{M}_x(t) = H_1[\chi'(\omega) \cos \omega t + \chi''(\omega) \sin \omega t] \tag{87}$$

By substituting this expression into Eq. (86), we find after the integration that only $\chi''(\omega)$ actually determines the spectral line shape:

$$\overline{P(\omega)} = \tfrac{1}{2}\omega\chi''(\omega)H_1^2 \tag{88}$$

When the lines are narrow, as is usually observed in ESR, ω varies very little within the linewidth and so the spectral shape is well represented by $\chi''(\omega)$. In the case of the two-level system discussed above, one finds

$$\chi''(\omega) = \tfrac{1}{2}\pi\omega_0 \chi_0 f(\omega) \tag{89}$$

where χ_0 is the static susceptibility and $f(\omega)$ is the Lorentzian function described by Eq. (9) and illustrated in Fig. 1.

It will be noticed that the shape function $f(\omega)$ is nothing else than the Fourier transform of the relaxation function of $\mathscr{M}_x(t)$ given in Eq. (84), describing the decay of the x component of the magnetization after the turning off of the rf field. These results are quite general in the framework of linear response theory (Abragam, 1961, p. 94). Relations like that given in Eq. (87) are expected to be obtained for the response of any system to monochromatic excitations, the excitation being in this case the rf field and the response the induced moment $\mathscr{M}_x$. Furthermore, the Fourier transformation is the general relation between the decay of the response measured in the time domain after the removal of the excitation, and the frequency dependence of the same physical property determined in the frequency domain under continuous excitation.

If the relaxation behavior of the system cannot be described by a single exponential decay, the absorption curve will be the superposition of Lorentzians of different widths (Freed and Fraenkel, 1963). The density matrix calculation relates the line-shape function to the microscopic structure of the spin system.

C. Relaxation Matrix and Correlation Functions

From the former discussion it is evident that the crucial point for the theoretical interpretation of the spectral line shapes is the computation of the relaxation matrix. We recall that Eq. (80) results from an approximate procedure that enable us to obtain the solutions for the density matrix of a system where a time-dependent perturbation $\mathscr{H}_1(t)$ is acting. Given the random nature of the time dependence of $\mathscr{H}_1(t)$, the effect of the perturbation on the system must be treated in a statistical way.

The general expression for $R_{\alpha\alpha'\beta\beta'}$ is found to be (Slichter, 1963)

$$\begin{aligned} R_{\alpha\alpha'\beta\beta'} &= 2J_{\alpha\beta\alpha'\beta'}(\omega_{\alpha\beta}) - \delta_{\alpha'\beta'} \sum_{\gamma} J_{\alpha\gamma\beta\gamma}(\omega_{\gamma\beta}) \\ &\quad - \delta_{\alpha\beta} \sum_{\gamma} J_{\gamma\alpha'\gamma\beta'}(\omega_{\beta'\gamma}) \\ J_{\alpha\alpha'\beta\beta'}(\omega) &= \frac{1}{2} \int_{-\infty}^{+\infty} G_{\alpha\alpha'\beta\beta'}(t) e^{-i\omega t}\, dt \\ G_{\alpha\alpha'\beta\beta'}(t) &= \hbar^{-2} \langle \mathscr{H}_1(0)_{\alpha\alpha'} \mathscr{H}_1(t)^*_{\beta\beta'} \rangle \end{aligned} \tag{90}$$

Here $\mathscr{H}_1(t)_{\alpha\beta}$ is a shorthand notation for the α,β matrix element of $\mathscr{H}_1(t)$, the brackets mean the average over the statistical ensemble, and $\omega_{\alpha\beta} = (E_\alpha - E_\beta)/\hbar$. $G(t)$ is called the *correlation function* of $\mathscr{H}_1(t)$. Even if the statistical average of $\mathscr{H}_1(t)$ is zero, the mean square interaction $\langle |\mathscr{H}_1(t)|^2 \rangle$ is not, and so $G(t)$ tells how the value of $\mathscr{H}_1$ at any time t_0 is correlated with values at later times $t_0 + t$. This function is independent of the time origin, due to the stationary nature of the random processes that cause the fluctuations of $\mathscr{H}_1$, as discussed later. The detailed time dependence of the correlation function depends on the physical model that is assumed to account for the main mechanism of modulation of $\mathscr{H}_1(t)$, e.g., anisotropic interactions modulated by rotational diffusion, but in general any statistical model leads to an exponential decay of the form

$$G(t) = \hbar^{-2} \langle |\mathscr{H}_1|^2 \rangle e^{-t/\tau_c} \tag{91}$$

where τ_c is the correlation time for the random motion.

We recall that $\mathscr{H}_1$ is a function of the Euler angles, and these are random functions of the time. Thus we need the definition of suitable probability distributions for the molecular orientations in order to be able to calculate the correlation function of the random function $\mathscr{H}_1(t)$.

The random processes that we shall encounter can always be described in terms of two probability distributions (Pedersen, 1972):

$P(x, t)\, dx$, which is the probability of finding x in the range $(x, x + dx)$ at time t;

$P(x_1 t_1, x_2 t_2)\, dx_1\, dx_2$, which is the joint probability of finding x in the range $(x_1, x_1 + dx_1)$ at time t_1 and in the range $(x_2, x_2 + dx_2)$ at time t_2.

It is useful to define a conditional probability $P(x_1 t_1 | x_2 t_2)$ as the probability that, given x_1 at time t_1, one finds x in the range $(x_2, x_2 + dx_2)$ at time t_2. The conditional probability is obviously related to the joint probability:

$$P(x_1 t_1, x_2 t_2) = P(x_1 t_1) P(x_1 t_1 | x_2 t_2)$$

For any function $f(x)$ of the random variable x we can now define the average value $\langle f(t)\rangle$ and the correlation function $\langle f(t_1)f^*(t_2)\rangle$:

$$\langle f(t)\rangle = \int f(x)P(x, t)\, dx \tag{92}$$

$$\langle f(t_1)f^*(t_2)\rangle = \iint f(x_1)f^*(x_2)P(x_1 t_1, x_2 t_2)\, dx_1\, dx_2 \tag{93}$$

The random processes that we are interested in are stationary in character, which means that they are invariant under translation of the time axis. It follows that the distribution probability $P(x, t)$ is independent of time, and that the joint or conditional probability depends on the times t_1 and t_2 only through their difference $t_2 - t_1 = t$. Under these circumstances we have

$$\langle f\rangle = \int f(x)P(x)\, dx \tag{94}$$

$$\begin{aligned}\langle f(0)f^*(t)\rangle &= \iint f(x_1)f^*(x_2)P(x_1 0, x_2 t)\, dx_1\, dx_2 \\ &= \int dx_1\, f(x_1)P(x_1) \int dx_2\, f^*(x_2)P(x_1 \mid x_2 t)\end{aligned}$$

$$\langle f(0)f^*(t)\rangle = \langle f(-t)f^*(0)\rangle = \langle f(0)f^*(-t)\rangle \tag{95}$$

The correlation time τ_c is defined by

$$\tau_c = \int_0^\infty \langle f(0)f^*(t)\rangle\, dt/\langle f^2\rangle \tag{96}$$

The probability distribution functions are normalized to unity:

$$\int P(x)\, dx = 1, \qquad \int P(x_1 \mid x_2 t)\, dx_2 = 1 \tag{97}$$

The following equalities result from the properties of $P(x)$ and $P(x_1 \mid x_2 t)$:

$$\int P(x_1)P(x_1 \mid x_2 t)\, dx_1 = P(x_2) \tag{98}$$

$$\lim_{t\to\infty} P(x_1 \mid x_2 t) = P(x_2) \tag{99}$$

Therefore the correlation function goes to $\langle f\rangle^2$ for $t \to \infty$ and reduces to $\langle f^2\rangle$ at $t = 0$.

D. Hamiltonian in Spherical Basis

The Hamiltonian given in Eq. (63) is a function of the Euler angles, which are in turn random functions of time. The dependence on the angles can be made explicit by transforming the Cartesian tensors from space to molecular axes. The expression so obtained is rather cumbersome, and for our purposes it would be desirable to have the functional dependence in a more tractable form. This can be achieved if the tensors are expressed in a spherical basis instead of in the usual Cartesian set.

A spherical tensor of rank l is a quantity represented by $2l + 1$ components that transform under rotation according to the law (Rose, 1957; Brink and Satchler, 1968)

$$T'^{(l,m)} = \sum_{m'} D^l_{m'm}(\alpha\beta\gamma)T^{(l,m')} \tag{100}$$

This transformation is analogous to the relation that gives the eigenfunctions of the angular momentum projection operator $L_{z'}$ in terms of the original eigenfunctions $|l, m\rangle$ of L_z after a rotation of the quantization axis.

Since m, m' can only assume the values $l, l - 1, \ldots, -l$, the coefficients $D^l_{m'm}$ form, for a given value of l, a matrix of dimension $2l + 1$, which is known as a Wigner rotation matrix. For the inverse rotation one has

$$\begin{aligned} T^{(l,m)} &= \sum_{m'} D^l_{m'm}(-\gamma - \beta - \alpha)T'^{(l,m')} \\ &= \sum_{m'} D^{l*}_{mm'}(\alpha\beta\gamma)T'^{(l,m')} \\ &= \sum_{m'} (-1)^{m-m'}D^l_{-m-m'}(\alpha\beta\gamma)T'^{(l,m')} \end{aligned} \tag{101}$$

The Wigner rotation matrix elements $D^l_{mm'}(\alpha\beta\gamma)$ are found to have the following functional dependence on the Euler angles:

$$D^l_{mm'}(\alpha\beta\gamma) = e^{-im\alpha}\, d_{mm'}(\beta)e^{-im'\gamma} \tag{102}$$

The reduced rotation matrices are real functions (Table III) and obey the symmetry relations

$$d^l_{mm'} = d^l_{-m'-m} = (-1)^{m-m'}\, d^l_{m'm}$$

The Wigner functions reduce in some special cases to spherical harmonics or to Legendre polynomials

$$\begin{aligned} D^l_{m0}(\alpha, \beta, 0) &= \left[\frac{4\pi}{2l+1}\right]^{1/2} Y^*_{lm}(\beta, \alpha) \\ D^l_{0m}(0, \beta, \gamma) &= (-1)^m\left[\frac{4\pi}{2l+1}\right]^{1/2} Y^*_{lm}(\beta, \gamma) \\ D^l_{00}(0, \beta, 0) &= P_l(\cos\beta) \end{aligned} \tag{103}$$

TABLE III

REDUCED WIGNER ROTATION MATRIX ELEMENTS $d^l_{mm'}(\beta)$ FOR $l = 0$, 1, AND 2

d^0_{00}	1
d^1_{11}	$\frac{1}{2}(1 + \cos\beta)$
d^1_{10}	$-\sin\beta$
d^1_{1-1}	$\frac{1}{2}(1 - \cos\beta)$
d^1_{00}	$\cos\beta$
d^2_{22}	$\frac{1}{4}(1 + \cos\beta)^2$
d^2_{21}	$-\frac{1}{2}(\sin\beta)(1 + \cos\beta)$
d^2_{20}	$\sqrt{\frac{3}{8}}\sin^2\beta$
d^2_{2-1}	$-\frac{1}{2}(\sin\beta)(1 - \cos\beta)$
d^2_{2-2}	$\frac{1}{4}(1 - \cos\beta)^2$
d^2_{11}	$\frac{1}{2}(2\cos^2\beta + \cos\beta - 1)$
d^2_{10}	$-\sqrt{\frac{3}{2}}\sin\beta\cos\beta$
d^2_{1-1}	$-\frac{1}{2}(2\cos^2\beta - \cos\beta - 1)$
d^2_{00}	$\frac{1}{2}(3\cos^2\beta - 1)$

They form a complete orthogonal set and the following orthogonality relation holds:

$$\int D^{l*}_{mk}(\alpha\beta\gamma)D^{l'}_{m'k'}(\alpha\beta\gamma)\, d\alpha \sin\beta\, d\beta\, d\gamma = \frac{8\pi^2}{2l+1}\delta_{ll'}\delta_{mm'}\delta_{kk'} \tag{104}$$

Finally, it should be remembered that the Wigner functions are the solutions of the Schrödinger equation for the symmetric rotator.

The relation between spherical components and Cartesian components of any tensor is easily found, beginning with the definition of the spherical components of a first-rank tensor and then by using the rule of the product of spherical tensors.

The spherical components of a first-rank tensor with Cartesian components T_x, T_y, and T_z are

$$\begin{aligned} T^{(1,1)} &= -(1/\sqrt{2})(T_x + iT_y) \\ T^{(1,0)} &= T_z \\ T^{(1,-1)} &= (1/\sqrt{2})(T_x - iT_y) \end{aligned} \tag{105}$$

Now a spherical tensor of rank l can be constructed from two tensors of ranks l_1 and l_2 according to the multiplication rule

$$T^{(l,m)}(A, B) = \sum_{m_1} C(l_1 l_2 l; m_1, m - m_1)A^{(l_1, m_1)}B^{(l_2, m-m_1)} \tag{106}$$

where $C(l_1 l_2 l; m_1 m_2)$ is a Clebsch–Gordan coefficient (Rose, 1957), which vanishes unless $|l_1 - l_2| \leqq l \leqq |l_1 + l_2|$. Note again that this relation is simply a generalization of the product of two angular momentum eigenfunctions $|l_1 m_1\rangle$ and $|l_2 m_2\rangle$ to give the eigenfunctions of the total angular momentum operator L^2. Numerical tables of the Clebsch–Gordan coefficients are found in the book by Heine (1964). By using Eq. (106) and the tabulated values of the C coefficients, we can build up successively, from the first rank components $T^{(1,m)}$, the spherical components of any rank. In this way we have obtained the irreducible spherical components of rank zero, one, and two corresponding to a general second-rank Cartesian tensor of components $A_i B_j$, and presented in the second column of Table IV.

Spherical tensors of the same rank can be *contracted* into an invariant according to the equation

$$A : B = \sum_m (-1)^m A^{(l,m)} B^{(l,-m)} \tag{107}$$

This corresponds to the scalar product of Cartesian tensors, which for second-rank tensors assumes the form

$$A : B = \sum_{ij} A_{ij} B_{ij} \tag{108}$$

The definitions used so far for Cartesian and spherical tensors can be extended directly to quantum mechanical operators. For example, the components S_x, S_y, and S_z can be regarded as the Cartesian components of a first-rank tensor, and the corresponding spherical components, written in terms of the "shift" operators $S_\pm$, are

$$\begin{aligned} S^{(1,1)} &= -(1/\sqrt{2})(S_x + iS_y) = -(1/\sqrt{2})S_+ \\ S^{(1,0)} &= S_z \\ S^{(1,-1)} &= (1/\sqrt{2})(S_x - iS_y) = (1/\sqrt{2})S_- \end{aligned} \tag{109}$$

Higher rank tensor operators are obtained in the same manner as just described. The third column of Table IV lists the components of rank zero, one, and two derived from the product of two first-rank spherical tensor operators. The two sets of expressions presented in Table IV are actually identical, but the spherical tensor operators are written for convenience in terms of the shift operators $A_\pm$ and $B_\pm$.

As already stated, the Hamiltonian must be invariant under rotation of the reference system and therefore it can be expressed as a sum of scalar products of tensors. In fact, the various terms of the magnetic Hamiltonian,

$$\mathbf{H} \cdot \mathbf{g} \cdot \mathbf{S} = \sum_{ij} H_i g_{ij} S_j, \qquad \mathbf{I} \cdot \mathbf{A} \cdot \mathbf{S} = \sum_{ij} I_i A_{ij} S_j$$

$$\mathbf{S} \cdot \mathbf{D} \cdot \mathbf{S} = \sum_{ij} S_i D_{ij} S_j$$

TABLE IV

GENERAL EXPRESSIONS FOR THE IRREDUCIBLE SPHERICAL COMPONENTS CORRESPONDING TO THE SECOND-RANK CARTESIAN TENSOR OF COMPONENTS $A_i B_j$

l, m	$T^{(l, m)}$	
0, 0	$-(1/\sqrt{3})(A_x B_x + A_y B_y + A_z B_z)$	$-(1/\sqrt{3})[A_z B_z + \frac{1}{2}(A_+ B_- + A_- B_+)]$
1, 1	$-\frac{1}{2}[A_x B_z - A_z B_x + i(A_y B_z - A_z B_y)]$	$-\frac{1}{2}(A_+ B_z - A_z B_+)$
1, 0	$-(i/\sqrt{2})(A_y B_x - A_x B_y)$	$-\frac{1}{2}(1/\sqrt{2})(A_+ B_- - A_- B_+)$
1, −1	$-\frac{1}{2}[A_x B_z - A_z B_x - i(A_y B_z - A_z B_y)]$	$-\frac{1}{2}(A_- B_z - A_z B_-)$
2, 2	$\frac{1}{2}[A_x B_x - A_y B_y + i(A_x B_y + A_y B_x)]$	$\frac{1}{2}A_+ B_+$
2, 1	$-\frac{1}{2}[A_x B_z + A_z B_x + i(A_y B_z + A_z B_y)]$	$-\frac{1}{2}(A_+ B_z + A_z B_+)$
2, 0	$\sqrt{\frac{2}{3}}[A_z B_z - \frac{1}{2}(A_x B_x + A_y B_y)]$	$\sqrt{\frac{2}{3}}[A_z B_z - \frac{1}{4}(A_+ B_- + A_- B_+)]$
2, −1	$\frac{1}{2}[A_x B_z + A_z B_x - i(A_y B_z + A_z B_y)]$	$\frac{1}{2}(A_- B_z + A_z B_-)$
2, 2	$\frac{1}{2}[A_x B_x - A_y B_y - i(A_x B_y + A_y B_x)]$	$\frac{1}{2}A_- B_-$

are the invariants obtained by the contraction of second-rank tensors with components g_{ij} and $H_i S_j$, of A_{ij} and $I_i S_j$, and of D_{ij} and $S_i S_j$, respectively. In the spherical basis, according to Eq. (107), the total Hamiltonian is expressible in the form

$$\mathscr{H} = \sum_{\mu} \sum_{lm} (-1)^m F'^{(l, -m)}_{\mu} A^{(l, m)}_{\mu} \tag{110}$$

where μ specifies any particular interaction (Zeeman, hyperfine, etc.), $F'^{(l, m)}_{\mu}$ are the irreducible spherical components related to the Cartesian tensor $\boldsymbol{\mu}$, and $A^{(l, m)}_{\mu}$ are the corresponding spin operators.

It is logical to have the spin operators expressed in the laboratory coordinate system; thus the spherical tensors in Eq. (110) will be specified in this system. However, the magnetic interaction tensors are naturally defined in the molecular axis system, and so we shall express the $F'^{(l, m)}$ components in terms of the components $F^{(l, m)}$ in the molecular system, according to the transformation law given in Eq. (101).

The magnetic Hamiltonian then takes the form

$$\mathscr{H} = \sum_{\mu} \sum_{l, m, m'} (-1)^{m'} F^{(l, -m')}_{\mu} D^{l}_{m, m'}(\alpha\beta\gamma) A^{(l, m)}_{\mu} \tag{111}$$

The interactions we have considered are represented by second-rank Cartesian tensors; hence the index l can only assume the values 0, 1, and 2. The explicit expressions for $F^{(l, m)}_{\mu}$ corresponding to a particular interaction μ can be derived from the second column of Table IV by setting $\mu_{ij} = A_i B_j$, whereas the most suitable choice for the operators $A^{(l, m)}$ are given by the third column, as already pointed out. From the table we see that all the $F^{(1, m)}_{\mu}$

terms vanish for symmetric tensors, which is our case. In addition, if the molecular axis system in which the magnetic interactions are expressed coincides with the principal axis system of the tensor $\boldsymbol{\mu}$, the off-diagonal elements μ_{ij} are zero and a simpler expression for $F_{\mu}^{(l,m)}$ results. A further simplification occurs if the tensor is axially symmetric because in this case only the $F_{\mu}^{(0,0)}$ and $F_{\mu}^{(2,0)}$ terms are left.

In solution, due to the molecular tumbling, the Wigner functions must be averaged over the probability distribution $P(\alpha\beta\gamma) \sin \beta \, d\alpha \, d\beta \, d\gamma$ of finding the molecular orientation defined within a volume element in the Euler space. For isotropic liquids this distribution is uniform and so all terms with $l \neq 0$ vanish because of the orthogonality properties of the rotation matrix components.

E. Rotational Diffusion Model

The functional form of the distribution functions needed to calculate the correlation function $G(t)$ in Eq. (90) can be obtained starting from a reasonable model for the random process under investigation. Thus the random reorientations of a molecule subjected to Brownian motion are usually described in terms of the rotational diffusion equation introduced by Debye (1945). If we denote the Euler angles that specify the orientation of the molecule relative to a fixed coordinate system by Ω, then the solution of the diffusion equation gives the probability $P(\Omega, t)$ of finding the molecule within the solid angle $d\Omega$ at time t.

In the case of axially symmetric molecules, the diffusion equation is formally analogous to the Schrödinger equation for the symmetric rotator, where the rotational constants are replaced by the principal values $D_{\parallel}$ and $D_{\perp}$ of the diffusion tensor (Freed, 1964). As already mentioned, the eigenfunctions of the symmetric rotator are the Wigner functions $D_{lm}^{j}(\Omega)$, and the corresponding eigenvalues are

$$\lambda_{jm} = -[j(j+1)D_{\perp} + (D_{\parallel} - D_{\perp})m^2] \tag{112}$$

The general solution of the diffusion equation is therefore

$$P(\Omega, t) = \sum_{jlm} c_{jlm}(0) D_{lm}^{j}(\Omega) \exp(\lambda_{jm} t) \tag{113}$$

The conditional probability is obtained by imposing the initial condition $P(\Omega, 0) = \delta(\Omega - \Omega_0)$ and by using the expansion of the Dirac delta function on the complete orthogonal set of the D_{lm}^{j} functions:

$$\delta(\Omega - \Omega_0) = \sum_{jlm} \frac{2j+1}{8\pi^2} D_{lm}^{j*}(\Omega_0) D_{lm}^{j}(\Omega) \tag{114}$$

We obtain therefore for $P(\Omega_0 \mid \Omega t)$

$$P(\Omega_0 \mid \Omega t) = \sum_{jlm} \frac{2j+1}{8\pi^2} D^{j*}_{lm}(\Omega_0) D^j_{lm}(\Omega) \exp(\lambda_{jm} t) \tag{115}$$

According to Eq. (99), the distribution probability $P(\Omega)$ is uniform and equal to $1/8\pi^2$, as expected. As we will see in next subsection, the relevant quantities to be evaluated are the correlation functions of the Wigner matrices, which in Eq. (111) carry out the transformation from the laboratory-fixed to the molecule-fixed coordinate system. We have then, using the orthogonality properties of the Wigner functions,

$$\begin{aligned}\langle D^{j'}_{l'm'}(0) D^{j*}_{lm}(t)\rangle &= \int d\Omega_0 D^{j'}_{l'm'}(\Omega_0) P(\Omega_0) \int d\Omega D^j_{lm}(\Omega) P(\Omega_0 \mid \Omega t) \\ &= \langle D^{j'}_{l'm'}(\Omega_0) D^{j*}_{lm}(\Omega_0)\rangle \exp(\lambda_{jm} t) \\ &= \delta_{jj'}\, \delta_{ll'}\, \delta_{mm'} [1/(2j+1)] \exp(-\mid t \mid / \tau_{jm})\end{aligned} \tag{116}$$

The correlation function decays exponentially with a characteristic time $\tau_{jm} = -\lambda_{jm}^{-1}$. The absolute sign for t is taken to satisfy the conditions expressed by Eq. (95). We can note that for an isotropic molecular diffusion, $D_{\parallel} = D_{\perp}$ and the correlation functions of all the Wigner rotation matrix components corresponding to the same value of j decay with the same characteristic time $\tau_j = [j(j+1)D]^{-1}$, where D is the diffusion constant. The values of the principal components of the diffusion tensor for the case of nonspherical molecules can be calculated from the molecular dimensions according to formulas reported by Freed (1964).

F. Linewidth Parameters

In this subsection we derive the linewidth parameters A, B, and C for a free radical in a liquid medium. We must calculate those elements $R_{\alpha\alpha'\beta\beta'}$ of the relaxation matrix that are involved in the determination of the x component of the magnetization. Equations (74) and (78) show that the indices α, α' and β, β' must correspond to states involved in allowed ESR transitions, the only ones that are connected by the operator S_x. It follows therefore from the condition $E_\alpha - E_{\alpha'} = E_\beta - E_{\beta'}$ that the relaxation matrix will be diagonal if there is no degeneracy in the ESR transitions. The simplest case for which this condition is satisfied is the case of an electron spin interacting with a single nuclear spin, and this will be treated in detail. We shall further assume that the paramagnetic molecules have an axially symmetric diffusion tensor, and the g factor and hyperfine anisotropies are the only (or the major) causes of spin relaxation. In this way we neglect fluctuating solvent interactions or molecular geometry variations, which can modify the instantaneous values of the magnetic parameters. Under these assumptions we have for the perturbing Hamiltonian $\mathscr{H}_1(t)$

$$\mathscr{H}_1(t) = \sum_{\mu} \sum_{m, m'} (-1)^{m'} F_\mu^{(-m')} D^2_{m, m'}(t) A_\mu^{(m)} \tag{117}$$

The correlation function of $\mathscr{H}_1(t)$ is then

$$G_{\alpha\alpha'\beta\beta'}(t) = \hbar^{-2} \sum_{\mu, \nu} \sum_{m, m'} (-1)^{m+m'} F_\mu^{(-m')} F_\nu^{(m')} g_{mm'}(t)$$
$$\times \langle \alpha | A_\mu^{(m)} | \alpha' \rangle \langle \beta | A_\nu^{(-m)} | \beta' \rangle \tag{118}$$

All the relevant interactions are described by second-rank tensors, and so the index 2 has been omitted for simplicity. In the above expression $g_{mm'}(t)$ stands for the correlation function of the rotation matrix component $D^2_{mm'}(\alpha\beta\gamma)$. Cross-correlation terms of the type $\langle D^2_{mm'}(0) D^{2*}_{nn'}(t) \rangle$ vanish according to Eq. (116).

Let $j_{mm'}(\omega)$ be the Fourier transform of the correlation function $g_{mm'}(t)$. From Eq. (116) we have

$$j_{mm'}(\omega) = \frac{1}{2} \int_{-\infty}^{+\infty} \tfrac{1}{5} \exp(-|t|/\tau_m) \exp(-i\omega t)\, dt = \tfrac{1}{5}\tau_m/(1 + \omega^2 \tau_m^2)$$
$$\tau_m = [6D_\perp + (D_\parallel - D_\perp) m^2]^{-1} \tag{119}$$

Finally we write for $J(\omega)$ of Eq. (90)

$$J_{\alpha\alpha'\beta\beta'}(\omega) = \hbar^{-2} \sum_{\mu\nu} \sum_{mm'} (-1)^{m+m'} F_\mu^{(-m')} F_\nu^{(m')} j_{mm'}(\omega)$$
$$\times \langle \alpha | A_\mu^{(m)} | \alpha' \rangle \langle \beta | A_\nu^{(-m)} | \beta' \rangle \tag{120}$$

The matrix elements of the spin operators on the basis of eigenfunctions of $\mathscr{H}_0$ are promptly calculated by standard methods. We note that the contributions to the relaxation matrix $R_{\alpha\alpha'\beta\beta'}$ can be conveniently divided into three groups of terms, which are called secular, pseudosecular, and nonsecular terms. The first come from those parts of $\mathscr{H}_1(t)$ that commute with the unperturbed Hamiltonian $\mathscr{H}_0$, and so have only diagonal matrix elements on the basis of the eigenfunctions of $\mathscr{H}_0$. Spin operators of the type $H_0 S_z$ and $I_z S_z$ satisfy this requirement, and for these terms $\omega_{\alpha\beta} = 0$. Pseudosecular terms result from spin operators, such as $I_\pm S_z$, that connect spin states whose energy difference is the hyperfine coupling constant a, or $\hbar\omega_{\rm hf}$ if $\omega_{\rm hf}$ is the hyperfine splitting in angular frequency units. The nonsecular part of $\mathscr{H}_1(t)$ contains the spin operators $S_\pm$ that have matrix elements between states with energy difference $\hbar\omega_0$, where ω_0 is the resonance frequency. This distinction is very important, because a is typically two orders of magnitude less than $g\beta H_0 = \hbar\omega_0$, and so, except for very rapid motions, it may happen that $\omega_0^2 \tau_m^2 \gg 1$ and $\omega_{hf}^2 \tau_m^2 \ll 1$. In this case we see from Eq. (119) that secular and pseudosecular terms give comparable contributions, and the nonsecular terms can be neglected.

After calculation for all the contributions to the relaxation matrix, the width of the spectral line corresponding to the value M of the nuclear projection quantum number is found to be

$$[T_2(M)]^{-1} = -R_{\alpha\alpha'\alpha\alpha'} = A + BM + CM^2 \tag{121}$$

$$\begin{aligned} A = {} & \tfrac{1}{6}H_0^2 \hbar^{-2} \sum_m (-1)^m F_g^{(m)} F_g^{(-m)} [4j_{0m}(0) + 3j_{1m}(\omega_0)] \\ & + \tfrac{1}{12} I(I+1) \sum_m (-1)^m F_A^{(m)} F_A^{(-m)} [j_{0m}(\omega_0) + 3j_{1m}(\omega_{hf}) + 6j_{2m}(\omega_0)] \\ B = {} & \tfrac{1}{6}H_0 \hbar^{-2} \sum_m (-)^m [F_A^{(m)} F_g^{(-m)} + F_A^{(-m)} F_g^{(m)}][4j_{0m}(0) + 3^j{}_{1m}(\omega_0)] \\ C = {} & \tfrac{1}{6}\hbar^{-2} \sum_m (-1)^m F_A^{(m)} F_A^{(-m)} \{4j_{0m}(0) + 3j_{1m}(\omega_0) \\ & - \tfrac{1}{2}[j_{0m}(\omega_0) + 3j_{1m}(\omega_{hf}) + 6j_{2m}(\omega_0)]\} \end{aligned} \tag{122}$$

The presence of a linear and a quadratic term in the expression for $[T_2(M)]^{-1}$ implies an asymmetric linewidth variation about the center of the spectrum. Examples of this behavior have been seen in the nitroxide free-radical spectra.

For moderately fast correlation time the nonsecular terms can be neglected, and $j_{lm}(\omega_{hf}) \simeq j_{lm}(0)$. In this case, if an approximate spherical symmetry of the diffusion tensor can be assumed, the following simplified expressions result for the linewidth parameters A, B, and C:

$$\begin{aligned} A/\tau_c &= (2/15)\beta_e^2 \hbar^{-2} H_0^2 (g' : g') \\ &\quad + (1/20)\hbar^{-2} I(I+1)(A' : A') \\ B/\tau_c &= (4/15)\beta_e \hbar^{-2} H_0 (g' : A') \\ C/\tau_c &= (1/12)\hbar^{-2}(A' : A') \\ \tau_c &= [6D]^{-1} \end{aligned} \tag{123}$$

where the scalar product of second-rank spherical tensors has been written in terms of the Cartesian tensors according to Eqs. (107) and (108). The expression for B and C used in Section IV.E are promptly derived from Eq. (123) after some simple manipulation.

The complete form of the relaxation matrix for the more general case of an electron spin interacting with several nonequivalent nuclei has been derived by Freed and Fraenkel (1963).

References

Abragam, A. (1961). "The Principles of Nuclear Magnetism." Oxford Univ. Press (Clarendon), London and New York.

Atherton, N. M. (1973). "Electron Spin Resonance." Ellis Horwood Ltd., Chichester, England.

Bersohn, M., and Baird, J. C. (1966). "An Introduction to Electron Paramagnetic Resonance." Benjamin, New York.

Briere, R., Lemaire, H., and Rassat, A. (1965). Nitroxydes XV: Synthèse et étude de radicaux libres stables pipéridiniques et pyrrolidinique. *Bull. Soc. Chim. Fr.* **1965**, 3273–3283.

Brink, D. M., and Satchler, G. R., (1968). "Angular Momentum." Oxford Univ. Press (Clarendon), London and New York.

Carrington, A., and McLachlan, A. D. (1967). "Introduction to Magnetic Resonance." Harper, New York.

Debye, P. (1945). "Polar Molecules." Dover, New York.

Edmonds, A. R. (1957). "Angular Momentum in Quantum Mechanics." Princeton Univ. Press, Princeton, New Jersey.

Freed, J. H. (1964). Anisotropic rotational diffusion and electron spin resonance linewidths, *J. Chem. Phys.* **41**, 2077–2083.

Freed, J. H., and Fraenkel, G. K. (1963). Theory of linewidths in electron spin resonance spectra, *J. Chem. Phys.* **39**, 326–348.

Griffith, O. H., Cornell, D. W., and McConnell, H. M. (1965). Nitrogen hyperfine tensor and g tensor of nitroxide radicals, *J. Chem. Phys.* **43**, 2909–2910.

Hameka, H. F. (1965). "Advanced Quantum Mechanics." Addison-Wesley, Reading, Massachusetts.

Heine, V. (1964). "Group Theory in Quantum Mechanics." Pergamon, Oxford.

Hudson, A., and Luckhurst, G. R. (1969). The electron resonance line shapes of radicals in solution, *Chem. Rev.* **69**, 191–225.

Kivelson, D. (1972). Electron spin relaxation in liquids, *In* "Electron Spin Relaxation in Liquids" (L. T. Muus and P. W. Atkins, eds.), pp. 213–277. Plenum, New York.

Pake, G. E. (1962). "Paramagnetic Resonance." Benjamin, New York.

Pedersen, J. B. (1972). Stochastic Processes, *In* "Electron Spin Relaxation in Liquids" (L. T. Muus and P. W. Atkins, eds.), pp. 25–70. Plenum, New York.

Poole, C. P., Jr., and Farach, H. A. (1972). "The Theory of Magnetic Resonance." Wiley, New York.

Redfield, A. G. (1966). The theory of relaxation processes, *Advan. Magn. Resonance* **1**, 1–32.

Rose, M. E. (1957). "Elementary Theory of Angular Momentum." Wiley, New York.

Slichter, C. P. (1963). "Principles of Magnetic Resonance." Harper, New York.

Van der Waals, J. H., and De Groot, M. S. (1959). Paramagnetic resonance in phosphorescent aromatic hydrocarbons. I: Napthalene, *Mol. Phys.* **2**, 333–340.

Weil, J. A., and Hecht, H. G. (1963). On the powder line shape of EPR spectra, *J. Chem. Phys.* **38**, 281–282.

Wertz, J. E., and Bolton, J. R. (1972). "Electron Spin Resonance." McGraw-Hill, New York.

3

Theory of Slow Tumbling ESR Spectra for Nitroxides

JACK H. FREED

DEPARTMENT OF CHEMISTRY
CORNELL UNIVERSITY
ITHACA, NEW YORK

I. INTRODUCTION

In this chapter we develop the theory for slow tumbling in ESR spectroscopy, with specific application to nitroxide free radical spectra. The slow tumbling region is that range of rotational reorientation times for which the

ESR spectrum can no longer be described as a simple superposition of Lorentzian lines, characteristic of the fast motional or motional narrowing region, for which the theory developed in the previous chapter applies. In this chapter we further take it to be the region for which the ESR spectrum still shows effects of the motion; i.e., the motion is not so slow as to yield a proper rigid-limit spectrum. For nitroxide free radicals, this usually means we are considering the range of rotational correlation times 10^{-9} sec $\leqq \tau_R \leqq 10^{-6}$ sec. This is an important range for nitroxide probes in viscous media or for nitroxide spin labels attached to large macromolecules.

In this region, the spectra are affected in a complicated way by both the motions and the magnetic spin interactions. As a result, a theory which can deal rigorously with describing this region must be both powerful and general. It is our objective to set forth this theory, which has been developed in the last few years, and to emphasize its general foundations. The general theory is presented in Section II.A. It is a characteristic of the slow motional region that it requires that we ask more intricate questions about the detailed nature of the molecular motions in order to properly analyze the spectra than is true for studies in the motional narrowing region. Thus in much of the remainder of Section II the dynamics of molecular motions is developed to an extent concomitant with the need to explain actual slow tumbling experiments.

It has also been our objective to demonstrate how the general theory can be applied to actual cases. Thus in Section III we give a detailed discussion of comparisons between the theoretical predictions and recent experimental studies. Wherever it appears reasonable, we have tried to indicate simplified approaches in the analysis of slow tumbling spectra short of running detailed computer simulations. Nevertheless, there are many cases where the researcher will need detailed simulations tailored to his specific needs. For that reason, we have, in Appendix B, supplied the computer program that is applicable to isotropic liquids.

The development of the theory and the specific examples given here draw most heavily on the recent work of the Cornell group, with which the author is most familiar. Detailed references to other work can be found in a recent review article (Freed, 1972b). Recently, Polnaszek (1975b) has reviewed our methods and compared them with the other approaches. It should be emphasized that the slow tumbling theory is in many ways based on a generalization of stochastic theories of jump models for magnetic resonance (Abragam, 1962; Kubo, 1962, 1969; Johnson, 1965) and we sometimes make use of the analogies when appropriate. A familiarity with motional narrowing theory, such as is discussed in Chapter 2, would be useful preparation for this chapter, and a familiarity with quantum mechanics and some statistical mechanics is assumed in Section II.

II. THEORY

A. General Method

A quantum mechanical wave function Ψ can be expanded in a complete set of orthonormal functions U_n as

$$\Psi(t) = \sum_n c_n(t) U_n \tag{1}$$

The density matrix is defined to be

$$\rho_{nm}(t) = \overline{c_n(t) c_m^*(t)} \tag{2}$$

where the bar indicates an average over a statistical ensemble. A useful property of the density matrix is the calculation of the expectation value of an operator O for a system described by the wave function Ψ. Thus

$$\langle \Psi | O | \Psi \rangle = \sum_{m,n} \langle \Psi | U_m \rangle \langle U_m | O | U_n \rangle \langle U_n | \Psi \rangle = \sum_{m,n} c_n(t) c_m^*(t) O_{mn} \tag{3a}$$

and for an average expectation value of an ensemble of such systems we have

$$\overline{\langle \Psi | O | \Psi \rangle} = \sum_{n,m} \rho_{nm} O_{mn} = \mathrm{Tr}\, \rho O \tag{3b}$$

where Tr is the trace. The trace is invariant to the choice of the complete orthonormal basis set. Thus all the information for calculating observable quantities is contained in the density matrix.

Since the wave function Ψ will vary with time, the coefficients $c_n(t)$ will be functions of time, and this time dependence can be obtained from the time-dependent Schrödinger equation. Then the density matrix equation of motion, assuming the same Hamiltonian $\hat{\mathscr{H}}(t)$ for all members of the ensemble, is given by the quantum mechanical Liouville equation

$$\partial \rho / \partial t = -i[\hat{\mathscr{H}}(t), \rho] \tag{4}$$

Now assume that the time dependence of the spin Hamiltonian $\hat{\mathscr{H}}(t)$ for a free radical arises from interactions with its environment such that $\hat{\mathscr{H}}(t)$ is fully determined by a complete set of random variables Ω. Also assume that this time dependence of Ω is described by a stationary Markov process, so that the probability of being in a state Ω_1 at time t, if in state Ω_2 at time $t - \Delta t$, is (a) independent of the value of Ω at any time earlier than $t - \Delta t$ and (b) depends only on Δt and not on t. A stationary Markov process can be described by a differential equation

$$\partial P(\Omega, t)/\partial t = -\Gamma_\Omega P(\Omega, t) \tag{5}$$

where $P(\Omega, t)$ is the probability of the free radical being in a state Ω at time t.

Since the process is assumed stationary, Γ_Ω is independent of time. The evolution operator Γ_Ω is an operator only on the random variables Ω and is independent of spin space. Γ_Ω includes such general Markov operators as the diffusion operators given by the Fokker–Planck equations and transition rate matrices among discrete states $\Omega_1, \Omega_2, \ldots, \Omega_n$. In this discussion, $\Omega \equiv \alpha, \beta, \gamma$ will represent Euler angles specifying orientation and Γ_Ω will be a rotational diffusion operator.

It is also assumed that the stochastic process has a unique equilibrium distribution $P_0(\Omega)$ characterized by

$$\Gamma_\Omega P_0(\Omega) = 0 \tag{6}$$

We can show (Kubo, 1969; Freed *et al.*, 1971; Freed, 1972a) that Eqs. (4)–(6) lead to the stochastic Liouville equation of motion

$$\partial\rho(\Omega, t)/\partial t = -i[\hat{\mathscr{H}}(\Omega), \rho(\Omega, t)] - \Gamma_\Omega \rho(\Omega, t) \tag{7}$$

where $\rho(\Omega, t)$ is now understood to be the value of ρ associated with a particular value of Ω, hence of $\hat{\mathscr{H}}(\Omega)$. Thus, instead of looking at the explicit time dependence of the spin Hamiltonian $\hat{\mathscr{H}}(t)$ involving the interaction with its environment, the spin Hamiltonian is written in terms of random angle variables Ω and its modulation (due to rotational motions) is expressed by the time dependence of Ω.

The steady-state spectrum in the presence of a single rotating microwave frequency field is determined by the power absorbed from this field. We find for the λth hyperfine line at an orientation specified by Ω

$$P_\lambda(\Omega) = 2N\hbar\omega d_\lambda Z_\lambda^{(1)''}(\Omega) \tag{8}$$

where P_λ is the power absorbed, N is the concentration of electron spins, d_λ is a "transition moment" given by $d_\lambda = \frac{1}{2}\gamma_e H_1 \langle \lambda^- | \hat{S}_- | \lambda^+ \rangle$ (where $\hat{S}_-$ is the electron spin-lowering operator and $\lambda^\pm$ are the $M_S = \pm\frac{1}{2}$ states of the electron spin for the λth transition), and $Z_\lambda^{(1)''}$ is defined by the series of equations

$$(\rho - \rho_0)_\lambda = \chi_\lambda \tag{9}$$

$$\chi_\lambda = \sum_{n=-\infty}^{\infty} [\exp(in\omega t)] Z_\lambda^{(n)} \tag{10}$$

and

$$Z_\lambda^{(n)} = Z_\lambda^{(n)'} + iZ_\lambda^{(n)''} \tag{11}$$

(Actually, the experimentally observed signal is proportional to $Z_\lambda^{(1)''}$ and not to $d_\lambda Z_\lambda^{(1)''}$.)

In Eq. (9), $\rho_0(\Omega)$ is the equilibrium spin density matrix. Equation (8) displays the fact that it is the $n = 1$ harmonic, i.e., the component rotating with the microwave field, that is directly observed. In the case of simple lines we can identify $Z_\lambda^{(1)'}$ and $Z_\lambda^{(1)''}$ with the magnetization components M_x and M_y for the λth line in the rotating frame.

The notation for a matrix element of an operator O is

$$O_{ab} = \langle a | O | b \rangle \tag{12a}$$

$$O_\lambda = \langle \lambda^- | O | \lambda^+ \rangle \tag{12b}$$

$$O_{\lambda\pm} = \langle \lambda^\pm | O | \lambda^\pm \rangle \tag{12c}$$

where a, b are eigenstates and λ^+, λ^- are, respectively, the upper and lower electron spin states between which the λth ESR transition occurs. For a nitroxide there are three allowed ESR transitions.

The total absorption is then obtained as the equilibrium average of Eq. (8) over all Ω. Thus averages are introduced such as

$$\overline{Z_\lambda^{(n)}} = \int d\Omega Z_\lambda^{(n)}(\Omega) P_0(\Omega) \tag{13}$$

so that

$$P_\lambda = 2N\hbar\omega d_\lambda \overline{Z_\lambda^{(1)''}} \tag{14}$$

where d_λ has been taken to be independent of orientation. The total spin Hamiltonian $\hat{\mathscr{H}}(t)$, expressed in angular frequency units, is now separated into three components,

$$\hat{\mathscr{H}}(\Omega) = \hat{\mathscr{H}}_0 + \hat{\mathscr{H}}_1(\Omega) + \hat{\varepsilon}(t) \tag{15}$$

In the high-field approximation

$$\hbar\hat{\mathscr{H}}_0 = g_s\beta_e H_0 \hat{S}_z - \hbar \sum_i \gamma_i \hat{I}_{z_i} H_0 - \hbar\gamma_e \sum_i a_i \hat{S}_z \hat{I}_{z_i} \tag{16}$$

yields the zeroth-order energy levels and transition frequencies (cf. Chapter 2). $\hat{\mathscr{H}}_1(\Omega)$ is the perturbation depending on the orientation angles Ω, and, being a scalar, can be expressed as the scalar product of two tensors. That is, in general, we write $\hat{\mathscr{H}}_1(\Omega)$ as [in the notation of Freed and Fraenkel (1963)]

$$\hat{\mathscr{H}}_1(\Omega) = \sum_{L,\, m,\, m'',\, \mu,\, i} \mathscr{D}^L_{-m,\, m''}(\Omega) F'^{(L,\, m)}_{\mu,\, i} A^{(L,\, m'')}_{\mu,\, i} \tag{17}$$

where the $F'^{(L,\, m)}_{\mu,\, i}$ and $A^{(L,\, m'')}_{\mu,\, i}$ are irreducible tensor components of rank L and component m, with the F' being spatial functions in molecule-fixed coordinates, while A consists only of spin operators quantized in the laboratory axis system. The subscripts μ and i refer to the type of perturbation and to the different nuclei, respectively. The Wigner rotation matrix elements $\mathscr{D}^{(L)}_{-m,m''}(\Omega)$ include the transformation from the molecule-fixed axis system (x', y', z') into the laboratory axis system (x, y, z). We shall be concerned with the **A** and **g** tensors, for which $L = 2$. [It has been found that effects of the ^{14}N quadrupole tensor is negligible (Goldman *et al.*, 1972a; Goldman, 1973).] In addition,

$$\hat{\varepsilon}(t) = \tfrac{1}{2}\gamma_e H_1[\hat{S}_+ \exp(-i\omega t) + \hat{S}_- \exp(i\omega t)] \tag{18}$$

is the interaction of the electron spin with the oscillating magnetic radiation field. [When more than one oscillating field, e.g. ELDOR or ENDOR, is present and/or when field modulation effects are to be explicitly incorporated then Eq. (18) may be appropriately modified to include these effects.]

When we take the $\langle\lambda^-|\ |\lambda^+\rangle$ matrix elements of Eq. (7) and utilize Eqs. (9)–(11), we find the steady-state equation for $Z_\lambda^{(n)}$ to be

$$(n\omega - \omega_\lambda)Z_\lambda^{(n)}(\Omega) + [\mathscr{H}_1(\Omega), Z^{(n)}(\Omega)]_\lambda - i[\Gamma_\Omega Z^{(n)}]_\lambda + d_\lambda[\chi_{\lambda+}^{(n-1)} - \chi_{\lambda-}^{(n-1)}] \\ = q\omega_\lambda d_\lambda \tag{19}$$

[The superscripts to $\chi_{\lambda\pm}$ refer to harmonics in the sense of Eq. (10). The harmonic components of any other oscillating fields present may be introduced in a similar manner.] For reasonable temperatures and typical ESR field strengths we may write $\rho_0 = N'^{-1} - q\mathscr{H}_0$, where N' is the number of spin eigenstates of $\mathscr{H}_0$ and $q = \hbar/N'kT$. Also, $\hbar\omega_\lambda = E_{\lambda+} - E_{\lambda-}$ and the $Z_\lambda^{(n)}(\Omega)$ are spin matrices defined by Eqs. (9)–(11), and the $E_{\lambda\pm}$ are the eigenenergies of $\mathscr{H}_0$ for the $\lambda^\pm$ states.

It is convenient at this point to introduce a "symmetrizing" transformation for the evolution operator. It is not needed for isotropic liquids but becomes useful for anisotropic liquids.
Thus

$$\tilde{\Gamma}_\Omega = P_0^{-1/2}\Gamma_\Omega P_0^{1/2} \tag{20}$$

and similarly

$$\tilde{Z}_\lambda^{(n)}(\Omega) = P_0^{-1/2}[Z_\lambda^{(n)}(\Omega)P_0] \tag{21}$$

This transformation usually renders Γ_Ω Hermitian. Now, Eqs. (6) and (20) combine to give

$$\tilde{\Gamma}_\Omega P_0^{1/2} = 0 \tag{22}$$

Equations (21) and (13) yield

$$\overline{Z_\lambda^{(n)}} = \int d\Omega P_0^{1/2}\tilde{Z}_\lambda^{(n)}(\Omega) \tag{23}$$

and Eq. (19) with Eqs. (20)–(21) becomes

$$(n\omega - \omega_\lambda)\tilde{Z}_\lambda^{(n)}(\Omega) + [\hat{\mathscr{H}}_1(\Omega), \tilde{Z}^{(n)}(\Omega)]_\lambda - i[\tilde{\Gamma}_\Omega Z^{(n)}]_\lambda + d_\lambda[\tilde{\chi}_{\lambda+}^{(n-1)} - \tilde{\chi}_{\lambda-}^{(n-1)}]$$
$$= q\omega_\lambda d_\lambda P_0^{1/2} \tag{24}$$

In order to solve for the absorption, Eq. (23), we first solve the diffusion equation (5). As in quantum mechanics, the solution of such a partial differential equation can be expressed in terms of a complete orthonormal set of eigenfunctions, call them $G_m(\Omega)$, such that

$$\tilde{\Gamma}_\Omega G_m(\Omega) = \tau_m^{-1} G_m(\Omega) \tag{25}$$

where τ_m^{-1} is the mth "eigenvalue." We generally find

$$G_0(\Omega) = P_0^{1/2} \tag{25a}$$

Then we expand matrix elements of $\tilde{Z}_\lambda^{(n)}(\Omega)$ in the complete orthonormal set $G_m(\Omega)$:

$$\tilde{Z}_\lambda^{(n)}(\Omega) = \sum_m [C_m^{(n)}(\omega)]_\lambda G_m(\Omega) \tag{26}$$

where the coefficient $C_m^{(n)}(\omega)$ is an operator in spin space and is a function of ω, but is independent of Ω.

Substituting Eq. (26) into Eq. (24), premultiplying the resulting equation by $G_m^*(\Omega)$, and then integrating over Ω and taking advantage of the orthonormal properties of $G_m(\Omega)$, we obtain

$$[(n\omega - \omega_\lambda) - i\tau_m^{-1}][C_m^{(n)}]_\lambda + \sum_{m'} \int d\Omega G_m^*(\Omega)[\mathscr{H}_1(\Omega), C_{m'}^{(n)}]_\lambda C_{m'}(\Omega)$$
$$+ d_\lambda([C_m^{(n-1)}]_{\lambda+} - [C_m^{(n-1)}]_{\lambda-}) = q\omega_\lambda d_\lambda \, \delta(m, 0)\, \delta(n, 1) \tag{27}$$

Since the absorption, Eq. (14), depends only on $Z_\lambda^{(1)}$, then solving Eq. (27) for $[C_0^{(1)}]_\lambda$ for all allowed transitions will give the spectral line shapes.

In the absence of saturation we can set $d_\lambda = 0$ on the left-hand side of Eq. (27) and then let $n = 1$, to obtain the needed expression.

A rotationally invariant Lorentzian linewidth T_2^{-1} can be included in Eq. (19) or Eq. (27) by letting

$$\omega_\lambda \rightarrow \omega_\lambda + iT_2^{-1} \tag{28}$$

In the near-rigid limit, this linewidth corresponds to the linewidth in a single crystal, or more precisely to a residual linewidth in a powder spectrum. More generally, an angular variation of the width can be introduced. For simplicity, this variation is allowed to take the form

$$T_2^{-1}(\theta) = \alpha + \beta \cos^2 \theta \tag{29}$$

where θ is the angle between the magnetic field axis and the molecular z axis.

The above equations yield coupled complex algebraic equations for the coefficients $[C_m^{(n)}]_\lambda$, and we attempt to solve for these equations utilizing only a finite number of coefficients. (The complete orthonormal set includes an infinite number of such eigenfunctions.) The convergence depends essentially on the ratio $|\mathscr{H}_1(\Omega)|/\tau_m^{-1}$. The larger the value of this ratio of off-diagonal to diagonal terms, the more terms $[C_m^{(n)}]_\lambda$ that are needed. The results obtained by relaxation theory valid in the fast motional limit (cf. Chapter 2) are recovered when only one order beyond $[C_0^{(1)}]_\lambda$ is included.

B. Rotational Modulation in Isotropic Liquids

When the general method of the previous section is applied to rotational modulation, Ω refers to the Euler angles for a tumbling molecular axis with respect to a fixed laboratory axis system. For a molecule undergoing many collisions, causing small random angular reorientations, the resulting isotropic Brownian rotational motion is a Markov process, which can be described by the rotational diffusion equation (Freed *et al.*, 1971; Freed, 1972a,b)

$$\partial P(\Omega, t)/\partial t = R\, \nabla_\Omega^2 P(\Omega, t) \tag{30}$$

where ∇_Ω^2 is the Laplacian operator on the surface of a unit sphere and R is the rotational diffusion coefficient. If the molecule is approximated by a rigid sphere of radius a rotating in a medium of viscosity η, then a rotational Stokes–Einstein relationship yields

$$R = kT/8\pi a^3 \eta \tag{31}$$

In an isotropic liquid, the equilibrium probability $P_0(\Omega)$ of Eq. (6) will be equal for all orientations, so that $P_0(\Omega) = 1/8\pi^2$. Here the Markov operator Γ_Ω, Eq. (5), for isotropic Brownian rotation, Eq. (30), is $-R\nabla_\Omega^2$, which is formally the Hamiltonian for a spherical top whose orthonormal eigenfunctions are the normalized Wigner rotation matrices or generalized spherical harmonics:

$$G_m \to \phi_{KM}^L(\Omega) = [(2L + 1)/8\pi^2]^{1/2} \mathscr{D}_{K,M}^L(\Omega) \tag{32}$$

with eigenvalues $RL(L + 1)$ (Freed, 1964, 1972a). Note that for $K = 0$, $\mathscr{D}_{0M}^L(\alpha, \beta, \gamma) = [4\pi/(2L + 1)]^{1/2} Y_{LM}(\beta, \gamma)$, where Y_{LM} is the well-known spherical harmonic (Rose, 1954; Edmonds, 1957).

Similarly, the Markov operator for axially symmetric Brownian rotation about a molecule-fixed z axis is formally the Hamiltonian for a symmetric top whose symmetry axis is the z axis. The orthonormal eigenfunctions are again the normalized Wigner rotation matrices with eigenvalues given by

$$\Gamma_\Omega \phi^L_{KM} = [R_\perp L(L+1) + (R_\parallel - R_\perp)K^2]\phi^L_{KM} \tag{33}$$

where $R_\perp$ and $R_\parallel$ are the rotational diffusion constants about the x, y axes and z axis, respectively (Freed, 1964; Favro, 1965). The "quantum numbers" K and M of the Wigner rotation matrices refer to projections along the body-fixed symmetry axis and along a space-fixed axis, respectively.

For completely asymmetric Brownian rotation (Freed, 1964, 1972a; Favro, 1965) $R_x \neq R_y \neq R_z$, where R_x, R_y, and R_z are the respective rotational diffusion constants about the x, y, and z axes, the Markov operator Γ_Ω has more complex solutions (see below). Also, for fast rotation about one axis (e.g., the z axis, so that $R_z \gg R_x, R_y$), the completely asymmetric rotation can be treated as axially symmetric with eigenvalues $R_+ L(L+1) + (R_z - R_+)K^2$, where $R_\pm = \frac{1}{2}(R_x \pm R_y)$.

In the fast rotational motion region, all rotational reorientation processes yield the same ESR line shapes, which, as predicted by earlier relaxation theories, yields Lorentzian lines (Abragam, 1961; Freed and Fraenkel, 1963). Naturally, in the very slow rotational motion region all models should tend to the rigid-limit powder line shape of the equilibrium distribution. However, in the intermediate slow rotational region $|\mathscr{H}_1(t)|\tau_R \gtrsim 1$, the ESR line shapes are found to be sensitive to the details of the molecular reorientation process.

A number of different models for rotational reorientation can be proposed. Some useful ones are: (a) Brownian rotational diffusion; (b) free diffusion in which a molecule rotates freely for time τ (i.e., inertial motion with $\tau = I/B$, with I the moment of inertia and B the friction coefficient) and then reorients instantaneously; and (c) jump diffusion in which a molecule has a fixed orientation for time τ and then "jumps" instantaneously to a new orientation (Goldman *et al.*, 1972; Egelstaff, 1970). For isotropic reorientation, we can summarize the results for these models as

(a)
$$\tau_L^{-1} = L(L+1)R \tag{34}$$

(b)
$$\tau_L^{-1} = L(L+1)R/[1 + L(L+1)R\tau]^{1/2} \tag{35}$$

and

(c)
$$\tau_L^{-1} = \tau^{-1}\left\{1 - (2L+1)^{-1}\int_0^\pi d\varepsilon W(\varepsilon)[\sin(L+\tfrac{1}{2})\varepsilon/\sin(\tfrac{1}{2}\varepsilon)]\right\} \tag{36}$$

where $W(\varepsilon)$ is the distribution function for diffusive steps by angle ε and is normalized so that

$$\int_0^\pi W(\varepsilon)\, d\varepsilon = 1 \tag{37}$$

One convenient form for $W(\varepsilon)$ is

$$W(\varepsilon) = A \sin(\tfrac{1}{2}\varepsilon) \exp(-\varepsilon/\theta) \tag{38}$$

where A is a normalization constant. For $\theta \ll \pi$, we obtain

$$\tau_L^{-1} = L(L + 1)R/[1 + R\tau L(L + 1)] \tag{39}$$

and

$$(\varepsilon^2)_{avg} = 6\theta^2 \tag{40}$$

where $R\tau$ is proportional to the size of a mean diffusive step when the diffusion coefficient is defined as

$$R = (\varepsilon^2)_{avg}/6\tau \tag{41}$$

Other possible choices for $W(\varepsilon)$ are summarized elsewhere (Goldman *et al.*, 1972). A good formal theoretical discussion of jump diffusion as has been employed here is given by Cukier and Lakatos-Lindenberg (1972); also the basic work of Ivanov on jump diffusion has recently been reviewed by Valiev and Ivanov (1973).

It should be noted that Eq. (35) is only an approximate expression for free diffusion. Since free diffusion includes inertial effects, the orientation of the molecule is not properly described as a simple Markovian process. A more accurate treatment of free diffusion must include angular momentum as well as orientational degrees of freedom, but recent work by Bruno and Freed (1974b) shows that the results for the more complete formulation of free diffusion are similar to those obtained using the simple model.

A comparison of Eqs. (34)–(36) shows that the L dependence of τ_L depends on the choice of reorientational model. Thus in the slow motional region, where the line shape is simulated in terms of an expansion in $\mathscr{D}^L_{K,M}$ with eigenvalues $\tau_{L,K}^{-1}$, the ESR spectra will be model sensitive. We can summarize these equations with the simple expression

$$\tau_L^{-1} = B_L L(L + 1)R \tag{42}$$

with the "model parameter" $B_L = 1$ for Brownian motion; $B_L = [1 + L(L + 1)]^{-1}$ for strong jump diffusion with $R\tau = 1$; and $B_L = [1 + L(L + 1)]^{-1/2}$ for free diffusion and $R\tau = 1$. For purposes of comparison, the definition of τ_R is generalized to

$$\tau_R \equiv (6B_2R)^{-1} \tag{43}$$

where B_2 is the appropriate model parameter for $L = 2$. Thus in the motionally narrowed region, where only the $L = 2$ term is important, it follows from Eqs. (42) and (43) that all models yield the same Lorentzian width for the same value of τ_R.

For anisotropic rotational diffusion, there are no convenient solutions for the jump and free diffusion models. The most straightforward course then is to generalize the equations for spherically symmetric rotation. Thus we let

$$\tau_{L,K}^{-1} = (B_L/R_\perp)[R_\perp L(L+1) + (R_\parallel - R_\perp)K^2] \tag{44}$$

where B_L is the same model parameter as in Eq. (42). In effect, it is assumed in Eq. (44) that although the L dependence of $\tau_{L,K}^{-1}$ is model dependent, the "quantum number" K plays the same role in all models (Goldman *et al.*, 1972a). Other interpretations of the model dependence in terms of more fundamental analyses of microscopic molecular dynamics have been discussed by Hwang *et al.* (1975).

C. Anisotropic Liquids

Suppose now that the liquid has a preferred axis of orientation, i.e., the director axis (cf. Chapter 8). We now write the perturbing Hamiltonian (17) as (Polnaszek *et al.*, 1973) [with the $(-1)^K$ which was included into the $F_{\mu,i}^{\prime(L,K)}$ of Eq. (17) now explicitly displayed]:

$$\mathscr{H}_1(\Omega, \Psi) = \sum_{\substack{L,M,M' \\ K,\mu,i}} (-1)^K \mathscr{D}_{-KM'}^L(\Omega) \mathscr{D}_{M'M}^L(\Psi) F_{\mu,i}^{\prime(L,K)} A_{\mu,i}^{(L,M)} \tag{45}$$

Equation (45) is based on two sets of rotations of the coordinate systems: first from the molecular axis system (x', y', z') into the director axis system (x'', y'', z'') with Euler angles $\Omega = (\alpha\beta\gamma)$; and then into the laboratory axis system (x, y, z) with Euler angles Ψ. The orientation of the director relative to the laboratory frame can be specified by the two polar angles θ' and ϕ' such that $\Psi = (0, \theta', \phi')$. More precisely, one means by the molecular coordinate system (x', y', z') the principal axis system for the orientation of the molecule in the mesophase. It may also be necessary to transform from the principal axis system of the magnetic interactions (x''', y''', z''') to the (x', y', z') system with Euler angles Θ, according to

$$F_{\mu,i}^{\prime(L,K)} = \sum_{K'} \mathscr{D}_{K,K'}^L(\Theta) F_{\mu,i}^{\prime\prime\prime(L,K')} \tag{46}$$

where $\Theta = (\alpha', \beta', \gamma')$.

The diffusion equation for a particle undergoing Brownian rotational diffusion in the presence of a potential V is given by (Favro, 1965; Polnaszek and Freed, 1975; Polnaszek, 1975a)

$$\frac{\partial P(\Omega, t)}{\partial t} = -\mathscr{M} \cdot \left[\mathbf{R} \cdot \frac{\mathscr{M} V(\Omega)}{kT} + \mathbf{R} \cdot \mathscr{M}\right] P(\Omega, t) \equiv -\Gamma_\Omega P(\Omega, t) \tag{47}$$

where $V(\Omega)$ can be taken to be the orienting pseudopotential for a liquid crystal, $\mathscr{M}$ is the vector operator, which generates an infinitesimal rotation, and is identified with the quantum mechanical angular momentum operator for a rigid rotator, and $\mathbf{R}$ is the diffusion tensor of the molecule. Both $\mathbf{R}$ and $\mathscr{M}$ are defined in the (x', y', z') molecular coordinate system. The angular momentum operator $\mathscr{M}$ is defined by

$$\mathscr{M}^2 \phi_{KM}^L(\Omega) = L(L+1)\phi_{KM}^L(\Omega) \tag{48a}$$

$$\mathscr{M}_{\pm} \phi_{KM}^L(\Omega) = [(L \mp K)(L \pm K + 1)]^{1/2} \phi_{K\pm 1, M}^L(\Omega) \tag{48b}$$

$$\mathscr{M}_{z'} \phi_{KM}^L(\Omega) = K\phi_{KM}^L(\Omega) \tag{48c}$$

where the $\phi_{KM}^L(\Omega)$ are the eigenfunctions of $\mathscr{M}^2$ and $\mathscr{M}_{z'}$, given by Eq. (32), and

$$\mathscr{M}_{\pm} = \mathscr{M}_{x'} \pm i\mathscr{M}_{y'} \tag{49}$$

When $V = 0$, Eq. (47) is simply the equation for (asymmetric) Brownian rotational diffusion in isotropic liquids. [Simpler expressions have been given for special cases by Nordio and Busolin (1971), Nordio *et al.* (1972), and Polnaszek *et al.* (1973).] Equation (47) is based on the assumption that the external torque $\mathbf{T}$ is derived from the potential $V(\Omega)$:

$$\mathscr{T} = -i\mathscr{M}V(\Omega) \tag{50}$$

[Cf. Eq. (30), where we have set $\nabla_\Omega^2 = -\mathscr{M}^2$.] The equilibrium solution to Eq. (47) is given by

$$P_0(\Omega) = \frac{\exp[-V(\Omega)/kT]}{\int d\Omega \exp[-V(\Omega)/kT]} \tag{51}$$

When the symmetrized forms of Eqs. (20)–(22) are used, we obtain the diffusion equation

$$\partial \tilde{P}(\Omega, t)/\partial t = -\tilde{\Gamma}_\Omega \tilde{P}(\Omega, t) \tag{52}$$

where

$$\tilde{\Gamma}_\Omega = \mathscr{M} \cdot \mathbf{R} \cdot \mathscr{M} + \frac{(\mathscr{M} \cdot \mathbf{R} \cdot V)}{2kT} + \frac{\mathscr{T} \cdot \mathbf{R} \cdot \mathscr{T}}{(2kT)^2} \tag{53}$$

The restoring potential for liquid crystals can be written in its most general form as

$$V(\Omega) = \sum_{L, K, M} \varepsilon_{KM}^L \mathscr{D}_{KM}^L(\alpha, \beta, \gamma) \tag{54}$$

The assumption of cylindrical symmetry about the director axis $\mathbf{n}$ implies that all averages taken over the angle γ vanish unless $M = 0$ (Glarum and

Marshall, 1966, 1967). The uniaxial property of nematic liquid crystals (i.e., $\mathbf{n} \equiv -\mathbf{n}$) implies that L must be even. It is useful to use the linear combinations of the $\mathscr{D}^L_{KM}$ that are of definite parity, i.e., the real linear combinations:

$$V(\Omega) = \sum_{\text{even } L}^{\infty} \left(\varepsilon_0^L \mathscr{D}^L_{00}(\Omega) + \sum_{K>0}^{L} \varepsilon^L_{K\pm}[\mathscr{D}^L_{K0}(\Omega) \pm \mathscr{D}^L_{-K0}(\Omega)] \right) \tag{55}$$

These have simpler properties for molecular symmetries less than cylindrical.

Usually we consider only the leading term $\varepsilon_0^2 \mathscr{D}^2_{00}(\Omega)$, i.e., the Meier–Saupe (1958) potential. The cylindrically symmetric case when $\varepsilon_0^4 \neq 0$ has also been considered (Polnaszek *et al.*, 1973) and it was shown that typical ESR spectral predictions are not very sensitive to having $\varepsilon_0^4 \neq 0$. In general, however, we expect the terms for $L > 2$ to be less important than those for $L = 2$, and we can approximate

$$V(\Omega) \simeq \varepsilon_0^2 \mathscr{D}^2_{00}(\Omega) + \sum_{K>0}^{2} \varepsilon^2_{K\pm}[\mathscr{D}^2_{K0}(\Omega) \pm \mathscr{D}^2_{-K0}(\Omega)] \tag{56}$$

The ε_0^2 and $\varepsilon^2_{K\pm} = \varepsilon^2_K \pm \varepsilon^2_{-K}$ (with the upper sign for $K > 0$, and the lower sign for $K < 0$) are themselves second-rank irreducible tensor components, so that, in the principal axis of molecular orientation system (x', y', z') their Cartesian components ε^2_{ij} are diagonalized, with $\mathrm{Tr}_i\, \varepsilon^2_{ii} = 0$, and complete specification is given by just ε_0^2 and ε^2_{2+}. [Equation (56) can be thought of as the scalar product of second-rank irreducible tensors.] The ordering tensor is defined by

$$\langle \mathscr{D}^L_{KM}(\Omega) \rangle = \int d\Omega P_0(\Omega) \mathscr{D}^L_{KM}(\Omega) \tag{57}$$

where $L = 2$ and $M = 0$. It is also a second-rank irreducible tensor whose symmetry properties are related to those of the $\varepsilon^2_{K\pm}$. Thus from Eqs. (51) and (56) and the orthogonality of the $\mathscr{D}^L_{KM}(\Omega)$ terms it follows that in the (x', y', z') system only $\langle \mathscr{D}^2_{00}(\Omega) \rangle$ and $\langle \mathscr{D}^2_{20} + \mathscr{D}^2_{-20} \rangle$ are nonzero, i.e., $\langle \mathscr{D}^2_{KM}(\Omega) \rangle$ is also diagonalized. Thus the "diagonalized" potential (retaining only $L = 2$ terms) becomes

$$V(\Omega) = \varepsilon_0^2 \mathscr{D}^2_{00}(\Omega) + \varepsilon^2_{2+}[\mathscr{D}^2_{20}(\Omega) + \mathscr{D}^2_{-20}(\Omega)] \tag{58}$$

or equivalently

$$V(\alpha, \beta) = \gamma_2 \cos^2 \beta + \varepsilon \sin^2 \beta \cos 2\alpha \tag{58$'$}$$

where $\varepsilon_0^2 = 2\gamma_2/3$ and $\varepsilon^2_{2+} = 2\varepsilon(6)^{-1/2}$. For molecules in which the molecular x' and y' axes are aligned to different extents, ε is nonzero. If we choose the orientation coordinate system such that the z' axis tends to align to a greater degree either parallel or perpendicular to the director than does the x' axis or the y' axis, we have $|\gamma_2| > |\varepsilon|$. The case $\varepsilon < 0$ corresponds to the y' axis being ordered preferential to the x' axis along the direction of **n** and/or to the x' axis being ordered to a greater degree perpendicular to the **n** than is the y' axis.

We can utilize Eqs. (50), (48b) and (48c) to obtain $\mathscr{T}$ from the potential in Eq. (58) in terms of its components in the (x', y', z') coordinate system:

$$\mathscr{T}_{\pm} \equiv \mathscr{T}_{x'} \pm i\mathscr{T}_{y'} = \pm i(\sin 2\beta)(\varepsilon e^{\mp i\alpha} - \gamma_2 e^{\pm i\alpha}) \tag{59a}$$

$$\mathscr{T}_{z'} = -2\varepsilon \sin^2 \beta \sin 2\alpha \tag{59b}$$

We assume axially symmetric rotation about z' such that $R_{x'x'} = R_{y'y'} = R_{\perp}$ and $R_{z'z'} = R_{\parallel}$. Further, we introduce the definitions

$$\lambda \equiv -\gamma_2/kT \tag{60a}$$

and

$$\rho \equiv -\varepsilon/kT \tag{60b}$$

Then the symmetrized Markov operator defined in Eq. (52) becomes (Polnaszek and Freed, 1975; Polnaszek, 1975a)

$$\tilde{\Gamma} = \mathscr{M} \cdot \mathbf{R} \cdot \mathscr{M} - f(R_{\perp}, R_{\parallel}, \lambda, \rho, \Omega) \tag{61a}$$

where

$$f(R_{\perp}, R_{\parallel}, \lambda, \rho, \Omega) = \sum_{L=0,2,4} [X_{00}^{L} \mathscr{D}_{00}^{L} + \sum_{0<K\leq L} X_{K0}^{L}(\mathscr{D}_{K0}^{L} + \mathscr{D}_{-K0}^{L})] \tag{61b}$$

with

$$X_{00}^{0} = -(2/15)[R_{\perp}(\lambda^2 + \rho^2) + 2R_{\parallel}\rho^2] \tag{62a}$$

$$X_{00}^{2} = 2\{R_{\perp}\lambda - [R_{\perp}(\lambda^2 + \rho^2) - 4R_{\parallel}\rho^2]/21\} \tag{62b}$$

$$X_{00}^{4} = 4[2R_{\perp}(\lambda^2 + \rho^2) - R_{\parallel}\rho^2]/35 \tag{62c}$$

$$X_{20}^{2} = 6^{1/2}\rho[\tfrac{1}{3}(R_{\perp} + 2R_{\parallel}) + \tfrac{2}{7}R_{\perp}\lambda] \tag{62d}$$

$$X_{20}^{4} = 4(10)^{1/2}R_{\perp}\rho\lambda/35 \tag{62e}$$

$$X_{40}^{4} = (8/35)^{1/2}R_{\parallel}\rho^2 \tag{62f}$$

and $\mathscr{M} \cdot \mathbf{R} \cdot \mathscr{M}$ is just the Γ_{Ω} of Eq. (33) with its associated eigenvalues.

We can also write the diffusion equation in terms of the general angular momentum operator $\mathscr{N}$ referred to the director frame (Favro, 1965). This is appropriate when **R** is diagonal in this frame, i.e., one has anisotropic viscosity. We then generate an analogous set of expressions (Polnaszek, 1975a;

Polnaszek and Freed, 1975). It is now assumed that $R_{xx} = R_{yy} = \hat{R}_{\perp}$, $R_{zz} = \hat{R}_{\parallel}$. The result is

$$\Gamma_{\Omega} = \mathcal{N} \cdot \mathbf{R}_n \cdot \mathcal{N} - \hat{R}_{\perp} \hat{f}(\lambda, \rho, \Omega) \tag{63}$$

where

$$\hat{f}(\lambda, \rho, \Omega) = \sum_{L=0,2,4} \left[\hat{X}_{00}^{L} \mathscr{D}_{00}^{L}(\Omega) + \sum_{0<K\leq L} \hat{X}_{K0}^{L}(\mathscr{D}_{K0}^{L}(\Omega) + \mathscr{D}_{-K0}^{L}(\Omega)) \right] \tag{64}$$

with

$$\hat{X}_{00}^{0} = -2(\lambda^2 + 3\rho^2)/15 \tag{65a}$$

$$\hat{X}_{00}^{2} = 2[\lambda - (\lambda^2 - 3\rho^2)/21] \tag{65b}$$

$$\hat{X}_{00}^{4} = 4(2\lambda^2 + \rho^2)/35 \tag{65c}$$

$$\hat{X}_{20}^{2} = 6^{1/2}\rho[1 + (2\lambda/7)] \tag{65d}$$

$$\hat{X}_{20}^{4} = 4(10)^{1/2}\lambda\rho/35 \tag{65e}$$

$$\hat{X}_{40}^{4} = (8/35)^{1/2}\rho^2 \tag{65f}$$

and

$$\mathcal{N} \cdot \mathbf{R} \cdot \mathcal{N} \cdot \phi_{KM}^{L} = [\hat{R}_{\perp} L(L+1) + (\hat{R}_{\parallel} - \hat{R}_{\perp})M^2]\phi_{KM}^{L} \tag{66}$$

It is shown in Appendix A how the slow tumbling equations for isotropic liquids can be simply modified to deal with anisotropic liquids in the simple case of a Meier–Saupe potential and with $\Psi = (0, 0, 0)$ and $\Theta = (0, 0, 0)$ (except that permutation of the labeling of the molecular axis system is permitted). The more complex expressions for the general cases are given by Polnaszek (1975a).

Finally, note that when $|\lambda| \gg 1$, and the eigenfunction expansion in $\Phi_{KM}^{L}(\Omega)$ is only slowly convergent for solutions of the diffusion equation, then there are other types of eigenfunction expansions, specifically tailored to these limiting cases, which become very useful (Polnaszek *et al.*, 1973).

D. Exchange and Slow Tumbling

In cases where the concentration of nitroxide spin probes is high, we also have to consider the effects of Heisenberg spin exchange. This phenomenon involves bimolecular collisions of radicals during which time an exchange integral J' is turned on, and, because of the Pauli principle, can be written as an added term in the spin Hamiltonian:

$$\mathscr{H}_{SS} = J(t)\mathbf{S}_1 \cdot \mathbf{S}_2 \tag{67}$$

where $J = 2J'$ and is time dependent due to the relative motion of the radical pairs. Its effect (viewed from an ESR point of view) is to cause the electron spins to exchange their nuclear environments.

A rigorous analysis of exchange (cf. Freed, 1967; Eastman *et al.*, 1969) shows that we can add to the left side of Eq. (19) for the allowed transitions λ, the terms:

$$-i\omega_{SS}\left(1 - \frac{2D_\lambda}{N'}\right)Z_\lambda + i\omega_{SS}\sum_{\eta \neq \lambda}\frac{2D_\eta}{N'}Z_\eta \tag{68}$$

where D_λ and D_η are the degeneracies of the λth and ηth allowed ESR transitions, respectively. That is, for nitroxides $D_\lambda = D_\eta = 1$ and $N' = 6$. The sum in Eq. (68) is over all allowed transitions η not equal to λ. Also, the effective exchange frequency ω_{SS} obeys

$$\omega_{SS} = \tau_2^{-1}[J^2\tau_1^2/(1 + J^2\tau_1^2)] \tag{69}$$

where τ_2 is the mean time between successive new bimolecular encounters of radicals, and τ_1 is the lifetime of the interacting pair. In the case of simple Brownian diffusion of uncharged radicals in solution we can write (Eastman *et al.*, 1969; Pedersen and Freed, 1973a, b):

$$\tau_2^{-1} = 4\pi dDN \tag{70}$$

$$\tau_1^{-1} = D/d\Delta r_J \tag{71}$$

where N is the density of radicals, d is the "interaction distance" for $J(r)$ which is nonzero (and equal to J) only in the range of $d < r < d + \Delta r_J$, and the diffusion coefficient in a Stokes–Einstein model is

$$D = kT/6\pi a\eta \tag{72}$$

Expression (68) is only appropriate for the allowed transitions, which then couple together by this mechanism. For each forbidden ESR transition (see below), we have instead of (68) to add the term to the left side of Eq. (19)

$$-i\omega_{SS}Z_\kappa \tag{73}$$

If we make the simplifying assumption that ω_{SS} is independent of any orientational effects [i.e., $J(r)$ is taken as a function only of the relative internuclear separation of the radical pair], then (68) and (73) can be added to the LHS of Eq. (27), where we let $Z_\lambda(\Omega) \rightarrow [C_m^{(n)}]_\lambda$, etc.

Expressions (69) and (70) are based on a simple contact-exchange model (Eastman *et al.*, 1969). More complex models of the motional modulation of $J(\mathbf{r}_1, \mathbf{r}_2)$ can also be dealt with (Pedersen and Freed, 1973a,b). Effects of radical charge and/or electrolyte concentration on Eqs. (70) and (71) must also be considered (Eastman *et al.*, 1970). Expressions (68) and (73) are also useful for two-dimensional motions, but Eqs. (69)–(72) would have to be modified. Also, in general, effects of intermolecular electron–spin dipolar interactions become important at higher concentrations as the motions slow down (Eastman *et al.*, 1969).

E. Nitroxides

For nitroxides, there are three allowed ESR transitions and six forbidden transitions which must be considered in a rigorous solution of Eq. (27). They are illustrated in Fig. 1a. The asymmetric g tensor $\mathbf{g}$ and the hyperfine tensor $\mathbf{A}$ yield an $\hat{\mathscr{H}}_1(\Omega)$ given by

$$\begin{aligned}\hat{\mathscr{H}}_1(\Omega) = {} & \mathscr{D}^2_{0,0}(\Omega)[F_0 + D'\hat{I}_z]\hat{S}_z + [\mathscr{D}^2_{-2,0}(\Omega) + \mathscr{D}^2_{2,0}(\Omega)] \\ & \times (F_2 + D^{(2)\prime}\hat{I}_z)\hat{S}_z + [\mathscr{D}^2_{0,1}(\Omega)\hat{I}_+ - \mathscr{D}^2_{0,-1}(\Omega)\hat{I}_-]D\hat{S}_z \\ & + [\mathscr{D}^2_{-2,1}(\Omega) + \mathscr{D}^2_{2,1}(\Omega)]D^{(2)}\hat{I}_+\hat{S}_z \\ & - [\mathscr{D}^2_{-2,-1}(\Omega) + \mathscr{D}^2_{2,-1}(\Omega)]D^{(2)}\hat{I}_-\hat{S}_z \end{aligned} \tag{74}$$

where

$$F_i = \sqrt{\tfrac{2}{3}}\, g^{(i)}\hbar^{-1}\beta_e H_0 \tag{75a}$$

$$g^{(0)} = 6^{-1/2}[2g_{z'} - (g_{x'} + g_{y'})] \tag{75b}$$

$$g^{(2)} = \tfrac{1}{2}[g_{x'} - g_{y'}] \tag{75c}$$

$$D = (|\gamma_e|/2\sqrt{6})(A_{x'} + A_{y'} - 2A_{z'}) \tag{76a}$$

$$D^{(2)} = \tfrac{1}{4}|\gamma_e|(A_{y'} - A_{x'}) \tag{76b}$$

with $D' = -(8/3)^{1/2}D$ and $D^{(2)\prime} = -(8/3)^{1/2}D^{(2)}$. The forbidden transitions 4–9 of Fig. 1a are coupled into the expressions for the allowed transitions because of the pseudosecular terms in Eq. (74), i.e., the terms involving $\hat{I}_\pm \hat{S}_z$ (where $\hat{I}_\pm$ are the nuclear spin raising and lowering operators). The Euler angles $\Omega \equiv \alpha, \beta, \gamma$ define the rotation between the laboratory coordinate system (x, y, z) and the principal axis system in the molecular frame in which the g and hyperfine tensors are diagonal. We assume they are diagonal in the same axis system [which is rigorously the x''', y''', z''' system, but

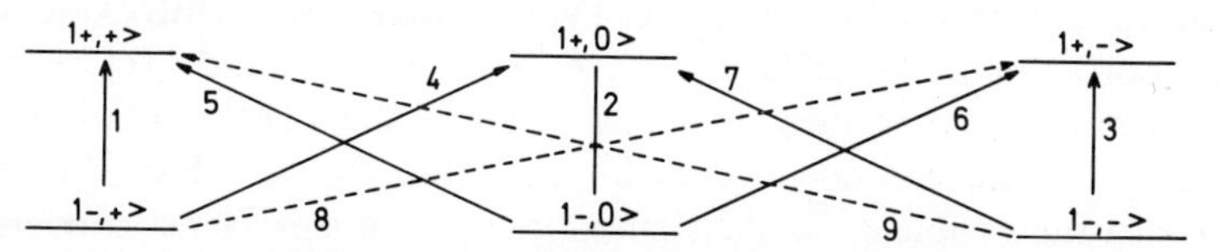

Fig. 1a. Energy levels and transitions for a nitroxide in high fields. Here $S = \frac{1}{2}$ and $I = 1$ and the notation is $|M_S, M_I\rangle$.

Eqs. (75a)–(75c) and (76a) and (76b) are written as though they correspond to the x', y', z' system].

It can be shown that the nuclear Zeeman term (which appears in the resonance frequencies of the forbidden transitions) makes a negligible contribution, so one can neglect it. Then it is only necessary to consider forbidden transitions 4 and 5, 6 and 7, and 8 and 9 in pairs, and this simplifies the expressions (see Appendix A).

We can now evaluate Eq. (27) (for $n = 1$) and the resulting expressions, neglecting saturation [i.e., set $d_\lambda = 0$ on the LHS of Eq. (27)], are given in Appendix A. They define an infinite set of coupled algebraic equations coupling the allowed and forbidden transitions. Only even values of the "quantum number" L appear for the allowed transitions for which $M = 0$. Also, one has the general restriction

$$0 \leqq K \leqq L \qquad \text{with } K \text{ even} \tag{77}$$

while $M = 1$ for the coefficients C^L_{KM} representing the "single forbidden" transition pairs (4, 5) and (6, 7) and $M = 2$ for the "doubly forbidden" pair (8, 9).

Approximations to the complete solution can be obtained by terminating the coupled equations at some finite limit by letting $C^L_{KM}(i) = 0$ for all $L > n_L$. While the number of equations needed to obtain a satisfactory convergent solution depends on the value of τ_R (the larger is τ_R, the greater is the value of n_L needed) the convergence also depends on the rotational reorientation model. The model that yields eigenvalues with the greatest dependence on L value in Eqs. (34)–(40) will have the fastest convergence. Therefore, in general, the convergence becomes poorer as one progresses from Brownian to simple free and intermediate jump, to strong collisional jump diffusion. For Brownian rotational diffusion with $\tau_R \sim 2 \times 10^{-8}$ sec, $n_L = 6$ is sufficient; with $\tau_R \sim 2 \times 10^{-7}$ sec, $n_L = 12$ is sufficient; and with $\tau_R \sim 2 \times 10^{-6}$ sec, $n_L = 24$ is sufficient. However, for simple free diffusion and $\tau_R \sim 2 \times 10^{-8}$ sec, $n_L = 10$ is needed and for a strong jump model $n_L = 16$ is needed. It is also often useful to terminate K at some value considerably less than n_L (i.e., for $n_K < n_L$). This would be especially applicable for isotropic or for axially symmetric reorientation about an axis parallel to the 2p π orbital of nitrogen, because in such a principal axis system, the **A** tensor is almost axially symmetric. Typical values for the nitroxide radical in this coordinate frame (cf. Table I) are $|F_0| \simeq 4.8$ G and $|D| \simeq 10.4$ G, while the asymmetric values are $|F_2| \simeq 1.6$ G and $|D^{(2)}| = 0.4$ G. These smaller terms are the coefficients for coupling the variable $\bar{C}^L_{KM}(i)$ with $K \neq 0$ into the problem, so their smallness guarantees faster convergence (cf. Appendix A). Also, if $R_\parallel \gg R_\perp$ or R_+, then this will greatly improve the convergence in K relative to that in L.

The computer program in Appendix B has provision for separate terminating values for K and L. The number of coupled equations is then found to be

$$r = 3 + \tfrac{9}{4}(n_L - \tfrac{1}{2}n_K + 1)n_K + \tfrac{3}{2}n_L, \qquad n_L \geq n_K, \quad n_K \geq 2 \qquad (78a)$$

or

$$r = 9 + 3(n_L - 2) \qquad \text{if } n_K = 0 \quad \text{and} \quad n_L \geqq 2 \qquad (78b)$$

A further reduction in the number of equations can be made by distinguishing between the allowed and forbidden transition coefficients. That is, if the coefficients for the allowed transitions $C_{K0}^{L}(i)$ are terminated at $L = n_L$, then the coefficients for the singly forbidden transitions $C_{K1}^{L}(j, k)$ are terminated at $L = n'$ and the doubly forbidden $C_{K2}^{L}(8, 9)$ at $L = n''$, where $n', n'' \leqq n_L$. We have found that for convergence usually $n' \simeq n_L$ or $n_L - 2$, but n'' may be truncated at values significantly below n_L (e.g., when $n_L = n' = 16$ was needed, $n'' = 8$ was sufficient). Also the terms of odd L, which exist only for the forbidden transitions, may be truncated at values of $L \ll n_L$. The computer program in Appendix B also provides for these truncations.

III. APPLICATIONS

A. Isotropic Liquids: Experiments

The validity of the slow tumbling theory has now been carefully confirmed in studies on model systems of PADS (peroxylamine disulfonate) and PD-TEMPONE (perdeuterated 2,2,6,6-tetramethyl-4-piperidone-1-oxyl) (cf. Fig. 1b) in viscous media (Goldman *et al.*, 1972a; Goldman, 1973; Hwang *et al.*, 1975).

One of the most important requirements in analyzing a slow tumbling spectrum is to have accurate values for the magnetic tensors **A** and **g**. These are best obtained from viscous solutions in the same solvent as is the slow tumbling spectrum. This is because, in general, nitroxides will exhibit magnetic parameters which are rather solvent dependent. Figure 2 shows one

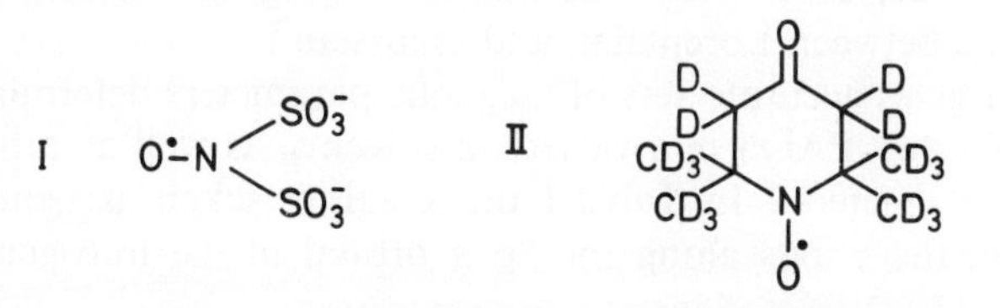

Fig. 1b. (I) Peroxylamine disulfonate anion (PADS). (II) Perdeuterated 2,2,6,6-tetramethyl-4-piperidone-1-oxyl (PD-TEMPONE).

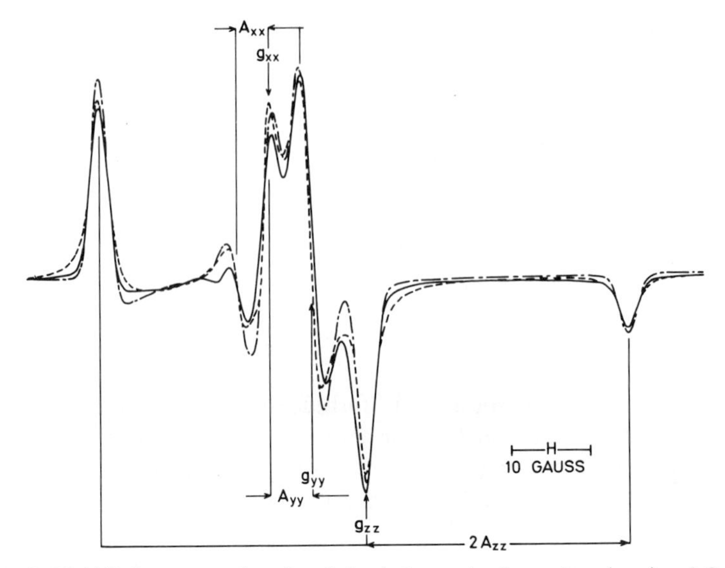

Fig. 2. Rigid limit spectrum (——) and simulations using Lorentizan (– – –) and Gaussian (- - - -) line shapes to the very well-resolved spectrum for PD-TEMPONE (cf. Fig. 1b) in toluene-d_8. See Table I for magnetic parameters. [Reprinted with permission from Hwang *et al.*, *J. Phys. Chem.* **79**, 489–511 (1975). Copyright by the American Chemical Society.]

such rigid-limit spectrum from PD-TEMPONE in toluene-d_8 (Hwang *et al.*, 1975). This is a convenient case, where the solution can be cooled sufficiently without any changes in phase until a rigid-limit spectrum is achieved. This is a particularly well-resolved spectrum because the effects of magnetic interactions of intra- and intermolecular protons on the ESR have been replaced with much weaker deuteron interactions. Note that while A_z and g_z can be read from the spectrum, accurate determination of the other principal values of the hyperfine tensor requires careful computer simulation. The rigid-limit computer simulations were performed according to standard methods; a program listing for nitroxides appears in Polnaszek's thesis (1975a). Note that a Lorentzian line shape gives an overall better fit in this case. (Other cases show line shapes ranging from near Lorentzian to intermediate between Lorentzian and Gaussian.)

Table I gives accurate sets of magnetic parameters determined for PD-TEMPONE and PADS in a variety of solvents, as well as other nitroxide magnetic parameters.† In Table I the x axis is taken as being along the N–O bond, the z axis along the 2p π orbital of the nitrogen, and the y

† Editor's note: See also Appendix II (page 564) and Chapter 6, Table II (p. 247) for additional **g** and **A** data.

TABLE I

MAGNETIC PARAMETERS[a]

	PADS in frozen D_2O[b]	PADS in 85% glycerol–H_2O[b]	DTBN[c]	PADS in $(KSO_3)_2NOH$[d]	^{17}O PADS in 85% glycerol–H_2O[e]
g_x	2.0081 ± 0.0002	2.00785 ± 0.0002	2.00872 ± 0.00005	2.0094 ± 0.0004	—
g_y	2.0057 ± 0.0002	2.00590 ± 0.0002	2.00616 ± 0.00005	2.0055 ± 0.0004	—
g_z	2.0025 ± 0.0001	2.00265 ± 0.0001	2.00270 ± 0.00005	2.0026 ± 0.0004	—
$\langle g \rangle$[f]	2.00543 ± 0.00017	2.00547 ± 0.00017	2.00586 ± 0.00005	2.0058 ± 0.0004	—
g_s[g]	2.00545 ± 0.00002	2.00548 ± 0.00001	—	—	—
$g(0)$[h] $z' = z$	$-3.61 \pm 0.15 \times 10^{-3}$	$-3.47 \pm 0.13 \times 10^{-3}$			
$g(0)$[h] $z' = y$	$3.1 \pm 1.5 \times 10^{-4}$	$5.3 \pm 1.4 \times 10^{-4}$	—	—	—
$g(2)$[i] $z' = z$	$1.2 \pm 0.1 \times 10^{-3}$	$9.8 \pm 1.0 \times 10^{-4}$			
$g(2)$[i] $z' = y$	$-2.8 \pm 0.1 \times 10^{-3}$	$-2.6 \pm 0.1 \times 10^{-3}$	—	—	—
A_x (G)	5.5 ± 0.5	5.5 ± 0.5	7.59 ± 0.05	7.7 ± 0.2	−8.7
A_y (G)	4.0 ± 0.5	5.0 ± 0.5	5.95 ± 0.05	5.5 ± 0.2	−8.7
A_z (G)	29.8 ± 0.3	28.7 ± 0.3	31.78 ± 0.05	27.5 ± 0.2	80.2 ± 0.8
$\langle A \rangle$[j] (G)	13.1 ± 0.4	13.1 ± 0.4	15.11 ± 0.05	13.6 ± 0.2	—
a_N[k] (G)	13.11 ± 0.03	13.03 ± 0.04	—	—	20.92 ± 0.06
$-D/2\pi$ $z' = z$	28.6 ± 0.6				
$-D/2\pi$ $z' = y$	−15.6 ± 0.9	−13.8 ± 0.9	—	—	—
$-D^{(2)}/2\pi$ $z' = z$	1.05 ± 0.7	0.35 ± 0.7			
$-D^{(2)}/2\pi$ $z' = y$	17.0 ± 0.6	16.2 ± 0.6	—	—	—

TABLE I (*cont.*)

	PD-TEMPONE[p] in toluene-d_8	PD-TEMPONE[p] in 85% glycerol-d_3–D_2O	PD-TEMPONE[p] in Acetone-d_6	PD-TEMPONE[p] in Ethanol-d_6	PD-TEMPONE[p] in phase V[q]
g_x	2.0096 ± 0.0002	2.0084 ± 0.0002	2.0095 ± 0.0003	2.0092 ± 0.0004	2.0097 ± 0.0002
g_y	2.0063 ± 0.0002	2.0060 ± 0.0002	2.0062 ± 0.0003	2.0061 ± 0.0004	2.0062 ± 0.0002
g_z	2.0022 ± 0.0001	2.0022 ± 0.0001	2.0022 ± 0.0002	2.0022 ± 0.0003	2.00215 ± 0.0001
$\langle g \rangle$[f]	2.0060 ± 0.00017	2.0055 ± 0.00017	2.0060 ± 0.00027	2.0058_5 ± 0.0037	2.0060 ± 0.00017
g_s[g]	2.00602 ± 0.00005	2.00570_5 ± 0.00005	2.00598 ± 0.00005	2.00589 ± 0.00005	2.00601 ± 0.00005
$g(0)$[h] $z' = z$	$-4.68 \pm 0.18 \times 10^{-3}$	$-4.29 \pm 0.18 \times 10^{-3}$	$-4.63 \pm 0.31 \times 10^{-3}$	$-4.52 \pm 0.43 \times 10^{-3}$	$-4.74 \pm 0.18 \times 10^{-3}$
$g(0)$[h] $z' = y$	$3.4 \pm 1.8 \times 10^{-4}$	$3.6 \pm 1.8 \times 10^{-4}$	$2.7 \pm 3.1 \times 10^{-4}$	$2.6 \pm 4.3 \times 10^{-4}$	$2.2 \pm 1.8 \times 10^{-4}$
$g(2)$[i] $z' = z$	$1.65 \pm 0.1 \times 10^{-3}$	$1.2 \pm 0.1 \times 10^{-3}$	$1.7 \pm 0.2 \times 10^{-3}$	$1.6 \pm 0.3 \times 10^{-3}$	$1.75 \pm 0.1 \times 10^{-3}$
$g(2)$[i] $z' = y$	$-3.7 \pm 0.1 \times 10^{-3}$	$-3.1 \pm 0.1 \times 10^{-3}$	$-3.7 \pm 0.2 \times 10^{-3}$	$-3.5 \pm 0.3 \times 10^{-3}$	$-3.77 \pm 0.1 \times 10^{-3}$
A_x (G)	4.1 ± 0.5	5.5 ± 0.5	4.8 ± 0.5	4.7_5 ± 0.6	5.61 ± 0.2
A_y (G)	6.1 ± 0.5	5.7 ± 0.5	5.4 ± 0.5	5.6_5 ± 0.6	5.01 ± 0.2
A_z (G)	33.4_5 ± 0.2	35.8 ± 0.3	34.0 ± 0.3	35.9 ± 0.4 (66%), 33.6_5 ± 0.4 (34%) } 35.1 ± 0.4	33.7 ± 0.3
$\langle A \rangle$[m] (G)	14.5_5 ± 0.4	15.7 ± 0.4	15.2 ± 0.4		14.77 ± 0.3
a_N[g] (G)	14.572 ± 0.015	15.740 ± 0.015	14.742 ± 0.015	15.173 ± 0.015	14.78 ± 0.02
α[l]	1.2,[n] 2.15	1.8, 2.9[n]	2.45, 3.1	1.54	2.2
β[l]	0.2, 0.0[o]	0.3, 0.5	0.5, 0.7	0.0	0.2
$-D/2\pi$ $z' = z$	32.4 ± 0.4	34.4 ± 0.4	33.1 ± 0.4	34.2 ± 0.7	32.5 ± 0.2
$-D/2\pi$ $z' = y$	−14.5 ± 0.9	−17.2 ± 0.9	−16.0 ± 0.9	−16.3 ± 1.1	−16.8 ± 0.5
$-D^{(2)}/2\pi$ $z' = z$	−1.39 ± 0.7	−0.105 ± 0.7	−0.441 ± 0.7	−0.631 ± 0.8	−0.420 ± 0.4
$-D^{(2)}/2\pi$ $z' = y$	20.6 ± 0.5	21.2 ± 0.6	20.5 ± 0.6	21.3 ± 0.7	19.7 ± 0.3
a_D (mG)	20.5 ± 0.2	16.0 ± 0.2	22.4 ± 0.2	20.2 ± 0.2	−21.5 ± 0.5
$A_{z,D}$ (mG)	—	—	—	—	−93 ± 7

	Tempone[r]	[s]	[t]	Tempol[u]	Tempane[v]
g_x	—	2.0089 ± 0.001	2.00783 ± 0.0002	2.0095	2.0103
g_y	—	2.0058 ± 0.001	2.00604 ± 0.0002	2.0064	2.0069
g_z	—	2.0021 ± 0.001	2.00270 ± 0.0002	2.0027	2.0030
$\langle g \rangle$	—	—	—	2.0062	2.0067
A_x (G)	5.2	5.8 ± 0.5	9.96 ± 1.07	—	—
A_y (G)	5.2	5.8 ± 0.5	7.94 ± 1.07	—	—
A_z (G)	31	30.8 ± 0.5	32.6 ± 1.07	—	—

[a] The magnetic parameters are given in their principal axis system (x''', y''', z'''), but the triple primes have been dropped for convenience. The (x', y', z') axes are the principal axes for the rotational diffusion.

[b] From Goldman *et al.* (1972a), PADS = peroxylamine disulfonate; cf. Fig. 1b.

[c] From Libertini and Griffith (1970).

[d] From Hamrick *et al.* (1972).

[e] From Goldman *et al.* (1973); entries give ^{17}O hyperfine entries. Note that it is assumed here that $A_x = A_y$.

[f] $\langle g \rangle = \frac{1}{3}(g_x + g_y + g_z)$.

[g] Measured in motionally narrowed region. Error limits reflect total range of observed values.

[h] $g(0) = 6^{-1/2}[2g_{z'} - (g_{x'} + g_{y'})]$

[i] $g^{(2)} = \frac{1}{2}(g_{x'} - g_{y'})$.

[j] $\langle A \rangle = \frac{1}{3}(A_x + A_y + A_z)$.

[k] $-D/2\pi = (|\gamma_e|/2\pi)(2\sqrt{6})^{-1}[2A_{z'} - (A_{x'} + A_{y'})]$ in MHz.

[l] $-D^{(2)}/2\pi = \frac{1}{4}(|\gamma_e|/2\pi)(A_{x'} - A_{y'})$ in MHz.

[m] Numbers are for Lorentzian and Gaussian fits, respectively.

[n] Lorentzian fit is best for PD-TEMPONE in toluene-d_8 and Gaussian fit is best for PD-TEMPONE in 85% glycerol–D_2O.

[o] Note that a small β would help the relative amplitude of $\tilde{M}(\pm 1)/\tilde{M}(0)$ but gives too much resolution in the central region.

[p] Perdeuterated 2,2,6,6-tetramethyl-4-piperidone-1-oxyl; cf. Fig. 1b, from Hwang *et al.* (1975).

[q] Cf. Fig. 1b, from Polnaszek and Freed (1975).

[r] 2,2,6,6-Tetramethyl-4-piperidone-1-oxyl in tetramethyl-1,3-cyclobutadione crystal, from Griffith *et al.* (1965).

[s] *N*-oxyl-4′,4′-dimethyloxazolidine derivative of 5-α-cholestan-3-one in cholesteryl chloride, from Hubbell and McConnell (1971).

[t] Nitroxide maleimide spin-labeled horse oxyhemoglobin crystal, from Ohnishi *et al.* (1966).

[u] 2,2,6,6-Tetramethyl-4-piperidinol-1-oxyl in crystal, from D. Bordeaux *et al.* (1973).

[v] 2,2,6,6-Tetramethyl piperidine-1-oxyl in crystal, from D. Bordeaux *et al.* (1973).

axis perpendicular to the other two. [These axes should more rigorously be written as x''', y''', z''', cf. Eq. (46), but the primes have been dropped for convenience, since there should be no confusion with the laboratory axes.] The x', y', z' axes are the principal axes of the diffusion tensor **R** and they are assumed in Table I to be either the same as the x, y, z axes or else to be a cyclic permutation of them. It should be clear from this tabulation that (1) nitroxides do exhibit significant solvent dependences in their magnetic parameters and (2) the different nitroxides will exhibit some difference in their magnetic parameters. One interesting observation in this context is the result for PD-TEMPONE (cf. Fig. 1b) in ethanol-d_6. It exhibits two distinct values of A_z, the larger one characteristic of the values in hydrogen-bonding solvents, while the smaller one is characteristic of the values in non-hydrogen-bonding solvents. In general, one must expect some variation in the magnetic parameters from site to site in a given solvent, and this will be an important source of the [orientation-dependent, cf. Eq. (29)] rigid-limit intrinsic width. These matters are discussed elsewhere (Hwang *et al.*, 1975).

Another source of valuable information for the analysis of the slow tumbling spectrum is the relaxation results from the fast motional spectrum in less viscous media, if it is at all available.

Figure 3 shows the results for PADS (cf. Fig. 1b) in D_2O (Goldman *et al.*, 1972a) where the derivative width $\delta(\tilde{M})$ is plotted as

$$\delta(\tilde{M}) = A + B\tilde{M} + C\tilde{M}^2 \tag{79}$$

The analysis of motional narrowing spectra is discussed by Nordio in Chapter 2. Suffice it to say here, that the motional narrowing theory, coupled with experimental values of B and C and *accurate* values of the magnetic parameters, is sufficient to determine $R_{\parallel}$ and $R_{\perp}$ at each temperature. It is important to note that in Fig. 3 the curves for B and C are very nearly parallel. This fact and the temperature insensitivity of a_N and g_s provide strong evidence against competing relaxation mechanisms affecting the interpretation. Figures 4a and 4b are plots of C versus B for PADS in glycerol–H_2O and for PD-TEMPONE in toluene-d_8, respectively. These results can be analyzed to yield

$$\tau_R \equiv (6\bar{R})^{-1} \tag{80}$$

$$\bar{R} \equiv (R_z R_{\perp})^{1/2} \tag{80'}$$

and

$$N \equiv R_z/R_{\perp} \tag{81}$$

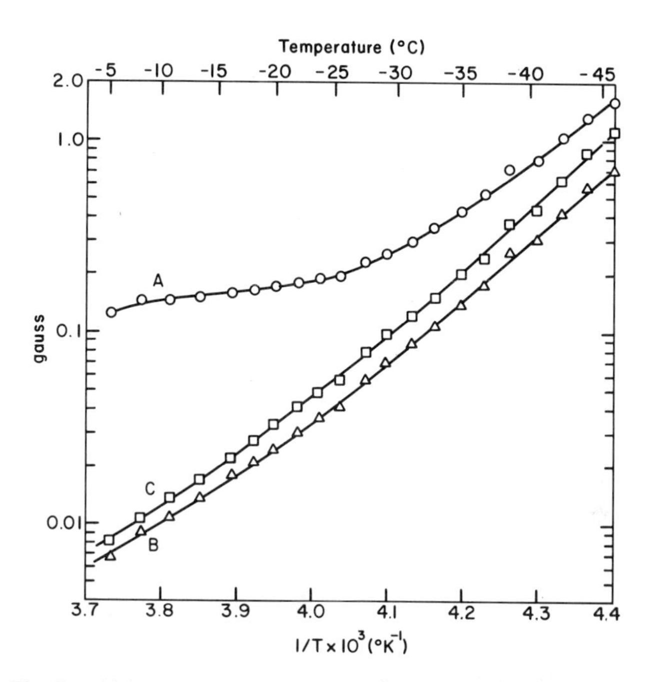

Fig. 3. The linewidth parameters A, B, and C [cf. Eq. (79)] versus $1/T$ for PADS (cf. Fig. 1b) in frozen water. [From Goldman *et al.* (1972a).]

It is found that the PADS system exhibits anisotropic rotational diffusion ($N = 4.7$ in aqueous glycerol solvents) where the z axis for the rotational diffusion tensor is parallel to the line through the two sulfur atoms, while the PD-TEMPONE system rotates isotropically (within experimental error). An interesting sidelight to this result is that, if we were to interpret the PD-TEMPONE motional narrowing results in glycerol solvent in terms of the magnetic parameters from toluene solvent, we would obtain $N \approx 3.6$, with fastest rotation about the molecular y axis, but when we use the correct magnetic parameters for glycerol, we again obtain $N = 1$.

The value of these motional narrowing results for the slow tumbling studies is that one can extrapolate the information obtained on τ_R and N into the slow motional region. That is, log τ_R is found to have a nearly linear dependence on $1/T$, as expected for an activation process. More precisely, for PD-TEMPONE in toluene, it is linear in η/T, as shown in Fig. 5. This is expected from Eq. (31) for Stokes–Einstein-type behavior. Also, N is found to be temperature independent. Figure 6 shows one such comparison of an experimental spectrum with computer simulated results utilizing the values of τ_R and N extrapolated from the motional narrowing region. This is the case of PADS in D_2O, where $N = 3$, and is one of incipient slow tumbling

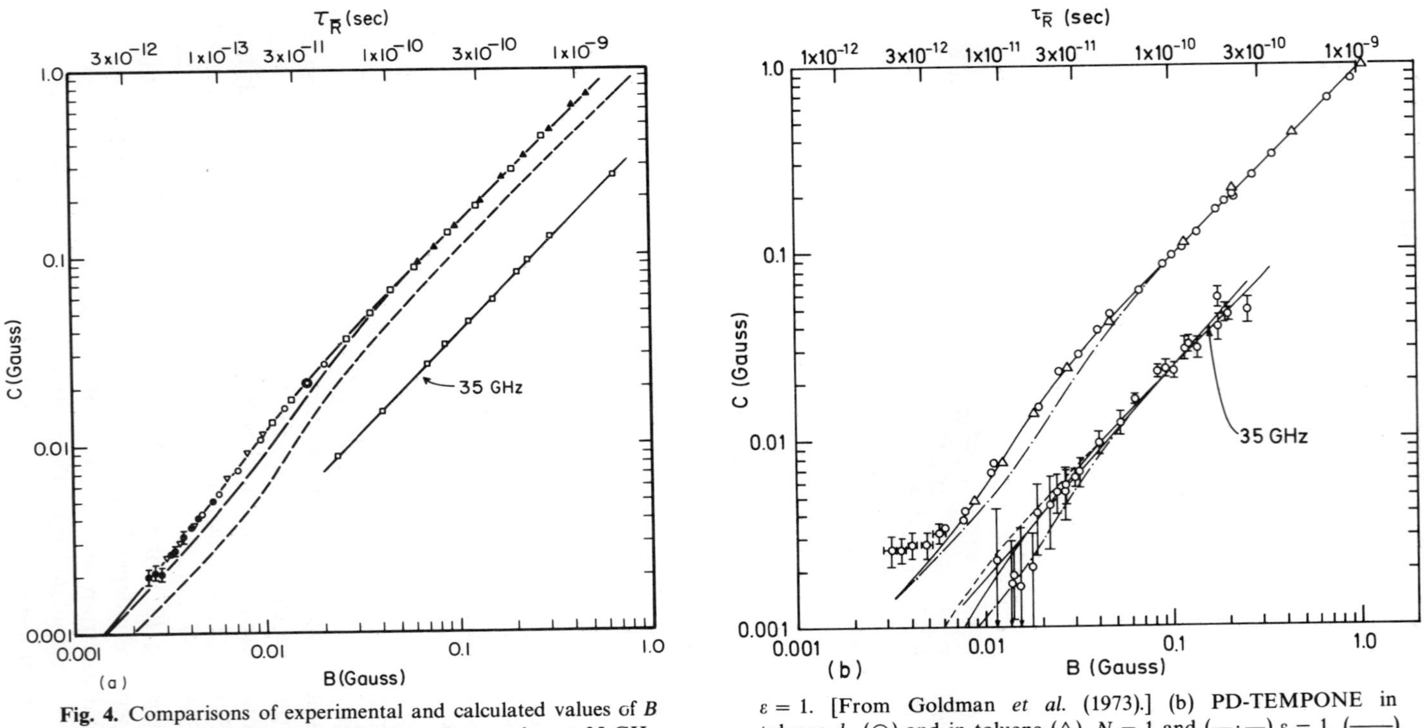

Fig. 4. Comparisons of experimental and calculated values of B vs. C. The τ_R values shown are for X-band results. τ_R values at 35 GHz correspond to the X-band results with the same value of C. (a) PADS in glycerol–H_2O: △, 85% glycerol; □, 50%; ○, 30%; ▽, 10%; ●, H_2O. (——) $N = 4.7$, $\varepsilon = 4$; (— —) $N = 4.7$, $\varepsilon = 1$; (- - -) $N = 1$, $\varepsilon = 1$. [From Goldman *et al.* (1973).] (b) PD-TEMPONE in toluene-d_8 (○) and in toluene (△). $N = 1$ and (—·—) $\varepsilon = 1$, (——) $\varepsilon = 5.4$, (---) $\varepsilon = 25$. [Reprinted with permission from Hwang *et al.*, *J. Phys. Chem.* **79**, 489–511 (1975). Copyright by the American Chemical Society.]

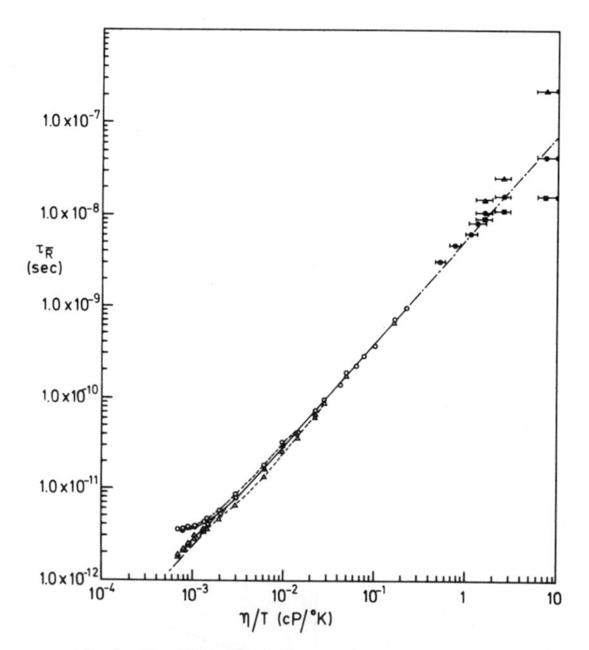

Fig. 5. τ_R versus $1/T$ for PD-TEMPONE in toluene-d_8. Motional narrowing results are designated by △ (τ_R from B values) and ○ (τ_R from C values). They are extrapolated to the slow tumbling region (-—-) and are compared to best fits for the different models: (●) free diffusion, (▲) Brownian diffusion, (■) strong jump diffusion. (——) $N = 1, \varepsilon = 5.4$. (---) $N = 1, \varepsilon = 1$. [Reprinted with permission from Hwang *et. al., J. Phys. Chem.* **79**, 489–511 (1975). Copyright by the American Chemical Society.]

($\tau_R = 4 \times 10^{-9}$ sec). A comparison is given for different values of N, and it clearly shows that the best fit is for $N = 3$. These simulations were performed for a Brownian diffusion model, with the model parameter $B_L = 1$. When slower motional spectra were obtained ($\tau_R \gtrsim 10^{-8}$ sec), it was found that a Brownian motion model was not yielding good simulations for the small spin probes PADS and PD-TEMPONE. Therefore other models were tried, i.e., the strong jump diffusion of Eq. (39) with $R\tau = 1$ and the free diffusion of Eq. (35) also with $R\tau = 1$. A typical comparison for the three models is shown in Fig. 7, from which it is clear that the best experimental fit is obtained with the free diffusion result. In this context, it is important to recognize that the values of B_L for this free diffusion model are not unique. For $\tau_R \lesssim 10^{-8}$ sec the simulated spectrum is determined mainly by coefficients of $L = 2, 4, 6$, and 8. Over this range of values of L, it is possible to reproduce the values of B_L of the free diffusion case reasonably well by the jump diffusion models given by Eq. (36) [and the special case of Eq. (39) with $R\tau \approx 0.13$ corresponding to an rms jump angle of 50°]. In fact, com-

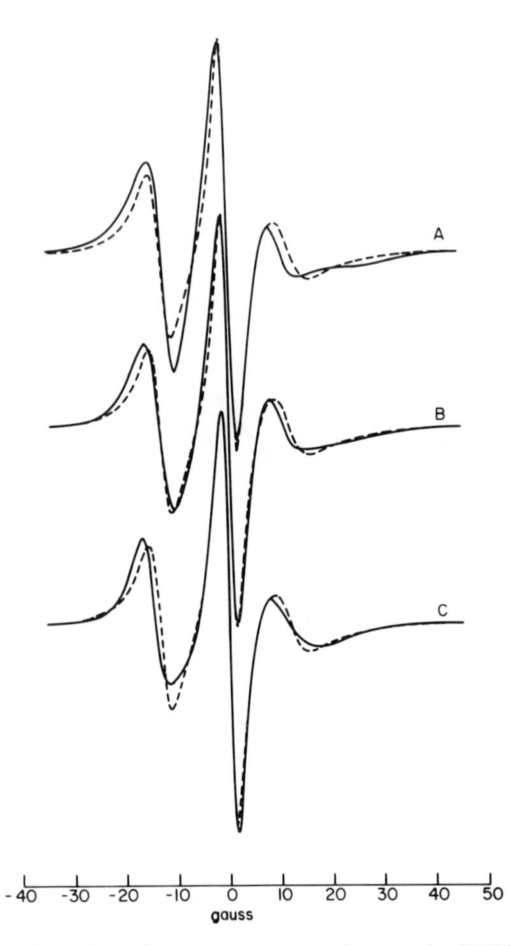

Fig. 6. A comparison of simulated and experimental spectra for PADS in frozen D_2O at $T = -50°C$: (---) experimental spectrum, (——) calculated for Brownian diffusion with $\tau_R = 4 \times 10^{-9}$ sec, $A' = 0.2$ G, and A, $N = 1$; B, $N = 3$; and C, $N = 6$. [From Goldman *et al.* (1972a).]

puter simulations of such jump models do yield virtually identical results. Figure 8 shows the results for PADS in D_2O in the model-sensitive region of τ_R for free diffusion and a range of values of N. This is also a case where free diffusion fits best.

Another way of testing the model dependence of the results is to extrapolate the τ_R values obtained in the motional narrowing region and to compare them with the "best" τ_R values obtained for each model. (This is associated with the S parameter discussed in Section III.B.) It is seen in Fig. 5 that the

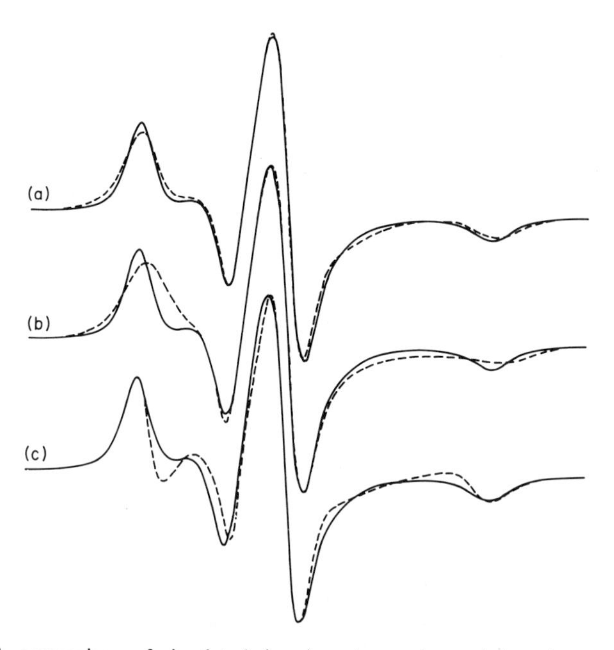

Fig. 7. A comparison of simulated (---) and experimental (——) spectra for PD-TEMPONE in toluene-d_8 at $\eta/T = 1.4$ cP/°K. The simulations are for: (a) the free diffusion model, $\tau_R = 1.0 \times 10^{-8}$ sec, $A' = 0.6$ G; (b) the strong jump diffusion model, $\tau_R = 8.7 \times 10^{-9}$ sec, $A' = 0.8$ G; (c) the Brownian diffusion model, $\tau_R = 1.4 \times 10^{-8}$ sec, $A' = 0.4$ G. [Reprinted with permission from Hwang *et al.*, *J. Phys. Chem.* **79**, 489–511 (1975). Copyright by the American Chemical Society.]

free diffusion model gives good agreement, but the others do not, especially for the slower values of $\tau_R > 10^{-8}$ sec.

The parameters τ_R and N are not the only ones that can be extrapolated from the motional narrowing region. Another is the parameter A' in G (cf. Fig. 7), which is that part of A in Eq. (79) that is not attributable to **g**- or **A**-tensor sources. (It is given as $T_{2,a}^{-1} = \frac{1}{2}\sqrt{3}\,|\gamma_e|\,A'$ in sec^{-1} in the computer programs.) The best A' for free diffusion again falls closest to the values extrapolated from the motional narrowing region, although the distinction here is not so clear (cf. Mason *et al.*, 1974).

An important point to emphasize at this stage is that the model-dependent studies summarized above were greatly aided by the very well-resolved spectra obtained from PADS and PD-TEMPONE in deuterated solvents. In general, the added intrinsic widths of typical spin labels due to the unresolved proton superhyperfine splitting will tend to obscure many of the spectral details in the slow motional and rigid-limit spectra. The analysis of the motional narrowing spectra would be particularly seriously affected.

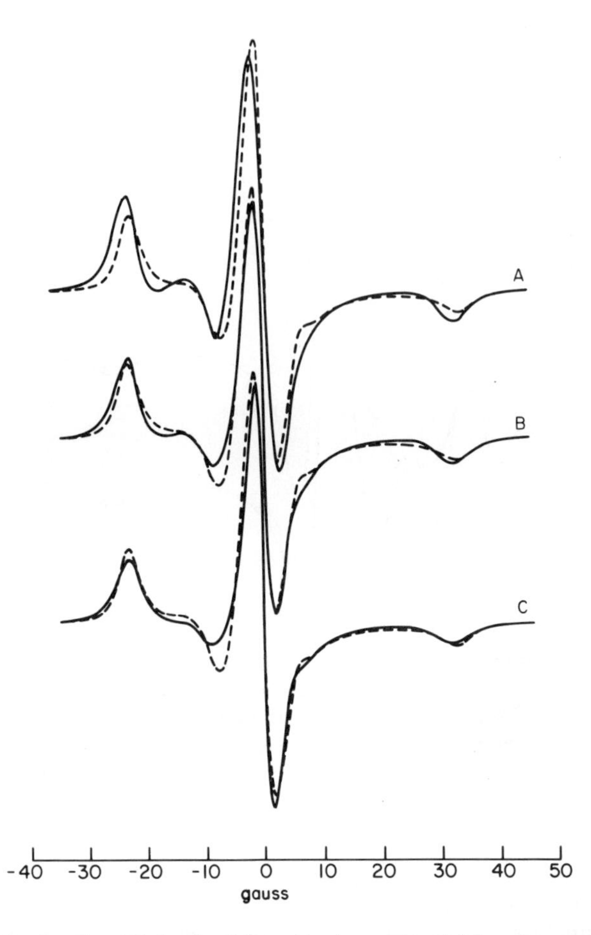

Fig. 8. A comparison of simulated (——) and experimental (- - -) spectra for PADS in frozen D_2O at $T = -60°C$. The simulated spectra are calculated for free diffusion with $\tau_R = 2 \times 10^{-8}$ sec, $A' = 0.6$ G, and A, $N = 1$; B, $N = 3$; and C, $N = 6$. [From Goldman *et al.* (1972a).]

It is interesting to note, however, that the ESR spectra obtained by McCalley *et al.* (1972)† from spin-labeled oxyhemoglobin in H_2O at $\tau_R \cong 2.6 \times 10^{-8}$ sec show many of the features that are characteristic of Brownian diffusion (cf. Fig. 7). This is an important confirmation of the theory, because one would expect that a macromolecule (unlike the small spin probes) would obey simple Brownian motion.

† Editor's note: Appendix I (p. 562) contains simulated spectra calculated by this model for a broad range of τ_R values.

B. Simplified Methods of Estimating τ_R

An important characteristic of a typical nitroxide slow motional spectrum is that it has two well-separated outer hyperfine extrema with an overlapped central region. It has been found that a useful parameter for describing these spectra is $S = A'_z/A_z$, where A_z has already been defined in Fig. 2 as one-half the separation of the outer hyperfine extrema, and A'_z is the slow tumbling value for the same spectral feature (Goldman *et al.*, 1972b). McCalley *et al.* (1972) have discussed the separate deviations of high-field and low-field positions from their rigid-limit values. Figure 9 compares the

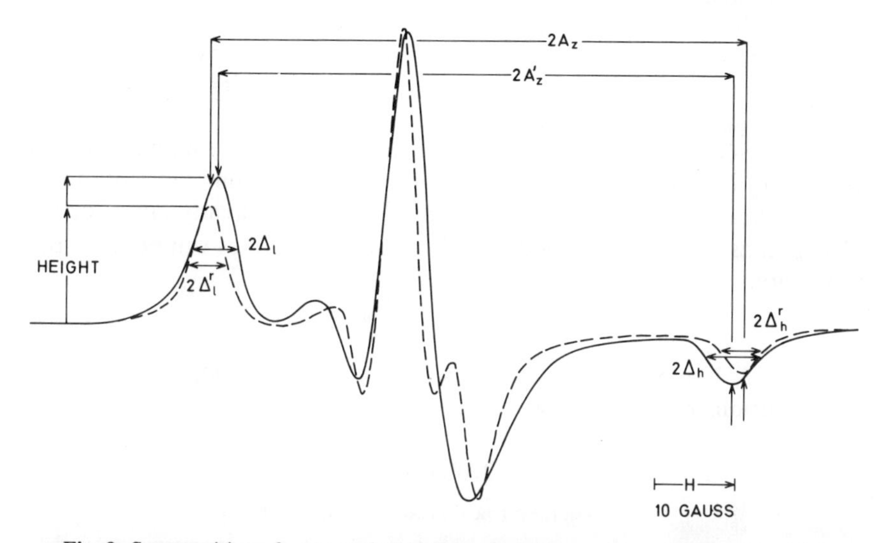

Fig. 9. Superposition of computed rigid-limit nitroxide spectrum with a computed slow-tumbling spectrum at $\tau_R = 5.0 \times 10^{-8}$ sec, demonstrating the measurements required for the parameters $S = A'_z/A_z$, $W_l = \Delta_l/\Delta_l^r$, and $W_h = \Delta_h/\Delta_h^r$. In an actual experiment, it is often necessary to estimate the Δ_i^{er} in place of the Δ_i^r as described in the text. The magnetic parameters utilized are $\delta = 3.0$ G, $g_x = g_y = 2.0075$, $g_z = 2.0027$, $A_x = A_y = 6.0$ G, $A_z = 32.0$ G, and $B_0 = 3.300$ G. [Reprinted with permission from Mason and Freed, *J. Phys. Chem.* **78**, 1321–1323 (1974). Copyright by the American Chemical Society.]

quantities A_z and A'_z. Note that A'_z decreases monotonically from its rigid-limit value of A_z as the motion becomes more rapid. Thus S is a sensitive, monotonically increasing function of τ_R. Furthermore, simulations performed for axial and asymmetric **A** and **g** tensors show that for a given value of A_z, the value of S is insensitive to changes in A_x, A_y, and the g-tensor components. Changes in the magnitude of A_z, however, do affect the value of S. This is expected, of course, since, as can be seen from Fig. 2, A_x, A_y, and **g** only contribute to the central regions of the rigid limit spectrum. This dependence can be approximately expressed in the functional form $S = S(\tau_R A_z)$, where S is simply dependent on the product $\tau_R A_z$. This functional dependence permits the scaling of results for one value of A_z to the range of values of A_z typical for nitroxides (27–40 G) with an error of less than 3%. Thus, if we know how S is affected by changes in the linewidth and rotational diffusion model, then it is possible to estimate τ_R without the necessity of making detailed line-shape calculations and comparisons. This is particularly useful for nitroxides that are broadened by inhomogeneous intramolecular or intermolecular (solvent) hyperfine and dipolar interactions. As already noted, this line broadening decreases the spectral resolution and obscures other τ_R-dependent line-shape changes.

The variation of S with τ_R is shown in Fig. 10 for Brownian, free, and strong jump diffusion models and isotropic diffusion, with $A_z = 32$ G and peak-to-peak derivative Lorentzian linewidths $[\delta = (2/\sqrt{3})|\gamma_e|^{-1}T_{2,a}^{-1}]$ of 0.3 and 3.0 G. It can be seen that S is model sensitive, and for an equivalent value of τ_R, S increases from a Brownian to a free to a jump reorientational model. (This is consistent with the analysis we have already given for the best τ_R fit as a function of model; cf. Fig. 7.) These curves can be fit to the expression

$$\tau_R = a(1 - S)^b \tag{82}$$

to within 2, 3, or 5% in the value of τ_R for a given S for jump, Brownian, or free diffusion, respectively, with the values of a and b given in Table II

TABLE II

PARAMETERS[a] FOR FITTING $\tau_R = a(1 - S)^b$

Diffusion model[b]	Linewidth[c] (G)	a	b	$\tau_R(S = 0.99)$[d](sec)
Brownian diffusion	0.3	2.57×10^{-10}	-1.78	9×10^{-7}
	3.0	5.4×10^{-10}	-1.36	3×10^{-7}
	5.0	8.52×10^{-10}	-1.16	2×10^{-7}
	8.0	1.09×10^{-9}	-1.05	2×10^{-7}
Free diffusion	0.3	6.99×10^{-10}	-1.20	3×10^{-7}
	3.0	1.10×10^{-9}	-1.01	1×10^{-7}
Strong diffusion	0.3	2.46×10^{-9}	-0.589	4×10^{-8}
	3.0	2.55×10^{-9}	-0.615	5×10^{-8}

[a] These values are calculated for an axial nitroxide with $A_\parallel = 32$ G, $A_\perp = 6$ G, $g_\perp - g_\parallel = 0.0041$, and isotropic reorientation. (From Goldman *et al.*, 1972b.)

[b] These models are discussed in detail in the text.

[c] Peak-to-peak derivative Lorentzian width: δ.

[d] For this τ_R value, $1 - S = 0.01$.

(Goldman *et al.*, 1972b; Goldman, 1973). The parameters for Brownian diffusion with linewidths of 5 and 8 G are also given. It should be noted that for $\tau_R < 7 \times 10^{-9}$ sec, S is undefinable since the outer lines begin to converge to the motionally narrowed spectrum. For longer τ_R's than shown in Fig. 10, the spectrum approaches the rigid limit, and the value of $1 - S$ become comparable to experimental uncertainties. The value of τ_R for which $1 - S = 0.01$ is given in Table II. The least squares fit to Eq. (82) was calculated for 7×10^{-9} sec $\leqq \tau_R \leqq \tau_R(S = 0.99)$.

The effect of linewidth on the value of S is also shown in Fig. 10. For Brownian and free diffusion models, S increases with increasing linewidth, while for jump diffusion a decrease in S is observed. The uncertainty in estimating τ_R due to an uncertainty in intrinsic linewidth, for a given value of S, increases for longer τ_R. Thus for a Brownian diffusion model and a 1.5 G uncertainty in the intrinsic width, the uncertainty in calculating τ_R for a given value of S increases from about 5% for $\tau_R \approx 1 \times 10^{-8}$ sec to about 50% for $\tau_R \approx 1 \times 10^{-7}$ sec to an order of magnitude for $\tau_R \gtrsim 1 \times 10^{-6}$ sec. Linear interpolations along the vertical line between the curves A and B (or

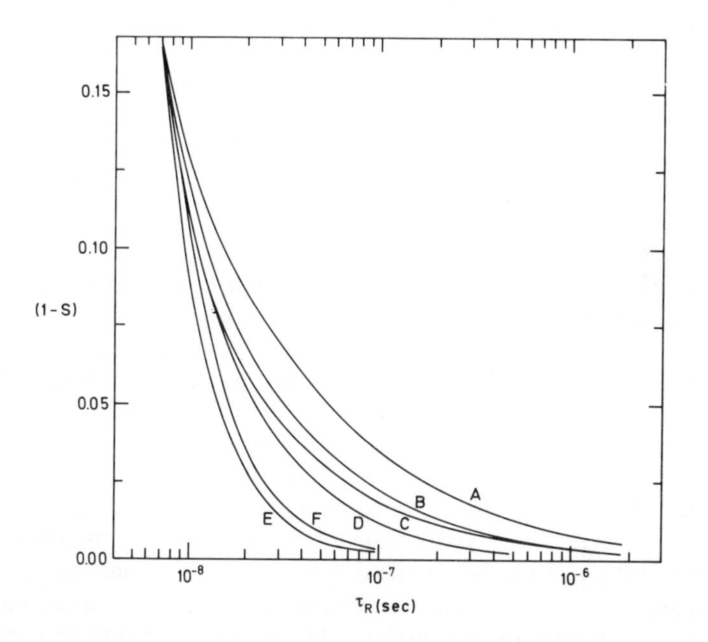

Fig. 10. Graph of $1 - S$ vs. τ_R. A, B: Brownian diffusion and derivative width $\delta = 0.3$ and 3.0 G, respectively; C, D: Free diffusion and $\delta = 0.3$ and 3.0 G, respectively; E, F: strong jump diffusion and $\delta = 0.3$ and 3.0 G, respectively. Values are calculated for isotropic reorientation with $A_z = 32$ G. See also Table II. [Reprinted with permission from Goldman *et al.*, *J. Phys. Chem.* **76**, 1858–1860 (1972b). Copyright by the American Chemical Society.]

C and D, or E and F) give the correct results for intermediate linewidth values.

The curves in Fig. 10 were calculated for isotropic rotational reorientation. For anisotropic diffusion about the z axis, with $R_{\parallel} > R_{\perp}$, the results are relatively straightforward. This type of rotation preserves the approximate axial symmetry of the spin parameters, and the observed value of S is the value expected for isotropic diffusion and $\tau_R = (6R_{\perp})^{-1}$. For relatively more rapid diffusion about the x or y axis, the results are more complicated. For small anisotropies about these axes, i.e., $R_{\parallel} = 3R_{\perp}$, the value of S is very slightly changed from the value calculated for isotropic diffusion and $\tau_R = (6\bar{R})^{-1} = \frac{1}{6}(R_{\parallel} R_{\perp})^{-1/2}$. This corresponds to a decrease of about 8% in the apparent value of τ_R obtained from Fig. 10. For larger anisotropies, a decrease in the value of S is observed (e.g., for $\tau_R = 3 \times 10^{-8}$ sec, Brownian diffusion, and an $A' = 0.3$ G, S decreases from 0.931 for $R_{\parallel}/R_{\perp} = 1$ to 0.897 for $R_{\parallel}/R_{\perp} = 20$, or an apparent decrease in τ_R obtained from Fig. 10 by a factor of two). The magnitude of this decrease is independent of whether the x or y axis is the symmetry axis. However, in general, if the axis of rotation is unknown or does not correspond to a molecular coordinate axis, or if the rotation is completely asymmetric, then estimates of the components of the rotational diffusion tensor can only be obtained from detailed spectral simulations.

Two criticisms of the general applicability of the method based on measuring S are (1) it becomes very insensitive when $\tau_R \gtrsim 10^{-7}$ sec and (2) in the region $\tau_R \gtrsim 10^{-8}$ sec the results are sensitive to the choice of residual width. A related simple technique has been proposed which may prove helpful in getting around these difficulties. Before we present it, it is useful to attempt a qualitative explanation of the observed behavior of the S parameter.

In the rigid limit (for an isotropic distribution of nitroxide spin labels), the outer hyperfine extrema arise from those nitroxide radicals for which the 2p π orbital of the nitrogen atom is nearly parallel to the applied field direction and for which the component of nuclear spin along the electron spin direction is $+1$ or -1. The Stanford group (McConnell and McFarland, 1970; Hubbell and McConnell, 1971) have shown that the derivative patterns of these outer hyperfine extrema are reasonably approximated as absorption curves with a shape function characteristic of the inhomogeneous broadening. Thus, when there are incipient motional effects in the near-rigid-limit spectrum, one can try to use the well-known magnetic resonance analogy of exchange occurring between distinct (and separated) resonance lines (Abragam, 1961; Johnson, 1965). In this well-known case, as the exchange rate increases, the lines are first observed to broaden and then to shift closer together. It is line shifts of just this type that cause $S < 1$. In fact, in the simple exchange case of two jump sites with equal probability, for which very simple equations exist (Johnson, 1965), we find that the separation of the lines decreases by the factor $[1 - 2(\tau s)^{-2}]^{1/2}$ [$\approx 1 - (\tau s)^{-2}$ when the line shifts are small], where $\tau/2$ and $2s$ are the lifetime in one state and the separation between peaks, respectively. If we now draw the analogy between the rotational motion (carrying the nitroxide radical between different orientations corresponding to substantially different ESR frequencies) at a rate of the order of τ_R^{-1} and the τ^{-1} of the two jump model, and we use the further analogy between A_z and s, then the result noted above, that $S = S(\tau_R A_z)$, is seen to follow. If we employ the two-jump expression for small shifts to the present case, then we would predict the form of Eq. (82) with $a = 1.78 \times 10^{-9}$ sec (for $A_z = 32$ G) and $b \approx -\frac{1}{2}$. We see from Table II that these results are of the correct order for strong jump diffusion, where the analogy is probably the best, but are substantially different than the Brownian diffusion results.

However, the analogy between τ_R^{-1} and τ^{-1} suggests that we examine the incipient line broadening of the outer hyperfine extrema, which should (1) roughly correspond to τ_R^{-1} and (2) be a more sensitive function of the motion than the shifts in position. Indeed, accurate computer simulations have confirmed these suggestions as being theoretically correct (Mason and Freed, 1974). In fact, the residual width is found to be given by τ_R^{-1} to within a factor of ~ 2 (or $\frac{1}{2}$) over most of the range of interest. This is about as good an agreement as we might hope for, when we recognize that the analogy is incomplete, because in the slow tumbling case, the rotational motion (a) modulates the ESR frequency over a continuous range, (and not between discrete values) and (b) induces nuclear spin flips as well because the quantization axis of the nuclear spins is orientation dependent; this is known as a nonadiabatic effect (Freed, 1972a). However, a recent approximate treatment, for small pseudosecular terms in Eq. (74), has shown that this latter effect is roughly proportional to τ_R^{-1} (Freed, 1974).

We give now a more quantitative discussion of this width effect (Mason and Freed, 1974). The average of the measured half-widths at half-heights, Δ, for the two outer extrema of a rigid-limit spectrum is found from simulations to be equal, to a very good approximation, to $\frac{1}{2}\sqrt{3}\,\delta = |\gamma_e|^{-1} T_{2,a}^{-1}$. The heights of the hyperfine extrema are measured from the true baseline (cf. Fig. 9). More precisely, we have

$$2\Delta_l^r = 1.59\delta \tag{83a}$$

$$2\Delta_h^r = 1.81\delta \tag{83b}$$

where the subscripts l and h refer to the low- and high-field lines, respectively, and the superscript r refers to the rigid-limit value. This result is found to be independent of δ over the range $1.0 \leqq \delta \leqq 4.0$ G and virtually independent of A_z over the range $27 \leqq A_z \leqq 40$ G. It is, of course, essentially independent of variations in the other nitroxide rigid-limit parameters. Equations (83a) and (83b) are valid for the assumption of Lorentzian inhomogeneous broadening. (No calculations for non-Lorentzian broadening have yet been performed for the method discussed.)

In the slow motional region, near the rigid limit, the linewidth Δ for Lorentzian line shapes can be decomposed into two contributions (cf. Abragam, 1961): (1) the Lorentzian inhomogeneous component given by Eqs. (83a) and (83b) and (2) the excess motional width (of order of magnitude τ_R^{-1}). (It is convenient to think in terms of this decomposition even though it is not necessary for the method.) A useful dimensionless parameter for describing these spectra is then

$$W_i \equiv \Delta_i/\Delta_i^r, \qquad W_i - 1 = (\Delta_i - \Delta_i^r)/\Delta_i^r \tag{84}$$

where $i = l$, h. In general, $W_i - 1$ is about an order of magnitude larger than $1 - S$ for a particular value of τ_R (cf. Fig. 11), and furthermore, it can be measured to *at least* comparable accuracy ($\sim 1\%$; cf. Fig. 9). The results in Fig. 11 were calculated utilizing the computer program in Appendix B. A study of how W_i is affected by changes in (1) the spin parameters, (2) linewidth, and (3) rotational diffusion model has been made. It was found that W_i, like S, is insensitive to deviations from axial **A** and **g** tensors, as well as to variations in $A_\perp$ and **g** typical of a nitroxide. However, in contrast to S, which is dependent on the product $\tau_R A_z$, W_i is virtually independent of A_z over the range $27 \leqq A_z \leqq 40$ G; (we have used $A_z = 32$ G in obtaining the results in Fig. 11), as expected from our simple analogy. However, W_i is found to depend on δ. Generally, a smaller δ implies a larger $\Delta_i - \Delta_i^r$ for a given τ_R. In particular, $\delta = 1$ G yields values of $\Delta_i - \Delta_i^r$ ranging from 1.3 to 2.5 times greater than those for $\delta = 3$ G. We can try to explain this observation qualitatively. The rigid-limit extrema of finite width Δ_i^r arise from those nitroxide radicals whose 2p π N-atom orbitals lie within a cone of angle Ω about the applied field direction, and the size of the cone increases rapidly with an increase in the rigid-limit δ (McConnell and McFarland, 1970). If we roughly identify the excess width $\Delta_i - \Delta_i^r$ with the rate at which radicals reorient out of the cone, then extrema from the larger cones (which result from greater values of δ) will be less broadened, since it takes longer for the radicals to leave the cone. The observation that $\Delta_h - \Delta_h^r$ is always significantly larger than $\Delta_l - \Delta_l^r$ at a given τ_R could be explained in a similar manner. It is known that the high-field resonance for a single-crystal spectrum changes with angle more rapidly than the low-field resonance; thus the range of Ω contained in the observed cone (from a polycrystalline sample) must be smaller for the high-field line. Reorientations out of the high-field cone thus occur at a more rapid rate, and, in general, W_h is a more sensitive function of τ_R than is W_l, as can be seen from Fig. 11.

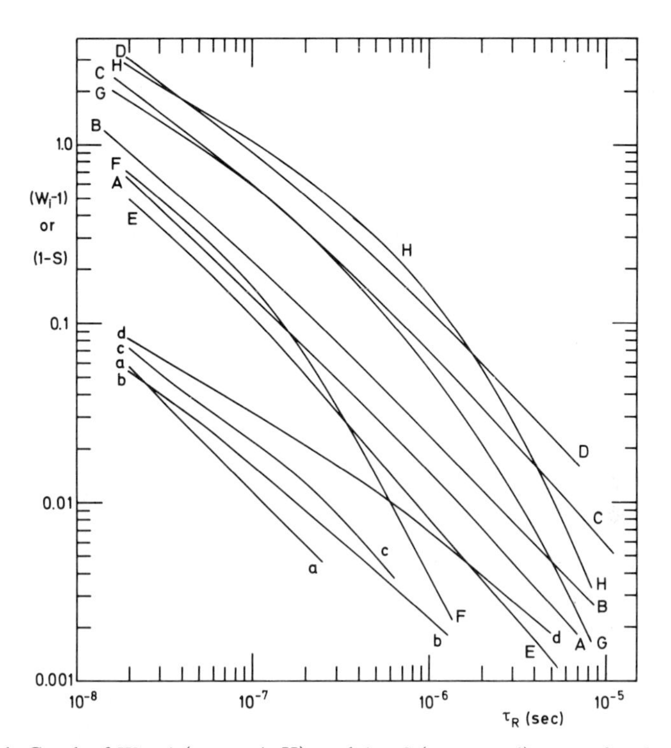

Fig. 11. Graph of $W_i - 1$ (curves A–H); and $1 - S$ (curves a–d) vs. τ_R for nitroxide isotropic rotational reorientation. Curves a and b are for free diffusion and $\delta = 3.0$ and 1.0 G, respectively; curves c and d are for Brownian diffusion and $\delta = 3.0$ and 1.0 G, respectively. Curves A and B are for free diffusion, $\delta = 3.0$ G, and the low- and high-field extrema, respectively, and curves C and D are the same as A and B, respectively, except $\delta = 1.0$ G. Curves E, F, G, and H correspond respectively to curves A, B, C, and D except that they are for Brownian diffusion. See also Table III. [Reprinted with permission from Mason and Freed, *J. Phys. Chem.* **78**, 1321–1323 (1974). Copyright by the American Chemical Society.]

The choice of a proper δ is clearly at the heart of the method. Near the rigid limit, an appropriate estimate of δ can be deduced from the Δ_i^r. The narrowest rigid-limit δ found in our laboratory is 1.5 G, which corresponds to $2\Delta_l^r = 2.4$ G. Hubbell and McConnell (1971) reported values of 4.6 and 5.5 G for $2\Delta_l^r$ for pseudoaxial "rigid-limit" spectra, corresponding to δ of 2.9 and 3.5 G, respectively. The rigid limit spectrum of the *N*-oxyl-4′,4′-dimethyloxazolidine derivative of 5α-androstan-3-one appears to have $2\Delta_l^r \approx 5.0$ G, which corresponds to a δ of 3.1 G (McConnell and McFarland, 1970). The motional broadening can easily double the widths of the outer extrema when the separation of the hyperfine extrema is not much different from the rigid-limit value (i.e., $S > 0.95$). Thus, very near the rigid limit, where δ can be determined from the rigid-limit extrema widths, two independent determinations of τ_R can be made using Fig. 11.

As has been noted, the major contributions to the Δ_i^{r} are electron nuclear hyperfine interactions between the electron and the protons of the spin label and host, while heterogeneity of the environment also contributes to Δ_i^{r}. These interactions will be quickly averaged with the onset of molecular motion, resulting in a decrease in the appropriate δ. When this is the case, it becomes necessary to estimate a single δ such that the rotational correlation times obtained from Fig. 11 for both the low- and high-field extrema are equal within experimental error. For this process we can define "effective inhomogeneous widths" Δ_i^{er}, which obey Eqs. (83a) and (83b) and which generally obey the relation $\Delta_i^{\mathrm{er}} \leqq \Delta_i^{\mathrm{r}}$. Then we should rewrite Eq. (84) as

$$W_i \equiv \Delta_i/\Delta_i^{\mathrm{er}} \tag{84'}$$

with the W_i given by Fig. 11. This procedure should then yield both τ_{R} and δ.

As noted, the uncertainty in δ can result in serious errors when τ_{R}'s of $> 3 \times 10^{-8}$ sec are determined from S. Once δ has been determined from the W_i, another estimate of τ_{R} may (when feasible) be obtained from a measurement of S. In other words, τ_{R} and δ could be obtained as a function of three experimental parameters, S and the W_i.

The model-dependent results shown in Fig. 11 were obtained for Brownian diffusion and free diffusion, as before. The free diffusion model results in a more nearly linear dependence (in a log–log plot) of $W_i - 1$ versus τ_{R} in Fig. 11 than the Brownian motion model. The plots in Fig. 11 have been fitted to the form

$$\tau_{\mathrm{R}} = a'(W_i - 1)^{-b'} \tag{85}$$

for the region $W_i - 1 > 0.01$, and the coefficients are given in Table III. The maximum variation between the curves and the results predicted from Eq. (85) is also given. It is clear that the use of Eq. (85) is a less accurate means of estimating τ_{R} than the curves. However, the fact that $b' \cong 1$ (except for the anomalous curve F, which is presumably affected by overlap) is consistent with the interpretation of $\Delta_i - \Delta_i^{\mathrm{r}}$ as a lifetime broadening.

When there is axially symmetric rotational diffusion about the molecular z axis with $R_{\parallel} > R_{\perp}$, it should mean that $\tau_{\mathrm{R}_{\perp}}$ is again obtained. Also, the introduction of an angle-dependent rigid-limit width given by Eq. (29) should have no effect on the outer hyperfine extrema, except that now

$$\delta = (2/\sqrt{3})\,|\gamma_{\mathrm{e}}|^{-1}(T_{2,a}^{-1} + T_{2,b}^{-1}) = (2/\sqrt{3})\,|\gamma_{\mathrm{e}}|^{-1}(\alpha + \beta) \tag{86}$$

It is very important in the experimental application of this method to avoid distortion of the true linewidth by overmodulation of the magnetic field and/or power saturation. Experimental applications of this method have not yet been reported.

TABLE III

PARAMETERS[a] FOR FITTING $\tau_R = a'(W_i - 1)^{b'}$

Curve	$a' \times 10^8$ sec	b'	Maximum deviation (%)[b]
A	1.29	1.033	3
B	1.96	1.062	6
C	5.32	1.076	18
D	7.97	1.125	18
E	1.15	0.943	5
F	2.12	0.778	18
G	5.45	0.999	30
H	9.95	1.014	55

[a] Table is based on approximate fit of Fig. 11 data to Eq. (85) for $W_i - 1 > 0.01$. See Fig. 11 for explanation of the different curves. [Reprinted with permission from Mason and Freed, *J. Phys. Chem.* **78**, 1321–1323 (1974). Copyright by the American Chemical Society.]

[b] Based on comparing values in Fig. 11 with Eq. (85).

The simplified methods discussed above are based on the relatively simple to analyze behavior of the outer extrema. While it is true that the central region of the nitroxide spectrum is very sensitive to motional effects, it is also very sensitive to the deviations of the nitroxide magnetic parameters from cylindrical symmetry and this can vary considerably (cf. Table I). Thus it would be difficult to develop general methods based on the central region; computer simulation with accurate magnetic parameters is probably required.

C. Very Anisotropic Rotational Reorientation

The simplified methods discussed in the previous section have an important failing in that they are not really applicable to spectra arising from highly anisotropic motion, especially when the molecular z axis is not itself a principal axis. The phenomenon of spin labels undergoing very rapid anisotropic rotational reorientation is a common one. In fact, the Stanford group (Hubbell and McConnell, 1969a,b, 1971; McConnell and McFarland, 1970; McFarland and McConnell, 1971) have developed a simple analysis in terms of an effective time-independent spin Hamiltonian $\mathscr{H}_{\text{eff}}$ to account for rapid anisotropic motion.

Suppose that there is very rapid motion about some molecular axis $\mathbf{v}$, while motion perpendicular to that axis is very slow. This is the case, for example, if the nitroxide spin label rotates about a single bond while the overall motion is that of the macromolecule to which it is attached. Then we can introduce effective $\mathbf{g}'$ and $\mathbf{A}'$ tensors that are axially symmetric about $\mathbf{v}$, and this yields an effective rigid-limit Hamiltonian to predict the spectrum. The use of such an effective Hamiltonian implies (1) a large enough $R_{\parallel}$ (representing rotational reorientation about $\mathbf{v}$) that residual time-dependent effects of the averaging process, which could lead to line broadening, are negligible; and (2) motion about axes perpendicular to $\mathbf{v}$, described by an effective $R_{\perp}$, is so slow that its effects on the spectrum are negligible.

Suppose that either of these conditions is not fulfilled, and that motional effects assert themselves in the spectrum. If condition 1 is fulfilled, but $R_{\perp}$ becomes fast enough to affect the spectrum, then one can simulate spectra using the program in Appendix B but with the effective axial tensors $\mathbf{A}'$ and $\mathbf{g}'$. The tensor $\mathbf{A}'$ is given in terms of the true $\mathbf{A}$ and the direction cosines α_i ($i = x'''$, y''', or z''', but we drop the triple primes for convenience in this section) of $\mathbf{v}$, in the molecular principal axis system as

$$A'_{\parallel} = \sum_i \overline{\alpha_i^2} A_i \tag{87}$$

$$A'_{\perp} = \frac{1}{2} \sum_i (1 - \overline{\alpha_i^2}) A_i \tag{88}$$

The simplified methods of the previous section would only apply in modified form if $A'_{\parallel}$ and $A'_{\perp}$ are not much different from typical nitroxide values (see below).

If, however, condition 1 is relaxed somewhat, then a motional narrowing theory can be applied to consider how the motion represented by $R_{\parallel}$ yields line broadening, etc., from the deviations between $\mathscr{H}_{\text{eff}}$ and the true $\mathscr{H}$. However, using our example above, the relaxation effects are a function of the orientation of the macromolecule, and we would have to compute such effects from $R_{\parallel}$ for each orientation.

However, the general theory given here can be rigorously applied to this example, including effects from both types of motion simultaneously. A series of simulations in which $R_{\perp}$ is just large enough to show incipient slow motional effects, namely $\tau_{R_{\perp}} = 5 \times 10^{-8}$ sec, were performed where $\tau_{R_{\parallel}}$ is allowed to vary from 5×10^{-8} to 6×10^{-11} sec (Mason *et al.*, 1974). This series of simulations, shown in Fig. 12b, was motivated by experimental results (cf. Fig. 12a) of Wee and Miller (1973) on a spin-labeled polybenzyl glutamate in DMF solution (cf. Fig. 13). These simulations required that the principal axes of $\mathbf{R}$ (i.e., x', y', and z') be tilted relative to the principal axes of the magnetic tensors (x''', y''', z'''). In this modified computer program, $\mathbf{A}$ and $\mathbf{g}$ are expressed in the x', y', z' coordinate system (cf. Polnaszek, 1975a). The spectra were calculated for a spatially isotropic distribution of spin labels with a tilt angle between respective z axes of 41.7°. The simulated spectra of

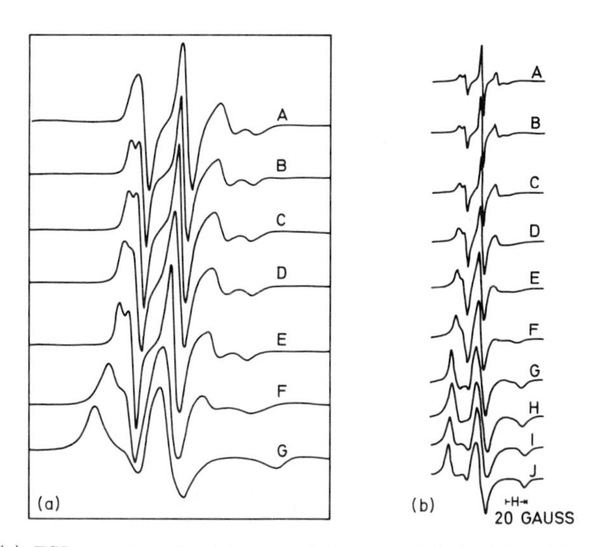

Fig. 12. (a) ESR spectra of polybenzyl glutamate labeled with 2,2,6,6-tetramethyl-4-aminopiperidine-1-oxyl (TEMP-NH_2) in DMF solutions of varying concentrations at room temperature. The polymer concentration (volume fraction) was A, 0.008; B, 0.0917; C, 0.128; D, 0.148; E, 0.200; F, 0.42 (0.5 weight fraction); and G, 1.0 (solid polymer). [Reprinted with permission from Wee and Miller, *J. Phys. Chem.* **77**, 182–189 (1973). Copyright by the American Chemical Society.] (b) Simulations computed with the magnetic parameters of the TEMP-NH_2 spin label: $A_z = 30.8$ G, $A_x = A_y = 5.8$ G, $g_x = 2.0089$, $g_y = 2.0058$, $g_z = 2.0021$. The symmetry axis of the rotational diffusion tensor is defined in the molecular axis system by the angles $\cos^{-1}(\overline{\alpha_x^2})^{1/2} = 48.3°$, $\cos^{-1}(\overline{\alpha_y^2})^{1/2} = 90°$, $\cos^{-1}(\overline{\alpha_z^2})^{1/2} = 41.7°$. $\tau_{R_\perp}$ is 5.0×10^{-8} sec and only $\tau_{R_\parallel}$ is varied in this series, except for the rigid-limit simulation J. See Table IV for values of $\tau_{R_\parallel}$ and δ used. [Reprinted with permission from Mason *et al.*, *J. Phys. Chem.*, **78**, 1324–1329 (1974). Copyright by the American Chemical Society.]

Fig. 12b for $\tau_{R_\parallel}$ of the same order of magnitude as $\tau_{R_\perp}$ appear similar to the isotropic and near-isotropic motional spectra already shown, and for which the simplified approaches already apply. But for $\tau_{R_\parallel} \ll \tau_{R_\perp}$ there are marked qualitative differences. (Note that this is an ideal case for K truncation, i.e., $n_K < n_L$.) The general progression of spectra in Fig. 12b from A to I, where $\tau_{R_\parallel}$ is increasing bears considerable resemblance to the progression of experimental spectra in Fig. 12a from A to G, where the polymer concentration, and hence the solution viscosity, is increasing. Rather close agreement is found between pairs 12b-I and 12a-G (the near-rigid limit), 12b-B and 12a-B (where $\tau_{R_\parallel}$ is very fast), and 12b-F and 12a-F (an intermediate case). These results provide evidence that there is fast motion of the piperidine ring about the NH–CH bond, and no observable motion of the overall polymer (i.e., $\tau_R > 10^{-7}$ sec).

The question now remains of the range of validity of the effective time-independent Hamiltonian in terms of $\mathbf{g}'$ and $\mathbf{A}'$. Its use by Hubbell and

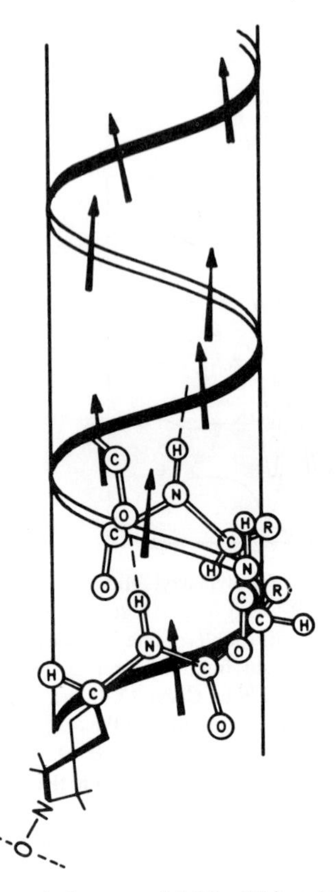

Fig. 13. End-labeled polybenzyl glutamate (cf. Fig. 12b), right-handed α-helix. [Reprinted with permission from Wee and Miller *J. Phys. Chem.* 77, 182–189 (1973). Copyright by the American Chemical Society.]

McConnell (1969a,b, 1971) is based by analogy with liquid crystalline spectra, on a so-called order parameter S_o given by

$$S_o = \tfrac{1}{2}(3\overline{\alpha_z^2} - 1) \approx (A'_{\parallel} - A'_{\perp})/(A_z - A_x) \tag{89}$$

where the bar implies a time average and the second equality is based on letting $A_x = A_y$ as is the case in many nitroxides. S_o is a measure of the extent of the motion leading to $\hat{\mathscr{H}}_{\text{eff}}$. We then have

$$A'_{\parallel} = A_o + \tfrac{2}{3}(A_z - A_x)S_o \tag{90a}$$

$$A'_{\perp} = \tfrac{1}{2}(3A_o - A'_{\parallel}) \tag{90b}$$

and

$$A_o = \tfrac{1}{3}\,\mathrm{Tr}\,A = \tfrac{1}{3}\,\mathrm{Tr}\,A' = A'_o \tag{91}$$

TABLE IV[a]

Spectrum index	$\tau_{R_\parallel}$ (nsec)	$A'_\parallel$ (G)	$A'_\perp$ (G)	$\Delta A'$ (G)	A'_o [g]	$\cos^{-1}(\alpha_z^2)^{1/2}$ (deg)[f]	$g'_\parallel - g'_\perp$	S_o [f]	$g'_\parallel$	$g'_\perp$	g'_o [h]	$\cos^{-1}(\alpha_x^2)^{1/2}$ (deg)[i]
A[b]	0.06	18.9	11.5	7.4	14.0	43.2	−.0008	0.296[k]	2.0051	2.0059	2.0056	55[j]
B[b]	0.10	18.9	11.3	7.6	13.8	42.9	−.0008	0.304	2.0051	2.0059	2.0056	55
C[b]	0.20	19.2	11.1	8.1	13.8	42.2	−.0010	0.324	2.0050	2.0060	2.0057	57
D[b]	0.40	20.2	10.7	9.5	13.9	40.0	−.0014	0.380	2.0047	2.0061	2.0056	60
E[b]	0.67	22.6	10.3_5	12.2_5	14.4	35.7	−.0025	0.490	2.0037	2.0062	2.0054	80
F[b]	1.00	24.1	10.9	13.2	15.3	34.1	−.0017	0.528	2.0033	2.0058	2.0050	47
G[b]	6.00	27.3	[d]	—	—	—	—	—	2.0030	[d]	—	—
H[c]	10.0	27.7	[d]	—	—	—	—	—	2.0028	[d]	—	—
I[c]	50.0	29.7_5	[d]	—	—	—	—	—	2.0025	[d]	—	—
J[c]	∞[e]	30.8	[d]	—	—	—	—	—	2.0021	[d]	—	—

[a] Reprinted with permission from Mason *et al.*, *J. Phys. Chem.* **78**, 1324–1329 (1974). Copyright by the American Chemical Society.
[b] Peak-to-peak residual derivative width δ of 1.0 G was used.
[c] Peak-to-peak residual derivative width δ of 3.0 G was used.
[d] The inner hyperfine extrema are not resolved.
[e] $R_\perp$ and $R_\parallel$ are zero.
[f] S_o and α_z are defined by Eq. (89).
[g] A_o is defined by Eq. (91).
[h] g'_o is defined as $\frac{1}{3}(g'_x + g'_y + g'_z)$.
[i] From construction $\cos^{-1}(\alpha_y^2)^{1/2} = \pi/2$.
[j] If $S_o = 0.336$ is used, $\cos^{-1}(\alpha_x^2)^{1/2}$ equals 50°; see text.
[k] If correction is made for the finite $\tau_{R_\perp} = 5 \times 10^{-8}$ sec, then $S_o = 0.338$ (cf. Mason *et al.*, 1974).

A check on the validity of using $\mathscr{H}_{\mathrm{eff}}$ for interpreting a spectrum in terms of the "pseudoaxial" rigid limit is that one must have Tr A = Tr A', which follows directly from the rotational invariance of the trace of a tensor. The simulated spectra of Fig. 12b have been analyzed just as though they were experimental results from anisotropically immobilized spin labels, and the results are summarized in Table IV (Mason *et al.*, 1974). These results are in fact very similar to those obtained by Wee and Miller (1973) and in spin label studies in membrane models and membranes (Hubbell and McConnell, 1969a,b, 1971; McConnell and McFarland, 1970; McFarland and McConnell, 1971). Note that (1) A_0' does vary slightly from $A_0 = 14.13$ G but (2) the apparent S_0 varies considerably even though the calculated spectra were from a tilt angle of 41.7°, corresponding to a constant true value of $S_0 = 0.336$. The results for A and B in Fig. 12b, where $\tau_{R_{\parallel}}$ is very fast, are reasonable; the only discrepancy with the true results probably arises from the residual motional effects on the spectrum of having $\tau_{R_{\perp}} = 5 \times 10^{-8}$ sec. However, for spectra such as E and F, where $\tau_{R_{\parallel}} \sim 5\text{–}10 \times 10^{-10}$ sec, the motional effects for the fast motion about a single bond are already slow enough for us to utilize the simple time-independent $\mathscr{H}_{\mathrm{eff}}$. Thus the invariance of S_0 is a necessary but not really sufficient condition for the simple approach.

Thus we note that a change in S_0 can arise from a real change in the angle between $\mathbf{v}$ and z or a change in the rotational rate about $\mathbf{v}$, as is the predominant phenomenon in the spectra of Fig. 12a (the spectrum A, however, has a shorter τ_R than the other spectra) (Mason *et al.*, 1974). In general, these two phenomena cannot be distinguished unless the rotational rate about $\mathbf{v}$ slows to where A_0' is clearly anomalous and/or the slowed motion manifests itself in the other spectral characteristics shown in Fig. 12a.

Note that the analysis in terms of a single $R_{\perp}$ and $R_{\parallel}$ represents a considerable simplification of the complex dynamics of polymer motion including localized bond motions and internal rotations. However, (1) as long as the internal rotation is much faster than the overall motion, it can be treated as uncoupled from the latter, and (2) if the overall motion is only showing marginal spectral effects, it would be difficult to obtain anything more precise than an effective $\tau_{R_{\perp}}$.

D. Anisotropic Liquids: Simulations

Some examples will now be given for nitroxides oriented in nematic liquid crystals (Polnaszek *et al.*, 1973; Polnaszek, 1975a). For convenience, the Meier–Saupe potential, coincidence of the magnetic and orientation principal axes, and axial magnetic parameters were used, unless otherwise noted. In all cases the director was assumed parallel to the static magnetic field.

First, the effect of keeping the rotational correlation time $\tau_R \equiv (6R)^{-1}$ constant but varying the ordering by changing the potential parameter λ is considered. Figure 14 shows such a case for an oblate nitroxide, i.e., one that

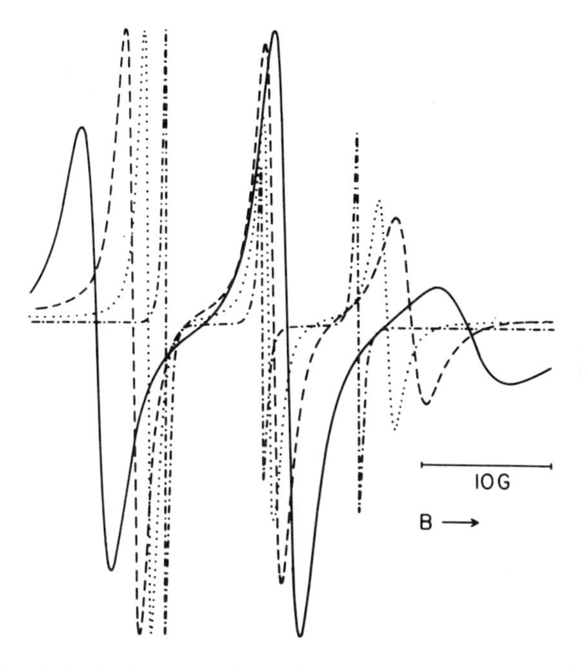

Fig. 14. First derivative line shapes for a nitroxide as a function of λ for Brownian diffusion. The different λ values are (——) 0, (- - -) -2.0, (···) -3.5, (-·-) -7.5. All correspond to $\tau_R = 1.84 \times 10^{-9}$, $g_{\parallel} = 2.0027$, $g_{\perp} = 2.0075$, $A_{\parallel} = 33.4$ G, $A_{\perp} = 5.42$ G, and $\delta = 0.1$ G. [From Polnaszek *et al.* (1973).]

tends to orient with its z axis perpendicular to the magnetic field. The rotational correlation time was held constant at a value of 1.84×10^{-9} sec, which is in the incipient slow motional region for nitroxides. One sees that the effect of increasing the absolute magnitude of the orienting potential is (1) to decrease linewidths considerably, and (2) to introduce larger shifts in the positions of the lines. The first effect is due to the fact that the effective $[\mathscr{H}_1(t) - \langle \mathscr{H}_1 \rangle]$ is reduced as the sample is being ordered, while $\langle \mathscr{H}_1 \rangle$, the average part of the perturbation, which causes the line shifts, departs more from its isotropic value of zero. It appears that the slow tumbling spectra begin to resemble motional narrowing results as $|\lambda|$ is increased at a constant τ_R, although the shift of line positions is characteristic of liquid crystals in the nematic range. However, the observed line shifts are not predicted correctly by expressions appropriate for the motional narrowing region (cf.

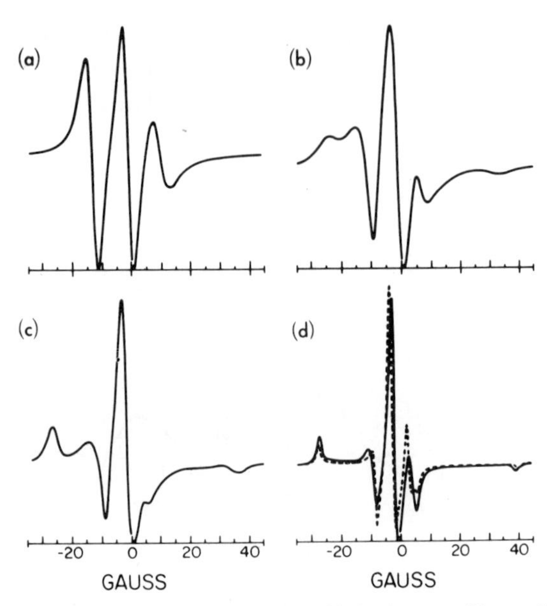

Fig. 15. First derivative line shapes for a nitroxide in the case of low ordering as a function of τ_R for Brownian diffusion: (a) $\tau_R = 3 \times 10^{-9}$ sec, IF = 5.08 (IF is the intensity factor, a measure of the relative integrated intensity); (b) $\tau_R = 10^{-8}$ sec, IF = 3.48; (c) $\tau_R = 3 \times 10^{-8}$ sec, IF = 4.38; (d) $\tau_R = 3 \times 10^{-7}$ sec, IF = 10.02; (---) rigid limit. Here $g_\parallel = 2.0024$, $g_\perp = 2.0078$, $A_\parallel = 33.75$ G, $A_\perp = 5.34$, and $\delta = 1.0$ G. All spectra are normalized to the same total height. [From Polnaszek *et al.* (1973).]

Chapter 10). We can use these shifts as an indication of slow motion in the mesophase. For $\lambda > 0$ a prolate top (a nitroxide that tends to orient with its z axis parallel to the field) similar linewidth behavior is observed, except that the spectrum spreads out and shifts to higher fields. In Fig. 15, λ is held constant at a value of -0.975, corresponding to a low degree of ordering typical of small nitroxides. Comparison with figures for isotropic liquids (cf. Figs. 6–8) shows that the trends are quite similar in both cases, but that there are distinct quantitative differences in the details of the line shapes at comparable τ_R's.

In Figs. 16a and b, λ is held constant (at -7.5 and 8.5, respectively) for large ordering parameters, with the correlation time varying over several orders of magnitude. They correspond respectively to large disklike and rodlike nitroxide molecules, which tend to be very well ordered. They show that spectra with correlation times $< 3 \times 10^{-8}$ sec will be fairly insensitive to changes in τ_R. In fact the observed ordering parameter for the oblate top nitroxide is nearly equal to the theoretical value for $\tau_R \lesssim 10^{-8}$ sec, and it is found that the rigid limit is not approached until $\tau_R \approx 10^{-6}$ (cf. Fig. 16c), compared to $\tau_R \approx 3 \times 10^{-7}$ for the low ordering case in Fig. 15. Thus for a very highly ordered nitroxide, we can extend the upper limit of rotational correlation times obtainable from the unsaturated slow motional line shapes.

For values of the potential parameter λ that lead to intermediate ordering (e.g., $\lambda = -3.5$, corresponding to $\langle \mathscr{D}^2_{00} \rangle = -0.30$), it has been found that the deviations from symmetric lines and from the theoretical ordering parameter begin to occur at somewhat longer values of τ_R in the incipient slow tumbling region than for isotropic liquids, but that the linewidth asymmetry starts to be appreciable at somewhat shorter τ_R's than those for which the apparent ordering parameter deviates significantly from the correct value. Therefore the linewidth asymmetry can be used as an indication of slow tumbling in nematics. This can also be seen from Fig. 16 for highly ordered nitroxides.

The effect of using different models of rotational reorientation has also been studied. For weakly ordered systems, one again sees the same qualitative behavior of spectral changes as the rotational model is changed, but there are qualitative differences between the spectral changes for the isotropic and nematic phases. As has been usually done for isotropic liquids, the correlation times for the non-Brownian models were determined to be those that gave the same values for the separation of the outer hyperfine extrema (i.e., S) as observed in the Brownian diffusion case. For liquid crystals S is expected to be a function of λ as well as of τ_R and A_z, thus complicating any attempts to use it as a quantitative measure of τ_R's in nematic phases. This was found to be true for weakly ordered systems. For the strongly ordered cases, the parameter S is meaningless since no outer extrema are observed.

Figure 17 gives an example in which the principal axis of orientation is permuted among the three principal axes of the nitroxide magnetic tensors. The hyperfine tensor is taken as axially symmetric, but the g tensor is asymmetric, as is typical for nitroxides. The rotational correlation time is 10^{-8} sec and λ was adjusted to make the S values nearly equal. The x- and y-axis spectra are cases where the molecule tends to align parallel to those axes, respectively, while the z axis tends to be perpendicular to the field. One sees significant changes in the line shapes as the principal axis of orientation is changed, even for this weakly ordered system. There are also "apparent" shifts in $\Delta g_{\parallel}$. The g shifts persist in the motional narrowing region for this case of three different orientations. Thus, especially when using any cylindrically symmetric potential, one must be careful to choose the principal axis of orientation correctly. However, when an asymmetric potential such as that given in Eq. (58) is used, the potential is invariant to permutation of the principal axes (i.e., the relabeling of x''', y''', z''' to obtain x', y', z'). Since the coefficients of the potential transform as the components of an irreducible tensor, we can transform the potential parameters for a principal axis system

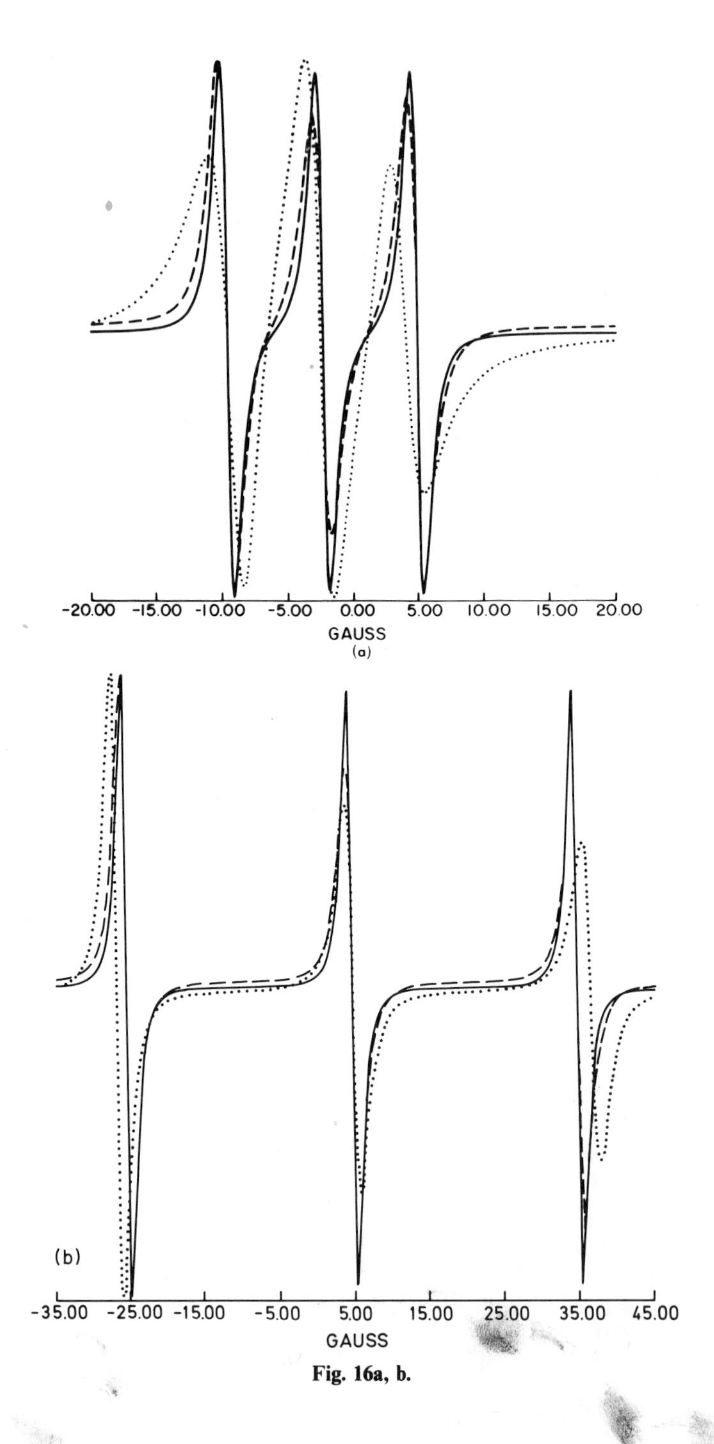
-20.00
-15.00
-10.00
-5.00
0.00
5.00
10.00
15.00
20.00
GAUSS
(a)
(b)
-35.00
-25.00
-15.00
-5.00
5.00
15.00
25.00
35.00
45.00
GAUSS

Fig. 16a, b.

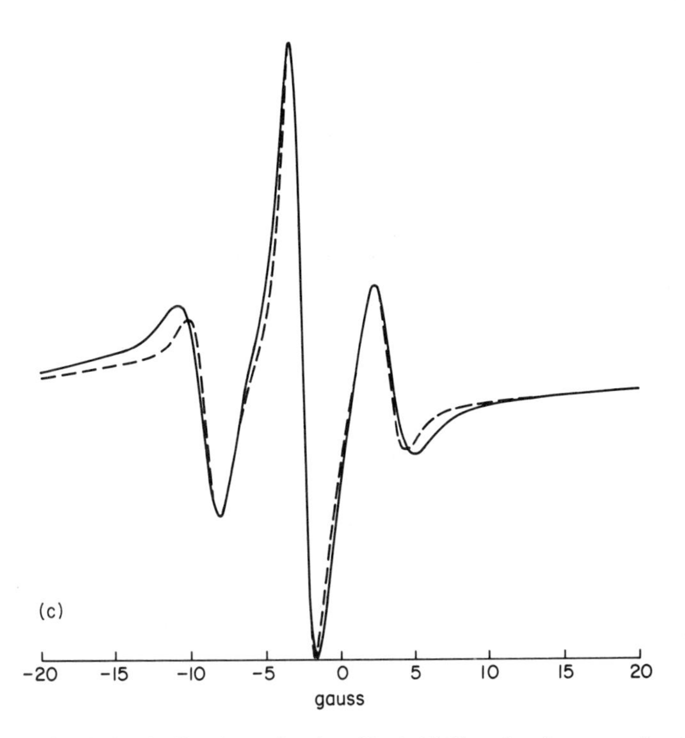

Fig. 16. First derivative line shapes for nitroxides in highly ordered cases as a function of τ_R for Brownian diffusion: (——) $\tau_R = 3 \times 10^{-10}$ sec, IF = 77.9; (---) $\tau_R = 3 \times 10^{-9}$ sec, IF = 52.9; (···) $\tau_R = 3 \times 10^{-8}$ sec, IF = 16.3. (a) $\lambda = -7.5$; (b) $\lambda = +8.5$. (c) $\lambda = -7.5$ and (——) $\tau_R = 3 \times 10^{-7}$ sec, IF = 29.8; (---) $\tau_R = 1 \times 10^{-6}$ sec, IF = 41.0. All other parameters as in Fig. 15. [From Polnaszek *et al.* (1973) and Polnaszek (1975a).]

in the molecule into those for another molecular axis system. For the potential given by Eq. (58), one has for a permutation of axes such that the y axis → z' axis

$$\lambda_y = -(\lambda_z - 3\rho_z)/2 \tag{92a}$$

and

$$\rho_y = -(\lambda_z + \rho_z)/2 \tag{92b}$$

where the subscripts refer to the principal axis of orientation of the molecule. We can also determine λ_x and ρ_x from the relation

$$\sum_i \lambda_i = \sum_i \rho_i = 0 \tag{93}$$

which follows from the fact that the ordering tensor is traceless.

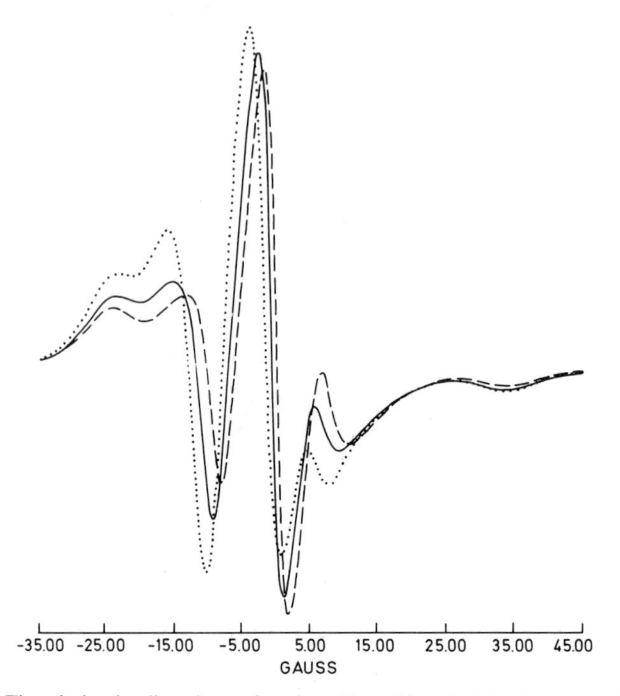

Fig. 17. First derivative line shapes for nitroxides with cylindrically symmetric potentials relative to the different principal axes of the magnetic tensor for Brownian diffusion. (——) The molecular z''' axis as the principal orientation axis with $\lambda = -0.975$; ($\cdots$) the x''' axis as the principal orientation axis with $\lambda = 1.6$; (– – –) the y''' axis as the principal orientation axis with $\lambda = 1.6$. In all cases $\tau_R = 1 \times 10^{-8}$ sec, $A_z = 33.75$ G, $A_x = A_y = 5.34$ G, $g_x = 2.0094$, $g_y = 2.0062$, $g_z = 2.0024$. [From Polnaszek (1975a).]

It is well known that liquid crystals exhibit an anisotropic viscosity when oriented in a magnetic field (Miesowicz, 1946). The effects of anisotropic viscosity on nitroxide slow tumbling spectra are shown in Fig. 18. In this figure $\tau_{R_\perp} = (6R_\perp)^{-1}$ is kept constant at 1×10^{-8} sec, the potential parameter λ is -0.975, and $\tau_{R_\parallel}$ is varied. The effect of keeping $R_\perp$ constant is to keep the value of S virtually constant. However, there are gross changes in the central region of the spectrum as $R_\parallel$ is increased relative to $R_\perp$. Note from Eq. (66) that the terms that contain the effect of anisotropic viscosity have $M \neq 0$. It is seen from Eq. (74) that such terms are the pseudosecular terms in $\mathscr{H}_1(\Omega)$. Thus we must include them in the Hamiltonian in order to see the effects of anisotropic viscosity. The effects of anisotropic viscosity on the slow motional line shapes are negligible when the ordering parameter is large or if the molecule tends to be aligned with its z axis, the axis of cylindrical symmetry of the hyperfine tensor, parallel to the field. Note that

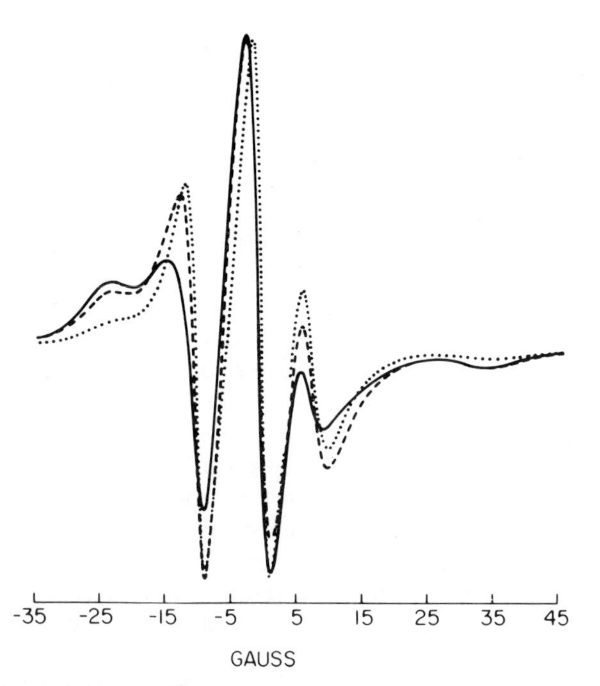

Fig. 18. First derivative line shapes for nitroxides as a function of the anisotropic viscosity parameter $\hat{N} = \hat{R}_{\parallel}/\hat{R}_{\perp}$ for Brownian diffusion. (——) $\hat{N} = 1$, IF = 3.48; (- - -) $\hat{N} = 6$, IF = 3.30; (···) $\hat{N} = 24$, IF = 5.33. All correspond to $\tau_{R_{\perp}} = 1 \times 10^{-8}$ sec. All other parameters as in Fig. 15. [From Polnaszek *et al.* (1973).]

the effects of anisotropic molecular reorientation on slow tumbling spectra from nitroxides are not as dramatic as the effects of anisotropic reorientation with respect to the director axis. Birrell *et al.* (1973) observed a system that can be thought of in terms of a highly anisotropic viscosity. Nitroxide free radicals were oriented in tubular cavities in inclusion crystals in which the molecule is free to rotate about the long axis but with its rotation hindered about the other two axes because of the cavity geometry. The system behaves as a highly ordered liquid crystal, which, as has already been noted, is fairly insensitive to the dynamics of the motion.

In all the preceding discussion, it has been assumed that **n** is fixed along the laboratory z axis, so that $\Psi = (0, 0, 0)$. When **n** is tilted relative to the z axis, then Eq. (45) must be used to expand $\mathscr{H}_1(\Omega, \Psi)$, but otherwise the same diffusion equations in terms of Ω are applicable. If there are random static distributions of directors, then, in principle, one must solve the problem for each value of Ψ and then integrate over the correct static distribution to predict the spectrum. When there is residual motion of the director, then the stochastic Liouville equation can be augmented to deal effectively with the simultaneous motions of Ψ and Ω (Polnaszek, 1975a; Polnaszek and Freed, 1975).†

E. Anisotropic Liquids: Experiments

Experiments have now been carried out to test the applicability of the slow tumbling theory to anisotropic liquids using the same general approach as has already been described for studies in isotropic liquids (Polnaszek, 1975a; Polnaszek and Freed, 1975). In particular, the PD-TEMPONE probe has been studied in several viscous nematic solvents. The study was limited by the fact that before very large viscosities could be reached, the nematics would freeze. However, in the case of phase V solvent, it was possible to reach the slow motional region (cf. Fig. 19). Note in Fig. 20 how the apparent $\langle \mathscr{D}^2_{00} \rangle$ changes markedly when the slow motional region is approached.

In the isotropic studies, the motional narrowing line shapes were carefully corrected for the residual inhomogenous broadening effects of the deuteron splittings given in Table I. In the nematic phase, the splittings were found to vary with temperature as a result of the increase of $\langle \mathscr{D}^2_{K0} \rangle$ with decreasing temperature. This added factor had to be corrected for in order to adequately deal with the line shapes.

Furthermore, careful measurements of the a_N and g shifts clearly demonstrated that the radical ordering required the use of the two-parameter potential of Eq. (58). For most cases (including phase V solvent), if the choice $x''' = x'$, $y''' = y'$, and $z''' = z'$ is made, then $|\lambda| > |\rho|$, with typical values for phase V being $\lambda_z = -0.8$ and $\rho = 0.3$. The higher temperature spectra from the isotropic phase again showed $N = 1$ for the anisotropic diffusion parameter, corresponding to isotropic rotation. However, in the nematic phase, only if the correct ordering potential involving *both* λ and ρ were used could the linewidth results be fit to isotropic rotational diffusion.

The appropriate values of τ_R, $N = 1$, λ, and ρ were then extrapolated to the slow motional region to obtain parameters to predict the slow motional spectrum. A typical comparison is shown in Fig. 21a, where $\tau_R = 3.6 \times 10^{-9}$ sec at $-25°$. There are serious discrepancies between experiment and prediction, unlike the good agreement (cf. Fig. 16) for isotropic liquids. This incipient slow tumbling region has been found to be rather model insensitive. However, careful analysis of the motional narrowing region for $\tau_R > 2 \times 10^{-10}$ sec showed that the discrepancy was developing

† Editor's note: Further and more expanded discussions of order parameters and the anisotropic motion of spin labels in liquid crystals and bilayer model membrane systems are found in Appendix IV and Chapters 10–12, respectively.

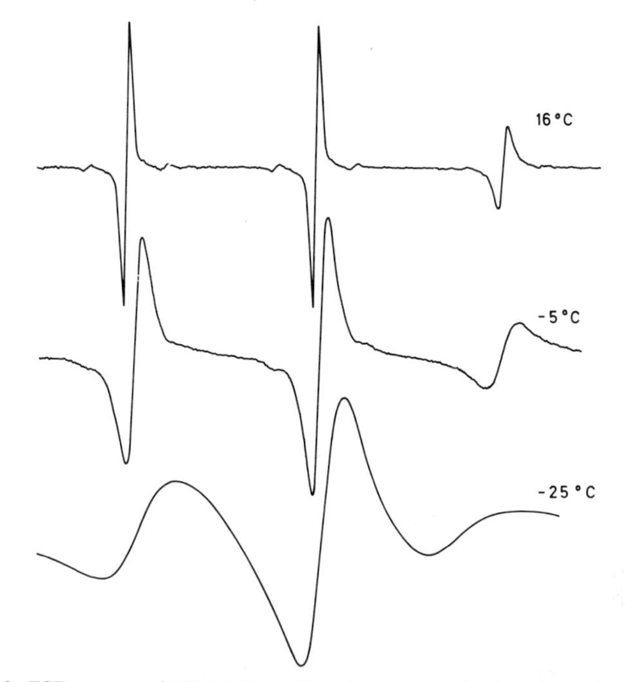

Fig. 19. ESR spectra of PD-TEMPONE in liquid crystal: phase V, at different temperatures. The scan range is 40 G. [From Polnaszek (1975a).]

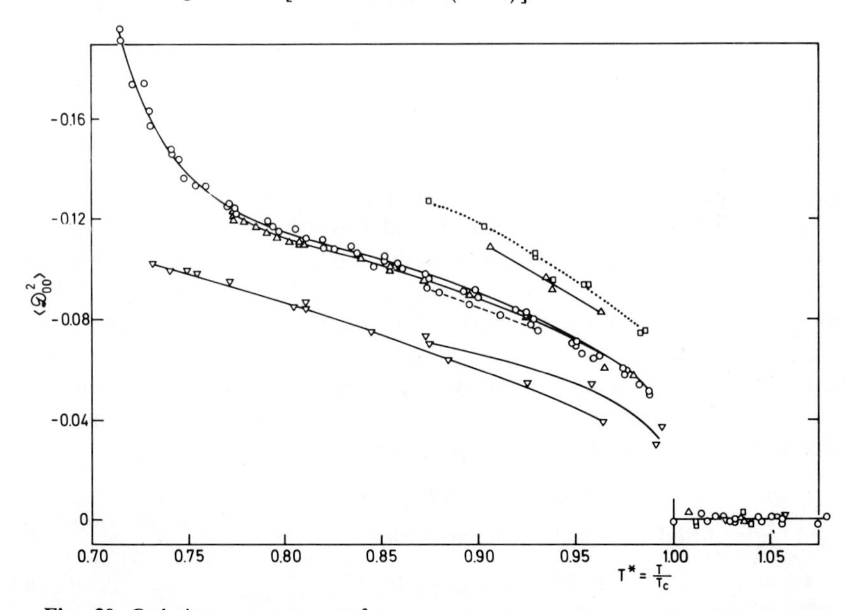

Fig. 20. Ordering parameter $\langle \mathscr{D}^2_{00} \rangle$ vs. reduced temperature $T^* \equiv T/T_c$ for PD-TEMPONE in several liquid crystals; ○, phase V solvent. [Reprinted with permission from Polnaszek and Freed, *J. Phys. Chem.* **79** (in press) (1975). Copyright by the American Chemical Society.]

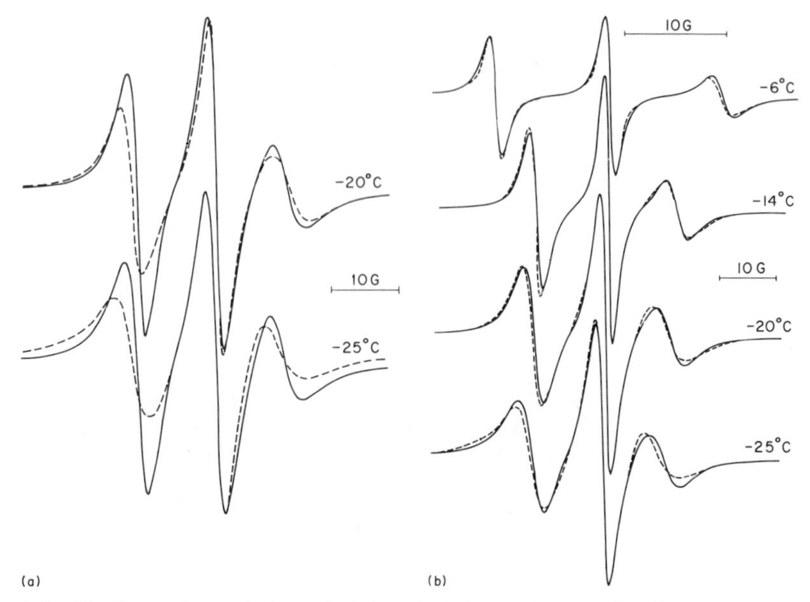

Fig. 21. Comparison of theoretical (——) and experimental (---) spectra for PD-TEMPONE in phase V liquid crystal at (a) $\varepsilon' = 1$ with $\tau_R = 2.5 \times 10^{-9}$ sec at $-20°C$ and $\tau_R = 3.6 \times 10^{-9}$ sec at $-25°C$ ($A' = 0$ G); (b) $\varepsilon'_{sec} = 1.2$, $\varepsilon'_{psec} = 20$. The values at $-6°$, $-14°$, $-20°$, and $-25°C$ are for τ_R: 0.9, 1.6, 2.5, and 3.6×10^{-9} sec, respectively, and for A': 0.55, 1.0, 1.45, and 1.75 G, respectively. The magnetic parameters are given in Table IV. [Reprinted with permission from Polnaszek and Freed, *J. Phys. Chem.* **79** (in press) (1975). Copyright by the American Chemical Society.]

there as well [i.e., the τ_R obtained from the B and C terms in Eq. (79) were no longer the same]. It was found possible to largely remove this discrepancy by the physically *unreasonable* model that anisotropic viscosity was developing such that while $\tau_{R_\perp}$ increased with η/T as is normal for a liquid, $\tau_{R_\parallel}$ remained virtually constant at $\sim 2 \times 10^{-10}$ sec. However, some alternative explanations have been proposed. To appreciate them, one must first examine Figs. 3 and 4 for isotropic liquids. There it is found that the nonsecular spectral densities for the rotational motion, which are expected to obey a Debye-type expression (cf. Chapter 2)

$$j(\omega_e) = \tau_R/(1 + \omega_e^2\tau_R^2) \tag{94}$$

were better fitted instead by the expression

$$j(\omega_e) = \tau_R/(1 + \varepsilon\omega_e^2\tau_R^2) \tag{94'}$$

with $\varepsilon \approx 5$. A similar correction was found to be the case in the work with nematic solvents both above and below the isotropic–nematic transition with a smaller one below the transition. Deviations from the simple Debye formula would be expected to arise if the rotational reorientation is significantly coupled to other degrees of freedom of the molecule or its surroundings (e.g., the angular momentum of the molecule or internal rotational degrees of freedom). A detailed statistical mechanical theory in terms of fluctuating torques experienced by the spin probe in its liquid environment has been proposed by Hwang *et al.* (1975).

The pseudosecular terms reflect the nuclear spin-flip transitions $\omega_{\pm} \approx a_N/2$ for isotropic liquids and $\langle a_N \rangle/2$ for nematics. Thus one might try

$$j(\omega_{\pm}) = \tau_R/(1 + \varepsilon' \omega_{\pm}^2 \tau_R^2) \tag{95}$$

It was found that a large value of ε' could account reasonably for the motional narrowing and slow tumbling spectra in the nematic (cf. Fig. 21b).

Actually, the analysis of this effect is fairly complex, involving incipient slow tumbling corrections from which it transpires that the secular spectral densities are now also characterized by nonzero frequencies and the pseudosecular spectral densities are modified. The proper analysis, which is complex, is discussed by Polnaszek and Freed (1975). It is found that the pseudosecular spectral densities require a correction of $\varepsilon'_{psec} \sim 15$–$20$ while for the secular spectral densities $\varepsilon'_{sec} \sim 1$–$2$. This yields satisfactory agreement with experiment (cf. Fig. 21b). Another approach to this problem, in terms of slowly fluctuating torques, has been introduced by Polnaszek and Freed (1975). It is based upon the concept of a local structure or ordering which is relaxing more slowly than the probe molecule. This structure is expected to result from the surrounding rodlike nematic solvent molecules, which reorient more slowly than the smaller spin probe. [Such effects are not included in the Meier–Saupe (1958) mean field analysis that leads to the effective pseudopotential.] A simple analysis of the spin relaxation effects of such a mechanism shows that it has many of the proper trends. We may let Ψ be the slowly relaxing set of Euler angles between the local structure and the lab frame, then for an axially symmetric potential of the spin probe relative to the local structure, rough estimates of $S_l^2 \equiv \langle \mathscr{D}_{00}^2(\Omega) \rangle^2 \sim 0.1$ and $\tau_x/\tau_R \sim 10$ are obtained from the simple model, where τ_x is the local structure relaxation time. If such a hypothesis were correct, further careful studies including larger spin probes of different shapes could shed further light on this phenomenon.

Some preliminary experimental spectra of the rodlike cholesteric spin label 3-doxyl-5α-cholestone in the viscous nematic phase of phase V are shown in Fig. 22 (Polnaszek, 1975a). The system is highly ordered and the apparent splitting constants do not change appreciably with temperature. The observed $\langle \mathscr{D}_{00}^2 \rangle$ values calculated from a motional narrowing theory analysis (cf. Chapter 8) are -0.35, -0.39, and -0.46 for $T = 26$, 3, and -26°C, respectively. The two splittings are not equal at the lowest tempera-

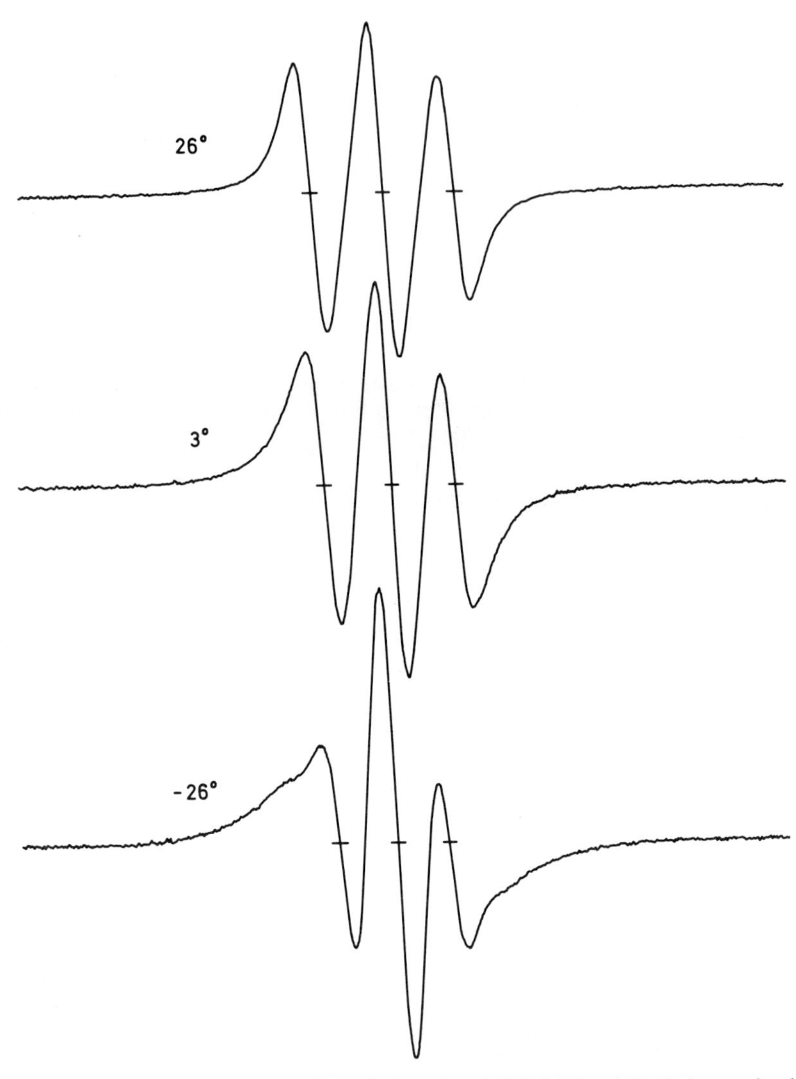

Fig. 22. Experimental spectra of the cholesteric spin label 3-doxyl-5α-cholestone in phase V at different temperatures. [From Polnaszek (1975a).]

tures, and, as already noted, this is indicative of slow motion (as is the exceedingly large value of $\langle \mathscr{D}_{00}^{2} \rangle$). The estimated τ_R for $-26°C$ is $\sim 4 \times 10^{-8}$ sec. These spectra compare favorably with the theoretical spectra simulated for high order ($\lambda = -7.5$) in Fig. 16a, although the actual ordering of the experimental spectra in Fig. 22 is somewhat less than that of the theoretical spectra.

F. Saturation and Nonlinear Effects

The general slow tumbling theory presented in Section I can also be applied to saturation phenomena (Freed *et al.*, 1971; Goldman *et al.*, 1973; Goldman, 1973; Bruno, 1973). A careful study of saturation and slow tumbling for PADS in viscous media has shown that the stochastic Liouville approach can be effectively employed to predict slow tumbling spectra (Goldman *et al.*, 1973). This is illustrated in Fig. 23. Note that the simulations were performed for axial nitroxide parameters, so the agreement should not be expected to be perfect.

In the discussion of Section II.B on simplified methods of estimating τ_R from features of the outer hyperfine extrema, it was found that in the region where $\tau_R > 10^{-8}$ sec they are quite insensitive to all parameters other than $\tau_{R_\perp}$, A_z, and A (the intrinsic width), while the central region is affected by the other magnetic tensor components. It is therefore reasonable to expect that simulations based on axial parameters should agree quite well with the saturation behavior of the outer extrema but not necessarily with the central extrema. Such a comparison is shown in Fig. 24, which shows the ratios of the $d_{\max}(\tilde{M}) = \frac{1}{2} | \gamma_e | H_{1,\max}(\tilde{M})$ [where $H_{1,\max}(\tilde{M})$ is the microwave field strength at which the $\tilde{M} = -1$, 0, or $+1$ regions of the spectra maximize, where these regions correspond, respectively, to the low field, central, and high-field extrema]. In particular, for $\tau_R > 10^{-8}$ sec, there is rather good agreement between experiment and prediction for $d_{\max}(+1)/d_{\max}(-1)$. It is possible to use such a ratio (which depends only on relative values of B_1 and not on its absolute magnitude) as a means of estimating τ_R, but this has not yet been studied in detail. In particular, we are adding the new parameter W_e, the electron-spin relaxation rate, into the analysis. Also note the rather small changes in the ratio for large changes in τ_R.

At the heart of all saturation and nonlinear phenomena is the fact that the eigenfunction expansion coefficients for the density matrix elements representing population differences {cf. Eq. (27), the terms $[C_m^{(0)}]_{\lambda^+} - [C_m^{(0)}]_{\lambda^-}$ where $n = 0$} are relaxed at rates that depend on

$$2W_e + B_L RL(L + 1) \tag{96}$$

when (1) an orientation-independent W_e is used and (2) isotropic reorientation is assumed. There is, of course, coupling among the different ESR transitions due to the nuclear spin-flip transitions induced by the pseudosecular terms in $\mathscr{H}_1(\Omega)$. Thus while the expansion coefficients for $L = 0$, representing the population difference for the proper isotropic average over orientations, is relaxed by $T_{1,e} = (2W_e)^{-1}$, the coefficients for $L > 0$ are relaxed by the rotational motion as well. Furthermore, when $R \gg W_e$ (typical values of W_e are $\sim 10^{-5}$ sec), then (1) the rotational motion is very

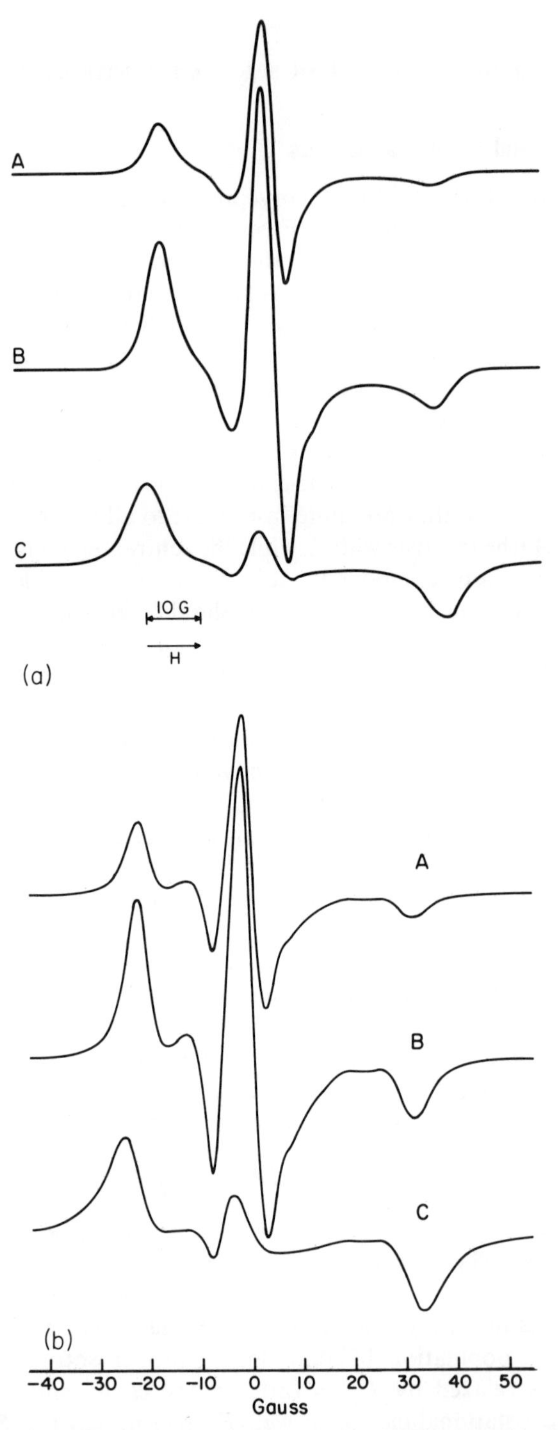

Fig. 23. (a) Experimental saturation spectra for PADS in 85% glycerol–H_2O at $T = -41°C$ and A, $d_e = 0.025$ G; B, $d_e = 0.079$ G; and C, $d_e = 0.45$ G. (b) Spectra simulated for free diffusion with $\tau_R = 2 \times 10^{-8}$ sec, $\delta = 1.2$ G, $W_e = 6.2 \times 10^4$ sec^{-1}, and A, $d_e = 0.03$ G; B, $d_e = 0.08$ G; and C, $d_e = 0.47$ G. Axial magnetic parameters were used in the simulations. $d_e = \frac{1}{2}\gamma_e H_1$. [From Goldman (1973).]

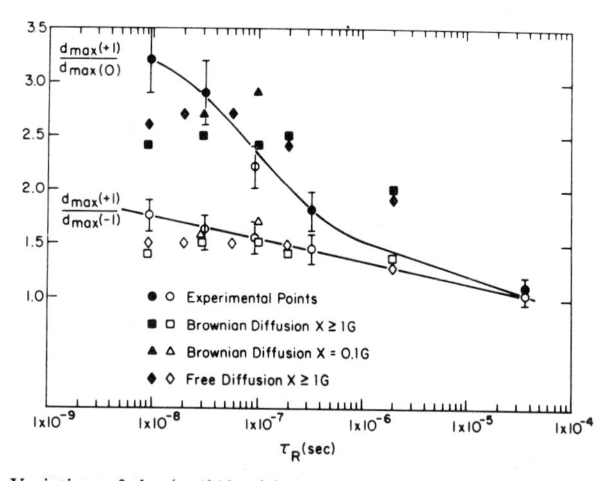

Fig. 24. Variation of $d_{max}(+1)/d_{max}(0)$ and $d_{max}(+1)/d_{max}(-1)$ with τ_R. Solid lines are drawn through the experimental points for the system PADS in 85% glycerol–H_2O. [From Goldman *et al.* (1973).]

effective in spreading the saturation throughout the spectrum, and (2) only the $L = 0$ component is effectively saturated. These ideas and the implications they have for pulsed saturation recovery studies are explored elsewhere (Freed, 1974). It is found in that work that the saturation recovery times (or at least the slowest relaxing mode) are often well approximated by $T_{1,e} \simeq (2W_e)^{-1}$ over a wide range of conditions, including the slow tumbling region.

Such considerations suggest that slow tumbling ELDOR experiments could be very interesting (Bruno, 1973; Bruno and Freed, 1974a; Smigel *et al.*, 1974). In such experiments, we saturate with a pump microwave field at a particular resonant frequency corresponding to a particular orientation (and nuclear spin) of the nitroxide and then observe at another position corresponding to a different orientation. Then the indirect saturation at the observing position will depend on the rotational motion of the molecule. We therefore expect that the ratio R/W_e will determine the relative importance of transmission of saturation to different environments (a form of "spin diffusion") versus simple ESR spin relaxation. The application of the stochastic Liouville theory to such experiments is discussed in detail elsewhere (Bruno, 1973; Bruno and Freed, 1974a; Smigel *et al.*, 1974).

Finally, we note the existence of a great variety of nonlinear phenomena, such as the modulation-frequency dependence of adiabatic rapid-passage effects explored by Hyde and Dalton (1972). This latter, more complex phenomenon can also be dealt with by stochastic Liouville methods and is discussed elsewhere (Dalton, 1973; Leniart, 1972; Thomas and McConnell, 1974). The rather striking spectral effects observed have been calibrated with typical samples by the use of the S parameter and extrapolation of τ_R with known values of η/T to yield estimates of $\tau_R \gtrsim 10^{-6}$ sec (Hyde and Dalton, 1972).

In the discussion of saturation and nonlinear effects there is an important distinction to be made with respect to intrinsic widths, which was not important in the theory for unsaturated spectra. Note that in Eqs. (28) and (29) intrinsic widths were introduced by replacing ω_λ by $\omega_\lambda + iT_2^{-1}$ with no distinction made as to whether these widths are due to homogeneous or inhomogeneous broadening. It is possible to show, from the general form of the solutions in terms of superpositions of "complex Lorentzians" (cf. Appendix A), that this is entirely adequate for the case of a Lorentzian distribution of inhomogeneous broadening and *no saturation* (Abragam, 1961; Goldman *et al.*, 1973). If the inhomogeneous widths may not be approximated as Lorentzian, then we must convolute the line shapes obtained from the slow tumbling theory given here with a more appropriate inhomogeneous line-shape function (e.g., a Gaussian). This can become a real problem when unresolved proton extra hyperfine structure is a dominant source of broadening (in which case rigorous line-shape simulations require adequate knowledge of the proton splittings and their orientation dependence). It is well known, however, that in the case of saturation, homogeneous and Lorentzian inhomogeneous lines are no longer formally equivalent (Abragam, 1961). Then Eqs. (28) and (29) are only appropriate for the homogeneous broadening. Again, inhomogeneous broadening could be accounted for by convolution methods, but for Lorentzian inhomogeneous broadening there is a simpler method. The solutions for saturated cases require both the coefficients $C_{KM}^L(i)$ of Eq. (27) and their complex conjugates $C_{KM}^L(i)^*$, to which they are coupled by the saturating terms. For homogenous broadening we replace $\omega_\lambda \to \omega_\lambda + iT_2^{-1}$ for the former, as already noted, but $\omega_\lambda \to \omega_\lambda - iT_2^{-1}$ for the latter. But for Lorentzian inhomogeneous broadening, we can merely let $\omega_\lambda \to \omega_\lambda + iT_2'^{-1}$ for both types of terms (Goldman *et al.*, 1973). The computer simulations of Fig. 23b were made for the case of homogeneous broadening.

APPENDIX A. GENERAL SOLUTIONS AND DISCUSSION OF THE COMPUTER PROGRAM FOR NITROXIDES

The equations upon which the computer program given in Appendix B for the ESR spectrum of a nitroxide in an isotropic fluid are based are (Bruno, 1973)

$$[(\omega - \omega_e + 2b) - i(T_{2,a}^{-1} + \tau_{L,K}^{-1})]\bar{C}_{K,0}^{L}(1)$$
$$- (F_0 + D' + iT_{2,b}^{-1}) \sum_{L'} N(L,L') \begin{pmatrix} L & 2 & L' \\ -K & 0 & K \end{pmatrix} \begin{pmatrix} L & 2 & L' \\ 0 & 0 & 0 \end{pmatrix} \bar{C}_{K,0}^{L'}(1)$$
$$- (F_2 + D^{(2)\prime}) \sum_{L'} N(L,L') \begin{pmatrix} L & 2 & L' \\ 0 & 0 & 0 \end{pmatrix}$$
$$\times \left[\begin{pmatrix} L & 2 & L' \\ -K & -2 & K+2 \end{pmatrix} \bar{C}_{K+2,0}^{L'}(1) \right.$$
$$\left. + \begin{pmatrix} L & 2 & L' \\ -K & 2 & K-2 \end{pmatrix} \bar{C}_{K-2,0}^{L'}(1) \right]$$
$$+ D \sum_{L'} N(L,L') \begin{pmatrix} L & 2 & L' \\ -K & 0 & K \end{pmatrix} \begin{pmatrix} L & 2 & L' \\ 0 & -1 & 1 \end{pmatrix}$$
$$\times \bar{C}_{K,1}^{L'}(4,5) + D^{(2)} \sum_{L'} N(L,L') \begin{pmatrix} L & 2 & L' \\ 0 & -1 & 1 \end{pmatrix}$$
$$\times \left[\begin{pmatrix} L & 2 & L' \\ -K & -2 & K+2 \end{pmatrix} \bar{C}_{K+2,1}^{L'}(4,5) \right.$$
$$\left. + \begin{pmatrix} L & 2 & L' \\ -K & 2 & K-2 \end{pmatrix} \bar{C}_{K-2,1}^{L'}(4,5) \right] = 2^{1/2} q\omega_1 d_1 \delta(L,0)\delta(K,0) \quad \text{(A.1)}$$

$$[(\omega - \omega_e) - i(T_{2,a}^{-1} + \tau_{L,K}^{-1})]\bar{C}_{K,0}^{L}(2)$$
$$- (F_0 + iT_{2,b}^{-1}) \sum_{L'} N(L,L') \begin{pmatrix} L & 2 & L' \\ -K & 0 & K \end{pmatrix} \begin{pmatrix} L & 2 & L' \\ 0 & 0 & 0 \end{pmatrix} \bar{C}_{K,0}^{L'}(2)$$
$$- F_2 \sum_{L'} N(L,L') \begin{pmatrix} L & 2 & L' \\ 0 & 0 & 0 \end{pmatrix} \left[\begin{pmatrix} L & 2 & L' \\ -K & -2 & K+2 \end{pmatrix} \bar{C}_{K+2,0}^{L'}(2) \right.$$
$$\left. + \begin{pmatrix} L & 2 & L' \\ -K & 2 & K-2 \end{pmatrix} \bar{C}_{K-2,0}^{L'}(2) \right] + D \sum_{L'} N(L,L') \begin{pmatrix} L & 2 & L' \\ 0 & -1 & 1 \end{pmatrix}$$
$$\times \begin{pmatrix} L & 2 & L' \\ -K & 0 & K \end{pmatrix} [\bar{C}_{K,1}^{L'}(4,5) + \bar{C}_{K,1}^{L'}(6,7)] + D^{(2)} \sum_{L'} N(L,L')$$
$$\times \begin{pmatrix} L & 2 & L' \\ 0 & -1 & 1 \end{pmatrix}$$
$$\times \left\{ \begin{pmatrix} L & 2 & L' \\ -K & -2 & K+2 \end{pmatrix} [\bar{C}_{K+2,1}^{L'}(4,5) + \bar{C}_{K+2,1}^{L'}(6,7)] \right.$$
$$\left. + \begin{pmatrix} L & 2 & L' \\ -K & 2 & K-2 \end{pmatrix} [\bar{C}_{K-2,1}^{L'}(4,5) + \bar{C}_{K-2,1}^{L'}(6,7)] \right\}$$
$$= 2^{1/2} q\omega_2 d_2 \delta(L,0)\delta(K,0) \quad \text{(A.2)}$$

$$
\begin{aligned}
&[(\omega - \omega_e - 2b) - i(T_{2,a}^{-1} + \tau_{L,K}^{-1})]\bar{C}_{K,0}^{L'}(3) \\
&\quad - (F_0 - D' + iT_{2,b}^{-1}) \sum_{L'} N(L, L') \begin{pmatrix} L & 2 & L' \\ -K & 0 & K \end{pmatrix} \begin{pmatrix} L & 2 & L' \\ 0 & 0 & 0 \end{pmatrix} \bar{C}_{K,0}^{L'}(3) \\
&\quad - (F_2 - D^{(2)'}) \sum_{L'} N(L, L') \begin{pmatrix} L & 2 & L' \\ 0 & 0 & 0 \end{pmatrix} \\
&\quad \times \left[\begin{pmatrix} L & 2 & L' \\ -K & -2 & K+2 \end{pmatrix} \bar{C}_{K+2,0}^{L'}(3) \right. \\
&\quad \left. + \begin{pmatrix} L & 2 & L' \\ -K & 2 & K-2 \end{pmatrix} \bar{C}_{K-2,0}^{L'}(3) \right] \\
&\quad + D \sum_{L'} N(L, L') \begin{pmatrix} L & 2 & L' \\ -K & 0 & K \end{pmatrix} \begin{pmatrix} L & 2 & L' \\ 0 & -1 & 1 \end{pmatrix} \\
&\quad \times \bar{C}_{K,1}^{L'}(6, 7) + D^{(2)} \sum_{L'} N(L, L') \begin{pmatrix} L & 2 & L' \\ 0 & -1 & 1 \end{pmatrix} \\
&\quad \times \left[\begin{pmatrix} L & 2 & L' \\ -K & -2 & K+2 \end{pmatrix} \right. \\
&\quad \left. \times \bar{C}_{K+2,1}^{L'}(6, 7) + \begin{pmatrix} L & 2 & L' \\ -K & 2 & K-2 \end{pmatrix} \bar{C}_{K-2,1}^{L'}(6, 7) \right] \\
&= 2^{1/2} q \omega_3 d_3 \delta(L, 0) \delta(K, 0)
\end{aligned} \tag{A.3}
$$

$$
\begin{aligned}
&[(\omega - \omega_e + b) - i(T_{2,a}^{-1} + \tau_{L,K}^{-1})]\bar{C}_{K,1}^{L}(4, 5) \\
&\quad + (F_0 + \tfrac{1}{2}D' + iT_{2,b}^{-1}) \sum_{L'} N(L, L') \begin{pmatrix} L & 2 & L' \\ -K & 0 & K \end{pmatrix} \\
&\quad \times \begin{pmatrix} L & 2 & L' \\ -1 & 0 & 1 \end{pmatrix} \bar{C}_{K,1}^{L'}(4, 5) \\
&\quad + (F_2 + \tfrac{1}{2}D^{(2)'}) \sum_{L'} N(L, L') \begin{pmatrix} L & 2 & L' \\ -1 & 0 & 1 \end{pmatrix} \\
&\quad \times \left[\begin{pmatrix} L & 2 & L' \\ -K & -2 & K+2 \end{pmatrix} C_{K+2,1}^{L'}(4, 5) \right. \\
&\quad \left. + \begin{pmatrix} L & 2 & L' \\ -K & 2 & K-2 \end{pmatrix} \bar{C}_{K-2,1}^{L'}(4, 5) \right] + D \sum_{L'} N(L, L') \begin{pmatrix} L & 2 & L' \\ -1 & 1 & 0 \end{pmatrix} \\
&\quad \times \begin{pmatrix} L & 2 & L' \\ -K & 0 & K \end{pmatrix} [\bar{C}_{K,0}^{L'}(1) + \bar{C}_{K,0}^{L'}(2)] \\
&\quad + D^{(2)} \sum_{L'} N(L, L') \begin{pmatrix} L & 2 & L' \\ -1 & 1 & 0 \end{pmatrix}
\end{aligned}
$$

(equation continues)

$$
\begin{aligned}
&\times \left\{ \begin{pmatrix} L & 2 & L' \\ -K & -2 & K+2 \end{pmatrix} [\bar{C}^{L'}_{K+2,0}(1) + \bar{C}^{L'}_{K+2,0}(2)] \right. \\
&+ \begin{pmatrix} L & 2 & L' \\ -K & 2 & K-2 \end{pmatrix} [\bar{C}^{L'}_{K-2,0}(1) \\
&\left. + \bar{C}^{L'}_{K-2,0}(2)] \right\} - 2^{-1/2} D \sum_{L'} N(L, L') \begin{pmatrix} L & 2 & L' \\ -K & 0 & K \end{pmatrix} \begin{pmatrix} L & 2 & L' \\ -1 & -1 & 2 \end{pmatrix} \\
&\times \bar{C}^{L'}_{K,2}(8, 9) - 2^{-1/2} D^{(2)} \sum_{L'} N(L, L') \begin{pmatrix} L & 2 & L' \\ -1 & -1 & 2 \end{pmatrix} \\
&\times \left[\begin{pmatrix} L & 2 & L' \\ -K & -2 & K+2 \end{pmatrix} \right. \\
&\left. \times \bar{C}^{L'}_{K+2,2}(8, 9) + \begin{pmatrix} L & 2 & L' \\ -K & 2 & K-2 \end{pmatrix} \bar{C}^{L'}_{K-2,2}(8, 9) \right] = 0 \qquad \text{(A.4)}
\end{aligned}
$$

$$
\begin{aligned}
&[(\omega - \omega_e - b) - i(T^{-1}_{2,a} + \tau^{-1}_{L,K})] C^{L}_{K,1}(6, 7) \\
&+ (F_0 - \tfrac{1}{2} D' + i T^{-1}_{2,b}) \sum_{L'} N(L, L') \begin{pmatrix} L & 2 & L' \\ -K & 0 & K \end{pmatrix} \\
&\times \begin{pmatrix} L & 2 & L' \\ -1 & 0 & 1 \end{pmatrix} \bar{C}^{L'}_{K,1}(6, 7) \\
&+ (F_2 - \tfrac{1}{2} D^{(2)\prime}) \sum_{L'} N(L, L') \begin{pmatrix} L & 2 & L' \\ -1 & 0 & 1 \end{pmatrix} \\
&\times \left[\begin{pmatrix} L & 2 & L' \\ -K & -2 & K+2 \end{pmatrix} \bar{C}^{L'}_{K+2,1}(6, 7) \right. \\
&\left. + \begin{pmatrix} L & 2 & L' \\ -K & 2 & K-2 \end{pmatrix} C^{L'}_{K-2,1}(6, 7) \right] + D \sum_{L'} N(L, L') \begin{pmatrix} L & 2 & L' \\ -K & 0 & K \end{pmatrix} \\
&\times \begin{pmatrix} L & 2 & L' \\ -1 & 1 & 0 \end{pmatrix} [\bar{C}^{L'}_{K,0}(2) + \bar{C}^{L'}_{K,0}(3)] \\
&+ D^{(2)} \sum_{L'} N(L, L') \begin{pmatrix} L & 2 & L' \\ -1 & 1 & 0 \end{pmatrix} \\
&\times \left\{ \begin{pmatrix} L & 2 & L' \\ -K & -2 & K+2 \end{pmatrix} [\bar{C}^{L'}_{K+2,0}(2) + \bar{C}^{L'}_{K+2,0}(3)] \right. \\
&\left. + \begin{pmatrix} L & 2 & L' \\ -K & 2 & K-2 \end{pmatrix} [\bar{C}^{L'}_{K-2,0}(2) + \bar{C}^{L'}_{K-2,0}(3)] \right\} \\
&- 2^{-1/2} D \sum_{L'} N(L, L') \begin{pmatrix} L & 2 & L' \\ -K & 0 & K \end{pmatrix} \begin{pmatrix} L & 2 & L' \\ -1 & -1 & 2 \end{pmatrix} \bar{C}^{L'}_{K,2}(8, 9)
\end{aligned}
$$

(equation continues)

$$
\begin{aligned}
&- 2^{-1/2} D^{(2)} \sum_{L'} N(L, L') \begin{pmatrix} L & 2 & L' \\ -1 & -1 & 2 \end{pmatrix} \\
&\times \left[\begin{pmatrix} L & 2 & L' \\ -K & -2 & K+2 \end{pmatrix} \bar{C}^{L'}_{K+2,\,2}(8, 9) \right. \\
&+ \left. \begin{pmatrix} L & 2 & L' \\ -K & 2 & K-2 \end{pmatrix} \bar{C}^{L'}_{K-2,\,2}(8, 9) \right] = 0
\end{aligned}
\tag{A.5}
$$

$$
\begin{aligned}
&[(\omega - \omega_e) - i(T_{2,a}^{-1} + \tau_{L,K}^{-1})] \bar{C}^{L}_{K,\,2}(8, 9) \\
&- (F_0 + i T_{2,b}^{-1}) \sum_{L'} N(L, L') \begin{pmatrix} L & 2 & L' \\ -K & 0 & K \end{pmatrix} \\
&\times \begin{pmatrix} L & 2 & L' \\ -2 & 0 & 2 \end{pmatrix} \bar{C}^{L'}_{K,\,2}(8, 9) \\
&- F_2 \sum_{L'} N(L, L') \begin{pmatrix} L & 2 & L' \\ -2 & 0 & 2 \end{pmatrix} \left[\begin{pmatrix} L & 2 & L' \\ -K & -2 & K+2 \end{pmatrix} \bar{C}^{L'}_{K+2,\,2}(8, 9) \right. \\
&+ \left. \begin{pmatrix} L & 2 & L' \\ -K & 2 & K-2 \end{pmatrix} \bar{C}^{L'}_{K-2,\,2}(8, 9) \right] - 2^{-1/2} D \sum_{L'} N(L, L') \\
&\times \begin{pmatrix} L & 2 & L' \\ -K & 0 & K \end{pmatrix} \\
&\times \begin{pmatrix} L & 2 & L' \\ -2 & 1 & 1 \end{pmatrix} [\bar{C}^{L'}_{K,\,1}(4, 5) + \bar{C}^{L'}_{K,\,1}(6, 7)] \\
&- 2^{-1/2} D^{(2)} \sum_{L'} N(L, L') \begin{pmatrix} L & 2 & L' \\ -2 & 1 & 1 \end{pmatrix} \\
&\times \left\{ \begin{pmatrix} L & 2 & L' \\ -K & -2 & K+2 \end{pmatrix} [\bar{C}^{L'}_{K+2,\,1}(4, 5) + \bar{C}^{L'}_{K+2,\,1}(6, 7)] \right. \\
&+ \left. \begin{pmatrix} L & 2 & L' \\ -K & 2 & K-2 \end{pmatrix} [\bar{C}^{L'}_{K-2,\,1}(4, 5) + \bar{C}^{L'}_{K-2,\,1}(6, 7)] \right\} = 0
\end{aligned}
\tag{A.6}
$$

Many of the terms in Eqs. (A.1)–(A.6) have been defined in Eqs. (75) and (76). ω_e is the electron–spin Larmor frequency and $b = -\frac{1}{2} a_N |\gamma_e|$.

The absorption is given by

$$
\overline{Z_1''} + \overline{Z_2''} + \overline{Z_3''} = \text{Im} \sum_{i=1}^{3} C^{0}_{0,\,0}(i) \tag{A.7}
$$

The following definitions have also been introduced:

$$
\bar{C}^{L}_{K,\,M}(i) = 2^{-1/2} [C^{L}_{K,\,M}(i) \pm C^{L}_{-K,\,M}(i)], \qquad K \geqq 0 \tag{A.8}
$$

where the plus sign is used for even L and the minus sign for odd L. For even L, i refers to all transitions; for odd L, i refers only to transitions 4, 5, 6, 7, 8, and 9;

$$\bar{C}^L_{K,1}(i,j) = 2^{-1/2}[\bar{C}^L_{K,1}(i) \mp \bar{C}^L_{K,-1}(j)] \tag{A.9}$$

where (i, j) are either (4, 5) or (6, 7), and the minus (plus) sign is for even (odd) L; and

$$\bar{C}^L_{K,2}(8,9) = 2^{-1/2}[\bar{C}^L_{K,2}(8) \pm \bar{C}^L_{K,-2}(9)] \tag{A.10}$$

where the plus (minus) sign is for even (odd) L. Also,

$$N(L, L') = [(2L+1)(2L'+1)]^{1/2} \tag{A.11}$$

and

$$T^{-1}_{2,a} = \alpha + \tfrac{1}{3}\beta \tag{A.12a}$$

$$T^{-1}_{2,b} = \tfrac{2}{3}\beta \tag{A.12b}$$

The quantity $\left(\begin{smallmatrix} L & 2 & L' \\ 0 & 0 & 0 \end{smallmatrix}\right)$ is a $3j$ symbol, whose values are tabulated or given by formulas (Rotenberg *et al.*, 1959; Edmonds, 1957). These are used to evaluate the integrals on the LHS of Eq. (27), utilizing

$$\int d\Omega \mathscr{D}^{L_1}_{m_1,m_1'}(\Omega)\mathscr{D}^{L_2}_{m_2,m_2'}(\Omega)\mathscr{D}^{L_3}_{m_3,m_3'}(\Omega) = 8\pi^2 \begin{pmatrix} L_1 & L_2 & L_3 \\ m_1 & m_2 & m_3 \end{pmatrix} \begin{pmatrix} L_1 & L_2 & L_3 \\ m_1' & m_2' & m_3' \end{pmatrix} \tag{A.13}$$

with the relationship

$$\mathscr{D}^{L*}_{m,m'}(\Omega) = (-)^{m-m'}\mathscr{D}^{L}_{-m,-m'}(\Omega) \tag{A.14}$$

There are a number of symmetry relations (Edmonds, 1957) in a $3j$ symbol $\left(\begin{smallmatrix} L_1 & L_2 & L_3 \\ m_1 & m_2 & m_3 \end{smallmatrix}\right)$. Among the more useful ones are: (1) The sum of the m values must be zero. (2) Naturally, L must be positive and the absolute value of m_i must not be greater than the corresponding L_i in a given column. (3) the columns can be permuted without changing the value of the $3j$ symbol if the sum of L values is even. If the sum of L values is odd, then a permutation of the columns results in a change in sign for the value of the $3j$ symbol. (4) If the sum of L values is even, then all the m values can change sign without changing the value of the $3j$ symbol. If the sum of L values is odd, then a change in sign for all m values results in a change in sign for the value of the $3j$ symbol. (5) The triangular property holds whereby the sum of any two L values must be equal to or greater than the third L value. Properties (4) and (5) result in the equations in L being coupled only to the equations in L and $L \pm 2$, for the three allowed transitions; but the forbidden transitions (only when there are asymmetric magnetic parameters) result in a coupling to equations in $L \pm 1$ as well. The coupling of odd L values is important only for slow motions when $\tau_R \geq 5 \times 10^{-8}$ sec. The required $3j$ symbols are automatically calculated in the computer program in Appendix B.

The solutions in Eqs. (A.1)–(A.6) can be expressed by the matrix equation

$$\mathscr{A}\mathbf{C} = \mathbf{U} \tag{A.15}$$

where $\mathbf{C}$ is an r-dimensional column vector consisting of the independent variables [i.e., the expansion coefficients $C^L_{K,M}(i)$], which are to be solved for, $\mathbf{U}$ is an r-dimensional column vector of constants given by the RHS of Eqs. (A.1)–(A.6), and $\mathscr{A}$ is an $r \times r$ square matrix formed by the coefficients of the $C^L_{KM}(i)$ variables in Eqs. (A.1)–(A.6). For purposes of computer programming efficiency, it is best that the coefficient matrix $\mathscr{A}$ be complex symmetric.

The above equations, except for saturation, are "naturally" symmetric because orthonormal expansions as well as a properly "normalized" linear combination of terms were used. [Actually, one must first replace the $\bar{C}^L_{0,M}(i)$ in Eqs. (A.1)–(A.6) by $C^L_{0,M}(i)$ to be able to render $\mathscr{A}$ symmetric.]

The diagonalization method for obtaining a spectrum is especially applicable, since Eq. (A.15) can be written in the form

$$(\mathscr{A}' + k\mathbf{1})\mathbf{C} = \mathbf{U} \tag{A.15$'$}$$

where $\mathscr{A}'$ does not contain the "sweep variable" (e.g., ω_e for a field-swept spectrum), $\mathbf{1}$ is the unit matrix, and k is a constant containing the sweep variable [i.e., $k = (\omega - \omega_e) - iT^{-1}_{2,a}$].

For a nondegenerate complex symmetric matrix $\mathscr{A}'$, a complex orthogonal matrix $\mathbf{O}$ exists such that

$$\mathbf{O}^{\mathrm{tr}}\mathscr{A}'\mathbf{O} = \mathscr{A}_{\mathrm{d}} \tag{A.16}$$

where $\mathscr{A}_{\mathrm{d}}$ is a diagonal matrix and $\mathbf{O}^{\mathrm{tr}}$ is the transpose of $\mathbf{O}$. Premultiplication of Eq. (A.15′) by $\mathbf{O}^{\mathrm{tr}}$, use of Eq. (A.16), and noting that $\mathbf{O}^{\mathrm{tr}} = \mathbf{O}^{-1}$ leads to

$$(\mathscr{A}_{\mathrm{d}} + k\mathbf{1})\mathbf{O}^{\mathrm{tr}}\mathbf{C} = \mathbf{O}^{\mathrm{tr}}\mathbf{U} \tag{A.17}$$

or

$$\mathbf{C} = \mathbf{O}(\mathscr{A}_{\mathrm{d}} + k\mathbf{1})^{-1}\mathbf{O}^{\mathrm{tr}}\mathbf{U} \tag{A.17$'$}$$

Note that the absorption $\bar{Z}''$ is given by

$$\bar{Z}'' = \mathrm{Im}[\mathbf{DC}] \tag{A.18}$$

where

$$\mathbf{D} = Y\mathbf{U} \tag{A.19a}$$

with

$$Y = [2^{-1/2} q \omega_e d_e]^{-1} \tag{A.19b}$$

where we use the fact that $\omega_1 \cong \omega_2 \cong \omega_3 \cong \omega_e$ in high fields and $d_1 = d_2 = d_3 = d_e$. It follows from Eqs. (A.17) and (A.18) that

$$\overline{Z''} = \mathrm{Im}[(\mathbf{DO})(\mathscr{A}_d + k\mathbf{1})^{-1}(\mathbf{O}^{tr}\mathbf{U})] \tag{A.20}$$

which can be written for the r dimensions as

$$\overline{Z''} = Y \,\mathrm{Im}\left[\sum_{i=1}^{r} \frac{(\mathbf{O}^{tr}\mathbf{U})_i^2}{(\mathscr{A}_d)_{ii} + (\omega - \omega_e) - iT_{2,a}^{-1}}\right] \tag{A.21}$$

and for the first derivative of an absorption field-swept spectrum

$$\frac{d\overline{Z''}}{d\omega_e} = Y \,\mathrm{Im}\left[\sum_{i=1}^{r} \frac{(\mathbf{O}^{tr}\mathbf{U})_i^2}{[(\mathscr{A}_d)_{ii} + (\omega - \omega_e) - iT_{2,a}^{-1}]^2}\right] \tag{A.22}$$

Thus only a single diagonalization is required to calculate an absorption line shape or the nth derivative of the absorption. Another advantage is that $\mathscr{A}_d$ and $\mathbf{O}$ are not functions of $T_{2,a}^{-1}$, so that spectral line shapes can be calculated for different values of $T_{2,a}^{-1}$ without performing additional diagonalizations.

The diagonalization subroutine used for the slow tumbling computer program in Appendix B is due to Gordon and Messenger (1972). This subroutine had the fastest execution time of all diagonalization subroutines tried. Besides its speed in diagonalizing a matrix, it has characteristics that make it especially useful in solving the slow motional equations. First, the subroutine takes advantage of the symmetry of $\mathscr{A}$, so that only the elements to one side of the diagonal are stored. Second, the subroutine retains the banded nature of the equations, so that only the subdiagonals containing nonzero elements are stored. Third, the subroutine performs the operation $(\mathbf{O}^{tr}\mathbf{U})$ "instantaneously" for each step of the diagonalization, so that only a single column vector $(\mathbf{O}^{tr}\mathbf{U})$ need be stored rather than the construction and storage of the entire $r \times r$ $\mathbf{O}^{tr}$ matrix. A modified version of this subroutine, which can be used to obtain all the eigenvectors, is given by Bruno (1973).

Anisotropic Liquids

Because $\tilde{\Gamma}_\Omega$ for anisotropic liquids is composed of a simple sum of terms in $\mathscr{D}_{K,M}^L(\Omega)$ plus the isotropic liquid Γ_Ω, there is much similarity between the equations for anisotropic liquids and those for isotropic liquids. In fact, the resulting equations for anisotropic liquids can be obtained by simple modifications of the isotropic liquid equations (A.1)–(A.6). These modifications can be specified by the following definitions for the simple case of a Meier–Saupe potential, in which **n** is completely aligned by the magnetic field (Polnaszek *et al.* (1973):

$$C_A(j) = -i\frac{2}{15}R\lambda^2 C^L_{K,M}(j) + i2R\lambda\left(1 - \frac{\lambda}{21}\right)(-)^{K-M} \times \sum_{L'} N(L, L')\begin{pmatrix} L & 2 & L' \\ -K & 0 & K \end{pmatrix}\begin{pmatrix} L & 2 & L' \\ -M & 0 & M \end{pmatrix} C^{L'}_{K,M}(j) + i\frac{8}{35}R\lambda^2 \times (-)^{K-M}\sum_{L'} N(L, L')\begin{pmatrix} L & 4 & L' \\ -K & 0 & K \end{pmatrix}\begin{pmatrix} L & 4 & L' \\ -M & 0 & M \end{pmatrix} C^{L'}_{K,M}(j) \tag{A.23}$$

and

$$U^L_A = q\omega_e d_e \delta(K, 0)\delta(M, 0)\frac{(2L+1)^{1/2}}{I''_0}\int_{-1}^{1} \mathscr{D}^L_{0,0}(0, \beta, 0)\exp\left(\frac{1}{2}x^2\lambda\right) dx \tag{A.24}$$

with $I''_0 = \int_{-1}^{1} \exp(\frac{1}{2}x^2\lambda)\, dx$ and λ given by Eq. (60a). The addition of the term $C_A(j)$ to the left-hand side and the replacement of the right-hand side by the term U^L_A for each of the respective isotropic equations (A.1)–(A.6) gives the desired set of equations for the anisotropic liquid.

The equations for Brownian axially symmetric rotational diffusion and for anisotropic (axially symmetric) viscosity are obtained by using Eqs. (61a), (61b), and (62a)–(62f) and Eqs. (63), (64), and (65a)–(65f), respectively.

The absorption is, from Eq. (23) or Eq. (13), proportional to

$$\overline{Z''} = \text{Im}\sum_{j=1}^{3}\sum_{L}\left\{\frac{(2L+1)^{1/2}}{I''_0}\left[\int_{-1}^{1} dx\, \mathscr{D}^L_{0,0}(0, \beta, 0)\exp\left(\frac{1}{2}x^2\lambda\right)\right] C^L_{0,0}(j)\right\} \tag{A.25}$$

If the anisotropic liquid equations are written in the matrix notation of Eq. (A.15), then from Eqs. (A.24) and (A.25) the absorption can again be written as proportional to

$$\overline{Z''} = Y\,\text{Im}[\mathbf{U}\cdot\mathbf{C}] \tag{A.26}$$

In the numerical evaluation of the terms in U^L_A the following recursion formulas are useful:

$$\mathscr{D}^L_{0,0}(x) = [(2L-1)x\mathscr{D}^{L-1}_{0,0}(x) - (L-1)\mathscr{D}^{L-2}_{0,0}(x)]/L \tag{A.27a}$$

and

$$I_n = \int_{-1}^{1} x^n \exp(x^2\lambda_2)\, dx = \lambda_2^{-1}[\exp(\lambda_2) - \tfrac{1}{2}(n-1)I_{n-2}] \tag{A.27b}$$

for n even. Appropriate expressions for the more complicated cases are given by Polnaszek (1975a).

Convergence for isotropic spectra has been discussed in Section II.E in terms of an n_L. The effect of having $\lambda \neq 0$ means that the ordering affects the convergence of the solutions. For example, in the motionally narrowed region where $n_L = 2$ is sufficient for isotropic liquids, then for $\lambda = -0.9$ (weak ordering) one needs $n_L = 4$ and for $\lambda = -3.5$ (moderately strong ordering) $n_L = 6$. It appears safe, for the slow motional region, to use as n_L the sum of the value required for isotropic liquids and $n_L' - 2$, where n_L' is required for convergence for that value of λ in the motional narrowing region, although usually smaller values of n_L may be used.

Computer programs are given by Polnaszek (1975a) for calculating nitroxide line shapes when (1) the asymmetric potential defined by Eq. (58′) describes the orientation of a nitroxide radical for which the principal magnetic (x''', y''', z''') and orientation (x', y', z') axes are coincident, and (a) Eqs. (61a), (61b), and (62a)–(62f) apply or (b) Eqs. (64) and (65a)–(65f) apply; (2) a Meier–Saupe potential is used; but the z' and z''' axes are titled by angle β'; (3) different reorientational models are used for a Meier–Saupe potential. All these programs contain the correction terms for nonsecular contributions to the resonant frequency shifts.

All the programs, including that given in Appendix B, have been written in FORTRAN IV language for an IBM 360/65 computer.

APPENDIX B. COMPUTER PROGRAM FOR SLOW TUMBLING NITROXIDES IN ISOTROPIC LIQUIDS

The following program was written in FORTRAN IV and is listed with 72 print positions per line. One of the subroutines in this program has been taken from Gordon and Messenger (1972), pp. 376–381, and has been reproduced with 60 print positions per line. The Gordon and Messenger subroutine is used with permission of Plenum Press.

```
                NITROXIDE FREE RADICALS

                        I=1,S=1/2
  -ASYMMETRIC  G  AND  A  TENSORS
  -ISOTROPIC OR AXIALLY SYMMETRIC ROTATIONAL REORIENTATION
  -MODELS OF ROTATIONAL REORIENTATION
  -K,L(ODD),L(DOUBLY FORBIDDEN TRANSITION)  TRUNCATION
```

THIS PROGRAM SIMULATES ESR SPECTRAL LINESHAPES FOR THE ROTATIONAL MODULATION OF AN UNPAIRED ELECTRON INTERACTING WITH A SINGLE NUCLEUS OF SPIN I=1. THE HYPERFINE LINES ARE BROADENED BY COMPLETELY ASYMMETRIC G-TENSOR AND A-TENSOR (ELECTRON-NUCLEAR DIPOLAR TENSOR). ROTATIONAL REORIENTATION MAY BE ISOTROPIC OR AXIALLY SYMMETRIC ABOUT A PRINCIPAL AXIS OF THE G AND A TENSOR (ASSUMED TO COINCIDE). OPTION OF ROTATIONAL REORIENTATION MODELS INCLUDE BROWNIAN DIFFUSION,SIMPLE 'FREE' DIFFUSION, OR JUMP REORIENTATION (STRONG OR INTERMEDIATE). OUTPUT MAY BE DISPERSION, ABSORPTION, OR THEIR FIRST DERIVATIVES.

IN THE DERIVATION OF THE EQUATIONS THAT ARE USED, THE FOLLOWING SIMPLIFICATIONS WERE MADE:
1. THE PRINCIPAL AXES FOR THE G AND A TENSORS WERE ASSUMED TO COINCIDE.
2. THE NUCLEAR ZEEMAN TERM WAS SET TO ZERO.
3. NON-SECULAR AND SATURATION TERMS WERE NEGLECTED.

THESE SIMPLIFICATIONS ARE USUALLY APPROPRIATE FOR NITROXIDE FREE RADICALS.

OPTION ON THE TERMINATING VALUES (KTERM,LDOUB,LODD) FOR DIFFERENT TYPES OF COEFFICIENT TERMS MAY BE MADE. THE ASYMMETRIC EXPANSION TERMS DUE TO THE DIFFERENCES IN THE X AND Y PRINCIPAL AXIS COMPONENTS OF THE A AND G TENSORS MAY BE TRUNCATED AT A SEPARATE VALUE (DESIGNATED BY KTERM). THE EXPANSION TERMS FOR THE 'DOUBLY FORBIDDEN' TRANSITIONS MAY BE TRUNCATED AT A SEPARATE VALUE (DESIGNATED BY LDOUB). ALSO THOSE EXPANSION TERMS WITH ODD L VALUES MAY BE TRUNCATED AT A SEPARATE VALUE (DESIGNATED BY LODD).

FOR THEORY, SEE FREED,BRUNO,POLNASZEK(J.PHYS.CHEM.75, 3385(1971)) AND GOLDMAN,BRUNO,POLNASZEK,FREED(J.CHEM.PHYS 56,716(1972)). THE DIAGONALIZATION METHOD IS USED TO SOLVE THE APPROPRIATELY SYMMETRIZED EQUATIONS GIVEN IN G. BRUNO THESIS,EQ.(II.C.37) .

LIST OF INPUT VARIABLES
(IN ORDER OF APPEARANCE IN PROGRAM)

```
 NPT = NUMBER OF PLOTS OR SETS OF PLOTS WHICH DIFFER IN
       G-TENSOR,A-TENSOR, OR ROTATIONAL REORIENTATION
       PARAMETERS (SPECIFICALLY  AX,AY,AZ,GX,GY,GZ,R,RZ,
       IROT,RTAU,BCEN,L,KTERM,LDOUB,LODD). PLOTS DIFFERING
       ONLY IN  T2IN  ARE NOT INCLUDED. (SEE BELOW FOR
       DEFINITIONS)

 LMAX = MAXIMUM  L  VALUE  TO BE USED FOR ANY PLOT

 NDIM = DIMENSION GIVEN IN THE DIMENSION STATEMENT OF THE
        MAIN PROGRAM FOR MATRICES  CSYM(NDIM,JDIM),
        CSYM1(NDIM,CSYM2(NDIM),VALUE(NDIM),AMP(NDIM).

 JDIM = DIMENSION GIVEN IN THE DIMENSION STATEMENT OF THE
        MAIN PROGRAM FOR MATRIX  CSYM(NDIM,JDIM)

 NSY = DIMENSION GIVEN IN DIMENSION STATEMENT OF MAIN
       PROGRAM FOR MATRIX  SY(NSY,8).

 NSY2 = DIMENSION GIVEN IN THE DIMENSION STATEMENT OF THE
        MAIN PROGRAM FOR MATRIX  SY2(NSY2,3).

 TITLE = TITLE FOR EACH OF THE  NPT  SETS OF PLOTS.

 AX,AY,AZ = VALUES OF THE A-TENSOR (ELECTRON-NUCLEAR
            DIPOLAR TENSOR PLUS FERMI-CONTACT TERM)
            COMPONENTS ALONG THE  X,Y,Z  PRINCIPAL AXES,
            RESPECTIVELY.
            -VALUE IN GAUSS.  (TYPICAL NITROXIDE SPIN
             LABEL VALUES FOR Z-AXIS ALONG THE NITROGEN
             2P-PI ORBITAL ARE  (6,5,32) RESPECTIVELY).

 GX,GY,GZ = G-TENSOR COMPONENTS ALONG THE  X,Y,Z
            PRINCIPAL AXES, RESPECTIVELY.

 R,RZ = ROTATIONAL DIFFUSION RATE CONSTANT ABOUT THE  X,Y
        AXES AND THE  Z  AXIS, RESPECTIVELY.
        -VALUE IS IN RADIANS PER SECOND.  (FOR ISOTROPIC
         BROWNIAN ROTATIONAL DIFFUSION, ROTATIONAL
         CORRELATION TIME IS  1/(6R) ).

 IROT = VALUE SPECIFIES ROTATIONAL REORIENTATION MODEL-
        1 BROWNIAN
        2 SIMPLE 'FREE'
        3 STRONG JUMP
        4 EGELSTAFF JUMP
        5 INTERMEDIATE JUMP WITH GAUSSIAN DISTRIBUTION
          FOR THE JUMP ANGLE
        6 INTERMEDIATE JUMP WITH EXPONENTIAL DISTRIBUTION
          FOR THE JUMP ANGLE

 RTAU = PRODUCT  R*TAU  TO BE GIVEN FOR THE INTERMEDIATE
        JUMP REORIENTATION OR SIMPLE 'FREE' DIFFUSION. R
        IS THE ROTATIONAL DIFFUSION CONSTANT (IN UNITS OF
        RADIANS PER SECOND) AND  TAU  IS EITHER THE TIME
        (IN UNITS OF SECONDS PER RADIAN) AT A FIXED
        ORIENTATION (INTERMEDIATE JUMP MODEL) OR THE TIME
        DURING WHICH THE MOLECULAE FREELY ROTATES (SIMPLE
        'FREE' DIFFUSION MODEL).

 BCEN = MAGNETIC FIELD VALUE AT CENTER OF SPECTRUM.
        -VALUE IN GAUSS.  (RESONANCE FIELD VALUE FOR THE
         ELECTRON ZEEMAN TERM WITH  G=(GX+GY+GZ)/3. FOR
         X-BAND TYPICAL VALUE IS 3300).

 L = TERMINATING  L  VALUE IN EXPANSION. L  MUST BE AN
     EVEN NUMBER. L=2  FOR THE FAST MOTIONAL LIMIT
     SPECTRUM. (L=10  IS TYPICAL FOR INTERMEDIATE SLOW
     MOTIONAL SPECTRUM WHILE  L=20  FOR A 'RIGID' LIKE
     SPECTRUM).

 TRUNC = .FALSE. -MAXIMUM TERMINATING VALUES SPECIFIED BY
                  L  ARE TO BE USED. THUS
                  KTERM = L; LDOUB = L ; LODD = L-1
         .TRUE. -TERMINATING VALUES  (KTERM,LDOUB,LODD)
                  ARE TO BE GIVEN

 KTERM = TERMINATING VALUE OF  K  FOR THE ASYMMETRIC
         EXPANSION TERMS DUE TO DIFFERENCES IN THE  X  AND
         Y  PRINCIPAL AXIS COMPONENTS OF THE  A  AND  G
         TENSORS. TO BE GIVEN WHEN  TRUNC  IS  .TRUE.
         VALUE MUST BE EVEN AND LESS THAN OR EQUAL TO  L.
         KTERM=0  IS FOR AN AXIALLY SYMMETRIC NITROXIDE
         FREE RADICAL.  (FOR Z AXIS ALONG THE 2P-PI
         NITROGEN ORBITAL, KTERM=4  IS TYPICAL FOR
         INTERMEDIATE SLOW MOTION AND  KTERM=8  FOR
         'RIGID' LIKE SPECTRUM).

 LODD = TERMINATING  L  VALUE FOR THOSE EXPANSION TERMS
        WITH ODD  L  VALUES. TO BE GIVEN WHEN  TRUNC  IS
        .TRUE.  VALUE MUST BE ODD, LESS THAN  L  AND
        GREATER OR EQUAL  1. IF  LODD=1  ALL ODD  L  TERMS
        ARE TRUNCATED.

 LDOUB = TERMINATING  L  VALUE FOR 'DOUBLY FORBIDDEN'
         TRANSITION EXPANSION TERMS. TO BE GIVEN WHEN
         TRUNC  IS  .TRUE.  VALUE MUST BE EVEN, NO GREATER
         THAN  L  AND NO LESS LESS THAN  2.

 NSET = NUMBER OF SPECTRA VARYING ONLY  T2IN  (OR  BI,BF,
        DISP) FOR A GIVEN SET OF VALUES FOR  TITLE,AX,AY,
        AZ,GX,GY,GZ,R,RZ,IROT,RTAU,BCEN,L,TRUNC,KTERM,
        LODD,LDOUB,WAMPEV,PAMPEV.

 WAMPEV = .TRUE. -PRINT OUT EIGENVALUES OF MATRIX  CSYM
                  AND COLUMN VECTOR  AMP.

 PAMPEV = .TRUE. -PUNCH OUT EIGENVALUES OF MATRIX  CSYM
                  AND THE COLUMN VECTOR  AMP.

 IPLOT = VALUE SPECIFIES TYPE OF PLOT-
         -1  ABSORPTION OR DISPERSION
          0  ABSORPTION OR DISPERSION AND ITS RESPECTIVE
             FIRST DERIVATIVE
         +1  FIRST DERIVATIVE OF THE ABSORPTION OR
             DISPERSION

 XPUNCH = .TRUE. -SPECTRAL POINTS PUNCHED OUT

 XPLOT = .TRUE. -SPECTRAL POINTS PRESENTED AS LINE PLOT

 DISP = .TRUE. -DISPERSION TYPE OUTPUT
        .FALSE. -ABSORPTION TYPE OUTPUT

 YLEN = FOR LINE PLOT OUTPUT, LENGTH OF ABSORPTION-
        DISPERSION/FIRST DERIVATIVE AXIS
        -VALUE IN INCHES

 XLEN = FOR LINE PLOT OUTPUT (XPLOT=.TRUE.), LENGTH OF
        FIELD SWEEP AXIS.
        -VALUE IN INCHES

 T2IN = ROTATIONALLY INVARIANT LORENTZIAN P-P FIRST
        DERIVATIVE LINEWIDTH
        -VALUE IN GAUSS.  (TYPICAL VALUE FOR NITROXIDE
         SPIN LABEL IS A FEW GAUSS).

 BI,BF = INITIAL AND FINAL FIELD SWEEP VALUES,
         RESPECTIVELY, WITH RESPECT TO CENTER OF SPECTRUM.
         -VALUE IN GAUSS

 NB = NUMBER OF POINTS FOR SPECTRUM, SPACED EQUIDISTANT
      APART ALONG THE SWEEP AXIS
```

THE STRUCTURE OF THE PROGRAM HAS AN INNER AND OUTER DO LOOP. THE OUTER ONE IS EXECUTED NPT TIMES WHILE THE INNER DO LOOP IS EXECUTED NSET TIMES FOR EACH CYCLE OF THE OUTER DO LOOP

```
      IMPLICIT REAL*8(A-H,O-Z)
      COMPLEX*16 CSYM,CSYM1,CSYM2,VALUE,AMP,ABSP,XAB
      REAL C,BFLD,DERAB,TITLE,YLEN,XLEN,ABMAX,ABMIN,DMAX,DMIN,X,Y,DXY
      LOGICAL XPUNCH,XPLOT,XPLEND,PAMPEV,WAMPEV,DISP,TRUNC
      DIMENSION SY(33,8),SY2(108,3),CSYM(108,34),VALUE(108),AMP(108),CSY
     1M1(108),CSYM2(108),C(401),DERAB(401),BFLD(401),TITLE(20),DROT(2)

                     DIMENSIONING OF MATRICES

      DIMENSION SY(NSY,8),SY2(NSY2,3),CSYM(NDIM,JDIM),VALUE(
     2NDIM),AMP(NDIM),CSYM1(NDIM),CSYM2(NDIM),C(NPNT),DERAB(
     2NPNT),BFLD(NPNT),TITLE(20),DROT(2)

 LIMITS IMPOSED BY THE DIMENSIONS ABOVE:

   1.MAXIMUM NUMBER OF POINTS FOR ONE SPECTRUM
     -GIVEN BY  NPNT

   2.MAXIMUM VALUE FOR  L (=LMAX,SEE ABOVE)
     -DETERMINED FROM  NSY,NSY2
      NSY = 3*(L + 1)
      NSY2 = (3*(L+2)*(L+2))/4
     NO VALUE OF  L  SHOULD YIELD  NSY,NSY2  GREATER
     THAN THOSE SPECIFIED IN DIMENSION STATEMENT

   3.COMBINATION OF VALUES FOR  L,KTERM,LODD,LDOUB  FOR A
     SINGLE PLOT
     -DETERMINED FROM  NDIM,JDIM

     A.FOR  TRUNC = .FALSE.  (KTERM=L,LDOUB=L,LODD=L-1)
       NDIM = ((9*L*L) + (30*L) + 24)/8
       JDIM = (9*(L/2)) + 7
       JDIM = 13                    WHEN  L=2

     B.FOR  LDOUB > LODD > KTERM
       NDIM=((20*L) + 24 + (10*L*KTERM) + (2*LDOUB*KTERM)
            + (6*LODD*KTERM) + (4*LDOUB) -(9*KTERM*KTERM)
            + (12*KTERM))/8
```

```
      JDIM = (9*(KTERM/2)) + 16
      JDIM = 10                    WHEN  KTERM=0
      JDIM = 7                     WHEN  KTERM=0  AND  L=2

   C.FOR  LDOUB > OR = KTERM > LODD
      NDIM=((20*L) + 21 + (10*L*KTERM) + (2*LDOUB*KTERM)
          + (3*LODD*LODD) + (4*LDOUB) - (6*KTERM*KTERM)
          + (12*KTERM))/8

      JDIM = (3*KTERM) + 16
      JDIM = (3*KTERM) + 15   WHEN  KTERM=LDOUB
      JDIM = (3*KTERM) + 10   WHEN  KTERM=L
      JDIM = 13               WHEN  L=2 AND KTERM=2
   OR
      JDIM = (9*((LODD-1)/2)) + 16
      JDIM = 13               WHEN  L=2 AND KTERM=2
     (JDIM  EQUALS THE GREATER OF THE TWO VALUES)

   D.FOR  LODD > LDOUB > OR = KTERM
      NDIM=((20*L) + 24 + (10*L*KTERM) + (4*LDOUB*KTERM)
          + (4*LODD*KTERM) + (4*LDOUB) -(9*KTERM*KTERM)
          + (10*KTERM))/8

      JDIM = (9*(KTERM/2)) + 16
      JDIM = (4*KTERM) + 15          WHEN  LDOUB=KTERM
      JDIM = (9*(KTERM/2)) + 7       WHEN  LDOUB=KTERM  AND  KTERM>16
      JDIM = 10                      WHEN  KTERM=0

   E.FOR  LODD> KTERM > LDOUB
      NDIM=((20*L) + 24 + (10*L*KTERM) + (4*LODD*KTERM)
          + (2*LDOUB*LDOUB) + (4*LDOUB)-(7*KTERM*KTERM)
          + (10*KTERM))/8

      JDIM = (7*(KTERM/2)) + 14
   OR
      JDIM = (4*LDOUB) + 15
      JDIM = (9*(LDOUB/2)) + 7       WHEN  LDOUB>16
     (JDIM  EQUALS THE GREATER OF THE TWO VALUES)

   F.FOR  KTERM > LDOUB > LODD
      NDIM=((20*L) + 21 + (10*L*KTERM) + (LDOUB*LDOUB)
          + (3*LODD*LODD) + (6*LDOUB) - (5*KTERM*KTERM)
          + (10*KTERM))/8

      JDIM = (5*(KTERM/2)) + 14
   OR
      JDIM = (3*LDOUB) + 15
   OR
      JDIM = (9*((LODD-1)/2)) + 16
     (JDIM  EQUALS THE GREATEST OF THE THREE VALUES)

   G.FOR  KTERM > LODD > LDOUB
      NDIM=((20*L) + 22 + (10*L*KTERM) + (2*LDOUB*LDOUB)
          + (2*LODD*LODD) + (4*LDOUB) - (5*KTERM*KTERM)
          + (10*KTERM))/8

      JDIM = (5*(KTERM/2)) + 14
      JDIM = (5*(KTERM/2)) + 9       WHEN  KTERM=L
   OR
      JDIM = (7*((LODD-1)/2)) + 14
   OR
      JDIM = (4*LDOUB) + 15
      JDIM = (9*(LDOUB/2)) + 10      WHEN  LDOUB>10
     (JDIM EQUALS THE GREATEST OF THE THREE VALUES)

      NO SET OF VALUES FOR  L,KTERM,LODD,LDOUB  SHOULD
      YIELD A VALUE FOR  NDIM,JDIM  GREATER THAN THOSE
      SPECIFIED IN DIMENSION STATEMENT


      EQUIVALENCE (CSYM1(1),CSYM(1,1)),(CSYM2(1),CSYM(1,2))
      XPLEND = .FALSE.

*************************************************************
* READ INPUT DATA - ONE READ STATEMENT                      *
*************************************************************
-NOTE THIS INPUT STATEMENT WILL BE READ ONLY ONCE

      READ(5,1) NPT,LMAX,NDIM,JDIM,NSY,NSY2
    1 FORMAT(I2,3X,I2,3X,I3,2X,I3,2X,I3,2X,I3)
      LMAX1 = LMAX + 1
      LMAX2 = (3*LMAX) + 1
      LMAX3 = 1 + LMAX
      LMAX4 = 3*(LMAX + 1)
      LMAX5 = (3*((LMAX + 2)**2))/4
      IF((NSY.LT.LMAX4).OR.(NSY2.LT.LMAX5)) GO TO 2

CALCULATION OF  3-J  SYMBOL VALUE.
VALUES STORED IN MATRICES  SY,SY2. FORMULAE FOR THESE
CALCULATIONS BASED UPON EQUATION 1.5 IN 'THE 3-J AND
6-J SYMBOLS' BY ROTENBERG,BIVINS,METROPOLIS,WOOTEN
AND FORMULAE GIVEN IN 'ANGULAR MOMENTUM IN QUANTUM
MECHANICS' BY EDMONDS.

      SY(1,1) = 0.D0
      SY(1,2) = 0.D0
      SY(2,1) = 1.D0/DSQRT(5.D0)
      SY(2,2) = 2.D-1
      DO 3 I=3,LMAX1,2
      XI = DFLOAT(I) - 1.D0
      XXI = 2.D0*XI
      SY(I,1) = -DSQRT((XI*(XI + 1.D0))/((XXI + 3.D0)*(XXI + 1.D0)*(XXI
     1 - 1.D0)))
      SY(I,2) = SY(I,1)*SY(I,1)
      SY(I+1,1) = DSQRT((3.D0*(XI + 2.D0)*(XI + 1.D0))/(2.D0*(XXI + 5.D0
     1)*(XXI + 3.D0)*(XXI + 1.D0)))
    3 SY(I+1,2) = SY(I+1,1)*SY(I+1,1)
      DO 4 I=3,8
      DO 4 J=1,3
    4 SY(J,I) = 0.D0
      SY(3,3) = -1.D0/DSQRT(5.D0)
      XODD = -1.D0
      DO 5 I=4,LMAX2,3
      XI = (DFLOAT(I) - 1.D0)/3.D0
      XXI = 2.D0*XI
      SY(I,3) = XODD*DSQRT(3.D0/(2.D0*(XXI + 3.D0)*(XXI + 1.D0)*(XXI - 1
     1.D0)))
      SY(I+1,3) = -XODD*DSQRT(XI/((XXI + 3.D0)*(XXI + 1.D0)*2.D0))
      SY(I+2,3) = -XODD*DSQRT(((XI + 1.D0)*(XI + 3.D0))/((XXI + 5.D0)*(X
     1XI + 3.D0)*(XXI + 1.D0)))
      SY(I,4) = XODD*(((XI*(XI + 1.D0)) - 3.D0)/DSQRT((XXI + 3.D0)*(XI +
     11.D0)*(XXI + 1.D0)*XI*(XXI - 1.D0)))
      SY(I+1,4) = XODD*DSQRT(3.D0/((XXI + 3.D0)*(XI + 1.D0)*(XXI + 1.D0)
     1))
      SY(I+2,4) = -XODD*DSQRT((3.D0*(XI + 3.D0)*XI)/(2.D0*(XXI + 5.D0)*(
     1XXI + 3.D0)*(XXI + 1.D0)))
      SY(I,5) = SY(I,3)
      SY(I+1,5) = -XODD*DSQRT((XI + 2.D0)/(2.D0*(XXI + 3.D0)*(XXI + 1.D0
     1)))
      SY(I+2,5) = XODD*DSQRT(((XI + 2.D0)*XI)/((XXI + 5.D0)*(XXI + 3.D0)
     1*(XXI + 1.D0)))
      SY(I,6) = -XODD*3.D0*DSQRT((3.D0*(XI + 2.D0)*(XI - 1.D0))/((XXI +
     1 3.D0)*(XI + 1.D0)*(XXI + 1.D0)*XXI*(XXI - 1.D0)))
      SY(I+1,6) = XODD*(XI - 2.D0)*DSQRT((XI + 3.D0)/((XXI + 3.D0)*(XI +
     1 1.D0)*(XXI + 1.D0)*XXI))
      SY(I+2,6) = XODD*DSQRT((XI*(XI + 4.D0)*(XI + 3.D0))/((XXI + 5.D0)*
     1(XI + 1.D0)*(XXI + 1.D0)*(XXI + 3.D0)))
      SY(I,7) = XODD*((12.D0 - (XI*(XI + 1.D0)))/DSQRT((XXI + 3.D0)*(XI
     1 + 1.D0)*(XXI + 1.D0)*XI*(XXI - 1.D0)))
      SY(I+1,7) = -XODD*2.D0*DSQRT((3.D0*(XI + 3.D0)*(XI - 1.D0))/((XI
     1 + 2.D0)*(XXI + 3.D0)*(XXI + 1.D0)*XI*(XI + 1.D0)))
      SY(I+2,7) = XODD*DSQRT(((XI + 4.D0)*(XI + 3.D0)*XI*(XI - 1.D0)*3.D
     10)/((XXI + 5.D0)*(XI + 2.D0)*(XXI + 3.D0)*(XXI + 2.D0)*(XXI + 1.D0
     2)))
      SY(I,8) = SY(I,6)
      SY(I+1,8) = XODD*(XI + 4.D0)*DSQRT((XI - 1.D0)/((XI + 2.D0)*(XXI +
     1 3.D0)*(XXI + 2.D0)*(XXI + 1.D0)))
      SY(I+2,8) = -XODD*DSQRT((XI*(XI + 3.D0)*(XI - 1.D0))/((XXI + 5.D0)
     1*(XI + 2.D0)*(XXI + 3.D0)*(XXI + 1.D0)))
    5 XODD = -XODD
      SY(4,7) = 0.D0
      K = 1
      XODD = 1.D0
      DO 6 I=1,LMAX3
      XI = DFLOAT(I) - 1.D0
      XXI = 2.D0*XI
      JK = (I+1)/2
      DO 7 J=1,JK
      XJ = (2.D0*DFLOAT(J)) - 2.D0
      IF((I+J).EQ.2) GO TO 8
      SY2(K,1) = XODD*(((3.D0*XJ*XJ) - (XI*(XI + 1.D0)))/DSQRT((XXI + 3.
     1D0)*(XI + 1.D0)*(XXI + 1.D0)*XI*(XXI - 1.D0)))
      SY2(K+1,1) = -XODD*XJ*DSQRT((3.D0*(XI + XJ + 1.D0)*(XI - XJ + 1.D0
     1))/((XI + 2.D0)*(XXI + 3.D0)*(XI + 1.D0)*(XXI + 1.D0)*XI))
      SY2(K,2) = XODD*DSQRT((6.D0*(XI - XJ - 1.D0)*(XI - XJ)*(XI + XJ +
     1 1.D0)*(XI + XJ + 2.D0))/((XXI + 3.D0)*(XXI + 2.D0)*(XXI + 1.D0)*X
     2XI*(XXI - 1.D0)))
      SY2(K+1,2) =  XODD*DSQRT(((XI + XJ + 1.D0)*(XI + XJ + 2.D0)*(XI +
     1XJ + 3.D0)*(XI - XJ))/((XI + 2.D0)*(XXI + 3.D0)*(XI + 1.D0)*(XXI +
     2 1.D0)*XXI))
      SY2(K,3) = XODD*DSQRT(( 6.D0*(XI + XJ - 1.D0)*(XI + XJ)*(XI - XJ +
     1 1.D0)*(XI - XJ + 2.D0))/((XXI + 3.D0)*(XXI + 2.D0)*(XXI + 1.D0)*X
     2XI*(XXI - 1.D0)))
      SY2(K+1,3) = -XODD*DSQRT(((XI - XJ + 1.D0)*(XI - XJ + 2.D0)*(XI -
     1XJ + 3.D0)*(XI + XJ))/((XI + 2.D0)*(XXI + 3.D0)*(XI + 1.D0)*(XXI +
     2 1.D0)*XXI))
    8 SY2(K+2,1) = XODD*DSQRT((6.D0*(XI + XJ + 2.D0)*(XI + XJ + 1.D0)*(X
     1I - XJ + 2.D0)*(XI - XJ + 1.D0))/((XXI + 5.D0)*(XXI + 4.D0)*(XXI +
     2 3.D0)*(XXI + 2.D0)*(XXI + 1.D0)))
      SY2(K+2,2) = XODD*DSQRT(((XI + XJ + 1.D0)*(XI + XJ + 2.D0)*(XI + X
     1J + 3.D0)*(XI + XJ + 4.D0))/((XXI + 5.D0)*(XXI + 4.D0)*(XXI + 3.D0
     2)*(XXI + 2.D0)*(XXI + 1.D0)))
      SY2(K+2,3) = XODD*DSQRT(((XI - XJ + 1.D0)*(XI - XJ + 2.D0)*(XI -XJ
     1 + 3.D0)*(XI - XJ + 4.D0))/((XXI + 5.D0)*(XXI + 4.D0)*(XXI + 3.D0)
     2*(XXI + 2.D0)*(XXI + 1.D0)))
    7 K = K + 3
    6 XODD = -XODD
      DO 9 I=1,2
      DO 9 J=1,3
    9 SY2(I,J) = 0.D0
      DO 10 LPT=1,NPT

*************************************************************
*  READ INPUT DATA - FIVE READ STATEMENTS                   *
*************************************************************
.-NOTE THAT THE FOLLOWING SET OF  5  INPUT CARDS IS
 FOLLOWED BY A SET OF  2*NSET  INPUT CARDS. THESE SETS
 OF INPUT CARDS ARE THEN REPEATED A TOTAL OF  NPT  TIMES

      READ(5,11) TITLE
   11 FORMAT(20A4)
      READ(5,12) AX,AY,AZ
   12 FORMAT(D10.3,5X,D10.3,5X,D10.3)
      READ(5,13) GX,GY,GZ
   13 FORMAT(F8.5,7X,F8.5,7X,F8.5)
      READ(5,14) R,RZ,IROT,RTAU,BCEN
   14 FORMAT(D10.3,5X,D10.3,5X,I1,4X,D10.3,5X,D10.3)
      READ(5,15) L,TRUNC,KTERM,LODD,LDOUB,NSET,PAMPEV,WAMPEV
   15 FORMAT(I2,3X,L1,4X,I2,3X,I2,3X,I2,3X,I2,3X,L1,4X,L1)
      DELG0 = GZ - ((GX + GY)/2.D0)
      DELG2 = GX - GY
      G = (GX + GY + GZ)/3.D0
      FOAB = +8.286D+07
      F0 = (DELG0*BCEN*2.D0*9.2731D+6)/(1.05443D0*3.D0)
      F2 = (DELG2*BCEN*9.2731D+6)/(DSQRT(6.D0)*1.05443D0)
      D = (AX + AY - (2.D0*AZ))/(2.D0*DSQRT(6.D0))
      D2 = ((AY - AX)/4.DU)
      B = -((AX + AY + AZ)/6.D0)
      GAMMA = (G*9.2731D+6)/1.05443D0
      IF(TRUNC) GO TO 16
      KTERM = L
      LODD = L - 1
      LDOUB = L
   16 KT = KTERM/2
      LO = (LODD - 1)/2
      LDB = LDOUB/2
      WRITE(6,17) TITLE,AX,AY,AZ,D,D2,B,GX,GY,GZ,F0,F2,R,RZ,L,KTERM,LODD
     1,LDOUB,BCEN
   17 FORMAT(1H1,20A4,///,' A(X), A-TENSOR COMPONENT ALONG  X  PRINCIPAL
     1 AXIS  ',D12.5,' GAUSS',/,' A(Y), A-TENSOR COMPONENT ALONG  Y  PRI
     2NCIPAL AXIS  ',D12.5,' GAUSS',/,' A(Z), A-TENSOR COMPONENT ALONG
     3Z  PRINCIPAL AXIS  ',D12.5,' GAUSS',/,9X,'CALCULATED VALUE OF  D(0
     4)  ',D12.5,' GAUSS',/,9X,'CALCULATED VALUE OF  D(2)  ',D12.5,' GAU
     5SS',/,9X,'CALCULATED VALUE OF  B  ',D12.5,' GAUSS',//,' G(X), G-TE
     6NSOR COMPONENT ALONG  X  PRINCIPAL AXIS  ',F8.5,/,' G(Y), G-TENSOR
     7 COMPONENT ALONG  Y  PRINCIPAL AXIS  ',F8.5,/,' G(Z), G-TENSOR COM
     8PONENT ALONG  Z  PRINCIPAL AXIS  ',F8.5,/,9X,'CALCULATED VALUE OF
     9 F(0)  ',D12.5,' RADIANS PER SECOND',/,9X,'CALCULATED VALUE OF  F(
```

```
*2)  ',D12.5,' RADIANS PER SECOND',//,' R,ROTATIONAL DIFFUSION CONS
*TANT ABOUT  X,Y  AXES  ',D12.5,' RADIANS PER SECOND',/,' R(Z), ROT
*ATIONAL DIFFUSION CONSTANT ABOUT  Z  AXIS  ',D12.5,' RADIANS PER S
*ECOND',//,' THE TERMINATING  L  VALUE  IS ',I3,//,' THE TERMINATIN
*G  K  VALUE IS  ',I3,//,' THE TERMINATING  LODD  VALUE ASSOCIATED
*WITH THE ODD  L  COEFFICIENTS IS ',I3,//,' THE TERMINATING  LDOUB
* VALUE ASSOCIATED WITH THE DOUBLY FORBIDDEN TRANSITION COEFFICIENT
*S IS  ',I3,//,' THE CENTER OF THE SPECTRUM  ',D12.5,' GAUSS',/)
   R=R/FOAB
   RZ = RZ/FOAB
   GO TO (18,19,20,21,22,23), IROT
18 CORRT = 1.D0/(6.D0*DSQRT(R*RZ)*FOAB)
   WRITE(6,24)
   GO TO 30
19 CORRT = DSQRT(7.D0)/(6.D0*DSQRT(R*RZ)*FOAB)
   WRITE(6,25)
   GO TO 30
20 CORRT = 1.D0/(DSQRT(R*RZ)*FOAB)
   WRITE(6,26)
   GO TO 30
21 CORRT = (1.D0 + (6.D0*RTAU))/(6.D0*DSQRT(R*RZ)*FOAB)
   WRITE(6,27) RTAU
   GO TO 30
22 CORRT = 5.D0/(6.D0*DSQRT(R*RZ)*FOAB*(1.D0/(1.D0 + (9.D0*RTAU*RTAU)
  1))*(4.D0/(1.D0 + (36.D0*RTAU*RTAU))))
   WRITE(6,28) RTAU
   GO TO 30
23 WRITE(6,29) RTAU
   CORRT = (5.D0*RTAU)/((2.D0*DSQRT(R*RZ)*FOAB)*(2.D0 - (DEXP(-(3.D0*
  1RTAU)) + DEXP(-(12.D0*RTAU)))))
24 FORMAT(1H ,'BROWNIAN ROTATIONAL DIFFUSION',/)
25 FORMAT(1H ,'SIMPLE''FREE'' ROTATIONAL DIFFUSION',/)
26 FORMAT(1H ,'RANDOM JUMP ROTATIONAL REORIENTATION',/)
27 FORMAT(1H ,'EGELSTAFF JUMP ROTATIONAL REORIENTATION ',/,' THE PROD
  1UCT  R*TAU  IS  ',D12.5,/)
28 FORMAT(1H ,'JUMP ROTATIONAL REORIENTATION WITH AN EXPONENTIAL DIST
  1RIBUTION OF JUMP ANGLES',/,' THE PRODUCT  R*TAU  IS  ',D12.5,/)
29 FORMAT(1H ,' JUMP ROTATIONAL REORIENTATION WITH A GAUSSIAN DISTRIB
  1UTION OF JUMP ANGLES',/,' THE PRODUCT  R*TAU  IS  ',D12.5,/)
30 WRITE(6,31) CORRT
31 FORMAT(1H ,/,' THE CALCULATED ROTATIONAL CORRELATION TIME IS  ',D1
  12.5,' SECONDS PER RADIAN',/)
   D = D*(GAMMA/FOAB)
   D2 = D2*(GAMMA/FOAB)
   B = B*(GAMMA/FOAB)
   DPR = -(D*DSQRT(8.D0/3.D0))
   D2PR = -(D2*DSQRT(8.D0/3.D0))
   F0 = F0/FOAB
   F2 = F2/FOAB
   IF((KTERM.GT.LODD).OR.(LODD.GT.LDOUB)) GO TO 301
   N = ((20*L) + 24 + (10*L*KTERM) + (2*LDOUB*KTERM) + (6*LODD*KTERM)
  1 + (4*LDOUB) - (9*(KTERM**2)) + (12*KTERM))/8
   M = (9*KT) + 16
   IF(KTERM.EQ.0) M=10
   GO TO 302
301 IF((LODD.GT.KTERM).OR.(KTERM.GT.LDOUB)) GO TO 303
   N = ((20*L) + 21 + (10*L*KTERM) + (2*LDOUB*KTERM) + (3*(LODD**2))
  1+ (4*LDOUB) - (6*(KTERM**2))+ (12*KTERM))/8
   M = (3*KTERM) + 16
   IF(KTERM.EQ.LDOUB) M = M - 1
   IF(KTERM.EQ.L) M = (3*KTERM) + 10
   M1 = (9*LO) + 16
   IF(M1.GT.M) M = M1
   GO TO 302
303 IF((KTERM.GT.LDOUB).OR.(LDOUB.GT.LODD)) GO TO 304
   N = ((20*L) + 24 + (10*L*KTERM) + (4*LDOUB*KTERM) + (4*LODD*KTERM)
  1 + (4*LDOUB) - (9*(KTERM**2))+ (10*KTERM))/8
   M = (9*KT) + 16
   IF(LDOUB.EQ.KTERM) M = (4*KTERM) + 15
   IF((LDOUB.EQ.KTERM).AND.(KTERM.GT.16)) M = (9*KT) + 7
   IF(KTERM.EQ.0) M=10
   GO TO 302
304 IF((LDOUB.GT.KTERM).OR.(KTERM.GT.LODD)) GO TO 305
   N = ((20*L) + 24 + (10*L*KTERM) + (4*LODD*KTERM) + (2*(LDOUB**2))
  1 + (4*LDOUB) - (7*(KTERM**2)) + (10*KTERM))/8
   M = (7*KT) + 14
   M1 = (4*LDOUB) + 15
   IF(LDOUB.GT.16) M1 = (9*LDB) + 7
   IF(M1.GT.M) M = M1
   GO TO 302
305 IF((LODD.GT.LDOUB).OR.(LDOUB.GT.KTERM)) GO TO 306
   N = ((20*L) + 21 + (10*L*KTERM) + (LDOUB**2) + (3*(LODD**2)) + (6*
  1LDOUB) - (5*(KTERM**2)) + (10*KTERM))/8
   M = (5*KT) + 14
   M1 = (3*LDOUB) + 15
   M2 = (9*LO) + 16
   IF(M1.GT.M) M = M1
   IF(M2.GT.M) M = M2
   GO TO 302
306 N = ((20*L) + 22 + (10*L*KTERM) + (2*(LDOUB**2)) + (2*(LODD**2)) +
  1 (4*LDOUB) - (5*(KTERM**2)) + (10*KTERM))/8
   M = (5*KT) + 14
   IF(L.EQ.KTERM) M = M - 5
   M1 = (7*LO) + 14
   M2 = (4*LDOUB) + 15
   IF(LDOUB.GT.10) M2 = (9*LDB) + 10
   IF(M1.GT.M) M = M1
   IF(M2.GT.M) M = M2
302 IF((L.EQ.2).AND.(KTERM.EQ.2)) M = 13
   IF((L.EQ.2).AND.(KTERM.EQ.0)) M = 7
   IF((N.GT.NDIM).OR.(M.GT.JDIM)) GO TO 814

 CONSTRUCTING THE COEFFICIENT MATRIX (SYMMETRIZED) OF THE
 C  VECTOR.

   DO 32 I=1,N
   DO 32 J=1,M
32 CSYM(I,J) = (0.D0,0.D0)
   CSYM(1,4) = (-1.D0,0.D0)*((F0 + DPR)*SY(2,2)*DSQRT(5.D0))
   CSYM(1,7) = (1.D0,0.D0)*(D*SY(2,1)*SY(3,3)*DSQRT(5.D0))
   CSYM(2,4) = (-1.D0,0.D0)*(F0*SY(2,2)*DSQRT(5.D0))
   CSYM(2,6) = CSYM(1,7)
   CSYM(2,7) = CSYM(2,6)
   CSYM(3,4) = (-1.D0,0.D0)*((F0 - DPR)*SY(2,2)*DSQRT(5.D0))
   CSYM(3,6) = CSYM(1,7)
   CSYM(1,1) = (1.D0,0.D0)*(2.D0*B)
   CSYM(3,1) = (-1.D0,0.D0)*(2.D0*B)
   IF(KTERM.EQ.0) GO TO 33
   CSYM(1,10) = (-1.D0,0.D0)*((F2 + D2PR)*SY(2,1)*SY2(3,2)*DSQRT(1.D+
  11))
   CSYM(1,13) = (1.D0,0.D0)*(D2*SY(3,3)*SY2(3,2)*DSQRT(1.D+1))
   CSYM(2,10) = (-1.D0,0.D0)*(F2*SY(2,1)*SY2(3,2)*DSQRT(1.D+1))
   CSYM(2,12) = CSYM(1,13)
   CSYM(2,13) = CSYM(2,12)
   CSYM(3,10) = (-1.D0,0.D0)*((F2 - D2PR)*SY(2,1)*SY2(3,2)*DSQRT(1.D+
  11))
   CSYM(3,12) = CSYM(2,13)
33 AL = 2.D0
   IEND = (L/2)
   J = 4
   DO 34 I=1,IEND
   K = (9*I) + 7
   IF((LDB.EQ.I).AND.(LO.GE.I).AND.(KT.GE.I)) K = (8*I) + 7
   IF((LDB.LT.I).AND.(LO.GE.I).AND.(KT.GE.I)) K = (7*I) + 6
   IF((LDB.GE.I).AND.(LO.LT.I).AND.(KT.GE.I)) K = (6*I) + 7
   IF((LDB.GT.I).AND.(LO.GE.I).AND.(KT.LT.I)) K = (9*KT) + 7
   IF((LDB.EQ.I).AND.(LO.GE.I).AND.(KT.LT.I)) K = (8*KT) + 7
   IF((LDB.LT.I).AND.(LO.LT.I).AND.(KT.GE.I)) K = (5*I) + 6
   IF((LDB.LT.I).AND.(LO.GE.I).AND.(KT.LT.I)) K = (7*KT) + 6
   IF((LDB.GE.I).AND.(LO.LT.I).AND.(KT.LT.I)) K = (3*KTERM) + 7
   IF((LDB.LT.I).AND.(LO.LT.I).AND.(KT.LT.I)) K = (5*KT) + 6
   KD = 0
   IF(I.GE.LDB) KD = 1
   LL1 = (2*I) + 1
   LL21 = LL1 + 1
   LL2 = (6*I) + 1
   LL12 = LL2 + 1
   LL22 = LL2 + 2
   L1L12 = LL2 + 3
   L1L22 = LL2 + 4
   L1L32 = LL2 + 5
   LL3 = (3*I*(I + 1)) + 1
   LL13 = LL3 + 1
   LL23 = LL3 + 2
   J1 = J + 1
   J2 = J + 2
   J3 = J + 3
   J4 = J + 4
   J5 = J + 5
   JCOUNT = 0
   SFLL = (2.D0*AL) + 1.D0
   SFL1L1 = (2.D0*AL) + 3.D0
   SFLL2 = DSQRT((4.D0*AL*AL) + (1.2D+1*AL) + 5.D0)
   SFLL1 = DSQRT((4.D0*AL*AL) + (8.D0*AL) + 3.D0)
   SFL1L2 = DSQRT((4.D0*AL*AL) + (1.6D+1*AL) + 1.5D+1)
   SFL1L3 = DSQRT((4.D0*AL*AL) + (2.D+1*AL) + 2.1D+1)
   ALL = AL
   LJUMP = 2*I
   DO 36 IR=1,2
   GO TO (37,38,39,40,41,42),IROT
37 DROT(IR) = R*ALL*(ALL + 1.D0)
   GO TO 36
38 DROT(IR) = R*((ALL*(ALL + 1.D0))/DSQRT(1.D0 + (ALL*(ALL + 1.D0))))
   GO TO 36
39 DROT(IR) = R
   GO TO 36
40 DROT(IR) = (R*ALL*(ALL + 1.D0))/(1.D0 + (RTAU*ALL*(ALL + 1.D0)))
   GO TO 36
41 DROT(IR) = 0.D0
   DO 43 IJUMP=1,LJUMP
43 DROT(IR) = DROT(IR) + ((DFLOAT(IJUMP)**2)/(1.D0 + ((3.D0*RTAU*DFLO
  1AT(IJUMP))**2)))
   DROT(IR) = DROT(IR)*((6.D0*R)/((2.D0*ALL) + 1.D0))
   GO TO 45
42 DROT(IR) = 0.D0
   DO 44 IJUMP=1,LJUMP
44 DROT(IR) = DROT(IR) + DEXP(-(3.D0*RTAU*DFLOAT(IJUMP)*DFLOAT(IJUMP)
  1))
   DROT(IR) = ((2.D0*R)/(((2.D0*ALL) + 1.D0)*RTAU))*(DFLOAT(LJUMP) -
  1DROT(IR))
45 LJUMP = LJUMP + 1
36 ALL = ALL + 1.D0
   IF(IEND.EQ.I) GO TO 46
   CSYM(J,K) = (-1.D0,0.D0)*((F0 + DPR)*SY(LL21,2)*SFLL2)
   CSYM(J,K+3) = (1.D0,0.D0)*(D*SY(LL21,1)*SY(LL22,3)*SFLL2)
   CSYM(J1,K) = (-1.D0,0.D0)*(F0*SY(LL21,2)*SFLL2)
   CSYM(J1,K+2) = CSYM(J,K+3)
   CSYM(J1,K+3) = CSYM(J1,K+2)
   CSYM(J2,K) = (-1.D0,0.D0)*((F0 - DPR)*SY(LL21,2)*SFLL2)
   CSYM(J2,K+2) = CSYM(J1,K+3)
   CSYM(J3,K) = (1.D0,0.D0)*((F0 + (DPR/2.D0))*SY(LL21,1)*SY(LL22,4)*
  1SFLL2)
   CSYM(J3,K-3) = (1.D0,0.D0)*(D*SY(LL21,1)*SY(LL22,5)*SFLL2)
   CSYM(J3,K-2) = CSYM(J3,K-3)
   CSYM(J4,K) = (1.D0,0.D0)*((F0 - (DPR/2.D0))*SY(LL21,1)*SY(LL22,4)*
  1SFLL2)
   CSYM(J4,K-3) = CSYM(J3,K-3)
   CSYM(J4,K-2) = CSYM(J4,K-3)
   IF(I.GT.LDB) GO TO 35
   CSYM(J5,K-2) = (-1.D0,0.D0)*((D/DSQRT(2.D0))*SY(LL21,1)*SY(LL22,8)
  1*SFLL2)
   CSYM(J5,K-1) = CSYM(J5,K-2)
   IF(I.GE.LDB) GO TO 35
   CSYM(J5,K) = (-1.D0,0.D0)*(F0*SY(LL21,1)*SY(LL22,7)*SFLL2)
   CSYM(J3,K+2) = (-1.D0,0.D0)*((D/DSQRT(2.D0))*SY(LL21,1)*SY(LL22,6)
  1*SFLL2)
   CSYM(J4,K+1) = CSYM(J3,K+2)
35 IF(KTERM.EQ.0) GO TO 47
   CSYM(J,K+6-KD) = (-1.D0,0.D0)*((F2 + D2PR)*SY(LL21,1)*SY2(LL23,2)*
  1SFLL2*DSQRT(2.D0))
   CSYM(J,K+9-KD) = (1.D0,0.D0)*(D2*SY(LL22,3)*SY2(LL23,2)*SFLL2*DSQR
  1T(2.D0))
   CSYM(J1,K+6-KD) = (-1.D0,0.D0)*(F2*SY(LL21,1)*SY2(LL23,2)*SFLL2*DS
  1QRT(2.D0))
   CSYM(J1,K+8-KD) = CSYM(J,K+9-KD)
   CSYM(J1,K+9-KD) = CSYM(J1,K+8-KD)
   CSYM(J2,K+6-KD) = (-1.D0,0.D0)*((F2 - D2PR)*SY(LL21,1)*SY2(LL23,2)
  1*SFLL2*DSQRT(2.D0))
   CSYM(J2,K+8-KD) = CSYM(J1,K+9-KD)
   CSYM(J3,K+6-KD) = (1.D0,0.D0)*((F2 + (D2PR/2.D0))*SY(LL22,4)*SY2(L
  1L23,2)*SFLL2*DSQRT(2.D0))
   CSYM(J3,K+3-KD) = (1.D0,0.D0)*(D2*SY(LL22,5)*SY2(LL23,2)*SFLL2*DSQ
  1RT(2.D0))
   CSYM(J3,K+4-KD) = CSYM(J3,K+3-KD)
   CSYM(J4,K+6-KD) = (1.D0,0.D0)*((F2 - (D2PR/2.D0))*SY(LL22,4)*SY2(L
  1L23,2)*SFLL2*DSQRT(2.D0))
   CSYM(J4,K+3-KD) = CSYM(J3,K+3-KD)
   CSYM(J4,K+4-KD) = CSYM(J4,K+3-KD)
   IF(I.GT.LDB) GO TO 100
   CSYM(J5,K+4-KD) = (-1.D0,0.D0)*(D2*SY(LL22,8)*SY2(LL23,2)*SFLL2)
```

```
      CSYM(J5,K+5-KD) = CSYM(J5,K+4-KD)

      CSYM(J5,K+6) = (-1.D0,0.D0)*(F2*SY(LL22,7)*SY2(LL23,2)*SFLL2*DSQRT
     1(2.D0))
      CSYM(J3,K+8) = (-1.D0,0.D0)*(D2*SY(LL22,6)*SY2(LL23,2)*SFLL2)
      CSYM(J4,K+7) = CSYM(J3,K+8)
  100 IF(I.GT.LO) GO TO 46
      KI = K - (3*I)
      IF((I.LT.LDB).AND.(I.GT.KT)) KI = K - (3*KT)
      IF((I.GE.LDB).AND.(I.LE.KT)) KI = K - (2*I)
      IF((I.GE.LDB).AND.(I.GT.KT)) KI = K - KTERM
      CSYM(J,KI) =  (1.D0,0.D0)*(D2*SY(LL12,3)*SY2(LL13,2)*SFLL1*DSQRT(2
     1.D0))
      CSYM(J1,KI-1) = CSYM(J,KI)
      CSYM(J1,KI) = CSYM(J,KI)
      CSYM(J2,KI-1) = CSYM(J,KI)
      CSYM(J3,KI-3) = (1.D0,0.D0)*((F2 + (D2PR/2.D0))*SY(LL12,4)*SY2(LL1
     13,2)*SFLL1*DSQRT(2.D0))
      CSYM(J4,KI-3) = (1.D0,0.D0)*((F2 - (D2PR/2.D0))*SY(LL12,4)*SY2(LL1
     13,2)*SFLL1*DSQRT(2.D0))
      IF(I.GT.LDB) GO TO 46
      CSYM(J5,KI-5) = (-1.D0,0.D0)*(D2*SY(LL12,8)*SY2(LL13,2)*SFLL1)
      CSYM(J5,KI-4) = CSYM(J5,KI-5)
      IF(I.GE.LDB) GO TO 46
      CSYM(J5,KI-3) = (-1.D0,0.D0)*(F2*SY(LL12,7)*SY2(LL13,2)*SFLL1*DSQR
     1T(2.D0))
      CSYM(J3,KI-1) = (-1.D0,0.D0)*(D2*SY(LL12,6)*SY2(LL13,2)*SFLL1)
      CSYM(J4,KI-2) = CSYM(J3,KI-1)
   46 IF(KTERM.EQ.0) GO TO 47
      KD = 0
      IF(I.GT.LDB) KD = 1
      CSYM(J,7-KD) = (-1.D0,0.D0)*((F2 + D2PR)*SY(LL1,1)*SY2(LL3,2)*SFLL
     1*DSQRT(2.D0))
      CSYM(J,10-KD) = (1.D0,0.D0)*(D2*SY(LL2,3)*SY2(LL3,2)*SFLL*DSQRT(2.
     1D0))
      CSYM(J1,7-KD) = (-1.D0,0.D0)*(F2*SY(LL1,1)*SY2(LL3,2)*SFLL*DSQRT(2
     1.D0))
      CSYM(J1,9-KD) = CSYM(J,10-KD)
      CSYM(J1,10-KD) = CSYM(J1,9-KD)
      CSYM(J2,7-KD) = (-1.D0,0.D0)*((F2 - D2PR)*SY(LL1,1)*SY2(LL3,2)*SFL
     1L*DSQRT(2.D0))
      CSYM(J2,9-KD) = CSYM(J1,10-KD)
      CSYM(J3,7-KD) = (1.D0,0.D0)*((F2 + (D2PR/2.D0))*SY(LL2,4)*SY2(LL3,
     12)*SFLL*DSQRT(2.D0))
      CSYM(J3,4-KD) = (1.D0,0.D0)*(D2*SY(LL2,5)*SY2(LL3,2)*DSQRT(2.D0)*S
     1FLL)
      CSYM(J3,5-KD) = CSYM(J3,4-KD)
      CSYM(J4,7-KD) = (1.D0,0.D0)*((F2 - (D2PR/2.D0))*SY(LL2,4)*SY2(LL3,
     12)*SFLL*DSQRT(2.D0))
      CSYM(J4,4-KD) = CSYM(J3,4-KD)
      CSYM(J4,5-KD) = CSYM(J4,4-KD)
      IF(I.GT.LDB) GO TO 47
      CSYM(J5,5) = (-1.D0,0.D0)*(D2*SY(LL2,8)*SY2(LL3,2)*SFLL)
      CSYM(J5,6) = CSYM(J5,5)
      IF(I.GE.LDB) GO TO 47
      CSYM(J5,7) = (-1.D0,0.D0)*(F2*SY(LL2,7)*SY2(LL3,2)*SFLL*DSQRT(2.D0
     1))
      CSYM(J3,9) = (-1.D0,0.D0)*(D2*SY(LL2,6)*SY2(LL3,2)*SFLL)
      CSYM(J4,8) = CSYM(J3,9)
   47 CSYM(J,4) = (1.D0,0.D0)*(D*SY(LL1,1)*SY(LL2,3)*SFLL)
      CSYM(J1,3) = CSYM(J,4)
      CSYM(J1,4) = CSYM(J,4)
      CSYM(J2,3) = CSYM(J,4)
      CSYM(J,1) = ((1.D0,0.D0)*((2.D0*B) - (SFLL*(F0 + DPR)*SY(LL1,2))))
     1 + ((0.D0,-1.D0)*DROT(1))
      CSYM(J1,1) = ((-1.D0,0.D0)*(SFLL*F0*SY(LL1,2))) + ((0.D0,-1.D0)*DR
     1OT(1))
      CSYM(J2,1) = ((1.D0,0.D0)*(-(2.D0*B) - (SFLL*(F0 - DPR)*SY(LL1,2))
     1)) + ((0.D0,-1.D0)*DROT(1))
      CSYM(J3,1) = ((1.D0,0.D0)*(B + (SFLL*(F0 + (DPR/2.D0))*SY(LL1,1)*S
     1Y(LL2,4)))) + ((0.D0,-1.D0)*DROT(1))
      CSYM(J4,1) = ((1.D0,0.D0)*(- B + (SFLL*(F0 - (DPR/2.D0))*SY(LL1,1)
     1*SY(LL2,4)))) + ((0.D0,-1.D0)*DROT(1))
      IF(I.GT.LDB) GO TO 101
      CSYM(J5,1) = ((-1.D0,0.D0)*(SFLL*F0*SY(LL2,7)*SY(LL1,1))) + ((0.D0
     1,-1.D0)*DROT(1))
      CSYM(J3,3) = (-1.D0,0.D0)*((D/DSQRT(2.D0))*SY(LL2,6)*SY(LL1,1)*SFL
     1L)
      CSYM(J4,2) = CSYM(J3,3)
      J = J + 1
  101 J = J + 5
      DO 48 II=1,I
      IF(II.GT.(KTERM/2)) GO TO 48
      J1 = J + 1
      J2 = J + 2
      J3 = J + 3
      J4 = J + 4
      J5 = J + 5
      JODD = J + (6*I) - (3*(II - 1))
      IF((I.LE.KT).AND.(I.EQ.LDB)) JODD = J + (6*I) - (4*II) + 4
      IF((I.LE.KT).AND.(I.GT.LDB)) JODD = J + (5*I) - (3*II) + 3
      IF((I.GT.KT).AND.(I.LT.LDB)) JODD = J + (3*KTERM) - (3*II) + 3
      IF((I.GT.KT).AND.(I.EQ.LDB)) JODD = J + (3*KTERM) - (4*II) + 4
      IF((I.GT.KT).AND.(I.GT.LDB)) JODD = J + (5*KT) - (3*II) + 3
      KD = 0
      IF(I.GT.LO) KD = (3*I)
      KLL20 = (9*I) + 7 - KD
      KLL21 = KLL20 - 6
      KLL22 = KLL20 + 6
      KLL10 = (6*I) - (3*II) + 4
      KLL11 = KLL10 - 3
      KLL12 = KLL10 + 3
      KL1L20 = 3*(I + II) + 7
      KL1L21 = KL1L20 - 6
      KL1L22 = KL1L20 + 6
      KL1L30 = (9*I) + 13
      KL1L31 = KL1L30 - 3
      KL1L32 = KL1L30 + 3
      IF((I.GT.KT).OR.(I.NE.LDB)) GO TO 102
      KD = I
      IF(I.GT.LO) KD = 0
      KLL20 = KLL20 - II - KD
      KLL21 = KLL20 - 5
      KLL22 = KLL20 + 5
      KLL10 = KLL10 - (II - 1)
      KLL11 = KLL10 - 2
      KLL12 = KLL10 + 2
      KL1L20 = KL1L20 - I - 1
      KL1L21 = KL1L20 - 5
      KL1L22 = KL1L20 + 5
      GO TO 103
  102 IF((I.GT.KT).OR.(I.LE.LDB)) GO TO 103
      KD = I
      IF(I.GT.LO) KD = 0
      KLL20 = KLL20 - I - 1 - KD
      KLL21 = KLL20 - 5
      KLL22 = KLL20 + 5
      KLL10 = KLL10 - I
      KLL11 = KLL10 - 2
      KLL12 = KLL10 + 2
      KL1L20 = KL1L20 - I - 1
      KL1L21 = KL1L20 - 5
      KL1L22 = KL1L20 + 5
  103 IF((I.GE.KT).OR.((I+1).NE.LDB)) GO TO 104
      KL1L30 = KL1L30 - II + 1
      KL1L31 = KL1L30 - 2
      KL1L32 = KL1L30 + 2
      GO TO 105
  104 IF((I.GE.KT).OR.((I+1).LE.LDB)) GO TO 106
      KL1L30 = KL1L30 - (2*I) - 2
      KL1L31 = KL1L30 - 2
      KL1L32 = KL1L30 + 2
      GO TO 105
  106 IF((I.LE.KT).OR.(I.GE.LDB)) GO TO 107
      KD = 0
      IF(I.GT.LO) KD = 3*KT
      KLL20 = (9*KT) + 7 - KD
      KLL21 = KLL20 - 6
      KLL22 = KLL20 + 6
      KLL10 = 3*(KTERM - II) + 4
      KLL11 = KLL10 - 3
      KLL12 = KLL10 + 3
      KL1L20 = (3*KT) + (3*II) + 7
      KL1L21 = KL1L20 - 6
      KL1L22 = KL1L20 + 6
      GO TO 108
  107 IF((I.LE.KT).OR.(I.NE.LDB)) GO TO 109
      KD = KTERM
      IF(I.GT.LO) KD = 0
      KLL20 = (3*KTERM) - II + 7 + KD
      KLL21 = KLL20 - 5
      KLL22 = KLL20 + 5
      KLL10 = (3*KTERM) - (4*II) + 5
      KLL11 = KLL10 - 2
      KLL12 = KLL10 + 2
      KL1L20 = KTERM + (3*II) + 6
      KL1L21 = KL1L20 - 5
      KL1L22 = KL1L20 + 5
      GO TO 108
  109 IF((I.LE.KT).OR.(I.LE.LDB)) GO TO 108
      KD = KTERM
      IF(I.GT.LO) KD = 0
      KLL20 = (5*KT) + 6 + KD
      KLL21 = KLL20 - 5
      KLL22 = KLL20 + 5
      KLL10 = (5*KT) - (3*II) + 4
      KLL11 = KLL10 - 2
      KLL12 = KLL10 + 2
      KL1L20 = KTERM + (3*II) + 6
      KL1L21 = KL1L20 - 5
      KL1L22 = KL1L20 + 5
  108 IF((I.LT.KT).OR.((I+1).GE.LDB)) GO TO 110
      KL1L30 = (9*KT) + 7
      KL1L31 = KL1L30 - 3
      KL1L32 = KL1L30 + 3
      GO TO 105
  110 IF((I.LT.KT).OR.((I+1).NE.LDB)) GO TO 111
      KL1L30 = (9*KT) - II + 8
      KL1L31 = KL1L30 - 2
      KL1L32 = KL1L30 + 2
      GO TO 105
  111 IF((I.LT.KT).OR.((I+1).LE.LDB)) GO TO 105
      KL1L30 = (7*KT) + 6
      KL1L31 = KL1L30 - 2
      KL1L32 = KL1L30 + 2
  105 LL3 = (3*I*(I + 1)) + (3*II) + 1
      LL13 = LL3 + 1
      LL23 = LL3 + 2
      L1L13 = LL3 + (3*I) + 3
      L1L23 = L1L13 + 1
      L1L33 = L1L13 + 2
      XK = DFLOAT(2*II)
      DROTK1 = DROT(1) + ((DROT(1)/(R*AL*(AL + 1.D0)))*(RZ - R)*XK*XK)
      DROTK2 = DROT(2) + ((DROT(2)/(R*(AL + 1.D0)*(AL + 2.D0)))*(RZ - R)
     1*XK*XK)
      IF(IEND.EQ.I) GO TO 49
      IF(II.EQ.(KTERM/2)) GO TO 50
      CSYM(J,KLL22) = (-1.D0,0.D0)*((F2 + D2PR)*SY(LL21,1)*SY2(LL23,2)*S
     1FLL2)
      CSYM(J,KLL22+3) = (1.D0,0.D0)*(D2*SY(LL22,3)*SY2(LL23,2)*SFLL2)
      CSYM(J1,KLL22)    = (-1.D0,0.D0)*(F2*SY(LL21,1)*SY2(LL23,2)*SFLL2)
      CSYM(J1,KLL22+2) = CSYM(J,KLL22+3)
      CSYM(J1,KLL22+3) = CSYM(J,KLL22+3)
      CSYM(J2,KLL22) = (-1.D0,0.D0)*((F2 - D2PR)*SY(LL21,1)*SY2(LL23,2)*
     1SFLL2)
      CSYM(J2,KLL22+2) = CSYM(J,KLL22+3)
      CSYM(J3,KLL22) = (1.D0,0.D0)*((F2 + (D2PR/2.D0))*SY(LL22,4)*SY2(LL
     123,2)*SFLL2)
      CSYM(J3,KLL22-3) = (1.D0,0.D0)*(D2*SY(LL22,5)*SY2(LL23,2)*SFLL2)
      CSYM(J3,KLL22-2) = CSYM(J3,KLL22-3)
      CSYM(J4,KLL22) = (1.D0,0.D0)*((F2 - (D2PR/2.D0))*SY(LL22,4)*SY2(LL
     123,2)*SFLL2)
      CSYM(J4,KLL22-3) = CSYM(J3,KLL22-3)
      CSYM(J4,KLL22-2) = CSYM(J4,KLL22-3)
      IF(I.GT.LDB) GO TO 112
      CSYM(J5,KLL22-2) = (-1.D0,0.D0)*((D2/DSQRT(2.D0))*SY(LL22,8)*SY2(L
     1L23,2)*SFLL2)
      CSYM(J5,KLL22-1) = CSYM(J5,KLL22-2)
      IF(I.GE.LDB) GO TO 112
      CSYM(J5,KLL22) = (-1.D0,0.D0)*(F2*SY(LL22,7)*SY2(LL23,2)*SFLL2)
      CSYM(J3,KLL22+2) = (-1.D0,0.D0)*((D2/DSQRT(2.D0))*SY(LL22,6)*SY2(L
     1L23,2)*SFLL2)
      CSYM(J4,KLL22+1) = CSYM(J3,KLL22+2)
  112 IF(I.GT.LO) GO TO 50
      IF(II.EQ.I) GO TO 51
      CSYM(J,KLL12) = (1.D0,0.D0)*(D2*SY(LL12,3)*SY2(LL13,2)*SFLL1)
      CSYM(J1,KLL12-1) = CSYM(J,KLL12)
      CSYM(J1,KLL12) = CSYM(J,KLL12)
```

```
      CSYM(J2,KLL12-1) = CSYM(J,KLL12)
      CSYM(J3,KLL12-3) = (1.D0,0.D0)*((F2 + (D2PR/2.D0))*SY(LL12,4)*SY2(
     1LL13,2)*SFLL1)
      CSYM(J4,KLL12-3) = (1.D0,0.D0)*((F2 - (D2PR/2.D0))*SY(LL12,4)*SY2(
     1LL13,2)*SFLL1)
      IF(I.GT.LDB) GO TO 51
      CSYM(J5,KLL12-5) = (-1.D0,0.D0)*((D2/DSQRT(2.D0))*SY(LL12,8)*SY2(L
     1L13,2)*SFLL1)
      CSYM(J5,KLL12-4) = CSYM(J5,KLL12-5)
      IF(I.GE.LDB) GO TO 51
      CSYM(J5,KLL12-3) = (-1.D0,0.D0)*(F2*SY(LL12,7)*SY2(LL13,2)*SFLL1)
      CSYM(J3,KLL12-1) = (-1.D0,0.D0)*((D2/DSQRT(2.D0))*SY(LL12,6)*SY2(L
     1L13,2)*SFLL1)
      CSYM(J4,KLL12-2) = CSYM(J3,KLL12-1)
   51 CSYM(JODD,KL1L22) = (1.D0,0.D0)*((F2 + (D2PR/2.D0))*SY(L1L22,4)*SY
     12(L1L23,2)*SFL1L2)
      CSYM(JODD,KL1L22-3) = (1.D0,0.D0)*(D2*SY(L1L22,5)*SY2(L1L23,2)*SFL
     11L2)
      CSYM(JODD,KL1L22-2) = CSYM(JODD,KL1L22-3)
      CSYM(JODD+1,KL1L22)  = (1.D0,0.D0)*((F2 - (D2PR/2.D0))*SY(L1L22,4)
     1*SY2(L1L23,2)*SFL1L2)
      CSYM(JODD+1,KL1L22-3) = CSYM(JODD,KL1L22-3)
      CSYM(JODD+1,KL1L22-2) = CSYM(JODD,KL1L22-3)
      IF(I.GT.LDB) GO TO 113
      CSYM(JODD+2,KL1L22-2) = (-1.D0,0.D0)*((D2/DSQRT(2.D0))*SY(L1L22,8)
     1*SY2(L1L23,2)*SFL1L2)
      CSYM(JODD+2,KL1L22-1) = CSYM(JODD+2,KL1L22-2)
      IF(I.GE.LDB) GO TO 113
      CSYM(JODD+2,KL1L22) = (-1.D0,0.D0)*(F2*SY(L1L22,7)*SY2(L1L23,2)*SF
     1L1L2)
      CSYM(JODD,KL1L22+2) = (-1.D0,0.D0)*((D2/DSQRT(2.D0))*SY(L1L22,6)*S
     1Y2(L1L23,2)*SFL1L2)
      CSYM(JODD+1,KL1L22+1) = CSYM(JODD,KL1L22+2)
  113 IF(I.GE.LO) GO TO 50
      CSYM(JODD,KL1L32)= (1.D0,0.D0)*((F2 + (D2PR/2.D0))*SY(L1L32,4)*SY2
     1(L1L33,2)*SFL1L3)
      CSYM(JODD+1,KL1L32) = (1.D0,0.D0)*((F2 - (D2PR/2.D0))*SY(L1L32,4)*
     1SY2(L1L33,2)*SFL1L3)
      IF(I.GT.LDB) GO TO 50
      CSYM(JODD+2,KL1L32-2) = (-1.D0,0.D0)*((D2/DSQRT(2.D0))*SY(L1L32,8)
     1*SY2(L1L33,2)*SFL1L3)
      CSYM(JODD+2,KL1L32-1) = CSYM(JODD+2,KL1L32-2)
      IF(I.GE.LDB) GO TO 50
      CSYM(JODD+2,KL1L32) = (-1.D0,0.D0)*(F2*SY(L1L32,7)*SY2(L1L33,2)*SF
     1L1L3)
      CSYM(JODD,KL1L32+2) = (-1.D0,0.D0)*((D2/DSQRT(2.D0))*SY(L1L32,6)*S
     1Y2(L1L33,2)*SFL1L3)
      CSYM(JODD+1,KL1L32+1) = CSYM(JODD,KL1L32+2)
   50 XZ = 1.D0
      IF(II.EQ.1) XZ = DSQRT(2.D0)
      CSYM(J,KLL20) = (-1.D0,0.D0)*((F0 + DPR)*SY(LL21,1)*SY2(LL23,1)*SF
     1LL2)
      CSYM(J,KLL21) = (-1.D0,0.D0)*((F2 + D2PR)*SY(LL21,1)*SY2(LL23,3)*S
     1FLL2*XZ)
      CSYM(J,KLL20+3) = (1.D0,0.D0)*(D*SY(LL22,3)*SY2(LL23,1)*SFLL2)
      CSYM(J,KLL21+3) = (1.D0,0.D0)*(D2*SY(LL22,3)*SY2(LL23,3)*SFLL2*XZ)
      CSYM(J1,KLL20) = (-1.D0,0.D0)*(F0*SY(LL21,1)*SY2(LL23,1)*SFLL2)
      CSYM(J1,KLL21) = (-1.D0,0.D0)*(F2*SY(LL21,1)*SY2(LL23,3)*SFLL2*XZ)
      CSYM(J1,KLL20+2) = CSYM(J,KLL20+3)
      CSYM(J1,KLL20+3) = CSYM(J,KLL20+3)
      CSYM(J1,KLL21+2) = CSYM(J,KLL21+3)
      CSYM(J1,KLL21+3) = CSYM(J,KLL21+3)
      CSYM(J2,KLL20) = (-1.D0,0.D0)*((F0 - DPR)*SY(LL21,1)*SY2(LL23,1)*S
     1FLL2)
      CSYM(J2,KLL21) = (-1.D0,0.D0)*((F2 - D2PR)*SY(LL21,1)*SY2(LL23,3)*
     1SFLL2*XZ)
      CSYM(J2,KLL20+2) = CSYM(J,KLL20+3)
      CSYM(J2,KLL21+2) = CSYM(J,KLL21+3)
      CSYM(J3,KLL20) = (1.D0,0.D0)*((F0 + (DPR/2.D0))*SY(LL22,4)*SY2(LL2
     13,1)*SFLL2)
      CSYM(J3,KLL21) = (1.D0,0.D0)*((F2 + (D2PR/2.D0))*SY(LL22,4)*SY2(LL
     123,3)*SFLL2*XZ)
      CSYM(J3,KLL20-3) = (1.D0,0.D0)*(D*SY(LL22,5)*SY2(LL23,1)*SFLL2)
      CSYM(J3,KLL20-2) = CSYM(J3,KLL20-3)
      CSYM(J3,KLL21-3) = (1.D0,0.D0)*(D2*SY(LL22,5)*SY2(LL23,3)*SFLL2*XZ
     1)
      CSYM(J3,KLL21-2) = CSYM(J3,KLL21-3)
      CSYM(J4,KLL20) = (1.D0,0.D0)*((F0 - (DPR/2.D0))*SY(LL22,4)*SY2(LL2
     13,1)*SFLL2)
      CSYM(J4,KLL21) = (1.D0,0.D0)*((F2 - (D2PR/2.D0))*SY(LL22,4)*SY2(LL
     123,3)*SFLL2*XZ)
      CSYM(J4,KLL20-3) = CSYM(J3,KLL20-3)
      CSYM(J4,KLL20-2) = CSYM(J4,KLL20-3)
      CSYM(J4,KLL21-3) = CSYM(J3,KLL21-3)
      CSYM(J4,KLL21-2) = CSYM(J3,KLL21-3)
      IF(I.GT.LDB) GO TO 114
      CSYM(J5,KLL20-2) = (-1.D0,0.D0)*((D/DSQRT(2.D0))*SY(LL22,8)*SY2(LL
     123,1)*SFLL2)
      CSYM(J5,KLL20-1) = CSYM(J5,KLL20-2)
      CSYM(J5,KLL21-2) = (-1.D0,0.D0)*((D2/DSQRT(2.D0))*SY(LL22,8)*SY2(L
     1L23,3)*SFLL2*XZ)
      CSYM(J5,KLL21-1) = CSYM(J5,KLL21-2)
      IF(I.GE.LDB) GO TO 114
      CSYM(J5,KLL20) = (-1.D0,0.D0)*(F0*SY(LL22,7)*SY2(LL23,1)*SFLL2)
      CSYM(J5,KLL21) =(-1.D0,0.D0)*(F2*SY(LL22,7)*SY2(LL23,3)*SFLL2*XZ)
      CSYM(J3,KLL20+2) = (-1.D0,0.D0)*((D/DSQRT(2.D0))*SY(LL22,6)*SY2(LL
     123,1)*SFLL2)
      CSYM(J3,KLL21+2) = (-1.D0,0.D0)*((D2/DSQRT(2.D0))*SY(LL22,6)*SY2(L
     1L23,3)*SFLL2*XZ)
      CSYM(J4,KLL20+1) = CSYM(J3,KLL20+2)
      CSYM(J4,KLL21+1) = CSYM(J3,KLL21+2)
  114 IF(I.GT.LO) GO TO 49
      CSYM(J,KLL10) = (1.D0,0.D0)*(D*SY(LL12,3)*SY2(LL13,1)*SFLL1)
      CSYM(J1,KLL10-1) = CSYM(J,KLL10)
      CSYM(J1,KLL10) = CSYM(J,KLL10)
      CSYM(J2,KLL10-1) = CSYM(J,KLL10)
      CSYM(J3,KLL10-3) = (1.D0,0.D0)*((F0 + (DPR/2.D0))*SY(LL12,4)*SY2(L
     1L13,1)*SFLL1)
      CSYM(J4,KLL10-3) = (1.D0,0.D0)*((F0 - (DPR/2.D0))*SY(LL12,4)*SY2(L
     1L13,1)*SFLL1)
      IF(I.GT.LDB) GO TO 115
      CSYM(J5,KLL10-5) = (-1.D0,0.D0)*((D/DSQRT(2.D0))*SY(LL12,8)*SY2(LL
     113,1)*SFLL1)
      CSYM(J5,KLL10-4) = CSYM(J5,KLL10-5)
      IF(I.GE.LDB) GO TO 115
      CSYM(J5,KLL10-3) = (-1.D0,0.D0)*(F0*SY(LL12,7)*SY2(LL13,1)*SFLL1)
      CSYM(J3,KLL10-1) = (-1.D0,0.D0)*((D/DSQRT(2.D0))*SY(LL12,6)*SY2(LL
     113,1)*SFLL1)
      CSYM(J4,KLL10-2) = CSYM(J3,KLL10-1)
  115 IF(II.EQ.1) GO TO 55
      CSYM(J,KLL11) = (1.D0,0.D0)*(D2*SY(LL12,3)*SY2(LL13,3)*SFLL1)
      CSYM(J1,KLL11-1) = CSYM(J,KLL11)
      CSYM(J1,KLL11) = CSYM(J,KLL11)
      CSYM(J2,KLL11-1) = CSYM(J,KLL11)
      CSYM(J3,KLL11-3) = (1.D0,0.D0)*((F2 + (D2PR/2.D0))*SY(LL12,4)*SY2(
     1LL13,3)*SFLL1)
      CSYM(J4,KLL11-3) = (1.D0,0.D0)*((F2 - (D2PR/2.D0))*SY(LL12,4)*SY2(
     1LL13,3)*SFLL1)
      IF(I.GT.LDB) GO TO 55
      CSYM(J5,KLL11-5) = (-1.D0,0.D0)*((D2/DSQRT(2.D0))*SY(LL12,8)*SY2(L
     1L13,3)*SFLL1)
      CSYM(J5,KLL11-4) = CSYM(J5,KLL11-5)
      IF(I.GE.LDB) GO TO 55
      CSYM(J5,KLL11-3) = (-1.D0,0.D0)*(F2*SY(LL12,7)*SY2(LL13,3)*SFLL1)
      CSYM(J3,KLL11-1) = (-1.D0,0.D0)*((D2/DSQRT(2.D0))*SY(LL12,6)*SY2(L
     1L13,3)*SFLL1)
      CSYM(J4,KLL11-2) = CSYM(J3,KLL11-1)
   55 CSYM(JODD,KL1L20) = (1.D0,0.D0)*((F0 + (DPR/2.D0))*SY(L1L22,4)*SY2
     1(L1L23,1)*SFL1L2)
      CSYM(JODD,KL1L20-3) = (1.D0,0.D0)*(D*SY(L1L22,5)*SY2(L1L23,1)*SFL1
     1L2)
      CSYM(JODD,KL1L20-2) = CSYM(JODD,KL1L20-3)
      CSYM(JODD+1,KL1L20) = (1.D0,0.D0)*((F0 - (DPR/2.D0))*SY(L1L22,4)*S
     1Y2(L1L23,1)*SFL1L2)
      CSYM(JODD+1,KL1L20-3) = CSYM(JODD,KL1L20-3)
      CSYM(JODD+1,KL1L20-2) = CSYM(JODD,KL1L20-3)
      IF(I.GT.LDB) GO TO 116
      CSYM(JODD+2,KL1L20-2) = (-1.D0,0.D0)*((D/DSQRT(2.D0))*SY(L1L22,8)*
     1SY2(L1L23,1)*SFL1L2)
      CSYM(JODD+2,KL1L20-1) = CSYM(JODD+2,KL1L20-2)
      IF(I.GE.LDB) GO TO 116
      CSYM(JODD+2,KL1L20) = (-1.D0,0.D0)*(F0*SY(L1L22,7)*SY2(L1L23,1)*SF
     1L1L2)
      CSYM(JODD,KL1L20+2) = (-1.D0,0.D0)*((D/DSQRT(2.D0))*SY(L1L22,6)*SY
     12(L1L23,1)*SFL1L2)
      CSYM(JODD+1,KL1L20+1) = CSYM(JODD,KL1L20+2)
  116 XZ = 1.D0
      IF(II.EQ.1) XZ = DSQRT(2.D0)
      CSYM(JODD,KL1L21) = (1.D0,0.D0)*((F2 + (D2PR/2.D0))*SY(L1L22,4)*SY
     12(L1L23,3)*SFL1L2*XZ)
      CSYM(JODD,KL1L21-3) = (1.D0,0.D0)*(D2*SY(L1L22,5)*SY2(L1L23,3)*SFL
     11L2*XZ)
      CSYM(JODD,KL1L21-2) = CSYM(JODD,KL1L21-3)
      CSYM(JODD+1,KL1L21) = (1.D0,0.D0)*((F2 - (D2PR/2.D0))*SY(L1L22,4)*
     1SY2(L1L23,3)*SFL1L2*XZ)
      CSYM(JODD+1,KL1L21-3) = CSYM(JODD,KL1L21-3)
      CSYM(JODD+1,KL1L21-2) = CSYM(JODD,KL1L21-3)
      IF(I.GT.LDB) GO TO 117
      CSYM(JODD+2,KL1L21-2) = (-1.D0,0.D0)*((D2/DSQRT(2.D0))*SY(L1L22,8)
     1*SY2(L1L23,3)*SFL1L2*XZ)
      CSYM(JODD+2,KL1L21-1) = CSYM(JODD+2,KL1L21-2)
      IF(I.GE.LDB) GO TO 117
      CSYM(JODD+2,KL1L21) = (-1.D0,0.D0)*(F2*SY(L1L22,7)*SY2(L1L23,3)*SF
     1L1L2*XZ)
      CSYM(JODD,KL1L21+2) = (-1.D0,0.D0)*((D2/DSQRT(2.D0))*SY(L1L22,6)*S
     1Y2(L1L23,3)*SFL1L2*XZ)
      CSYM(JODD+1,KL1L21+1) = CSYM(JODD,KL1L21+2)
  117 IF(I.GE.LO) GO TO 49
      CSYM(JODD,KL1L30) = (1.D0,0.D0)*((F0 + (DPR/2.D0))*SY(L1L32,4)*SY2
     1(L1L33,1)*SFL1L3)
      CSYM(JODD+1,KL1L30) = (1.D0,0.D0)*((F0 - (DPR/2.D0))*SY(L1L32,4)*S
     1Y2(L1L33,1)*SFL1L3)
      IF(I.GT.LDB) GO TO 118
      CSYM(JODD+2,KL1L30-2) = (-1.D0,0.D0)*((D/DSQRT(2.D0))*SY(L1L32,8)*
     1SY2(L1L33,1)*SFL1L3)
      CSYM(JODD+2,KL1L30-1) = CSYM(JODD+2,KL1L30-2)
      IF(I.GE.LDB) GO TO 118
      CSYM(JODD+2,KL1L30) = (-1.D0,0.D0)*(F0*SY(L1L32,7)*SY2(L1L33,1)*SF
     1L1L3)
      CSYM(JODD,KL1L30+2) = (-1.D0,0.D0)*((D/DSQRT(2.D0))*SY(L1L32,6)*SY
     12(L1L33,1)*SFL1L3)
      CSYM(JODD+1,KL1L30+1) = CSYM(JODD,KL1L30+2)
  118 IF(II.EQ.1) GO TO 49
      CSYM(JODD,KL1L31) = (1.D0,0.D0)*((F2 + (D2PR/2.D0))*SY(L1L32,4)*SY
     12(L1L33,3)*SFL1L3)
      CSYM(JODD+1,KL1L31) = (1.D0,0.D0)*((F2 - (D2PR/2.D0))*SY(L1L32,4)*
     1SY2(L1L33,3)*SFL1L3)
      IF(I.GT.LDB) GO TO 49
      CSYM(JODD+2,KL1L31-2) = (-1.D0,0.D0)*((D2/DSQRT(2.D0))*SY(L1L32,8)
     1*SY2(L1L33,3)*SFL1L3)
      CSYM(JODD+2,KL1L31-1) = CSYM(JODD+2,KL1L31-2)
      IF(I.GE.LDB) GO TO 49
      CSYM(JODD+2,KL1L31) = (-1.D0,0.D0)*(F2*SY(L1L32,7)*SY2(L1L33,3)*SF
     1L1L3)
      CSYM(JODD,KL1L31+2) = (-1.D0,0.D0)*((D2/DSQRT(2.D0))*SY(L1L32,6)*S
     1Y2(L1L33,3)*SFL1L3)
      CSYM(JODD+1,KL1L31+1) = CSYM(JODD,KL1L31+2)
   49 IF((II.EQ.I).OR.(II.EQ.KT)) GO TO 56
      KD = 0
      IF(I.GT.LDB) KD = 1
      CSYM(J,7-KD) = (-1.D0,0.D0)*((F2 + D2PR)*SY(LL1,1)*SY2(LL3,2)*SFLL
     1)
      CSYM(J,10-KD) = (1.D0,0.D0)*(D2*SY(LL2,3)*SY2(LL3,2)*SFLL)
      CSYM(J1,7-KD) = (-1.D0,0.D0)*(F2*SY(LL1,1)*SY2(LL3,2)*SFLL)
      CSYM(J1,9-KD) = CSYM(J,10-KD)
      CSYM(J1,10-KD) = CSYM(J,10-KD)
      CSYM(J2,7-KD) = (-1.D0,0.D0)*((F2 - D2PR)*SY(LL1,1)*SY2(LL3,2)*SFL
     1L)
      CSYM(J2,9-KD) = CSYM(J,10-KD)
      CSYM(J3,7-KD) = (1.D0,0.D0)*((F2 + (D2PR/2.D0))*SY(LL2,4)*SY2(LL3,
     12)*SFLL)
      CSYM(J3,4-KD) = (1.D0,0.D0)*(D2*SY(LL2,5)*SY2(LL3,2)*SFLL)
      CSYM(J3,5-KD) = CSYM(J3,4-KD)
      CSYM(J4,7-KD) = (1.D0,0.D0)*((F2 - (D2PR/2.D0))*SY(LL2,4)*SY2(LL3,
     12)*SFLL)
      CSYM(J4,4-KD) = CSYM(J3,4-KD)
      CSYM(J4,5-KD) = CSYM(J3,4-KD)
      IF(I.GT.LDB) GO TO 119
      CSYM(J3,9) = (-1.D0,0.D0)*((D2/DSQRT(2.D0))*SY(LL2,6)*SY2(LL3,2)*S
     1FLL)
      CSYM(J4,8) = CSYM(J3,9)
      CSYM(J5,7) = (-1.D0,0.D0)*(F2*SY(LL2,7)*SY2(LL3,2)*SFLL)
      CSYM(J5,5) = (-1.D0,0.D0)*((D2/DSQRT(2.D0))*SY(LL2,8)*SY2(LL3,2)*S
     1FLL)
      CSYM(J5,6) = CSYM(J5,5)
  119 IF(I.GT.LO) GO TO 56
      KD = 0
      IF(I.GE.LDB) KD = 1
```

```
      CSYM(JODD,4-KD) = (1.D0,0.D0)*((F2 + (D2PR/2.D0))*SY(L1L12,4)*SY2(
     1L1L13,2)*SFL1L1)
      CSYM(JODD+1,4-KD) = (1.D0,0.D0)*((F2 - (D2PR/2.D0))*SY(L1L12,4)*SY
     12(L1L13,2)*SFL1L1)
      IF(I.GE.LDB) GO TO 56
      CSYM(JODD,6) = (-1.D0,0.D0)*((D2/DSQRT(2.D0))*SY(L1L12,6)*SY2(L1L1
     13,2)*SFL1L1)
      CSYM(JODD+1,5) = CSYM(JODD,6)
      CSYM(JODD+2,4) = (-1.D0,0.D0)*(F2*SY(L1L12,7)*SY2(L1L13,2)*SFL1L1)
      CSYM(JODD+2,2) = (-1.D0,0.D0)*((D2/DSQRT(2.D0))*SY(L1L12,8)*SY2(L1
     1L13,2)*SFL1L1)
      CSYM(JODD+2,3) = CSYM(JODD+2,2)
   56 CSYM(J,4) = (1.D0,0.D0)*(D*SY(LL2,3)*SY2(LL3,1)*SFLL)
      CSYM(J1,3) = CSYM(J,4)
      CSYM(J1,4) = CSYM(J,4)
      CSYM(J2,3) = CSYM(J,4)
      CSYM(J,1) = ((1.D0,0.D0)*((2.D0*B) - (SFLL*(F0 + DPR)*SY(LL1,1)*SY
     12(LL3,1)))) + ((0.D0,-1.D0)*DROTK1)
      CSYM(J1,1) =((-1.D0,0.D0)*(SFLL*F0*SY(LL1,1)*SY2(LL3,1))) + ((0.D0
     1,-1.D0)*DROTK1)
      CSYM(J2,1) = ((1.D0,0.D0)*(-(2.D0*B) - (SFLL*(F0 - DPR)*SY(LL1,1)*
     1SY2(LL3,1)))) + ((0.D0,-1.D0)*DROTK1)
      CSYM(J3,1) = ((1.D0,0.D0)*(B + (SFLL*(F0 + (DPR/2.D0))*SY(LL2,4)*S
     1Y2(LL3,1)))) + ((0.D0,-1.D0)*DROTK1)
      CSYM(J4,1) = ((1.D0,0.D0)*(-B + (SFLL*(F0 - (DPR/2.D0))*SY(LL2,4)*
     1SY2(LL3,1)))) + ((0.D0,-1.D0)*DROTK1)
      IF(I.GT.LDB) GO TO 120
      CSYM(J3,3) = (-1.D0,0.D0)*((D/DSQRT(2.D0))*SY(LL2,6)*SY2(LL3,1)*SF
     1LL)
      CSYM(J4,2) = CSYM(J3,3)
      CSYM(J5,1) = ((-1.D0,0.D0)*(SFLL*F0*SY(LL2,7)*SY2(LL3,1))) + ((0.D
     10,-1.D0)*DROTK1)
      J = J + 1
  120 J = J + 5
      IF(I.GT.LO) GO TO 48
      CSYM(JODD,1) = ((1.D0,0.D0)*(B + (SFL1L1*(F0 + (DPR/2.D0))*SY(L1L1
     12,4)*SY2(L1L13,1)))) + ((0.D0,-1.D0)*DROTK2)
      CSYM(JODD+1,1) = ((1.D0,0.D0)*(-B + (SFL1L1*(F0 - (DPR/2.D0))*SY(L
     11L12,4)*SY2(L1L13,1)))) + ((0.D0,-1.D0)*DROTK2)
      IF(I.GE.LDB) GO TO 121
      CSYM(JODD,3) = (-1.D0,0.D0)*((D/DSQRT(2.D0))*SY(L1L12,6)*SY2(L1L13
     1,1)*SFL1L1)
      CSYM(JODD+1,2) = CSYM(JODD,3)
      CSYM(JODD+2,1) = ((-1.D0,0.D0)*(SFL1L1*F0*SY(L1L12,7)*SY2(L1L13,1)
     1)) + ((0.D0,-1.D0)*DROTK2)
      JCOUNT = JCOUNT + 1
  121 JCOUNT = JCOUNT + 2
   48 CONTINUE
      J = J + JCOUNT
   34 AL = AL + 2.D0
      DO 60 I=1,N
      AMP(I) = (0.D0,0.D0)
   60 VALUE(I) = (0.D0,0.D0)
      AMP(1) = (1.D0,0.D0)
      AMP(2) = (1.D0,0.D0)
      AMP(3) = (1.D0,0.D0)

 SUBROUTINES  CSQZ  AND  CQRT  DIAGONALIZE A COMPLEX
 SYMMETRIC MATRIX. THESE SUBROUTINES WERE DEVELOPED BY
 GORDON AND MESSINGER,'ESR RELAXATION IN LIQUIDS',ED.
 L.T.MUUS AND P.W.ATKINS(PLENEM,NEW YORK,1972).

      CALL CSQZ(CSYM,AMP,NDIM,JDIM,N,M,SQTOL)
      CALL CQRT(CSYM1,CSYM2,AMP,NDIM,N,20,TOL,VALUE)
      IF(.NOT.PAMPEV) GO TO 821
      WRITE(7,822) GAMMA,FOAB,N,(CSYM1(I),AMP(I),I=1,N)
  822 FORMAT(D20.13,5X,D20.13,5X,I3,/,(6D13.6))
  821 IF(.NOT.WAMPEV) GO TO 823
      WRITE(6,824)(CSYM1(I),AMP(I),I=1,N)
  824 FORMAT(1H ,//,10X,'EIGENVALUE',26X,'AMP',25X,'EIGENVALUE',26X,'AMP
     1',//,(1X,2D14.5,4X,2D14.5,4X,2D14.5,4X,2D14.5))
  823 IF(NSET.LT.1) GO TO 10
      DO 10 ISET=1,NSET
      IF(ISET.GT.1) WRITE(6,825)
  825 FORMAT(1H1,'INPUT VALUES ARE THE SAME AS FOR PRECEDING SET OF INPU
     1T VALUES ',/,' EXCEPT FOR THE FOLLOWING:',//)

 ***************************************************************
 *   READ INPUT DATA - TWO READ STATEMENTS                      *
 ***************************************************************
 -NOTE THESE TWO INPUT CARDS ARE REPEATED UNTIL THEY ARE
  READ A TOTAL OF  NSET  TIMES.

      READ(5,87) IPLOT,XPUNCH,XPLOT,DISP,YLEN,XLEN
   87 FORMAT(I2,3X,L1,4X,L1,4X,L1,4X,E10.3,5X,E10.3)
      READ(5,88) T2IN,BI,BF,NB
   88 FORMAT(D10.3,5X,D10.3,5X,D10.3,5X,I3)
      WRITE(6,89) T2IN,BI,BF,NB
   89 FORMAT(1H ,//,' ROTATIONALLY INVARIANT P-P FIRST DERIVATIVE LINEWI
     1DTH ',D12.5,' GAUSS',//,' INITIAL FIELD SWEEP VALUE RELATIVE TO CE
     2NTER  ',D12.5,' GAUSS',/,' FINAL FIELD VALUE RELATIVE TO CENTER ',
     3D12.5,' GAUSS',//,' NUMBER OF POINTS FOR SPECTRUM ',I3,/)
      T2IN = T2IN*(DSQRT(3.D0)/2.D0)*(GAMMA/FOAB)
      BI = BI*(GAMMA/FOAB)
      BF = BF*(GAMMA/FOAB)
      DEL = (BF - BI)/DFLOAT(NB-1)
      XAB = (0.D0,-1.D0)
      IF(DISP) XAB = (1.D0,0.D0)
      IF(IPLOT.GT.0) GO TO 241

 CALCULATION OF THE ABSORPTION-DISPERSION VALUES. NOTE
 THE ABSORPTION-DISPERSION VALUES ARE STORED IN  C  MATRIX
 AND THE CORRESPONDING FIELD VALUES ARE IN  BFLD  MATRIX.

      RES = BI
      DO 235 IB=1,NB
      ABSP = (0.D0,0.D0)
      DO 223 I=1,N
  223 ABSP = ABSP + ((AMP(I)*AMP(I))/(CSYM1(I) + ((-1.D0,0.D0)*RES) + ((
     10.D0,-1.D0)*T2IN)))
      C(IB) = XAB*(ABSP/(3.D0*3.14159D0))
      BFLD(IB) = RES*(FOAB/GAMMA)
      RES = RES + DEL
  235 CONTINUE
      IF(DISP) WRITE(6,906)
  906 FORMAT(1H ,' DISPERSION PLOT',//)
      WRITE(6,242)(C(I),BFLD(I),I=1,NB)
  242 FORMAT(1H ,///,47X,'COORDINATES FOR ABSORPTION PLOT',//,4X,'ABSORP
     1TION',6X,'FIELD',11X,'ABSORPTION',6X,'FIELD',11X,'ABSORPTION',6X,'
     2FIELD',11X,'ABSORPTION',6X,'FIELD',//,(1X,2E14.5,4X,2E14.5,4X,2E14
     4.5,4X,2E14.5))
      IF(.NOT.XPUNCH) GO TO 243

 PUNCH OUTPUT - COORDINATES OF ABSORPTION-DISPERSION
                SPECTRUM PUNCHED ON CARDS

      WRITE(7,244)(C(I),BFLD(I),I=1,NB)
  244 FORMAT(E13.6,2X,E13.6,7X,E13.6,2X,E13.6)
  243 IF(.NOT.XPLOT) GO TO 241

 LINE PLOT FOR ABSORPTION-DISPERSION SPECTRUM (CORNELL
 U. PLOT SUBROUTINES)

      ABMAX = MAXIMUM ABSORPTION-DISPERSION VALUE
      ABMIN = MINIMUM ABSORPTION-DISPERSION VALUE

 SUBROUTINES FOR PLOTTING-
      AXIS = DRAWS AND LABELS AXIS
      PLOT = MOVES PEN TO A SPECIFIED POINT (COORDINATES IN
             INCHES).
      PLTEND = SPECIFIES END OF ALL PLOTTING

      ABMAX = C(1)
      DO 245 I=1,NB
      IF(ABMAX.LT.C(I)) ABMAX = C(I)
  245 CONTINUE
      ABMIN = C(1)
      DO 246 I=1,NB
      IF(ABMIN.GT.C(I)) ABMIN = C(I)
  246 CONTINUE
      DXY = (BFLD(NB) - BFLD(1))/XLEN
      CALL AXIS(0.0,1.0,5HGAUSS,-5,XLEN,0.0,BFLD(1),DXY)
      Y = 1.0 + (YLEN*((C(1) - ABMIN)/(ABMAX - ABMIN)))
      X = 0.0
      CALL PLOT(X,Y,3)
      DO 247 I=1,NB
      Y = 1.0 + (YLEN*((C(I) - ABMIN)/(ABMAX - ABMIN)))
      X = XLEN*((BFLD(I) - BFLD(1))/(BFLD(NB) - BFLD(1)))
  247 CALL PLOT(X,Y,2)
      XPLEND = .TRUE.
      Y = 0.0
      X = X + 3.
      CALL PLOT(X,Y,-3)
  241 IF(IPLOT.LT.0) GO TO 10

 CALCULATION OF FIRST DERIVATIVE VALUES.
      THE FIRST DERIVATIVE VALUES ARE STORED IN THE  DERAB
      MATRIX AND THE CORRESPONDING MAGNETIC FIELD VALUES ARE
      IN THE MATRIX  BFLD.

      RES = BI
      DO 260 IB=1,NB
      ABSP = (0.D0,0.D0)
      DO 261 I=1,N
  261 ABSP = ABSP + ((AMP(I)*AMP(I))/((CSYM1(I) + ((-1.D0,0.D0)*RES) + (
     1(0.D0,-1.D0)*T2IN))**2))
      DERAB(IB) = XAB*ABSP
      BFLD(IB) = RES*(FOAB/GAMMA)
  260 RES = RES + DEL
      IF(DISP) WRITE(6,906)
      WRITE(6,248)(DERAB(I),BFLD(I),I=1,NB)
  248 FORMAT(1H ,///,41X,'COORDINATES FOR FIRST DERIVATIVE SPECTRUM',//,
     14X,'1ST DERIV',7X,'FIELD',11X,'1ST DERIV',7X,'FIELD',11X,'1ST DERI
     2V',7X,'FIELD',11X,'1ST DERIV',7X,'FIELD',//,(1X,2E14.5,4X,2E14.5,4
     3X,2E14.5,4X,2E14.5))
      IF(.NOT.XPUNCH) GO TO 249

 PUNCH OUTPUT - COORDINATES OF FIRST DERIVATIVE PUNCHED
                ON CARDS

      WRITE(7,250)(DERAB(I),BFLD(I),I=1,NB)
  250 FORMAT(E13.6,2X,E13.6,7X,E13.6,2X,E13.6)
  249 IF(.NOT.XPLOT) GO TO 10

 LINE PLOT FOR A FIRST DERIVATIVE SPECTRUM.
      SAME AS FOR ABSORPTION-DISPERSION PLOTS EXCEPT  DMAX,
      DMIN  REFER TO MAXIMUM AND MINIMUM VALUES,
      RESPECTIVELY, OF FIRST DERIVATIVE SPECTRUM.

      DMAX = DERAB(1)
      DO 251 I=1,NB
      IF(DMAX.LT.DERAB(I)) DMAX = DERAB(I)
  251 CONTINUE
      DMIN = DERAB(1)
      DO 252 I=1,NB
      IF(DMIN.GT.DERAB(I)) DMIN = DERAB(I)
  252 CONTINUE
      DXY = (BFLD(NB) - BFLD(1))/XLEN
      CALL AXIS(0.0,1.0,5HGAUSS,-5,XLEN,0.0,BFLD(1),DXY)
      Y = 1.0 + (YLEN*((DERAB(1) - DMIN)/(DMAX - DMIN)))
      X = 0.0
      CALL PLOT(X,Y,3)
      DO 253 I=1,NB
      Y = 1.0 + (YLEN*((DERAB(I) - DMIN)/(DMAX - DMIN)))
      X = XLEN*((BFLD(I) - BFLD(1))/(BFLD(NB) - BFLD(1)))
  253 CALL PLOT(X,Y,2)
      XPLEND = .TRUE.
      Y = 0.0
      X = X + 3.
      CALL PLOT(X,Y,-3)
   10 CONTINUE
      GO TO 813
    2 WRITE(6,812) LMAX4,NSY,LMAX5,NSY2
  812 FORMAT(1H1,' JOB IS TERMINATED BECAUSE DIMENSIONS OF  SY  OR  SY2
     1  ARE LESS THAN REQUIRED BY LMAX',/,' VALUE OF DIMENSION  N  FOR
     2MATRIX  SY(N,8)  SHOULD BE ',I3,/,' VALUE SPECIFIED IS',2X,I3,//,'
     3 VALUE OF DIMENSION  M  FOR MATRIX  SY2(M,3)  SHOULD BE  ',I3,/,'
     4 VALUE SPECIFIED IS  ',I3)
      GO TO 813
  814 WRITE(6,815) N,NDIM,M,JDIM
  815 FORMAT(1H1,' JOB TERMINATED BECAUSE BANDWIDTH OR ORDER OF MATRIX
     1 CSYM  IS GREATER THAN SPECIFIED',//,' VALUE FOR THE ORDER  N  FOR
     2 THE MATRIX  CSYM(N,M)  SHOULD BE ',I3,/,' VALUE SPECIFIED IS ',I3
     3,//,' VALUE FOR THE DIMENSION(BANDWIDTH)  M  OF MATRIX  CSYM(N,M)
     4 SHOULD BE ',I3,/,' VALUE SPECIFIED IS ',I3)
  813 IF(XPLEND) CALL PLTEND
      STOP
      END
```

```
C           GORDON AND MESSENGER SUBROUTINES
C
      SUBROUTINE CSQZ(A,AMP,IDIM,JDIM,N,M,SQTOL)
      IMPLICIT REAL*8(A-H,O-Z)
C
C  SUBROUTINE  CSQZ  TRANSFORMS, BY A SERIES OF COMPLEX
C  JACOBI ROTATIONS, AN  N BY N  COMPLEX BAND MATRIX,  A, OF
C  BAND WIDTH  2M-1  INTO A CCMPLEX SYMMETRIC TRIDIAGONAL
C  MATRIX,  T = TRANSPOSE(R)*A*R
C
C  ENTERING THE SUBROUTINE,  A(I,1)  CONTAINS THE  ITH
C  DIAGONAL ELEMENT AND  A(I,J+1)  CONTAINS THE  JTH
C  ELEMENT FRCM THE DIAGONAL IN THE  ITH ROW.  N  IS THE
C  LENGTH OF THE COLUMNS OF  A, AND  M  IS THE HALF BAND
C  WIDTH PLUS ONE.
C
C  LEAVING THE SUBROUTINE  A(I,1)  CONTAINS THE DIAGONAL
C  ELEMENTS AND  A(I,J)  CONTAINS THE NEW OFF-DIAGONAL
C  ELEMENTS CRDERED AS BEFORE.
C  ON ENTERING THE VECTOR TO BE POTATED IS IN  AMP. AT EXIT
C  AMP  CONTAINS THE ROTATED VECTOR.
C
      CCMPLEX*16 A(IDIM,JCIM),AMP(IDIM),
     1 B, G, S, SX, C,CX,CS,U,V,TEMP
      DIMENSICN RG(2)
      EQUIVALENCE (G,RG(1))
      SQTOL = 0.DO
      GRASS = 0.DO
      DO 101 LJM=1,N
      GRASS = GRASS + 1.DO
  101 SQTOL = CCABS(A(LJM,1)) + SQTOL
      DO 102 KID=2,M
      LOW = N+1-KID
      DO 102 LJM=1,LOW
      GRASS = GRASS + 2.DO
  102 SQTCL=2.DO*CDABS(A(LJM,KID))+SQTOL
      SQTOL = 1.D-15*SQTOL/DSQRT(GRASS)
      MX = M - 2
C  EACH PASS THROUGH THIS LOOP WILL DESTROY A SET OF OFF-
C  DIAGONAL ELEMENTS  A(I,NQ)
      DO 20 L=1,MX
      NR=M-L
      NQ=NR+1
      NX=N-NR
C  THIS STATEMENT SAVE CCMPUTATION WHEN THE BAND IS LARGER
C  THAN THE MATRIX
      IF(NX) 20,20,5
C  EACH PASS THRCUGH THIS DESTROYS ONE OFF-DIAGONAL ELEMENT
C  A(K,NQ)
    5 DC 15 K=1,NX
C  THE FIRST PASS THROUGH THIS LCOP DESTROYS AN OFF-DIAGONAL
C  ELEMENT  A(K,NQ)  AND CREATES ANOTHER ELEMENT  G  BY THE
C  JACOBI ROTATION OF COLUMN AND ROWS  I  AND  I+1.
C  SUBSEQUENT PASSES DESTROY THE ELEMENT  G  AND CREATE
C  ANCTHER UNTIL THE LAST PASS WHICH DESTROYS WITHOUT
C  CREATING.
      DO 25 J=K,NX,NR
      IF(J-K) 4C,50,40
   50 IF(CCABS(A(J,NQ)) - SQTCL) 15,15,6C
   60 B = -A(J,NR)/A(J,NQ)
      GO TO 80
   40 IF(RG(1)**2+RG(2)**2-SQTOL**2) 15,15,70
   70 B= -A(J-1,NQ)/G
C  THESE FIVE STATEMENTS CCMPUTE THE CORRECT SINE S AND
C  COSINE  C.
   80 SX = 1.DO/(1.DO + B**2)
      S = CDSQRT(SX)
      C=B*S
      CX = C**2
      CS = C*S
      I= J+NR - 1
C  THESE THREE STATEMENTS ROTATE  AMP
      TEMP = AMP(I)
      AMP(I) = C*TEMP - S*AMP(I+1)
      AMP(I+1) = C*AMP(I+1) + S*TEMP
C  THESE FIVE STATEMENTS ROTATE THE CROSS ELEMENTS
      U = A(I,1)
      V=A(I+1,1)
      A(I,1) = CX*U-2.DO*CS*A(I,2)+SX*V
      A(I+1,1) = SX*U + 2.DO*CS*A(I,2) + CX*V
      A(I,2) = CS*(U-V) + (CX-SX)*A(I,2)
   65 IZ = I - 1
C  THIS LOOP TRANSFORMS THE APPROPRIATE COLUMNS
      DO 90 IX=J,IZ
      IK = I-IX+1
      U = A(IX,IK)
      V = A(IX,IK+1)
      A(IX,IK) = C*U - S*V
   90 A(IX,IK+1) = S*U + C*V
   75 IF(J-K) 95,85,95
   95 A(J-1,NQ) = C*A(J-1,NQ) - S*G
   85 IR = MINO(NR,N-I)
      IF(IR-2)55,45,45
C  THIS LOOP TRANSFORMS THE APPROPRIATE ROWS
   45 DO 10 IX=2,IR
      U = A(I,IX+1)
      A(I,IX+1) = C*U-S*A(I+1,IX)
   10 A(I+1,IX)=S*U+C*A(I+1,IX)
   55 IF(I+NR-N) 30,25,25
   30 G = -S*A(I+1,NQ)
      A(I+1,NQ) = C*A(I+1,NC)
   25 CCNTINUE
C  END OF THE LOOP OVER  J. ONE OFF-DIAGONAL ELEMENT
C  DESTROYED
   15 CCNTINUE
C  END OF THE LOOP OVER  K. ONE SET OF OFF-DIAGONAL
C  ELEMENTS DESTROYED.
   20 CCNTINUE
C  END OF THE LCOP CVER L. MX SETS OF OFF-DIAGONAL
C  ELEMENTS DESTROYED
      RETURN
      END
      SUBROUTINE CQRT(A,B,AMP,IDIM,N,M,TOL,EIGVAL)
      IMPLICIT REAL*8(A-H,O-Z)
CCQRT
C
C  CQRT IS INTENDED TO PRODUCE THE EIGENVALUE OF A COMPLEX
C  SYMMETRIC (NOT HERMITIAN) TRIDIAGONAL MATRIX BY ITERATIVE
C  QR  TRANSFORMS AS DESCRIBED IN WILKINSCN 'THE ALGEBRAIC
C  EIGENVALUE PROBLEM' , P.565.
C
C  ENTERING THE SUBROUTINE THE ARRAY  A  OF DIMENSION  N
C  CONTAINS THE DIAGONAL ELEMENTS, B  THE OFF-DIAGONAL
C  ELEMENTS. THE VECTOR TO BE ROTATED IS  AMP. N  IS THE
C  DIMENSICN OF THE MATRIX AND THE LENGTH OF  AMP.
C  ITERATICN ARE CONTINUED UNTIL ALL OFF-DIAGONAL ELEMENTS
C  HAVE BEEN REDUCED TO LESS THAN  'TOL'  IN MAGNITUDE.
C  M  IS THE MAXIMUM NUMBER OF ITERATIONS ALLOWED PER
C  EIGENVALUE. M=20  HAS ALWAYS BEEN MORE THAN SUFFICIENT.
C  THE NUMBER OF ITERATICNS REQUIRED MAY BE REDUCED IF
C  ACCURATE ESTIMATES OF THE EIGENVALUES ARE ALREADY KNOWN.
C  THESE INITIAL GUESSES SHOULD BE PLACED IN THE ARRAY
C  'EIGVAL' . IF NO GUESSES ARE FURNISHED, EIGVAL(1)
C  SHOULD BE SET EQUAL TO A CCMPLEX ZERO, (0.DO,0.DO), IN
C  THE CALLING PROGRAM. CQRT  WILL GENERATE ITS OWN GUESSES.
C
C  THE MATRICES  A,B,AMP,  ARE TREATED AS REAL ARRAYS OF
C  DIMENSION  N*2  INSIDE THE SUBROUTINE. OUTSIDE THEY
C  ARE TREATED AS CCMPLEX ARRAYS OF DIMENSICN  N.
      INTEGER HELL
      DIMENSICN A(1),B(1),AMP(1),SM(2),EIGVAL(1)
      CCMPLEX*16 SC
C  THE EQUIVALENCING OF THE ARRAY  SM  TO THE COMPLEX NUMBER
C  SC  PERMITS THE USE OF THE COMPLEX SQUARE ROOT ROUTINE
C  CSQRT.
      EQUIVALENCE (SC,SM(1))
C  NX  WILL BE THE UPPER LIMIT FOR THE ITERATION, NI  WILL
C  BE THE LOWER LIMIT AND BOTH MAY VARY BETWEEN ITERATIONS
      NX = 2*N
      NI=2
C  THE TOLERANCE--THAT IS, THE SIZE TO WHICH OFF-DIAGONAL
C  ELEMENTS ARE REDUCED--IS SET IN  CQRT  ON THE BASIS OF
C  THE  NORM OF THE TRIDIAGCNAL MATRIX.
      TOL = 0.DO
      GRASS = DFLOAT(N)
      DO 200 KID=1,NX,2
  200 TOL = DSQRT(A(KID)**2+A(KID+1)**2)+TOL+ 2.DO*DSQRT(B(K
     C  ID)**2 + B(K
     1ID+1)**2)
      TOL = 1.D-15*TOL/DSQRT(3.DO*GRASS)
C  SHR  AND SHI  CONTAIN THE TOTAL SHIFT
      SHR = 0.DO
      SHI = 0.DO
C  K  WILL CCUNT THE NUMBER OF ITERATIONS PER EIGENVALUE
      K = 0
C  IF A SET OF 'EIGVALS' IS AVAILABLE FOR USE AS INITIAL
C  SHIFTS --I.E. IF THE FIRST 'EIGVAL' IS NOT (0.DC,0.DC)--
C  THEN WE USE THE GUESSES IN 'EIGVAL' AS SHIFTS BY SETTING
C  'HELL'  EQUAL TO  101.
      ASSIGN 1CC TO HELL
      IF(DABS(EIGVAL(1))+DABS(EIGVAL(2)).NE.0.DO)ASSIGN 101
     C  TO HELL
      GO TO HELL,(100,101)
C  EACH NEW ITERATION BEGINS AT THIS STATEMENT OR AT
C  STATEMENT 101 --WHICHEVER  'HELL' IS.
  100 K=K+1
C  THESE SEVEN STATEMENTS SOLVE THE TWO BY TWO MATRIX IN
C  THE LOWER LEFT CCRNER AND USE CNE RCOT AS THE SHIFT
      ATR = A(NX-1) + A(NX-3)
      ATI=A(NX) + A(NX-2)
      SM(1)=ATR**2-ATI**2 -4.DO*(A(NX-1)*A(NX-3)-A(NX)*A(NX-
     C  2)-
     1  B(NX-3)**2+B(NX-2)**2)
      SM(2)=2.DO*ATR*ATI-4.DO*(A(NX-1)*A(NX-2)+A(NX)*A(NX-3)
     C  -
     1   2.DO*B(NX-3)*B(NX-2))
      SC = CDSQRT(SC)
      STR=(ATR-SM(1))*C.5DO
      STI=(ATI-SM(2))*0.5DO
      STTR=(ATR+SM(1))*.5DO
      STTI=(ATI+SM(2))*.5DO
      TEMP = CDABS(DCMPLX(A(NX-1)-STR,A(NX)-STI))
      TEMPO = CDABS(DCMPLX(A(NX-1)-STTR,A(NX)-STTI))
      IF(TEMPO.GT.TEMP) GO TO 102
      STR = STTR
      STI = STTI
      GO TO 102
  101 K=K+1
      IF(K .GE. M)  GO TO 999
      STR = EIGVAL(NX-1) - SHR
      STI=EIGVAL(NX) - SHI
  1C2 IF(K.LE.M) GO TO 104
  999 WRITE(6,1C3) M
  103 FORMAT(26H CQRT HAS NOT CONVERGED IN,I3,11H ITERATIONS
     C  )
      CALL EXIT
C  THESE TWO STATEMENTS INCREASE THE SHIFT BY THE
C  TEMPORARY SHIFT
  104 SHR = SHR + STR
      SHI = SHI+STI
C  THIS LOOP SUBTRACTS THE TEMPORARY SHIFT FROM THE DIAGONAL
C  ELEMENTS.
      DC 20 I=2,NX,2
      A(I-1) = A(I-1)-STR
   20 A(I) = A(I) -STI
C  THESE FOUR STATEMENTS SUPPLY INITIAL VALUES FOR THE
```

```
C     ITERATION.
      CR=1.0D0
      CI=0.0D0
      CSR =0.D0
      CSI=0.D0
      UR =0.0D0
      UI=0.0D0
C     THIS LOOP COMPLETES ONE ITERATION, THAT IS, ONE  QR
C     TRANSFORM. I  IS DOUBLE THE  I  IN THE OTHER CQRT  OR IN
C     QRTR
      DO 10 I=NI,NX,2
C     G  IS THE GAMMA IN WILKINSON.
      GR = A(I-1)-UR
      GI = A(I) - UI
C     Q  IS THE  P  IN WILKINSON.
      QR=CR*A(I-1)-CI*A(I)-CSR*B(I-3)+CSI*B(I-2)
      QI=CI*A(I-1)+CR*A(I)-CSI*B(I-3)-CSR*B(I-2)
C     THIS BRANCH AVOIDS UNNECESSARY COMPUTATION AT THE END
   16 IF(I.EQ.NX) GO TO 10
      SM(1) = QR**2-QI**2+B(I-1)**2-B(I)**2
      SM(2) =2.D0*(QR*QI+B(I-1)*B(I))
      SC= CDSQRT(SC)
      RR = SM(1)
      RI=SM(2)
      IF(I.EQ.NI) GO TO 18
C     THESE TWO STATEMENTS ROTATE AN OFF-DIAGONAL ELEMENT
      B(I-3) = SR*RR-SI*RI
      B(I-2) = SR*RI+SI*RR
   18 RS =1.D0/(RR**2+RI**2)
C     THESE TWO STATEMENTS COMPUTE THE NEW SINE FOR THE
C     JACOBI ROTATION.
      SR =(RR*B(I-1)+B(I)*RI)*RS
      SI=(RR*B(I)-B(I-1)*RI)*RS
C     THESE TWO STORE THE PRODUCT OF THE NEW SINE AND THE OLD
C     COSINE.
      CSR = CR*SR-CI*SI
      CSI = CR*SI+CI*SR
C     THESE TWO COMPUTE THE NEW COSINE.
      CR=(RR*QR+RI*QI)*RS
      CI=(RR*QI-RI*QR)*RS
C     THESE SIX COMPUTE A NEW  U.
      TR = GR + A(I+1)
      TI=GI+A(I+2)
      SXR = SR**2-SI**2
      SXI = 2.D0*SR*SI
      UR = SXR*TR-SXI*TI
      UI=SXR*TI+SXI*TR
C     THESE TWO ROTATE THE DIAGONAL ELEMENT.
      A(I-1) = GR + UR
      A(I) = GI + UI
C     THESE NINE ROTATE  AMP.
      TAR = AMP(I-1)
      TAI = AMP(I)
      TBR = AMP(I+1)
      TBI = AMP(I+2)
      AMP(I-1)=TAR*CR-TAI*CI+TBR*SR -TBI*SI
      AMP(I) = TAI*CR+TAR*CI+TBI*SR+TBR*SI
      AMP(I+1) = TBR*CR-TBI*CI-TAR*SR+TAI*SI
      AMP(I+2)=TBI*CR+TBR*CI-TAI*SR-TAR*SI
   10 CONTINUE
C     THIS ENDS ONE ITERATION.
C     THESE TWO COMPUTE THE LAST OFF-DIAGONAL ELEMENT.
      B(NX-3) = SR*QR -SI*QI
      B(NX-2) = SR*QI+SI*QR
C     THESE TWO COMPUTE THE LAST DIAGONAL ELEMENT
      A(NX-1) = GR
      A(NX) = GI
C     AT THIS POINT WE BEGIN CHECKING UPWARD THROUGH THE OFF-
C     DIAGONAL ELEMENTS TO FIND THOSE LESS THAN  TOL.
   85 IT=NX
C     THESE THREE STATEMENTS CONSTITUTE AN EFFECTIVE BACKWARD
C     LOOP
   30 IT= IT-2
C     THIS LOOP IS LEFT WHEN AN ELEMENT LESS THAN  TOL  IS
C     FOUND.
      IF(DABS(B(IT-1)) + DABS(B(IT)).LE.TOL) GO TO 40
      IF(IT-NI) 100,100,30
C     IF NO OFF-DIAGONAL ELEMENT LESS THAN  TOL  ARE FOUND, WE
C     PERFORM ANOTHER ITERATION.
C     THIS CONDITIONAL BRANCHES ACCORDING TO WHETHER THE
C     MATRIX ISOLATED BY THE SMALL OFF-DIAGONAL ELEMENT IS OF
C     DIMENSION  ONE, TWO, OR GREATER THAN TWO
   40 IF(NX-IT-4) 50,60,70
C     THESE TWO STATEMENTS EXTRACT THE EIGENVALUE OF A ONE BY
C     ONE MATRIX, ADDING BACK THE SHIFT.
   50 A(NX-1) = A(NX-1) + SHR
      A(NX) = A(NX) + SHI
C     THIS DECREASES THE SIZE OF THE PORTION OF THE MATRIX
C     AFFECTED BY LATER ITERATIONS.
      NX = NX - 2
C     THIS RESETS THE ITERATION COUNTER.
      K = 0
      GO TO 80
C     THIS SECTION EXTRACTS THE EIGENVALUE FROM A TWO BY TWO
C     MATRIX SECTION WHICH HAS BECOME ISOLATED FROM THE REST.
C     IT ALSO PERFORMS THE CORRESPONDING ROTATIONS ON  AMP.
C     THE FIRST TWENTY-NINE STATEMENTS COMPUTE THE PROPER
C     VALUES FOR THE SINE AND COSINE.
   60 ALR=-B(NX-3)
      ALI=-B(NX-2)
      AMR= 0.5D0*(A(NX-3) - A(NX-1))
      AMI = 0.5D0*(A(NX-2) - A(NX))
      SM(1) = ALR**2-ALI**2+AMR**2-AMI**2
      SM(2) =2.D0*(ALR*ALI+AMR*AMI)
      SC = CDSQRT(SC)
      ANR=SM(1)
      ANI= SM(2)
      TAR = AMR + ANR
      TAI = AMI+ANI
      TBR=ANR-AMR
      TBI=ANI-AMI
      SIG = 1.0D0
C     THIS BRANCH CHOOSES THE ROOT OF THE IMPLICITLY SOLVED
C     QUADRATIC SO THE COSINE HAS THE LARGER ABSOLUTE VALUE.
      IF(TAR**2+TAI**2.GT.TBR**2+TBI**2) GO TO 65
      TAR = TBR
      TAI = TBI
      SIG = -1.0D0
C     THESE SIX COMPUTE THE COSINE.
   65 TBR = 0.5D0/(ANR**2+ANI**2)
      SM(1) = (ANR*TAR+ANI*TAI)*TBR
      SM(2) = (ANR*TAI-ANI*TAR)*TBR
      SC = CDSQRT(SC)
      CR=SM(1)
      CI = SM(2)
C     THESE FIVE COMPUTE THE SINE.
      TAR = ANR*CR-ANI*CI
      TAI=ANR*CI+ANI*CR
      TBR=0.5D0/(TAR**2+TAI**2)
      SR=SIG*(TAR*ALR+TAI*ALI)*TBR
      SI= SIG*(TAR*ALI-ALR*TAI)*TBR
C     THESE SIXTEEN STORE DATA NEEDED IN THE ROTATION OF THE
C     MATRIX ELEMENTS.
      TAR = A(NX-3)
      TAI= A(NX-2)
      TBR=A(NX-1)
      TBI=A(NX)
      TCR=B(NX-3)
      TCI=B(NX-2)
      CXR=CR**2-CI**2
      CXI = 2.D0*CR*CI
      SXR=SR**2-SI**2
      SXI = 2.D0*SR*SI
      CSR=CR*SR-CI*SI
      CSI=CR*SI+CI*SR
      TR=CXR-SXR
      TI= CXI-SXI
      UR = 2.D0*(TCR*CSR - TCI*CSI)
      UI = 2.D0*(TCI*CSR + TCR*CSI)
C     THESE FOUR ROTATE THE DIAGONAL ELEMENTS
      A(NX-3) =TAR*CXR-TAI*CXI+TBR*SXR-TBI*SXI-UR+SHR
      A(NX-2)=TAR*CXI+TAI*CXR+TBR*SXI+TBI*SXR-UI+SHI
      A(NX-1)=TAR*SXR-TAI*SXI+TBR*CXR-TBI*CXI+UR+SHR
      A(NX) =TAR*SXI+TAI*SXR+TBR*CXI+TBI*CXR+UI+SHI
C     THESE TWO ROTATE THE OFF-DIAGONAL ELEMENTS
      B(NX-3) = 2.D0*(AMR*CSR-AMI*CSI)+TCR*TR-TCI*TI
      B(NX-2) = 2.D0*(AMR*CSI+AMI*CSR) + TCR*TI+TCI*TR
      I=NX-2
C     THESE NINE ROTATE  AMP.
C     NOTICE THAT THE SENSE OF ROTATION IS OPPOSITE TO THAT
C     OF THE FIRST ROTATION.
      TAR=AMP(I-1)
      TAI=AMP(I)
      TBR=AMP(I+1)
      TBI=AMP(I+2)
      AMP(I-1) = TAR*CR-TAI*CI-TBR*SR+TBI*SI
      AMP(I)=TAI*CR+TAR*CI-TBI*SR-TBR*SI
      AMP(I+1)=TBR*CR-TBI*CI+TAR*SR-TAI*SI
      AMP(I+2)=TBI*CR+TBR*CI+TAI*SR+TAR*SI
C     THIS RESETS THE ITERATION COUNT.
      K=0
C     THIS DECREASES THE SIZE OF THE PORTION OF THE MATRIX
C     AFFECTED BY THE LATER ITERATIONS.
      NX=NX-4
      GO TO 80
C     THIS STATEMENT IS REACHED WHEN THE PORTION OF THE MATRIX
C     ISOLATED IS GREATER THAN TWO BY TWO. IT CHANGES THE
C     LOWER LIMIT OF THE ITERATION SO THAT ONLY THIS PORTION
C     WILL BE  AFFECTED BY SUBSEQUENT ROTATIONS UNTIL ALL ITS
C     EIGENVALUES ARE FOUND.
   70 NI=IT+2
C     THIS STATEMENT TRANSFERS TO THE BEGINNING OF ANOTHER
C     ITERATION.
      GO TO HELL,(100,101)
C     THIS STATEMENT IS REACHED AFTER EITHER ONE OR TWO
C     EIGENVALUES HAVE JUST BEEN FOUND. IT TRANSFERS IF ALL
C     THE EIGENVALUES IN THIS PORTION OF THE MATRIX HAVE BEEN
C     FOUND.
   80 IF(NX.LT.NI) GO TO 90
C     THIS BRANCH TRANSFERS ACCORDING TO WHETHER ONE,TWO, OR
C     MORE EIGENVALUES REMAIN TO BE FOUND IN THIS PORTION.
   95 IF(NX-NI-2) 50,60,85
C     THIS CONDITIONAL IS REACHED WHEN ALL THE EIGENVALUES IN
C     THIS PORTION OF THE MATRIX HAVE BEEN FOUND. IT RETURNS IF
C     THIS IS THE LAST PORTION OF THE MATRIX
   90 IF(NI.EQ.2) RETURN
C     THIS ENLARGES THE PORTION OF THE MATRIX BEING TREATED TO
C     INCLUDE THE BEGINNING OF THE MATRIX.
      NI=2
      GO TO 95
      END
```

Acknowledgments

Support of this work by grants from the National Science Foundation, the Petroleum Research Fund (Grant No. 6818-AC6) administered by the American Chemical Society, and the Cornell University Materials Science center are gratefully acknowledged. The efforts of the past and present members of the Cornell Chemistry Department ESR group have been the basis for this chapter and their critical comments are greatly appreciated. Special thanks are due Dr. G. V. Bruno for developing the computer program given here and Dr. C. F. Polnaszek for his considerable help with it. This chapter was completed while the author was a guest professor at the Department of Physical Chemistry of Aarhus University (Spring semester, 1974), and he greatly appreciates the facilities made available to him.

References

Abragam, A. (1961). "The Principles of Nuclear Magnetism," pp. 49–51, Oxford Univ. Press, London and New York.

Bordeaux, O. *et al.* (1973). *J. Org. Mag. Res.* **5**, 47.

Bruno, G. V. (1973). Application of the Stochastic Liouville Method in Calculating ESR Line Shapes in the Slow Tumbling Region and An ESR-ELDOR Study of Exchange, Ph.D. Thesis, Cornell Univ.

Bruno, G. V., and Freed, J. H. (1974a). ESR lineshapes and saturation in the slow motional region: ELDOR, *Chem. Phys. Lett.* **25**, 328–332.

Bruno, G. V., and Freed, J. H. (1974b). Analysis of inertial effects on ESR spectra in the slow tumbling region, *J. Phys. Chem.* **78**, 935–940.

Birrell, G. B., Van, S. P., and Griffith, O. H. (1973). Electron spin resonance of spin labels in organic inclusion crystals. Models for anisotropic motion in biological membranes, *J. Amer. Chem. Soc.* **95**, 2451–2464.

Cukier, R. I., and Lakatos-Lindenberg, K. (1972). Rotational relaxation of molecules in fluids for reorientations of arbitrary angle, *J. Chem. Phys.* **57**, 3427–3435.

Dalton, L. (1973). Private communication.

Eastman, M. P., Kooser, R. G., Das, M. R., and Freed, J. H. (1969). Studies of Heisenberg spin exchange in ESR spectra, I. Linewidth and saturation effects, *J. Chem. Phys.* **51**, 2690–2709.

Eastman, M. P., Bruno, G. V., and Freed, J. H. (1970). Studies of Heisenberg spin exchange in ESR spectra II. Effects of radical charge and size, *J. Chem. Phys.* **52**, 2511–2522.

Edmonds, A. R. (1957). "Angular Momentum in Quantum Mechanics." Princeton Univ. Press, Princeton, New Jersey.

Egelstaff, P. A. (1970). Cooperative rotation of spherical molecules, *J. Chem. Phys.* **53**, 2590–2598.

Favro, L. D. (1965). Rotational Brownian motion, *In* "Fluctuation Phenomena in Solids" (R. E. Burgess, ed.), pp. 79–101. Academic Press, New York.

Freed, J. H. (1964). Anisotropic rotational diffusion and electron spin resonance linewidths, *J. Chem. Phys.* **41**, 2077–2083.

Freed, J. H. (1967). Theory of saturation and double resonance effects in electron spin resonance spectra. II. Exchange vs. dipolar mechanisms, *J. Phys. Chem.* **71**, 38–51.

Freed, J. H. (1972a). Ch. VIII, ESR relaxation and lineshapes from the generalized cumulant and relaxation matrix viewpoint, and Ch. XIV, ESR lineshapes and saturation in the slow motional region. The stochastic Liouville approach, *In* "Electron Spin Relaxation in Liquids" (L. T. Muus and P. W. Atkins, eds.), pp. 165–191, 387–409. Plenum, New York.

Freed, J. H. (1972b). Electron spin resonance, *Ann. Rev. Phys. Chem.* **23**, 265–310.

Freed, J. H. (1974). Theory of saturation and double resonance in ESR spectra VI: Saturation recovery, *J. Phys. Chem.* **78**, 1155–1167.

Freed, J. H., and Fraenkel, G. K. (1963). Theory of linewidths in electron spin resonance spectra, *J. Chem. Phys.* **39**, 326–348.

Freed, J. H., Bruno, G. V., and Polnaszek, C. F. (1971). Electron spin resonance lineshapes and saturation in the slow motional region, *J. Phys. Chem.* **75**, 3385–3399.

Glarum, S. H., and Marshall, J. H. (1966). ESR of the perinaphthenyl radical in a liquid crystal, *J. Chem. Phys.* **44**, 2884–2890.

Glarum, S. H., and Marshall, J. H. (1967). Paramagnetic relaxation in liquid-crystal solvents, *J. Chem. Phys.* **46**, 55–62.

Goldman, S. A. (1973). An ESR Study of Rotational Reorientation and Spin Relaxation in Liquid and Frozen Media, Ph.D. Thesis, Cornell Univ.

Goldman, S. A., Bruno, G. V., Polnaszek, C. F., and Freed, J. H. (1972a). An ESR study of anisotropic rotational reorientation and slow tumbling in liquid and frozen media, *J. Chem. Phys.* **56**, 716–735.

Goldman, S. A., Bruno, G. V., and Freed, J. H. (1972b). Estimating slow motional rotational correlation times for nitroxides by electron spin resonance, *J. Phys. Chem.* **76**, 1858–1860.

Goldman, S. A., Bruno, G. V., and Freed, J. H. (1973). ESR studies of anisotropic rotational reorientation and slow tumbling in liquid and frozen media II. Saturation and non-secular effects, *J. Chem. Phys.* **59**, 3071–3091.

Gordon, R. G., and Messenger, T. (1972). Ch. XIII, Magnetic resonance line shapes in slowly tumbling molecules, *In* "Electron Spin Relaxation in Liquids" (L. T. Muus and P. W. Atkins, eds.), pp. 341–381. Plenum, New York.

Griffith, O. H., *et al.* (1965). *J. Chem. Phys.* **43**, 2909.

Hamrick, P. J., *et al.* (1972). *J. Chem. Phys.* **57**, 5029.

Hubbell, W. L. and McConnell, H. M. (1969a). Motion of steroid spin labels in membranes, *Proc. Nat. Acad. Sci. U.S.* **63**, 16–22.

Hubbell, W. L., and McConnell, H. M. (1969b). Orientation and motion of amphiphilic spin labels in membranes, *Proc. Nat. Acad. Sci. U.S.* **64**, 20–27.

Hubbell, W. L., and McConnell, H. M. (1971). Molecular motion in spin labeled phospholipids and membranes, *J. Amer. Chem. Soc.* **93**, 314–326.

Hyde, J. S., and Dalton, L. (1972). Very slowly tumbling spin labels: Adiabatic rapid passage, *Chem. Phys. Lett.* **16**, 568–572.

Hwang, J. S., Mason, R., Hwang, L. P., and Freed, J. H. (1975). ESR studies of anisotropic rotational reorientation and slow tumbling in liquid and frozen media III: Perdeuterotempone and an analysis of fluctuating torques, *J. Phys. Chem.* **79**, 489–511.

Johnson, C. S. Jr. (1965). Chemical rate processes and magnetic resonance, *Advan. Magn. Res.* **1**, 33–102.

Kubo, R. (1962). A stochastic theory of line shape and relaxation, *In* "Fluctuation, Relaxation, and Resonance in Magnetic Systems" (D. ter Haar, ed.), pp. 23–67. Oliver and Boyd, London.

Kubo, R. (1969). A stochastic theory of lineshape, *In* "Stochastic Processes in Chemical Physics, Advances in Chemical Physics" (K. E. Shuler, ed.), Vol. XVI, pp. 101–127. Wiley, New York.

Leniart, D. S. (1972). Private Communication.

Libertini, L. J., and Griffith, O. H. (1970). *J. Chem. Phys.* **53**, 1359.

Mason, R., and Freed, J. H. (1974). Estimating microsecond rotational correlation times from lifetime broadening of nitroxide ESR spectra near the rigid limit, *J. Phys. Chem.* **78**, 1321–1323.

Mason, R., Polnaszek, C. F., and Freed, J. H. (1974). Comments on the interpretation of ESR spectra of spin labels undergoing very anisotropic rotational reorientation, *J. Phys. Chem.* **78**, 1324–1329.

McCalley, R. C., Shimshick, E. J., and McConnell, H. M. (1972). The effect of slow rotational motion on paramagnetic resonance spectra, *Chem. Phys. Lett.* **13**, 115–119.

McConnell, H. M., and McFarland, B. G. (1970). Physics and chemistry of spin labels, *Quart. Rev. Biophys.* **3**, 91–136.

McFarland, B. G., and McConnell, H. M. (1971). Bent fatty acid chains in lecithin bilayers, *Proc. Nat. Acad. Sci. U.S.* **68**, 1274–1278.

Meier, V. W., and Saupe, A. (1958). Eine einfache molekulare theorie des nematischen kristallinflüssigen zustandes, *Z. Naturforsch.* **13a**, 564–566.

Miesowicz, M. (1946). The three coefficients of viscosity of anisotropic liquids, *Nature* (*London*) **158**, 27.

Nordio, P. L., and Busolin, P. (1971). Electron spin resonance line shapes in partially oriented systems, *J. Chem. Phys.* **55**, 5485–5490.

Nordio, P. L., Rigatti, G., and Segre, U. (1972). Spin relaxation in nematic solvents, *J. Chem. Phys.* **56**, 2117–2123.

Ohnishi, S., Boeyens, J. C. A., and McConnell, H. M. (1966). Spin-labeled hemoglobin crystals, *Proc. Nat. Acad. Sci. U.S.* **56**, 809–813.

Pedersen, J. B., and Freed, J. H. (1973a). Theory of chemically-induced dynamic electron polarization. I., *J. Chem. Phys.* **58**, 2746–2762.

Pedersen, J. B., and Freed, J. H. (1973b). Theory of chemically-induced dynamic electron polarization. II., *J. Chem. Phys.* **59**, 2869–2885.

Polnaszek, C. F. (1975a). An ESR Study of Rotational Reorientation and Spin Relaxation in Liquid Crystal Media, Ph.D. Thesis, Cornell Univ.

Polnaszek, C. F. (1975b). The analysis and interpretation of electron spin resonance spectra of nitroxide spin probes in terms of correlation times for rotational reorientation and of order parameters, *Quart. Rev. Biophys.* (to be published).

Polnaszek, C. F., and Freed, J. H. (1975). ESR studies of anisotropic ordering, spin relaxation and slow tumbling in liquid crystalline solvents, *J. Phys. Chem.* **79** (in press).

Polnaszek, C. F., Bruno, G. V., and Freed, J. H. (1973). ESR lineshapes in the slow-motional region: Anisotropic liquids, *J. Chem. Phys.* **58**, 3185–3199.

Rose, M. E. (1957). "Elementary Theory of Angular Momentum." Wiley, New York.

Rotenberg, M., Bivins, R., Metropolis, N., and Wooten, J. K. (1959). "The 3*j*- and 6*j*-Symbols," Technology Press, M.I.T., Boston, Massachusetts.

Smigel, M., Dalton, L. R., Hyde, J. S., and Dalton, L. A. (1974). Investigation of very slowly tumbling spin labels by nonlinear spin response techniques: Theory and experiment for stationary electron–electron double resonance, *Proc. Nat. Acad. Sci. U.S.* **71**, 1925–1929.

Thomas, D. D., and McConnell, H. M. (1974). Calculation of paramagnetic resonance spectra sensitive to very slow rotational motion, *Chem. Phys. Lett.* **25**, 470–475.

Valiev, K. A., and Ivanov, E. N. (1973). Rotational Brownian motion, *Sov. Phys.-Usp.* **16**, 1–16.

Wee, E. L., and Miller, W. G. (1973). Studies on nitroxide spin-labeled poly-γ-benzyl-α,L-glutamate, *J. Phys. Chem.* **77**, 182–189.

4

Biradicals as Spin Probes

GEOFFREY R. LUCKHURST

DEPARTMENT OF CHEMISTRY
THE UNIVERSITY, SOUTHAMPTON, ENGLAND

I. INTRODUCTION

The obvious success of the spin probe technique stems from the ability of the environment to influence the appearance of the probe's electron resonance spectrum. It is important therefore to understand the parameters that determine the form of this spectrum and how these can be modified by the interaction of the probe with its surroundings. For the purpose of the discussion it is convenient, but not essential, to assume that any molecular motion is sufficiently fast to give a motionally narrowed spectrum. The positions of the spectral lines will then be determined by the scalar magnetic interactions provided the system is isotropic. If, however, some space-fixed orientations are preferred, then the line positions will depend on the anisotropic as well as the scalar couplings. In general the linewidths are affected by a variety of spin relaxation processes, but for organic free radicals the dominant process is invariably associated with the anisotropic interactions coupled to the molecular rotation. There are then four variables that determine the spectral shape and each can be influenced by the interaction between the probe and its environment. Such interactions have been described in earlier chapters and so we shall mention just three examples encountered with nitroxide monoradicals. The formation of a hydrogen bond with the nitroxide group increases the nitrogen coupling constant and this change has been employed to probe the polarity of the environment. The spacing between the hyperfine lines is altered by the partial alignment of the spin probe because of the anisotropy in the nitrogen hyperfine interaction. This behavior has been employed to considerable advantage in the study of membranes and other liquid crystals. Finally, the rate of molecular reorientation is profoundly influenced by the macroscopic and the local viscosity of the environment; changes in these viscosities are reflected in the linewidths because of the anisotropy in the g and nitrogen hyperfine tensors.

Nitroxide monoradicals would therefore appear to satisfy the particular requirements for a spin probe and so it is proper to ask what advantages, if any, are provided by biradical probes. Of course any magnetic interaction present in the monoradical will also occur in the corresponding biradical. But, in addition, there is a new interaction, which is the coupling between the two electron spins; this, like all magnetic interactions, has an isotropic and an anisotropic part. The scalar component or exchange interaction is found to depend critically on the conformation of the biradical, whereas the anisotropic part is determined by the electron–electron separation. For flexible biradicals both of these quantities will be influenced by their surroundings and so biradical probes provide a way of examining another facet of the environment. Further, the rate at which the environment modifies the conformation of the probe can also be inferred from the widths of the spectral lines. The anisotropy in the electron–electron interaction provides an alternative method for studying those features already investigated with the aid of the anisotropy in the g and hyperfine interactions.

In the following sections we shall consider, in some detail, how the introduction of these electron–electron interactions influences the appearance of the electron resonance spectrum. We shall confine our attention to biradicals, largely because no new principles are encountered for polyradicals but also because only biradicals appear to have been employed as spin probes. At various stages in the development of the theory we shall pause to consider those specific applications of biradicals as spin probes that illustrate this theory. We shall not pay particular attention to systems in which the molecular motion is slow, although we shall comment on the difficulties encountered in such systems. Finally, we shall return to the problem of the relative merits of mono- and biradicals as spin probes in Section V.

II. THE TRIPLET STATE

A. Isotropic Solutions

We shall be predominantly concerned in this chapter with those biradicals formed by joining two nitroxides together with saturated linkages. Consequently, the spin Hamiltonian will contain contributions from the hyperfine interactions between the electron and nuclear spins; it is, however, convenient to begin our discussion by ignoring such terms. The orientation-independent or scalar Hamiltonian therefore contains the electron Zeeman interaction and the electron exchange coupling:

$$\mathscr{H}^0 = g\beta B(S_z^{(1)} + S_z^{(2)}) + J S^{(1)} \cdot S^{(2)} \tag{1}$$

where J is the exchange integral.† The biradical is taken to be symmetric and so the g factors for the two electrons have been set equal. The three triplet spin functions

$$|1\rangle \equiv |\alpha\alpha\rangle, \qquad |0_a\rangle \equiv (1/\sqrt{2})(|\alpha\beta\rangle + |\beta\alpha\rangle), \qquad |-1\rangle \equiv |\beta\beta\rangle \tag{2}$$

and the singlet

$$|0_e\rangle = (1/\sqrt{2})(|\alpha\beta\rangle - |\beta\alpha\rangle) \tag{3}$$

are eigenfunctions of the spin Hamiltonian with energies

$$E_1 = g\beta B + J/4, \qquad E_{0_a} = J/4, \qquad E_{-1} = -g\beta B + J/4 \tag{4}$$

and

$$E_{0_e} = -3J/4 \tag{5}$$

The subscripts a and e on the spin states denote symmetric and antisymmetric functions, respectively. The allowed spin transitions are within the triplet manifold and satisfy the selection rule

$$\Delta M_s = \pm 1 \tag{6}$$

† Editor's note: Note that here the symbol B, instead of H, is used for magnetic field. Both symbols are common in the literature.

where M_s is the total spin quantum number. The observed transitions are then

$$|-1\rangle \leftrightarrow |0_a\rangle \qquad (\equiv |-\rangle) \tag{7}$$

and

$$|0_a\rangle \leftrightarrow |1\rangle \qquad (\equiv |+\rangle) \tag{8}$$

These are degenerate and occur when

$$B = h\nu_0/g\beta \tag{9}$$

where ν_0 is the operating frequency of the spectrometer. The position of this single line is independent of the exchange integral, although it can influence the intensity $\mathscr{I}$ of the spectrum. This intensity is proportional to the population of the triplet state and, provided the sample is not saturated, is given by

$$\mathscr{I} \propto \frac{1}{T}\frac{3e^{J/RT}}{1 + 3e^{J/RT}} \tag{10}$$

When J is comparable to the thermal energy the spectral intensity will increase with increasing temperature if the exchange integral is positive, corresponding to a singlet ground state, but decrease if it is negative. Under favorable conditions the sign and often the magnitude of J can be obtained from the temperature dependence of the signal intensity.

The width of the single line in the solution spectrum of a triplet is often found to be rather large and in certain cases the line is undetectably broad. The linewidths are considerable because the anisotropic electron–electron interaction coupled to the molecular reorientation modulates the energies of the spin levels and induces transitions between them (Weissman, 1958). The calculation of the linewidth resulting from this powerful spin relaxation process can be accomplished with the aid of Redfield's relaxation theory, which is outlined in Chapter 2. The first step in the calculation is the formulation of the dynamic spin Hamiltonian, which is that for the anisotropic electron–electron coupling:

$$\mathscr{H} = 2(S_Z^{(1)}D_{ZZ}S_Z^{(2)} + S_X^{(1)}D_{XX}S_X^{(2)} + S_Y^{(1)}D_{YY}S_Y^{(2)}) \tag{11}$$

where $\mathbf{D}$ is the zero-field splitting tensor and $\mathscr{H}$ is written in a molecule-fixed coordinate system XYZ. It is important to note that the zero-field splitting spin Hamiltonian takes the same form when written in terms of the total spin operator

$$\mathscr{H} = D_{ZZ}S_Z^2 + D_{XX}S_X^2 + D_{YY}S_Y^2 \tag{12}$$

but now the coefficients are just one-half the value required for the Hamiltonian involving individual spin operators (Carrington and McLachlan,

1967). The tensor **D** is referred to as the zero-field splitting because this interaction removes the degeneracy of the three triplet spin levels even in the absence of a magnetic field.

As the triplet rotates, the spin operators fluctuate in time and this time dependence is conveniently handled using the irreducible operator techniques developed in Chapter 2. The dynamic perturbation is therefore written as

$$\mathscr{H}'(t) = \sum_p (-1)^p D^{(2,p)} T^{(2,-p)}(t) \tag{13}$$

where $T^{(2,-p)}(t)$ denotes the time-dependent spin operators. This time dependence can be removed by transforming to a laboratory coordinate system where the operators are independent of time. This transformation is accomplished with a Wigner rotation matrix and gives

$$\mathscr{H}'(t) = \sum_{p,q} (-1)^p D^{(2,p)} \mathscr{D}^{(2)}_{q,-p}(t) T^{(2,q)} \tag{14}$$

Redfield's relaxation matrix contains four elements because the transitions given in Eqs. (7)–(8) are doubly degenerate; these four elements are readily calculated, from Eq. (100) of Chapter 2, with the result

$$\begin{array}{c|cc|} & |+\rangle & |-\rangle \\ \langle + | & A & B \\ \langle - | & B & A \end{array} \tag{15}$$

where

$$A = -\tfrac{1}{10} \sum_p (-1)^p D^{(2,p)} D^{(2,-p)} \{3j_0 + 3j_1 + 2j_2\} \tag{16}$$

and

$$B = -\tfrac{1}{5} \sum_p (-1)^p D^{(2,p)} D^{(2,-p)} j_1 \tag{17}$$

The spectral densities j_n are defined by

$$j_n = \tau/(1 + 4\pi^2 n^2 \nu_0^2 \tau^2) \tag{18}$$

where τ is the correlation time characterizing the isotropic reorientation process. The linewidths are simply minus one times the eigenvalues of the relaxation matrix and for this problem both transitions are found to have the same width

$$T_2^{-1} = \tfrac{1}{10} \sum_p (-1)^p D^{(2,p)} D^{(2,-p)} \{3j_0 + 5j_1 + 2j_2\} \tag{19}$$

and so the line shape is Lorentzian (Carrington and Luckhurst, 1964).

Consequently, if rotational modulation of the zero-field splitting constitutes the dominant relaxation process, the width of the single line can be used to estimate the correlation time (Michon and Rassat, 1974). However, since there is only one line, it may prove difficult to make an unambiguous assignment of the relaxation process. There has been one attempt to assign the relaxation process for the biradical **I** dissolved in ethyl alcohol (Rassat,

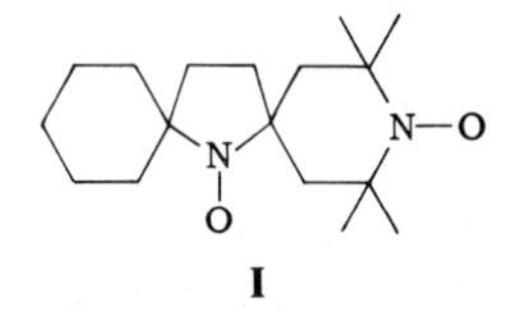

1972). The peak-to-peak linewidth ΔB_{pp} was measured as a function of temperature, with the results given in Fig. 1, where ΔB_{pp} is plotted against η/T. The linearity of the plot can be understood in the following way. For a

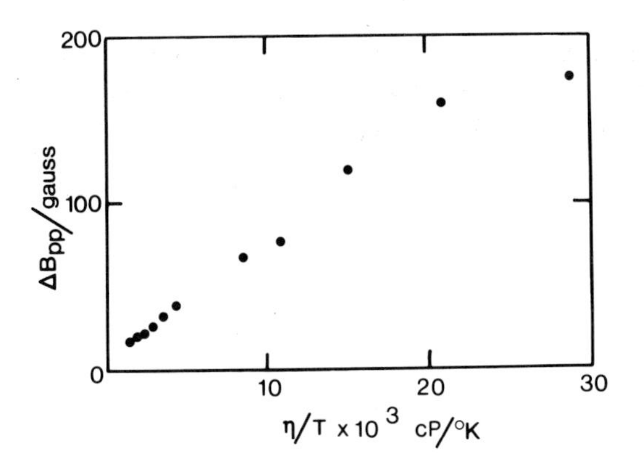

Fig. 1. The temperature dependence of the peak-to-peak linewidth ΔB_{pp} for the biradical **I** dissolved in ethyl alcohol. (Rassat, 1972.)

Lorentzian line shape the linewidth T_2^{-1} is directly proportional to ΔB_{pp} and for moderately viscous systems the spectral densities j_1 and j_2 may be ignored in comparison with j_0, since $n^2 v_0^2 \tau^2 > 1$. The peak-to-peak separation is then directly proportional to τ, which, according to the Debye model of rotation, is related to the viscosity and size of the molecule by

$$\tau = \tfrac{4}{3}\pi r^3 \eta / kT \tag{20}$$

Although the Debye model is admittedly crude, it has been found to work rather well for molecules as large as biradical **I**. The only necessary modification is the introduction of an anisotropic interaction parameter κ, which reflects the anisotropy in the solute–solvent intermolecular potential and is apparently independent of temperature (Hwang *et al.*, 1973); the modified Debye equation is

$$\tau = \tfrac{4}{3}\pi r^3 \kappa\eta/kT \tag{21}$$

Accordingly, the peak-to-peak separation should be proportional to η/T, in agreement with experiment; the deviation from linearity at high viscosities is probably associated with the failure of Redfield's theory. It is important to note that the linewidth will also be proportional to the correlation time, and hence η/T, if the molecular reorientation is fast in the sense $n^2\nu_0^2\tau^2 < 1$, for now all of the spectral densities are equal to τ.

The validity of the expression for the linewidth of a triplet state rotating rapidly in fluid solution can be tested in quite a different way. This involves the measurement of the linewidth, at constant temperature, but for different microwave frequencies ν_0. In practice such experiments are restricted to three readily available frequencies: 3 GHz (S band), 9.5 GHz (X band), and 35 GHz (Q band); although these are usually sufficient. The frequency dependence of T_2^{-1} given in Eq. (19) is caused by the occurrence of ν_0 in the spectral densities j_1 and j_2, as we can see from Eq. (18). Consequently, provided $n^2\nu_0^2\tau^2$ is comparable to unity, changing the frequency will modify the spectral densities and hence T_2^{-1}; in general the linewidth should decrease on increasing the microwave frequency. Clearly, if the zero-field splitting **D** is known, then measurement of the linewidth at different frequencies will permit independent estimates of the correlation time; these should, of course, be equal if the theory is correct. Alternatively, if **D** is not known, then such variable-frequency experiments should enable us to determine the correlation time together with the inner product $\sum_p (-1)^p D^{(2,p)} D^{(2,-p)}$. This procedure has yet to be applied to the triplet state, although it has been used with some success in studies of spin relaxation in states of higher multiplicity, such as iron(III) and manganese(II) (Levanon *et al.*, 1970). However, it must be realized that such experiments are essentially tests of one aspect of the theory, namely the assumption of an exponential decay of the correlation function for the rotation matrices.

The expression for the linewidth assumes that all of the components of the diffusion tensor **R** are the same and in fact equal to $(6\tau)^{-1}$. Although this assumption may be reasonable for biradical **I** it can hardly be realistic for **II**,

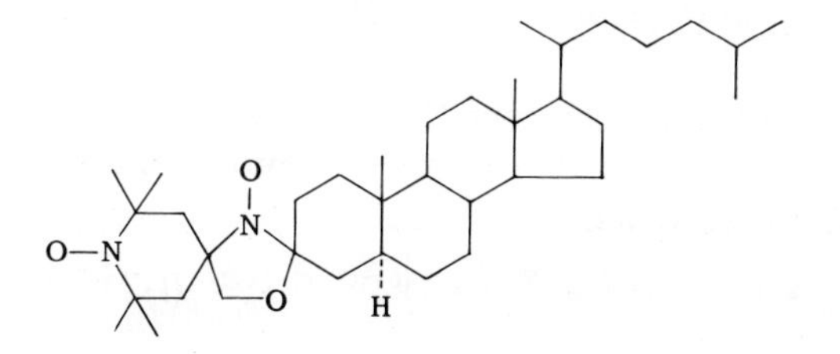

II

where the diffusion tensor should possess cylindrical symmetry about the long molecular axis. The linewidth predicted (Luckhurst *et al.*, 1975) for such a biradical is

$$T_2^{-1} = \tfrac{1}{10} \sum_p (-1)^p D^{(2,\,p)} D^{(2,\,-p)} \{3j_{0p} + 5j_{1p} + 2j_{2p}\} \tag{22}$$

where the spectral densities are

$$j_{np} = \tau_p/(1 + 4\pi^2 n^2 \tau_p^2 \nu_0^2) \tag{23}$$

Now the correlation times depend on p as well as on the components of the diffusion tensor parallel ($R_{\parallel}$) and perpendicular ($R_{\perp}$) to the symmetry axis:

$$\tau_p = [6R_{\perp} + (R_{\parallel} - R_{\perp})p^2]^{-1} \tag{24}$$

When applying Eq. (22) for the linewidth, it is important to remember that the coordinate system employed to calculate the irreducible components of the zero-field splitting must also diagonalize the diffusion tensor. Since the observed linewidth depends on two components of the diffusion tensor, it will never be possible to determine them from this single width and so triplet spin probes are not of value in studying rapid anisotropic rotational diffusion.

B. Anisotropic Environments

Although, as we have seen, the zero-field splitting has a pronounced effect on the linewidth when the triplet is tumbling rapidly in an isotropic solvent, it does not influence the line position. This is not the case when the spin probe is dissolved in a macroscopically anisotropic system such as a liquid crystal (Falle *et al.*, 1966); the properties of such systems are described more fully in Chapter 10. Since all orientations are no longer equivalent, **D** is not averaged to zero by reorientation and its retention in the static spin Hamiltonian removes the degeneracy of the two spin transitions. The spectrum will therefore contain two lines and we shall now seek to calculate their separation. Our starting point is the spin Hamiltonian for a particular orientation of the spin probe; this is the sum of the scalar Hamiltonian given in Eq. (1) and the anisotropic Hamiltonian in Eq. (13):

$$\mathscr{H}(t) = \mathscr{H}^0 + \sum_p (-1)^p D^{(2,\,p)} T^{(2,\,-p)} \tag{25}$$

As we have seen, the spin operators $T^{(2,\,-p)}$ fluctuate in time as the molecule rotates, but this time dependence can be removed by transforming to a laboratory coordinate system:

$$\mathscr{H}(t) = \mathscr{H}^0 + \sum_{p,\,q} (-1)^p D^{(2,\,p)} \mathscr{D}^{(2)}_{q,\,-p}(t) T^{(2,\,q)} \tag{26}$$

The static Hamiltonian is obtained from this expression by taking a time or ensemble average of the Wigner rotation matrix. In an isotropic environment the ensemble average of any rotation matrix $\mathscr{D}^{(L)}_{p,p}$ is just

$$\overline{\mathscr{D}^{(L)}_{q,p}} = \delta_{L0}\delta_{qp} \tag{27}$$

and so only the scalar terms in Eq. (26) contribute to the static spin Hamiltonian since $\overline{\mathscr{D}^{(2)}_{q,-p}}$ is zero. This is not the case for an anisotropic environment; however, to evaluate the ensemble average $\overline{\mathscr{D}^{(2)}_{q,-p}}$ and hence the form of the spectrum it is necessary to know the symmetry of both the spin probe and the potential that is responsible for its partial alignment. The optical properties of nematic and smectic A liquid crystals show that this ordering potential is cylindrically symmetric and it is convenient and reasonable to adopt the same symmetry for other anisotropic systems. With this assumption $\overline{\mathscr{D}^{(2)}_{q,-p}}$, and indeed the averages of all even rotation matrices, vanish unless q is zero; the static spin Hamiltonian then reduces to

$$\overline{\mathscr{H}} = \mathscr{H}^0 + \sum_p (-1)^p D^{(2,p)} \overline{\mathscr{D}^{(2)}_{0,-p}} T^{(2,0)} \tag{28}$$

in a coordinate system containing the direction of the ordering potential as one of the axes. This is entirely equivalent to the spin Hamiltonian for a triplet state with a cylindrically symmetric zero-field splitting tensor with the symmetry axis parallel to the ordering potential. The components of the effective or partially averaged zero-field splitting are denoted by $\tilde{D}_{\parallel}$ and $\tilde{D}_{\perp}$. The electron resonance spectra expected for such a Hamiltonian are well understood as a result of numerous studies involving transition metal ions (Low, 1960). Indeed the energy levels for a comparable spin Hamiltonian are shown in Fig. 4 of Chapter 2 together with the allowed spin transitions. Thus the spectrum is found to contain two lines whose separation d changes as the orientation of the sample with respect to the magnetic field is varied. The angular dependence for our problem is

$$d = 3(\tilde{D}_{\parallel}/g\beta)(3\cos^2\gamma - 1)/2 \tag{29}$$

provided $\tilde{\mathbf{D}}$ is sufficiently small compared with the Zeeman splitting that the nonsecular terms may be ignored. Here γ is the angle between the magnetic field and the symmetry axis of the environment. The result in Eq. (29) can be obtained by extending the following argument. When the magnetic field is parallel to the symmetry axis of $\tilde{\mathbf{D}}$ the static spin Hamiltonian in Eq. (28) reduces to

$$\overline{\mathscr{H}} = g\beta BS_z + \tfrac{1}{2}J(S^2 - \tfrac{3}{2}) + \tfrac{1}{2}\tilde{D}_{\parallel}(3S_z^2 - S^2) \tag{30}$$

where S is the total spin operator. The triplet functions given in Eq. (2) are still eigenfunctions of $\mathscr{H}$ but with eigenvalues

$$E_1 = g\beta B + \tfrac{1}{4}J + \tfrac{1}{2}\tilde{D}_{\parallel}, \qquad E_{0_a} = \tfrac{1}{4}J - \tilde{D}_{\parallel}$$

$$E_{-1} = -g\beta B + \tfrac{1}{4}J + \tfrac{1}{2}\tilde{D}_{\parallel} \tag{31}$$

The two allowed transitions are again $|+\rangle$ and $|-\rangle$ but they are no longer degenerate:

$$B_+ = (h\nu_0/g\beta) - (3\tilde{D}_{\parallel}/2g\beta) \tag{32}$$

$$B_- = (h\nu_0/g\beta) + (3\tilde{D}_{\parallel}/2g\beta) \tag{33}$$

and the separation between the two spectral lines is $3\tilde{D}_{\parallel}/g\beta$, in agreement with Eq. (29).

We shall now consider two examples that illustrate these ideas. The first is for the biradical dispiro((dimethyl-4,4-oxazolidine-3-oxyl)-2,1′-cyclohexane-4′,2‴-(dimethyl-4‴,4‴-oxazolidine-3‴-oxyl)) **III** dissolved in a

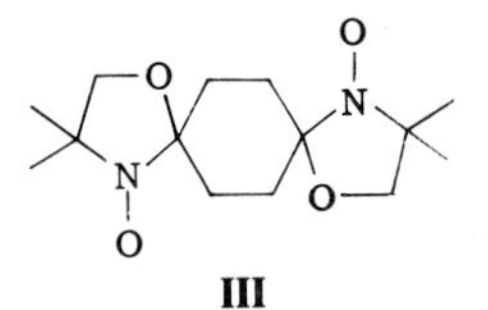

III

nematogen. The spectrum found for the isotropic phase is shown in Fig. 2a and contains the expected single broad line; the three sharp lines originate from a monoradical impurity. In the nematic mesophase this single line is split into two, as the spectrum in Fig. 2b shows, because the liquid crystal is aligned with its symmetry axis parallel to the magnetic field (cf. Chapter 8). These observations confirm the theoretical analysis and also provide a dramatic confirmation of the anisotropic environment provided by a liquid crystal. The magnitude of the line separation in this experiment is just $3\tilde{D}_{\parallel}$, which is related to the total zero-field splitting tensor by

$$\tilde{D}_{\parallel} = \tfrac{2}{3}\sum_p (-1)^p D^{(2,p)}\overline{\mathscr{D}^{(2)}_{0,-p}} \tag{34}$$

where the averages $\overline{\mathscr{D}^{(2)}_{0,-p}}$ are a measure of the partial alignment of the spin probe.

These order parameters can also be described in terms of the direction cosines of the environmental symmetry axis in a molecular coordinate system, as we shall see in Chapter 10. This description follows naturally when the various magnetic interactions are expressed as Cartesian rather than spherical tensors and leads directly to the concept of the ordering matrix which was originally introduced by Saupe (1964). The relationship between this ordering matrix and the order parameters $\overline{\mathscr{D}^{(2)}_{0,-p}}$ has been

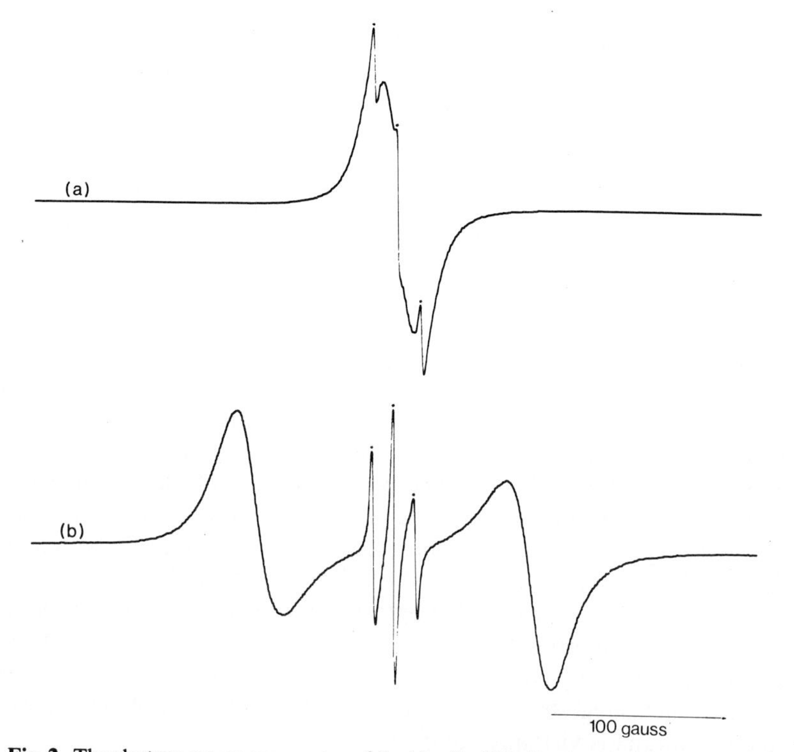

Fig. 2. The electron resonance spectra of the biradical **III** dissolved in (a) the isotropic and (b) the nematic mesophase of a liquid crystal. The three lines marked with an asterisk come from a monoradical impurity.

shown to be relatively straightforward, as we might have expected (Falle and Luckhurst, 1970). The determination of these parameters is of some importance because they provide a valuable insight into the anisotropy of the solute–solvent intermolecular potential (Humphries *et al.*, 1971). In general there are five independent order parameters and it is clearly not possible to determine all of them using a biradical spin probe if the anisotropic interactions are restricted to the electron–electron coupling. However, when the biradical is cylindrically symmetric only one order parameter $\overline{\mathscr{D}^{(2)}_{0,0}}$ is required and such uncommon biradicals could prove useful in studying the extent of solute alignment.

In the second example (Keana and Dinerstein, 1971) the cholestane biradical **II** was used to study a dipalmitoyl lecithin bilayer whose structure is described in Chapters 10 and 11. An attempt was made to obtain a homogeneous sample by spreading the spin-doped membrane onto a glass plate since the symmetry axis of the bilayer has been found to be oriented perpendicular to the surface. The electron resonance spectra are shown in

Fig. 3 for three orientations of the glass plate with respect to the magnetic field; the central line should be ignored since it originates from a monoradical impurity. Provided the rotational motion of the biradical is fast, the spectrum should contain just two lines with a separation given by Eq. (29). Two lines are observed when the magnetic field is either parallel or perpendicular to the glass surface, although the line shape is not symmetric. This asymmetry is probably caused by the imperfect alignment of the membrane but it could result if the molecular motion was slow, as we discovered in Chapter 3 (Brooks *et al.*, 1971; Norris and Weissmann, 1969). Both interpretations are consistent with the observation of four peaks in the spectra for intermediate orientations. Similar, although less pronounced, asymmetry in the line shape is also found for the liquid crystal spectrum shown in Fig. 2b. Such observations present rather a problem in electron resonance studies of liquid crystals because it is extremely difficult to devise ways of distinguishing unambiguously between the two possible origins of the asymmetric line shapes. The separation d between the lines is difficult to estimate because of the asymmetry; however, d is about 450 G when γ, the angle between the field and the symmetry axis, is 0° and decreases to 250 G when γ is 90°. These numbers are in reasonable agreement with Eq. (29), which predicts the modulus of the ratio to be 2 : 1. Finally, the observed value for $\tilde{D}_{\parallel}$ of 420 MHz is extremely close to the largest component of the total zero-field splitting, which suggests that the alignment of the long axis of the spin probe in the membrane is virtually complete.

The dominant spin relaxation process for a triplet reorienting in an anisotropic environment almost certainly results from the anisotropy in the electron–electron interaction coupled to this rotation. There are, however, two essential differences between the operation of this relaxation mechanism in isotropic and in anisotropic solvents (Luckhurst *et al.*, 1975). The first of these results from the partial alignment of the spin probe, which means that not all of the zero-field splitting contributes to the dynamic spin Hamiltonian:

$$\mathscr{H}'(t) = \sum_{p} (-1)^p D^{(2,p)}(\mathscr{D}^{(2)}_{q,-p}(t) - \overline{\mathscr{D}^{(2)}_{0,-p}}\, \delta_{0q})T^{(2,q)} \tag{35}$$

Second, the widths of the two lines, like their positions, now depend on the orientation of the sample with respect to the magnetic field. The angular dependence of the linewidths is found to be

$$T_2^{-1}(\gamma) = X_0 + X_2 P_2(\cos\gamma) + X_4 P_4(\cos\gamma) \tag{36}$$

where $P_L(\cos\gamma)$ is the Lth Legendre polynomial. The magnitudes of the angular linewidth coefficients X_L depend on the model adopted for the

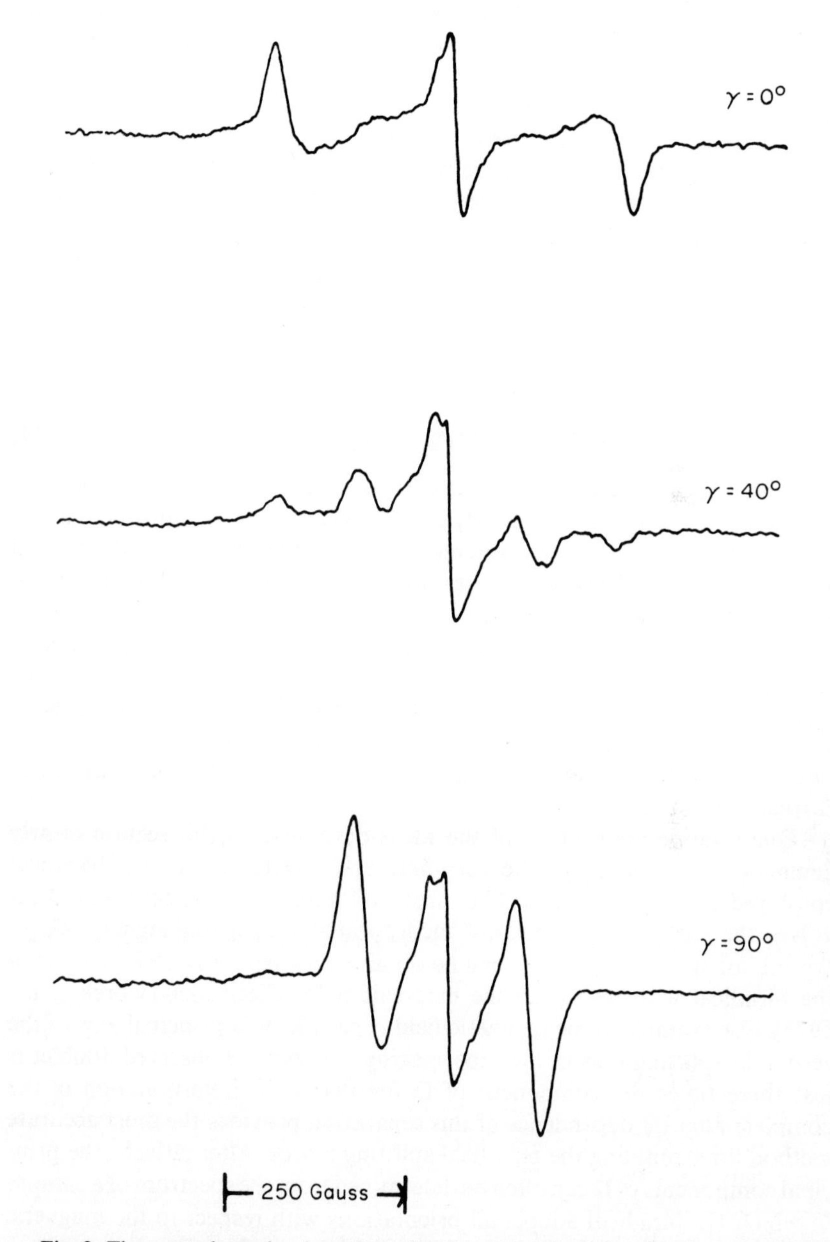

Fig. 3. The spectral angular dependence for the spin probe **II** dissolved in a dipalmitoyl lecithin bilayer. The central peak comes from a monoradical impurity. [Reprinted with permission from Keana and Dinerstein, *J. Amer. Chem. Soc.* **93**, 2808–2810 (1971). Copyright by the American Chemical Society.]

reorientation process. The strong collision model is mathematically attractive, for according to this model reorientation proceeds via collisions with no correlation between the molecular orientation before and after a collision. This model can be modified to allow for molecular anisotropy by the introduction of several correlation times; for example, if the molecule is axially symmetric, there would be two such times (Luckhurst and Sanson, 1972). One, τ_0, would be associated with collisions resulting in reorientation of the symmetry axis, whereas the other, τ_2, would relate to reorientation about this axis. The angular coefficients for such an axially symmetric spin probe are found to be

$$X_0 = (3/10)\{D^{(2,0)2}(1 - \bar{P}_2^2)\tau_0 + 2D^{(2,2)2}\tau_2\} \tag{37}$$

$$X_2 = (3/7)\{D^{(2,0)2}(\bar{P}_2 - \bar{P}_2^2)\tau_0 - 2D^{(2,2)2}\bar{P}_2\tau_2\} \tag{38}$$

$$X_4 = (9/35)\{3D^{(2,0)2}(\bar{P}_4 - \bar{P}_2^2)\tau_0 + D^{(2,2)2}\bar{P}_4\tau_2\} \tag{39}$$

provided the nonsecular terms in $\mathscr{H}'(t)$ are negligible. These coefficients depend on the order parameters $\bar{P}_2$ and $\bar{P}_4$, where $\bar{P}_2$ is the equivalent of the order parameter $\overline{\mathscr{D}^{(2)}_{0,p}}$ encountered for molecules with less than axial symmetry; it can, of course, be determined from the line positions. There are few techniques capable of yielding the other parameter $\bar{P}_4$, and so such linewidth studies may prove to be invaluable in the investigation of anisotropic systems. The three linewidth coefficients depend on three unknowns $\bar{P}_4$, τ_0, and τ_2; consequently each of these can be determined if the angular dependence of the linewidth is known. This contrasts with the situation for isotropic systems, where, as we have seen, it is impossible to obtain both correlation times.

Quantitative application of the ideas developed in this section clearly demands a knowledge of the zero-field splitting tensor for the biradical employed as the spin probe. The most satisfactory determination of **D** involves the study of the biradical incorporated into a diamagnetic single crystal, for then the spectrum can be obtained for particular orientations of the biradical with respect to the magnetic field (Ciecierska-Tworek *et al.*, 1973). For example, if the magnetic field is parallel to a principal axis of the zero-field splitting tensor, then the spacing between the observed doublet is just three times the component of **D** for that axis. Determination of the complete angular dependence of this separation provides the most accurate method for estimating the zero-field splitting tensor. Alternatively, the principal components of **D** can often be determined from the spectrum of a sample in which the biradical adopts all orientations with respect to the magnetic field. The spectra of such polycrystalline samples are found to contain three pairs of lines with separations equal to three times the principal components of the zero-field splitting tensor. An example of a polycrystalline spectrum is

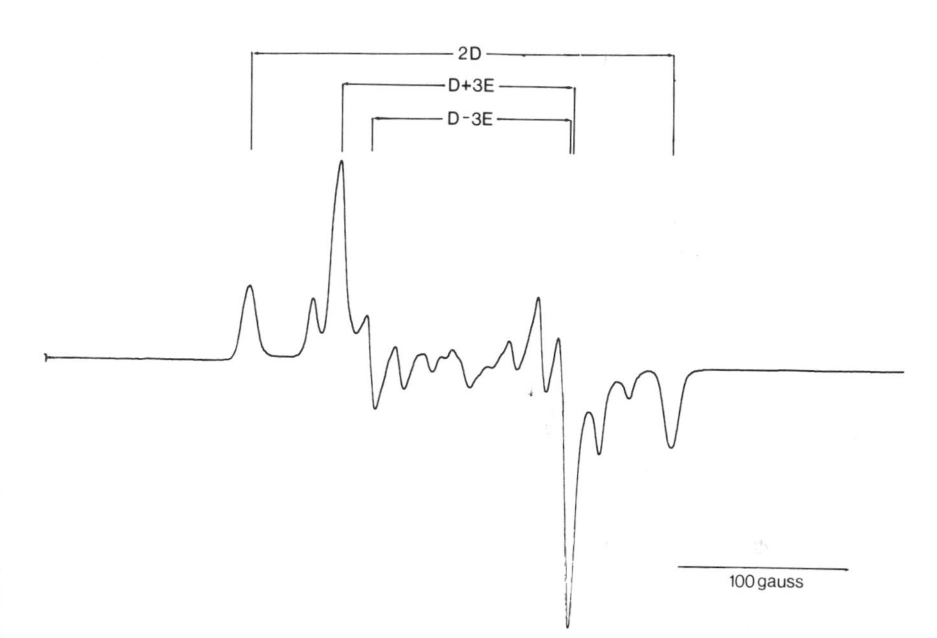

Fig. 4. The electron resonance spectrum of biradical **III** in a toluene glass at −196°C.

shown in Fig. 4 for biradical **III** dissolved in frozen toluene. Nitrogen hyperfine structure is only observed on one of the pairs of lines because the components of the hyperfine tensor associated with the other two directions are so small. The trace of the zero-field splitting is zero and so only two quantities are required to define the principal components of the tensor; it is conventional to choose these as

$$D = \tfrac{3}{2}D_{ZZ} \tag{40}$$

and

$$E = \tfrac{1}{2}(D_{XX} - D_{YY}) \tag{41}$$

where the Z direction corresponds to the largest component of **D**. These parameters are directly related to the irreducible components of **D**:

$$D^{(2,\,0)} = (2/3)^{1/2}D \tag{42}$$

and

$$D^{(2,\,\pm 2)} = \pm E \tag{43}$$

The values of D and E for the nitroxide biradicals **I–III** and for biradicals **IV** and **V** are listed in Table I.

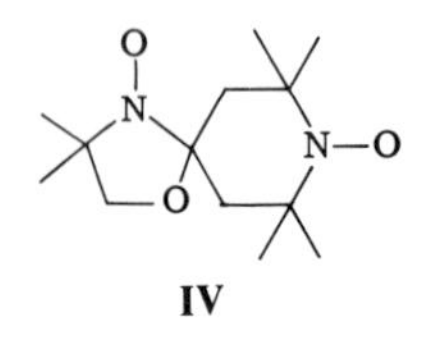

IV

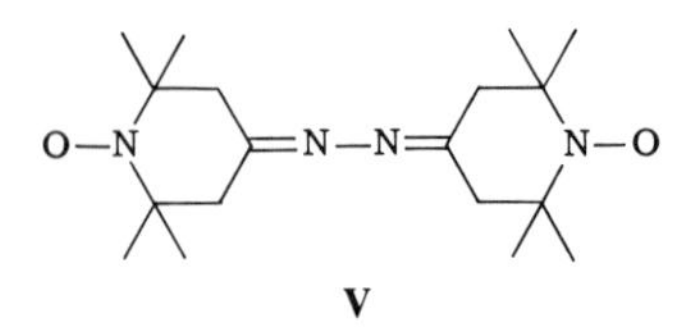

V

TABLE I

THE ZERO-FIELD SPLITTING PARAMETERS FOR SEVERAL NITROXIDE BIRADICALS

Biradical	D (MHz)	E (MHz)	Matrix	Reference
I	644	50	Ethyl alcohol (−120°C) or Nujol oil (−20°C)	Rassat (1972)
II	636	—	Ethyl ether/isopentane/ethyl alcohol (−190°C)	Keana *et al.* (1971)
III	350	—	Dimethyl formamide (−100°C)	Michon (1970)
	336	9	Toluene (−150°C)	Luckhurst and Poupko (1974)
IV	605	14	Single crystal of corresponding diamine (19°C)	Ciecierska-Tworek *et al.* (1973)
	613	—	Ethyl ether/isopentane/ethyl alcohol (−196°C)	Ciecierska-Tworek *et al.* (1973)
V	143	9.5	Single crystal of corresponding diamine	Nakajima (1973)

C. The Zero-Field Splitting

We must now consider what factors determine the magnitude of the zero-field splitting parameters D and E. In general the calculation of the zero-field splitting tensor is a difficult task in quantum mechanics because it demands a precise knowledge of the spatial electronic wave functions for the molecule. Given these functions, **D** is then obtained from the matrix elements

$$\mathbf{D} = \tfrac{1}{2}g^2\beta^2\langle - \,|\, \mathbf{D}^{\mathrm{op}} \,|\, - \rangle \tag{44}$$

where $|\,-\rangle$ is the antisymmetric wave function

$$|\,-\rangle = (1/\sqrt{2})(\,|\,a(1)b(2)\rangle - \,|\,a(2)b(1)\rangle) \tag{45}$$

in which $|a\rangle$ and $|b\rangle$ are the molecular orbitals containing the unpaired electrons. The spatial operators are defined in terms of the coordinates of these two electrons as

$$D^{op}_{\alpha\beta} = \{(r^2\, \delta_{\alpha\beta} - 3\alpha_{12}\beta_{12})/r^5\} \tag{46}$$

where r is the electron–electron separation and α_{12} is the difference in the α coordinates of the electrons. Unless the principal axes of **D** are determined by the molecular symmetry, the total tensor must be calculated in some arbitrary coordinate system and the principal components obtained by diagonalization.

This entire procedure is considerably simpler when the size of the molecular orbitals $|a\rangle$ and $|b\rangle$ is small in comparison with their separation, for then the electrons may be assumed to be localized. When the electrons are confined to specific locations, the matrix elements in Eq. (44) simply reduce to $D^{op}_{\alpha\beta}$ with the localized coordinates of electrons in place of the spatial operators. The obvious choice of coordinate system for this system is with the Z axis parallel to the interelectron vector, although the assignment of the X and Y axes is immaterial. If one electron is found at the origin of this system then the coordinates of the other electron are

$$Z_2 = r, \qquad X_2 = 0, \qquad Y_2 = 0 \tag{47}$$

Calculation of the appropriate differences from these coordinates then gives the components of the zero-field splitting tensor as

$$D_{ZZ} = -g^2\beta^2/r^3, \qquad D_{XX} = D_{YY} = -g^2\beta^2/2r^3, \qquad D_{XY} = D_{XZ} = D_{YZ} = 0 \tag{48}$$

The tensor is therefore diagonal and cylindrically symmetric within this coordinate system; of course this result follows immediately from the symmetry of two localized electrons. The zero-field splitting parameters are then

$$D = 3g^2\beta^2/2r^3 = -1.949 \times 10^4 g^2/r^3 \quad \text{MHz} \tag{49}$$

when r is expressed in angstroms and $E = 0$. Consequently, a knowledge of D should enable the electron–electron separation r to be determined and occasionally the molecular configuration to be assigned to the biradical. Strictly a necessary but not sufficient condition for these calculations is that E is zero or at least small in comparison with D and this is the case for the biradicals listed in Table I. Consider, as an example, the biradical **III**, which may exist in either a cis or a trans configuration. However, the electron–electron separation estimated from D is about 6 Å and this separation is only compatible with the trans configuration, where the separation is found, from molecular models, to be 6.2 Å (Michon, 1970; Rassat, 1971). Similarly, the zero-field splittings for the biradicals **I**, **II**, and **IV** correspond to an electron–electron separation of about 5 Å, which is in accord with a twist-boat configuration.

Of course the assumption of localized electrons cannot explain the non-zero values of E observed for certain biradicals and is not strictly valid for nitroxide biradicals where the electrons are delocalized over their respective nitroxide groups. Consequently it is necessary to extend the theory if a more precise analysis of the experimental data is required and a more rigorous theory is available (Pullman and Kochanski, 1967). This theory starts with Eq. (44) but now the molecular orbitals are written as a sum of atomic 2p orbitals, for example,

$$|a\rangle = \sum_i C_i^a |i\rangle \tag{50}$$

where $|i\rangle$ denotes an orbital on atom i with a coefficient C_i^a, which could be calculated from Hückel molecular orbital theory. With this approximation, which would be valid for most organic biradicals, the zero-field splitting tensor is found to be

$$\mathbf{D} = \tfrac{1}{2}g^2\beta^2 \sum_{i<j} (C_i^a C_i^b - C_j^a C_j^b)^2 \langle ii | \mathbf{D}^{\text{op}} | jj \rangle \tag{51}$$

where $\langle ii | \mathbf{D}^{\text{op}} | jj \rangle$ is a two-center integral for the two electrons in orbitals on atoms i and j. The problem is therefore reduced to the evaluation of these integrals; this calculation is particularly simple, and yet accurate, if the 2p orbitals are represented as two half charges (McWeeny, 1961). The calculation proceeds in analogous manner to that for localized electrons which was described previously; however, there is now an arbitrary parameter corresponding to the separation between the two half charges. The most appropriate value of this separation is found to be 1.4 Å by comparing experimental and theoretical values of the zero-field splitting (Pullman and Kochanski, 1967). The complete evaluation of **D** is now reduced, in essence, to a geometrical calculation and has been applied with some success to a conjugated nitroxide biradical (Calder *et al.*, 1969). The calculation is still simpler for aliphatic nitroxide biradicals because the electrons are localized on their respective nitroxide groups and so the coefficients required in Eq. (51) only involve the spin densities on the nitrogen and oxygen atoms.

III. BIRADICALS WITH HYPERFINE INTERACTIONS

A. Spectral Analysis

The previous sections contained examples of nitroxide biradicals and in each of these there must be a coupling between the electron and nuclear spins. It was possible to ignore such hyperfine interactions because their effect on the spectra was obscured by the large linewidths. In this section we shall be concerned with biradicals for which the hyperfine structure can be resolved; however, before considering biradicals that exhibit such structure we shall find it helpful to deal with a biradical where the electron spins are associated with different g factors. The scalar spin Hamiltonian is then

$$\mathscr{H} = g^{(1)}\beta B S_z^{(1)} + g^{(2)}\beta B S_z^{(2)} + J S^{(1)} \cdot S^{(2)} \tag{52}$$

which is equivalent to that for an AB system in nuclear magnetic resonance spectroscopy (cf. Carrington and McLachlan, 1967). When the g factors are identical the eigenfunctions of $\mathscr{H}$ are just the triplet and singlet spin states. However, the difference in the g factors is responsible for a competition for the two spins and the result of this competition depends on the ratio $\Delta g\beta B/J$, where

$$\Delta g = g^{(1)} - g^{(2)} \tag{53}$$

At one extreme, when this ratio is infinite, the two electrons are unaware of each other and the spectrum contains two lines at resonance fields $h\nu_0/g^{(1)}\beta$ and $h\nu_0/g^{(2)}\beta$. The other extreme corresponds to a strong coupling between the electron spins and so the ratio is zero; as a consequence they share a common resonance field, which is $2h\nu_0/(g^{(1)} + g^{(2)})\beta$. These two extreme situations are frequently described as the strong ($\Delta g\beta B/J \ll 1$) and weak ($\Delta g\beta B/J \gg 1$) exchange limits. It is not uncommon to see them referred to as the fast and slow exchange limits, but this terminology is quite misleading, for no physical motion is involved. For intermediate values of $\Delta g\beta B/J$ the spectrum is more complex and therefore more informative; we shall now determine the form of such spectra. The choice of the basis set clearly depends on the value of the ratio and in our calculation we shall find it convenient to employ the triplet and singlet spin functions. With this basis the Hamiltonian matrix is

$$\begin{array}{c|cccc|} & |1\rangle & |0_a\rangle & |0_e\rangle & |-1\rangle \\ \langle 1| & \dfrac{g^{(1)}+g^{(2)}}{2}\beta B + \dfrac{J}{4} & 0 & 0 & 0 \\ \langle 0_a| & 0 & \dfrac{J}{4} & \dfrac{\Delta g\beta B}{2} & 0 \\ \langle 0_e| & 0 & \dfrac{\Delta g\beta B}{2} & \dfrac{-3J}{4} & 0 \\ \langle -1| & 0 & 0 & 0 & -\dfrac{g^{(1)}+g^{(2)}}{2}\beta B + \dfrac{J}{4} \end{array} \tag{54}$$

where only states $|0_a\rangle$ and $|0_e\rangle$ are mixed by the g-factor difference. The matrix can therefore be cast in diagonal form by taking linear combinations of these two states; the appropriate combinations are

$$|0+\rangle = \cos\phi\,|0_a\rangle + \sin\phi\,|0_e\rangle \tag{55}$$

and

$$|0-\rangle = \cos\phi\,|0_e\rangle - \sin\phi\,|0_a\rangle \tag{56}$$

provided

$$\tan 2\phi = \Delta g\beta B/J \tag{57}$$

The eigenvalues for the two new levels are then

$$E_{0+} = \tfrac{1}{4}J + \tfrac{1}{2}\Delta g\beta B \tan\phi \tag{58}$$

and

$$E_{0-} = -\tfrac{3}{4}J - \tfrac{1}{2}\Delta g\beta B \tan\phi \tag{59}$$

The allowed transitions are now

$$\begin{aligned} |0+\rangle &\leftrightarrow |1\rangle & h &= 1 + \cos 2\phi \\ |0-\rangle &\leftrightarrow |1\rangle & h &= 1 - \cos 2\phi \\ |-1\rangle &\leftrightarrow |0+\rangle & h &= 1 + \cos 2\phi \\ |-1\rangle &\leftrightarrow |0-\rangle & h &= 1 - \cos 2\phi \end{aligned} \tag{60}$$

where h is the transition probability obtained from the square of the matrix element of $(S_x^{(1)} + S_x^{(2)})$ between the relevant states. The resonance fields for the two transitions involving $|0+\rangle$ are determined by

$$h\nu_0(T) = \tfrac{1}{2}(g^{(1)} + g^{(2)})\beta B \pm \tfrac{1}{2}\Delta g\beta B \tan\phi \tag{61}$$

and those involving $|0-\rangle$ are

$$h\nu_0(S) = \tfrac{1}{2}(g^{(1)} + g^{(2)})\beta B \pm J \pm \tfrac{1}{2}\,\Delta g\beta B \tan\phi \tag{62}$$

In general the spectrum will contain two pairs of lines, each centered on the mean resonance field. When the ratio $\Delta g\beta B/J$ is comparable to or smaller than one, it is possible to label these transitions and spectral lines in a particularly convenient manner. For this region the value of $\cos\phi$ is approximately one and so the state $|0+\rangle$ is largely composed of the triplet function $|0_a\rangle$, whereas $|0-\rangle$ closely resembles the singlet function. The transitions involving $|0+\rangle$ may therefore be thought of as triplet (T) transitions while those involving $|0-\rangle$ are essentially singlet (S) transitions. The spacing between the triplet lines is $\Delta g\beta B \tan\phi$ and that between the singlet

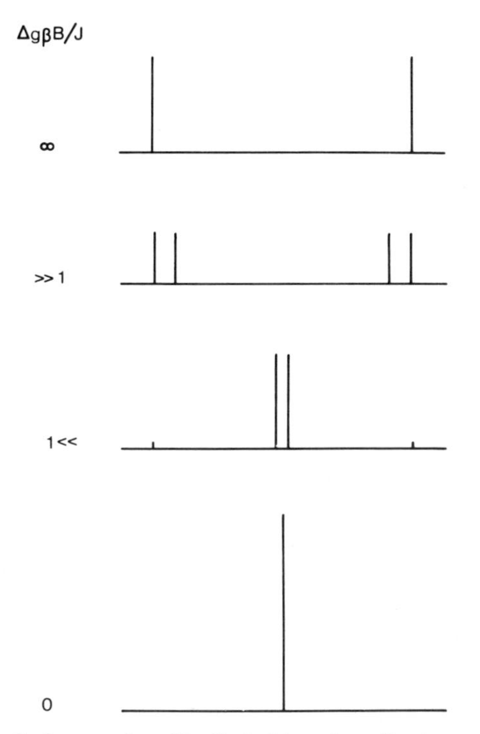

Fig. 5. The theoretical spectra for a biradical without hyperfine interactions calculated for various values of the ratio $\Delta g\beta B/J$.

pair is $2J + \Delta g\beta B \tan \phi$. The spectra calculated for such a biradical are shown in Fig. 5 for a range of values for the ratio $\Delta g\beta B/J$.

We are now in a position to develop the relevant theory for a biradical composed of two essentially independent halves, which could, for example, be joined by a saturated linkage. The spin Hamiltonian for the system is usually taken to be the sum of the Hamiltonians for the isolated fragments together with the familiar exchange interaction:

$$\mathcal{H} = g^{(1)}\beta BS_z^{(1)} + \sum_i a_i^{(1)}I^{(i)} \cdot S^{(1)} + g^{(2)}\beta BS_z^{(2)} + \sum_j a_j^{(2)}I^{(j)} \cdot S^{(2)} + JS^{(1)} \cdot S^{(2)} \tag{63}$$

Here $a_i^{(1)}$ is the coupling constant for the ith nucleus in fragment 1 (Slichter, 1955). The conditions under which the hyperfine contribution to the spin Hamiltonian takes this form were derived by Reitz and Weissman (1960); their derivation is outlined in Appendix A. When the hyperfine interactions are small compared with the electron Zeeman splitting, the nonsecular hyperfine terms may be ignored and $\mathscr{H}$ reduces to

$$\mathscr{H} = g^{(1)}\beta B S_z^{(1)} + \sum_i a_i^{(1)} I_z^{(i)} S_z^{(i)} + g^{(2)}\beta B S_z^{(2)} + \sum_i a_j^{(2)} I_z^{(j)} S_z^{(2)} + J S^{(1)} \cdot S^{(2)} \tag{64}$$

The basis set must now be extended to include the nuclear spin states and the simplest way to achieve this is to employ product spin functions of the form $|M_s; \prod_i m_i^{(1)}, \prod_j m_j^{(2)}\rangle$, where $m_i^{(1)}$ is the quantum number for nucleus i in fragment 1. Because the nuclear components of the spin states are not mixed by any term in $\mathscr{H}$, we can obtain an effective spin Hamiltonian that operates solely on the electron spin states. The effective operator is found by evaluating the matrix elements of the nuclear spin operators between the product spin states, which gives

$$\mathscr{H}_{\text{eff}} = \left\{g^{(1)}\beta B + \sum_i a_i^{(1)} m_i^{(1)}\right\} S_z^{(1)} + \left\{g^{(2)}\beta B + \sum_j a_j^{(2)} m_j^{(2)}\right\} S_z^{(2)} + J S^{(1)} \cdot S^{(2)} \tag{65}$$

This has exactly the same form as the spin Hamiltonian in Eq. (52) for a biradical without hyperfine interactions and we can therefore define effective g factors by

$$g_{\text{eff}}^{(1)} = g^{(1)} + (1/\beta B) \sum_i a_i^{(1)} m_i^{(1)} \tag{66}$$

with a similar expression for fragment 2. It is possible therefore to determine the transition fields and probabilities for even the most complex biradical simply by substituting the effective g factors into Eqs. (60)–(62).

However, our objectives are far less ambitious, for we shall consider a symmetric nitroxide biradical in which the only important hyperfine interaction is with the nitrogen nuclei. The effective g factors are

$$g_{\text{eff}}^{(1)} = g + am^{(1)}/\beta B \tag{67}$$

and

$$g_{\text{eff}}^{(2)} = g + am^{(2)}/\beta B \tag{68}$$

where the subscript on the nuclear quantum number is omitted because there is only one nucleus in each fragment; the labels on the coupling constant have been dropped for similar reasons. The spectral behavior is, of course, determined by $\Delta g_{\text{eff}}\,\beta B/J$, which is $a(m^{(1)} - m^{(2)})/J$. The line positions are readily calculated from the expressions for the effective g factors and the results embodied in the single equation

$$h\nu_0\begin{pmatrix}\text{T}\\ \text{S}\end{pmatrix} = g\beta B + \frac{a}{2}M_I + \begin{pmatrix}\pm 1\\ \mp 1\end{pmatrix}\frac{a\,\Delta m}{2}\tan\phi + \begin{pmatrix}0\\ \mp 1\end{pmatrix}J \tag{69}$$

where M_I is the total nuclear quantum number and Δm is the difference $m^{(1)} - m^{(2)}$. In the strong exchange limit tan ϕ vanishes and so the triplet transitions occur when

$$h\nu_0(T) = g\beta B + \tfrac{1}{2}aM_I \tag{70}$$

The singlet transitions are, of course, forbidden in this limit. Since the spin of the common nitrogen isotope is 1, the total spin quantum number takes all integer values from 2 to -2. The spectrum therefore contains five lines and, as we can see from Table II, they have intensities 1 : 2 : 3 : 2 : 1 with a

TABLE II

THE LINE POSITIONS FOR A NITROXIDE BIRADICAL

			Line positions relative to the central peak	
M_I	$m^{(1)}$	$m^{(2)}$	$a/J \ll 1$	$a/J < 1$
2	1	1	$a/g\beta$	$a/g\beta$
1	1	0	$a/2g\beta$	$a/2g\beta \pm a^2/4Jg\beta$
	0	1	$a/2g\beta$	$a/2g\beta \pm a^2/4Jg\beta$
0	0	0	0	0
	1	-1	0	$\pm a^2/Jg\beta$
	-1	1	0	$\pm a^2/Jg\beta$
-1	-1	0	$-a/2g\beta$	$-a/2g\beta \pm a^2/4Jg\beta$
	0	-1	$-a/2g\beta$	$-a/2g\beta \pm a^2/4Jg\beta$
-2	-1	-1	$-a/g\beta$	$-a/g\beta$

spacing of $a/2g\beta$, which is just one-half the coupling constant for the corresponding monoradical. The biradical bis(2,2,6,6-tetramethyl piperidin-4-one-1-oxyl)azide (V) exhibits strong exchange, as we can see from the spectrum shown in Fig. 6b. There are five hyperfine lines, with the expected relative intensities, each separated by 7.5 G; this gives a value for the nitrogen coupling constant a of 42 MHz, which is equal to that found for the relevant monoradical 2,2,6,6-tetramethyl piperidin-4-one-1-oxyl (Nakajima *et al.*, 1972). Of course it is not possible to determine the magnitude of J from the electron resonance spectrum in the strong exchange limit. However, as the exchange interaction is reduced the degeneracy of the spin transitions is removed and it is possible to estimate J. The magnitude of the splitting of the triplet transitions can be obtained from Eq. (69) by expanding tan ϕ for small ϕ and this gives

$$h\nu_0(T) = g\beta B + \tfrac{1}{2}aM_I \pm (a^2\,\Delta m^2/4J) \tag{71}$$

Accordingly, those transitions for which Δm is zero are unshifted but all other lines do move. Their new positions are given in Table II, and, as we

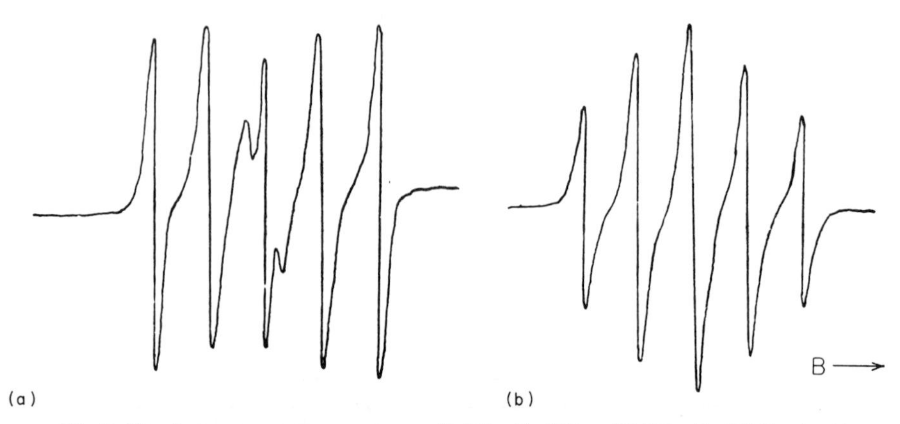

Fig. 6. The electron resonance spectrum of (a) biradical **VI** and (b) biradical **V** dissolved in tetrahydrofuran. (Nakajima *et al.*, 1972.)

can see, the central peak is split into three lines with a spacing between adjacent lines of $a^2/Jg\beta$. The degeneracy of the peaks with $M_I = \pm 1$ is also removed but here the splitting between the two lines is only $a^2/2Jg\beta$. When these component lines can be resolved it is then possible to estimate the exchange integral J, although the accuracy of such measurements is not high. An example of this situation is provided by bis(2,2,6,6-tetramethyl piperidin-4-ol-1-oxyl) sulfite (**VI**), whose spectrum is shown in Fig. 6a,

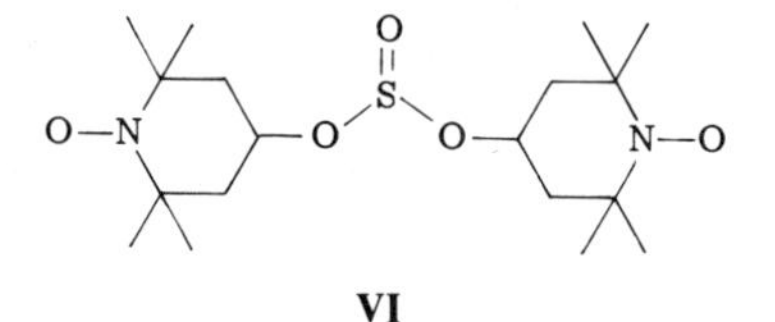

VI

where only the splitting of the central peak is large enough to be observed; in this case a/J was found to be 0.33 (Nakajima *et al.*, 1972). When the splitting is insufficient to allow the various components to be resolved the removal of the degeneracy will produce an inhomogeneous broadening of the central peak. In addition there will be a smaller broadening of the two lines with $M_I = \pm 1$. However, the use of the widths of the spectral lines to gauge the magnitude of J is fraught with difficulties, for, as we shall see in Section IV.B, the widths of just these lines are strongly influenced by spin relaxation processes. Consequently when the spectrum of a nitroxide biradical contains just five lines the only sound conclusion which may be drawn is that the strong exchange limit obtains. It is unfortunate that several pioneering studies with biradicals as spin probes failed to appreciate this point, thus invalidating their conclusions (Calvin *et al.*, 1969; Ferruti *et al.*, 1969; Hsia *et al.*, 1969).

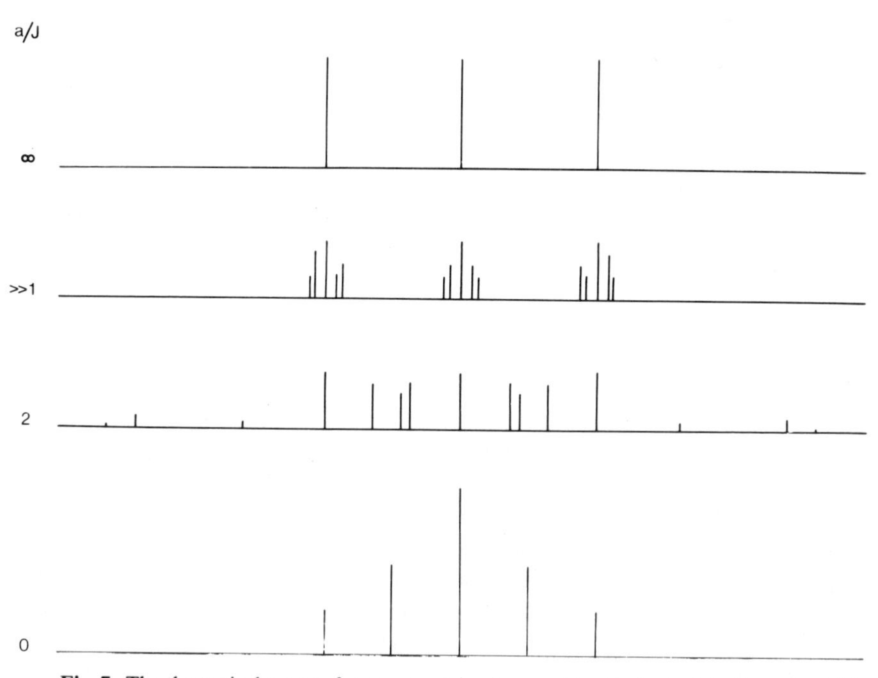

Fig. 7. The theoretical spectra for a symmetric nitroxide biradical calculated from Eq. (69) for various values of a/J.

The exchange integral can be determined with the greatest precision when it is comparable to the hyperfine interaction. This situation is illustrated in Fig. 7, which shows the spectra to be expected for a nitroxide biradical, calculated from Eq. (69), for various values of a/J (Briere *et al.*, 1965). When this ratio is one-half, the singlet transitions can be observed as satellite lines, and, as we noted earlier, their separation is essentially $2J$. An example of a spectrum with the singlet lines is shown in Fig. 9 of the following section. However, as we shall discover, these singlet lines can often be unobservably broad and then the exchange interaction must be extracted from the spacing between the triplet peaks. The theoretical spectra also show that as the ratio a/J is increased still further the transitions are arranged into three groups of lines and finally collapse in the weak exchange limit to the three-line spectrum characteristic of a nitroxide monoradical.

B. The Exchange Integral

Now that we know how to determine the magnitude of the exchange interaction in biradicals we shall require a theoretical basis with which to rationalize these results. A variety of mechanisms may contribute to the exchange integral and a quantitative prediction of their magnitudes is often difficult. However, in the case of unconjugated nitroxide biradicals there are just two contributions to J and the qualitative behavior of these is readily appreciated. Let us consider a particularly simple system containing two electrons, which may occupy the two molecular orbitals $|a\rangle$ and $|b\rangle$. The spatial Hamiltonian for the system will be composed of the Hamiltonians for the individual components together with the term e^2/r, which allows for electron–electron repulsion. It is straightforward to show that J, which is equal to the energy separation between the bonding and antibonding combinations of $|a\rangle$ and $|b\rangle$, is given by

$$J = 2e^2\langle a(1)b(2)|r^{-1}|a(2)b(1)\rangle \tag{72}$$

The symbol $|a(1)\rangle$ implies that electron 1 occupies orbital $|a\rangle$; these orbitals are discussed in greater detail in Appendix A. The magnitude of the matrix element clearly depends on the direct overlap between $|a\rangle$ and $|b\rangle$; indeed it is suggested that this mechanism makes a negligible contribution to J if the molecular orbitals are separated by more than 14 Å. In contrast if the separation is less than 10 Å the contribution is said to be large, although the basis of these statements is obscure (Calvin *et al.*, 1969). This direct contribution to J is positive and clearly sensitive to the configuration of the biradical. It is, of course, just this configurational dependence which makes flexible biradicals potentially useful as environmental spin probes.

The second mechanism by which the spins are made aware of each other is not immediately obvious from the form of Eq. (72), for it involves the spin polarization of the σ core in the biradical. An entirely analogous mechanism was described in Chapter 2 to account for the scalar hyperfine interaction when the unpaired electron did not appear to occupy an orbital with some s character. McConnell (1960) has employed perturbation theory to estimate the magnitude of J caused by this mechanism; although we shall not be concerned with the details, the final result is of interest. When the orbitals containing the unpaired electrons are linked by n σ bonds the spin polarization contribution to J is approximately

$$J = (-1)^n 3 \times 10^{6-n} \quad \text{MHz} \tag{73}$$

As expected, J decreases rapidly with the number of σ bonds separating $|a\rangle$ and $|b\rangle$; in addition, the calculation makes the interesting prediction that the sign of the exchange integral should alternate.

We shall now see how these ideas can be used to help us understand the temperature dependence of J for the biradical bis(2,2,6,6-tetramethyl piperidin-4-ol-1-oxyl) oxalate (**VII**) dissolved in carbon disulfide (Glarum

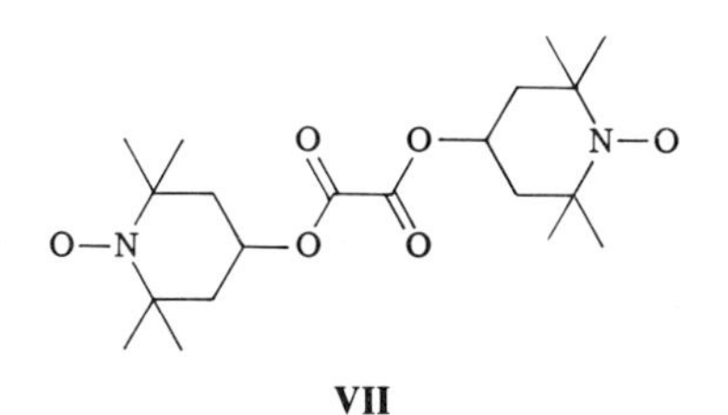

VII

and Marshall, 1967). This temperature dependence is shown in Fig. 8 and is typical of that found for structurally related biradicals. At low temperatures the exchange integral reaches a limiting value of about 14 MHz. The fluctuations in molecular geometry are presumably quenched at these temperatures and the biradical then adopts a configuration in which the direct contribution to J is negligible. The observed limiting value is then attributable to the indirect mechanism and, according to Eq. (73), should only be about

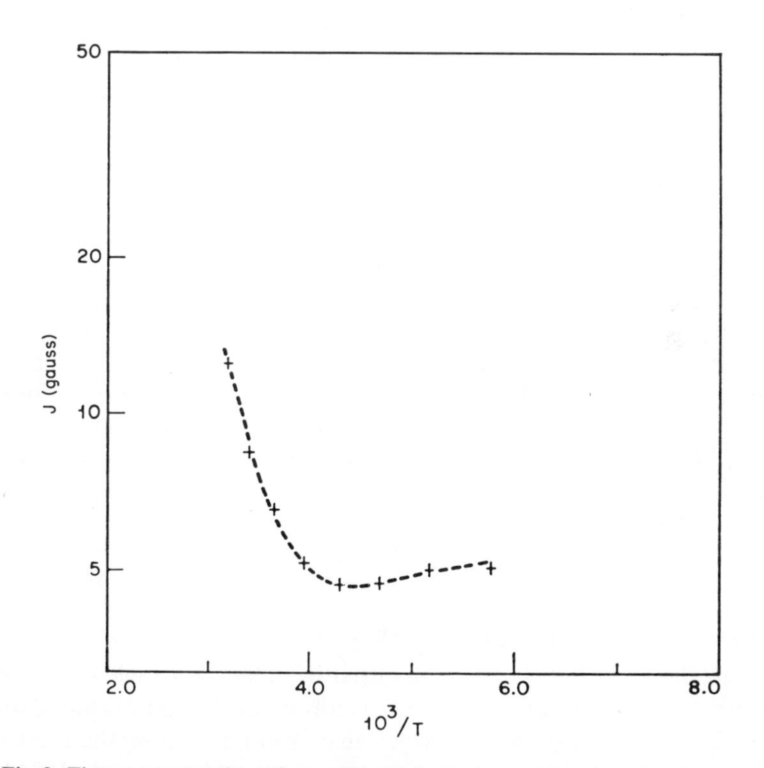

Fig. 8. The temperature dependence of the exchange integral for biradical **VII** dissolved in carbon disulfide. (Glarum and Marshall, 1967.)

3×10^{-5} MHz. This enormous discrepancy between theory and experiment probably results from relatively minor errors in the integrals involved in McConnell's calculation. Indeed, the agreement is considerably improved if the exchange integral is attenuated by a factor of $\frac{1}{6}$ rather than $\frac{1}{10}$ by each σ bond as suggested in Eq. (73). As the temperature is increased there is a slight decrease in J, followed by an exponential increase, which may be attributed to the population of high-energy configurations with the nitroxide groups in close proximity. The slope of the log J–$1/T$ plot corresponds to an activation energy of about 12 kJ mole^{-1} but it is not possible to use this value to identify the nature of the process responsible for the configurational changes. These changes might be controlled simply by the barriers to rotation about single bonds in the biradical or they could be determined by positional changes in the solvent sheath. Indeed, the exchange integral does exhibit a solvent dependence; for example, the low-temperature limiting values of J for bis(2,2,6,6-tetramethyl-piperidin-4-ol-1-oxyl) carbonate (**VIII**) range from 125 MHz in chloroform through 95 MHz in carbon

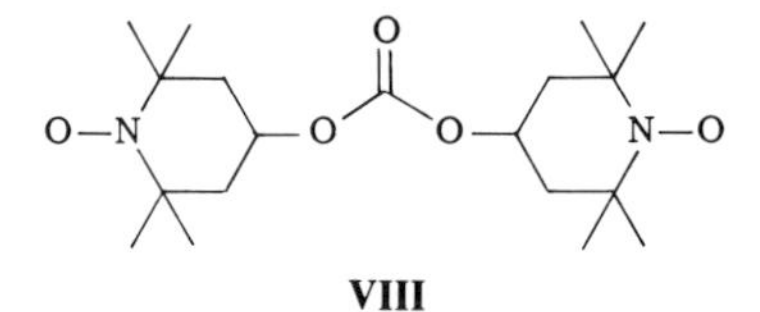

VIII

disulfide to 74 MHz in *n*-hexane. There is an apparent correlation with the anisotropy of the solvent molecules which may force the spin probe to adopt less favorable configurations for the direct contribution to J even at low temperatures.

The sensitivity of the exchange integral to the biradical's environment has been employed in a study of micelle formation in aqueous solutions of sodium dodecyl sulfate (Ohnishi *et al.*, 1970). The biradical used was *N*,*N'*-bis[4-(1-oxyl-2,2,6,6-tetramethyl piperidyl)] urea (**IX**), which gives the

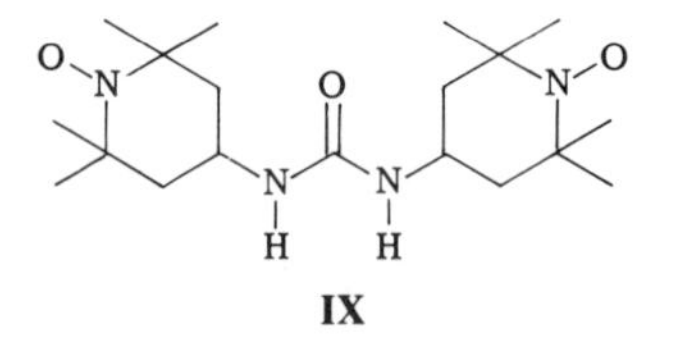

IX

spectrum shown in Fig. 9a when dissolved in water. The singlet transitions are observed, the spectrum is therefore readily analyzed, and J is found to be 65 MHz. Addition of sodium dodecyl sulfate had no effect on the spectrum until the critical micelle concentration of 8×10^{-3} M was reached. At this point additional lines were observed in the spectrum and these are indicated by arrows in Fig. 9b. Further increase in the concentration of the surfacant

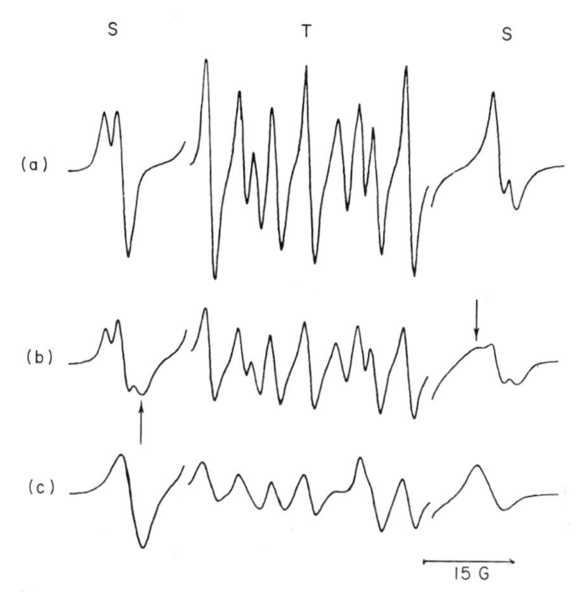

Fig. 9. The electron resonance spectrum of the spin probe **IX** dissolved in aqueous solutions of sodium dodecyl sulfate (a) below, (b) at, and (c) above the critical micelle concentration. (Ohnishi *et al.*, 1970.)

yields the single spectrum shown in Fig. 9c; the singlet lines can still be discerned and give a value for the exchange integral of 56 MHz. This is close to the value found when the biradical is dissolved in hydrocarbon solvents. The interpretation of these results is quite straightforward. Below the critical micelle concentration the spin probe experiences an aqueous environment. Far above this concentration the probe is associated entirely with the hydrocarbon region of the micelles. The relatively sharp spectral lines observed for the spin probe in this environment imply a highly fluid state for the interior of the micelle. At intermediate concentrations the biradical is partitioned between the micelles and the surrounding aqueous phase. The observation of spectra from both regions shows that the rate of exchange between the two environments is slow. Similar conclusions have also been reached with the aid of monoradical spin probes (Atherton and Strach, 1972; Oakes, 1972).

In these experiments the molecular configuration and hence the exchange integral were modified by the weak van der Waals interactions with the solvent. However, in an idealized application of biradicals as spin probes there should be a much stronger interaction with the environment (Calvin *et al.*, 1969). For example, we might hope to attach the spin probe via covalent or ionic links to the system under investigation, which could be a membrane or a nucleic acid. Then changes in the conformation of the system, induced by an external stimulus, would induce corresponding structural variations in the biradical and these could be detected by changes in the exchange integral. The success of such experiments clearly demands a rather special spin probe, for not only must it bind specifically to the system, but, in addition, the exchange integral must be comparable to the hyperfine interaction in order for the magnitude of J to be determined from the spectrum (Ferruti *et al.*, 1970).

C. Anisotropic Environments

In this section we shall consider the spectral analysis for a biradical dissolved in a macroscopically anisotropic environment with particular reference to nematic liquid crystals. As we saw in Section II.B, the difference between isotropic and anisotropic systems is the retention of angle-dependent contributions in the static spin Hamiltonian $\overline{\mathscr{H}}$. A straightforward extension of Eq. (28) to a biradical for which there are several angle-dependent terms shows that the formal expression for $\overline{\mathscr{H}}$ is

$$\overline{\mathscr{H}} = \mathscr{H}^0 + \sum_{\mu;\, p} (-1)^p F_\mu^{(2,\, p)} \overline{\mathscr{D}_{0,\, -p}^{(2)}}\, T_\mu^{(2,\, 0)} \tag{74}$$

when the symmetry axis of the system is parallel to the magnetic field. These various magnetic interactions are denoted by μ and the interaction tensors $F^{(2,\, p)}$ are expressed in a molecular coordinate system. If the contributions are restricted to the Zeeman, hyperfine, and electron–electron interactions, then this spin Hamiltonian is

$$\begin{aligned}\overline{\mathscr{H}} &= \tilde{g}_\parallel^{(1)} \beta B S_z^{(1)} + \tilde{g}_\parallel^{(2)} \beta B \tilde{S}_z^{(2)} + \tilde{A}_\parallel^{(1)} I_z^{(1)} S_z^{(1)} + \tilde{A}_\parallel^{(2)} I_z^{(2)} S_z^{(2)} \\ &\quad + (J - \tilde{D}_\parallel) S^{(1)} \cdot S^{(2)} + 3\tilde{D}_\parallel S_z^{(1)} S_z^{(2)}\end{aligned} \tag{75}$$

provided the nonsecular hyperfine terms are negligible. The parallel components of the partially averaged g and hyperfine tensors are defined in an analogous manner to that for the zero-field splitting tensor $\tilde{\mathbf{D}}$. The two g tensors $\mathbf{g}^{(1)}$ and $\mathbf{g}^{(2)}$ are not equal even in a symmetric biradical unless the principal axes of the total g tensors are parallel; when this condition is satisfied the two electrons are said to be completely equivalent. Similarly the partially averaged hyperfine tensors for the two nitrogen nuclei will only be identical if the nitrogens are completely equivalent. The contribution of the zero-field splitting to $\overline{\mathscr{H}}$ is identical to that given in Eq. (30) but has been rewritten in terms of the individual electron spin operators.

According to Eq. (75), the partial alignment of the biradical shifts both the g factor and the coupling constant from their isotropic values; however, the major spectral change results from the contribution of the anisotropic electron–electron coupling (Falle *et al.*, 1966). In the strong exchange limit this coupling removes the degeneracy of the electron spin transitions, as we

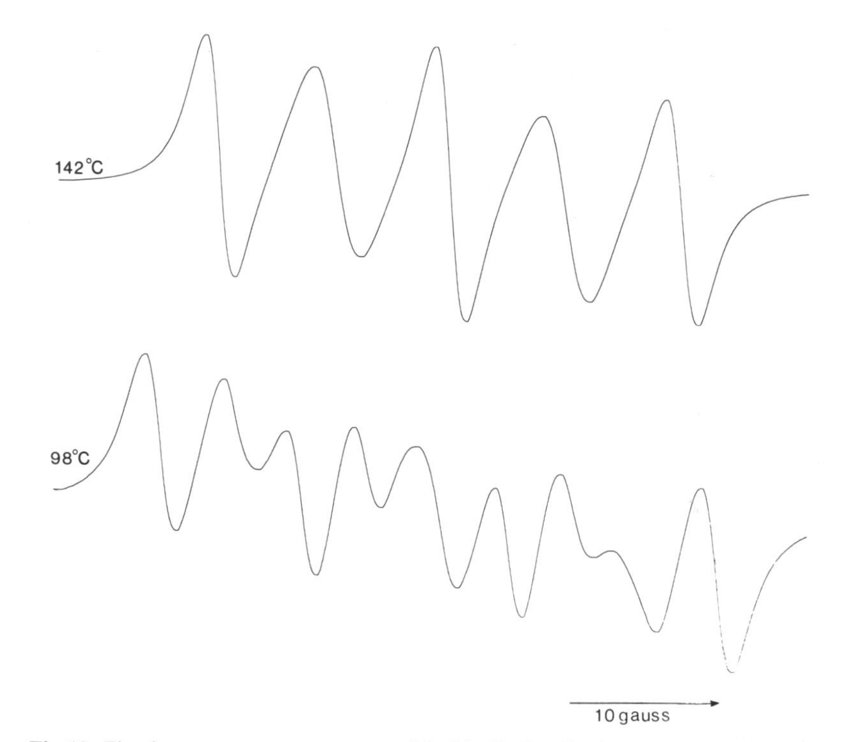

Fig. 10. The electron resonance spectrum of the biradical **X** dissolved in the isotropic phase (142°C) and nematic mesophase (98°C) of 4,4′-dimethoxyazoxybenzene. (Luckhurst, 1970.)

saw in Section II.B. Consequently the number of spectral lines is doubled and the spacing between these doublets is $3\tilde{D}_{\parallel}$. The splitting of the spectral lines is illustrated in Fig. 10 for bis(2,2,6,6-tetramethyl piperidin-4-ol-1-oxyl) glutarate (**X**) dissolved in the nematogen 4,4′-dimethoxyazoxybenzene. The

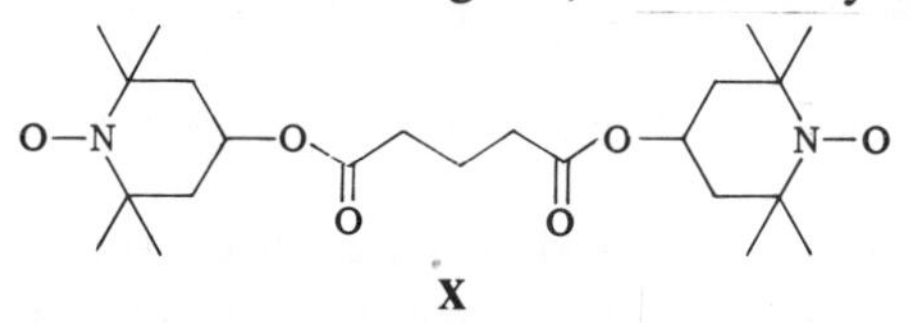

X

spectrum in the isotropic phase (142°C) contains five lines indicative of strong exchange, although the lines do not have their expected relative heights. This deviation is a result of symmetric line broadening, which is described in Section IV.B. In the liquid crystal phase each of the lines is indeed split into a doublet, although only nine of the ten lines are observed, because of overlap in the center of the spectrum. Determination of this additional splitting together with a knowledge of the total zero-field splitting tensor would then provide a value for the order parameter of the spin probe.

The spectral changes produced by partial alignment are much more complex for intermediate values of spin exchange (Lemaire, 1967; Lemaire *et al.*, 1968; Corvaja *et al.*, 1970). The problem is further complicated by the contribution of the zero-field splitting to $\overline{\mathscr{H}}$, which has the same form as the scalar electron–electron coupling and as a consequence influences the extent of triplet–singlet mixing. The effect of this mixing can best be appreciated by returning to an effective spin Hamiltonian, for the Hamiltonian matrix is readily found to be

$$\begin{array}{c|cccc|} & |1\rangle & |0_a\rangle & |0_e\rangle & |-1\rangle \\ \langle 1| & \begin{array}{c}\frac{1}{2}(\tilde{g}_{\parallel}^{(1)}+\tilde{g}_{\parallel}^{(2)})\beta B\\ \frac{1}{4}J'+\frac{3}{4}\tilde{D}_{\parallel}\end{array} & 0 & 0 & 0 \\ \langle 0_a| & 0 & \frac{1}{4}J'-\frac{3}{4}\tilde{D}_{\parallel} & \frac{1}{2}\Delta\tilde{g}_{\parallel}\beta B & 0 \\ \langle 0_e| & 0 & \frac{1}{2}\Delta\tilde{g}_{\parallel}\beta B & -\frac{3}{4}J'-\frac{3}{4}\tilde{D}_{\parallel} & 0 \\ \langle -1| & 0 & 0 & 0 & \begin{array}{c}-\frac{1}{2}(\tilde{g}_{\parallel}^{(1)}+\tilde{g}_{\parallel}^{(2)})\beta B\\ +\frac{1}{4}J'+\frac{3}{4}\tilde{D}_{\parallel}\end{array} \end{array} \tag{76}$$

where

$$J' = J - \tilde{D}_{\parallel} \tag{77}$$

This matrix is identical in form to Eq. (54) and so the eigenfunctions are still given by Eqs. (55) and (56), but now

$$\tan 2\phi = \Delta\tilde{g}_{\parallel}\beta B/J' \tag{78}$$

Evaluation of the eigenvalues shows that the four allowed spin transitions occur when

$$h\nu_0(\mathrm{T}) = (\tilde{g}_{\parallel}^{(1)} + \tilde{g}_{\parallel}^{(2)})(\beta B/2) \pm \Delta\tilde{g}_{\parallel}(\beta B/2)\tan\phi \pm \tfrac{3}{2}\tilde{D}_{\parallel} \tag{79}$$

and

$$h\nu_0(\mathrm{S}) = (\tilde{g}_{\parallel}^{(1)} + \tilde{g}_{\parallel}^{(2)})(\beta B/2) \pm J' \pm \Delta\tilde{g}_{\parallel}(\beta B/2)\tan\phi \pm \tfrac{3}{4}\tilde{D}_{\parallel} \tag{80}$$

The partially averaged zero-field splitting makes the anticipated contribution to the spacing between the triplet lines, but in addition there is also a contribution to the separation between the singlet lines. Since these two separations depend in different ways on J and $\tilde{D}_{\parallel}$, it is possible to determine both from the spectrum of the biradical in the ordered phase. Such determinations are potentially important because the molecular configurations should change on passing from a disordered to an ordered system and these changes will be reflected in J. However, environmental changes in the exchange integral should be negligible for the rigid biradical azoxy-5,5′-tetraethyl-1,1,3,3 isoindoline oxyl-2 **(XI)** synthesized by Giroud

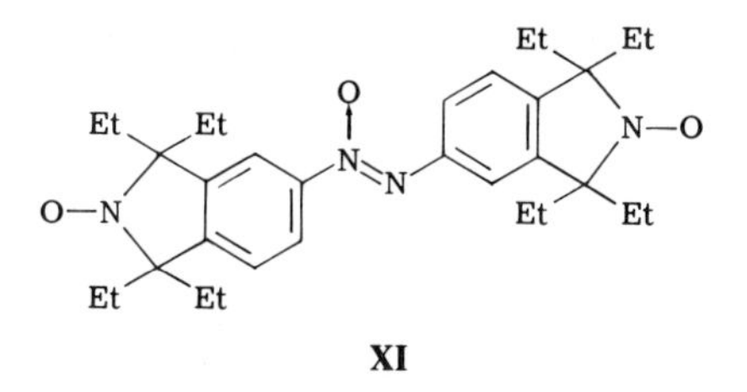

XI

et al. (1974). This biradical could be a particularly valuable spin probe because its structure closely resembles that of the many liquid crystals containing the azoxybenzene nucleus. In addition, the determination of the partially averaged g factor, nitrogen hyperfine splitting, and zero-field splitting from the spectrum in the liquid crystal mesophase should permit all of the order parameters $\overline{\mathscr{D}^{(2)}_{0,-p}}$ to be obtained.

IV. SPIN RELAXATION

An appreciation of the various factors which may influence the linewidths in a spectrum is important not only as an aid to a correct analysis but also because the linewidths can often afford a valuable insight into the molecular dynamics. There are two distinct relaxation processes which contribute to the linewidths for a biradical exhibiting hyperfine structure. One results in an asymmetric variation in the widths of the spectral lines and in the other case the symmetric appearance of the spectrum is preserved.

A. Asymmetric Line Broadening

When the solvent viscosity is sufficiently high to impede molecular rotation the modulation of the anisotropic interactions by this reorientation provides the dominant spin relaxation process (Lemaire, 1967; Luckhurst and Pedulli, 1971). Except in a few rare cases, these interactions are second rank and so the dynamic perturbation required in the linewidth calculation can be written formally as

$$\mathscr{H}'(t) = \sum_{\mu;\, p,\, q} (-1)^p F_\mu^{(2,\, p)} \mathscr{D}^{(2)}_{q,\, -p}(t) T_\mu^{(2,\, q)} \tag{81}$$

where the Wigner rotation matrix connects the molecular and space-fixed coordinate systems. For nitroxide biradicals the subscript μ denotes the anisotropic Zeeman coupling, nitrogen hyperfine interaction, and electron–electron coupling. The calculation of the Redfield relaxation matrix for these three interactions is tedious; however, the bookkeeping can be reduced by assuming that the two electrons and two nuclei are completely equivalent as well as by ignoring the nonsecular terms in the dynamic perturbation. This can then be written as

$$\begin{aligned}\mathscr{H}'(t) = \sum_{p} (-1)^p \{ & g^{(2,p)}\beta B \mathscr{D}^{(2)}_{0,-p}(t)(2/3)^{1/2}(S_z^{(1)} + S_z^{(2)}) \\ & + A^{(2,p)}[\mathscr{D}^{(2)}_{0,-p}(t)(2/3)^{1/2}(I_z^{(1)}S_z^{(1)} + I_z^{(2)}S_z^{(2)}) \\ & + \mathscr{D}^{(2)}_{-1,-p}(t)(1/2)(I_+^{(1)}S_z^{(1)} + I_+^{(2)}S_z^{(2)}) \\ & + \mathscr{D}^{(2)}_{1,-p}(t)(1/2)(I_-^{(1)}S_z^{(1)} + I_-^{(2)}S_z^{(2)})] \\ & + D^{(2,p)}\mathscr{D}^{(2)}_{0,-p}(t)(8/3)^{1/2} \\ & \times [S_z^{(1)}S_z^{(2)} - \tfrac{1}{4}(S_+^{(1)}S_-^{(2)} + S_-^{(1)}S_+^{(2)})]\} \end{aligned} \tag{82}$$

Neglect of the nonsecular terms will, of course, only be valid when the rotational correlation time τ satisfies the inequality $4\pi^2 \nu_0^2 \tau^2 \gg 1$.

The size of the relaxation matrix is determined by the degeneracy of the transition and this, for a biradical, is a sensitive function of the ratio a/J. For example, in the strong exchange limit the central component of the five-line multiplet observed from a nitroxide biradical is sixfold degenerate because of the triple degeneracy of the nuclear spin states and the double degeneracy of the electron spin transitions. In contrast, when a is comparable to J each transition is only doubly degenerate. For those transitions between states with $m^{(1)}$ equal to $m^{(2)}$ this degeneracy is associated with the electron spin. However, for the other transitions the electron spin degeneracy is removed and the double degeneracy comes from the nuclear spin states. The detailed calculation of the elements of the relaxation matrix, whatever its size, is somewhat involved and since no new principles are introduced, we shall not be concerned with these details. In addition, the results of the calculation are complicated and so have been relegated to Appendix B. It is difficult to draw qualitative conclusions simply by inspection of these results, especially since the widths of the component lines for a doubly degenerate transition may differ and so cause the overall line shape to deviate from a Lorentzian.

In view of these difficulties we shall illustrate the theory by simulating spectra for the typical biradical **VIII** with the configuration shown, for various correlation times. The other parameters required by the linewidth calculations are the g and hyperfine tensors, which have been determined from single-crystal studies of the relevant monoradicals. The zero-field splitting tensor has not been measured for this biradical but can be estimated with the aid of the point-dipole approximation. The simulations are shown in Fig. 11 for three values of the rotational correlation time; when the motion is fast the spectrum exhibits a slight broadening of the spectral lines which is asymmetric with respect to the center of the spectrum. As the correlation time is increased the asymmetric linewidth variation becomes more pro-

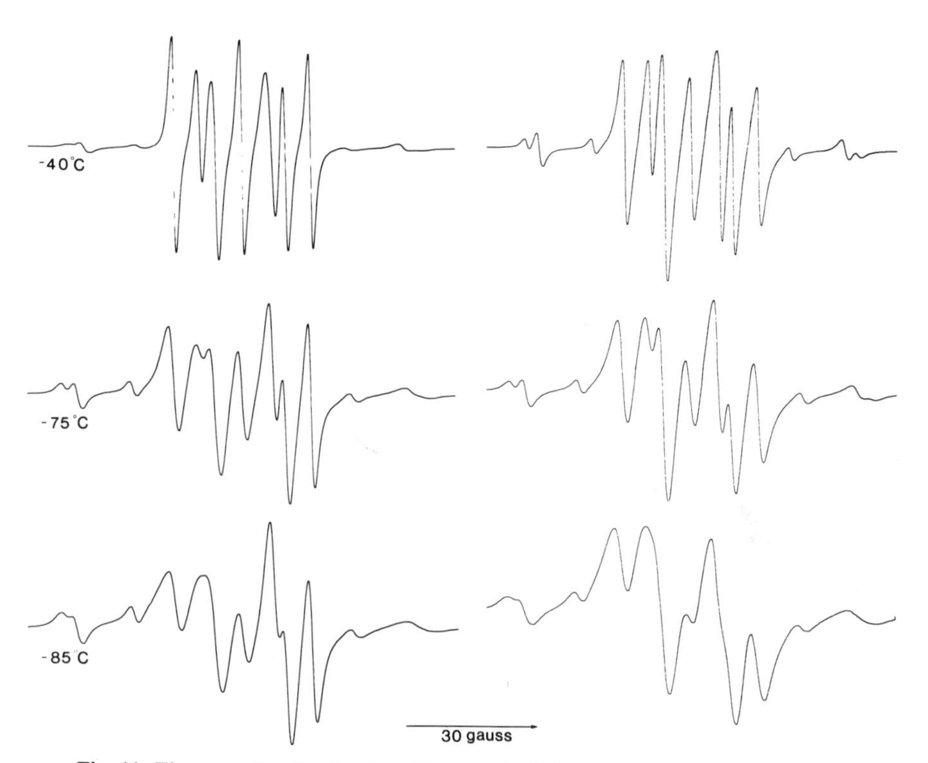

Fig. 11. Three spectra simulated to illustrate the influence of the asymmetric linewidth variation on the spectrum; also included are three spectra measured for the biradical **VIII**. (Luckhurst and Pedulli, 1971.)

nounced. Indeed the sharp features of the triplet spectrum rapidly disappear, although in contrast the singlet transitions become more important because their widths are less affected by this relaxation process. Figure 11 also includes, for comparison, the electron resonance spectrum of biradical **VIII** dissolved in toluene at three different temperatures. The general agreement between theory and experiment is good and clearly demonstrates the value of the theory in analyzing the spectra of biradical spin probes in viscous environments.

When the spin probe is rigid, the only unknown in the calculation is the rotational correlation time and so a comparison between theory and experiment simply yields τ. A more interesting situation obtains when the spin probe is nonrigid, for then the configuration and, hence, the zero-field splitting may be influenced by the environment and so are not known. Consequently analysis of the asymmetric line broadening should enable both $\mathbf{D}$ and τ to be determined even though the spin probe is in fluid solution. The sensitivity of the spectral appearance to the zero-field splitting is demonstrated by the following calculation. Biradical **VIII** is assumed to adopt a hypothetical configuration in which the piperidine rings are parallel and with the largest component of **D** parallel to that of the nitrogen hyperfine tensor. In other words the principal components of the zero-field splitting are unchanged although the orientation of the principal coordinate system with respect to that for the hyperfine tensor has been rotated through 90°. The spectrum, calculated for a correlation time of 7.23×10^{-10} sec, is shown in Fig. 12; it is quite unlike any of those obtained for the real configuration and so confirms the dramatic dependence of the spectrum on the zero-field splitting.

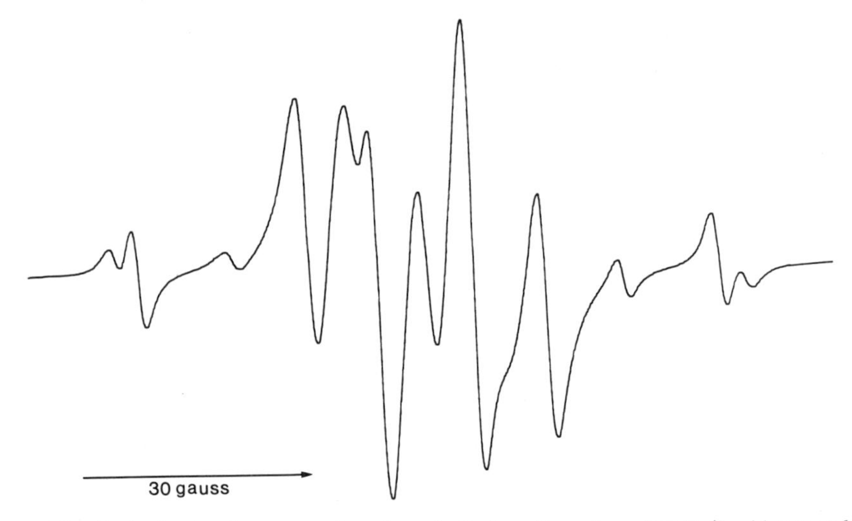

Fig. 12. A simulated spectrum for a hypothetical configuration of **VIII**. (Luckhurst and Pedulli, 1971.)

B. Symmetric Linewidth Variations

According to Eq. (72), the exchange integral is sensitive to the separation between the orbitals containing the unpaired electrons and so in a flexible biradical J will vary with the molecular configuration. This modulation of J by the intramolecular motion constitutes the dominant spin relaxation process for a nonrigid biradical dissolved in a low-viscosity solvent (Luckhurst, 1966). This behavior is not entirely unexpected, for as we saw in Section III.A, the positions of the spectral lines are dependent on J. Further, according to the theoretical spectra in Fig. 7, the singlet lines are extremely sensitive to the value of the exchange integral, whereas the triplet lines are less sensitive and indeed certain of these lines are independent of J altogether. When the modulation of the exchange integral is an important relaxation process we must therefore expect the singlet lines to be considerably broader than the triplet lines, while certain of these will remain unbroadened. We shall now seek to quantify these predictions.

It will again prove advantageous to consider a biradical described by an effective spin Hamiltonian in which the hyperfine interactions have been incorporated into an effective g factor. This approach is valid because the dynamic perturbation involves just the electron–electron interaction and so does not mix the nuclear spin states. The rate of J modulation is assumed to be fast in the sense that the static spin Hamiltonian, which determines the line positions, can be obtained by taking an average of the exchange integral over all molecular configurations. This time-averaged spin Hamiltonian is

$$\overline{\mathscr{H}} = g^{(1)}\beta B S_z^{(1)} + g^{(2)}\beta B S_z^{(2)} + \overline{J} S^{(1)} \cdot S^{(2)} \tag{83}$$

where the bar indicates a time or ensemble average. The dynamic perturbation is obtained by subtracting the static from the instantaneous spin Hamiltonian; this gives

$$\mathscr{H}'(t) = (J(t) - \overline{J}) S^{(1)} \cdot S^{(2)} \tag{84}$$

which has a zero time average. In general there are four allowed spin transitions for our effective biradical and they are listed in Eq. (60). These transitions are nondegenerate and so the relevant relaxation matrices contain a single element for which the expression given in Chapter 2 reduces to

$$R_{\alpha\alpha', \alpha\alpha'} = 2J_{\alpha\alpha, \alpha'\alpha'}(0) - \sum_{\gamma} \{J_{\alpha\gamma, \alpha\gamma}(\omega_{\alpha\gamma}) + J_{\alpha'\gamma, \alpha'\gamma}(\omega_{\alpha'\gamma})\} \tag{85}$$

for the transition between states $|\alpha\rangle$ and $|\alpha'\rangle$. In this problem the spectral density $J_{\alpha\beta, \alpha'\beta'}(\omega_{\alpha\alpha'})$ is

$$J_{\alpha\beta, \alpha'\beta'}(\omega_{\alpha\beta}) = \langle\alpha | S^{(1)} \cdot S^{(2)} | \beta\rangle\langle\alpha' | S^{(1)} \cdot S^{(2)} | \beta'\rangle^* j(\omega_{\alpha\beta}) \tag{86}$$

and $j(\omega_{\alpha\beta})$ is the Fourier transform of the autocorrelation function for $(J(t) - \overline{J})$, that is,

$$j(\omega_{\alpha\beta}) = \frac{1}{2}\int_{-\infty}^{\infty} \overline{(J(0) - \overline{J})(J(t) - \overline{J})} e^{-i\omega_{\alpha\beta}t}\, dt \tag{87}$$

This calculation gives the widths of the two triplet transitions as

$$T_2^{-1}(\mathrm{T}) = j(0) \sin^4 \phi + j(\Delta\omega)\tfrac{1}{4} \sin^2 2\phi \tag{88}$$

where

$$\Delta\omega = J + \Delta g\beta \tan \phi \tag{89}$$

which is the frequency separating states $|0_+\rangle$ and $|0_-\rangle$. There is an analogous expression for the widths of the singlet lines:

$$T_2^{-1}(\mathrm{S}) = j(0)\cos^4\phi + j(\Delta\omega)\tfrac{1}{4}\sin^2 2\phi \tag{90}$$

It is now possible to see why the singlet satellite lines may be difficult to detect for flexible biradicals (Glarum and Marshall, 1967). Consider the situation when $\Delta g\beta B/J$ is one-half; then the widths of the triplet and singlet lines are calculated to be

$$T_2^{-1}(\mathrm{T}) = 0.0022j(0) + 0.038j(\Delta\omega) \tag{91}$$

and

$$T_2^{-1}(\mathrm{S}) = 0.62j(0) + 0.038j(\Delta\omega) \tag{92}$$

Since $j(\Delta\omega)$ can never be greater than $j(0)$, the widths of the singlet lines could well be orders of magnitude greater than those of the triplet lines. The absolute linewidths will depend on the magnitude of the spectral density $j(0)$, which will, in turn, be a function of the rate at which the exchange integral is modulated as well as the values which it adopts. It is, however, difficult to calculate $j(\omega)$ in any rigorous fashion because the exact dependence of J on the molecular configuration is not available and in addition the nature of the motion between configurations is unknown. The qualitative features of $j(\omega)$ can be obtained by appealing to a mathematically tractable but possibly naive model in which the biradical is assumed to exist in just two configurations, each with its own lifetime and exchange integral. An entirely similar problem was considered in Chapter 2 and we can use these results to write the spectral density as

$$j(\omega) = (J_1 - J_2)^2 \frac{\tau^3}{(\tau_1 + \tau_2)^2} \frac{1}{1 + \omega^2\tau^2} \tag{93}$$

where τ is the geometric mean of the individual lifetimes. The magnitude of the spectral density therefore depends on the difference in the exchange integral for the two configurations and is essentially proportional to the lifetime. The model could, in principle, be made more realistic by increasing the number of configurations, although this would increase the number of adjustable parameters in the theory. From a pragmatic point of view such extensions are not warranted, for, at best, the spectrum in the limit of rapid modulation only provides us with the two spectral densities $j(0)$ and $j(\Delta\omega)$. Finally, we consider the width of the triplet line in the strong exchange limit when $\Delta g\beta B/\bar{J}$ is small. This limiting width is of considerable interest, for many flexible biradicals are found to exhibit strong exchange. By taking the limit of Eqs. (88) and (89) for small $\Delta g \beta B/\bar{J}$ the width of the triplet line is found to be

$$T_2^{-1} = (\Delta g^2 \beta^2 B^2 / 4\bar{J}^2) j(\bar{J}) \tag{94}$$

We can now proceed to consider the linewidths for nitroxide biradicals, which exhibit hyperfine structure, simply by replacing the g-factor difference in Eqs. (88), (90), and (94) by the difference in the effective g factors. As an example of this procedure, we consider a symmetric nitroxide biradical exhibiting strong spin exchange with intramolecular motions modulating the exchange integral. The spectrum therefore contains five lines with intensities 1 : 2 : 3 : 2 : 1 and their widths are obtained by substituting the effective g factors defined by Eqs. (67) and (68) into Eq. (94), as

$$T_2^{-1}(m^{(1)}, m^{(2)}) = [a^2(m^{(1)} - m^{(2)})^2 / 4\bar{J}^2] j(\bar{J}) \tag{95}$$

The linewidths calculated from this expression for the nine transitions corresponding to the nine combinations of $m^{(1)}$ and $m^{(2)}$ are listed in Table III. Since the linewidth depends on the square of $(m^{(1)} - m^{(2)})$, lines whose total quantum numbers are related by a reversal of sign will have the same width and consequently this broadening mechanism preserves the symmetry about the center of the spectrum; this is readily seen from the linewidths listed in Table III. The three lines for which $m^{(1)}$ is equal to $m^{(2)}$ are not broadened by

TABLE III

SYMMETRIC LINEWIDTH VARIATIONS

M_I	$m^{(1)}$	$m^{(2)}$	$T_2^{-1}/[a^2 j(\bar{J})/4\bar{J}^2]$
2	1	1	0
1	1	0	1
	0	1	1
0	0	0	0
	1	−1	4
	−1	1	4
−1	−1	0	1
	0	−1	1
−2	−1	−1	0

modulation of the exchange integral. However, although one component of the central line is unbroadened, the widths of the other two are increased considerably and we can now see why the width of this line may not provide a reliable method for determining the exchange integral. The lines with $M_I = \pm 1$ are broadened by the relaxation process and so the widths of the spectral lines will alternate. An example of such linewidth alternation is

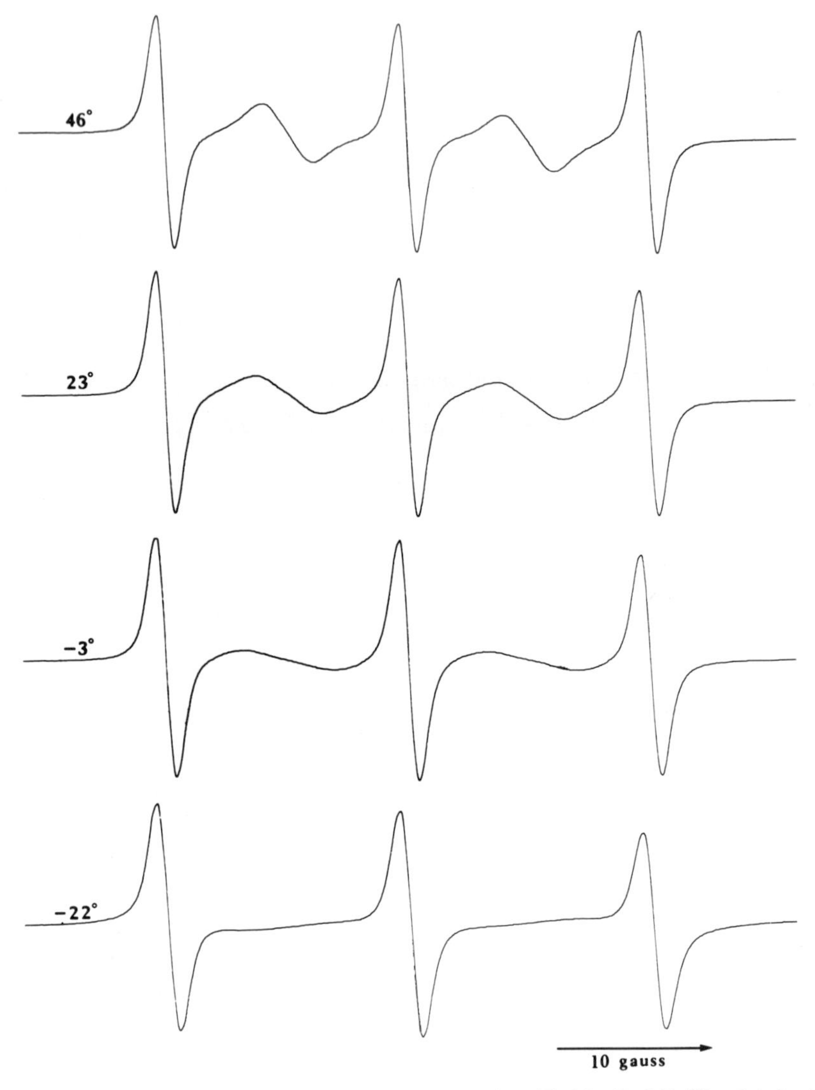

Fig. 13. Symmetric linewidth variations for the nitroxide biradical **X**. [Reprinted with permission from Luckhurst and Pedulli, *J. Amer. Chem. Soc.* **92**, 4738–4739 (1970). Copyright by the American Chemical Society.]

provided by the spectra, shown in Fig. 13, for the nitroxide biradical **X** dissolved in toluene (Luckhurst and Pedulli, 1970). Even at 46°C the relative line heights are not in the ratio of 1 : 2 : 3 : 2 : 1 expected when the linewidths are equal; indeed only one component of the central line can be observed. The extreme spectral lines corresponding to $M_I = \pm 2$ are also unbroadened, while the peaks with $M_I = \pm 1$ are rather broad. As the temperature is reduced the widths of the $M_I = \pm 1$ lines increase still further, although those of the other peaks remain unchanged. At $-22°C$ only the three lines with $m^{(1)}$ equal to $m^{(2)}$ are observed and this spectrum is identical to that expected for a biradical exhibiting weak exchange. Clearly, considerable care must be exercised when interpreting a biradical spectrum that is identical to that for the corresponding monoradical (Lemaire *et al.*, 1968). According to Eq. (95), these spectral changes reflect the temperature dependence of the ratio $j(\bar{J})/\bar{J}^2$ and it is not possible to associate them with either variations in $j(\bar{J})$ or $\bar{J}$ alone.

The preceding linewidth analysis is based on Redfield's relaxation theory and so will only be valid when the exchange interaction is modulated rapidly by the intramolecular motion. If this motion is not fast, then theoretical techniques analogous to those described in Chapter 3 must be employed to calculate not only the widths of the spectral lines but also their positions. However, a rigorous calculation is not possible with a flexible biradical, for exactly the same reasons that the spectral density $j(\omega)$, defined by Eq. (87), cannot be evaluated. We are therefore forced to perform calculations based on the simple two-site model of intramolecular motion. The line-shape theory is relatively straightforward (Parmon and Zhidomirov, 1974) and can be simplified by the use of an effective spin Hamiltonian. The details of the theory is outside the scope of this chapter and we must be content to quote some of the more pertinent results. These can be illustrated by reference to the calculation for a nitroxide biradical interconverting between two conformations with equal lifetimes; in one configuration the exchange integral is zero and for the other it is 30 times the nitrogen hyperfine interaction. The spectra shown in Fig. 14 were calculated for a range of lifetimes τ, which are listed in the figure as the dimensionless parameter $a\tau_1$. The average exchange integral $\bar{J}$ is $15a$ and so when the intramolecular motion is fast the spectrum corresponds to the strong exchange limit; this situation can be seen in Fig. 14g. As the lifetime is increased certain lines in the simulated spectra are observed to broaden, in complete accord with the predictions of Eq. (95). It is also of interest to note the similarity between the theoretical spectra and the experimental spectra shown in Fig. 13. The spectrum given in Fig. 14c contains essentially three lines of equal intensity because of the extreme widths of the other peaks; however, as the lifetime is increased still further new lines appear in the spectra, as we can see from Figs. 14b and 14a. The origin of these new peaks is readily understood, for in the limit of slow conversion the observed spectrum will be the sum of the spectra from the two configurations. For one of these the exchange integral is zero and it therefore has a three-line spectrum, whereas the other configuration corresponds to strong spin exchange and has a five-line spectrum. It is the peaks

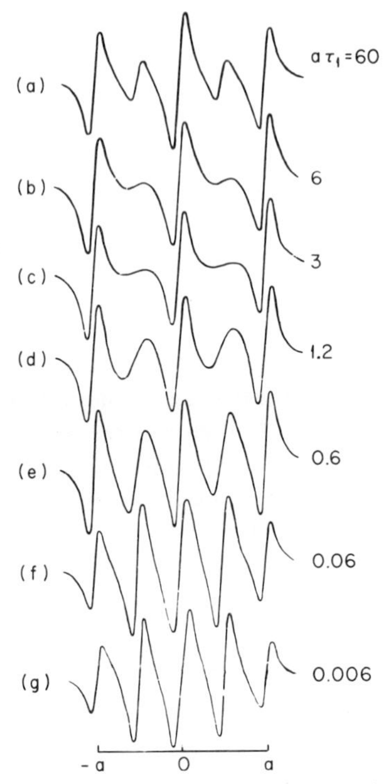

Fig. 14. The theoretical spectrum of a nitroxide biradical capable of existing in two configurations calculated for both fast and slow rates of intramolecular motion. (Parmon and Zhidomirov, 1974.)

from the latter spectrum that are responsible for the reappearance of the spectral lines. These model calculations therefore imply that for flexible biradicals apparently similar spectra may be obtained for two quite different situations; in one the rate of J modulation is fast and in the other it is slow. Consequently considerable caution must be exercised in the analysis of a solitary biradical spectrum and ideally spectra should be obtained for a range of temperatures to assist in the assignment of the rate for the molecular dynamics.

V. CONCLUSIONS

In this chapter we have been concerned with those magnetic interactions peculiar to a biradical and have seen how they can influence the appearance of its electron resonance spectrum. From our analysis it is clear that there are at least four ways in which a biradical could prove valuable as a spin probe. The first of these is to determine the rate of molecular reorientation from linewidth measurements. Since the correlation time that can be determined by such techniques depends on the anisotropy in the magnetic interactions, nitroxide biradicals are more useful than the corresponding monoradical. This advantage accrues because the zero-field splitting in a biradical can be varied simply by changing the separation between the two nitroxide groups, whereas the g and hyperfine tensors for a nitroxide are essentially independent of the molecular structure.

The other applications as spin probes would seem to be unique to biradicals and so there is no direct conflict with the use of monoradicals. The ability to investigate the molecular configuration of the biradical from a knowledge of the exchange integral is important. Of course the detailed molecular geometry cannot be inferred from the magnitude of J but changes in the exchange integral will reflect conformational variations induced by an external stimulus. In addition the rate and nature of the intramolecular motion can often be deduced from the linewidths; consequently changes in the dynamics caused by a specific perturbation to the system can also be examined. Finally, the ability to determine the zero-field splitting tensor from the line broadening in moderately viscous systems is important because the interelectron separation can be determined and so the effect of the environment on the molecular geometry investigated.

Nitroxide biradicals would therefore appear to possess unique properties which make them valuable as environmental spin probes and yet their applications to real problems are rare. It is not too difficult to see why they should be so unpopular in comparison with their monoradical counterparts. In the first instance the additional magnetic interactions in the biradical often make its electron resonance spectrum complex and the lines poorly resolved. Further, the analysis of such a spectrum need not be straightforward or even unambiguous. The lack of suitable biradicals for which the exchange integrals can be determined from their spectra may also be a reason for their unpopularity. Nonetheless, it is to be expected that as the theory demanded for the spectral analysis becomes more appreciated and as the range of biradicals is extended their obvious merits as spin probes will be exploited.

APPENDIX A. THE SPIN HAMILTONIAN

Here we seek to justify the form of the spin Hamiltonian given in Eq. (63) for a biradical exhibiting hyperfine structure. The scalar coupling between the electron and nuclear spins is represented in the Hamiltonian by the Fermi contact interaction, which was described in Chapter 2. For a biradical this takes the form

$$\mathscr{H}_c = (8\pi/3)g\beta g_N\beta_N \sum_i [\delta(\mathbf{r}_1 - \mathbf{r}_i)S^{(1)} + \delta(\mathbf{r}_2 - \mathbf{r}_i)S^{(2)}] \cdot I^{(i)} \quad (A.1)$$

where the summation is restricted to nuclei in one-half of the biradical; there is a comparable summation, which has been omitted, for the other half. Here g_N is the nuclear g factor, β_N is the nuclear magneton, and $\delta(\mathbf{r}_1 - \mathbf{r}_i)$ vanishes unless the coordinates $\mathbf{r}$ of electron 1 and nucleus i coincide. This Hamiltonian contains both spatial and spin operators; since we require a spin Hamiltonian, the spatial operators must be removed by evaluating their matrix elements. The molecular orbitals $|a\rangle$ and $|b\rangle$ containing the unpaired electrons can be combined to form symmetric,

$$|+\rangle = (1/\sqrt{2})(|a(1)b(2)\rangle + |a(2)b(1)\rangle) \tag{A.2}$$

and antisymmetric,

$$|-\rangle = (1/\sqrt{2})(|a(1)b(2)\rangle - |a(2)b(1)\rangle) \tag{A.3}$$

functions. The total wave function is obtained by combining these spatial functions with spin functions in such a way that the total function is antisymmetric with respect to the interchange of the electron labels. Consequently the symmetric state $|+\rangle$ can only combine with the antisymmetric singlet spin function and the antisymmetric state $|-\rangle$ must combine with the triplet spin functions. To evaluate the matrix element of the hyperfine spin Hamiltonian between the triplet spin functions we have first to obtain the matrix element of the spatial operators between the $|-\rangle$ states:

$$\langle -|\mathscr{H}_c|-\rangle = (8\pi/3)g\beta g_N\beta_N \sum_i \tfrac{1}{2}[\langle a|\delta(\mathbf{r}_i)|a\rangle + \langle b|\delta(\mathbf{r}_i)|b\rangle$$
$$-2\langle a|\delta(\mathbf{r}_i)|b\rangle\langle a|b\rangle]I^{(i)}\cdot(S^{(1)} + S^{(2)}) \tag{A.4}$$

together with a corresponding term for the other half of the biradical. Here the symbol $\langle a|\delta(\mathbf{r}_i)|a\rangle$ implies that the square of the wave function is evaluated at nucleus i. Before considering the matrix elements for the other states we shall simplify the expression in Eq. (A.4). If the orbitals containing the unpaired electrons are orthogonal, then the last term in the square brackets vanishes. In addition, if these orbitals are spatially separated, then when the summation is over nuclei in the fragment associated with orbital $|a\rangle$ the value of the wave function $|b\rangle$ at nucleus i will be negligibly small. The matrix element reduces to

$$\langle -|\mathscr{H}_c|-\rangle = \sum_i \tfrac{1}{2}a^{(i)}I^{(i)}\cdot(S^{(1)} + S^{(2)}) \tag{A.5}$$

where

$$a^{(i)} = (8\pi/3)g\beta g_N\beta_N\langle a|\delta(\mathbf{r}_i)|a\rangle \tag{A.6}$$

and is identical to the coupling constant for the ith nucleus in the corresponding monoradical.

Similar arguments show that the operator required for evaluating the matrix element for the singlet state is also given by Eq. (A.5). In contrast the spin operator employed in calculating the matrix element between the triplet and singlet spin functions is

$$\langle + | \mathscr{H}_c | - \rangle = \sum_i \tfrac{1}{2} a^{(i)} I^{(i)} \cdot (S^{(1)} - S^{(2)}) \tag{A.7}$$

Since the electron spin operator $(S^{(1)} - S^{(2)})$ only connects triplet with singlet spin levels, we can obtain an equivalent hyperfine Hamiltonian to operate on any spin function simply by adding Eqs. (A.5) and (A.7). This gives

$$\mathscr{H}_c = \sum_i a^{(i)} I^{(i)} \cdot S^{(1)} \tag{A.8}$$

with a corresponding term for the nuclei in the fragment associated with orbital $|b\rangle$. We have achieved our objective, for this expression is identical to the hyperfine operator given in Eq. (63).

In conclusion we note that when the conditions concerning the overlap and extent of the molecular orbitals are relaxed, the coefficients in Eqs. (A.5) and (A.7) are different and not equal to the coupling constant for the appropriate monoradical.

APPENDIX B. THE LINEWIDTHS FOR A NITROXIDE BIRADICAL

We list here the equations for the linewidths of a nitroxide biradical when the dominant spin relaxation process results from the molecular rotation coupled to the anisotropy in the magnetic interactions.

In the strong exchange limit the linewidth for the transition

$$|M_s; m^{(1)} m^{(2)}\rangle \leftrightarrow |0_a; m^{(1)} m^{(2)}\rangle \tag{B.1}$$

where M_s takes the values ± 1, is

$$\begin{aligned} T_2^{-1} = \frac{\tau}{5}\bigg\{ & \frac{2}{3}\frac{\beta^2 B^2}{h^2}(g:g) + \frac{M_s \beta B}{h}(g:D) + \frac{3}{8}(D:D) \\ & + \frac{2M_I \beta B}{3h}(g:A) + \frac{M_I M_s}{2}(A:D) + \frac{M_I^2}{6}(A:A)\bigg\} \\ & + \frac{(A:A)}{10}\bigg\{\frac{\Delta m^2}{3} j(J) \\ & + \tfrac{1}{2}[2I(I+1) - (m^{(1)2} + m^{(2)2})]\left[j\left(\frac{a}{2}\right) + j(J)\right]\bigg\} \end{aligned} \tag{B.2}$$

The symbol $(X:Y)$ denotes the inner product of the two second-rank tensors **X** and **Y**:

$$(X:Y) = \sum_q X^{(2,q)} Y^{(2,q)*} \tag{B.3}$$

The spectral density $j(\omega)$ is defined by

$$j(\omega) = \tau/(1 + \omega^2\tau^2) \tag{B.4}$$

where τ is the rotational correlation time; when applying Eq. (B.2) the argument ω is usually assumed to be small compared with τ and the spectral densities are set equal to this correlation time.

For intermediate values of the exchange integral the contribution of the secular terms in the dynamic spin Hamiltonian to the widths of the triplet and singlet lines is

$$\begin{aligned}
T_2^{-1(\mathrm{sec})}\begin{pmatrix}\mathrm{T}\\ \mathrm{S}\end{pmatrix} = \frac{\tau}{5}\bigg\{&\frac{2\beta^2B^2}{3h^2}(g:g) + \frac{\beta BM_s}{3h}\begin{pmatrix}2+\cos 2\phi\\ 2-\cos 2\phi\end{pmatrix}(g:D) \\
&+ \begin{pmatrix}(2+\cos 2\phi)^2\\ (2-\cos 2\phi)^2\end{pmatrix}\frac{(D:D)}{24} \\
&+ \frac{2\beta B}{3h}\begin{pmatrix}M_I - M_s\,\Delta m \sin 2\phi\\ M_I + M_s\,\Delta m \sin 2\phi\end{pmatrix}(g:A) \\
&+ \begin{pmatrix}M_IM_s(2+\cos 2\phi) - \Delta m \sin 2\phi(2+\cos 2\phi)\\ M_IM_s(2-\cos 2\phi) + \Delta m \sin 2\phi(2-\cos 2\phi)\end{pmatrix}\frac{(A:D)}{6} \\
&+ \begin{pmatrix}(M_I - M_s\,\Delta m \sin 2\phi)^2\\ (M_I + M_s\,\Delta m \sin 2\phi)^2\end{pmatrix}\frac{(A:A)}{6}\bigg\} \\
&+ \frac{1}{5}j(J\sec 2\phi)\bigg\{\sin^2\phi\,\frac{(D:D)}{24} \\
&+ \Delta m \sin 4\phi\,\frac{(A:D)}{12} + \Delta m^2\cos^2 2\phi\,\frac{(A:A)}{6}\bigg\}
\end{aligned} \tag{B.5}$$

The pseudosecular terms in $\mathscr{H}'(t)$ make the same contribution to the widths of both triplet and singlet lines:

$$\begin{aligned}
T_2^{-1(\mathrm{ps\text{-}sec})} = \tfrac{1}{20}(A:A)\Big[&\tfrac{1}{2}\{2I(I+1) - (m^{(1)2} + m^{(2)2})\} \\
&\times [j(\tfrac{1}{2}a) + \tfrac{1}{2}\{\sin^2(\phi+\phi')j(\omega'_-) \\
&+ \sin^2(\phi+\phi'')j(\omega''_+)\}] - \tfrac{1}{4}M_I\{\sin^2(\phi+\phi')j(\omega'_-) \\
&- \sin^2(\phi+\phi'')j(\omega''_-) + \cos^2(\phi+\phi')j(\omega'_+) \\
&- \cos^2(\phi+\phi'')j(\omega''_+)\}\Big]
\end{aligned} \tag{B.6}$$

The new symbols in this equation are

$$\tan 2\phi' = a(\Delta m + 1)/J \tag{B.7}$$

$$\tan 2\phi'' = a(\Delta m - 1)/J \tag{B.8}$$

$$\omega'_{\pm} = \tfrac{1}{2}J(\sec 2\phi' \pm \sec 2\phi) \tag{B.9}$$

with analogous expressions for $\omega''_{\pm}$. For nitroxide biradicals both a and J are usually small compared with the inverse of the rotational correlation time and so the expression in Eq. (B.6) can be reduced to

$$T_2^{-1(\text{ps-sec})} = \tfrac{1}{20}\tau\{2I(I+1) - (m^{(1)2} + m^{(2)2})\}(A:A) \tag{B.10}$$

The total linewidth is obtained by adding Eq. (B.10) to Eq. (B.5) after replacing $j(J \sec 2\phi)$ by τ.

References

Atherton, N. M., and Strach, S. J. (1972). E.s.r. study of aqueous sodium dodecyl sulphate solutions using di-*t*-butyl nitroxide as a probe, *J. C. S. Faraday II* **68**, 374–381.

Briere, R., Dupeyre, R., Lemaire, H., Morat, C., Rassat, A., and Rey, P. (1965). Nitroxydes XVII: biradicaux stables du type nitroxyde, *Bull. Soc. Chim. Fr.* **1965**, 3290–3297.

Brooks, S. A., Luckhurst, G. R., and Pedulli, G. F. (1971). Thermal fluctuations in the nematic mesophase, *Chem. Phys. Lett.* **11**, 159–162.

Calder, A., Forrester, A. R., James, P. G., and Luckhurst, G. R. (1969). Nitroxide radicals. Part V. *N,N'*-Di-*t*-butyl-*m*-phenylenebinitroxide, a stable triplet, *J. Amer. Chem. Soc.* **91**, 3724–3727.

Calvin, M., Wang, H. H., Entine, G., Gill, D., Ferruti, P., Harpold, M. A., and Klein, M. P. (1969). Biradical spin labelling for nerve membranes, *Proc. Nat. Acad. Sci. U.S.* **63**, 1–8.

Carrington, A., and Luckhurst, G. R. (1964). Electron spin resonance line widths of transition metal ions in solution. Relaxation through zero-field splitting. *Mol. Phys.* **8**, 125–132.

Carrington, A., and McLachlan, A. D. (1967). "Introduction to Magnetic Resonance." Harper, New York.

Ciercierska-Tworek, Z., Van, S. P., and Griffith, O. H. (1973). Electron–electron dipolar splitting anisotropy of a dinitroxide oriented in a crystalline matrix, *J. Mol. Struct.* **16**, 139–148.

Corvaja, C., Giacometti, G., Kopple, K. D., and Ziauddin. (1970). Electron spin resonance studies of nitroxide radicals and biradicals in nematic solvents, *J. Amer. Chem. Soc.* **92**, 3919–3924.

Falle, H. R., and Luckhurst, G. R. (1970). The electron resonance spectra of partially oriented radicals, *J. Magn. Res.* **3**, 161–199.

Falle, H. R., Luckhurst, G. R., Lemaire, H., Marechal, Y., Rassat, A., and Rey, P. (1966). The electron resonance of ground state triplets in liquid crystal solutions, *Mol. Phys.* **11**, 49–56.

Ferruti, P., Gill, D., Klein, M. P., and Calvin, M. (1969). Correlation between conformation and pairwise spin exchange in flexible biradicals in solution. Control of conformation by pH-dependent ionic forces, *J. Amer. Chem. Soc.* **91**, 7765–7766.

Ferruti, P., Gill, D., Klein, M. P., Wang, H. H., Entine, G., and Calvin, M. (1970). Synthesis of mono-, di-, and polynitroxides. Classification of electron spin resonance spectra of flexible dinitroxides dissolved in liquids and glasses, *J. Amer. Chem. Soc.* **92**, 3704–3713.

Giroud, A. M., Rassat, A., and Sieveking, H. U. (1974). Nitroxydes LXV. Mono et biradicaux derives de l'isoindoline, *Tetrahedron Lett.* **8**, 635–638.

Glarum, S. H., and Marshall, J. H. (1967). Spin exchange in nitroxide biradicals, *J. Chem. Phys.* **47**, 1374–1378.

Hsia, J. C., Kosman, D. J., and Piette, L. H. (1969). Organophosphate spin-label studies of inhibited esterases, α-chymotrypsin and cholinesterase, *Biochem. Biophys. Res. Commun.* **36**, 75–78.

Humphries, R. L., James, P. G., and Luckhurst, G. R. (1971). A molecular field treatment of liquid crystalline mixtures. *Symp. Faraday Soc.* **5**, 107–118.

Hwang, J., Kivelson, D., and Plachy, W. (1973). Esr linewidths in solution VI. Variation with pressure and study of functional dependence of anisotropic interaction parameter, κ, *J. Chem. Phys.* **58**, 1753–1765.

Keana, J. F. W., and Dinerstein, R. J. (1971). A new highly anisotropic dinitroxide ketone spin label. A sensitive probe for membrane structure, *J. Amer. Chem. Soc.* **93**, 2808–2810.

Lemaire, H. (1967). Nitroxydes XX. Résonance paramagnétique d'un biradical nitroxyde; détermination du signe de l'echange, *J. Chim. Phys.* **64**, 559–571.

Lemaire, H., Rassat, A., Rey, P., and Luckhurst, G. R. (1968). Le téréphthalate de di(tétraméthyl-2,2,6,6 pipéridinyl-4 oxyle-1) est-il un biradical a echange fort ou a echange faible? *Mol. Phys.* **14**, 441–447.

Levanon, H., Stein, G., and Luz, Z. (1970). E.s.r. study of complex formation and electronic relaxation of Fe^{3+} in aqueous solutions, *J. Chem. Phys.* **53**, 876–887.

Low, W. (1960). "Paramagnetic Resonance in Solids." Academic Press, New York.

Luckhurst, G. R. (1966). Alternating linewidths. A novel relaxation process in the electron resonance of biradicals, *Mol. Phys.* **10**, 543–550.

Luckhurst, G. R. (1970). Electron resonance in anisotropic solvents, *R.I.C. Rev.* **3**, 61–84.

Luckhurst, G. R., and Pedulli, G. R. (1970). Interpretation of biradical electron resonance spectra, *J. Amer. Chem. Soc.* **92**, 4738–4739.

Luckhurst, G. R., and Pedulli, G. F. (1971). Asymmetric line broadening in the electron resonance spectra of biradicals, *Mol. Phys.* **20**, 1043–1055.

Luckhurst, G. R., and Poupko, R. (1974). Unpublished results.

Luckhurst, G. R., and Sanson, A. (1972). Angular dependent linewidths for a spin probe dissolved in a liquid crystal, *Mol. Phys.* **24**, 1297–1311.

Luckhurst, G. R., Poupko, R., and Zannoni, C. (1975). Spin relaxation for biradical spin probes in anisotropic environments, *Mol. Phys.* (in press).

McConnell, H. M. (1960). Theory of singlet–triplet splittings in large biradicals, *J. Chem. Phys.* **33**, 115–121.

McWeeny, R. (1961). Zero-field splitting of molecular Zeeman levels, *J. Chem. Phys.* **34**, 399–401.

Michon, P. (1970). Thesis, Univ. of Grenoble, France.

Michon, J., and Rassat, A. (1974). Nitroxides LIX. Rotational correlation time determination of nitroxide biradical application to solvation studies, *J. Amer. Chem. Soc.* **96**, 335–337.

Nakajima, A. (1973). Magnetic properties of some iminoxyl polyradicals. V. Epr studies of the TEMPAD biradical, *Bull. Chem. Soc. Japan* **46**, 1129–1134.

Nakajima, A., Ohya-Nishiguchi, H., and Deguchi, Y. (1972). Magnetic properties of some iminoxyl polyradicals. III. Exchange interaction in iminoxyl biradicals, *Bull. Chem. Soc. Japan* **45**, 713–716.

Norris, J. R., and Weissman, S. I. (1969). Studies of rotational diffusion through electron–electron dipolar interaction, *J. Phys. Chem.* **73**, 3119–3124.

Oakes, J. (1972). Magnetic resonance studies in aqueous systems. Part 1. Solubilization of spin probes by micellar solution, *J. C. S. Faraday II* **68**, 1464–1471.

Ohnishi, S., Cyr, T. J. R., and Fukushima, H. (1970). Biradical spin-labeled micelles, *Bull. Chem. Soc. Japan* **43**, 673–676.

Parmon, V. N., and Zhidomirov, G. M. (1974). Calculation of the e.s.r. spectrum shape of a dynamic biradical system, *Mol. Phys.* **27**, 367–375.

Pullman, A., and Kochanski, E. (1967). *Int. J. Quant. Chem.* **1**, 251.

Rassat, A. (1971). Application of electron spin resonance to conformational analysis, *Pure Appl. Chem.* **25**, 623–634.

Rassat, A. (1972). Private communication.

Reitz, D. C., and Weissman, S. I. (1960). Spin exchange in biradicals, *J. Chem. Phys.* **33**, 700–704.

Saupe, A. (1964). Kernresonanzen in kristallinen Flussigkeiten und in kristallinflussigen Losungen, Teil I., *Z. Naturforsch.* **19a**, 161–171.

Slichter, C. P. (1955). Spin resonance of impurity atoms in silicon, *Phys. Rev.* **99**, 479–480.

Weissman, S. I. (1958). On detection of triplet molecules in solution by electron spin resonance, *J. Chem. Phys.* **29**, 1189–1190.

5

The Chemistry of Spin Labels

BETTY JEAN GAFFNEY

DEPARTMENT OF CHEMISTRY
THE JOHNS HOPKINS UNIVERSITY
BALTIMORE, MARYLAND

Much of the organic chemistry of nitroxides as stable free radicals has been covered in several books and articles (Buchachenko, 1965; Forrester *et al.*, 1968; Nelson, 1973; Rozantsev, 1970; Rozantsev and Scholle, 1971). The stability of the free radical portion of nitroxides under synthetic conditions that are aimed at the non-free radical portion of the molecule and also in biological samples is probably the most important aspect of nitroxide chemistry for spin-label studies. This subject is covered in Section I of this chapter. The synthesis of the spin labels often occupies a major part of the time devoted to a spin-label investigation of a problem of biological or biophysical interest. Section II of this chapter summarizes the types (i.e., spin-labeled protein modification reagents, spin-labeled nucleotides, carbohydrate spin labels, etc.) of spin labels that have been synthesized and pertinent considerations for the syntheses. Section III presents the experimental procedures for preparation of a representative sampling of spin labels. Many of these procedures have not been published previously.

I. THE STABILITY OF THE PARAMAGNETIC NITROXIDE GROUP IN SPIN LABELS

Although nitroxides are extraordinarily stable free radicals, they may be destroyed, with loss of paramagnetism, by components of some biological systems and under some of the experimental conditions employed in synthetic steps aimed at the nonparamagnetic portions of a spin-label molecule. The reactions, which involve the unpaired electron of nitroxides, have been discussed in detail (Buchachenko, 1965; Forrester *et al.*, 1968; Nelson, 1973; Rozantsev, 1970; and Rozantsev and Scholle, 1971; Corker *et al.*, 1966). In summary, nitroxides have been observed to undergo the following types of reactions:

(1) Disproportionation (Dupeyre and Rassat, 1966)
(2) Free-radical reaction at oxygen (Keana *et al.*, 1971)
(3) Carbon–nitrogen bond cleavage (Keana and Baitis, 1968)
(4) One-electron oxidation (Golubev *et al.*, 1965)
(5) One-electron reduction (Golubev *et al.*, 1965; Neimann *et al.*, 1964)
(6) Reaction with strong acids (Neimann *et al.*, 1964).

$$2\ -\overset{|}{\underset{|}{C}}-\overset{}{\underset{\underset{O}{|\cdot}}{N}}-\overset{|}{\underset{\underset{H}{|}}{C}}- \longrightarrow -\overset{|}{\underset{|}{C}}-\underset{\underset{HO}{|}}{N}-\overset{|}{\underset{\underset{H}{|}}{C}}- + -\overset{|}{\underset{|}{C}}-\underset{\underset{O_-}{|}}{\overset{+}{N}}=C\Big< \tag{1}$$

Disproportionation

$$-\overset{|}{\underset{|}{C}}-\underset{\underset{O}{|\cdot}}{N}-\overset{|}{\underset{|}{C}}- + R\cdot \longrightarrow -\overset{|}{\underset{|}{C}}-\underset{\underset{\underset{R}{|}}{O}}{|}{N}-\overset{|}{\underset{|}{C}}- \tag{2}$$

Free radical reaction

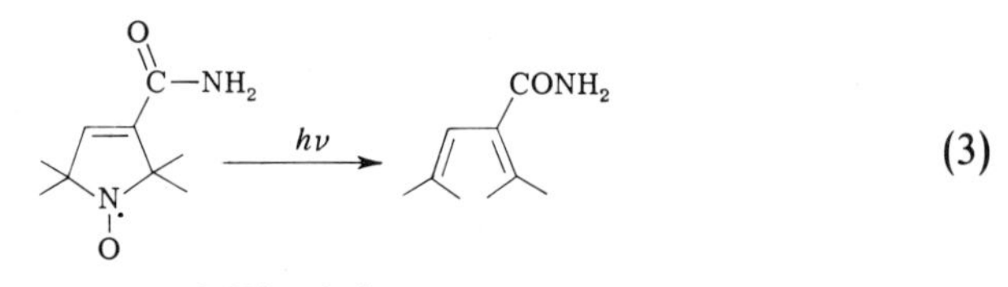

C-N bond cleavage

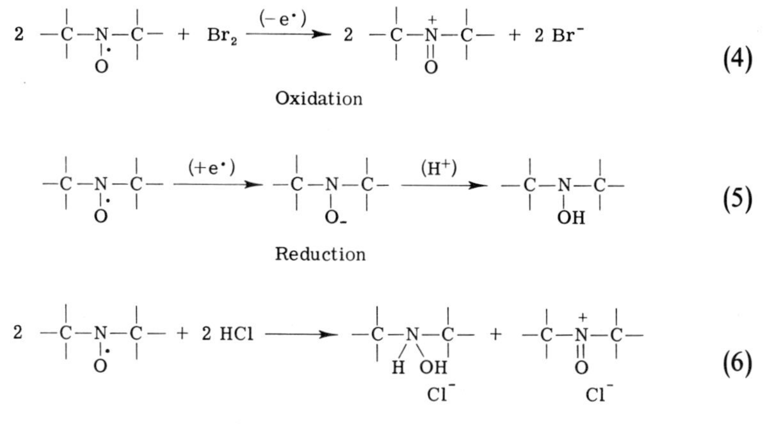

Reaction (1) may be easily disposed of as a problem for spin labeling by using only nitroxides that bear tertiary carbon atoms adjacent to the nitroxide nitrogen. [The paramagnetic resonance spectra of the unstable nitroxides bearing α hydrogens have additional characteristic splittings, however, and this gives rise to the interesting subject of "spin trapping" whereby transient free radicals are identified (Janzen, 1971; Adevih and Lagercrantz, 1970).] Reactions (2)–(4) are primarily of concern in the synthesis of spin labels in that they may compete with the nonradical portion of the nitroxide molecule for consumption of reagents and may lead to unwanted products. Reaction (3) is a special case of photolysis of stable nitroxides (Keana *et al.*, 1971; Keana and Baitis, 1968). In the nitroxide ring structures that have no unsaturation, reactions of type (2) account for the predominant products in photochemical reactions (Keana *et al.*, 1971).

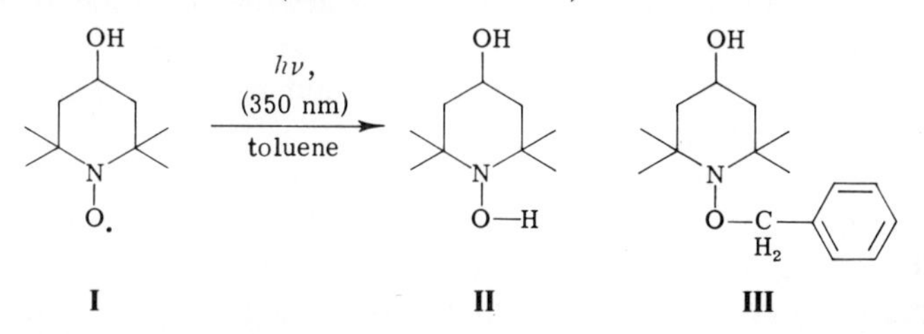

From the spin-labeling point of view, reactions (5) and (6) present the most serious complications, although the use of specific reductants in biological systems to produce a signal decay can be used to good advantage, as will be discussed at the end of this section. Nitroxides are readily reduced to hydroxylamines by hydrogen and platinum catalyst and further to secondary amines when palladium on carbon is used as a catalyst (Rozantsev, 1970). Lithium aluminum hydride reductions of other functional groups in a nitroxide molecule, without appreciable reduction of the nitroxide group, can be accomplished by use of a stoichiometric amount of hydride. Nitroxides are stable during borohydride reductions (Rozantsev, 1970). A common source of signal loss in spin-labeled samples of biological origin is the reduction of nitroxides by sulfhydryl groups (Morrisett and Drott, 1969). The reaction is usually relatively slow and very sensitive to the presence of traces of oxygen and iron. Thus, the slow reduction of nitroxides by mercaptans is not an insurmountable problem in spin-label studies. For instance, when spin-labeled aspartate transcarbamylase was stored at 4°C in the presence of 2-mercaptoethanol, the paramagnetism was not completely lost for several weeks (Buckman, 1970).

The half-wave potential ($E_{\frac{1}{2}}$), relative to the saturated calomel electrode, has been measured polarographically for reduction of nitroxides at a dropping mercury electrode (Neimann *et al.*, 1964). The $E_{\frac{1}{2}}$ for nitroxides **IV** and **V** is constant between pH 6 and 8 at a value of about -150 mV (the precise

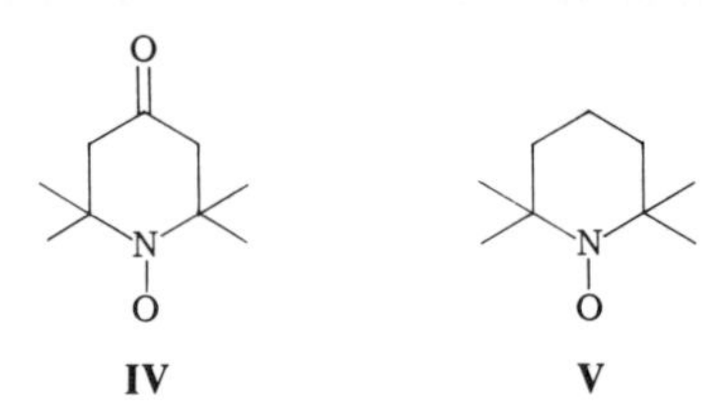

values are dependent on the structure of the nitroxide) in citrate–phosphate and borate buffer solutions. Outside of this pH region, there is a strong dependence of $E_{\frac{1}{2}}$ on pH: $\Delta E_{\frac{1}{2}}/\Delta\text{pH} \sim -60$ mV. The one-electron oxidation and reduction potentials for di-*t*-butyl nitroxide, in acetonitrile, are $+550$ mV and -163 mV, respectively (Hoffman and Henderson, 1961). It is clear from the reduction potentials for nitroxides that a number of components of an intact cell are capable of reducing nitroxides. The fairly rapid reduction of spin labels by lymphocytes could be reversed by including 1 m*M* potassium ferricyanide in the incubation medium (Kaplan *et al.*, 1973). However, if ferricyanide is not in considerable excess, the ferrous ion formed by reduction of ferricyanide may itself catalyze reduction of spin labels. Therefore, if high concentrations of ferricyanide are undesirable, it is preferable to carry out spin-label studies on biological samples in an iron-free buffer containing EDTA to prevent ferrous ion catalysis of nitroxide reduction.

The reaction of nitroxides with strong acids [reaction (6)] places the restriction on the design of a scheme for spin-label synthesis so that steps requiring strong acids must be performed before the secondary amine or the hydroxyl amine is oxidized to the nitroxide. It is possible to do quick extractions of organic solutions of nitroxides with cold 1 *M* HCl without appreciable loss of paramagnetism. The reaction of nitroxides with acids probably proceeds via disproportionation of the protonated nitroxide, a species that has been observed under anhydrous conditions (Hoffman and Eames, 1969).

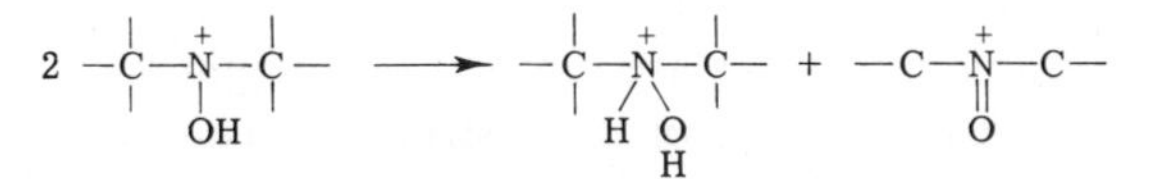

Nitroxides are reduced rapidly by aqueous solutions of sodium ascorbate (in excess). This reduction has been used effectively to differentiate labels on the inside and outside surfaces of phospholipid membranes and to measure the rate of motion of lipids from one side of the membrane to the other (Kornberg and McConnell, 1971).

The signal decay of nitroxides in the presence of biological systems lends itself to kinetic analysis and has proven very useful in some cases (Corker *et al.*, 1966; Stier and Sackmann, 1973; Stier and Reitz, 1971).

Di-*t*-butyl nitroxide has been used to trap free-radical intermediates in photosynthesis (Corker *et al.*, 1966). Rapid and irreversible loss of the spin-label signal in the presence of chloroplasts and photosynthetic organisms was observed. The signal decay was faster in light than dark. This study also lists a number of organic materials capable of free-radical or oxidation–reduction reactions that react with the paramagnetic nitroxide.

The enzyme system consisting of cytochrome P_{450} : cytochrome P_{450} reductase (12 : 1) may be used to oxidize 2,2,6,6-tetramethylpiperidine to 1-oxyl-2,2,6,6-tetramethylpiperidine (TEMPO)† or, in the presence of NADPH, to reduce TEMPO (Stier and Sackmann, 1973; Stier and Reitz, 1971). This enzyme system is a component of microsomal membranes. The NADPH-dependent reduction has been employed as a new method of measuring phase transitions in biological membranes (Stier and Sackmann, 1973). For this measurement, a lipophilic spin label is used. The rate of reduction of the spin label by the enzyme depends on the lateral mobility of the label and of the membrane components within the plane of the membrane. A plot of the rate of reduction of the lipophilic label versus temperature showed a sharp break at about 32°C. As a control, a similar plot for a nonlipophilic spin label showed no break.

II. THE SYNTHESIS OF SPIN LABELS

In this section, representative syntheses of different classes of spin labels are discussed. The numerous reviews on spin labeling (Berliner, 1974; Feher, 1969; Gaffney, 1974; Griffith and Wagonner, 1969; Hamilton and McCon-

† *Editor's note:* The use of this acronym is found rather frequently in the literature and refers to the basic piperidine nitroxide moiety of a spin-label structure.

nell, 1968; Jost and Griffith, 1972; Jost *et al.*, 1971; Kalmanson and Grigoryan, 1973; McConnell and McFarland, 1970; Ohnishi, 1968; and Smith, 1972), as well as Chapter 8 of this book, augment the list of labels presented here. In addition, because of the limitation of the discussion to "spin labels," a number of stable nitroxides are omitted that might very well be used for biophysical studies. A survey of these structures can be obtained from reviews of the chemistry of nitroxide free radicals (Buchachenko, 1965; Forrester *et al.*, 1968; Nelson, 1973; Rozantsev, 1970; Rozantsev and Scholle, 1971). A notable example of the omissions from this chapter is the molecule TEMPO (V), which binds nonspecifically to the fluid lipid regions of biological membranes and has been very useful for investigating phase separations (Shimshick and McConnell, 1973; Linden *et al.*, 1973). Section III of

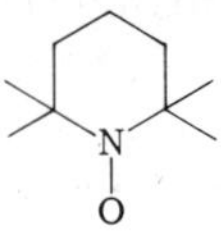

N
O
V

this chapter presents the experimental procedures for a representative selection of spin labels. Throughout the discussion in Section II, those molecules for which the detailed synthetic procedures will be given in Section III are indicated by an asterisk. The synthetic procedures for most of the nitroxide precursors that are used in spin-label synthesis have been collected in Rozantsev's book (Rozantsev, 1970).

In general, nitroxides may be treated as normal organic molecules for structure determination by physical measurements (Gaffney, 1974). The primary exception to this generality, of course, is that the paramagnetism broadens NMR signals considerably.† However, NMR signals of nitroxides have been investigated (Kreilick, 1967), and nitroxides may be reduced, for instance by addition of dithionite to the NMR tube for an aqueous sample to give nonparamagnetic molecules. The mass spectral fragmentations (Morrison and Davies, 1970) and infrared spectra (Forrester *et al.*, 1968) of nitroxides have been well characterized. The visible absorption of nitroxides has a very low extinction coefficient ($\varepsilon < 20$). The ultraviolet absorption is more intense [$\varepsilon(230 \text{ nm}) \cong 3000$] (Forrester *et al.*, 1968). Although there are some characteristic differences between the paramagnetic resonance spectra of the 5- and 6-membered ring nitroxides, EPR spectra are of limited value in making structural assignments.

† *Editor's note:* This phenomenon can be used to great advantage in quantitatively mapping macromolecular structures as discussed in Chapter 9.

A. Isotopically Labeled Nitroxides

The synthesis of isotopically labeled nitroxides is discussed here as an illustrative example of the synthetic procedures used for preparation of the basic structures from which spin labels are derived. Isotopic substitution of nitroxides (Chiarelli and Rassat, 1973) is also of interest because it offers several relatively unexplored possibilities for future spin-label experiments. Labels that are both paramagnetic and radioactive would have numerous applications in the study of biological problems—for instance, for quantitative measurement of label concentration. The syntheses that have been developed for introduction of deuterium (Chiarelli and Rassat, 1973; McFarland and McConnell, 1971) and ^{13}C, (Briere *et al.*, 1968, 1971) may equally well be used as a means of preparing tritium and ^{14}C substituted spin labels. Another reason for preparing isotopically substituted spin labels is to achieve a varied paramagnetic resonance spectrum. The replacement of nitroxide ^{14}N (nuclear spin quantum number $I = 1$) by the isotope ^{15}N ($I = \frac{1}{2}$) is of particular interest because the paramagnetic resonance spectrum is simplified from three spectral lines to two (Briere *et al.*, 1965; Antsiferova *et al.*, 1970; Stryukov and Rozantsev, 1973; Williams *et al.*, 1971). In certain cases where detailed theoretical analysis of resonance spectra is required, the ^{15}N nitroxide spectrum could simplify the analysis. In addition, the existence of spin labels with two fundamentally different resonance spectra allows one to consider double label experiments and measurement of relative spectral intensities.

The first stable nitroxide was prepared by oxidation of triacetoneamine (Lebedev and Kazarnovskii, 1959). This synthesis of triacetoneamine has been reinvestigated (Chiarelli and Rassat, 1973), in order to optimize yield and conditions for work with isotopic precursors and is illustrated here for the preparation of triacetoneamine-d_{16} (**VIII**-d_{16}):

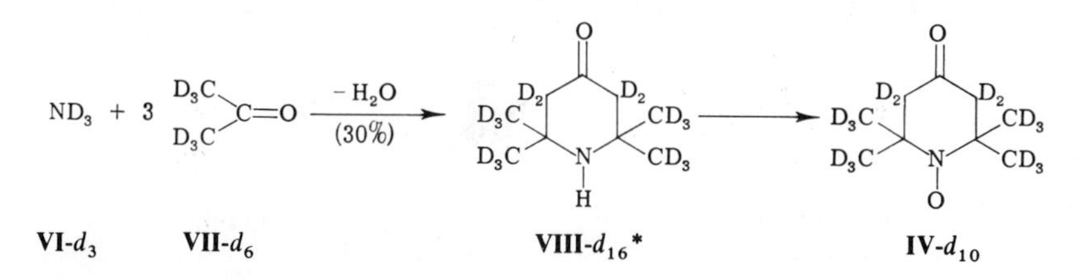

Oxidized triacetoneamine, often referred to as TEMPone, has been labeled with ^{15}N (Briere *et al.*, 1965; Williams *et al.*, 1971) and deuterium (Chiarelli and Rassat, 1973).

Deuterium-substituted di-*t*-butylnitroxide (**XIV**-d_{18}) (Chiarelli and Rassat, 1973) has been prepared by a multistep synthesis that begins with labeled acetone and ends with a step identical to the original method used by Hoffman *et al.* (Kaplan *et al.*, 1973) for preparation of di-*t*-butylnitroxide from *t*-nitrobutane:

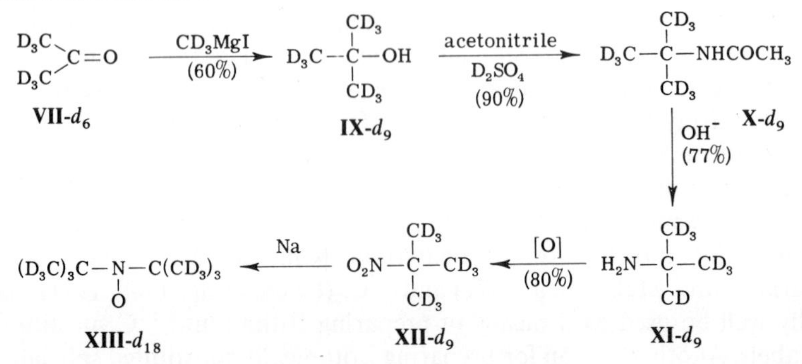

A third general class of nitroxides, the *N*-oxyloxazolidines, (Keana *et al.*, 1967), has been prepared with deuterium substitution, (Chiarelli and Rassat, 1973). This type of label is prepared from a ketone precursor and 2-amino-2-methylpropanol. Deuterium-substituted *N*-oxyloxazolidine derivatives of acetone-d_6 (Chiarelli and Rassat, 1973) and 5-ketopalmitic acid-4,4,6,6-d_4 (McFarland and McConnell, 1971) have been prepared. The syntheses require, in addition to deuterium exchange of the protons adjacent to the ketone carbonyl, the preparation of 2-amino-2-methylpropanol-d_8 as outlined:

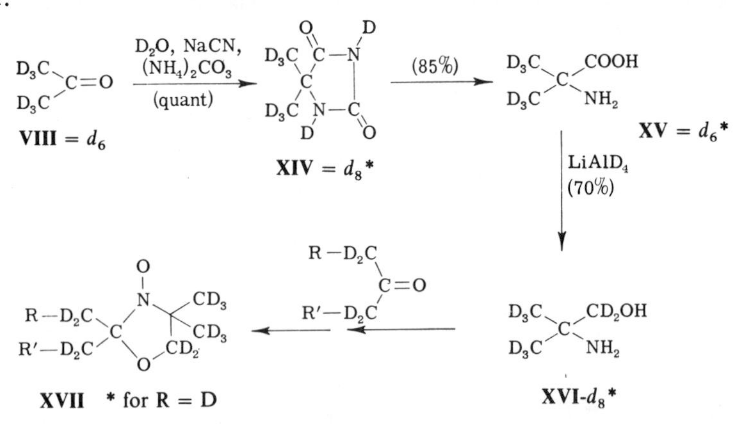

B. Synthetic Precursors

The majority of spin labels that have been prepared are derivatives of a few cyclic nitroxide precursors that are themselves derivatives of triacetone-amine. The reaction schemes leading to spin-label precursors derived from

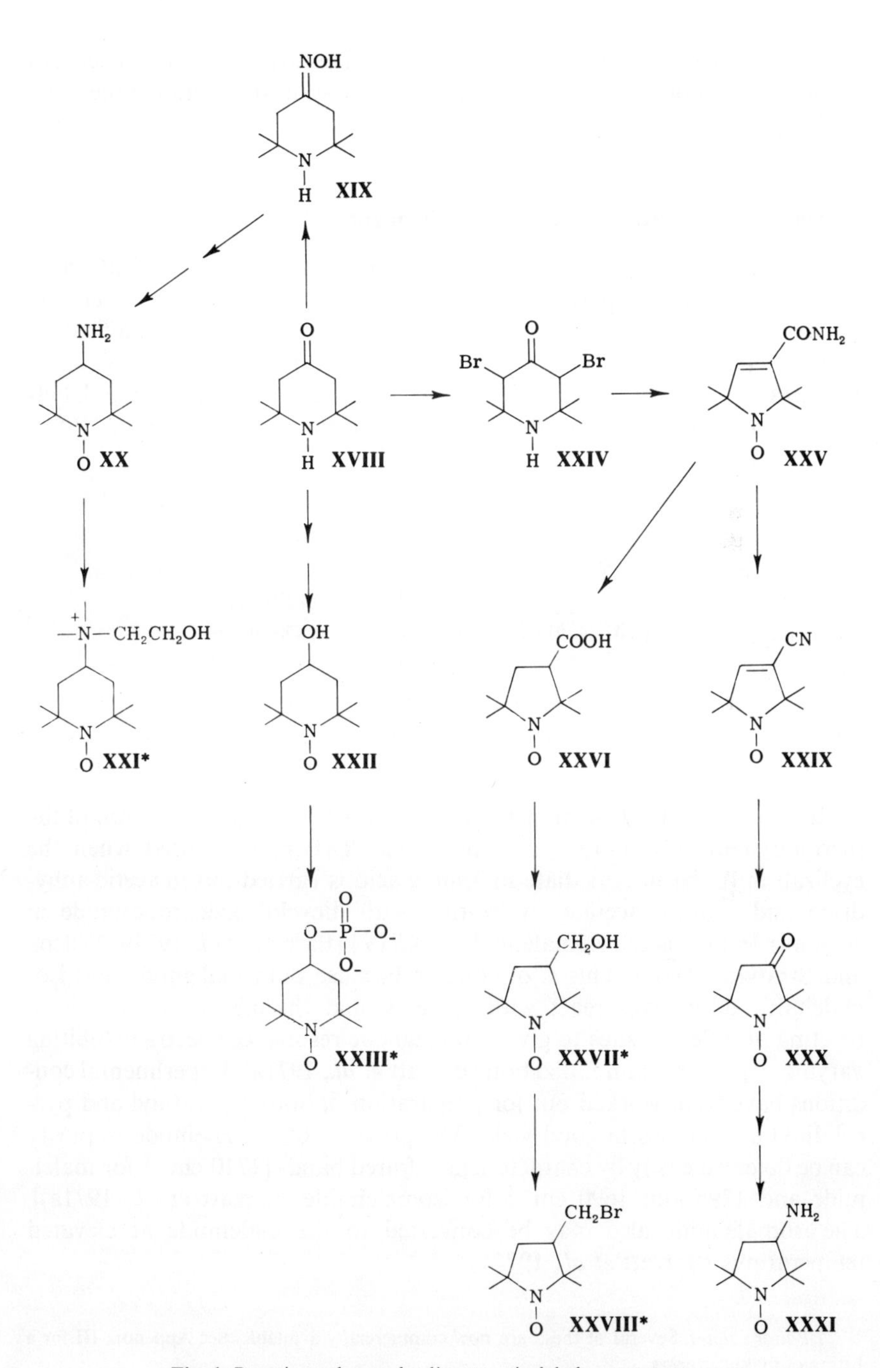

Fig. 1. Reaction schemes leading to spin-label precursors.

triacetoneamine are summarized in Fig. 1.† The experimental procedures for preparation of most of these molecules are described in detail in the book by Rozantsev (1970). The syntheses of the remaining precursors are given in Section III of this chapter.

C. Spin-Labeled Protein Modification Reagents

Spin-labeled derivatives of many of the known protein modification reagents have been prepared (Hirs, 1972; Barker, 1971). These labels are enumerated in Chapter 8 of this book. The spin-label protein modification reagents that have received by far the most use are the maleimide **XXXII**(a, b) (Griffith and McConnell, 1966) and iodoacetamide labels **XXXIII***a*, *b* (Ogawa and McConnell, 1967; McConnell and Hamilton, (1968).

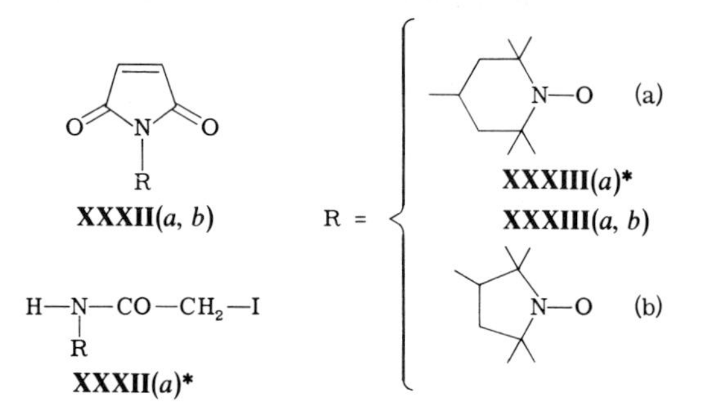

In the synthesis of the maleic acid derivative, although formation of the thermodynamically more stable maleimide **XXII**(*b*) is favored when the cyclization of the intermediate maleamic acid is carried out in acetic anhydride and sodium acetate, cyclization with dicyclohexylcarbodiimide in acetonitrile affords the isomaleimide (**XXXIV**) (Barratt *et al.*, 1971a; Wilcox and Stratigos, 1967). This is of concern because both maleimide and isomaleimide derivatives react with proteins, and the labels are capable of reacting at different sites to give paramagnetic resonance spectra exhibiting varying degrees of immobilization (Barratt *et al.*, 1971a). Experimental conditions have been worked out for preparation of both piperidinyl and pyrrolidinyl maleimides in good yield. The presence of isomaleimide impurity can be detected easily by characteristic infrared bands [1710 cm^{-1} for maleimide and 1795 and 1690 cm^{-1} for isomaleimide (Barratt *et al.*, 1971a)]. The isomaleimide also may be converted to the maleimide at elevated temperatures (Barratt *et al.*, 1971a).

† *Editor's note:* Several of these are now commercially available. See Appendix III for a listing of manufacturers.

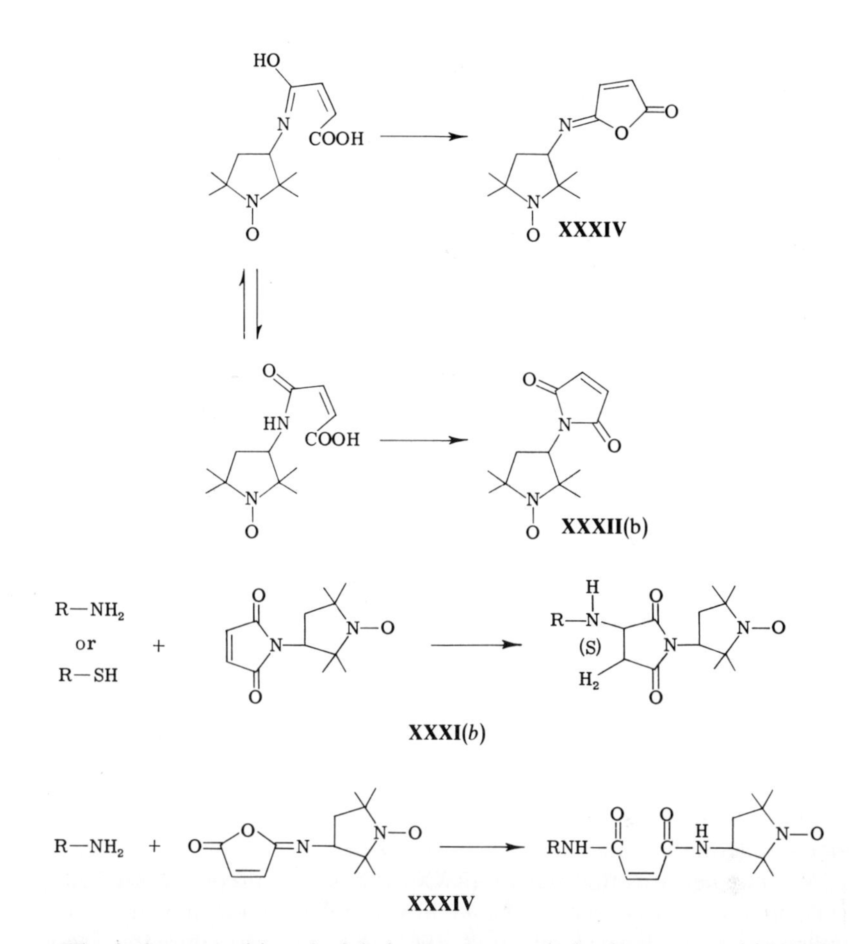

The iodoacetamide spin labels are prepared via the more stable bromo- or chloroacetamides (**XXXV**) (McConnell *et al.*, 1969). In general, the iodoacetamides react with sulfhydryls somewhat more slowly than the maleimide labels and in some cases, this may influence the specificity achieved in labeling sulfhydryl versus amino groups. Thus, the number of *S*-carbomethylated cysteines resulting from the reaction of the iodoacetamide **XXXIII** with hemoglobin (two reactive sulfhydryls per tetramer) was found

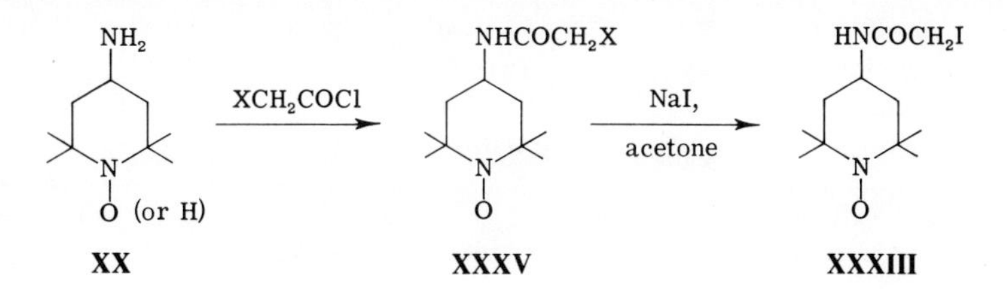

to be 2.0 ± 0.25 moles of labeled sulfhydryls per mole of protein after reaction for 28 hr at pH 7.8 and 2°C (Ogawa, 1967). Other groups on the protein could only be labeled after prior blocking with —SH reagents and prolonged reaction times. On the other hand, in maleimide spin-labeled hemoglobin, both the β-93 sulfhydryls and other sites are labeled in a ratio of 5 to 1 (Ohnishi *et al.*, 1966).

In the case of reaction with ribonuclease A, which has no sulfhydryl groups, bromoacetamide labels (**XXXV**; X = Br) may react with histidinyl and lysyl residues when the protein is denatured in 8 *M* urea at pH 5.5 (Smith, 1968).

The following series of iodoacetamide, bromoacetamide, and maleimide spin labels of varying length is commercially available (Syva Associates, 1970):

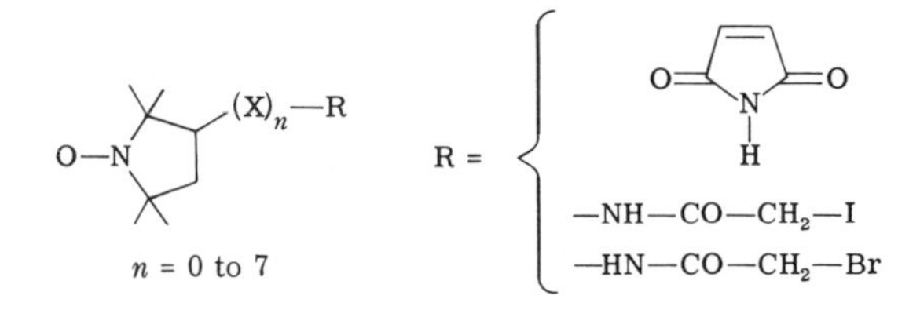

In addition to protein-labeling studies, the iodoacetamide and maleimide labels have been very useful synthetic intermediates in reactions with small molecule mercaptans. They have led to syntheses of spin-labeled ATP (Cooke and Duke, 1971), and to a thiogalactoside (see Section III.X) and a thiocholesterol derivative (Huang *et al.*, 1970).

A spin-labeled isothiocyanate (**XXXVI**) has been prepared (see Section III.L) in order to overcome the lability problems encountered with the similar isocyanate label (Stone *et al.*, 1965). A simple procedure (see Section III.L) involving addition of freshly distilled thiophosgene to a basic aqueous solution of TEMPOamine (**XX**) avoided the experimental difficulties, including destruction of paramagnetism, which accompanied the standard isothiocyanate synthesis involving (1) carbon disulfide and (2) chloroethyl carbonate (Crossley, 1955). The isothiocyanate is insoluble in the aqueous reaction solvent and may be collected simply by filtration.

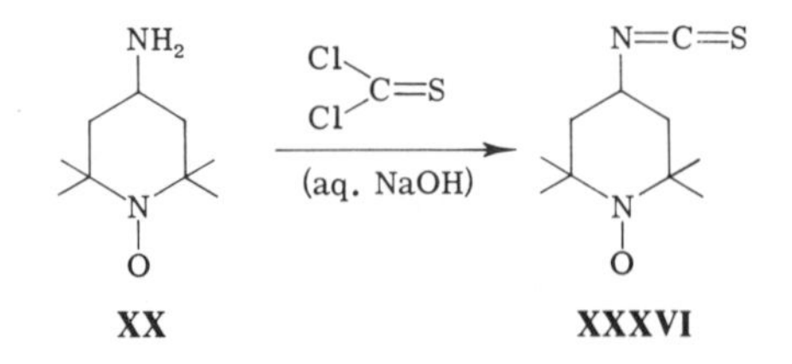

The reactions of the isothiocyanate spin label have not been investigated in detail, but in studies of the molecular control of muscle contraction, it was noted that the isothiocyanate label reacted with groups (presumably $-NH_2$) quite different from those labeled by iodoacetamide and maleimide spin labels (Tonomura *et al.*, 1969; Cooke and Morales, 1969). In reaction with troponin as well as with myoglobin (Shimshick, 1973), lysozyme, (Shimshick, 1973), and hemoglobin (Shimshick, 1973), the isothiocyanate label produced weakly immobilized signals, implying reaction with peripheral amino groups. The isothiocyanate label probably resembles phenyl isothiocyanate in reaction with N-terminal amino groups (Edman, 1956).

A series of isothiocyanate spin labels with varying numbers of atoms between the spin label ring and isothiocyanate groups have been prepared and are commercially available (Syva Associates, 1970).

Several fluorophosphonate labels have been prepared by the reaction of methyl phosphonodifluoridate (Morrisett *et al.*, 1969) or phosphorous oxydichlorofluoride (Hsia *et al.*, 1969; Hoff *et al.*, 1971) with TEMPO alcohol (2,2,6,6-tetramethylpiperidin-4-ol-1-oxyl). The labels are selective for the active site of serine enzymes such as acetylcholinesterase.

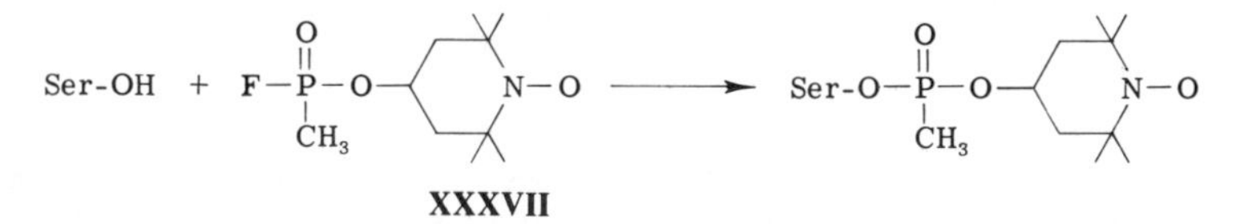

Work with these reagents should be carried out with *extreme care* because fluorophosphonates are potent cholinesterase inhibitors (Saunders and Stacey, 1948; Chapman and Saunders, 1948). Diisopropyl fluorophosphonate is a more potent cholinesterase inhibitor than eserine. One part per million of this fluorophosphonate produced noticeable effects lasting at least seven days on human volunteers (Saunders and Stacey, 1948).

Several spin-label acylating reagents have been prepared. For example, the spin-label hydroxysuccinimide ester **XXXVIII** reacts with valyl-*t*-RNA

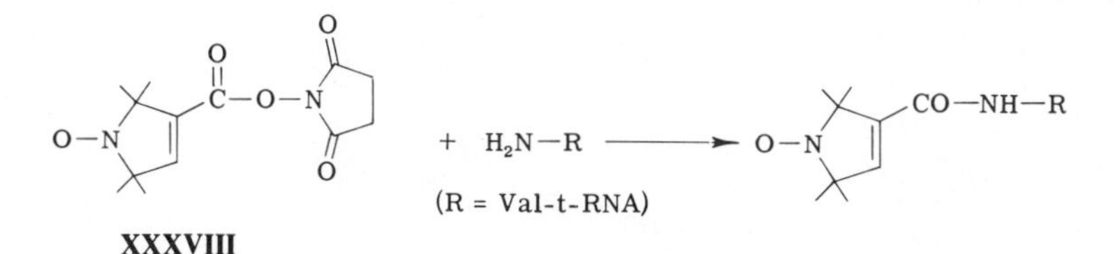

to give a spin labeled aminoacyl-*t*-RNA (Hoffman *et al.*, 1970). Mixed carboxylic–carbonic anhydride acylating reagents (**XXXIX, XL**) have similar reactivity (Griffith *et al.*, 1967; Barratt *et al.*, 1971b).

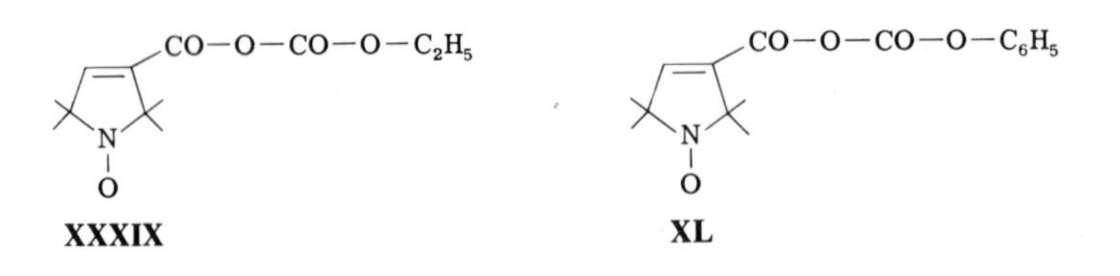

D. Lipid Spin Labels

Spin-labeled derivatives of steroids (Huang *et al.*, 1970; Hubbell and McConnell, 1969; Keana *et al.*, 1967), fatty acids (Hubbell and McConnell, 1971; Seelig, 1970; Waggoner *et al.*, 1969), and phospholipids (Aneja and Davies, 1970; Devaux and McConnell, 1974; Hubbell and McConnell, 1971) have been prepared. In the case of the steroid labels, all but one are derived via the procedure developed by Keana for conversions of ketones to nitrox-

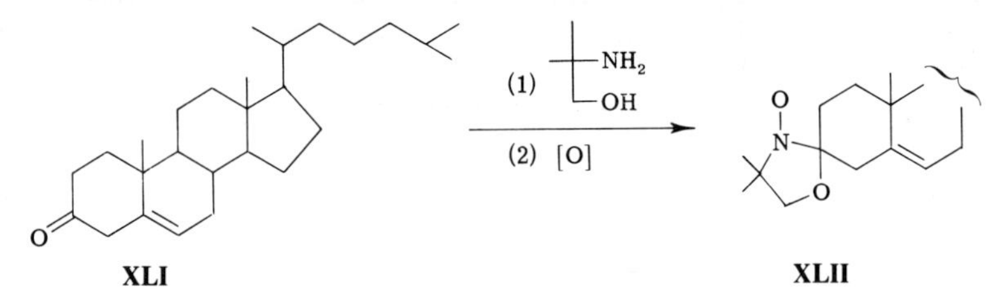

ides (Keana *et al.*, 1967). Other steroid nitroxides which have been prepared by Keana's procedure are **XLIII**, **XLIV**, and **XLV** (Hubbell and McConnell,

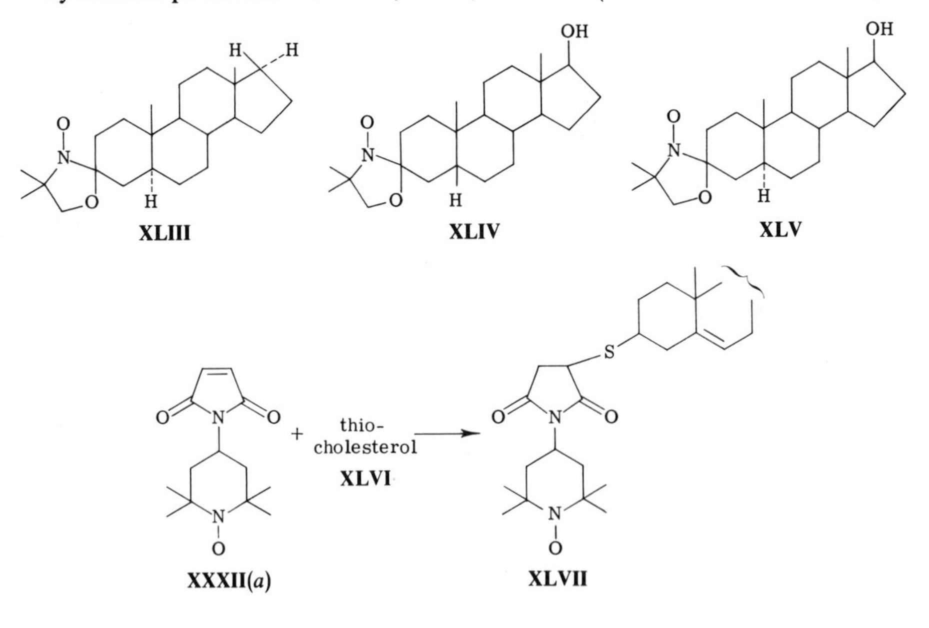

1969). An alternative procedure has been used to prepare a thiocholesterol spin label (Huang *et al.*, 1970). The oxazolidine nitroxides are prepared in three steps, illustrated below.

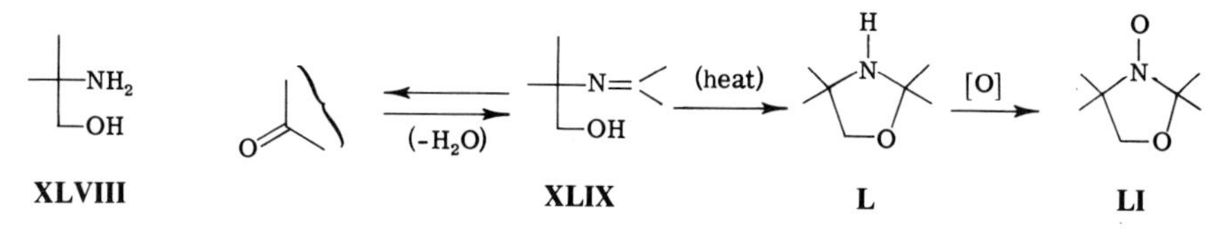

In general, the first two steps (water removal and heating) are accomplished by heating the appropriate components for several days in refluxing toluene with water removal via a Dean–Stark trap. However, this reaction can also be accomplished by heating the reaction components in a sealed tube in the absence of solvent (Chiarelli and Rassat, 1973). The latter approach was used in preparation of the deuteriated dimethyloxazolidine derivative of acetone-d_6. The oxazolidine was isolated in 61% yield in this case. The maximum yield of oxazolidine is limited by the equilibrium between ring-opened (**XLIX**) and closed (**L**) structures (Elderfield, 1957). Oxidation of the oxazolidine **L** to the nitroxide **LI** may be accomplished by hydrogen peroxide–sodium tungstate if the material is water soluble (Chiarelli and Rassat, 1973). The oxidation may also be achieved with *m*-chloroperbenzoic acid (Keana *et al.*, 1967).

A recent variation on the oxazolidine spin-label preparation has been used to create a spin-label cholesterol biradical in good yield (Keana and Dinerstein, 1971).

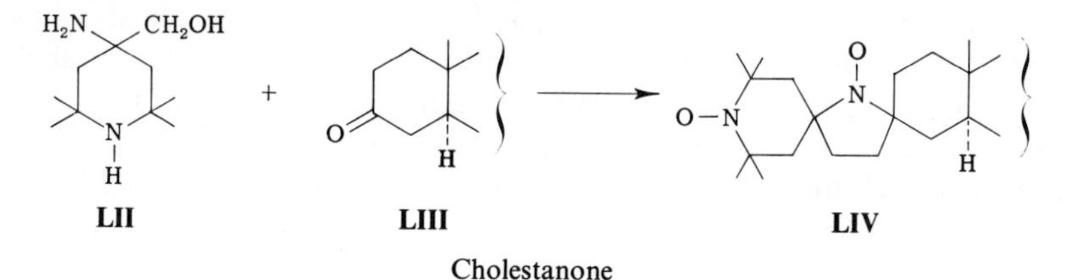

The fatty acid spin labels **LV**(m, n) have been synthesized in most combinations of $m + n + 3 = 16$ or 18 and are prepared by the Keana procedure for converting ketones to *N*-oxyloxazolidines (Hubbell and McConnell, 1971; Waggoner *et al.*, 1969; Seelig, 1970).

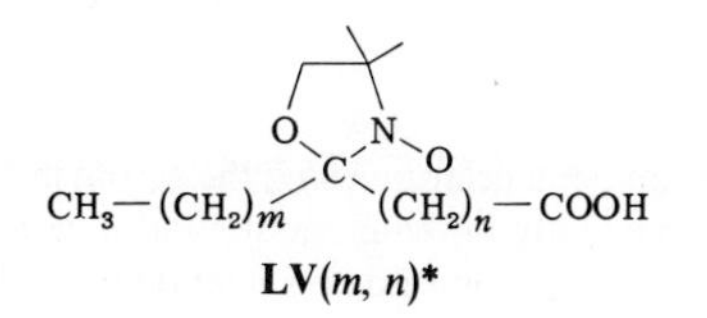

LV(m, n)*

The ketoester precursors are synthesized by the following general scheme:

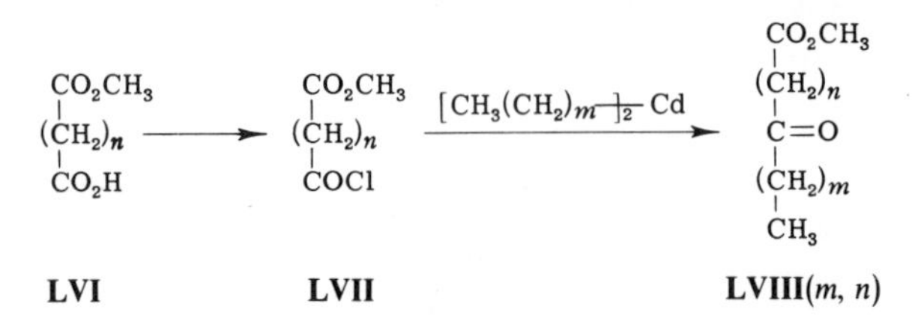

For n less than 7, the half acid ester **LVI** results from acid-catalyzed equilibration of the corresponding diacid and diester (Hubbell and McConnell, 1971). For n greater than 7, the diester is half hydrolyzed in methanolic barium hydroxide to give **LVI** (Durham *et al.*, 1963). The ketoester **LVIII**(5, 10)† may be prepared somewhat more easily by Raney nickel oxidation of the natural product, 12-hydroxystearic acid (Freedman and Applewhite, 1966).

The synthesis of a new type of spin-labeled fatty acid has recently been reported (Williams *et al.*, 1971).

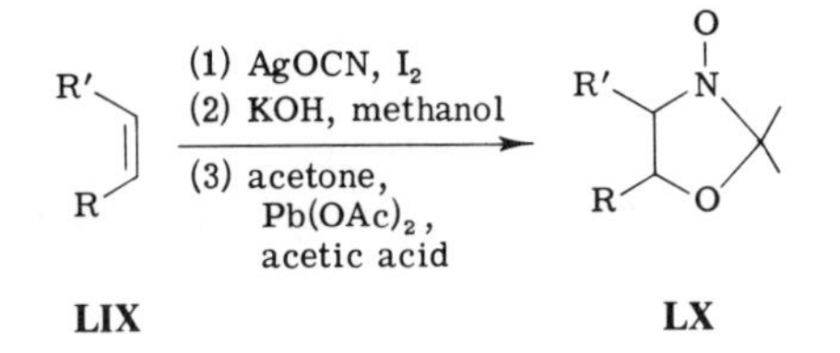

Many of the techniques developed for preparation of phospholipids (Slotboom and Bonsen, 1970) may be applied to the synthesis of spin-labeled phospholipids. Derivatives of phospholipids that are spin labeled in the head group region have been prepared from two spin-label precursors. That is, precursors **XXI** and **XXVII** have been used to prepare spin-labeled lecithin

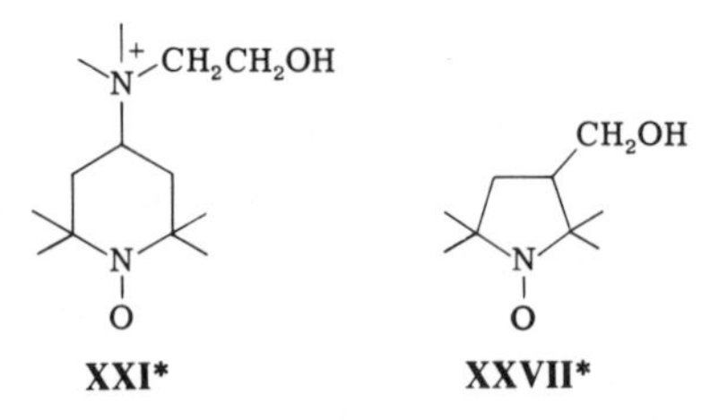

† *Editor's note:* A common nomenclature using the acronym "doxyl" for the oxazolidine-*N*-oxyl moiety is now used frequently in the literature and in later chapters. For example, the nitroxide from the 12-keto stearic acid is 12-doxyl stearate instead of **LV**(5, 10).

and phosphatidic acid, respectively. In both syntheses, the spin-labeled alcohol was coupled to phosphatidic acid in the presence of 2,4,6-triisopropylbenzenesulfonyl chloride (TPS). Lecithins bearing spin-labeled

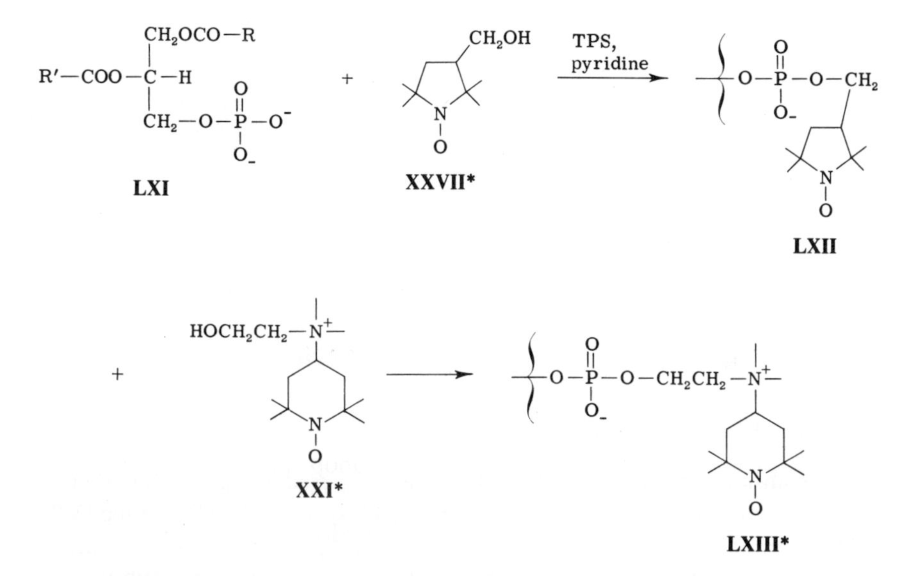

fatty acid chains have been prepared by the reaction of the fatty acid anhydrides with lysolecithin in the absence of solvent (Hubbell and McConnell,

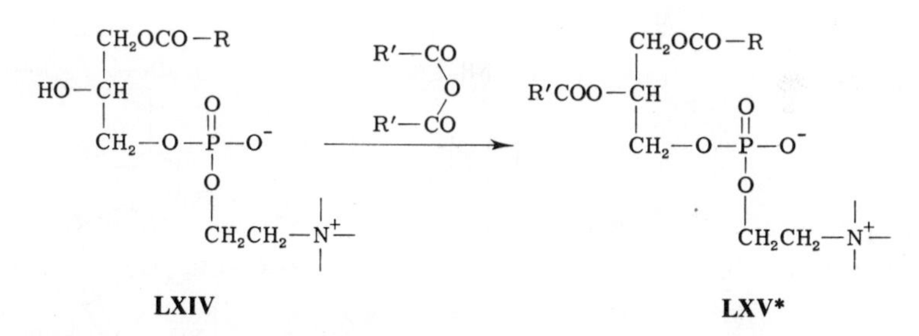

1971; Robles and Van den Berg, 1969). One of these spin-labeled derivatives (**LXV**) also served as a starting point for synthesis of spin-labeled phosphatidylethanolamine (**LXVII**) (Devaux and McConnell, 1974; Barzilay and Lapidot, 1971).

The biosynthetic incorporation of spin-labeled fatty acids has been studied as a method of preparing spin-labeled phospholipids. Preparation of spin-labeled mycoplasma lipids polar lipids (Tourtellotte, 1970), phosphatidic acid (Stanacev *et al.*, 1972), lecithin (Colbeau *et al.*, 1972), and phosphotidylethanolamine (Colbeau *et al.*, 1972) have been reported.

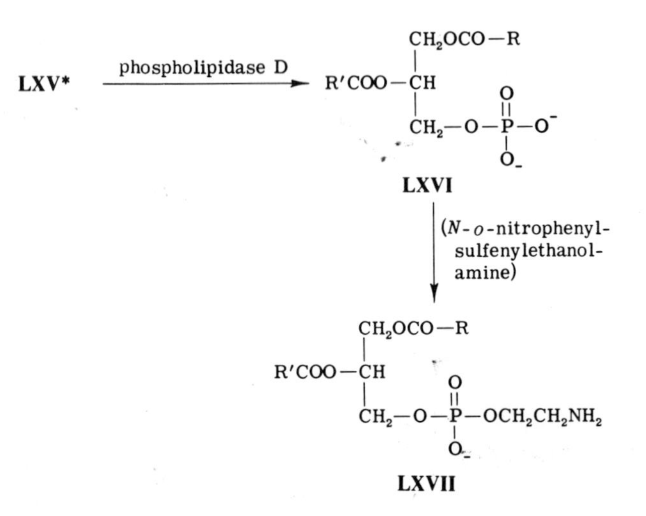

E. Spin-Labeled Nucleotides

A spin-label derivative of AMP has been prepared in high yield (74%) by the reaction of 6-chloropurine ribotide (**LXVIII**) with TEMPOamine (**XX**) (Atkinson *et al.*, 1969). Conversion of the monophosphate to the triphosphate was achieved by DCC coupling of the AMP label and tributylammonium pyrophosphate under rigorously anhydrous conditions.

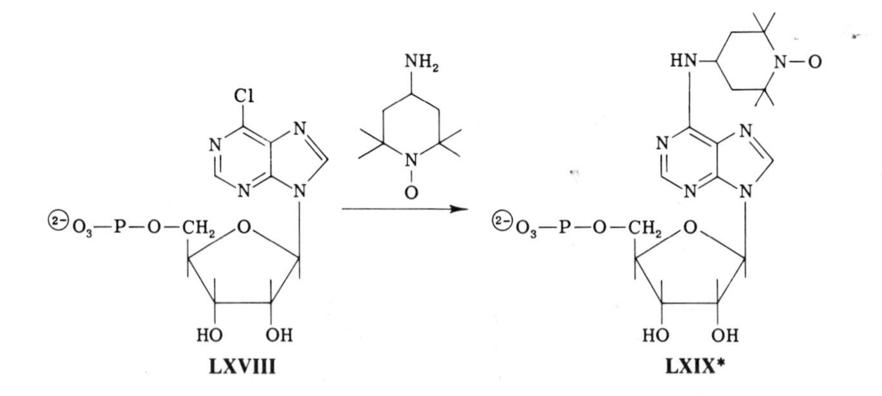

Another derivative of ATP has been prepared by the reaction of TEMPO iodoacetamide with 6-mercapto ATP (Cooke and Duke, 1971). In a similar reaction, the pyrrolidinyl iodoacetamide **XXXIII**(*a*) was coupled to 4-thiouridine 2',(3')-phosphate as well as to the thiouridine of several *E. coli* *t*-RNAs (Hara *et al.*, 1970).

A spin label has also been coupled to the phosphate end of a nucleotide diphosphate to give a spin-labeled analog of NAD (see Section II.F).

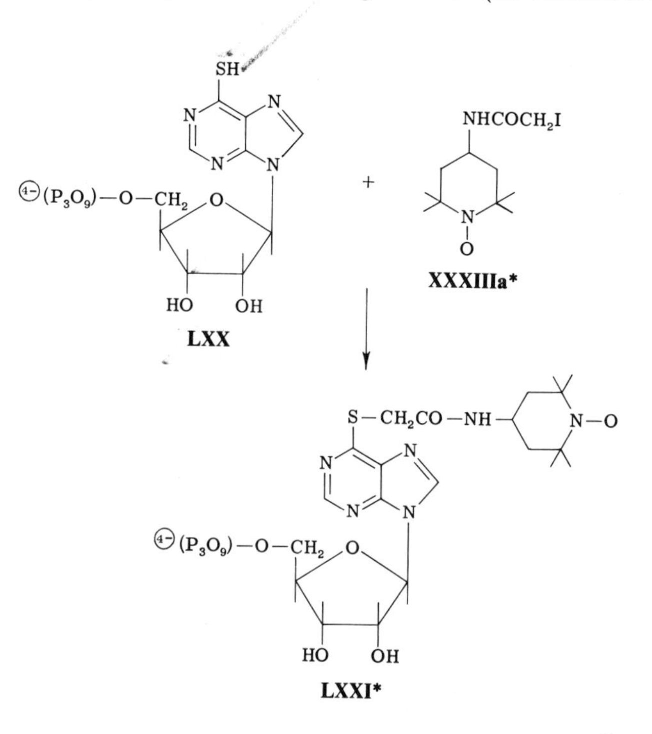

F. Spin-Labeled Cofactors and Prosthetic Groups

Spin-labeled derivatives of (1) NAD (Weiner, 1969), (2) corrinoids and vitamin B_{12} (Buckman *et al.*, 1969; Law *et al.*, 1971), and of the hemes of (3) cytochrome *c* peroxidase, myoglobin, hemoglobin, and horseradish peroxidase (Asakura *et al.*, 1969, 1971), and of (4) cytochrome *c* (Raykhman *et al.*, 1972), have been prepared. These syntheses are outlined below.

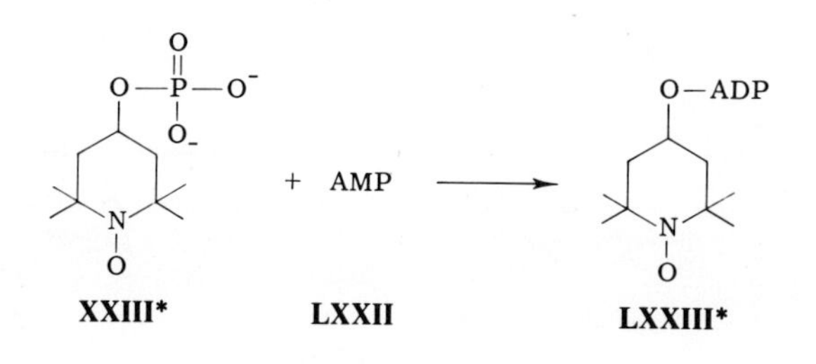

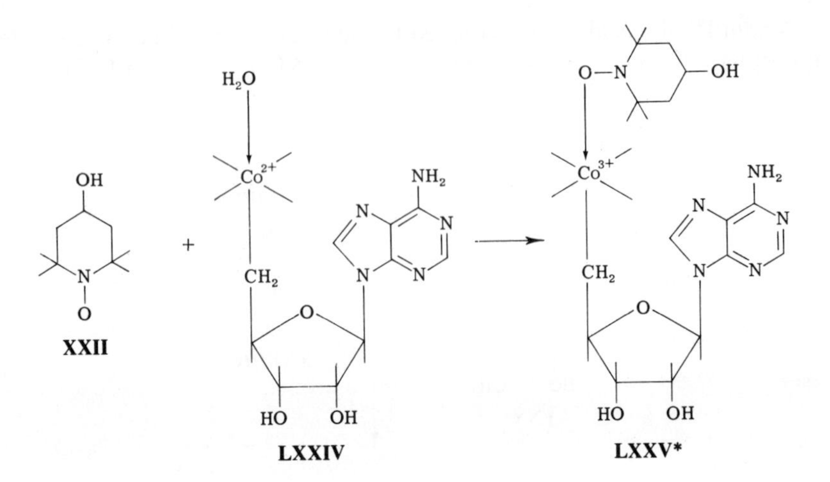
H$_2$O
OH
Co^{2+}
NH$_2$
N
N
+
CH$_2$
N
N
N
O
O
XXII
HO
OH
LXXIV
O—N
OH
Co^{3+}
NH$_2$
N
N
CH$_2$
N
N
O
HO
OH
LXXV*

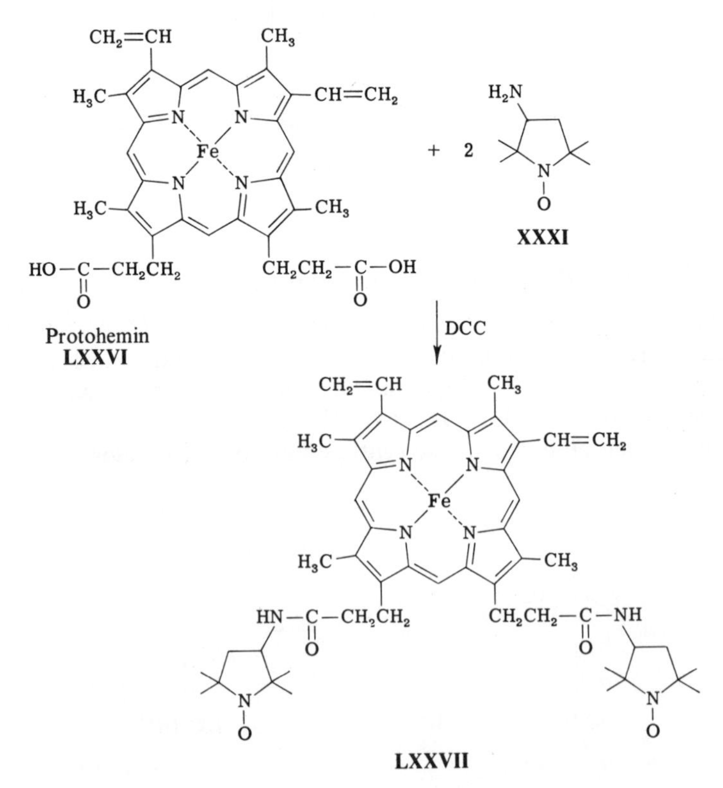
CH$_2$=CH
CH$_3$
H$_3$C
CH=CH$_2$
H$_2$N
N
N
Fe
N
N
+ 2
N
O
H$_3$C
CH$_3$
XXXI
HO—C—CH$_2$CH$_2$
O
CH$_2$CH$_2$—C—OH
O
Protohemin
LXXVI
DCC
CH$_2$=CH
CH$_3$
H$_3$C
CH=CH$_2$
N
N
Fe
N
N
H$_3$C
CH$_3$
HN—C—CH$_2$CH$_2$
O
CH$_2$CH$_2$—C—NH
O
N
O
N
O
LXXVII

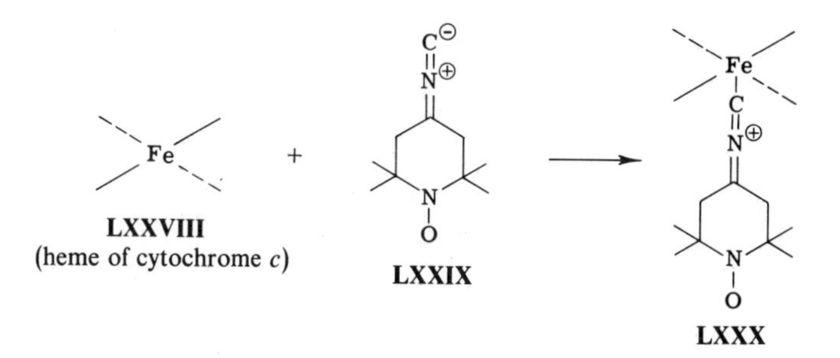

G. Spin-Labeled Sugars

A tritiated galactoside spin label (Struve and McConnell, 1972), two *N*-acetyl glycosides (Wien *et al.*, 1972), and two thioglycosides (see Section III) have been prepared by employing reactions well known in carbohydrate chemistry. The syntheses of these labeled glycosides are summarized below.

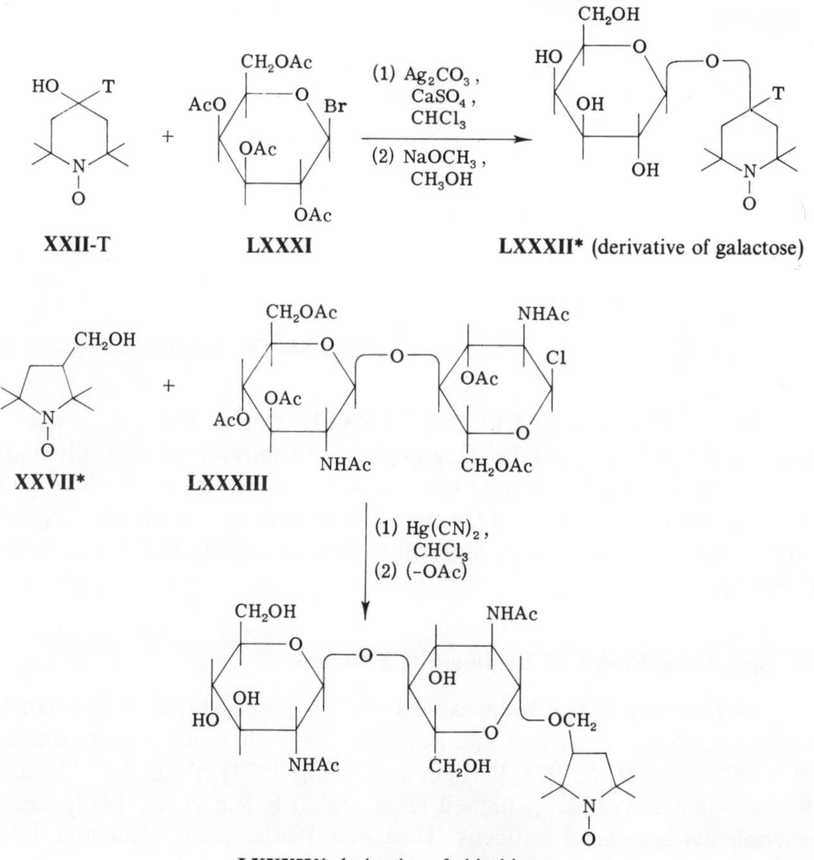

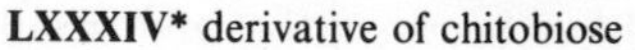
LXXXIV* derivative of chitobiose

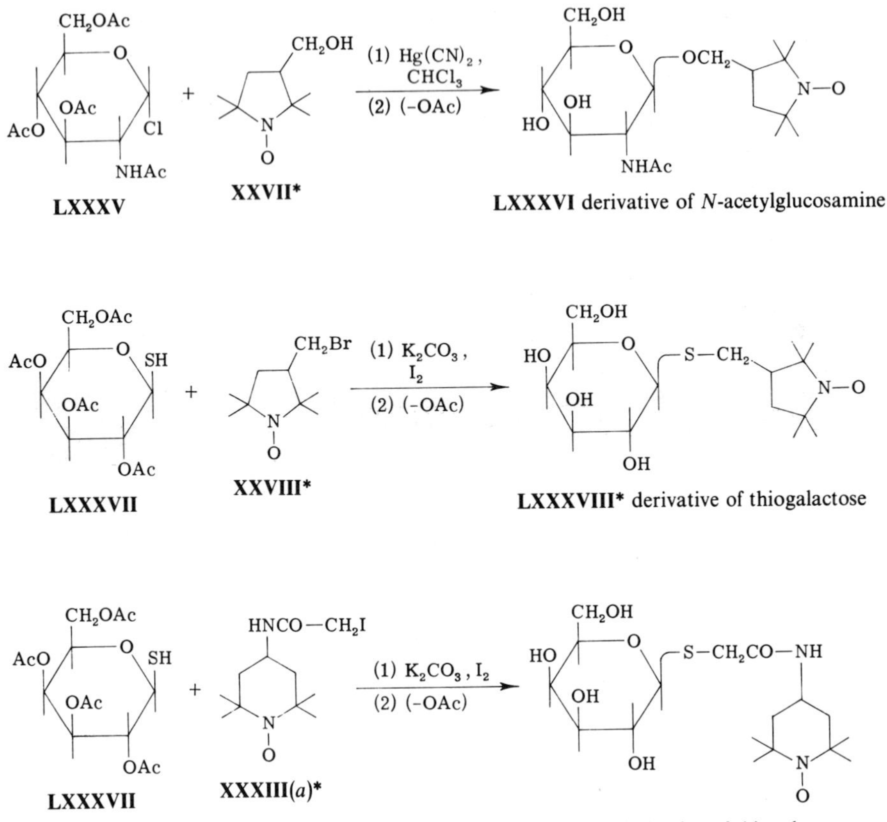

Labels **LXXXIV**, **LXXXVI**, and **LXXXVIII** exist as two optical isomers because the added spin-label ring contains an unresolved optically active center. Optically pure derivatives could be prepared by employing the reported optical resolution of the pyrrolidine carboxylic acid (**XXVI**) (Flohr and Kaiser, 1972) or by using a pyrroline alcohol in place of the pyrrolidinol (**XXVII**).

H. Spin-Label Probes of Binding-Site Structure

Extensive work has been carried out on preparations of spin-labeled nitrophenyl ester substrates and inhibitors for α-chymotrypsin (Kosman *et al.*, 1969; Wong *et al.*, 1974; Berliner and Wong, 1974), of sulfonamide labels for carbonic anhydrase (Chignell *et al.*, 1972; Erlich *et al.*, 1973), and of nitrophenyl spin-label antigens (Hsia and Piette, 1969; Hsia and Little, 1971). The labels that have been synthesized in all of these cases provide good examples of syntheses of a series of reagents with a variable chain length between the "reporter" group and the group bound by the protein. Several of these series as well as a series of protein modification reagents (Syva Associates, 1970), are shown below.

Substrates for α-chymotrypsin

XC*a–c* (Kosman *et al.*, 1969).

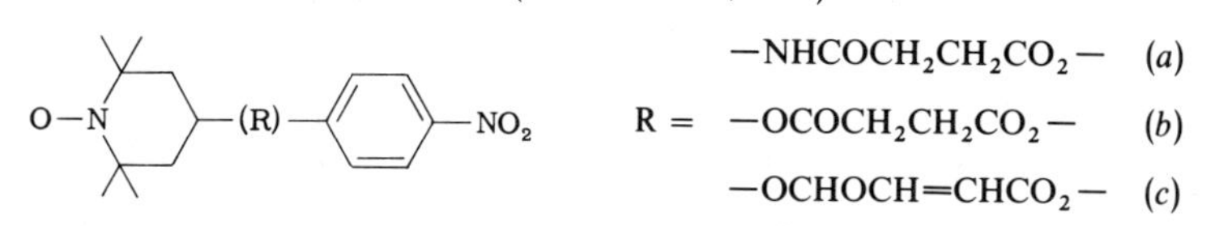

R = $-NHCOCH_2CH_2CO_2-$ (*a*)
$-OCOCH_2CH_2CO_2-$ (*b*)
$-OCHOCH{=}CHCO_2-$ (*c*)

XCI *a–d* Wong *et al.*, 1974

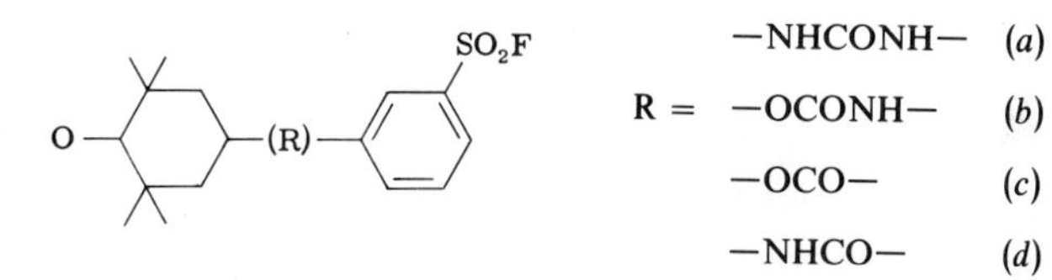

R = $-NHCONH-$ (*a*)
$-OCONH-$ (*b*)
$-OCO-$ (*c*)
$-NHCO-$ (*d*)

Substrates for carbonic anhydrase

XCII *a–e* (Erlich *et al.*, 1973)

R = $-CONH-$ (*a*)
$-CONHCH_2-$ (*b*)
$-CONHCH_2CONH-$ (*c*)
$-NHCO(CH_2)_2CONH-$ (*d*)
$-NHCO(CH_2)_3CONH-$ (*e*)

Protein modification reagents

XCIII *a–e* (Syva Associates, 1970)

R = (*a*)
$-CH_2-$ (*b*)
$-CH_2CH_2-$ (*c*)
$-(CH_2)_3-$ (*d*)
$-CH_2CH_2OCH_2CH_2-$ (*e*)

I. Spin Labels in Immunochemistry

Spin-labeled nitrophenyl haptens have been the subject of many spin label studies involving antibodies (Hsia and Piette, 1969; Hsia and Little, 1971; Stryer and Griffith, 1965). Some of these structures were summarized in Section II.H.

A spin-labeled morphine derivative (**XCIV**) is the basis of a new general immunoassay technique (Leute *et al.*, 1972). This label produces a broad, immobilized paramagnetic resonance spectrum when bound to antibodies to carbomethylmorphine–BSA conjugate, but an intense, sharp signal when it is freely rotating in solution. The assay procedure depends on the competition of unlabeled morphine with **XCIV** for the antibody-binding site.

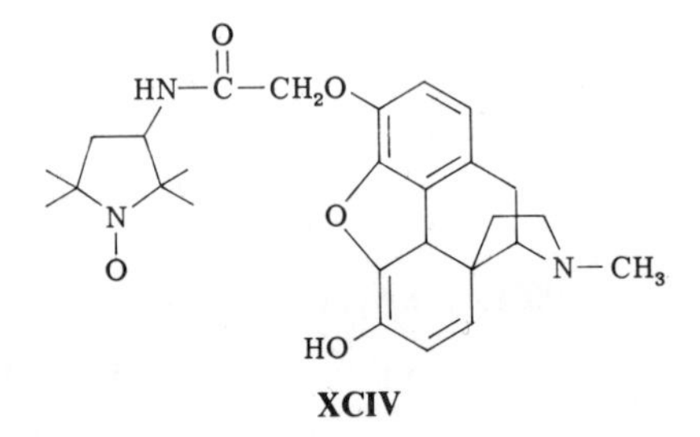

XCIV

A novel use of spin labels in immunochemistry has recently been reported (Humphries and McConnell, 1974). In this case, the spin label is contained in high concentration in the aqueous interior of red blood cell ghosts. These spin-label-loaded ghosts are subject to complement-mediated immune lysis in the presence of antibodies. When the highly concentrated spin label is contained within the ghosts, the paramagnetic resonance signal is very broad. Upon lysis of a ghost, spin label becomes diluted in the medium with a resulting increase in sharp signal intensity. This approach constitutes a new spin-label immunoassay technique by which as little as 10^{-11} antigen in 20 μl may be detected. A particular virtue of this immunoassay technique is that no separation of ghosts from supernate is required, and thus, the assay may be performed quite rapidly.

III. EXPERIMENTAL PROCEDURES FOR PREPARATION OF SPIN LABELS

A. Triacetoneamine-d_{16} (**VIII-d_{16}**) (Chiarelli and Rassat, 1973, translated by B. Gaffney)

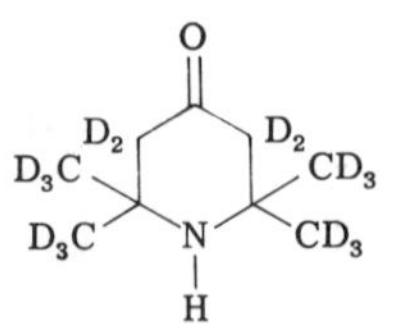

Liquid ammonia-d_3 (3.5 ml, 0.12 mole) was added to 25 ml acetone-d_6 (99.7% *d*, 0.31 mole) and 8 g $CaCl_2$ in an autoclave cooled in liquid nitrogen. The mixture was sealed and stirred and heated in the autoclave at

50°C for 24 hr. The reaction products were placed in a 50 ml flask and heated for 5 hr at 70°C to drive off ammonia. To the remaining red-brown liquid was added 0.5 ml of water, and the mixture was cooled to −15°C and stirred continually, whereupon the hydrate of triacetoneamine-d_{16} crystallized. The product was collected by filtration, recrystallized from ether, and then sublimed under vacuum. The yield was 5.1 g (30%) of triacetoneamine-d_{16} (**VIII**-d_{16}, m.p. 58°C).

B. α-Aminoisobutyric acid-d_6 (**XV**-d_6) (Chiarelli and Rassat, 1973)

$$D_3C-\overset{\overset{\displaystyle NH_2}{|}}{\underset{\underset{\displaystyle CD_3}{|}}{C}}-COOH$$

A solution of 15 gm NaCN in 50 ml of deuterium oxide was added drop by drop to a solution of 20 ml acetone-d_6 (16 gm, 0.25 mole) and 60 gm NH_4CO_3 (0.60 mole) in 150 ml deuterium oxide. The reaction mixture was kept at 50°C for 48 hr. After continuous extraction with ether for 48 hr, 33 gm of 5,5-dimethylhydantoin-d_8 (**XIV**-d_8, quantitative yield) was obtained.

The d_8-hydantoin (25 gm) and 250 gm of $Ba(OH)_2$ were dissolved in 500 ml water. The solution was heated at 130°C for 3 days. After cooling, 1 *N* H_2SO_4 was added until the pH of the solution was 5–6. Barium sulfate precipitated and was removed by filtration. The filtrate was evaporated in a rotary evaporator and the resulting amino acid was recrystallized from ethanol–water (80–20). The yield of α-aminobutyric acid-d_6 (**XV**-d_6) was 18.4 gm (85% yield), mp 337°C (sublimes from 260°C).

C. 2-Methyl-2-aminopropanol-d_8 (**XVI**-d_8) (Chiarelli and Rassat, 1973)

$$D_3C-\overset{\overset{\displaystyle NH_2}{|}}{\underset{\underset{\displaystyle CD_3}{|}}{C}}-CD_2OH$$

α-Aminobutyric acid-d_6 (**XV**-d_6) (10 gm) was added in small amounts to a suspension of 8 gm of lithium aluminum deuteride in 600 ml tetrahydrofuran. The suspension boiled gently during addition. When boiling ceased, the solution was heated under reflux for 5 days. The reaction was terminated by addition of 8 ml deuterium oxide, 8 ml 15% sodium deuteroxide and 24 ml deuterium oxide. Tetrahydrofuran was evaporated from the organic phase and the residue was distilled. The fraction boiling between 162° and 166°C was collected to give 5.8 gm of 2-amino-2-methylpropanol-d_8 (**XVI**-d_8) (70% yield).

D. 1-Oxyl-2,2,4,4-tetramethyloxazolidine-d_{14} (**XVII**, $R + R' = D$) (Chiarelli and Rassat, 1973)

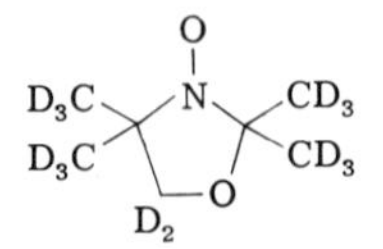

A mixture of 4 gm 2-amino-2-methylpropanol-d_8 (**XVI**-d_8), 200 mg *p*-toluenesulfonic acid, and 25 ml acetone-d_6 was sealed in a tube, and the tube was heated in an autoclave at 60°C for 5 days. The reaction mixture was dried over Na_2SO_4 and taken up in 100 ml petroleum ether. A stream of dry HCl was bubbled into the solution and the resulting hydrochloride was removed by filtration. The hydrochloride was dissolved in water and neutralized with dilute NaOH and extracted with ether. The ether was distilled slowly through a Vigreux column. The residue was distilled yielding 3.1 gm (61% yield) of 2,2,4,4-tetramethyloxazolidine-d_{14} (bp 122°–126°C).

In order to oxidize the oxazolidine, 0.5 ml 5% sodium tungstate solution and 40 mg disodium EDTA were added to a 20 ml flask containing 0.5 gm 2,2,4,4-tetramethyloxazolidine-d_{14} and cooled in an ice bath. To the cooled mixture was added 0.5 ml of 10% H_2O_2. After 6 hr, the mixture was extracted with ether, the ether layers dried over Na_2SO_4, and the ether evaporated slowly. The residue was purified by chromatography on alumina (50 gm, Woelm, activity III). A mixture of 90% pentane and 10% ether was used to elute the product, **XVII**-d_{14} 143 mg (28% yield).

E. 1-Oxyl-2,2,5,5-tetramethyl-3-hydroxymethylpyrrolidine (**XXVII**) (B. J. Gaffney, unpublished)

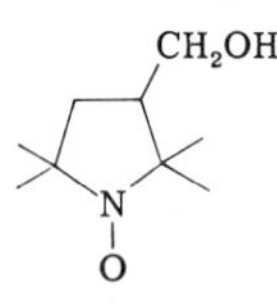

Three grams (0.016 mole) of 1-oxyl-2,2,5,5-tetramethyl-3-pyrrolidine carboxylic acid (**XXVI**) (Rozantsev, 1970) was dissolved in 40 ml dry tetrahydrofuran, and the solution was cooled in an ice-water bath. Lithium aluminum hydride (0.70 gm, 0.024 equivalents; freshly opened can) was added to the cooled solution in small portions. The addition of lithium aluminum hydride required 1 hr. The reaction mixture was stirred at room temperature for 4 hr. Excess lithium aluminum hydride was decomposed by careful addition of a saturated aqueous solution of Na_2SO_4. The Na_2SO_4

solution was added until the white salts coagulated. The tetrahydrofuran layer was decanted and dried over Na_2SO_4. The dried solution was filtered and evaporated to give 2.48 gm (89% yield) of 1-oxyl-2,2,5,5-tetramethyl-3-hydroxypyrrolidine (**XXVII**) (mp 112°C).

F. 1-Oxyl-2,2,5,5-tetramethyl-3-bromomethylpyrrolidine (**XXVIII**) (B. J. Gaffney, unpublished)

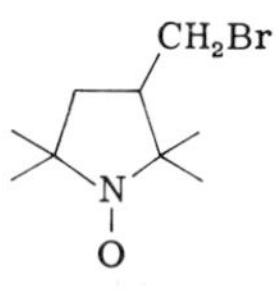

The alcohol above (**XXVII**, 1.72 gm, 0.01 mole) was dissolved in 3 ml (0.037 mole) pyridine and the solution was cooled in an ice bath. To the cooled solution was added 1 ml methanesulfonyl chloride. The reaction mixture was allowed to stand at room temperature for 18 hr. Water (10 ml) was added to the reaction mixture, and the resulting solution was extracted three times with 10 ml portions of ether. The combined ether layers were extracted once with cold dilute HCl and once with saturated NaCl solution. After it was dried over Na_2SO_4, the ether solution was filtered and evaporated to give the methane sulfonate as an oil. The methane sulfonate was used for the next step without further purification.

The crude methane sulfonate (2.4 gm) was dissolved in 20 ml dimethylformamide. Lithium bromide (2.6 gm, dried prior to use for 12 hr at 120°C) was added to the solution of the methane sulfonate, and the reaction mixture was protected from moisture by Drierite. The reaction was carried out at 70°C for 2 hr. Water (10 ml) was added to the cooled mixture. The aqueous layer was extracted four times with ether, and the ether layers were extracted once with saturated NaCl solution. The ether solution was dried over Na_2SO_4 and evaporated to give a yellow oil. Thin-layer chromatography (Eastman prepared silica gel plate, benzene : chloroform, 1 : 4) showed two spots. The components of the mixture were separated by column chromatography on 30 gm silica gel, eluting with 50% chloroform in benzene. The second of two fast-moving bands was collected and evaporated (without heating) to give 0.85 gm of the bromide (**XXVII**) as an oil that crystallized after storage for 2 days in a desiccator over P_2O_5 (mp 43–45°C).

Anal. calcd. for $C_9H_{17}NOBr$:
C, 46.00; H, 7.23; N, 5.95; Br, 34.01
Found: C, 45.95; H, 7.20; N, 5.69; Br, 34.08.

G. TEMPOmonophosphate (**XXIII**) (*4-phospho-2,2,6,6-tetramethylpiperidine-1-oxyl*) (Weiner, 1969)

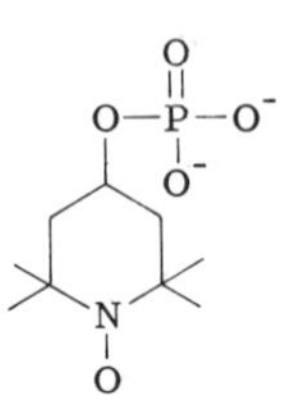

The phosphorylation of TEMPO alcohol (**XXII**) was based on the procedure outlined by Tener (1961). To 1 mmole **XXII** (172 mg) was added 4 ml 1 *M* cyanoethyl phosphate in aqueous pyridine. The solution was evaporated to dryness and reconstituted with 5 ml pyridine. This was repeated four times to remove the water. To the final mixture, in 5 ml pyridine, was added 1.5 gm of *N,N'*-dicyclohexylcarbodimide. After 3 days at room temperature, the reaction was terminated by adding 2 ml water. The urea formed was removed after 1 hr by filtration and washed with 25 ml water. The wash was combined with the yellow solution, and the resulting solution was made 0.4 *M* in LiOH. The basic solution was refluxed for 1 hr to hydrolyze the cyanoethyl group, then filtered. To the yellow supernate was added 6 ml water and additional precipitate was removed. The yellow solution was passed through a 2 × 28 cm Dowex 50X-8 ion-exchange column to remove the lithium. The phosphorylated product was made basic with pyridine and used without further purification.

H. TEMPO triphosphate (Ogata, (1971))

TEMPO monophosphate (~ 10 mmoles) was concentrated at 40°C to 25 ml using a rotary evaporator. Tributylamine (2.4 ml) and 100 ml anhydrous pyridine were then added, and the solution was concentrated as before to ~ 10 ml. Another 100 ml anhydrous pyridine were added, and the solution was again concentrated to 10 ml. Forty milliliters anhydrous *N,N*-dimethylformamide (DMF) were then added, and the solution was again concentrated to 10 ml. After repeating the last step once, 50 ml anhydrous DMF and 8 gm (50 mmoles) of fresh 1,1'-carbonyldiimidazole (dissolved in 100 ml dry DMF) were added to the residue. The reaction flask was then tightly stoppered, shaken vigorously for 30 min, then placed in a desiccator and allowed to stand at room temperature for 1–2 days. Methanol (3.3 ml) was then added. After standing for 30 min, 40 mmoles tributylammonium pyrophosphate dissolved in 200 ml dry DMF were added and the mixture shaken vigorously for 5 min.

The tightly stoppered flask was again placed in a desiccator and the mixture was filtered and the residue washed with two 10 ml portions of dry DMF. To the combined filtrates was added an equal volume of methanol, and this was concentrated on a rotary evaporator to a heavy dark yellow oil. The oil was dissolved in a small volume of 0.01 *M* triethylammonium bicarbonate (TEABC) applied to a 3.4 × 40 cm column of DEAE-Sephadex (HCO_3^-) and eluted with a linear gradient of TEABC 0.01 to 0.4 *M* in 4 liters. The progress of the column could be followed visually (the nitroxide is yellow) and four very well separated bands were collected. Material from the first band was chromatographically indistinguishable from the starting nitroxide monophosphate [Whatmann no. 3MM paper developed in ethanol : 1 *M* ammonium acetate (5 : 2)]. The second band exhibited a five-line paramagnetic resonance spectrum characteristic of a nitroxide biradical and is thus likely to be a dimer of the monophosphate. Band three was assumed to be the diphosphate, and band four (the principal band) the triphosphate. Band four was concentrated to 5 ml, 250 ml ethanol was added, and the solution was concentrated again to remove TEABC. This was repeated twice. Methanol (~ 50 ml) and a solution of 20.5 gm sodium perchlorate in 250 ml acetone were then added to the residue. The precipitated sodium salt of the triphosphate was collected by centrifugation, washed with four 20 ml portions of acetone, and then dried in a desiccator over P_2O_5. The solid obtained was then dissolved in 50 ml water and the solution adjusted to pH 8.5 with 0.1 *M* $Ba(OH)_2$. The precipitated barium pyrophosphate was carefully and completely removed by centrifugation, and the light orange supernate was then passed through a column (~ 100 ml bed volume) of AG 50X2 ion exchange resin. The eluent was made basic with triethylamine, concentrated to 5 ml, freed of triethylamine by repeated addition and evaporation of absolute ethanol, and precipitated with sodium perchlorate as before. The precipitate was then dried exhaustively over P_2O_5 at 0.1 Torr.

In the first synthesis, the resulting bright red product had a total phosphate to labile phosphate ratio of 3 : 2.04 and a spin (measured by comparing the paramagnetic resonance spectral amplitude of a solution of the triphosphate with that of a standard solution of 2,2,6,6-tetramethyl-4-piperidinol) to phosphate ratio of 0.15 (expected for a nitroxide triphosphate is 0.33). The total phosphorus found was 18.3%; calculated for $C_9H_{19}O_{11}NP_3Na_4$ is 18.6%.

I. TEMPOcholine chloride (**XXI**) (Kornberg, 1972; Kornberg and McConnell, 1971)

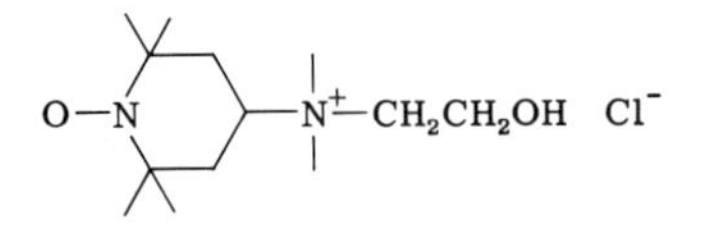

(a) 4-(*N*,*N*-Dimethylamino)-2,2,6,6-tetramethylpiperidine was prepared by the method of Icke and Wisegarver (1955). Twenty-five grams 4-amino-2,2,6,6-tetramethylpiperidine (**XX**) (Aldrich Chemical Co.) was added with magnetic stirring to 38 gm 98% formic acid in an ice bath. The mixture was removed from the ice bath, combined with 28.5 gm 37% formaldehyde, refluxed for about 10 hr, cooled in an ice bath, and combined with 18.2 gm concentrated HCl. Formic acid, formaldehyde, and water were removed under reduced pressure. The residue was combined with about 100 ml of water and about 50 gm NaOH. The lower phase was extracted twice with diethyl ether. The upper phase and ether extracts were dried over BaO. The ether was removed by distillation at atmospheric pressure. Vacuum distillation (at 29–36°C and less than or about 1 Torr) yielded 18 gm (67%) of clear, colorless liquid.

Anal. calcd. for $C_{11}H_{24}N_2$: C, 71.67; H, 13.13; N, 15.20.
Found: C, 71.55; H, 12.99; N, 14.95.

(b) 2-Bromoethyl acetate was prepared by the reaction of 2-bromoethanol with acetic anhydride. Excess acetic anhydride (about 100 gm) was added with magnetic stirring to 100 gm of 2-bromoethanol (Aldrich Chemical Co.) in an ice bath. The mixture was heated on a steam bath for 1 hr, cooled in an ice bath, and repeatedly extracted with cold 1% $NaHCO_3$ until the pH of the upper phase was about 7. The lower phase was then dried over anhydrous Na_2SO_4. Vacuum distillation yielded a clear, colorless liquid: IR (film), band at 1740 cm^{-1} (carboxylic ester).

(c) A mixture of 8.52 gm 4-(*N*,*N*-dimethylamino)-2,2,6,6-tetramethylpiperidine and 15.5 gm 2-bromoethyl acetate was kept in the dark at room temperature for 3 days. The white solid product was dispersed in anhydrous diethyl ether, filtered and washed with anhydrous diethyl ether by suction, and dried to 11.6 gm (71%) under vacuum over P_2O_5. A mixture of 6.15 gm of the dry solid, 1.70 gm EDTA, 0.93 gm NaOH, 2 gm sodium tungstate dihydrate, 25 ml 30% H_2O_2 were added. After 2 days, the remaining H_2O_2 was destroyed by vigorous magnetic stirring. The water was removed under reduced pressure. Benzene was added and then removed under reduced pressure. The residual solids were dispersed in about 500 ml of absolute ethanol and filtered. The filtrate was evaporated under reduced pressure and redissolved in 20 ml water. Fifteen milliliters of this solution was applied to a column containing 60 ml AG 50W-X8 (Bio-Rad Laboratories, ammonium ion form), and eluted with 250 ml 0.25 *M* ammonium bicarbonate and 400 ml 1 *M* ammonium bicarbonate. The first 300 ml of 1 *M* ammonium bicarbonate eluate was evaporated under reduced pressure and the residual solids were dispersed in 100 ml of absolute ethanol and filtered. The filtrate was evaporated under reduced pressure and the residual solids were twice dispersed in 100 ml absolute ethanol and evaporated under reduced pressure. The pH of a solution of the residual solids in 50 ml water was adjusted to 2.9 by the addition of 6 ml 1.2 *M* HCl. The water was removed under reduced pressure. The orange solid was dried to 2.1 gm under vacuum over P_2O_5. It appeared as a single spot on a cellulose thin-layer chromatogram that was developed in propanol ammonium hydroxide (14.8 *M*)–water (6 : 3 : 1, v/v/v) and stained with iodine vapor.

Anal. calcd. for $C_{13}H_{28}N_2O_2Cl$:
C, 55.80; H, 10.09; N, 10.01; Cl, 12.67.
Found: C, 54.76; H, 9.97; N, 9.95; Cl, 12.48.

J. N-(1-Oxyl-2,2,6,6-tetramethyl-4-piperidinyl)maleimide **[XXXII(*a*)]** (C. L. Hamilton and H. M. McConnell, unpublished)

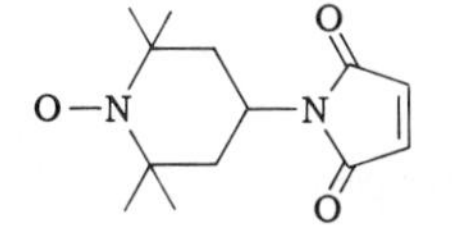

Maleic anhydride (0.33 gm, 3.4 mmole) was dissolved in 20 ml anhydrous ether in a 50 ml round bottomed flask equipped with a magnetic stirrer. 2,2,6,6-Tetramethyl-4-aminopiperidine-1-oxyl (0.58 gm, 3.4 mmole) was dissolved in a few ml anhydrous ether; this solution was added dropwise to the stirred maleic anhydride solution from a dropping funnel. The mixture was then stirred at room temperature for an hour, and the product (0.75 gm, 82%) was collected by filtration and washed with several portions of ether. Further purification was unnecessary.

A mixture of the above *N*-nitroxylmaleamic acid (0.96 gm, 3.6 mmole), 0.15 gm (1.8 mmole) of anhydrous sodium acetate, and 15 ml of acetic anhydride were heated for 2–3 hr in a boiling water bath. The acetic anhydride was removed by distillation *in vacuo*. The residue was dissolved in benzene as much as possible; the solids remaining were removed by filtration. The residue obtained after removing the solvent was distilled at 5×10^{-3} mm

(140°–160°C) to yield solid material (72%), mp 99°C. Thin-layer chromatography (Eastman chromogram, chloroform) indicated normal maleimide (IR; 1690 cm^{-1}) plus small amounts of isomaleimide (IR; 1790, 1690 cm^{-1}).

K. N-(1-Oxyl-2,2,6,6-tetramethyl-4-piperidinyl)iodoacetamide **(XXXIII)**

Procedure (1) (McConnell *et al.*, 1969)

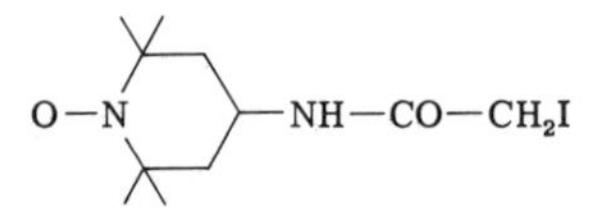

15.6 gm 2,2,6,6-tetramethyl-4-aminopiperidine (Aldrich Chemical Co.) was acetylated with an equivalent amount of chloroacetyl chloride. The reaction mixture was extracted with dilute HCl, and the extract was neutralized with aqueous NaOH. The white solid that precipitated was collected on a filter and suspended in about 600 ml water; 3 gm sodium tungstate, 3 gm disodium EDTA, and 60 ml 30% H_2O_2 were added to the suspension, and the resulting mixture was stirred until all solid material dissolved. The solution was allowed to stand for 2 hr, and then the resulting orange solution was saturated with NaCl and extracted with ether. The ether phase was washed with water to remove excess peroxide. The dark orange crystals obtained after drying over $MgSO_4$ and removal of the ether *in vacuo* were dissolved in acetone, and a slight excess of NaI dissolved in acetone was added. After standing overnight, the mixture was filtered and the acetone was removed *in vacuo*. The dark oil remaining was dissolved in 200 ml hot toluene, and after 3 days, the product **(XXXIII)** crystallized as dark orange needles that melted at 114°–117°C.

Anal. calcd. for $C_{11}H_{20}IN_2O_2$: C, 39.0; H, 5.9; I, 37.4.
Found: C, 39.3; H, 6.2; I, 37.5.

Procedure (2) (Procedure of B. J. Gaffney)

To a solution of 2,2,6,6-tetramethyl-4-aminopiperidine-1-oxyl **(XX)** (5 gm, 0.029 mole) and NaOH (1.17 gm) in 50 ml water was added chloroacetyl chloride (3.33 gm, 0.029 mole) over a period of 1 hr. The reaction mixture was stirred for 1 hr, and the orange precipitate was removed from the mixture by filtration and dried overnight over P_2O_5 to give 2,2,6,6-tetramethyl-4-chloroacetamidopiperidine-1-oxyl (2.00 gm, 28% yield). One recrystallization from benzene yielded red prisms (mp 120°–121°C):

Anal. calcd. for $C_{11}H_{20}ClN_2O_2$: C, 53.22; H, 8.14; N, 11.31.
Found: C, 53.68; H, 8.32; N, 10.96

A solution of 2,2,6,6-tetramethyl-4-chloroacetamidopiperidine-1-oxyl (1.95 gm, 0.0079 mole) and NaI (1.18 gm, 0.0079 mole) in 20 ml acetone was allowed to stand for 48 hr at 15°–20°C. The white precipitate of NaCl was removed by filtration and washed with acetone and dried (0.35 gm). To the remaining acetone solution was added an additional 3 gm NaI. After 48 hr, the acetone was removed under vacuum at 20°C. To the remaining red oil was added 5 ml water, and the mixture was rapidly extracted three times with ether. The ether solution was dried (Na_2SO_4) and evaporated to give a red oil that solidified on trituration with benzene. The red crystals were collected by filtration and dried (2.10 gm, 78% yield; mp 121°–123°C). This sample was recrystallized twice from benzene to give an analytical sample.

Anal. calcd. for $C_{11}H_{20}IN_2O_2$: C, 38.95; H, 5.94; N, 8.26.
Found: C, 40.69; H, 5.95; N, 8.55.

L. 1-Oxyl-2,2,6,6-tetramethyl-4-isothiocyanatopiperidine **(XXXVI)** (B. J. Gaffney, unpublished)

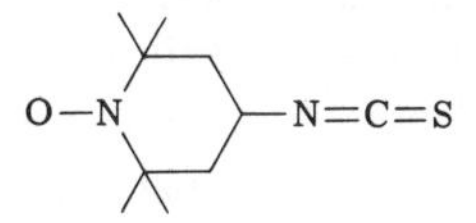

To 150 ml of a 5% NaOH solution of 1-oxyl-2,2,6,6-tetramethyl-4-aminopiperidine-1-oxyl (6.42 gm, 0.0375 mole) at 0°C was added freshly distilled thiophosgene (4.32 gm, 0.0375 mole, 3.1 ml; bp 73°-75°C/758 Torr) over a period of 15 min. The resulting mixture was stirred for 10 min, and the orange precipitate was removed by filtration, washed three times with water, and dried under vacuum over P_2O_5 to give a red powder (mp 107°–112°C, 5.7 gm, 71% yield). This powder was recrystallized from ethanol and water to give red prisms (1.6 gm, mp 124°–125°C). Two further recrystallizations from benzene afforded an analytical sample (mp 127°–128°C):

Anal. calcd. for $C_{10}H_{17}N_2OS$: C, 56.30; H, 8.03; N, 13.13.
Found: C, 56.24; H, 8.06; N, 12.97.

M. TEMPOphosphatidylcholine **(LXIII)** (Kornberg and McConnell, 1971; Kornberg, 1972)

H_2C—O—CO—R
R—CO—O—CH
H_2C—O—PO—O—CH_2CH_2—$\overset{+}{N}(CH_3)_2$—(TEMPO ring) N—O
O^-

(a) Two grams dipalmitoylphosphatidylcholine (derived from egg phosphatidylcholine by the methods of Brockeroff and Yurkowski (1965) and Baer and Buchnea (1959)) were dispersed in 50 ml water with a Bronwill 20 kHz sonicator and intermittent cooling in an ice bath. The sonication was continued until further sonication did not produce further clearing of the homogeneous opalescent dispersion. A mixture of 3.3 ml 1 *M* pH 5.6 acetic acid–calcium acetate buffer, 100 ml Savoy cabbage supernate [freshly prepared by the method of Davidson and Long (1958)], 100 ml alcohol-free diethyl ether (freshly distilled from P_2O_5), and 20 ml water was shaken with the dipalmitoylphosphatidylcholine dispersion for 13 hr at room temperature, then shaken with 35 gm citric acid (monohydrate) and 170 ml water and centrifuged. The aqueous phase was extracted three times with 500 ml portions of chloroform. The combined ethereal phase and chloroform extracts were evaporated under reduced pressure. The residue was dissolved in absolute ethanol, and then the ethanol and last traces of water were removed under pressure. The residue was dissolved in 15 ml chloroform and added dropwise to a magnetically stirred solution of 20 gm of barium acetate (monohydrate) and 400 ml methanol–water (1 : 1, v/v). The precipitate was collected by centrifugation, shaken for about 10 min at room temperature with 250 ml 0.5 *N* H_2SO_4 and 250 ml chloroform, and centrifuged. The lower phase was extracted once more with 250 ml 0.5 *N* H_2SO_4, once with 500 ml methanol–water (1 : 1, v/v), combined with an extract of the upper phase from the first 0.5 *N* H_2SO_4 extraction (the upper phase was extracted once with 250 ml 0.5 *N* H_2SO_4 and once with 250 ml water), and evaporated under reduced pressure. Traces of water were removed from the residue by repeated evaporation of absolute ethanol, and traces of ethanol were removed by repeated evaporation of toluene, yielding 1.83 gm (103%) white solid which was dissolved in 10 ml anhydrous chloroform. An oxalic acid-impregnated silica thin-layer chromatogram of this solution was developed in chloroform–methanol–HCl (12 *M*) (87 : 13 : 0.5, v/v/v). It showed one spot with the same R_f value as phosphatidic acid derived from egg phosphadidylcholine by the method of Davidson and Long (1958).

(b) One-half of the anhydrous chloroform solution that resulted from the procedure in (a) was transferred to a two-necked flask, equipped for mechanical overhead stirring, which contained about 45 gm of glass beads (roughly 2 mm in diameter), 2.5 ml anhydrous pyridine, 0.55 gm TEMPOcholine chloride (**XXI**) and 1.21 gm 2,4,6-triisopropylbenzenesulfonyl chloride. The mixture was vigorously stirred for 5 hr at room temperature. Ten milliliters of water was added with cooling and then the mixture was stirred for 3 hr at room temperature, shaken with 250 ml chloroform and 150 ml of 0.5 *N* H_2SO_4, and centrifuged. The lower phase was washed successively with 150 ml 0.5 *N* H_2SO_4, 150 ml water, 225 ml 2% $NaHCO_3$ in methanol–water (1 : 2, v/v), and 225 ml methanol–water (1 : 2, v/v), evaporated under reduced pressure, dissolved in 50 ml chloroform–methanol (1 : 1, v/v), applied to a column containing 600 ml alumina [E. Merck, activity I, washed extensively with chloroform–methanol (1 : 1, v/v) for the removal of fines, activated for 3 hr at 110°C, and packed in the column as a slurry in chloroform], and eluted with chloroform–methanol (1 : 1, v/v). The first 25 ml of orange eluate were discarded. The next 250 ml of orange eluate were concentrated to a small volume under reduced pressure, diluted with absolute ethanol, and centrifuged for removal of alumina particles. The ethanol was removed under reduced pressure. The orange solids were dissolved in 20 ml chloroform. This spin-labeled phosphatidylcholine solution contained 914 μmoles of phosphorus (78% of the total phosphorus in the solution that was applied to the alumina column). Silica thin-layer chromatograms of the solution were developed in chloroform–methanol–NH_4OH (14.8 *M*)–water (70 : 30 : 4 : 1, v/v), chloroform–methanol–water (65 : 25 : 4, v/v), and chloroform–methanol–acetic acid–water (50 : 25 : 7 : 3, v/v). In each case there was only one spot. There were, however, two spots (R_f values 0.42 and 0.47) on a silica-impregnated paper chromatogram developed in diisobutyl ketone–acetic acid–water (8 : 5 : 1, v/v) and sprayed with the phosphate reagent of Dittmer and Lester (1964). The spot corresponding to an R_f value of 0.42 was about 10% as intense as the other spot. A 12% contamination of the spin-labeled phosphatidylcholine solution by some nonparamagnetic phospholipid would explain why the ratio of nitroxide/phosphorus was lower than expected. *Analysis*: Nitroxide/phosphorus, 0.88; palmitic acid/phosphorus, 2.00. The nonparamagnetic impurity was removed by silica column chromatography. Four milliliters of the chloroform solution of spin-labeled phosphatidylcholine were applied to a bed of silica (Mallinckrodt 100 mesh silicic acid, sieved to 60–140 mesh, activated for 2 days at 110°C) in chloroform, roughly 50 cm high, and eluted successively with 300 ml chloroform, 800 ml 5% methanol in chloroform, 400 ml 10%, 800 ml 12.5%, 800 ml 15%, and 2.5 liters of 17.5%. The second liter of 17.5% methanol in chloroform contained about 125 μmoles of phosphorus. The spot on the silica-impregnated paper chromatogram corresponding to an R_f value of 0.42 was about 1% as intense as the other spot. *Analysis*: Nitroxide/phosphorus, 1.04. Palmitic acid/phosphorus, 2.00.

The quantitative hydrolysis of spin-labeled phosphatidylcholine—0.5 ml alcohol-free diethyl ether, 0.02 ml 1 *M* pH 5.6 acetic acid–calcium acetate buffer, and 0.5 ml of cabbage supernate were shaken for 15 hr at room temperature—yielded only phosphatidic acid [identified on an oxalic acid-impregnated silica thin-layer chromatogram developed in chloroform–

methanol–HCl (12 *M*) (87 : 13 : 0.5, v/v)] by comparison with phosphatidic acid derived from egg phosphatidylcholine by the method of Davidson and Long (1958) and TEMPOcholine [identified on a cellulose thin-layer chromatogram developed in propanol–NH_4OH (14.8 *M*)–water (6 : 3 : 1, v/v) by comparison with TEMPOcholine chloride]. The R_f values of spin-labeled phosphatidylcholine on silica chromatograms in basic and acidic solvents were greater than the corresponding R_f values of phosphatidylcholine [R_f values of 0.67 and 0.40 for spin-labeled phosphatidylcholine and phosphatidylcholine, respectively, on a silica thin-layer chromatogram developed in chloroform–methanol–NH_4OH (14.8 *M*)–water (70 : 30 : 4 : 1, v/v), and R_f values of 0.47 and 0.42 for spin-labeled phosphatidylcholine and phosphatidylcholine, respectively, on a silica-impregnated paper chromatogram developed in diisobutyl ketone–acetic acid–water (8 : 5 : 1, v/v)]; that is, spin-labeled phosphatidylcholine was less polar than phosphatidylcholine by the criterion of silica chromatography.

N. The 1-oxyl-2,2-dimethyloxazolidine derivative of methyl-12-keto stearate (**LVIII**, *5*, *10*) (Procedure of Waggoner *et al.*, 1969)

$$CH_3—(CH_2)_m—C(—O—CH_2—C(CH_3)_2—N(\to O)—)—(CH_2)_n—COOCH_3$$

To form the oxazolidine compound, 3.0 gm (10 mmole) of methyl-12-keto stearate, 20 ml (200 mmole) of 2-amino-2-methyl-3-propanol (Aldrich), and 35 mg (0.2 mmole) of *p*-toluenesulfonic acid monohydrate were dissolved in 200 ml xylene. The mixture was refluxed for 10 days using a Dean–Stark trap containing anhydrous $CaSO_4$ for continuous water removal. After refluxing, the solution was cooled and washed four times with 50 ml portions of saturated $NaHCO_3$ solution, once with 50 ml saturated NaCl solution, then dried over Na_2SO_4. The xylene was removed with a rotary evaporator leaving 2.7 gm of thick colorless liquid that contained about 25% ketone starting material, traces of other impurities, and about 75% oxazolidine as estimated by thin-layer chromatography using silica gel G as an adsorbent, hexane–diethyl ether (7 : 3) as the developer, and H_2SO_4 spray followed with charring for detection of the spots. The ketone has R_f of about 0.7 and the oxazolidine an R_f of about 0.3. The $—CH_2—$ protons on the oxazolidine ring are found at 3.43 ppm (TMS) in the nmr spectrum. The oxazolidine was not isolated but was oxidized directly to nitroxide.

The mixture containing the oxazolidine was dissolved in 300 ml anhydrous diethyl ether containing 1.5 gm *m*-chloroperoxybenzoic acid (Aldrich Chemical Co.) (87% pure) was added dropwise to the oxazolidine solution over a period of 1 hr. The number of moles of *m*-chloroperoxybenzoic acid added was 1.50 times the number of moles of oxazolidine present (estimated by thin-layer chromatography as described above) in the 2.7 gm mixture of ketone and oxazolidine. As the peroxide was added, the solution slowly turned pale yellow. The ether solution was allowed to come to room temperature, and 48 hr later it was washed four times with 50 ml saturated $NaHCO_3$, once with water, dried over Na_2SO_4, evaporated to half its former volume, and placed in dry ice to precipitate excess ketone starting material. After evaporating the ether, the viscous, yellow, oily nitroxide was purified by preparative thin-layer chromatography again using hexane–ether (7 : 3). The yellow nitroxide band lies between the starting material and the oxazolidine. The total nitroxide yield based on the starting material was about 35%.

Anal. calcd. for $C_{23}H_{44}NO_4$: C, 69.3; H, 11.1; N, 3.5.
Found in two separate determinations:
C, 69.3 (69.4); H, 11.0 (10.9); N, 3.5 (3.5).

The ir spectrum of the nitroxide in chloroform shows a typical ester carbonyl band at 5.78 μm. The number of unpaired electrons per mole of methyl stearate nitroxide was found to be $5.0 \times 10^{23} \pm 0.5 \times 10^{23}$ based on two determinations using nitroxides of known purity.

O. The 1-oxyl-2,2-dimethyloxazolidine derivative of 5-keto palmitic acid (**LVIII**, *10, 3*) (Hubbell and McConnell, 1971)

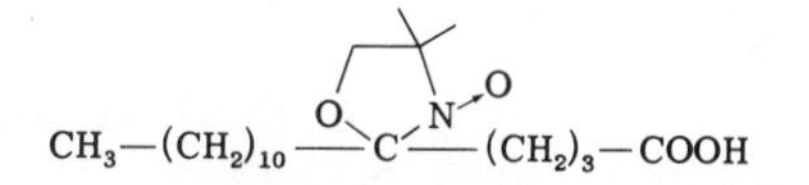

A solution of 68.2 gm undecyl bromide (0.29 mole) in 350 ml of dry diethyl ether was added in the usual way to 6.94 gm (0.29 gm-atom) of magnesium turnings in 100 ml of dry diethyl ether. After the addition, the reaction was refluxed until almost complete disappearance of the magnesium.

The Grignard solution was cooled in an ice bath and 27.5 gm (0.15 mole) finely powdered anhydrous $CdCl_2$ was added in one portion. The ice bath was removed, and after a period of 30 min the ether was almost completely removed by distillation. Dry benzene (350 ml) was added and 28 gm (0.23 mole) methyl-4-(chloroformyl)butyrate (Aldrich Chemical Co.) in 100 ml dry benzene was added over a period of 10 min with rapid stirring. The mixture was then heated to reflux for 1 hr, after which time the mixture was cooled in an ice bath and approximately 100 ml water was added slowly with stirring. Then a large excess of 0.1 *N* H_2SO_4 was added until two distinct phases formed. The benzene was removed under reduced pressure and the solid twice recrystallized from pentane, yielding 40 gm methyl 5-ketopalmitate: mp (uncorr) 49°–50°C; ir (KBr) bands at 1735 (carboxylic ester) and 1715 cm^{-1} (ketone).

To 500 ml toluene were added 30 gm (0.1 mole) of the methyl 5-ketopalmitate, 100 ml (1.0 mole) of 2-amino-2-methyl-1-propanol, and 100 mg *p*-toluenesulfonic acid monohydrate. The mixture was refluxed for 6 days using a Dean–Stark trap for water removal. The toluene phase was then washed with six 200 ml portions of saturated $NaHCO_3$ solution and four 200 ml portions of water and dried with anhydrous Na_2SO_4. The toluene was removed under reduced pressure, yielding a colorless, viscous liquid. The 4′,4′-dimethyloxazolidine derivative of the methyl 5-ketopalmitate was not purified at this stage, but oxidized directly to the *N*-oxyl-4′,4′-dimethyloxazolidine derivative.

The 4′,4′-dimethyloxazolidine was dissolved in 500 ml diethyl ether and cooled to 0°C in an ice bath, and 100 ml diethyl ether containing 22.4 gm *m*-chloroperbenzoic acid (85%, Aldrich Chemical Co.) was added over a period of 2 hr. The mixture was allowed to stand for 12 hr, at which time the ether phase was washed four times with 200 ml portions of saturated $NaHCO_3$ and four times with 200 ml portions of water and dried with anhydrous Na_2SO_4. The ether was removed under reduced pressure, and the yellow, viscous oil chromatographed on silica gel, eluting with benzene–ether (7 : 3 v/v). The center portion of the fast-moving yellow band was collected, yielding 10 gm of the *N*-oxyl-4′,4′-dimethyloxazolidine derivative of methyl 5-ketopalmitate: ir (film) band at 1735 cm^{-1} (carboxylic ester).

Anal. calcd for $C_{21}H_{40}NO_4$: C, 68.07; H, 10.88; N, 3.78.
Found: C, 68.19; H, 10.78; N, 3.73.

The nitroxide ester was dissolved in 100 ml of dioxane, and 4% NaOH solution was added until an oil began to form. When the oil had dissolved, more of the NaOH solution was added until another oily phase formed. This was repeated until no precipitate formed on addition of NaOH solution. The solution was brought to pH 1 with concentrated HCl, 100 ml water was added, and the solution was extracted with ethyl acetate. The ethyl acetate phase was dried over anhydrous Na_2SO_4 and the ethyl acetate removed under reduced pressure to yield the corresponding *N*-oxyl-4′,4′-dimethyloxazolidine derivative of 5-ketopalmitic acid: mp (uncorr) 45–47°, ir (KBr) band at 1710 cm^{-1} (carboxylic acid).

P. Acylation of lysolecithin (Hubbell and McConnell, 1971)

(a) **LV**(*m*, *n*) anhydrides. The anhydrides were prepared by the following general method of Selinger and Lapidot (1966). To 75 ml dry CCl_4

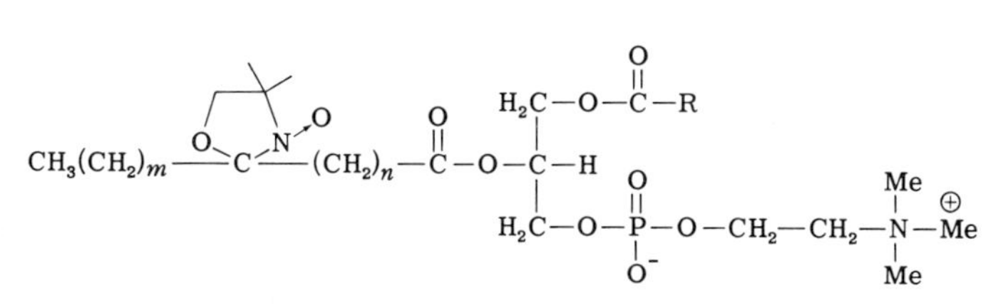

containing 10 mmole **LV**(m, n) was added 1.03 gm (5 mmole) dicyclohexylcarbodiimide in 25 ml of dry CCl_4. After 12 hr, the precipitated dicyclohexylurea was removed by filtration through sintered glass and the CCl_4 removed under reduced pressure. The anhydrides were used without further purification: ir (CCl_4) bands at 1810 and 1750 cm^{-1} (carboxylic anhydride).

(b) Egg lysolecithin. To 1 gm egg yolk lecithin in 300 ml ether was added 50 mg lyophilized *Crotalus adamanteus* venom (Pierce Chemical Co.) in 4 ml 5 mM $CaCl_2$ solution. At the end of 1.5 hr, 100 ml of ethanol was added and the volume reduced under vacuum to about 25 ml. The remaining solution was centrifuged at low speed to remove the precipitated enzyme. The precipitate was washed once with 2 : 1 v/v chloroform–methanol and the extract combined with the original supernate. The volume was reduced to approximately 2 ml and added to 40 ml diethyl ether. The precipitated lysolecithin was collected by centrifugation and washed twice with 40 ml of diethyl ether.

(c) Acylation of egg lysolecithin with **LV**(m, n) anhydrides. The method used here is that of Cubero Robles and Van den Berg (1969) and is illustrated by the acylation of egg lysolecithin with **LV**(10, 3) anhydride. To 50 mg (approximately 0.1 mmole) of egg lysolecithin in a 10 ml pear-shaped flask was added 0.27 gm (0.39 mmole) of **LV**(10, 3) anhydride and 3 mg (0.05 mmole) of finely powdered Na_2O. The flask was sealed and slowly rotated in an oil bath at 60°C. Periodically, the reaction mixture was examined by thin-layer chromatography. A gradual increase in the amount of lecithin was observed with a concomitant decrease in the lysolecithin, which had completely disappeared after 24 hr. A small amount (0.5 ml) of chloroform was then added to the reaction mixture and resulting solution applied to 7 gm Unisil (Clarkson Chemical Co., Williamsport, Pa.) that had been activated for 12 hr at 120°C. The fatty acids and excess anhydride were eluted with 2% methanol in chloroform and the yellow band collected. Evaporation of the solvent yielded 40 mg of a waxy yellow solid. The thin-layer chromatographic behavior of the lecithins prepared by this method [**LXV**(10, 3), **LXV**(7, 6), and **LXV**(5, 10) lecithins] is nearly identical with that of natural egg lecithin.

The spin-labeled lecithins hydrolyzed by phospholipase A to yield the corresponding fatty acid and a compound that cochromatographed with natural lysolecithin. Phospholipase D degradation gave a new compound that behaved on thin-layer chromatography like authentic phosphatidic acid. It should be noted that while the **LXV**(5, 10) lecithin hydrolyzed by both enzymes at approximately the same rate as natural lecithin, the **LXV**(7, 6) and **LXV**(10, 3) lecithins were hydrolyzed at a much slower rate, the effect being largest with phospholipase A. The ir spectra of the lecithins are consistent with published spectra of synthetic lecithins, with bands at 1735, 1075, and 1250 cm^{-1} (solid film).

Q. 6-Amino-TEMPO-ATP (6-amino[N-(1-oxyl-2,2,6,6-tetramethyl-4-piperidinyl)]-9-β-D-ribofuranosylpurine-5′-triphosphate) (Atkinson *et al.*, 1969).

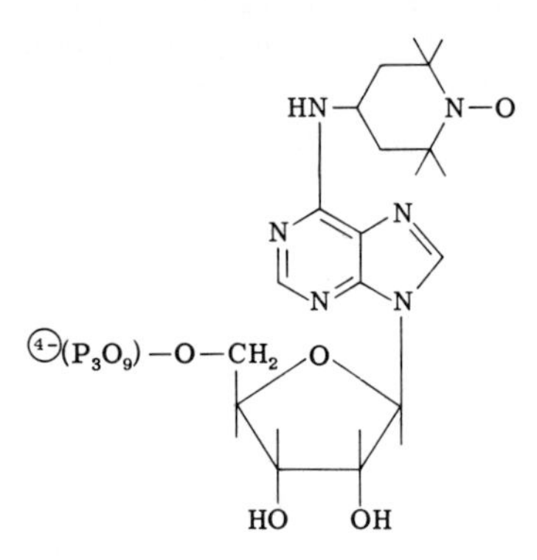

(a) Preparation of 6-chloropurine riboside. A mixture of 2′,3′,5′-tri-*o*-acetylinosine (50 gm, 0.127 mole), dimethylaniline (13.25 ml), and phosphoryl chloride (62.5 ml, 0.68 mole) in 62.5 ml chloroform was heated under reflux for 15 min. The reaction mixture turned light brown. The reaction mixture was poured into 500 ml ice water and the excess phosphoryl chloride was hydrolyzed. The organic layer was separated and washed twice with 500 ml 1 *N* HCl and with water until the pH of the aqueous layer was over 6. This extract was evaporated to a thick syrup and dissolved in 600 ml methanolic NH_3 and let stand over night at 5°C. The mixture was evaporated under reduced pressure and methanol added. The crude 6-chloropurine riboside was crystallized and yielded 20.03 gm, 55%, mp 176.5°–177°C. The paper chromatogram of the crude product showed traces of adenosine, which were removed by repeating the crystallization from aqueous methanol.

(b) Phosphorylation of 6-chloropurine riboside. To a mixture of phosphorus oxychloride (1.38 ml, 15 mmole) in trimethyl phosphate, 6-chloropurine riboside (1.25 gm, 5 mmole) was added at 0°C and the mixture was kept for 2 hr with stirring at 0°C. The mixture was poured into 50 ml of ice water and was neutralized to pH 8 with 6 *N* NaOH. Trimethyl phosphate was removed by extraction twice with 50 ml of chloroform. The aqueous layer was concentrated to 10 ml at a temperature below 50°C, and methanol (10 ml) was added. (Disodium 6-chloropurine ribotide is soluble in 60% aqueous methanol.) Inorganic salts were separated on a filter and the filtrate was evaporated to dryness.

The residue was dissolved in 10 ml water and an excess of aqueous barium acetate was added. The precipitates of inorganic salts and a small amount of Ba_2IMP (side product) were separated on a filter. The barium salt of 6-chloropurine ribotide was crystallized by adding 20 ml ethanol. Barium acetate is soluble in 80% aqueous ethanol. Recrystallization was repeated in the same way. 1.39 g of pure barium salt was obtained. Yield was 52% based on the nucleoside. Absorbance 250 nm/260 nm = 0.80, 280 nm/260 nm = 0.164.

(c) Preparation of TEMPO-AMP from 6-chloropurine ribotide. The barium salt of 6-chloropurine ribotide was converted to the free acid by passing through a Dowex 50 column (H^+ form). Then 200 μmoles of 6-chloropurine nucleotide in 2.6 ml was adjusted to pH 8.5 with TEMPO-amine and evaporated to dryness. 0.1 ml TEMPOamine and 3 ml *t*-butanol were added and heated to the boiling point to give a clear solution. The reaction was checked by electrophoresis in 0.05 *M* citrate pH 3.5. The slower component (TEMPO-AMP, $R_f = 0.48$ relative to 6-chloropurine ribotide) had $\lambda_{max} = 267$ nm and the reaction went to 95% conversion.

2.4 ml of reaction mixture containing 162 μmoles was diluted to 100 ml with 5 m*M* TEABC (triethyl ammonium bicarbonate) pH 8.2 and was applied to a DEAE-Sephadex column (10 cm × 1 cm^2 DEAE-TEABC). The column was washed with 5 m*M* TEABC and then the product eluted with a 200 ml gradient (5–200 m*M* TEABC). TEMPO-AMP eluted at 50 m*M*. The pooled fractions were reduced to dryness four times from water, and dissolved in 2 ml water. Yield was 119 μmoles or 74% from 6-chloropurine ribotide. Pentose : organic phosphate : orthophosphate = $1.00 \pm 0.04 : 0.99 \pm 0.03 : 0.02$; $\lambda_{max} = 267$ nm, $E_{267} = 20700$.

(d) Pyrophosphorylation of TEMPO-AMP. 119 μmoles TEMPO-AMP in 2 ml water was passed through Dowex-50 in the pyridine form. Five column volumes were collected and evaporated to near dryness. Then 119 μmoles of tributylamine (redistilled and anhydrous) in 3 ml pyridine was added to the residue. The residue was evaporated three times from pyridine and three times from acetetonitrile (spectroquality, anhydrous). The residue was dissolved in 2 ml HMP (hexamethylphosphoric triamide, redistilled and anhydrous). 600 μmoles of carbonyldiimidazole was added in

2 ml HMP. The reaction was allowed to proceed overnight at room temperature. The reaction was checked by electrophoresis on 0.05 *M* NH_4HCO_3 pH 8.0. The imidazole derivative had about 0.75 the mobility of the ribotide at pH 3.0. 1070 μmoles methanol was added in 0.5 ml HMP, and the mixture was allowed to stand for 4 hr. Then 1.19 ml HMP containing 600 μmoles of pyrophosphate (tributylamine salt) was added and the reaction proceeded at room temperature. After 4 hr, 1 ml of 1.2 *M* TEABC pH 8.2 and several grams of ice were added. The mixture was then diluted with 500 ml cold water and absorbed onto a DEAE-Sephadex column (as above). The product was eluted with a 200 ml, 0–450 m*M* TEABC gradient, pH 8.2. TEMPO-ATP eluted at 250 m*M*. The pooled fractions were evaporated to dryness two times from water and two times from methanol. Chromatography in *n*-propanol–NH_3–water (55 : 20 : 25) showed that the material was 97% TEMPO-ATP, with both a faster and a slower component (2.0% and 0.6%, respectively).

R. 6-Thioacetamido-TEMPO-ATP (**LXXI**) (*6-mercapto[N-(1-oxyl-2,2,6,6-tetramethyl-4-piperidinyl)]-acetamido-9-β-*D*-ribofuranosylpurine-5′-triphosphate*) (Cooke and Duke, 1971).

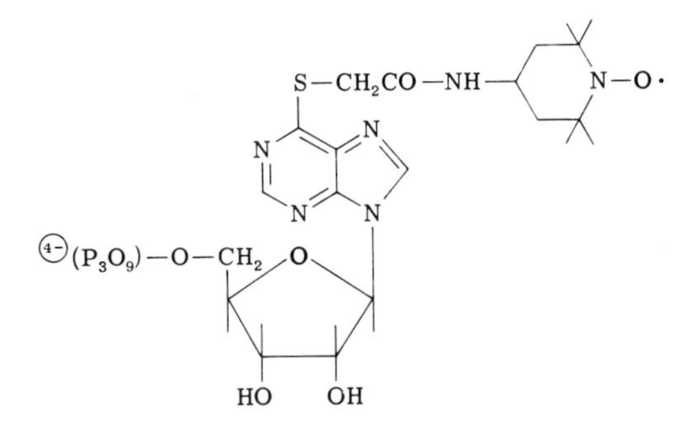

A typical preparation involved reaction of 10 mg (3.14×10^{-5} moles) of spin label dissolved in 10 ml of tris buffer (10^{-3} *M*) at pH 8.5 with 22 mg (3.42×10^{-5} moles) of 6-mercaptoATP. The mixture was stored in an oxygen-free atmosphere in the dark at 25°C. The progress of the reaction was followed by monitoring the disappearance of the absorption peak at 322 nm and the appearance of the peak at 282 nm. Upon completion of the reaction (72 hr), the reaction mixture was placed on a DEAE-cellulose (Bio-Rad, Richmond, California) column (9 × 2 cm) in the bicarbonate form at pH 7.6 and eluted with a 2 liter linear gradient (0.0–0.2 *M*) of tetraethylammonium bicarbonate. The eluent was monitored at 282 nm for the product that appeared in a well-defined peak separated from both unreacted SH-ATP and unreacted spin label. The fraction containing the product was evaporated, and the residue, dissolved in methanol, was precipitated with acetone containing a twentyfold molar excess of sodium perchlorate. The yellow precipitate was washed with acetone, then ether, and finally dried over P_2O_5. The yield was customarily 50% of the theoretical. The product had an extinction coefficient at 282 nm of 2.04×10^4 in a solvent consisting of 0.05 *M* potassium acetate (pH 4.8) and exhibited an R_f value of 0.78 on thin-layer cellulose plates in an isopropanol–ammonium bicarbonate system (3 : 2, v/v).

S. Preparation of coordinate nitroxide corrinoids (**LXXV**) (Buckman *et al.*, 1969)

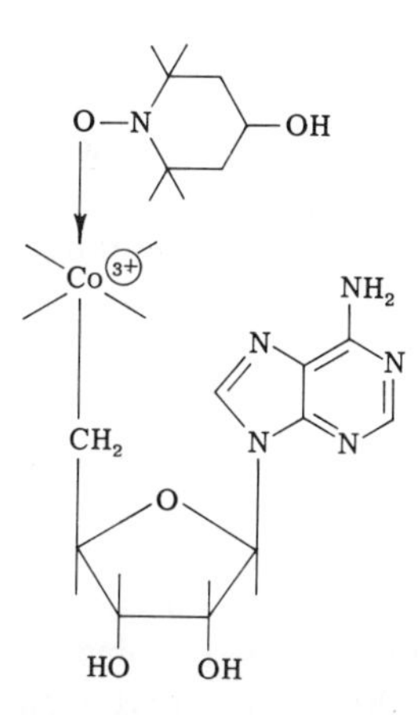

Aquo-Co(III)-nitroxylcobinamide was prepared by allowing a tenfold excess of either 2,2,6,6-tetramethylpiperidine-*N*-oxyl or 4-hydroxy-2,2,6,6-tetramethylpiperidine-*N*-oxyl with 10.0 mg diaquocobinamide to stand at room temperature for 10 days in 20 ml ethanol. Diaquocobinamide was synthesized by exhaustive aerobic photolysis of methylcobinamide. Methylcobinamide was synthesized and purified by the method described by Wood and Wolfe (1966). Aquo-Co(III)-nitroxycobinamide was crystallized by addition of excess acetone, followed by a 2 day growth period at 4°C. The crystals (7.0 mg) were centrifuged and Soxhlet-extracted for 48 hr with diethyl ether at a rate of 5 ml/min. When the colorless ether layer, dried over $CaSO_4$, was evaporated to dryness, no residue remained. The coordinate complex was stored at −15°C over $CaSO_4$. The ultraviolet-visible spectrum of this derivative indicates that water was not displaced from the upper axial ligand.

T. Spin-labeled protohemin and protoporphyrin (**LXXVII**) (Asakura *et al.*, 1969)

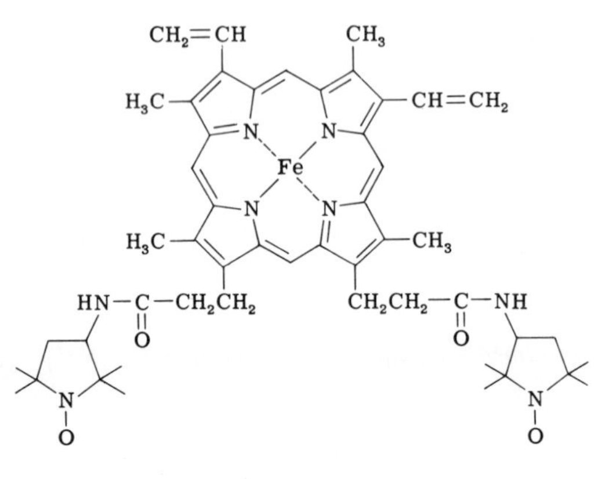

Spin-labeled protohemin, labeled at positions 6 and 7 of the porphyrin ring, was prepared from protohemin and 2,2,5,5-tetramethyl-3-aminopyrrolidine-1-oxyl as follows. Protohemin (13.2 mg) dissolved in 20 ml of tetrahydrofuran was mixed with the nitroxide in a molar ratio 4 to 6 radicals per hemin in the presence of *N*,*N*-dicyclohexylcarbodiimide (50 mg, Eastman Kodak Company). The reaction mixture was kept in a dark room for 24 hr at 23°C. The progress of the reaction was traced by thin-layer chromatography (2,6-lutidine–water, 10 : 1, v/v). Protohemin that did not move in the lutidine–water system gradually changed to mononitroxide compounds and finally to spin-labeled protohemin that has nitroxyl free radicals at positions 6 and 7 of the porphyrin ring. After the reaction, the solution was mixed with 200 ml of ethyl acetate and washed twice with water, once with 2 *N* Na_2CO_3 to remove unreacted protohemin, and again several times with water. By these washings, more than 95% of the unreacted spin labels were removed. The solution was then evaporated to dryness. The dried material was further purified by chromatography on Sephadex LH-20. Spin-labeled protoporphyrin was prepared by the same technique as mentioned above. Spin-labeled protohemin gave a clear single spot in thin-layer chromatography with an R_f value of 1.0, corresponding to protohemin dimethyl ester. The spin-labeled protohemin was soluble in organic solvent, but completely insoluble in aqueous alkaline solution, indicating that no free carboxyl groups remained. By optical and epr measurements, the ratio of hemin to the nitroxyl free radicals was calculated as approximately 1 : 2. The pyridine hemochromogen spectrum of the spin-labeled protohemin had the same absorption maximum at 555 mμm as either protohemin or protohemin dimethyl ester, suggesting that positions 2 and 4 of the porphyrin ring were intact. All these results suggest that the nitroxyl free radicals were bound at positions 6 and 7 of the porphyrin ring.

U. Spin-labeled NAD (adenosine 5′-diphosphate 4(2,2,6,6-tetramethylpiperidine-1-oxyl) (**LXXIII**) (Weiner, 1969)

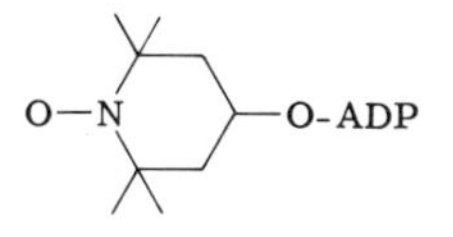

The solution of 1-oxyl-2,2,6,6-tetramethyl-4-phosphopiperidine (**XXIII**) in pyridine was evaporated to dryness many times. To the flask was added 5 ml dry pyridine and 600 mg AMP morpholidate (Calbiochemical Co.); the resulting solution was evaporated to dryness and 5 ml dry pyridine was added and the evaporation was repeated five more times. The reaction proceeded at room temperature in 5 ml pyridine for 5 days and was terminated by the addition of 15 ml water. The solvents were removed under vacuum on a rotary vacuum evaporator, keeping the temperature below 40°C. More water was added and the evaporation was continued until the odor of pyridine was absent. Finally the precipitate was dissolved in 20 ml water and the insoluble material was removed by filtration. To the aqueous solution was added 350 mg barium acetate and 100 ml ethyl alcohol and the solution was left at -15°C for 12 hr. The white precipitate was collected and dissolved in pH 6, $\mu = 0.006$, phosphate, and the barium salts were removed by centrifugation. The product was separated from AMP by chromatography on a DEAE-cellulose column using a linear gradient between $\mu = 0.006$ and 0.2 phosphate.†

The product was homogeneous on paper chromatography using ethanol–1.0 *M* ammonium acetate (pH 7.5, 7 : 3) or *n*-butanol–acetone–acetic acid–5% NH_4OH (4.5 : 1.5 : 1 : 1 : 2). The migration of the spin-labeled ADP was identical with that of commercial ADP ribose (Sigma Chemical Co.). The molar extinction coefficient for spin-labeled ADP is ε_{260} 17,000.

† Recent work indicates that the work-up time can be accelerated by separating the products on a 2 m Sephadex G-10 column rather than DEAE. Thus, the reaction is terminated with water as before; the water–pyridine is evaporated and the residue is dissolved in ca. 1 m*M* pH 6 pyridine–acetate buffer. Insoluble material is removed by centrifugation and the products are separated on a G-10 column. (H. Weiner, personal communication.)

V. (1-Oxyl-2,2,6,6-tetramethyl-4[³H]-piperidinyl)-β-D-galactoside **(LXXXII)** (Struve and McConnell, 1972)

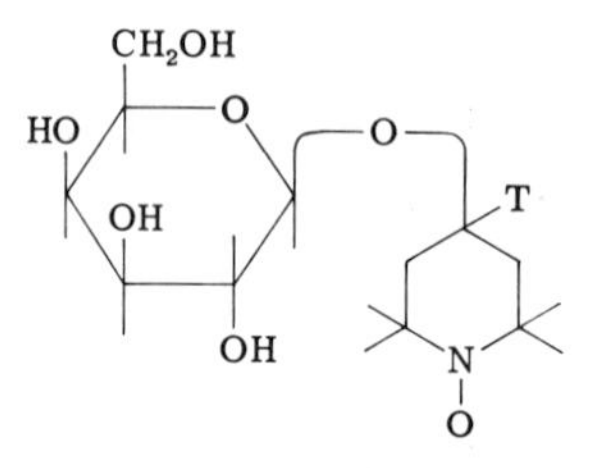

The tritiated alcohol (prepared by $NaBT_4$ reduction of 1-oxyl-triactoneamine) (1.89 gm, ~60 mCi), 3.04 gm silver carbonate, 10 gm Drierite, and 10 ml dry, alcohol-free chloroform were placed in a three-necked round-bottom flask equipped with a sealed mechanical stirrer, a drying tube, and a dropping funnel. The flask was wrapped in black paper and stirred for 1 hr. Then 0.41 gm anhydrous silver perchlorate was added as a catalyst. Acetobromo-α-D-galactose (4.13 gm), synthesized from D-galactose (Lemieux, 1963), was dissolved in 15 ml dry, alcohol-free chloroform, and the solution added to the stirred mixture over a period of 1 hr. Stirring was continued for 4 hr and the mixture filtered. The residue was washed with 10 ml chloroform and the combined filtrates were concentrated under reduced pressure. The compound was purified by chromatography on a silica gel column (Grace, grade 62, 60–200 mesh) 3.5 cm × 31.6 cm, eluted with dichloromethane–ethyl ether (8 : 2). Thin-layer chromatography in this solvent system separates the bromide, alcohol, and product from each other. The first orange band (270–330 ml) from the column was collected and evaporated to dryness. Thin layer chromatography in the same solvent showed only one spot ($R_f = 0.5$) by uv absorption, radioactivity, or by ashing with 50% H_2SO_4. The optical rotation was $[\alpha]_D^{24} = -9.9 \pm 0.2°C$ ($c = 6.7$, methanol).

The residue was dissolved in 10 ml of anhydrous methanol and 40 μl of 1 *M* sodium methoxide in methanol was added. After 3 hr at room temperature, the sodium methoxide was neutralized by addition of 20 mg anhydrous Dowex 50 (H^+ form). The mixture was filtered after stirring for $\frac{1}{2}$ hr. The filtrate was evaporated to a small volume and applied to Whatman 3MM paper. The chromatogram was developed by descending chromatography with ethyl acetate–*n*-propanol–water (5 : 3 : 2 v/v). The uv absorbing band nearest the origin was eluted with water. The eluate was evaporated to dryness and the residue was dissolved in a small amount of absolute ethanol. A slight precipitate that formed was removed by filtration and the filtrate evaporated to dryness. Yield: 0.74 mole (7.4%).

W. (1-Oxyl-2,2,5,5-tetramethyl-3-pyrrolidinyl)methyl-β-chitobiose (**LXXXIV**) (Wien *et al.*, 1972)

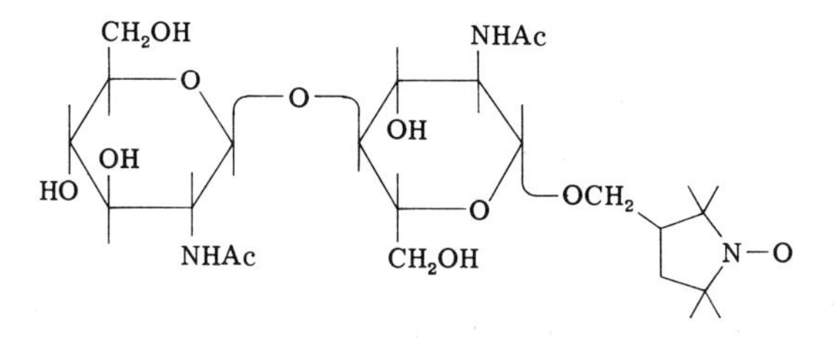

The procedure for the preparation of this compound is a modification of the method described by Kuhn and Kirschenlohr (1953). Acetochlorochitobiose (314 mg) prepared by the method of Dahlquist and Raftery (1969) was intimately mixed with 344 mg of (1-oxyl-2,2,5,5-tetramethyl-3-pyrrolidinyl)carbinol and 504 mg of $Hg(CN)_2$ using a mortar and pestle. This mixture was transferred to a 25 ml flask and dried *in vacuo* for 5 hr. The reaction was initiated by adding 3 ml of chloroform (freshly distilled from P_2O_5). The reaction mixture was stirred at room temperature for 72 hr, then evaporated to dryness under reduced pressure at less than 40°C. The residue was redissolved in chloroform and excess mercuric cyanide removed by filtration. The yellow filtrate was washed exhaustively with water to remove excess **IV**, dried over $MgSO_4$, and evaporated to dryness yielding a yellow glass. This was applied to a column (0.6 × 10 cm) of SilicAR, CC-4, 100–200 mesh (Mallinckrodt Chemical Works) equilibrated with methylene chloride. The column was developed by eluting successively with methylene chloride, methylene chloride–chloroform (1 : 1), chloroform, and chloroform–methanol (10 : 1). The desired blocked glycoside was eluted last. Evaporation to dryness afforded a yellow glass that could not be crystallized. The infrared spectrum exhibited carbonyl bands at 1745 cm^{-1} (acetate) and 1660 cm^{-1} (acetamide and no hydroxyl absorption. This material was deacylated by dissolving in 20 ml methanol saturated with NH_3 at 5°C. After 24 hr, the reaction mixture was evaporated to dryness at less than 30°C. The yellow syrup was completely soluble in water; lyophilization afforded 91 mg of a yellow hygroscopic powder. The infrared spectrum exhibited carbonyl absorption at 1655 cm^{-1} (acetamide) but not at 1730–1750 cm^{-1} (acetate).

Anal. calcd. for $C_{25}H_{44}N_3O_{12}NH_3$:
C, 50.41; H, 7.95; N, 9.40.
Found: C, 50.56; H, 8.07; N, 10.04.

X. (1 - Oxyl - 2,2,6,6 - tetramethyl - 4 - amidopiperidinyl)methyl - β - D - thiogalactoside **(LXXXIX)** (S. J. Opella and B. J. Gaffney, unpublished)

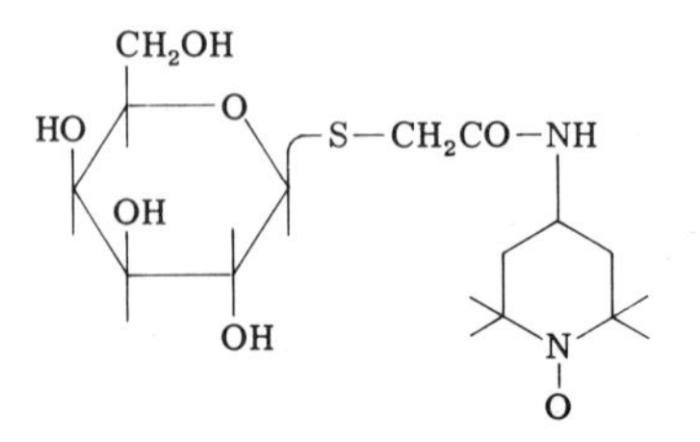

Tetra-*O*-acetyl-α-D-galactosyl bromide (Lemieux, 1963) was converted to tetra-*O*-acetyl-*β*-D-thiogalactose and then to the thiogalactoside by the method of Černý and Pacák (1959). To a solution of 0.76 gm (0.002 mole) acetyl-*β*-D-thiogalactoside and 0.7 gm (0.002 mole) 1-oxyl-2,2,6,6-tetramethyl-4-iodoacetamidopiperidine in 14 ml acetone was added 2 ml water and 0.27 gm K_2CO_3. The mixture was stirred at room temperature for 4 hr, at which time thin-layer chromatography [chloroform–ether, 1 : 1, Keisegel plates (Merck)] showed no remaining tetraacetylthiogalactose. The reaction mixture was filtered and evaporated to complete dryness. The anhydrous, acetylated reaction product was dissolved in 5 ml anhydrous ethanol, and 20 mg sodium methoxide was added to this mixture to achieve deacetylation. After 5 min, 5 ml water was added. The cation exchange resin IR-120 (H^+ form) was added to the deacetylation mixture to remove cations, and after 5 min, the suspension was filtered. The eluate was evaporated to dryness, taken up in 2 ml water, extracted three times with ether, and again evaporated to dryness. Thin-layer chromatography on Kieselgel plates [5 × 20 cm, solvent: isopropanol–ethyl acetate–water (7 : 1 : 2); charred with 50 % H_2SO_4] gave single spots for each of the following: the deacetylated thiogalactoside spin label, the product of the reaction of iodoacetamide spin label with tetraacetylthiogalactose, and the fully acetylated spin-labeled thiogalactoside (acetylated with acetic anhydride–sodium acetate) (R_f values: 0.61, 0.66, and 0.71).

Anal. calcd. for $C_{17}H_{31}N_2SO_7$: C, 50.10; H, 7.67; N, 6.87.
Found: C, 49.46; H, 8.33; N, 5.53.

Y. (1-Oxyl-2,2,5,5-tetramethylpyrrolidinyl)-β-D-thiogalactoside **(LXXXVIII)** (B. J. Gaffney, unpublished)

This procedure follows that of Černý and Pacák (1959) for preparation of *β*-D-thioglucosides. To a solution of 0.32 gm (0.89 × 10^{-3} moles) tetra-*O*-acetylthio-*β*-D-galactose in 1.5 ml acetone was added 0.23 gm (0.98 × 10^{-3}

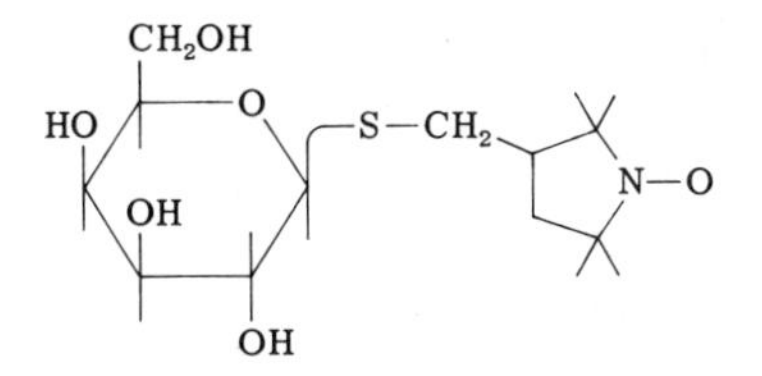

moles) of 2,2,5,5-tetramethyl-3-bromomethylpyrrolidinylbromomethane and 0.12 gm K_2CO_3 dissolved in 1 ml water. A crystal of NaI was added to the reaction mixture, and the mixture was heated at 35°C for 3 hr in a stoppered flask. After addition of 10 ml water and 2 ml chloroform, the organic and aqueous layers were separated and the aqueous layer was extracted three times with 1 ml chloroform. The organic layers were dried over Na_2SO_4 and evaporated to give 0.45 gm of a yellow oil. A portion (0.23 gm) of this oil was purified by preparative thin-layer chromatography (three plates, 20 × 20 cm, 0.75 mm thick, Merck Kieselgel HF_{254}, dried 30 min at 120°C and allowed to stand 20 hr in air) using chloroform–methanol (180 : 1) as solvent R_f values for the product, tetraacetylthio-β-D-galactoside, and nitroxide bromide were 0.17, 0.33, 0.47). The yield was 0.14 gm (59%).

Deacetylation was achieved by dissolving 0.14 gm of the spin-labeled acetyl sugar in 1 ml anhydrous methanol containing a trace of sodium methoxide. The solution was filtered through charcoal and evaporated to give 0.084 gm of (1-oxyl-2,2,5,5-tetramethylpyrrolidinyl)thio-β-D-galactoside.

Z. 1 - Oxyl - 2,2,5,5 - tetramethyl - 3 - [3 - [(p - sulfamoylphenyl)carbamoyl] - propionamide]-1-pyrrolidine [**XCII**(e)] (Procedure of Erlich *et al.*, 1973)

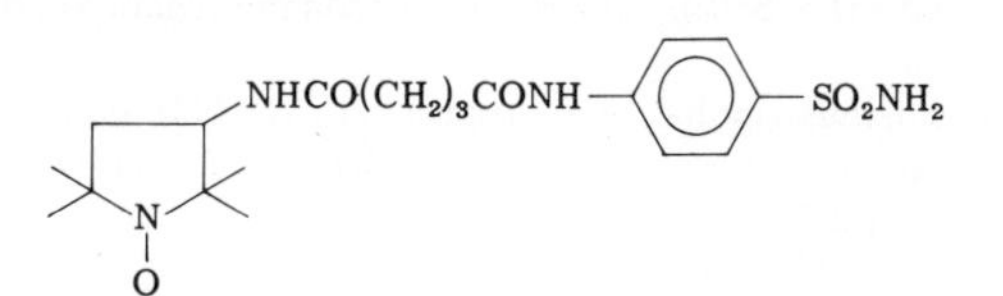

Sulfanilamide (2.0 gm; 12 mmoles) and succinic anhydride (1.16 gm; 12 mmoles) was dissolved in acetone (10 ml), and the reaction mixture was allowed to stand overnight. The precipitate (2.5 gm) was removed by filtration recrystallized from methanol to give colorless needle crystals, mp 208°–210°C.

Calculated: C, 44.11; H, 44.44; N, 10.29.
Found: C, 43.77; H, 4.47; N, 10.03.

Freshly distilled isobutylchloroformate (0.48 ml; 3.7 mmoles) was added dropwise to a solution of 4-succinamidobenzenesulfonamide (1.0 gm; 3.7 mmoles) in dry tetrahydrofuran (20 ml) cooled in an ice–salt bath. After the mixture had been stirred for 2 hr, triethylamine (0.52 ml; 3.7 mmoles) was added, followed by a solution of 1-oxyl-3-amino-2,2,5,5-tetramethyl-1-pyrrolidine (0.58 gm; 3.7 mmoles) in dry tetrahydrofuran (3 ml). The reaction mixture was then allowed to warm to room temperature and was stirred for 18 hr. After the reaction mixture had been evaporated to dryness, the residue was dissolved in water (10 ml) and then acidified to pH 3.0 with aqueous HCl and immediately extracted with three 10 ml portions of ethyl acetate. The ethyl acetate extracts were combined, dried over Na_2SO_4, and then evaporated to dryness. The residue (1.2 gm) was recrystallized from methanol to give **XCII**(*e*) as pale orange needle crystals, mp 208°–209°C.

Calculated: C, 52.54; H, 6.62; N, 13.62.
Found: C, 52.55; H, 6.75; N, 13.63.

Acknowledgments

I would like to thank Professor H. M. McConnell and the colleagues with whom I have worked during a number of years in his laboratory for the numerous inspiring conversations which have brought to my attention many of the papers cited in this chapter.

I am very grateful to Professor A. Kornberg for permission to include the unpublished synthesis of 6-amino-TEMPO-ATP which he, D. Brutlag, and the late M. Atkinson developed.

Some of the synthetic procedures included were developed in the laboratory of Professor K. Nakanishi, Tohoku University, Sendai, Japan, while I was a Varian Associates Postdoctoral Fellow (1967–1968).

The assistance of Robert S. Schepp and Sandi Hanson in preparation of the manuscript is gratefully acknowledged.

Preparation of this manuscript has been supported by the NSF under grant GB33501X-1, and by the NIH under grant NS-08058-06 (to H. M. McConnell) and by the NIH under grant 1 R01 CA 15997-01 (to myself).

References

For reviews of nitroxide chemistry, see:

Buchachenko, A. L. (1965). "Stable Radicals," Consultants Bureau, New York.

Forrester, A. R., Hay, J. M., and Thomson, R. H. (1968). "Organic Chemistry of Stable Free Radicals." Academic Press, New York.

Nelson, S. F. (1973). *In* "Free Radicals" (J. K. Kochi, ed.), Vol. II, Chapter 21, pp. 527–594. Wiley, New York.

Rozantsev, E. G. (1970). "Free Nitroxyl Radicals" (translated by B. J. Hazzard). Plenum Press, New York.

Rozantsev, E. G., and Scholle, V. D. (1971). Synthesis and reactions of stable nitroxyl radicals, *Synthesis* 190–202.

For reviews of spin labeling, see:

Berliner, L. J. (1974). Applications of spin labeling to structure-conformation studies of enzymes, *Progr. Bioorg. Chemi.* **3**, 1–80.

Feher, G. (1969). Spin-labeled biomolecules, *In* "Electron Paramagnetic Resonance," pp. 108–139. Gordon and Breach, New York.

Gaffney, B. J., and McNamee, C. M. (1974). Spin label measurements in membranes, *Methods Enzymol.* **32**, 161–198.

Griffith, O. H., and Waggoner, A. S. (1969). Nitroxide free radicals: Spin labels for probing biomolecular structure, *Accounts. Chem. Res.* **2**, 17–24.

Hamilton, C. L., and McConnell, H. M. (1968). Spin labels, *In* "Structural Chemistry and Molecular Biology" (A. Rich and N. Davidson, eds.), pp. 115–149. Freeman, San Francisco, California.

Jost, P., and Griffith, O. H. (1972). Electron spin resonance and the spin labeling method, *Methods Pharmacol.* **2**, 223–276.

Jost, P., Waggoner, A. S., and Griffith, O. H. (1971). Spin labeling and membrane structure, *In* "Structure and Function of Biological Membranes" (L. I. Rothfield, ed.), pp. 83–144. Academic Press, New York.

Kalmanson, A. E., and Grigoryan, G. L. (1973). Spin labels in E.P.R. investigation of biological systems, *In* "Experimental Methods in Biophysical Chemistry" (C. Nicolau, ed.), pp. 589–612. Wiley, New York.

McConnell, H. M., and Gaffney-McFarland, B. (1970). Physics and chemistry of spin labels, *Quart. Rev. Biophys.* **3**, 91–136.

Ohnishi, S. (1968). The spin-label technique, *Seibutsu Butsuri* **8**, 118–129.

Smith, I. C. P. (1972). The spin-label method, *In* "Biological Applications of Electron Spin Resonance Spectroscopy" (J. R. Bolton, D. Borg, and H. Schwartz, eds.). Wiley (Interscience), New York.

General References

Adevih, G., and Lagercrantz, C. (1970). Spin trapping of the radicals formed by α-irradiation of sodium phosphite in the solid state, *Acta Chem. Scand.* **24**, 2253–2255.

Aneja, R., and Davies, A. P. (1970). The synthesis of a spin-labeled glycero-phospholipid, *Chem. Phys. Lipids* **4**, 60–71.

Antsiferova, L. I., Lazarev, A. V., and Stryukov, V. B. (1970). Effect of rotational diffusion on the shape of a paramagnetic resonance line, *Pis'ma Zh. Eksp. Theor. Fiz.* **12**, 108–12.

Asakura, T., Drott, H. R., and Yonetani, T. (1969). Studies on cytochrome *c* peroxidase, *J. Biol. Chem.* **244**, 6626–6631.

Asakura, T., Leigh, J. S., Jr., Drott, H. R., Yonetani, T., and Chance, B. (1971). Structural measurements in hemoproteins: use of spin-labeled protoheme as a probe of HEME environment, *Proc. Nat. Acad. Sci. U.S.* **68**, 861–865.

Atkinson, M. R., Brutlag, D. L., and Kornberg, A. (1969). Unpublished.

Baer, E., and Buchnea, D. (1959). Synthesis of saturated and unsaturated L-α-lecithins, *Can. J. Biochem. Physiol.* **37**, 953–959.

Barker, R. (1971). "Organic Chemistry of Biological Compounds," Table 4.3, p. 59. Prentice Hall, Englewood Cliffs, New Jersey.

Barrat, M. D., Davies, A. P., and Evans, M. T. A. (1971a). Maleimide and isomaleimide pyrrolidine-nitroxide spin labels, *Eur. J. Biochem.* **24**, 280–283.

Barrat, M. D., Leslie, R. B., and Scanu, A. M. (1971b). Protein-protein and protein–lipid interactions in human serum high-density lipoprotein: an analysis by a spin-label method, *Chem. Phys. Lipids* **7**, 345–355.

Barzilay, I., and Lapidot, Y. (1971). The use of *O*-nitrophenylsulfenyl group as amino protecting group in the synthesis of phosphatidylethanolamine, *Chem. Phys. Lipids* **7**, 93–97.

Berliner, L. J., and Wong, S. S. (1974). Spin-labeled sulfonyl fluorides as active site probes of protease structure. I. Comparison of the active site environments in α-chymotrypsin and trypsin, *J. Biol. Chem.* **249**, 1668–1677.

Briere, R., Dupeyre, R. M., Lemaire, H., Norat, C., Rassat, A., and Rey, P. (1965). Nitroxydes XVII: Biradicaux stables du type nitroxyde, *Bull. Soc. Chim. Fr.* 3290–3297.

Briere, R., Lemaire, H., and Rassat, A. (1968). Nitroxides, XXV. Isotropic splitting constant for ^{13}C in the tertiary carbon position of di-*tert*-butyl nitroxide, *J. Chem. Phys.* **48**, 1429–1430.

Briere, R., Chapelet-Letourneux, G., Lemaire, H., and Rassat, A. (1971). Nitroxydes. XXXV étude de l'interaction hyperfine electron-carbon-13., radicaux nitroxydes selectivement marques en α de 1′ azote, *Mol. Phys.* **20**, 211–224.

Brockerhoff, H., and Yurkowski, M. (1965). Simplified preparation of L-α-glyceryl phosphoryl choline, *Can. J. Biochem.* **43**, 1777.

Buckman, T. (1970). Spin-labeling studies of asparate transcarbamylase. I. Effects of nucleotide binding and subunit separation, *Biochemistry* **9**, 3255–3265.

Buckman, T., Kennedy, F. S., and Wood, J. M. (1969). Spin labeling of vitamin B_{12}, *Biochemistry* **8**, 4437–4442.

Černý, M., and Pacák, J. (1959). Eine neue methode zur darstellung acetylierter β-D-thioglucopyranoside, *Coll. Czech. Chem. Commun.* **24**, 2566–2570.

Chapman, N. B., and Saunders, B. C. (1948). Esters containing phosphorus. Part VI. Preparation of esters of fluorophosphonic acid by means of phosphorus oxydichlorofluoride, *J. Chem. Soc.* 1010–1014.

Chiarelli, R., and Rassat, A. (1973). Syntheses de radicaux nitroxydes deuteries, *Tetrahedron* **29**, 3639–3647.

Chignell, C. F., Starkweather, D. K., and Erlich, R. H. (1972). The interaction of some spin-labeled sulfonomides with bovine erythrocyte carbonic anhydrase B, *Biochim. Biophys. Acta.* **271**, 6–15.

Colbeau, A., Vignais, P. M., and Piette, L. H. (1972). Biosynthetic synthesis and isolation of phospholipid spin-labels from rat liver microsomes and mitochondria, *Biochem. Biophys. Res. Commun.* **48**, 1495–1503.

Cooke, R., and Duke, J. (1971). The synthesis and some properties of a spin label analogue of adenosine-5′-triphosphate, *J. Biol. Chem.* **246**, 6360–6369.

Cooke, R., and Morales, M. F. (1969). Spin-label studies of glycerinated muscle fibers, *Biochemistry* **8**, 3188–3194.

Corker, G. A., Klein, M. P., and Calvin, M. (1966). Chemical trapping of a primarily quantum conversion product in photosynthesis, *Proc. Nat. Acad. Sci. U.S.* **56**, 1365–1369.

Crossley, F. S. (1955). Methyl isothiocyanate, *Org. Syn. Coll.* **3**, 599–600.

Dahlquist, F. W., and Raftery, M. A. (1969). Some properties of contiguous binding subsites on lysozyme as determined by proton magnetic resonance spectroscopy, *Biochemistry* **8**, 713–724.

Davidson, F. M., and Long, C. (1958). The structure of the naturally occurring phosphoglycerides, *Biochem. J.* **69**, 458–466.

Devaux, P., and McConnell, H. M. (1974). Equality of the rates of lateral diffusion of phosphatidyl ethanolamine and phosphatidylcholine spin labels in rabbit sarcoplasmic reticulum, *Ann. N.Y. Acad. Sci.* **222**, 489–496.

Dittmer, J. C., and Lester, R. C. (1964). A simple, specific spray for the detection of phospholipids on thin-layer chromatograms, *J. Lipid Res.* **5**, 126–127.

Dupeyre, R. M., and Rassat, A. (1966). "Nitroxides. XIX, Norpseudopelletierine-*N*-oxyl; A new, stable, unhindered free radical, *J. Amer. Chem. Soc.* **88**, 3180–3181.
Durham, L. J., McLeod, D. J., and Cason, J. (1963). Methyl hydrogen hendecandioate, *Org. Syn. Coll.* **4**, 635–638.
Edman, P. (1956). On the mechanism of the phenyl isothiocyanate degradation of peptides, *Acta Chem. Scand.* **10**, 761–768.
Elderfield, R. C. (1957). "Heterocyclic Compounds," Vol. 5, 391–392. Wiley, New York.
Erlich, R. H., Starkweather, D. K., and Chignell, C. F. (1973). A spin label study of human erythrocyte carbonic anhydrases B and C, *Mol. Pharmacol.* **9**, 61–73.
Flohr, K., and Kaiser, E. T. (1972). Enantiomeric specificity in the chymotrypsin-catalysed hydrolysis of 3-carboxy-2,2,5,5-tetramethylpyrrolidin-1-oxy-*p*-nitrophenyl ester, *J. Amer. Chem. Soc.* **94**, 3675–3676.
Freedman, T. H., and Applewhite, T. H. (1966). Dehydrogenation of methyl 12-hydroxystearate to methyl 12 ketostearate with Raney nickel, *J. Amer. Oil. Chem. Soc.* **43**, 125–127.
Golubev, V. A., Rozantsev, E. G., and Neiman, M. B. (1965). Some reactions of free iminoxyl radicals with the participation of the unpaired electron, *Akad. Nauk. SSSR (Eng.)* 1898–1904.
Griffith, O. H., and McConnell, H. M. (1966). A nitroxide-maleimide spin label, *Proc. Nat. Acad. Sci. U.S.* **55**, 8–11.
Griffith, O. H., Keana, J. F. W., Noall, D. L., Ivey, J. L. (1967). Nitroxide mixed carboxylic–carbonic acid anhydrides—A new class of versatile spin labels, *Biochim. Biophys. Acta* **148**, 583–585.
Hara, H. Horiuchi, T., Saneyoshi, M., and Nishimura, S. (1970). 4-thiouridine-specific spin-labeling of *E. coli* transfer RNA, *Biochem. Biophys. Res. Commun.* **38**, 305–311.
Hirs, E. H. W., and Timasheff, S. N. (eds.) (1972). Modification reagents, *Methods Enzymol.* **25**, Part B, Sect. VIII, 387–672.
Hoff, A. J., Oosterbaan, R. A., and Deen, R. (1971). Specific and non-specific binding of an organophosphate spin label to some serine esterases, *FEBS Lett.* **14**, 17–21.
Hoffman, A. K., and Henderson, A. T. (1961). A new stable free radical: Di-*t*-butylnitroxide, *J. Amer. Chem. Soc.* **83**, 4671–4672.
Hoffman, B. M., and Eames, T. B. (1969). Protonated nitroxide free radical, *J. Amer. Chem. Soc.* **91**, 2169–2170.
Hoffman, B. M., Schofield, P., and Rich, A. (1970). Spin labeling studies of aminoacyl transfer ribonucleic acid, *Biochemistry* **9**, 2525–2532.
Hsia, J. C., and Piette, L. H. (1969). Spin labeling as a general method in studying antibody active site, *Arch. Biochem. Biophys.* **129**, 296–303.
Hsia, J. C., and Little, J. R. (1971). Alterations of antibody binding properties and active-site dimensions in the primary and secondary immune response, *Biochemistry* **10**, 3742–3748.
Hsia, J. C., Kosman, D. J., and Piette, L. H. (1969). Organophosphate spin-label studies of inhibited esterases, α-chymotrypsin and cholinesterase, *Biochem. Biophys. Res. Commun.* **36**, 75–78.
Huang, C., Charlton, J. P., Shyr, C. I., and Thompson, J. E. (1970). Studies on phosphatidylcholine vesicles with thiocholesterol and a thiocholesterol-linked spin label incorporated in the vesicle wall, *Biochemistry* **9**, 3422–3426.
Hubbell, W. L., and McConnell, H. M. (1969). Motion of steroid spin labels in membranes, *Proc. Nat. Acad. Sci. U.S.* **63**, 16–22.
Hubbell, W. L., and McConnell, H. M. (1971). Molecular motion in spin-labeled phospholipids and membranes, *J. Amer. Chem. Soc.* **93**, 314–326.
Humphries, G. K., and McConnell, H. M. (1974). *Proc. Nat. Acad. Sci. U.S.* **71**, 1691–1694.
Icke, R. N., and Wisegarver, B. B. (1955). β-phenylethyl dimethylamine, *Org. Syn. Coll.*, **3**, 723–725.
Janzen, E. G. (1971). Spin trapping, *Accounts Chem. Res.* **4**, 31–40.

Kaplan, J., Canonico, P. G., and Caspary, W. J. (1973). Electron spin-labeled mammalian cells by detection of surface-membrane signals, *Proc. Nat. Acad. Sci. U.S.* **70**, 66–70.

Keana, J. F. W., and Baitis, F. (1968). Photolysis of the stable nitroxide, 3-carbamoyl-2,2,5,5-tetramethylpyrrolidine-1-oxyl, *Tetrahedron Lett.* 365–368.

Keana, J. F. W., and Dinerstein, R. J. (1971). A new highly anisotropic dinitroxide ketone spin label. A sensitive probe for membrane structure, *J. Amer. Chem. Soc.* **93**, 2809–2810.

Keana, J. F. W., Keana, S. B., and Beetham, D. (1967). A new versatile ketone spin label, *J. Amer. Chem. Soc.* **89**, 3055–3056.

Keana, J. F. W., Dinerstein, R. J., and Baitis, F. (1971). Photolytic studies on 4-hydroxy-2,2,6,6-tetramethylpiperidine-1-oxyl, a stable nitroxide free radical, *J. Org. Chem.* **36**, 209–211.

Kornberg, R. D. (1972). The Diffusion of Phospholipids in Membranes. Thesis, Stanford Univ.

Kornberg, R. D., and McConnell, H. M. (1971). Inside-outside transitions of phospholipids in vesicle membranes, *Biochemistry* **10**, 1111–1120.

Kosman, D. J., Hsia, J. C., and Piette, L. H. (1969). ESR probing of macromolecules: Function and operation of structural units within the active site of α-chymotrypsin, *Arch. Biochem. Biophys.* **133**, 29–37.

Krielich, R. W. (1967). NMR studies of a series of aliphatic nitroxide radicals, *J. Chem. Phys.* **46**, 4260–6264.

Kuhn, R., and Kirschenlohr, W. (1953). β-Glycosides of *N*-acetyl-β-D-glucosamine, *Ber.* **86**, 1331–1333.

Law, P. W., Brown, D. G., Lien, E. L., Babior, B. M., and Wood, J. M. (1971). Synthesis and catalytic activity of spin-labeled cobinamide coenzymes, *Biochemistry* **10**, 3428–3435; also *Biochemistry* **8**, 4437–4442 (1969).

Lebedev, O. L., and Kazarnovskii, S. N. (1959). Intermediate products of oxidation of amines by pertungstate, *Papers Chem. Chem. Technol.* **2**, 649–665 [*Chem. Abstr.* **56**, 15479f (1962)].

Lemieux, R. V. (1963). *In* "Methods in Carbohydrate Chemistry" (R. L. Whistler and M. L. Wolfrom, eds.), Vol. II, pp. 221–222. Academic Press, New York.

Leute, R. K., Ullman, E. F., Goldstein, A., and Herzenberg, L. A. (1972). Spin immunoassay technique for determination of morphine, *Nature (London) New Biol.* **236**, 93–94.

Linden, C., Wright, K., McConnell, H. M., and Fox, C. F. (1973). Phase separations and glucoside uptake in *E. coli* fatty acid auxotrophs, *Proc. Nat. Acad. Sci. U.S.* **76**, 2271–2275.

McConnell, H. M., and Hamilton, C. L. (1968). Spin-labeled hemoglobin derivatives in solution and in single crystals, *Proc. Nat. Acad. Sci. U.S.* **60**, 776–781.

McConnell, H. M., Deal, W., and Ogata, R. T. (1969). Spin-labeled hemoglobin derivatives in solution, polycrystalline suspensions and single crystals, *Biochemistry* **8**, 2580–2585.

McFarland, B. G., and McConnell, H. M. (1971). Bent fatty acid chains in lecithin bilayers, *Proc. Nat. Acad. Sci.* **68**, 1274–1278.

Morrisett, J. D., and Drott, H. R. (1969). Oxidation of the sulfhydryl and disulfide groups by the nitroxyl radical, *J. Biol. Chem.* **244**, 5083–5086.

Morrisett, J. D., Broomfield, C. A., and Hackley, B. E. Jr. (1969). A new spin label specific for the active site of serine enzymes, *J. Biol. Chem.* **244**, 5758–5761.

Morrison, A., and Davies, A. P. (1970). Mass spectrometry of piperidine nitroxides—A class of stable free radicals, *Org. Mass. Spectrosc.* **3**, 353–366.

Neimann, M. B., Mairanovskii, S. G., Kovarskaya, B. M., Rozantsev, E. G., and Gintsberg, E. G. (1964). Polarographic study of certain *N*-oxide free radicals, *Akad. Nauk. SSSR Bull. (Eng)* 1424–1426.

Ogata, R. T. (1971). The Binding of Triphosphate Spin Labels to Hemoglobin. Thesis, Stanford Univ.

Ogawa, S. (1967). Spin Label Studies of Conformational Changes in Hemoglobin. Thesis, Stanford Univ.

Ogawa, S., and McConnell, H. M. (1967). Spin-label study of hemoglobin conformations in solution, *Proc. Nat. Acad. Sci. U.S.* **58**, 19–26.

Ohnishi, S., Boeyens, J. C. A., and McConnell, H. M. (1966). Spin-labeled hemoglobin crystals, *Proc. Nat. Acad. Sci. U.S.* **56**, 809–813.

Raykhman, L. M., Annayev, B., and Rozantsev, E. G. (1972). Studies of conformational transitions of cytochrome *c* by the spin label method, *Mol. Biol. (SSSR)* **6**, 553–558 (translated by NASA, Washington, D.C.).

Robles, E. C., and Van den Berg, O. (1969). Synthesis of lecithins by acylation of *O*-(SN-glycero-3-phosphoryl)choline with fatty acid anhydrides, *Biochim. Biophys. Acta* **187**, 520–525.

Saunders, B. C., and Stacey, G. J. (1948). Esters containing phosphorus, Part IV. Diisopropyl fluorophosphonate, *J. Chem. Soc.* 695–699.

Seelig, J. (1970). Spin label studies of oriented smectic liquid crystals (A model system for bilayer membranes), *J. Amer. Chem. Soc.* **92**, 3881–3887.

Selinger, Z., and Lapidot, Y. (1966). Synthesis of fatty acid anhydrides by reaction with dicyclohexylcarbodiimide, *J. Lipid Res.* **7**, 174–175.

Shimshick, E. J. (1973). Part I. Rotational Diffusion of Spin Labeled Proteins. Part II. Lateral Phase Separations in Lipid Membranes. Thesis, Stanford Univ.

Shimshick, E. J., and McConnell, H. M. (1973). Lateral phase separations in phospholipid membranes, *Biochemistry* **12**, 2351–2360.

Slotboom, A. J., and Bonsen, P. P. M. (1970). Recent developments in the chemistry of phospholipids, *Chem. Phys. Lipids* **5**, 301–398.

Smith, I. C. P. (1968). A study of the conformational properties of bovine pancreatic ribonuclease by electron paramagntic resonance, *Biochemistry* **7**, 745–757.

Stanacev, N. Z., Stuhne-Sekalec, L., Schreier-Mucillo, S., and Smith, I. C. P. (1972). Biosynthesis of spin labeled phospholipids. Enzymatic incorporation of spin labeled stearic acid into phosphatidic acid, *Biochem. Biophys. Res. Commun.* **46**, 114–119.

Stier, A., and Sackmann, E. (1973). Spin labels as enzyme substrates: Heterogenous lipid distribution in liver microsomal membranes, *Biochim. Biophys. Acta* **311**, 400–408.

Stier, A., and Reitz, I. (1971). Radical production in amine oxidation by liver microsomes, *Xenobiotica* **1**, 499–500.

Stone, T. J., Buckman, T., Nordio, R. L., and McConnell, H. M. (1965). Spin labeled biomolecules, *Proc. Nat. Acad. Sci. U.S.* **54**, 1010–1017.

Struve, W. G., and McConnell, H. M. (1972). A new spin labeled substrate for β-galactosidase and β-galactoside permease, *Biochem. Biophys. Res. Commun.* **49**, 1631–1637.

Stryer, L., and Griffith, O. H. (1965). A spin-labeled hapten, *Proc. Nat. Acad. Sci.* **54**, 1785–1791.

Stryukov, V. B., and Rozantsev, E. G. (1973). The anisotropy parameters of the iminoxyl radical containing ^{15}N, *Theor. Exp. Chem. (USSR) (Eng.)* **7**, 209–212.

Syva Associates (1970). A Catalog Introduction To Spin Labels, 3221 Porter Drive, Palo Alto, California 94304.

Tener, G. M. (1961). 2-Cyanoethyl phosphate and its use in the synthesis of phosphate esters, *J. Amer. Chem. Soc.* **83**, 159–168.

Tonomura, Y., Watanabe, S., and Morales, M. (1969). Conformational changes in molecular control of muscle contraction, *Biochemistry* **8**, 2171–2176.

Tourtellotte, M. E., Branton, D., and Keith, A. (1970). Membrane structure: Spin labeling and freeze etching of *mycoplasma laidlawii*, *Proc. Nat. Acad. Sci. U.S.* **66**, 909–916.

Waggoner, A. S., Kingzett, T. J., Rottschaefer, S., Griffith, O. H., and Keith, A. D. (1969). A spin-labeled lipid for probing biological membranes, *Chem. Phys. Lipids* **3**, 245–253.

Weiner, H. (1969). Interaction of a spin-labeled analog of nicotinamide-adenine dinucleotide with alcohol dehydrogenase. I. Synthesis, kinetics and electron paramagnetic resonance studies, *Biochemistry* **8**, 526–533.

Wien, R. W., Morisett, J. D., and McConnell, M. (1972). Spin label-induced nuclear relaxation. Distances between bound saccharides, histidine-15, and tryptophan-123 on lysozyme in solution, *Biochemistry* **11**, 3707–3716.

Wilcox, R. D., and Stratigos, H. P. (1967). Varian Tech. Rep. Varian Ass., Palo Alto, California.

Williams, J. C., Mehlhorn, R., and Keith, A. D. (1971). Syntheses and novel uses of nitroxide motion probes, *Chem. Phys. Lipids.* **7**, 207–230.

Wong, S. S., Quiggle, K., Triplett, C., and Berliner, L. J. (1974). Spin-labeled sulfonyl fluorides as active site probes of protease structure. II. Spin label syntheses and enzyme inhibition, *J. Biol. Chem.* **249**, 1678–1682.

Wood, J. M., and Wolfe, R. S. (1966). Propylation and purification of a B_{12} enzyme involved in methane formation, *Biochemistry* **5**, 3598–3603.

6

Molecular Structures of Nitroxides

J. LAJZEROWICZ-BONNETEAU

LABORATORY OF PHYSICAL SPECTROMETRY
UNIVERSITY OF SCIENCE AND MEDICINE OF GRENOBLE
GRENOBLE, FRANCE

I. X-RAY ANALYSIS OF SINGLE NITROXIDE CRYSTALS

Appropriately substituted nitroxyl derivatives of piperidine, pyrroline, pyrrolidine, and oxazolidine are solids at room temperature and crystallize quite readily. The crystal and molecular structures of a number of these derivatives have been determined by X-ray diffraction methods (Stout and Jensen, 1968; Woolfson, 1970). The dozen structures that have been determined by this method are listed in Fig. 1.

There are generally two, four, or eight molecules per elementary crystal unit cell derived from each other by symmetry operations. This leads to the determination of the coordinates of approximately thirty atoms. These data allow determination of molecular stereochemistry and conformation (interatomic angles and distances) as well as the relative orientation of different molecules within the crystal lattice. These structures have been determined

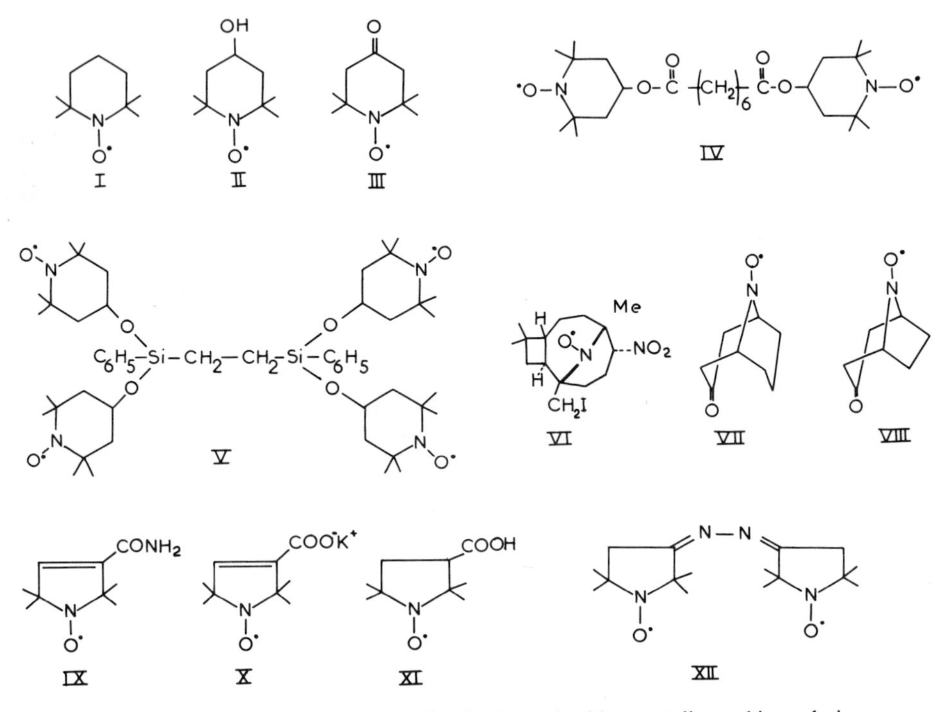

Fig. 1. Structures of nitroxide molecules determined by crystallographic analysis.

I 2,2,6,6-Tetramethylpiperidine-1-oxyl (Bordeaux *et al.*, 1973; Capiomont *et al.*, 1972b)
II 2,2,6,6-Tetramethylpiperidine-1-oxyl-4-ol (Lajzerowicz, 1968; Berliner, 1970)
III 2,2,6,6-Tetramethyl-4-piperidinone-1-oxyl (Shibaeva *et al.*, 1972; Bordeaux and Lajzerowicz, 1974a)
IV Di(2,2,6,6-tetramethyl-4-piperidinyl-1-oxyl) suberate (Capiomont, 1972)
V Iminoxyl organosilicon tetra radical (Shibaeva *et al.*, 1973)
VI Caryophylene iodonitrosite (Hawley *et al.*, 1968)
VII 9-Azabicyclo[3.3.1]nonan-3-one-9-oxyl (Capiomont *et al.*, 1971)
VIII 1,5-Dimethyl-8-azabicyclo[3.2.1]octane-3-one-8-oxyl (Capiomont, 1973)
IX 2,2,5,5-Tetramethyl-3-carbamidopyrroline-1-oxyl (Turley and Boer, 1972)
X (+)-3-Carboxy-2,2,5,5-tetramethyl-1-pyrrolidinyl-oxyl (Ament *et al.*, 1973)
XI Potassium 2,2,5,5-tetramethyl-3-carboxy pyrroline-1-oxyl (Boeyens and Kruger, 1970)
XII 2,2,5,5-Tetramethyl-1-aza-3-cyclo pentanone-1-oxyl-3-azine (Chion *et al.*, 1972)

to various degrees of accuracy; the standard deviation for bond distances between carbon, nitrogen, and oxygen atoms varies from 3×10^{-3} to 2×10^{-2} Å. Bond angles vary from 0.2° to 1°. The positions of hydrogen atoms have not been determined for all of the twelve molecules in Fig. 1. For those molecules whose hydrogen atom positions are known, the standard deviations are often as much as ten times greater than those for the larger atoms. These standard deviations will be quoted in parentheses immediately after the number to which they refer: e.g. C(1)—C(2) = 1.536(5).

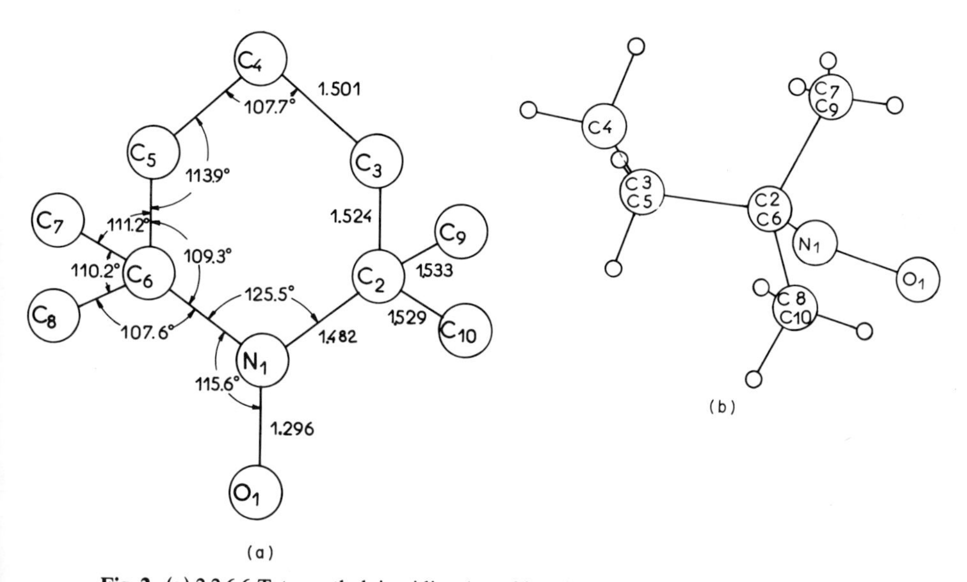

Fig. 2. (a) 2,2,6,6-Tetramethylpiperidine-1-oxyl bond angles and distances [Bordeaux *et al.* (1973), with permission of Heyden & Son, Ltd.]. (b) Projection of the molecule on its symmetry plane (hydrogen atoms are represented by small circles).

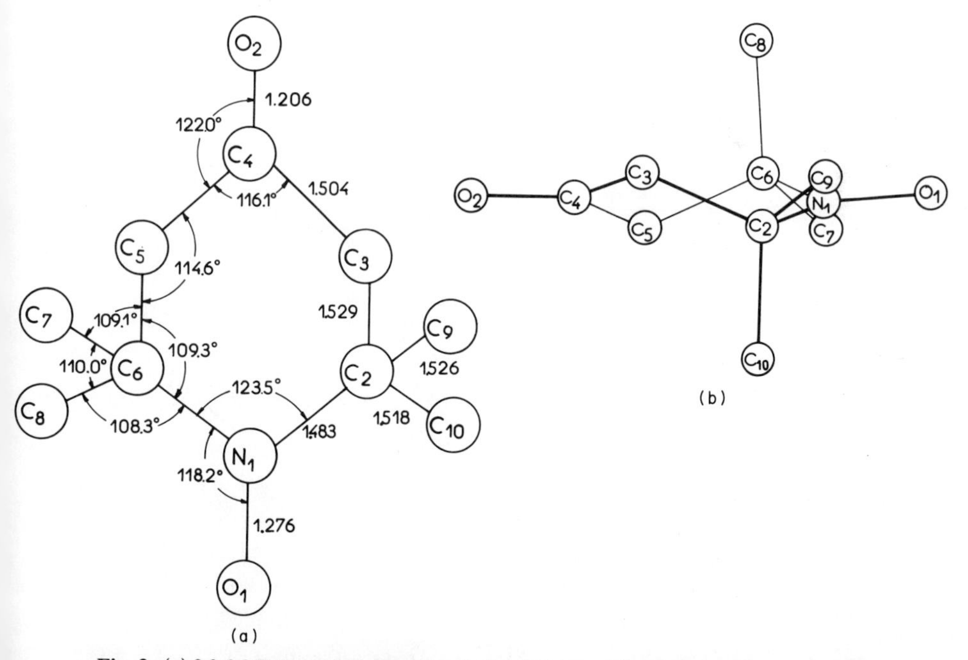

Fig. 3. (a) 2,2,6,6-Tetramethylpiperidine-4-piperidinone-1-oxyl bond angles and distances. (b) Projection of the molecule parallel to its mean plane; the binary axis $O_1N_1C_4O_2$ is in the projection plane. (Bordeaux and Lajzerowicz, 1974a.)

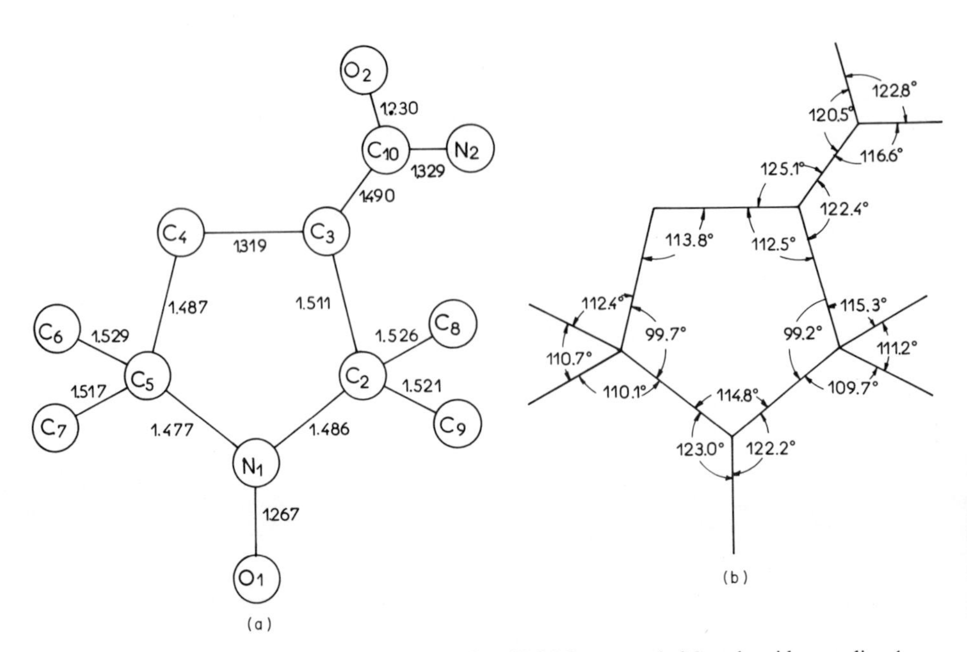

Fig. 4. (a and b) Bond distances and angles of 2,2,5,5-tetramethyl-3-carbamidopyrroline-1-oxyl. The pyrroline ring and the nitroxide group are coplanar. (Turley and Boer, 1972.)

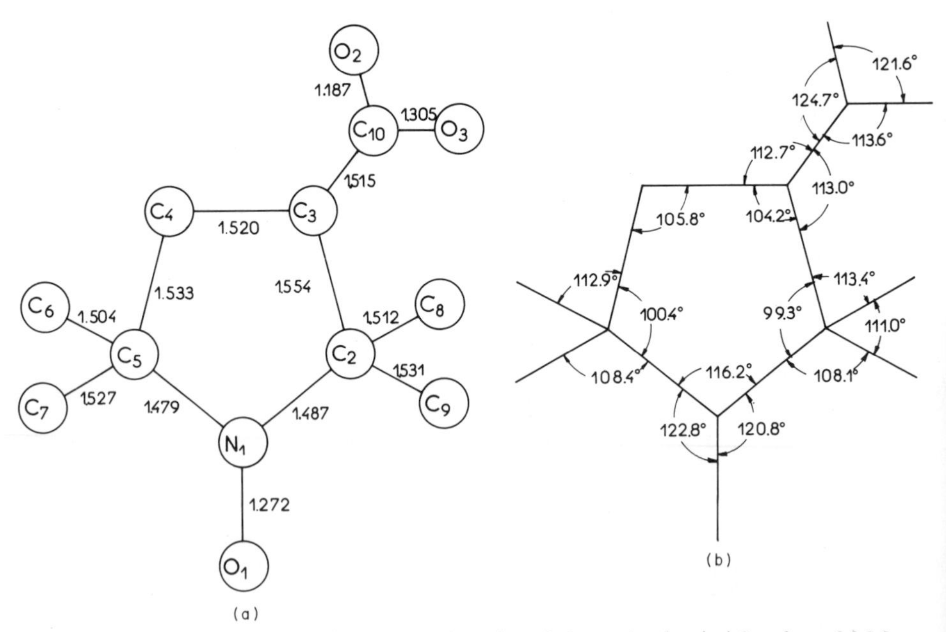

Fig. 5. (a and b) Bond distances and angles of the molecules (+)-3-carboxy-2,2,5,5-tetramethyl-1-pyrrolidinyl-oxyl.

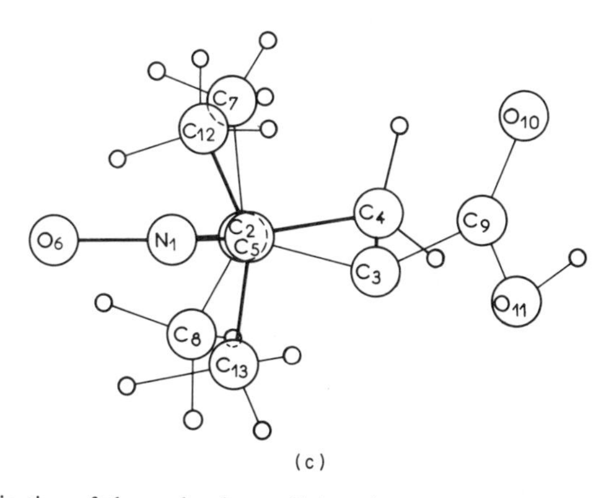

Fig. 5c. Projection of the molecule parallel to its mean plane. The pseudobinary axis of the cycle (O_2N_1) is in the projection plane. (Ament *et al.*, 1973.)

TABLE I

CONFORMATION OF THE GROUP $\begin{matrix} C \\ C \end{matrix}\!\!>\!N{-}O\cdot$ [a]

Nitroxide	Symmetry of the ring	N—O bond length (Å)	Angle C–N–C (deg)	Angle α between the N—O· bond and the CNC plane (deg)
I	Chair form, mirror plane	1.296 (5)	125.5 (3)	19.4
II	Chair form, mirror plane	1.291 (7)	125.4 (5)	15.8
III	Twist form, twofold axis	1.276 (3)	123.5 (2)	0
IV	Chair form, mirror plane	1.276 (5)	123.9 (5)	18.2
V	Chair form	1.31	126	20
VI		1.31 (2)	121 (1)	24
VII	Mirror plane	1.289 (4)	114.2 (3)	30.5
VIII	Mirror plane	1.274 (4)	107.4 (3)	24.9
IX	Planar	1.267 (5)	114.8 (5)	0
X	Twist form	1.272 (3)	116.2 (2)	3.3
XI	Planar	1.277 (8)	114.8 (9)	0
		1.263 (13)	114.7 (7)	
XII	Planar	1.266 (7)	112	0

[a] References to the structures are found in the text.

The conformation obtained is that of the molecule in the crystal. The interactions between molecules are generally of the van der Waals type, and it is reasonable to consider that the conformation of the molecule in the crystal differs little from that of the molecule in solution. Apart from the actual conformation of the molecule, there are several other noteworthy features:

(1) In molecules containing groups that may participate in hydrogen bonding (e.g., OH, NH_2), the oxygen atom in the N—O· group participates in a weak intermolecular hydrogen bond (**II**, **IX**, **X**), a phenomenon that might occur in solution as well.

(2) When studying the crystal of a compound containing one or more centers of asymmetry, it is possible to obtain the absolute configuration of the molecule; such is the case for the nitroxide **X** (Ament *et al.*, 1973).

(3) Some nitroxides have been found to be polymorphic; i.e., they have various crystalline forms. This is the case with compound **I** (Capiomont *et al.*, 1972). In the stable phase between 20° and 40°C, the molecules are in chair–chair inversion movement; this inversion, which also exists in solution, must be capable of occurring when the nitroxide is bound to a macromolecule.

II. EXPERIMENTAL RESULTS

The detailed results for four nitroxides of different structural types are:

Nitroxide I	Piperidine chair-form ring
Nitroxide II	Piperidine twisted-form
Nitroxide IX	Pyrroline
Nitroxide X	Pyrrolidine

For each compound we give a projection of the molecule on a particular plane (Figs. 2b, 3b, 5c) and diagrams with interatomic angles and distances (Figs. 2a, 3a, 4a, 4b, 5a, 5b). In addition, Table I gives all the data on the
C
>N—O· nitroxide group: N—O· bond distance, CNC angle, and angle α
C
between the N—O· bond and the CNC plane.

III. CONCLUSIONS ON THE CONFORMATIONS OF NITROXIDES

All the five- and six-membered rings have the expected form and symmetry, but are nevertheless very deformed. The dimensions of the group
C
>N—O· are variable. This group is planar when the ring is either planar or
C
contains a binary axis of symmetry; otherwise the group is pyramid shaped.

Angle CNC is particularly sensitive to the constraints of the ring; its value changes from 125° for piperidine rings to 107° for the bicyclo compound **VIII**. Diagrammatically, two models for the nitroxides can be proposed (Fig. 6).

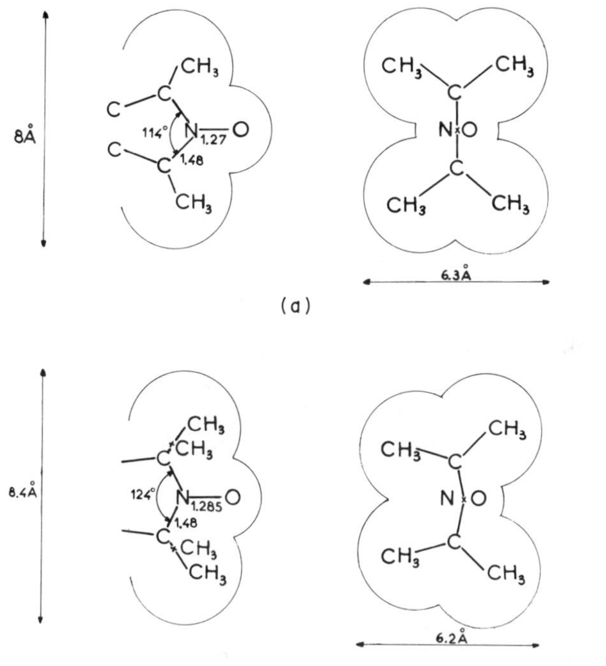

Fig. 6. Diagrams of the nitroxide group and its overall van der Waals thickness. (a) Five-membered ring, the nitroxide group is planar; (b) six-membered ring, the nitroxide group is bent.

Nonempirical methods have been used for calculating wave functions of the elementary radical H_2NO (Salotto and Burnelle, 1970; Ellinger *et al.*, unpublished results). These calculations lead to an aminoxyl group without any well-defined intrinsic geometry. In particular, the total electronic energy varies only slightly with the torsion of the group (two nominal minima are found for $\alpha = \pm 17°$). The theoretical model predicts the N—O bond length to be 1.27 Å and the CNC angle to be 123°. Both values are in excellent agreement with the experimental values found for nitroxides in which the N—O group is not highly constrained (e.g., six-membered rings).

Semiempirical calculations have been done on compound **I** (PCILO method—perturbative configuration interaction using localized orbitals). Here, too, the calculated geometry agrees well with the experimentally determined structure. The energy difference between the planar form and the pyramidal configuration is 1.05 kcal/mole (Ellinger *et al.*, unpublished results).

From studies on polarization parameters and spin density distribution in the N—O· group, Hayat and Silver (1973) find that bending is necessary to account for the observed values. They obtain an angle $\alpha = 17°$ for the nitroxide **III**, which is found planar in the crystalline state.

IV. ESR SPECTRA OF SINGLE NITROXIDE CRYSTALS

A. Role of the Exchange Interaction

The ESR studies of single nitroxide crystals are useful for obtaining information about molecular orientations in the unit cell and therefore facilitate crystal structure determinations. In addition, these studies allow precise measurements of the principal values of the g tensors. ESR spectra of single nitroxide crystals contain only a single line, which is Lorentzian in shape and about 10 G in width. These spectral properties are exhibited by all nitroxide single crystals regardless of the number of radicals per unit cell or the orientation of the crystal in the magnetic field (Bordeaux *et al.*, 1973). In these crystals the distances between the (C)(C)>N—O· groups are about 6 Å. Consequently, the exchange interaction between neighboring spins is very large compared to the difference of absorption energies of these spins themselves. Thus, for each orientation, the measured experimental g factor (g_{exp}) is the mean of the different individual nitroxide g factors (Capiomont *et al.*, 1974).

B. Experimental g Tensor Values

From these ESR studies, principal values of the g tensors g_{XX}, g_{YY}, g_{ZZ} have been found and are shown in Table II. The estimated accuracy is ± 0.0002. These components cover a very wide range of values, and there does not seem to be a relation between them and the molecular conformation. The lower mean values for $\frac{1}{3}(g_{XX} + g_{YY} + g_{ZZ})$ are for the nitroxides **II** and **IX**, which exhibit hydrogen bonding in the crystal.

C. Relative Orientation of the g Tensor and the (C)(C)>N—O· Group

When the nitroxide group is planar, the principal axes X, Y, and Z coincide respectively with the N—O· bond, the C—C direction, and the perpendicular to axes X and Y, which passes through the long axis of the

TABLE II

DATA RELATIVE TO THE g TENSOR FOR SOME NITROXIDES[a,b]

Compound	g_{XX}	g_{YY}	g_{ZZ}	$\frac{1}{3}(g_{XX} + g_{YY} + g_{ZZ})$	Relative orientation $\text{C}_2\text{N—O}\cdot/g$ tensor
I	2.0103	2.0069	2.0030	2.0067	Coincidence
II	2.0096	2.0064	2.0027	2.0062	Deviation
III	2.0104	2.0074	2.0026	2.0068	Coincidence
IV	2.0112	2.0063	2.0039	2.0071	Coincidence
VIII	2.0104	2.0066	2.0038	2.0069	Deviation
IX	2.0086	2.0066	2.0032	2.0061	Coincidence
XII	2.0101	2.0068	2.0028	2.0066	Coincidence

[a] A. Capiomont *et al.*, 1974.

[b] *Editor's note:* A complete table of all nitroxide g and hyperfine tensors reported to date is compiled in Appendix II, p. 564, and in Chapter 3, Table I, p. 73.

2p π orbital. In other cases it is sometimes necessary to assume a noncoincidence of the X axis and the N—O bond in order to interpret the ESR results.

The examples of compound **I** and **II** are instructive (Bordeaux *et al.*, 1973). The crystallographic structures are isomorphic with only one molecule in the unit cell. The symmetry plane of the crystals is the molecular mirror plane, so the orientation of the principal g axes is easy to find. In compound **I** the X axis and N—O· bond coincide. In compound **II** a displacement of 8.5° is found between these directions (Fig. 7). The intermolecular hydrogen bond present in compound **II** is probably responsible for this effect.

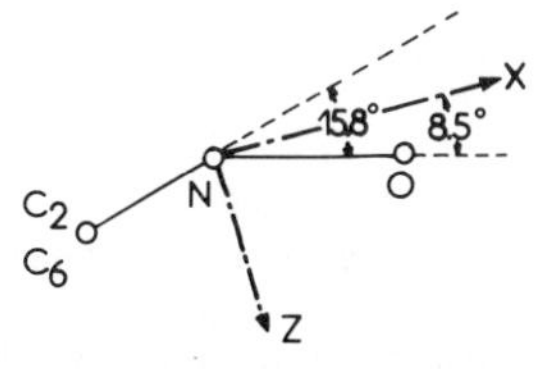

Fig. 7. 2,2,6,6-tetramethyl-4-piperidinol-1-oxyl deviation of the principal axes of the g tensor (X and Z) from the N—O bond. [Bordeaux *et al.* (1973), with permission of Heyden & Son, Ltd.]

Crystal and molecular structures of two N-oxyl oxazolidine nitroxides have been determined: **XIII**, 1,4-bis(4′,4′-diméthyloxazolidine-N-oxyl)cyclohexane (Gleason, 1973); **XVIV**, spiro(1,2′-(4′,4′-dimethyloxazolidine)cyclohexane) (Bordeaux and Lajzérowicz, 1974b). In both compounds the

C⟩N—O· group is planar (strictly for **XIV**). The NO bond lengths are 1.25 (1) **XIII** and 1.259 (4) **XIV**. The principal values of the *g* tensor for the compound XIV are 2.0089, 2.0076, and 2.0036.

ACKNOWLEDGMENTS

Thanks are due to Dr. S. Ament, J. Wetherington, J. Moncrief, K. Flohr, M. Mochizuki, E. Kaiser, Y. Ellinger, J. Douady, A. Rassat, and M. Subra for communicating their results prior to publication, and to Drs. D. Bordeaux, A. Capiomont, and B. Chion and Professor A. Rassat and associates for helpful discussions.

REFERENCES

Ament, S., Wetherington, J., Moncrief, J., Flohr, K., Mochizuki, M., and Kaiser, E. (1973). Determination of the absolute configuration of (+)-3-carboxy-2,2,5,5-tetramethyl-1-pyrrolidinyloxy, *J. Amer. Chem. Soc.* **95**, 7896–7897.

Berliner, L. J. (1970). Refinement and location of the hydrogen atoms in the nitroxide 2,2,6,6-tetramethyl-4-piperidinol-1-oxyl, *Acta Cryst.* **B26**, 1198–1202.

Boeyens, J. C. A., and Kruger, G. J. (1970). Remeasurements of the structure of potassium 2,2,5,5-tetramethyl-3 carboxypyrroline-1-oxyl, *Acta Cryst.* **B26**, 668–672.

Bordeaux, D., Lajzerowicz, J., Briere, R., Lemaire, H., and Rassat, A. (1973). Détermination des axes propres et des valeurs principales du tenseur *g* dans deux radicaux libres nitroxydes par étude de monocristaux, "Organic Magnetic Resonance," Vol. 5, pp. 47–52. Heyden and Son Ltd.

Bordeaux, D., and Lajzerowicz, J. (1974a). Polymorphism du radical libre 2,2,6,6-tetra-methylpiperidinone oxyl-1. Affinement de la structure de la phase orthorhombique, *Acta Cryst.* **B30**, 790–793.

Bordeaux, D., and Lajzerowicz, J. (1974b). Structure cristalline du spiro [cyclohexane-1,2′-(dimethyl-4′,4′ oxazolidine)], *Acta Cryst.* **B30**, 2130–2132.

Capiomont, A., Chion, B., and Lajzerowicz, J. (1971). Affinement de la structure du radical nitroxyde dimérisé bicyclo[3.3.1]-nonanone-3-aza-9-oxyl-9, *Acta Cryst.* **B27**, 322–326.

Capiomont, A. (1972a). Structure cristalline du radical nitroxide: suberate de ditétraméthyl-2,2,6,6 piperidinyl-4-oxyl-1, *Acta Cryst.* **B28**, 2298–2301.

Capiomont, A., Bordeaux, D., and Lajzerowicz, J. (1972). Structure cristallographique de la forme quadratique du nitroxyde tétramethyl 2,2,6,6-piperidineoxyl-1, *C. R. Acad. Sci. Paris* **275C**, 317–320.

Capiomont, A. (1973). Structure cristalline du nitroxyde 1,5-dimethyl-8-aza-bicyclo(3.2.1)octane-3-one-8-oxyl, *Acta Cryst.* **B29**, 1720–1722.

Capiomont, A., Chion, B., Lajzerowicz, J., and Lemaire, H. (1974). Interpretation and utilization for crystal structure determination of ESR spectra of nitroxide free radicals in single crystals, *J. Chem. Phys.* **60**, 2530–2535.

Chion, B., Capiomont, A., and Lajzerowicz, J. (1972). Structure du biradical tétramethyl-2,2,5,5-aza-1-cyclopentanone-3-azine-3-oxyl, *Acta Cryst.* **B28**, 618–621.

Gleason, W. B. (1973). 1,4-bis-(4′,4-dimethyloxazolidine-*N*-oxyl)cyclohexane, *Acta Cryst.* **B29**, 2959–2960.

Hayat, H., and Silver, B. L. (1973). Oxygen-17 and nitrogen-14 σ–π polarization parameters and spin density distribution in the nitroxyl group, *J. Phys. Chem.* **77**, 72–78.

Hawley, D. M., Ferguson, G., and Robertson, J. M. (1968). The crystal structure and absolute stereochemistry of caryophyllene "iodonitrosite," a stable nitroxide radical, *J. Chem. Soc. B* 1255–1261.

Lajzerowicz, J. (1968). Structure du radical nitroxyde tétraméthyl 2,2,6,6-piperidinol-4-oxyl-1, *Acta Cryst.* **B24**, 196–199.

Salotto, A. W., and Burnelle, L. (1970). *Ab initio* calculation of H_2NO geometry and hyperfine splittings, *J. Chem. Phys.* **53**, 333–340.

Shibaeva, R. P., Atovmian, L. D., and Nejgauv, M. G. (1972). The crystal structure of the nitroxide free radical 2,2,6,6-tetramethyl-4-piperidinone-1-oxyl, *Zh. Strukt. Khim.* **13**, 887–891.

Shibaeva, R. P., Atovmian, L. O., Rozenberg, L. P., and Stryukov, B. (1973). Crystal and molecular structure of the stable iminoxyl-organo-silicon tetraradical, *Dokl. Akad. Nauk. SSSR.* **210**, 833–836.

Stout, G. H., and Jensen, L. H. (1968). "X-ray Structure Determination." Macmillan, New York.

Turley, J. W., and Boer, F. P. (1972). The crystal structure of the nitroxide free radical 2,2,5,5-tetramethyl 3-carbamidopyrroline-1-oxyl, *Acta Cryst.* **B28**, 1641–1644.

Woolfson, M. M. (1970). "X-ray Crystallography." Cambridge Univ. Press, London and New York.

7

Instrumental Aspects of Spin Labeling†

PATRICIA JOST and O. HAYES GRIFFITH

INSTITUTE OF MOLECULAR BIOLOGY
AND
DEPARTMENT OF CHEMISTRY
UNIVERSITY OF OREGON
EUGENE, OREGON

I. ESR INSTRUMENTATION AND INSTRUMENTAL ARTIFACTS

A. ESR Spectrometers

A number of books are available which describe the design, construction, and sensitivity of ESR spectrometers (Ayscough, 1967; Poole, 1967; Alger, 1968; Wilmhurst, 1968). Wertz and Bolton (1972) give a good description of

† Adapted from P. Jost and O. H. Griffith, Electron spin resonance and the spin labeling method, *in* "Methods in Pharmacology" (C. Chignell, ed.), Vol. 2: Physical methods, pp. 242–257, © 1972 by Meredith Corporation. Reprinted by permission of Prentice-Hall, Inc., Englewood Cliffs, N.J.

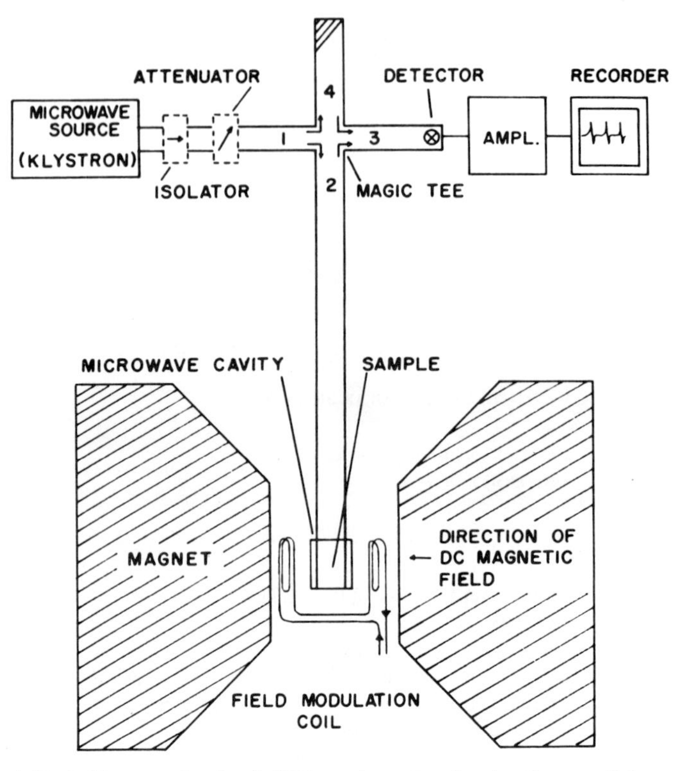

Fig. 1. A block diagram of a simple ESR spectrometer showing the essential components.

the basic instrumentation of electron spin resonance. These are useful to the expert in the field but most potential users have a commercial ESR spectrometer available and must use it as it is. Our aim here is to provide the nonexpert with background to help in the use of the ESR spectrometer as a black box. Particular emphasis is placed on instrument settings, since most operator errors are made here. Any comparison among the many commercial spectrometers is deliberately avoided because the various spectrometers are constantly being upgraded and because the authors have not made a careful comparison of the various instruments available. All spectra displayed in this chapter were recorded on Varian ESR spectrometers, most of them on a Varian E-3.†

A schematic diagram of a simple ESR spectrometer is shown in Fig. 1. The basic parts are the klystron, magic tee, cavity, detector-amplifier

† Editor's note: See Appendix III, p. 566, for a compendium of commercial equipment and supplies.

combination, and a recording device. (Modern commercial instruments use a microwave circulator instead of the magic tee to accomplish the same purpose.) The klystron, in conjunction with the fixed isolator and adjustable attenuator, provides the operator with a variable source of microwave energy. The magic tee is the microwave equivalent of the familiar Wheatstone bridge. The spectrometer is constructed so that the impedances of arms 2 and 4 are nearly matched and very little power reaches the detector on arm 3. The sample is situated in the microwave cavity, a region of high microwave field H_1. The klystron frequency is fixed during the experiment and the magnetic field is scanned. As the resonance condition is approached, the sample absorbs microwave energy, creating an imbalance in the bridge. This imbalance is detected as increased power in arm 3 and is amplified and displayed on a recorder. The modulation coils also play an important role. They are needed in the conversion of the dc signal to an ac voltage, which can then be easily amplified using the powerful methods of phase sensitive detection. The modulation coils are frequently imbedded in plastic and fastened to the sides of the microwave cavity, out of sight of the operator. The use of modulation causes the spectra to appear as a first derivative of the microwave absorption, and provides the operator with one more control that must be adjusted properly, the modulation amplitude control.

B. The Microwave Cavity and the Sample

The cavity and sample are of special importance. Many cavity designs are available, including rectangular cavities, cylindrical cavities, and special slow-wave structures. Rectangular cavities are the most common type at present and all spectra presented here were recorded using a standard Varian rectangular cavity. However, any high-quality (high-Q) cavity can be used in spin labeling studies. The user should be aware of which type of cavity is available since this determines the geometry of the sample holder. A typical cavity will have an opening 1 cm in diameter, but the actual liquid or solid sample is much smaller than this opening. As one might expect, the optimum sample geometry for a rectangular cavity is a flat, rectangular cell and the best geometry for a cylindrical cavity is a cylindrical sample tube. More care must be exercised when dealing with aqueous samples than organic materials because the dielectric loss of water tends to lower the spectrometer sensitivity by lowering the cavity Q. Commercial quartz flat aqueous sample cells have approximate dimensions 6 cm × 1 cm × 0.3 mm i.d. These useful cells are equipped with stoppered ports at the top and the bottom. When dealing with moderate concentrations of spin label, 10^{-3} M to 10^{-6} M, less expensive cells may be used. For example, the aqueous samples may be drawn into 0.4-mm-i.d. thin-walled capillary tubes made of ordinary flint (soft) glass or Pyrex. The capillary tubes are available from standard laboratory supply companies. After filling, these capillary tubes may be sealed at the dry end with a flame or with a soft wax such as Universal Red Wax available from Central Scientific Company (note that some nitroxides are soluble in wax). The filled capillaries are dropped into standard 2–4-mm-i.d. quartz ESR sample tubes. The combination is then placed in the cavity. After the experiment the capillary tube is discarded and the quartz tube is ready for reuse. The quartz tubes are easily made from stock quartz tubing, but special care must be taken since standard quartz tubing frequently has an intrinsic ESR signal. It is more convenient to avoid this problem by purchasing the quartz ESR sample tubes.

Dry tissues, powders, hydrocarbon solvents, and other samples that do not contain water are less troublesome. Large-diameter flint glass or Pyrex capillaries make useful sample tubes. Disposable pipettes sealed at the small end make inexpensive tubes that are easy to degas. Of course, the quartz flat cells and ESR sample tubes may be used with organic samples as well as aqueous samples.

Many special sample holders have been constructed. These range from a simple glass cover slip used to support phospholipid multilayers, to the complex arrangement shown in Fig. 2. The sample holder of Fig. 2 was designed to permit simultaneous nerve excitation and ESR signal recording. Note that the sample itself (the lobster nerve) is quite small and most of the supporting structure is made of thin plastic and other dielectric materials.

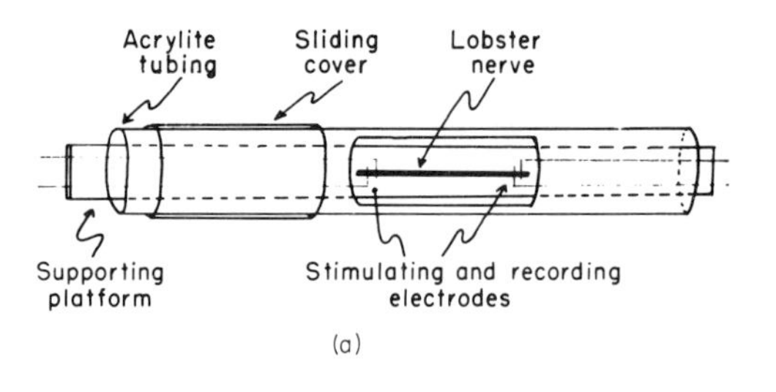

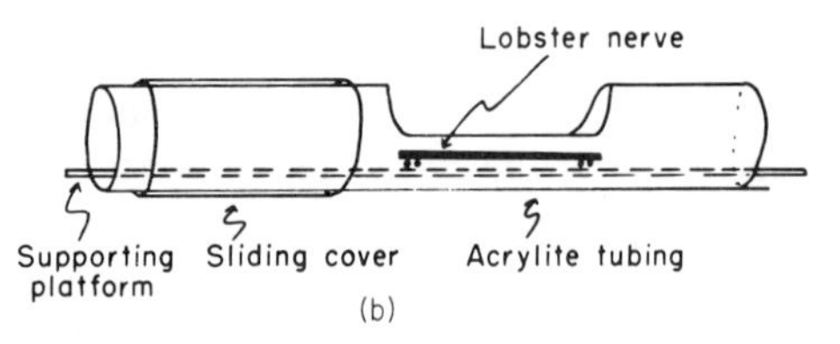

Fig. 2. One example of a specialized ESR sample cell. A nerve chamber for stimulating and recording the action potential in the ESR cavity. This sample configuration permits the observation of spin-labeled nerves during nerve stimulation. (a) Top view, (b) side view. [From Calvin *et al.* (1969).]

With reasonable care, electrodes and other small objects may be placed in the cavity. The presence of moderate quantities of metal can drastically alter the characteristics of the microwave cavity. Another kind of sample holder, designed for reproducible sample insertion into the cavity, is described by Gaffney and McNamee (1974).

Some experimental applications require controlling the humidity of the sample. This is readily accomplished by placing the sample in a chamber over a reservoir containing an appropriate saturated salt solution for which the equilibrium relative humidity has been reported (see "Handbook of Chemistry and Physics," CRC Press, any edition). A simple chamber can be improvised from a 1-cm Pyrex tube approximately 20 cm long, and a small salve jar, as shown in Fig. 3a. The middle 8–10 cm of the Pyrex tube is etched with hydrofluoric acid to reduce the amount of glass in the cavity. The sample is supported in the tube either on glass wool (Fig. 3b) or on a thin glass cover slip held in a split tubing, the other end of which is slipped over a plastic rod with a rubber stopper at the top to close the chamber (Fig. 3a). The Pyrex tube, with sample inserted, is joined to the salve jar containing the salt solution with a one-hole rubber stopper, and the sample is allowed to equilibrate. The whole assembly is then inserted into the cavity from the bottom, supported so that the sample, surrounded by the thinned glass wall, is in the center of the cavity. Standard taper fittings can also be modified for this use (see Fig. 3b). This amount of glass reduces the cavity Q somewhat, but for many spin labeling purposes this much loss of sensitivity can be tolerated. A similar chamber made of polyethylene or quartz may be preferable when higher sensitivity is important.

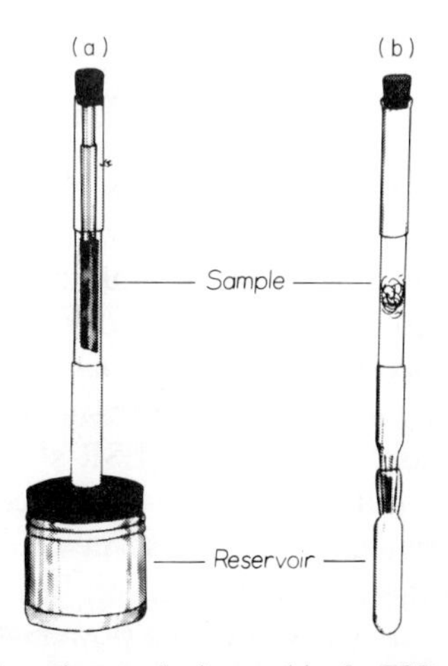

Fig. 3. Example of chambers that can be inserted in the ESR cavity when it is desirable to maintain a spin-labeled sample in equilibrium with a vapor phase.

C. Receiver Gain, Field Modulation Amplitude, and Microwave Power

The gain, modulation amplitude, and microwave power controls are grouped together because they strongly influence the signal amplitude. If any or all three of these controls are set at zero, no signal will be observed. As each variable is increased, the signal amplitude increases, passes through an optimal range, and then either decreases or becomes distorted. The optimal range will depend on the sample and, to a lesser extent, on the ESR spectrometer. The receiver gain needs no elaboration since this control is encountered in all chemical instruments. The receiver gain (also called amplifier gain or signal level control) may be increased until the amplifier becomes unstable and the signal-to-noise ratio decreases. The effect of modulation amplitude is more dramatic, as illustrated in Fig. 4. As the modulation amplitude increases, the ESR lines first increase in height, then broaden, and finally become greatly distorted. A useful rule of thumb is to set the modulation amplitude equal to or less than the ESR linewidth in gauss. In Fig. 4 the peak-to-peak linewidth is 0.5–1.0 G. As seen in Fig. 4, the best choice for the modulation amplitude is also 0.5 G. (For convenience, spectra for this chapter were usually recorded at 1.0 G modulation amplitude. For very accurate line shape studies somewhat lower settings are usually optimal.) A word of caution is needed here. This rule of thumb and other generalizations mentioned in this chapter can be misused. If careful line shape measurements are being made, the investigator should always decrease the modulation amplitude (or other instrument setting) and determine if this has an effect on the line shape. Only if the change in line shape or relative peak heights is insignificant can the setting be considered appropriate.

The effects of microwave power on nitroxide ESR spectra have not been carefully investigated. In any system, relaxation processes are present that allow the spins to return to the ground state after absorption of microwave energy. If the power becomes too large, however, the relaxation processes are unable to return the spin system to equilibrium, and saturation takes place. The power required to saturate depends on the spin–lattice relaxation time. Saturation and relaxation effects will play an important role in future spin labeling experiments involving electron–nuclear and electron–electron double resonance. In most conventional ESR studies, however, saturation should be avoided. It is difficult to determine precisely the point at which saturation becomes significant. One test involves measuring the signal height as a function of the microwave power. Below saturation, the amplitude of the signal varies linearly with the power. As saturation sets in, the amplitude increases at a lower rate and eventually either flattens out or decreases as the power is increased. Some effects of saturation are illustrated

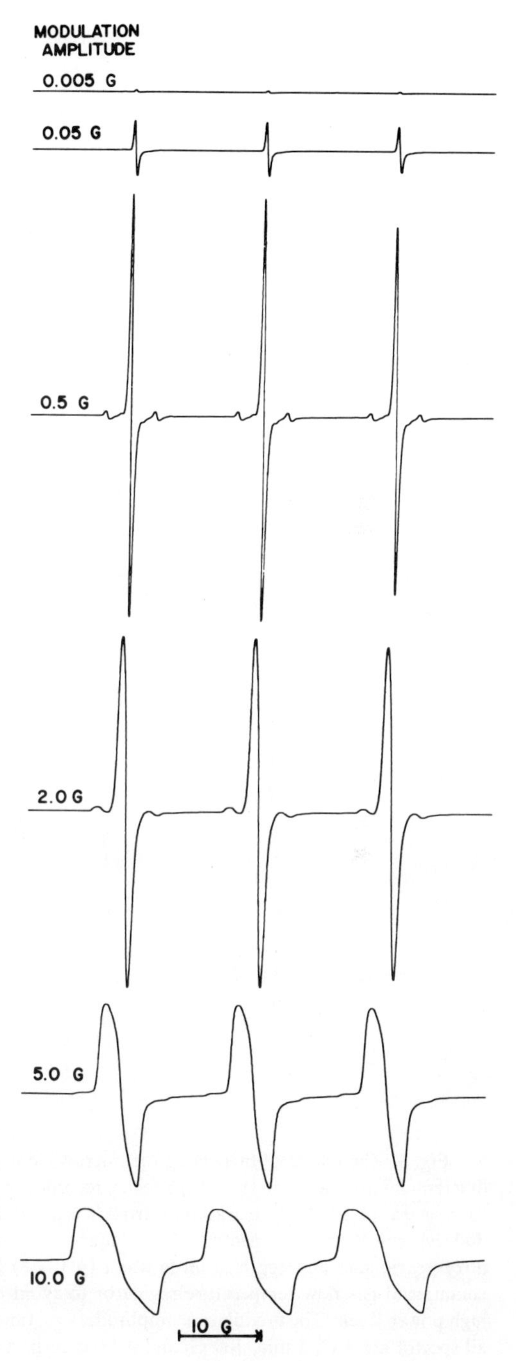

Fig. 4. The ESR spectrum of a typical nitroxide as a function of field modulation amplitude. The sample is an aqueous solution containing 5×10^{-4} *M* piperidone nitroxide (2,2,6,6-tetramethyl-piperidine-1-oxyl-4-one) at room temperature. The microwave power, scan time, scan range, and filter time constant are 5 mW, 4 min, 100 G, and 0.3 sec, respectively.

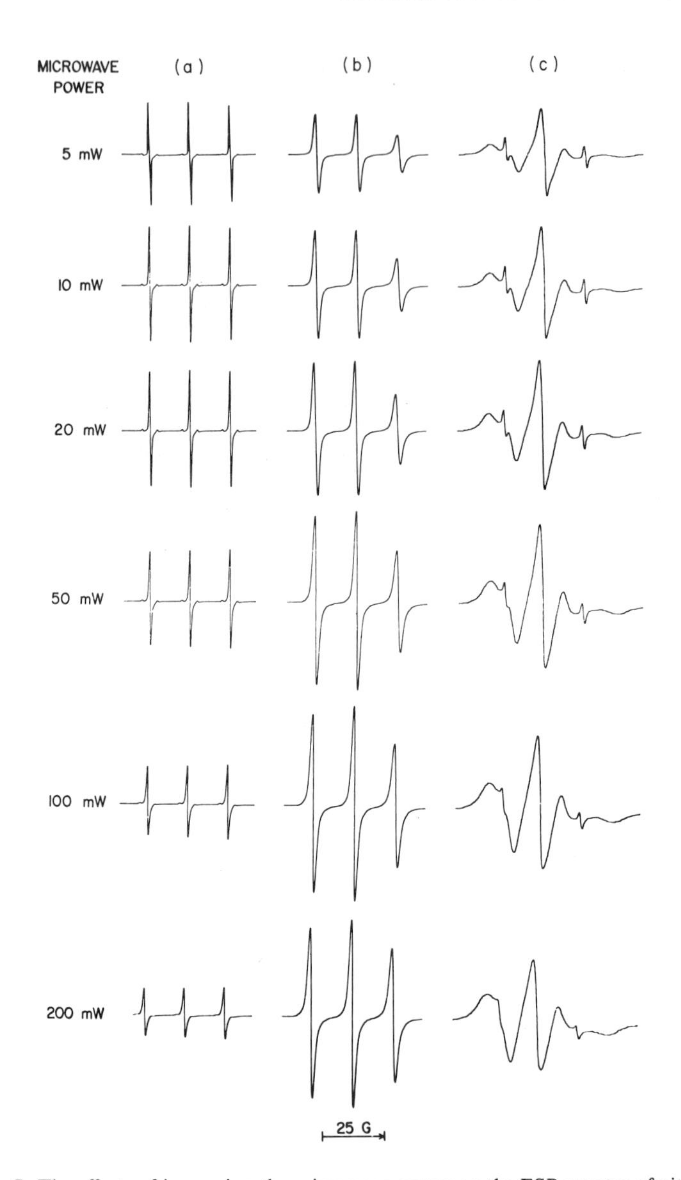

Fig. 5. The effects of increasing the microwave power on the ESR spectra of nitroxides in different environments at 18°C. (a) Spectra recorded for 10^{-4} *M* piperidone nitroxide in water. (b) Spectra for 10^{-4} *M* piperidinol nitroxide ester of dodecanoic acid in 5% aqueous sodium dodecyl sulphate. (c) Spectra of an aqueous dispersion containing 1.5×10^{-3} *M* 12-doxylstearic acid and egg lecithin in water (pH = 7). The temperature was stabilized using a commercial gas flow temperature regulator to avoid possible effects of microwave heating at high power levels. The modulation amplitude, scan time, scan range, and filter time constant for all spectra are 1 G, 4 min, 100 G, and 0.3 sec, respectively.

in Fig. 5. Figures 5a and 5b represent nitroxides at room temperature in water and in an aqueous sodium dodecyl sulfate solution, respectively. The relative peak heights change with increasing microwave power. Rotational correlation times are calculated from peak heights (or line shapes) and will be in error if saturation occurs. Figure 5c illustrates another possible pitfall. These spectra were recorded for a doxylstearic acid in an aqueous phospholipid dispersion. The concentration and pH of the solution were adjusted so that the nitroxide was present in both the aqueous phase (sharp lines) and in the phospholipid vesicles (broad lines). As the microwave power is increased, the sharp lines saturate more rapidly than the broad lines. Equilibrium constants based on relative line heights at higher powers would underestimate the nitroxide concentration in the aqueous phase. Microwave power of 1–5 mW appears to be acceptable for room-temperature spin labeling studies. Some saturation may be occurring at these power levels, but the effects evidently are not serious. For spectra of nitroxide free radicals at liquid nitrogen temperatures the power level should be reduced to 1 mW or lower to avoid saturation effects. Once again, the investigator should check the results by decreasing the power and looking for changes in the ESR spectrum.

D. Scan Time, Filter Time Constant, and Field Inhomogeneity

Improper adjustments in these three experimental parameters produce similar line shape distortions. The scan time or sweep time is the time required to vary the dc magnetic field slowly over a specified interval (the scan range). The effect of varying the scan time on a typical nitroxide ESR spectrum with a 100-G scan range is illustrated in Fig. 6. The 4.0- and 8.0-min scans are almost superimposable, whereas the 0.5-min scan is badly distorted. There is a tradeoff here between the quality of the spectrum and the amount of time required to record it. In this example a 4.0-min scan appears to be a good choice, but if the spin-labeled sample is stable, the investigator may choose the 8.0-min scan time. It is interesting to note that the same effect is evident in the much broader spectrum of a membrane model system (Fig. 6b). Of course, in both samples the distortion caused by shortening the scan time can also be obtained by increasing the scan range while holding the scan time constant.

Nearly all chemical instruments have filter networks to increase the signal-to-noise ratio, and an ESR spectrometer is no exception. The effect of varying the filter time constant, given a fixed scan time and scan range, is shown in Fig. 7. In this example distortion occurs in all but the top spectrum. The rule to follow is that the time constant must be much shorter than the time required to sweep through the ESR line. In this example the scan

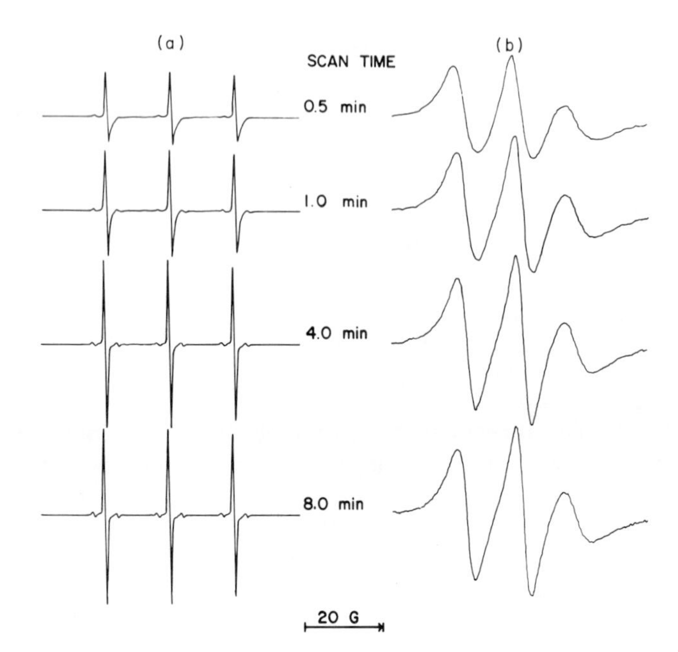

Fig. 6. The dependence of nitroxide ESR on scan time. (a) Spectra for 5×10^{-4} M piperidone nitroxide in water at room temperature. (b) Spectra for a room-temperature aqueous dispersion containing 10^{-4} M 12-doxylstearate, methyl ester, and 1 wt % egg lecithin (pH 7). For all spectra the microwave power, modulation amplitude, scan range, and filter time constant are 5 mW, 1 G, 100 G, and 0.3 sec, respectively. The gain setting is the same for all spectra.

time was 4 min and the scan range was 100 G. One ESR line covers about 2 G. Therefore it takes about $(4/100) \times 60 = 2.4$ sec to scan through each ESR line. It is easy to see why the 0.1-sec time constant was the only reasonable choice. To avoid distorted spectra, it is good practice to check the time constant, scan time, and scan range experimentally. The bottom spectrum of Fig. 7 represents a special case of signal distortion. In this example the time constant of the filter network is so large that the line shape approximates the integral of the first derivative curve. This behavior is expected of a simple filter circuit (consider, for example, a resistor in series with a capacitor). However, as discussed below, the digital computer provides a better method of integration.

The distortion caused by field inhomogeneity is illustrated in Fig. 8. The inhomogeneity of the dc magnetic field increases from the top spectrum to the bottom spectrum. The inhomogeneity was created by moving the cavity various distances from the center of the magnet. Field inhomogeneity of this magnitude is not ordinarily observed in commercial 9.5-GHz instruments.

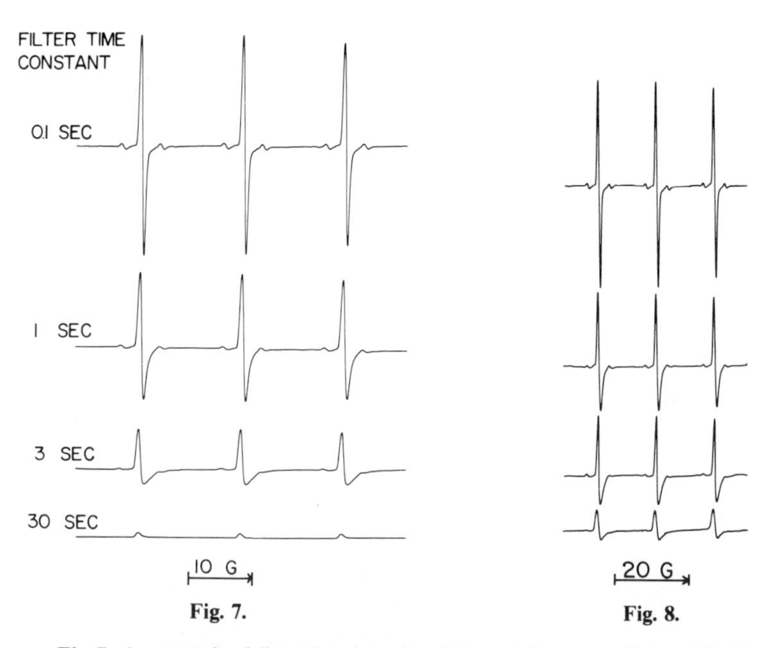

Fig. 7. **Fig. 8.**

Fig. 7. An example of distortions introduced through improper choice of the time constant. The sample is the same as for Fig. 3. The microwave power, modulation amplitude, scan time, and scan range for all spectra are 5 mW, 1 G, 4 min, and 100 G, respectively.

Fig. 8. Nitroxide spectra distorted by an inhomogeneous magnetic field. The sample is the same as for Fig. 3. The microwave power, modulation amplitude, scan time, scan range, and filter time constant are 5 mW, 1 G, 4 min, 100 G, and 0.3 sec, respectively. The gain setting is the same for all spectra, and the field inhomogeneity increases from top to bottom.

In 35-GHz spectrometers, special dual-cavity applications, or when using older magnet systems, the problem is common. It can usually be corrected by adjusting the position of the microwave cavity. The reason for including this effect is to point out that the bottom spectrum of Fig. 8 is almost identical to the top left spectrum of Fig. 6 and the third spectrum of Fig. 7. Fortunately, this is one of the few examples of a line distortion that has several possible causes.

E. Degassing and Concentration Effects

Anyone performing spin labeling experiments soon encounters oxygen–nitroxide interactions and nitroxide–nitroxide interactions. The ground state of molecular oxygen is a triplet and oxygen is therefore paramagnetic. It is rarely observed directly by ESR, but oxygen does interact with the spin label through exchange and dipolar mechanisms. The result is the well-known oxygen broadening, illustrated in Fig. 9. The top spectra are of the

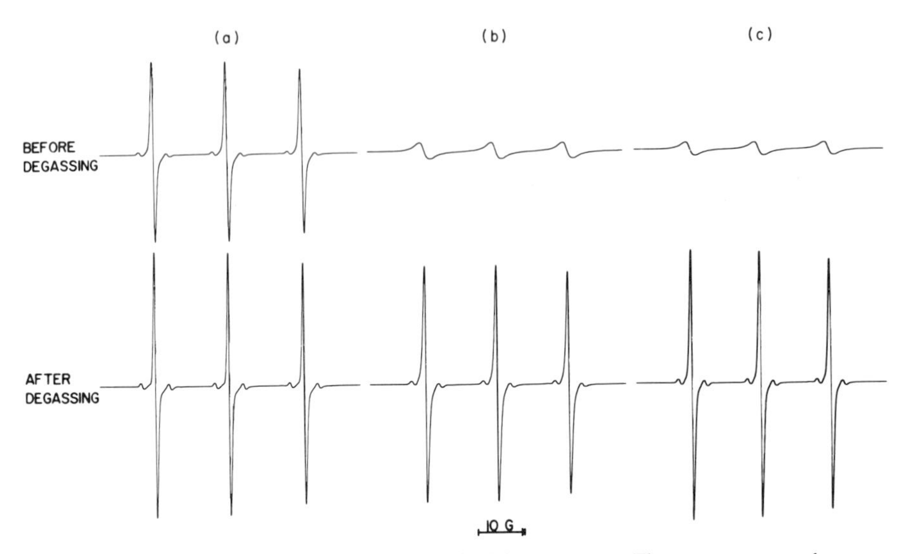

Fig. 9. Effects of oxygen broadening on the ESR spectrum. The spectra were taken on 10^{-4} M solutions of piperidone nitroxide in (a) water, (b) methanol, and (c) chloroform, respectively. The top spectra were recorded on samples equilibrated with the atmosphere. The bottom spectra were recorded after nitrogen was bubbled through the solutions at room temperature for 30 sec. The microwave power, modulation amplitude, scan time, scan range, and filter time constant are 5 mW, 1 G, 4 min, 100 G, and 0.3 sec, respectively. All spectra were recorded at room temperature and at the same gain setting. Reversibility was tested by bubbling air through the sample to verify that concentrations were not altered by the degassing procedure.

piperidone nitroxide dissolved in water, methanol, or chloroform at room temperature. Although it is not obvious from the spectra, the three solutions contain approximately equal concentrations of the nitroxide. Considerably more oxygen broadening is observed in methanol and chloroform (and other hydrocarbon solvents) than in aqueous solutions. The oxygen may be removed either by the freeze–thaw method or by bubbling a good grade of nitrogen or argon gas through the samples. The bottom spectra of Fig. 9 were recorded after nitrogen gas was passed through the solutions

---

Fig. 10. Concentration effects on nitroxide ESR spectra. (a) Spectra recorded on varying concentrations of a long-chain nitroxide, piperidinol nitroxide ester of dodecanoic acid, in dodecane (degassed) at room temperature. (b) Spectra taken using varying concentrations of the same nitroxide in an aqueous solution containing 0.5 wt % sodium dodecyl sulfate at room temperature. The nitroxide concentration is the same for the two spectra in each row and the actual value is written between the two spectra. The microwave power, modulation amplitude, scan time, scan range, and filter time constant for all spectra are 5 mW, 1 G, 4 min, 100 G, and 0.3 sec, respectively, with variable gain settings.

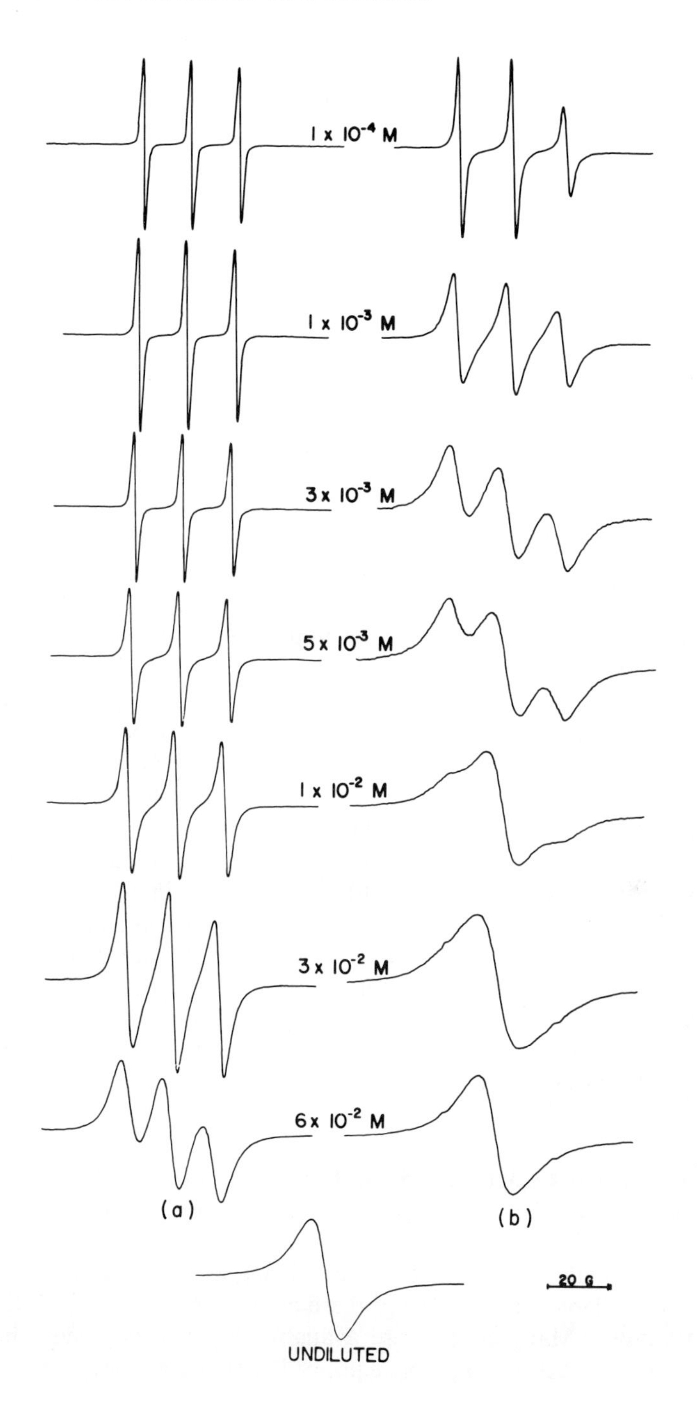
1 x 10⁻⁴ M
1 x 10⁻³ M
3 x 10⁻³ M
5 x 10⁻³ M
1 x 10⁻² M
3 x 10⁻² M
6 x 10⁻² M
(a)
(b)
UNDILUTED
20 G

for approximately 30 sec. The effects are completely reversible. After passing air through these same samples, the spectra are indistinguishable from the top row of Fig. 9.

Nitroxide–nitroxide interactions also occur via the familiar dipolar and exchange mechanisms. The effect of increasing the nitroxide concentration of piperidinol nitroxide ester of dodecanoic acid in degassed dodecane at room temperature is shown in Fig. 10a. The sharp three-line spectrum remains essentially unchanged until the concentration exceeds 10^{-3} *M*. As the concentration is increased beyond this point, the three lines gradually broaden and move together. At 6×10^{-2} *M*, the three lines are still evident. The exchange-narrowed, single-line spectrum of the pure nitroxide is shown at the bottom of the figure for comparison.

The values given in the center of Fig. 10 represent average concentrations. Local concentrations can exceed these average values, producing unusual results. For example, if the nitroxide were not soluble to the extent of 1×10^{-3} *M*, solid flakes of nitroxide would be present and a superposition of three sharp lines and a single broad line would result. A more subtle example is given in Fig. 10b. In this case the same water-insoluble nitroxide is introduced into an aqueous solution containing small clusters (micelles) of sodium dodecyl sulfate. The equilibrium concentration of nitroxide molecules is very much higher in the micelles than in the aqueous phase. The net result is that the local concentration of nitroxide is much greater than the average bulk concentration, and nitroxide–nitroxide broadening is observed at lower bulk concentrations. At 1×10^{-4} *M* the spectrum consists of three sharp lines. The heights are unequal due to slower rates of tumbling but this effect is not of primary concern here. At 1×10^{-3} *M* all three lines are already greatly broadened. At 5×10^{-3} *M* the three lines are disappearing, and the 3×10^{-2} *M* solution gives a spectrum very similar to that of the pure nitroxide. The point is that it is important to watch for nitroxide–nitroxide interactions even when the average concentration of nitroxide is low. With reasonable care, errors in interpretation can be avoided and nitroxide–nitroxide interactions will provide valuable information about the system under investigation.

II. THE USE OF COMPUTERS TO PROCESS SPIN LABELING DATA

Small computers appropriate for interfacing with spectrometers are commercially available, and the use of these computers in spin labeling can be very profitable. Many choices are available, but our experience has been with the Varian 620/i computer, equipped with 8K memory, teletype, and general interface, connected directly to the ESR spectrometer, and more recently with the Varian 620/L-100 16K system used in a time-sharing mode among three spectrometers. Since this section was originally written, many other systems have been developed and marketed for digitizing and manipulating spectroscopic data. Given the memory to store several sizable arrays in addition to the necessary program and the software capability for redefining the start and end of arrays (operationally this amounts to horizontally shifting one spectrum into register with another spectrum), in principle the data manipulations described here are quite straightforward. They can be performed either with a dedicated minicomputer or with a central computer with interactive graphics. In the University of Oregon system, developed by Prof. Charles Klopfenstein, the computer controls the ESR spectrometer through commands from the teletype and digitizes the data as the spectrum is collected. Various data treatments described below can be performed, either as the points are collected or immediately after collection. The data can also be transferred to paper or magnetic tape and reentered into the computer for later processing. With ESR spectra reduced to digital form, it is possible to perform a number of operations that are tedious or impossible to perform by hand. Among these are time averaging, correction for a slanting baseline, scaling spectra, integration, and spectral titration.

A. Time Averaging

The earliest application was time averaging and was first used in ESR by Klein and Barton (1963). In this application, several approaches are possible, and all improve the signal-to-noise ratio in a very noisy spectrum by repeated scanning. The change in signal-to-noise ratio is proportional to $n^{1/2}$, where n is the number of scans. We have found most useful the procedure of adding successive scans, dividing each point by the number of scans, and instructing the computer to list the number of scans on the teletype. The resulting spectrum can be plotted out at any time during the repeated scans, either on an external recorder or on the spectrometer recorder itself. It is preferable to increase the signal-to-noise ratio by introducing more label or concentrating the sample, but, if this is not feasible, time averaging can be very useful in improving the quality of the spectrum and revealing otherwise inaccessible details.

B. Scaling Experimental Spectra

One simple application that we have routinely found useful is the scaling of experimental spectra. This is illustrated in Fig. 11, where three spectra were collected using the same sample. The sample is the spin label 5-doxylstearic acid in an aqueous dispersion of lecithin : cholesterol (molar

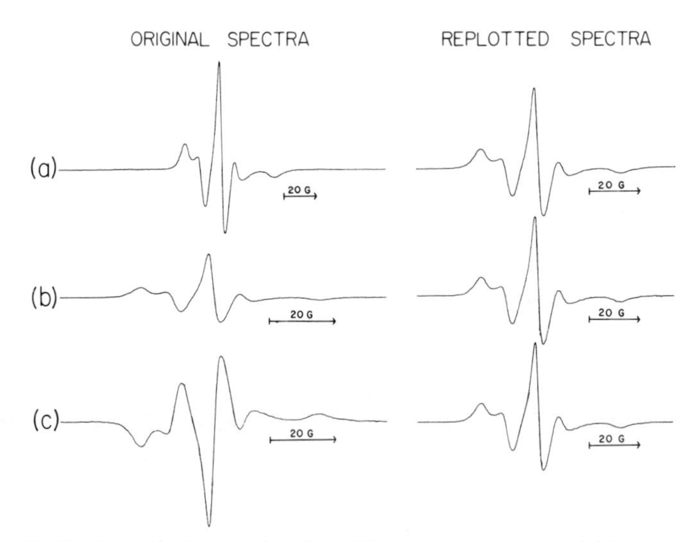

Fig. 11. Horizontal and vertical scaling of first derivative spectra. (a) Spectrum recorded with a scan range of 200 G. Spectra (b) and (c) were recorded with a scan range of 100 G, with spectrum (c) recorded with reversed phase, and spectrum (b) recorded using a lower gain setting. Each of the original spectra has been replotted on the right to have the same horizontal and vertical dimensions. [From Klopfenstein *et al.* (1972).]

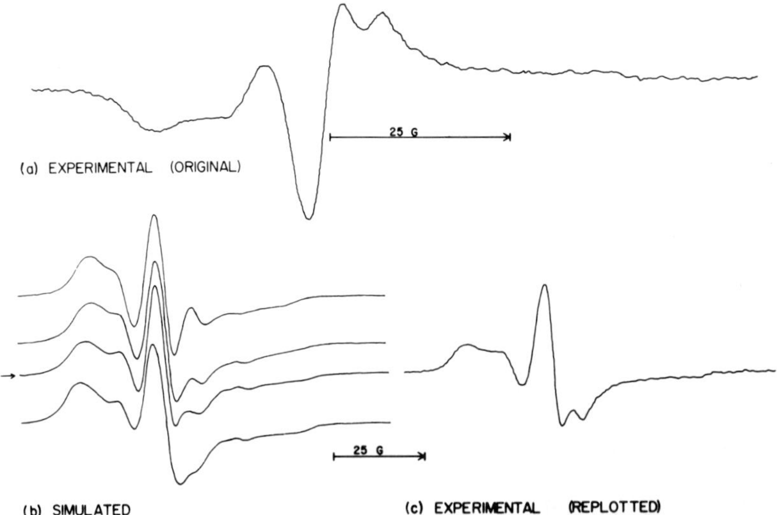

Fig. 12. Replotting experimental data. (a) The original experimental spectrum from a dehydrated oriented lecithin preparation containing 5-doxylstearic acid; (b) simulations from large computer to be compared; (c) experimental spectrum (a) replotted to match horizontal and vertical scales and phase of spectra (b). The arrow in (b) indicates the best fit to the experimental spectrum in both splittings and line shape. [From Jost *et al.* (1971).]

ratio 1 : 1). The instrumental variables of scan range, gain, and phase reversal contribute to the apparent differences in the experimental spectra (left column, Fig. 11). In the right column these spectra have been replotted to the same horizontal scale and normalized to an arbitrary vertical scale. It is now clear that all three experimental spectra are identical. In practice, such large variations may not be encountered, but when comparing closely related spectra, this procedure makes clear small differences (conversely, small apparent differences frequently disappear). Scaling can also be proportional to the relative spin label concentrations. This involves integration of the spectra and examples will be given in Section C.

Frequently experimental spectra are compared to computer-calculated theoretical line shapes, in order to estimate molecular motion or order in the samples. In principle, the calculated line shapes can be scaled individually to each experimental spectrum. In practice, it is simpler to scale the experimental spectra to the calculated spectra. A typical example is given in Fig. 12.

C. Integration of the Experimental Spectrum

Figure 13 illustrates integration of first derivative ESR spectra. We include this example to show how difficult it is to estimate by eye the relative concentrations from peak heights. The nitroxide concentration present varies as the product of the linewidth squared times the height of the first derivative ESR lines. Thus, if the ESR linewidth increases by a factor of 10 upon binding of the spin label, the height of the bound spectrum will be reduced 100-fold and small concentrations of rapidly tumbling spin label can almost obscure a much larger concentration of the bound spin label present. The bottom row of Fig. 13 consists of the familiar first derivative ESR spectra, representing examples of three different mobilities of nitroxide spin labels. Integrating once yields the absorption spectra, analogous to the familiar absorption spectra of optical absorption or nuclear magnetic resonance spectroscopy (middle row of Fig. 13). A second integration (i.e., integrating the absorption spectrum) has been performed to determine relative concentrations, and the results are shown in the top spectra of Fig. 13. The heights I_a, I_b, and I_c are proportional to the concentrations of the probe, and $I_a : I_b : I_c$ is approximately 1 : 19 : 40. For purposes of illustration, the middle line of each of the first derivative spectra in Fig. 13 has been arbitrarily adjusted to the same height before integration. When this height is held constant, the relationship between ESR linewidth and concentration is dramatically evident. A small concentration of freely tumbling nitroxide has a relatively large first derivative peak height. It can be readily seen from Fig. 13 that quite a low concentration of unbound spin label in the presence of bound label may dominate the composite spectrum obtained.

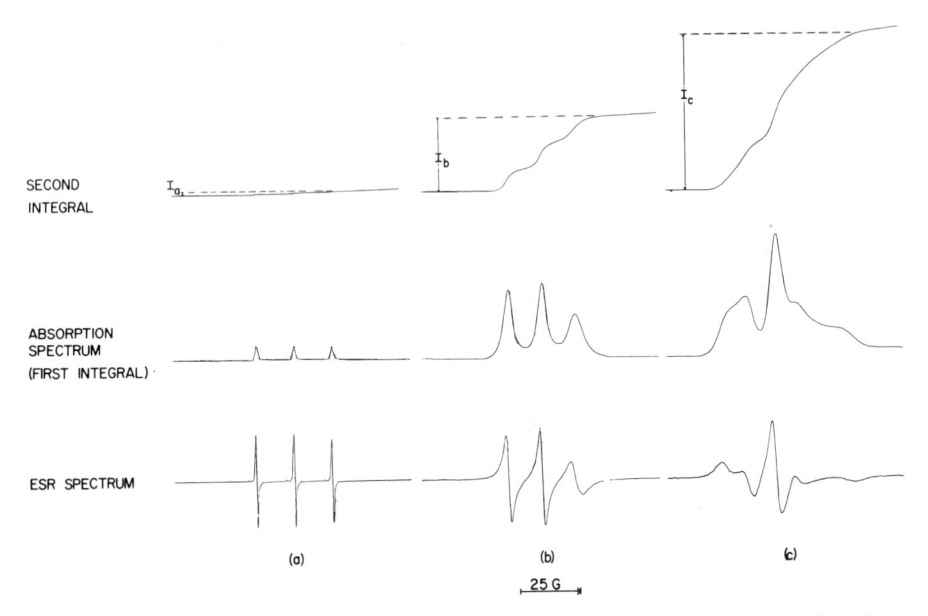

Fig. 13. Integration by computer. (a) Freely tumbling nitroxide: the same sample as for Fig. 3; (b) moderately immobilized nitroxide: 16-doxylstearic acid in an aqueous dispersion of lecithin; (c) strongly immobilized nitroxide: 5-doxylstearic acid in an aqueous dispersion of lecithin : cholesterol, molar ratio 2 : 1. All spectra were recorded at room temperature. The middle lines of all three first derivative ESR spectra are of the same height.

D. Spectral Titration

Another problem frequently encountered in spin labeling experiments is the occurrence of composite spectra. This happens whenever the nitroxide partitions between two or more environments. For example, when a lipid spin label is diffused into an aqueous suspension of membranes a complex spectrum is often obtained. It may be due to the sum of the two spectra arising from equilibrium concentrations of aqueous and membrane associated probes. Composite spectra are also encountered when the lipid spin label partitions between the fluid bilayer phase and sites on the hydrophobic protein surface within the membrane. Other examples occur in antibody–antigen studies and enzyme spin labeling experiments. A set of composite spectra is often obtained as a result of some experimental variable such as concentration, temperature, or pH. The related spectra may exhibit common points of intersection when the spectra are superimposed. These isoclinic points are usually the result of isosbestic points in the absorption spectra and may well indicate the presence of a simple equilibrium, although some caution must be used in this kind of interpretation (Marriott and Griffith, 1974).

If the data are digitized, combinations of spectral subtraction and subsequent double integration can be used to determine relative concentrations of label in the two environments. It is necessary to obtain or simulate the spectrum of at least one component. For example, the spectrum of the spin label in the aqueous phase can be recorded separately, or samples with and without the protein or lipid-depleted samples can be used. The next step is to adjust the spectra for slanting baseline and offset so that the baseline value is zero. The spectra are then shifted horizontally relative to each other so that they are in register. Finally, incremental subtractions are carried out until an obvious endpoint is reached (e.g., a negative baseline or phase reversal of peaks). An example is given in Fig. 14. To simulate a typical experimental

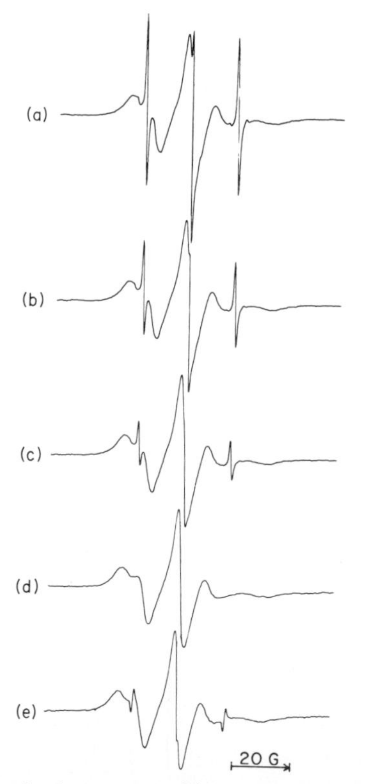

Fig. 14. Spectral subtraction by computer. (a) Is a room-temperature, composite spectrum. Two capillary tubes were placed in the cavity, one containing an aqueous dispersion of 1.5×10^{-3} *M* 5-doxylstearic acid in lecithin, the second containing 5×10^{-5} *M* piperidone nitroxide in water. (b)–(e) Are traces produced by subtracting increasing amounts of the sharp three-line spectrum from the composite spectrum, with total subtraction occurring in spectrum (d). In spectrum (e), too much of the sharp three-line spectrum has been subtracted (note the three sharp lines of this spectrum are phase-reversed). [From Jost *et al.* (1971).]

sample, two capillary samples representative of two different mobilities were placed together in the ESR cavity (see top spectrum, Fig. 14). The composite spectrum is then titrated with the sharp three-line spectrum from the aqueous sample. In the bottom spectrum oversubtraction has occurred and the sharp lines are phase-reversed. In this case the endpoint is relatively precise. When the line shapes are related more closely, the endpoint becomes progressively harder to estimate. Depending on the relative linewidths of the two (or more) components, the range between obvious undersubtraction and oversubtraction may be as much as 10–15% of the absorption of one of the contributing spectra.

As a result of such a titration, the spectral components are resolved. The line shapes can be examined and relative concentrations determined for the two environments. The next step is to integrate each spectrum twice and compare the double integrals. This is illustrated in Fig. 15 for the composite spectrum and the two separate components. The quantities I_c, I_f, and I_b are the areas under the ESR absorption curves for the composite, free, and bound line shapes, respectively. These quantities are proportional to the

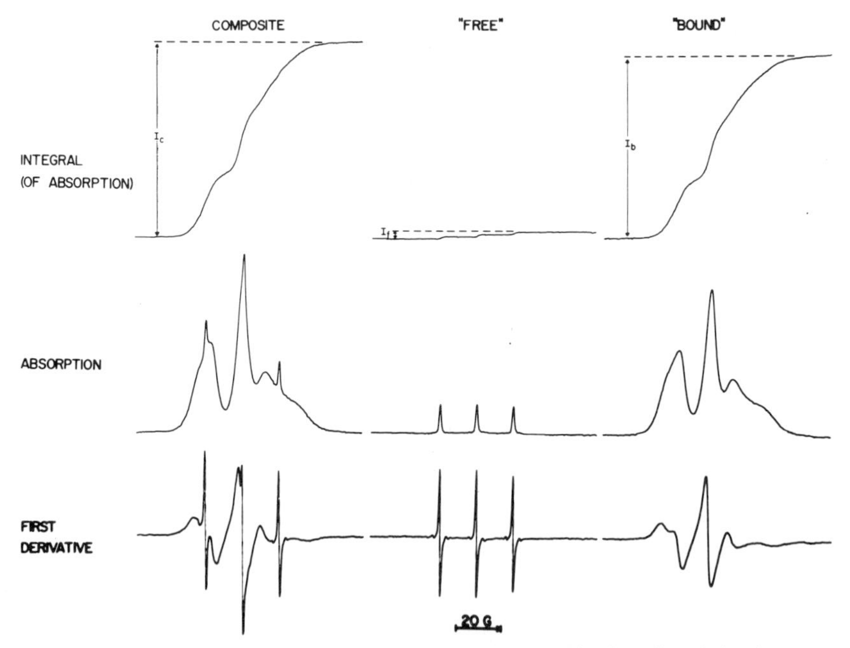

Fig. 15. Integration and double integration by computer. The three first derivative spectra in the bottom row are, respectively, the composite spectrum, the aqueous sample, and the phospholipid dispersion of Fig. 14. Single integration of these ESR spectra yields the three absorption spectra shown. The corresponding double integrals of the first derivative spectra are shown in the top row. [From Jost *et al.* (1971).]

concentrations of the spin label present. In Fig. 15, $I_c = 236$, $I_f = 7$, and $I_b = 222$ in arbitrary units. Note that I_c very nearly equals $I_f + I_b$. Since this figure was made we have added enough memory to use double precision arrays to store the integrals (i.e., each point in the array can be used without prior division by a constant) and the accuracy has been somewhat improved. A common source of much larger error is, of course, the presence of a nonzero baseline in the first derivative spectrum, which is easily detected by displaying or plotting the first integral. If the last 5–10 G of the absorption curve gives a nonzero average value, this way of determining concentration would be inappropriate.

The actual concentrations of nitroxides in the two capillary samples used in this example were $C_f = 5 \times 10^{-5}\ M$ and $C_b = 1.5 \times 10^{-3}\ M$. As another check the ratio $I_b/I_f = 32$ is in good agreement with the known ratio of concentrations $C_b/C_f = 30$.

We have discussed our general system and the software we use elsewhere (Klopfenstein *et al.*, 1972; Jost *et al.*, 1971), and the interested reader can consult these for more detail and other applications.

ACKNOWLEDGMENTS

We are grateful to Miss Dee Brightman for technical assistance and to Professor Charles Klopfenstein for helpful discussions. This investigation was supported by Public Health Service Research Grant No. CA-10337 from the National Cancer Institute.

REFERENCES

Alger, R. S. (1968). "Electron Paramagnetic Resonance: Techniques and Applications." Wiley (Interscience), New York.

Ayscough, P. B. (1967). "Electron Spin Resonance in Chemistry." Methuen, London.

Calvin, M., Wang, H. H., Entine, G., Gill, D., Ferruti, P., Harpold, M. A., and Klein, M. P. (1969). Biradical spin labeling for nerve membranes, *Proc. Nat. Acad. Sci. U.S.* **63**, 1–8.

Gaffney, B. J., and McNamee, C. M. (1974). Spin label measurements in membranes, *Methods Enzymol.* **32B**, 161–198.

Jost, P. C., and Griffith, O. H. (1972). Electron spin resonance and the spin labeling method, *In* "Methods in Pharmacology" (C. Chignell, ed.), Vol. II, pp. 223–276. Appleton, New York.

Jost, P. C., Waggoner, A. S., and Griffith, O. H. (1971). Spin labeling and membrane structure, *In* "Structure and Function of Biological Membranes" (L. Rothfield, ed.), pp. 84–144. Academic Press, New York.

Klein, M. P., and Barton, G. W., Jr. (1963). Enhancement of signal-to-noise ratio by continuous averaging: applications to magnetic resonances, *Rev. Sci. Instrum.* **34**, 754–759.

Klopfenstein, C., Jost, P., and Griffith, O. H. (1972). The dedicated computer in electron spin resonance spectroscopy, *In* "Computers in Chemical and Biochemical Research" (C. Klopfenstein and C. Wilkins, eds.), pp. 175–221. Academic Press, New York.

Marriott, T. B., and Griffith, O. H. (1974). Comments on isosbestic and isoclinic points in spin labeling studies, *J. Mag. Res.* **13**, 45–52.

Poole, C. P., Jr. (1967). "Electron Spin Resonance." Wiley (Interscience), New York.

Wertz, J. E., and Bolton, J. R. (1972). "Electron Spin Resonance: Elementary Theory and Practical Applications." McGraw-Hill, New York.

Wilmshurst, T. H. (1968). "Electron Spin Resonance Spectrometers." Plenum, New York.

8

The Use of Spin Labels for Studying the Structure and Function of Enzymes

JOEL D. MORRISETT

DEPARTMENT OF MEDICINE
BAYLOR COLLEGE OF MEDICINE
HOUSTON, TEXAS

DEFINITIONS

Doxyl	Used to describe a molecule containing the 4′,4′-dimethyloxazolidinyl-*N*-oxyl heterocyclic group
Piperidinyl	Used loosely to describe a molecule containing the 1-oxyl-2,2,6,6-tetramethylpiperidinyl group
Pyrrolidinyl	Used loosely to describe a molecule containing the 1-oxyl-2,2,5,5-tetramethylpyrrolidinyl group
Pyrrolinyl	Used loosely to describe a molecule containing the 1-oxyl-2,2,5,5-tetramethylpyrrolinyl group
Phosph*or*yl	Used to describe organophosphorus compounds containing P—O bonds but not P—C bonds
Phosph*on*yl	Used to describe organophosphorus compounds containing P—O bonds and P—C bonds

I. INTRODUCTION

Enzymes are very interesting molecules. Much of the interest that surrounds these molecules derives from their great complexity. This complexity is due to the infinite number of chain lengths (molecular weights) and sequences (primary structures) that can be generated from about twenty naturally occurring amino acids. This complexity is increased further by several special ways in which amino acids in a local region are sterically disposed (secondary structure), the numerous ways in which these local regions can be spatially interrelated (tertiary structure), and finally, the large number of ways these separate chains may interrelate three dimensionally (quaternary structure). Determining the precise interatomic relationships that completely define the structure of an enzyme represents an enormous task. Yet physical, organic, and biological chemists have met this challenging task with much enthusiasm and a host of techniques that have permitted the achievement of an impressive series of accomplishments in this area. One of the techniques that has made highly significant contributions to our understanding of enzyme structure is spin labeling.

Many reagents for the chemical modification of specific side chains in proteins had been developed by the time the first spin-label experiment was reported (Ohnishi and McConnell, 1965). Several of these reagents were quickly and easily altered to include the nitroxyl moiety, thereby generating a series of covalent spin-label reagents. Some of these reagents were found to react with several side chains, giving rise to mixed EPR spectra, which could not always be interpreted unambiguously. This problem was solved in part by the refinement of site-specific methodology (Singer, 1967), which has culminated in the elegant technique of affinity labeling.

Although the successful application of spin labeling to protein structure determination has depended heavily on the state of the art of protein chemistry, the single most important factor responsible for the widespread use of this method in any area has been the intrinsic magnetic properties of the nitroxyl radical itself. These magnetic properties of nitroxides have, in turn, allowed the measurement of a variety of important structural properties of enzymes, a description of which is the subject of this chapter.

II. REAGENTS FOR SPIN LABELING ENZYMES

A. Covalently Binding Spin Labels

A wide selection of covalently binding spin labels has been synthesized, and many of these have been used to study various enzyme systems. The more common ones are commerically available from such companies as Syva, Aldrich, or Frinton.† Procedures or references for the synthesis of many of these labels are given in Chapter 5. Covalently binding spin labels may be divided into four general categories according to the type of reaction they undergo: alkylation (Fig. 1), acylation (Fig. 2), sulfonylation (Fig. 3), and phosphor(n)ylation (Fig. 4).

1. Alkylating Spin-Label Reagents

The reaction of proteins with spin-labeled alkyl halides is conveniently monitored by measuring the release of halide ion with a halogen electrode. The analogues of bromoacetamide (**I**, **VIII** in Fig. 1) are especially useful because acid hydrolysis of a protein labeled with either reagent yields the carboxymethyl derivative of the alkylated amino acid(s) (see Section III.A.1). The most common sites of reaction for these reagents are the

† Editor's note: Addresses for these and other manufacturers may be found in Appendix III, p. 566.

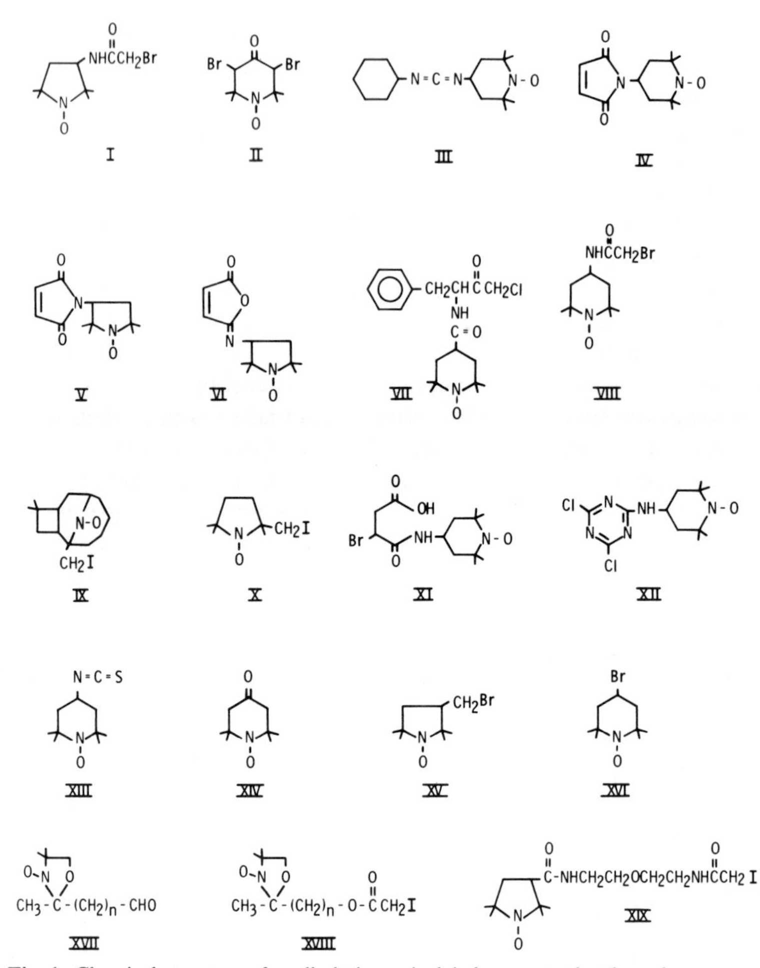

Fig. 1. Chemical structures for alkylating spin label reagents that have been or may be useful for probing the structure and function of enzymes. A few pertinent references to either the synthesis and/or the use of these reagents are given below. **I.** Smith (1968); Morrisett (1969); Jackson *et al.* (1973); Daniel *et al.* (1973). **II.** Rozantsev (1970). **III.** Azzi (1973). **IV.** Smith (1968). **V.** Smith (1968); Barratt *et al.* (1971). **VI.** Barratt *et al.* (1971). **VII.** Kosman (1972). **VIII.** McConnell and Hamilton (1968); Jones *et al.* (1972, 1973); Spallholz and Piette (1972). **IX.** Hawley *et al.* (1967). **X.** Motherwell and Roberts (1972). **XI.** Smith (1968). **XII.** Likhtenshtein and Bobodzhanov (1969). **XIII.** McConnell and McFarland (1970). **XIV.** Wagner and Hsu (1970). **XV.** McConnell and McFarland (1970). **XVI.** Rozantsev (1970). **XVII.** Balthasar (1971). **XVIII.** Balthasar (1971). **XIX.** Spallholz and Piette (1972).

sulfhydryl group of cysteine, the imidazole group of histidine, the ε-amino group of lysine, and the thioether of methionine. The carboxymethyl derivatives of each of these amino acids have been isolated and characterized (Gundlach *et al.*, 1959). These derivatives exhibit characteristic elution times on the amino acid analyzer. Their quantitation allows estimation of the site and extent of spin-label incorporation.

The dibromoketone (**II**, Fig. 1) may be useful for cross-linking two reactive groups which are near each other (e.g., the two active site histidine residues (His 12 and His 119) of bovine pancreatic ribonuclease (Wyckoff *et al.*, 1970) or the adjacent thiol (Cys 25) and imidazole (His 159) in papain (Drenth *et al.*, 1970). Another attractive possibility is the use of this reagent to join two sulfhydryl groups resulting from the reductive cleavage of a specific disulfide bond in a protein (Sperling *et al.*, 1969).

The carbodiimide spin label (**III**, Fig. 1) has been used by Azzi *et al.* (1973) to inhibit membrane-bound ATPase. Presumably, an unusually nucleophilic group at the active site of this enzyme adds across one of the double bonds of the carbodiimide. This recently developed reagent may also prove useful in studies of enzymes involved in nucleic acid metabolism, since the reaction of a similar carbodiimide with uridine and guanosine bases in nucleic acids has been reported (Metz and Brown, 1969).

The piperidinyl and pyrrolidinyl derivatives of *N*-ethylmaleimide (**IV** and **V**, Fig. 1) have been employed extensively. However, serious problems can attend the use of these labels. The nucleophilic groups (sulfhydryl and amino) that normally react with these reagents can either add across the carbon–carbon double bond or attack one of the carbonyl groups. In the latter case, the imide ring opens, thereby increasing the number of "spacer" atoms between the nitroxyl ring and the bond which connects the reagent to the protein. If *both* reaction mechanisms occur during labeling of a *single* group on the protein, the resulting EPR spectrum will most likely be a mixed one containing two components. In some cases, workers have interpreted such spectra as binding of the label at two or more different sites on the protein. In aqueous solutions, the pyrrolidinyl derivative (**V**, Fig. 1) has been found to hydrolyze rapidly to an unreactive product (Smith, 1968). Furthermore, Barratt *et al.* (1971) have shown that unless the reaction conditions are carefully controlled during the synthesis of maleimide **V** (Fig. 1), the isomaleimide **VI** will also be produced. Albumin spin-labeled with maleimide **V** (Fig. 1) gave an EPR spectrum quite different from that obtained when the isomaleimide (**VI**, Fig. 1) was used.

Reagent **VII** (Fig. 1) is an active-site-directed analogue of *p*-toluene sulfonylphenylalanine chloromethyl ketone (TPCK), which selectively alkylates His 57 of chymotrypsin (Kosman, 1972). While several noncovalently binding affinity spin labels have been synthesized (e.g., Fig. 5), few such covalent derivatives have been made. These specific reagents are of great value in obtaining high levels of labeling at a single site, an important consideration in distance-measuring experiments.

Caryophylene iodonitrosite (**IX**, Fig. 1) contains a very bulky alkyl substituent that might be useful for studying spacious active or binding sites (Hawley *et al.*, 1967). However, due to extensive proton coupling, the paramagnetic resonance lines of this reagent are rather broad (3.5 G), thereby limiting its usefulness (Morrisett, unpublished data). At the other extreme is **X** (Fig. 1), which represents one of the smallest covalent spin labels yet synthesized (Motherwell and Roberts, 1972). This molecule, like **XV** (Fig. 1), enjoys the advantage of having only an interposed methylene group between the nitroxyl ring and the bond joining it to the protein. Such spin labels are often more sensitive to conformational changes of proteins than compounds that have several intervening atoms between the ring and the protein. Spin labels that are closely attached to a protein will often have their motion strongly impeded. In such cases, these labels can be used to determine the rotational correlation time of the protein molecule (Shimshick and McConnell, 1972).

Cyanuric chloride has been derivatized with 1-oxyl-2,2,6,6-tetramethyl-4-aminopiperidine to give **XII** (Fig. 1). This reagent has been used by Likhtenshtein and Bobodzhanov (1969) to tag histidine and lysine side chains of albumin. It, like the dibromoketone (**II**, Fig. 1), possesses two potentially reactive halogens and may be useful as a cross-linking spin label.

The isothiocyanate nitroxide (**XIII**, Fig. 1) may be used to carbamylate amino groups (McConnell and McFarland, 1970). This reagent is considerably more stable than the corresponding isocyanate originally prepared by Stone *et al.* (1965).

1-Oxyl-2,2,6,6-tetramethylpiperidone-4 (**XIV**, Fig. 1) reacts with a free amino group to form a Schiff base, which upon reduction by sodium borohydride, yields a stable *N*-alkyl derivative (Wagner and Hsu, 1970). There is no apparent reduction of the nitroxide under the conditions used by these workers (excess sodium borohydride, several hours at pH 9 and 0°C).

Compound **XVIII** (Fig. 1) has been used to alkylate the active site SH group of D-glyceraldehyde-3-phosphate dehydrogenase (Balthasar, 1971). This reagent, like **I** and **VIII** (Fig. 1), is a derivative of haloacetic acid, and therefore will give a stable carboxymethylamino acid that is quantifiable by amino acid analysis. The aldehyde **XVII** (Fig. 1) has been used to alkylate the same enzyme. Both of these spin labels might be useful for studying the active sites of lipid hydrolyzing enzymes such as phospholipase A, lecithin : cholesterol acyltransferase, or lipoprotein lipase.

2. Acylating Spin-Label Reagents

The acylating spin-label reagents have been used extensively to tag reactive hydroxyl, thiol, or amino functionalities of proteins. These compounds generally exhibit greater specificity than the alkylating reagents but less than the sulfonylating or phosphorylating reagents. They do not allow an advantage afforded by many of the alkylating spin labels, namely, the formation of amino acid derivatives that are stable to conditions used for total protein hydrolysis (i.e., 6 *N* HCl, 110°C, 24 hr).

The succinic amide ester, succinic diester, and maleic diester (**I**, **II**, **III** in Fig. 2) have been used by Kosman *et al.* (1969) to study the active site geometry of α-chymotrypsin. These reagents, along with **IV**, **V**, **IX**, and **X** (Fig. 2) contain the nitrophenyl leaving group. The release of this chromophore may be monitored spectrophotometrically (Flohr *et al.*, 1972), providing a convenient method for studying the kinetics of the labeling reaction.

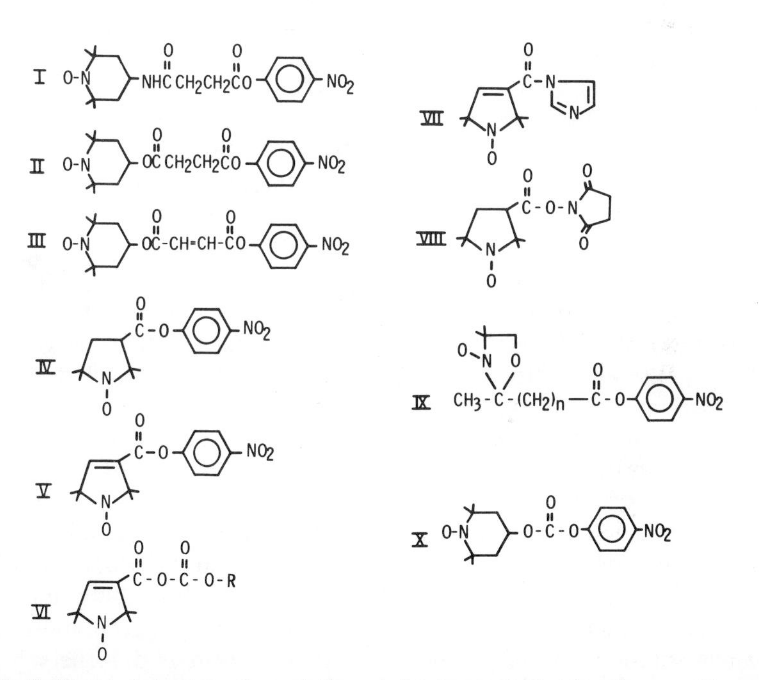

Fig. 2. Chemical structures for acylating spin label reagents that have been used to probe the structure and function of enzymes. Some pertinent references to either the synthesis and/or use of these molecules are: **I.** Kosman *et al.* (1969). **II.** Kosman *et al.* (1969). **III.** Kosman *et al.* (1969). **IV.** Berliner and McConnell (1966); Flohr *et al.* (1971); Flohr and Kaiser (1972); Ament *et al.* (1973). **V.** Berliner (1972). **VI.** Griffith *et al.* (1967). **VII.** Barratt *et al.* (1969). **VIII.** Hoffman *et al.* (1969). **IX.** Balthasar (1971). **X.** Sparrow (1975).

The mixed carboxylic–carbonic acid anhydride **VI** (Fig. 2) is a highly reactive reagent that has been used to acylate the ε-amino groups of poly-L-lysine (Griffith *et al.*, 1967). Apparently, the acylation reaction successfully competes with the hydrolysis reaction in aqueous solution. A possible disadvantage of this reagent is that the nonparamagnetic R group, instead of the paramagnetic nitroxyl moiety, may become attached to the protein or peptide, in which case undetectable urethane formation would occur. A spin-labeled analog of *N*-acetylimidazole (**VII**, Fig. 2) has been synthesized by Barratt *et al.* (1969) and used to tag the phenolic hydroxyl groups of poly-L-tyrosine. This reagent appears to exhibit some degree of specificity since treatment of an equimolar mixture of poly-L-tyrosine and poly-L-lysine resulted in preferential labeling of tyrosine residues. This reagent might be useful for studying the active site of carboxypeptidase A, which bears a tyrosine side chain at this locus (Lipscomb *et al.*, 1970).

The *N*-hydroxysuccinimide ester of a spin-labeled carboxylic acid (**VIII**, Fig. 2) has been used to acylate the α-amino group of valyl-tRNA. The attached label was used to monitor thermal unfolding of the amino acid–nucleic acid complex (Hoffman *et al.*, 1969).

The spin-labeled carbonate **X** (Fig. 2) is a very convenient acylating agent, which upon reaction with amino groups yields a urethan. This reagent has been used to label the deblocked amino terminus of peptide polymers obtained during the solid phase synthesis of apolipoprotein fragments (Sparrow, 1975). The accessibility of the amino terminus of a growing peptide (on the polymer) has been estimated from the EPR spectrum and has been studied as a function of the swelling solvent and the type of polymer support. An obvious advantage of this reagent is that the group attached to the amino terminus can be removed by standard deblocking procedures and the synthesis continued.

3. Sulfonylating Spin-Label Reagents

An array of fifteen different ortho-, meta-, and para-substituted spin-labeled sulfonylating agents has been synthesized by Berliner and colleagues to study and compare the active sites of trypsin and α-chymotrypsin (Berliner and Wong, 1974; Wong *et al.*, 1974). Several of these compounds were insoluble in water alone and also underwent hydrolytic decomposition in aqueous solutions, making it necessary to add them as dioxane solutions and portionwise over a period of time. These technical problems prevented carrying out accurate kinetic measurements of the inhibition reaction. The release of the paramagnetic center from the reagent before sulfonylation of the enzyme resulted in a mixture of sulfonylated enzyme molecules that did and did not carry a spin label. Once the intact reagent became attached to the protein, there appeared to be no release of the nitroxide from the aromatic moiety. There was, however, slow desulfonylation, a process that had been documented earlier by Fahrney and Gold (1963). In spite of these difficulties, Berliner and Wong (1974) were able to obtain very meaningful data by use of these labels (Fig. 3).

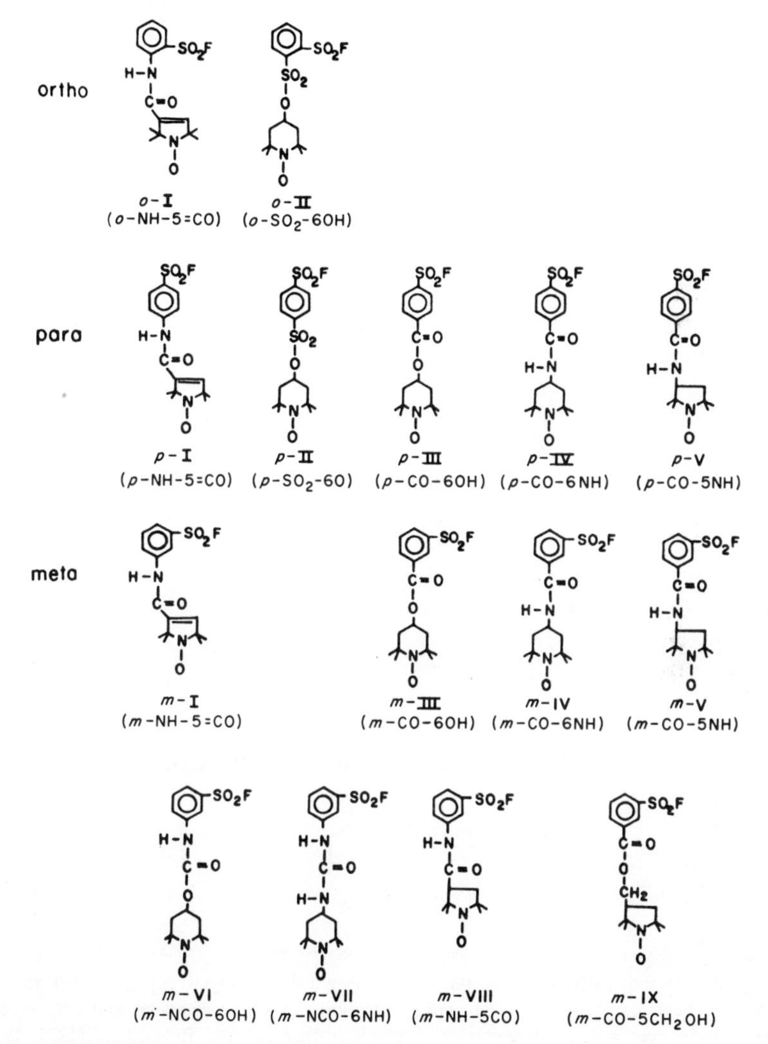

Fig. 3. Chemical structures of sulfonylating spin-label reagents used by Berliner and Wong (1974) and Wong *et al.* (1974) in studies comparing the active site geometries of α-chymotrypsin and trypsin.

4. Phosphor(n)ylating Spin-Label Reagents

The organophosphorus spin labels were first introduced by Hsia *et al.* (1969) and Morrisett *et al.* (1969). These reagents are highly specific for the nucleophilic serine residue at the active site of many enzymes. However, there are at least two known exceptions to this specificity. Reagent **I** (Fig. 4) has been used to spin label the unusually reactive tyrosine residues of papain and lysozyme (Morrisett, unpublished experiments). The nonparamagnetic reagent diisopropylphosphorofluoridate (DFP) has been shown to react with Tyr 123 of papain (Chaiken and Smith, 1969) and with Tyr 20 and Tyr 23 of lysozyme (Kato and Murachi, 1971). Retention of activity by the

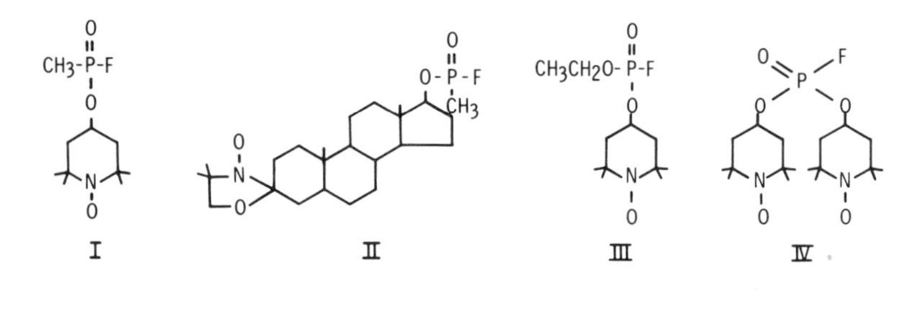

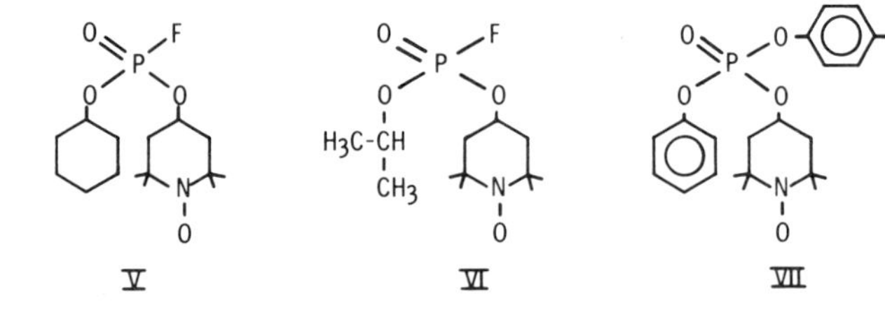

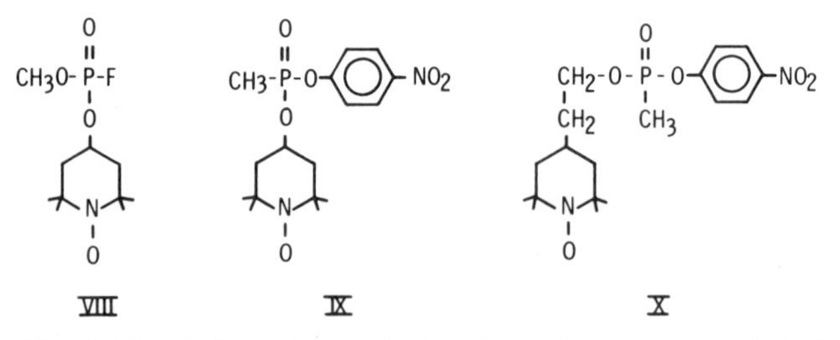

Fig. 4. Chemical structures of phosphorylating (phosphonylating) spin-label reagents which have been used to study the active site region of serine esterases and/or proteases. Pertinent references are: **I.** Morrisett *et al.* (1969); Morrisett and Broomfield (1971, 1972); Berliner and Wong (1973); Hoff *et al.* (1971). **II.** Morrisett and Broomfield (1971); Shimshick and McConnell (1972). **III.** Berliner and Wong (1973). **IV.** Hsia *et al.* (1972); Kosman and Piette (1972). **V.** Hsia *et al.* (1969). **VI.** Hsia *et al.* (1969). **VII.** Struve and Goldstein (1971). **VIII.** Hsia *et al.* (1972); Kosman and Piette (1972). **IX.** Grigoryan *et al.* (1973). **X.** Grigoryan *et al.* (1973).

phosphorylated enzymes and X-ray crystallographic data of the native molecules (Drenth *et al.*, 1970; Phillips, 1967) indicate that none of these tyrosine residues is involved in catalysis.

Most of the phosphor(n)ylating agents shown in Fig. 4 are extremely toxic and should be handled with rubber gloves in a well-ventilated hood. It is wise to keep 1.0 *M* NaOH nearby for rinsing glassware exposed to the agent and hydrolyzing any spilled material. Generally, these reagents react

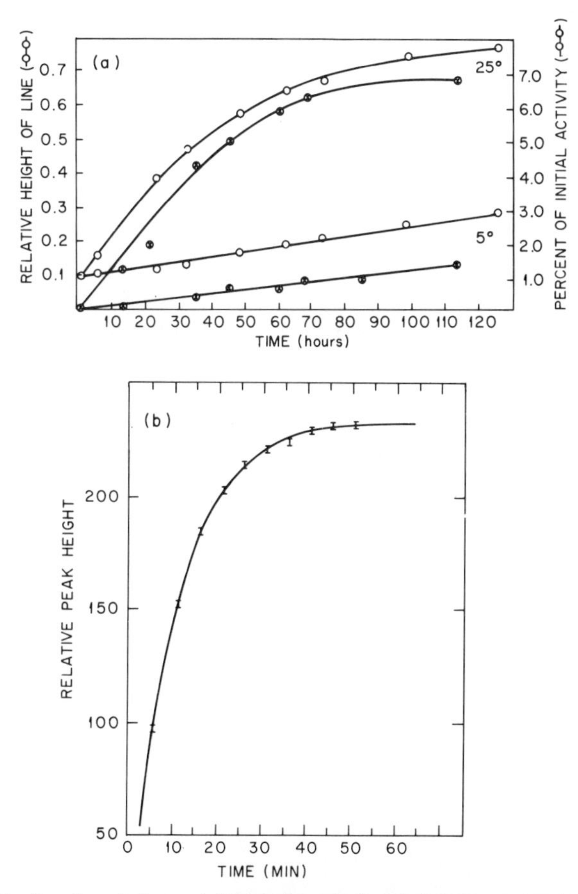

Fig. 5. Dephosphonylation and deacylation of spin-labeled chymotrypsin. (a) Spontaneous release of chymotrypsin spin-labeled with **I** (Fig. 4) at pH 6.6 at two different temperatures as measured by the time-dependent increase in spectral amplitude and enzymatic activity. [Reprinted with permission from Morrisett and Broomfield *J. Amer. Chem. Soc.* **93**, 7297 (1971). Copyright by the American Chemical Society.] (b) Typical plot of relative peak height versus time for the paramagnetic product arising from the deacylation of chymotrypsin spin labeled with **IV** (Fig. 2). Conditions: 0.05 *M* phosphate, pH 6.8. The smooth curve represents a theoretical first order decay process for 1.6×10^{-3} sec^{-1}. [From Berliner and McConnell (1966).]

very rapidly. For example, 125 mg of chymotrypsin (5 μmoles) in 5 ml buffer was inhibited by 50 μmoles of **I** (Fig. 4) to the extent of about 95% after 5 min (Morrisett and Broomfield, 1972).

The phosphor(n)ylating and sulfonylating reagents give spin-labeled protein derivatives that are usually more resistant to hydrolysis (i.e., hydrolysis that results in regeneration of active enzyme) than the acylating reagents. This is graphically illustrated in Fig. 5, which compares the time dependence of nitroxide released at pH 6.6–6.8 (as determined by increase in amplitude of the narrow line component) from α-chymotrypsin spin labeled with piperidinyl phospho*n*ate **I** (Fig. 4) and pyrrolidinyl carboxylate **IV** (Fig. 2). According to the model of Steitz *et al.* (1969), this stability of the phosphor(n)yl and sulfonyl analogues of acyl enzymes is due to the steric exclusion of an activated water molecule in the vicinity of the atom (phosphorus or sulfur) which is normally attacked.

Phosphor(n)ylated enzymes can undergo a process known as "aging," whereby an alkyl or alkoxy group is released from the enzyme while the phosphorus moiety remains attached. This process may be acid or base catalyzed. In the case of acid catalysis, a carbonium ion is released (Fig. 6), whereas when the process is base catalyzed, an alkoxide ion or an olefin is

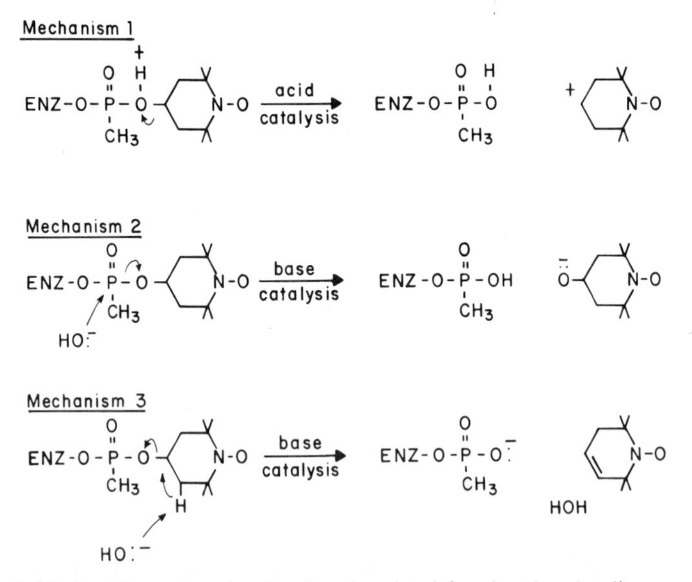

Fig. 6. Mechanistic routes whereby phospho*n*ylated (or phospho*r*ylated) enzymes might undergo release of the nitroxyl moiety without dephosphonyiation (dephosphorylation). This process (known as aging) is not accompanied by recovery of enzymatic activity. In the case of acid catalysis, this process is highly dependent on the stability of the incipient carbonium ion. [From Morrisett *et al.* (1970a).]

released. The aging phenomenon can give rise to mixed EPR spectra that contain narrow line components which are *not* the result of hydrolytic dephosphor(n)ylation, since the aging is *not* accompanied by return of enzymatic activity. This problem was essentially ignored by Hoff *et al.* (1971) in a spin-label study of atropinesterase, subtilisin, and chymotrypsin. Aging has been shown to be an important consideration in studies on acetylcholinesterase (Morrisett *et al.*, 1970a,b) inhibited with phospho*n*ylating reagents **I** or **II** (Fig. 4). The effect of pH, ionic strength, and temperature on the aging of phospho*n*ylated cholinesterases is well known (Keijer *et al.*, 1974).

Reagents **I** and **II** (Fig. 4) have been used by Morrisett and Broomfield in studies on the mechanism of guanidine-induced denaturation of α-chymotrypsin (Morrisett and Broomfield, 1971) and in studies comparing the active site geometries of trypsin, chymotrypsin, subtilisin, elastase, and acetylcholinesterase (Morrisett and Broomfield, 1972). Phospho*n*ylating reagent **I** and phospho*r*ylating reagent **III** (Fig. 4) have been employed by Berliner and Wong (1973) to elucidate a subtle autolysis effect observed with spin-labeled trypsin. The biradical phospho*r*ate **IV** and mono-radical phospho*r*ate **VIII** (Fig. 4) have been used by Hsia *et al.* (1972) and Kosman and Piette (1972) in active-site studies of cholinesterase, chymotrypsin, trypsin, elastase, and thrombin. Phosphor(n)ylating reagents for which the leaving group is *p*-nitrophenoxide (rather than fluoride) have been employed by Grigoryan *et al.* (1973) to investigate the urea denaturation of α-chymotrypsin (**IX**, **X**, Fig. 4) and by Struve and Goldstein (1971) to spin label serine esterases in synaptic membranes (**VII**, Fig. 4).

B. Noncovalently Binding Spin Labels†

Spin-labeled ligands that are bound to proteins by forces other than covalent chemical bonds (e.g. hydrophobic, ionic, and hydrogen bonding) undergo chemical exchange with unbound species. The strength of these interactions will govern the residence time of the ligand and may affect the motional freedom of the nitroxyl group attached to that ligand. If the affinity of the protein for the ligand is great enough, an EPR spectrum of only the bound species may be obtained (e.g., Chignell *et al.*, 1972). For lesser affinities, a mixed spectrum of bound and unbound species may result (e.g., Ogata and McConnell, 1972). When the association constant is quite low, the spectrum of the bound species will be masked by that of the unbound species. In favorable cases where the spectral amplitude of the bound label does not significantly increase the amplitude of the unbound species,

† Editor's note: The name "spin probe" is also used to distinguish noncovalent from (covalently bound) spin labels.

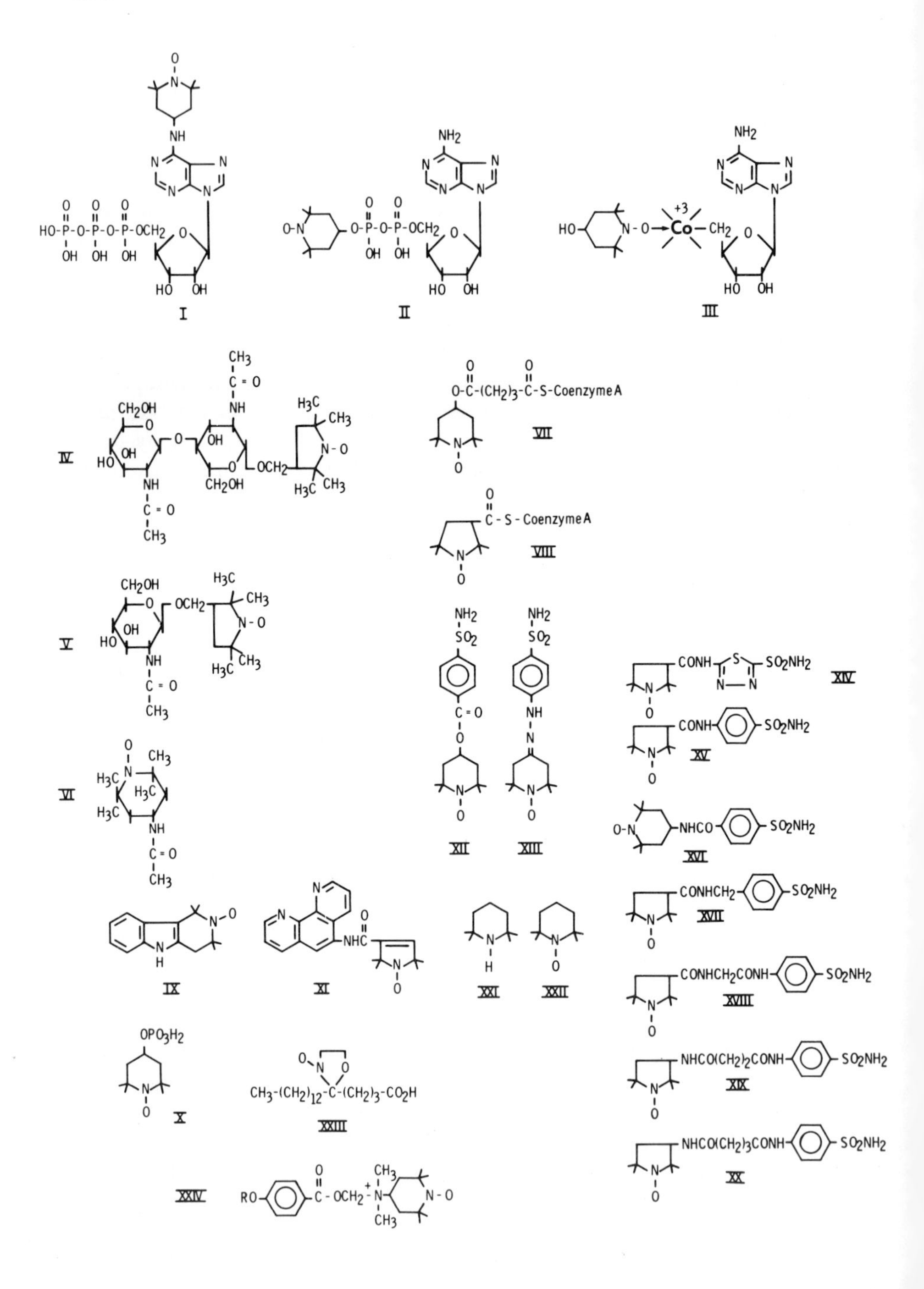
I
II
III
IV
V
VI
VII
VIII
IX
X
XI
XII
XIII
XIV
XV
XVI
XVII
XVIII
XIX
XX
XXI
XXII
XXIII
XXIV
O-C-(CH2)3-C-S-CoenzymeA
C-S-CoenzymeA
CH3-(CH2)12-C-(CH2)3-CO2H

the latter quantity may be used to calculate an association constant for the ligand (Wien *et al.*, 1972). The affinity and specificity requirements of noncovalently binding spin labels are major considerations in designing experiments that employ these molecules. Both of these parameters must be accurately known for distance-measuring experiments to be meaningfully interpreted (Wien *et al.*, 1972).

A number of spin-labeled coenzymes have been prepared and gainfully utilized. A nitroxyl derivative of ATP (**I**, Fig. 7) has been used by Krugh (1971) to measure distances between groups in the ATP and AMP binding sites on DNA polymerase. A phosphate group of ADP has been esterified with 1-oxyl-2,2,6,6-tetramethylpiperidin-4-ol giving an analogue of NAD (**II**, Fig. 7) which was used in a study of alcohol dehydrogenase (Weiner, 1969; Mildvan and Weiner, 1969a,b). These workers were able to measure the distance from the coenzyme binding site to the sites occupied by substrates or inhibitors such as ethanol, acetaldehyde, and isobutyramide. By coordination of the nitroxyl oxygen to the cobalt atom of cobinamides, spin-labeled derivatives of methylcobinamide and 5′-deoxyadenosylcobinamide (**III**, Fig. 7) have been prepared (Law *et al.*, 1971). An analogue of vitamin B_{12} has been prepared by alkylation of Co^{3+} 5,6-dimethylbenzimidazolylcobamide with the piperidinylbromoamide **VIII** (Fig. 1). Homolytic cleavage of the cobalt–carbon bond could be brought about by photolysis, and EPR could be used to follow the kinetics of this process (Buckman *et al.*, 1969). Two analogues of coenzyme A (**VII** and **VIII**, Fig. 7) have been synthesized and used to study the active site of citrate synthase (Weidman *et al.*, 1973).

Glycosides of *N*-acetylglucosamine (**V**) and di-*N*-acetylglucosamine (**IV**) with 1-oxyl-2,2,5,5-tetramethyl-4-hydroxymethylpyrrolidine as the aglycone have been prepared by Wien *et al.* (1972). These spin-labeled sugars were used to measure the distance from subsite D to His 15 on lysozyme. The pseudosugar piperidinylacetamide **VI** (Fig. 7) was originally expected to bind at one of the six subsites in the catalytic region, but X-ray crystallographic data indicated that its highest occupancy was at an anomalous binding site near Trp 123 (Berliner, 1971).

Fig. 7. Chemical structures of noncovalently binding spin-labeled ligands that have been used to elucidate the structural and functional properties of enzymes. References to the synthesis or use of these molecules are: **I.** Krugh (1971). **II.** Weiner (1969); Mildvan and Weiner (1969a,b). **III.** Law *et al.* (1971). **IV–VI.** Wien *et al.* (1972). **VII–VIII.** Weidman *et al.* (1973). **IX.** Rozantsev and Shapiro (1964); Rosnati and Palazzo (1945); Morrisett *et al.* (1975). **X.** Roberts *et al.* (1969); Mushak *et al.* (1972). **XI.** Spallholz and Piette (1972). **XII.** Hower *et al.* (1971). **XIII.** Mushak and Coleman (1972). **XIV–XX.** Chignell *et al.* (1972); Erlich *et al.* (1973). **XXI–XXII.** Stier and Reitz (1971).

A spin-labeled derivative of indole (**IX**, Fig. 7) was prepared some time ago (Rozantsev and Shapiro, 1964; Rosnati and Palazzo, 1945). This molecule binds in one of the NAD binding sites of alcohol dehydrogenase (McConnell, 1967) and in the indole binding site of bovine serum albumin (Morrisett *et al.*, 1975). This nitroxide is not very water soluble, but counterbalancing that drawback are several advantages. The nitroxyl ring is almost coplanar with the aromatic ring, providing an essentially flat molecule that might intercalcate between base pairs of bihelical nucleic acids and therefore might be useful for observing structural changes in such systems. The molecule is only weakly fluorescent, but the diamagnetic precursor (the secondary amine) is strongly fluorescent, and its emission and absorption maxima are similar to those of indole (Pownall and Morrisett, unpublished data). Since the amine and nitroxide are so structurally similar, they comprise an ideal spin label–fluorophore pair that allows one to measure several different parameters by two independent techniques using essentially the same probe molecule.

1,10-Phenanthroline strongly coordinates divalent cations. A spin-labeled derivative of this chelating agent (**XI**, Fig. 7) has been prepared by Spallholz and Piette (1972). It rigidly binds to two of the four zinc ions normally present in liver alcohol dehydrogenase.

Aromatic sulfonamides are well-known specific inhibitors of carbonic anhydrase. Chignell *et al.* (1972) have synthesized a number of spin-labeled sulfonamides (**XIV–XX**, Fig. 7) in which the distance between the nitroxyl and the aromatic sulfonamide group varied from 9.6 to 15.2 Å. These compounds were used to determine the depth of the active site cleft in bovine erythrocyte carbonic anhydrase B (Chignell *et al.*, 1972) and in human erythrocyte carbonic anhydrase B and C (Erlich *et al.*, 1973). Carbonic anhydrase isozymes from various species were studied using the spin-labeled hydrazide **XIII** (Mushak and Coleman, 1972). A sulfoamide ester (**XII**, Fig. 7) has been used by Hower *et al.* (1971) to probe the active site of a mixture of bovine carbonic anhydrase isoenzymes.

Ribonuclease strongly binds a single mole of phosphate. 1-Oxyl-2,2,6,6-tetramethylpiperidine-4-phosphate (**X**, Fig. 7) has been used by Roberts *et al.* (1969) to establish that the phosphate-binding site is in the active site and very near His 12 and His 119. The same probe molecule is a substrate for alkaline phosphatase. The rate of catalyzed hydrolysis has been monitored by observing the loss of proton splitting and the shift in g value of the EPR spectrum of the reactant (Mushak *et al.*, 1972).

Stier and Reitz (1971) have used piperidine **XXI** (Fig. 7) as a substrate for oxidizing enzymes in liver microsomes that convert this compound to **XXII** (Fig. 7), the nitroxyl derivative.

III. LABORATORY TECHNIQUES FOR SPIN LABELING ENZYMES

The success of any spin-label study is highly dependent on careful experimental design and execution. This is especially true of studies designed to yield information about the structure of an enzyme and to correlate that information with what is known about the enzyme's biological function. With all probe techniques that involve the use of an extrinsic reporter group such as a spin label, it is essential not only to anticipate but to evaluate the effect of that group on the structure and function of the system under study. Only when such an evaluation has been made can the results of probe experiments be placed in full perspective and their meaning correctly interpreted. In this section are outlined several techniques that have proven useful in minimizing ambiguity and maximizing information content of results from spin-labeled enzyme experiments.

A. Labeling the Enzyme

1. Covalent Labeling

The enzyme of interest may be labeled by either covalent attachment or noncovalent binding. The usefulness of covalent labeling depends on the presence and reactivity of amino acid side chains at or near the active site. Covalent labeling experiments usually fall into one of two different categories: those in which the label is attached *near* the active site and does not severely impair biological function (e.g., Cohn *et al.*, 1971; Jones *et al.*, 1973) and those in which the label is attached to a group *in* the active site and destroys most or all catalytic activity (e.g., Berliner and Wong, 1974; Morrisett and Broomfield, 1972).

In the ideal case, a single functional group at a desirable location is labeled 100%, and no other groups in the protein are labeled at all. The attainment of such specificity may require special conditions of pH, ionic strength, temperature, reaction time, or reagent concentration. Often this high level of specificity can be achieved by using a spin-label reagent whose structure closely resembles that of a substrate or pseudosubstrate (e.g. Kosman, 1972). A working knowledge of affinity labeling (Singer, 1967) is very useful in the design (Means and Feeney, 1971; Baker, 1967) and synthesis (Rozantsev, 1970; Forrester *et al.*, 1968) of such molecules. Reaction conditions for covalent binding of spin labels will generally be the same as those used for any protein-modifying reagent. However, there are a few buffer constituents that should be carefully considered before use. Thiols such as mercaptoethanol or glutathione (Morrisett and Drott, 1969) can be oxidized by spin labels with concomitant reduction of the nitroxyl moiety and loss of paramagnetism. Nitroxide reduction is virtually instantaneous with ascorbic acid (Kornberg and McConnell, 1971). High hydrogen ion concentration (pH < 2) for extended periods leads to destruction of the paramagnetic center. Paramagnetic ions can give rise to EPR spectra that overlap the nitroxide spectrum. Such ions can also enhance electronic relaxation of the spin label, resulting in diminution of spectral amplitude (Taylor *et al.*, 1969; Leigh, 1970). Cobaltous ion has been shown to complex to the nitroxyl group (Beck *et al.*, 1967; Buckman *et al.*, 1969; Law *et al.*, 1971). In cases where it is necessary to remove divalent cations from spin-labeled proteins, this can be done by adding excess ethylenediaminetetraacetate (EDTA) and removing the EDTA–Co(II) complex. In experiments where the spin concentration is of critical importance, it is not sufficient merely to add excess EDTA, since EDTA alone can cause anomalous changes in spectral amplitude.

Upon completion of the labeling reaction, excess unbound label can be removed by any one of several convenient techniques. Dialysis (Craig and Chen, 1969) involves minimal equipment but has several disadvantages, including possible loss of protein due to adsorption to the dialysis tubing or escape through the pores, and slow or incomplete removal of unbound label. These difficulties can be minimized by use of Spectrapor™ tubing (Spectrum Medical Industries), which can be obtained with molecular weight cutoffs of 3500, 6000–8000, and 12,000–14,000. Gel filtration (Reiland, 1971) over Sephadex G-10 (Pharmacia) or Bio-gel P-2 (Bio-Rad) removes unbound label much more rapidly and completely than dialysis. In cases where a small amount of unbound label is present due to either incomplete removal or hydrolytic release from the protein (Berliner and McConnell, 1966; Morrisett and Broomfield, 1971), a mixed EPR spectrum containing narrow and broad components may be obtained. In such cases, the narrow spectral component can be removed by computer subtraction techniques (Jost *et al.*, 1971).

At this point in the experiment it is necessary to compare the physical, chemical, and biological properties of the labeled and native enzyme. Changes in secondary, tertiary, or quaternary structure may be assessed by such techniques as ultracentrifugation (Schachman and Edelstein, 1973), circular dichroism (Adler *et al.*, 1973), and intrinsic fluorescence (Brand and Witholt, 1967). The sequence position of modified residues may be determined from tryptic (Canfield and Anfinsen, 1963) or cyanogen bromide (Gross and Witkop, 1962) fragmentation. Which amino acid side chain(s) has actually been tagged can sometimes be determined from a change in the amino acid composition (Spackman *et al.*, 1958). In cases where the bond joining the amino acid and the spin label is labile in 6 *N* HCl, a change in the analysis will not be observed. In some favorable cases, however, a derivative amino acid will result which appears on the analysis chromatogram at a position not masked by other species. For example when ribonuclease was tagged with pyrrolidinylbromoacetamide **I** (Fig. 1), its amino acid composition decreased by one histidine residue and increased by one carboxymethylhistidine residue (Morrisett, 1969; Daniel *et al.*, 1973), which eluted immediately after proline. While the extent of labeling may be determined from the amino acid analysis, values so obtained are not as reliable as those obtained by using a radioactive spin label or by counting the number of incorporated spins. In the latter case, a known amount of accurately determined spin-labeled protein is hydrolyzed in 1.0 *M* NaOH for 24 hr at 60°C, a portion of the hydrolysate is transferred to a Corning 100 μl capillary sealed at one end, its EPR spectrum recorded under a standard condition, and the spectral amplitude compared to that of a standard solution of the corresponding spin label of known concentration (Wien *et al.*, 1972). The number of incorporated spin labels may also be obtained by double integration of the first-derivative spectrum. Measurements of the biological activity of the labeled protein by at least one and preferably several different assay methods should be carried out and the results of these measurements compared to those obtained from the unlabeled protein.

Some reactions may give rise to several differently labeled proteins, in which case the interpretation of spectral changes can be difficult if not impossible. Such a mixture can often be resolved into its components by use of ion-exchange chromatography (Himmelhoch, 1971), isoelectric focusing (Vesterberg, 1971), or polyacrylamide gel electrophoresis (Shuster, 1971). By using covalently binding spin-label reagents, one exposes himself to the same pitfalls encountered by anyone involved in the general area of protein modification. These include: (a) failing to perform *complete* amino acid analyses on the modified protein; (b) assuming that retention or loss of enzyme activity toward one class of substrate implies similar behavior toward all classes, and (c) concluding that demonstrated modification or survival of particular residues in model peptides or other proteins may be extrapolated to the case in question. Such pitfalls deserve careful consideration before proceeding to detailed studies on the labeled enzyme.

2. Noncovalent Labeling

Many of the techniques described above for covalently spin-labeled enzymes are equally useful for enzymes to which labels are noncovalently bound. While the spin labels used in this latter case need not carry reactive functionalities, they must bear moieties that will confer on the total molecules sufficient affinity for the binding site of interest on the protein. Such spin labels are exemplified by the analogues of ATP (**I**, Fig. 7; Krugh,

1971), *N*-acetylglucosamine (**V**, Fig. 7; Wien *et al.*, 1972), and NAD (**II**, Fig. 7; Weiner, 1969). Unless the association constants of such ligands are very high, the EPR spectrum of the bound label may be partially or totally obscured by the unbound species. In such cases computer methods can be useful for removing the narrow line spectral component due to the unbound label (Jost *et al.*, 1971).† In cases where the "bound" spectrum does not make a significant contribution to the total spectrum, diminution of the signal amplitude upon adding the enzyme to the spin-label solution can be used to calculate an association constant for that label (Wien *et al.*, 1972). There is an upper limit on the concentration of unbound label that can be present when this type of measurement is made. At concentrations of unbound nitroxide $\geqq 10^{-3}$ *M*, spin exchange can occur. This may shorten the electronic spin relaxation time (τ_s) and diminish the paramagnetic resonance peak intensity due to exchange broadening.

B. Properties of Spin-Labeled Enzymes

The contribution of a nitroxyl group to the ultraviolet absorption of the protein bearing it is dependent on the structure of the group and the wavelength of measurement. As shown in Fig. 8, the λ_{max} of the oxazolidinyl-, piperidinyl-, and pyrrolidinyl-1-oxyl rings are in the region of 230–340 nm, well removed from the λ_{max} of most proteins, which occurs at about 280 nm. The molar extinction of a nitroxide at this latter wavelength is in the range of 200–800 as compared to 1100 for tyrosine and 5200 for tryptophan. Hence, for a protein such as chymotrypsin, which contains four tyrosines and eight tryptophans (Matthews *et al.*, 1967) and bears one spin label at the active site (Morrisett and Broomfield, 1971), the contribution of the label's absorbance to that of the total molecule is less than 1%.

The ability of nitroxides to induce fluctuating magnetic fields that enhance nuclear relaxation is a well-known phenomenon. The resonance lines of nuclei on a protein become greatly broadened when a spin label is covalently attached. Hence, this effect makes it difficult if not impossible to measure distances between a covalently bound spin label and a nucleus on the same protein molecule. However, the effect is extremely useful for measuring distances to binding sites of chemically exchanging nuclei (Wien *et al.*, 1972; Morrisett *et al.*, 1973).

In some cases it is desirable to concentrate the spin-labeled protein in order to obtain a required optical density or resonance signal amplitude. This is achieved by such techniques as lyophilization, pressure ultrafiltration

† Editor's note: The preceeding chapter covers these methods in some detail.

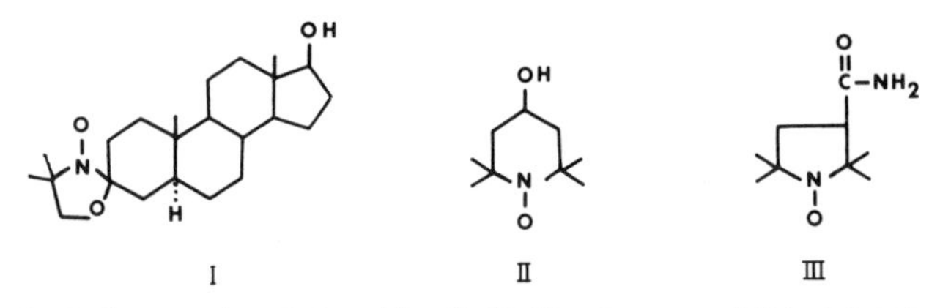

Fig. 8. Optical properties of nitroxides. (J. D. Morrisett, unpublished data.)

Compound	Solvent	λ_{max}	$E_{\lambda_{max}}(M^{-1}\ cm^{-1})$	$E_{280}(M^{-1}\ cm^{-1})$
I	EtOH	227	2.41×10^3	7.6×10^2
II	EtOH	240	1.44×10^3	3.42×10^2
III	EtOH	230	2.58×10^3	2.16×10^2

(Amicon or Millipore), suction ultrafiltration using collodion bags (Schleicher and Schuell), or simply adding an appropriate amount of dry Sephadex to the enzyme solution. While these methods for concentrating samples may be useful, they may also lead to aggregation of the protein. Aggregation, in turn, may give rise to significant changes in the correlation time τ_c of the spin label and/or particular nuclei on the protein.

C. Sample Measurement

There are a variety of cells, both commercial† and home made, that can be used to contain spin-labeled protein solutions. The uniform size of Corning calibrated capillaries makes them ideal for spin-count determinations where the sample must be reproducibly oriented in the spectrometer cavity. For routine measurements at room temperature, disposable Pasteur pipets sealed at the smaller end are adequate. For experiments that require use of the temperature controller and maximum signal level, the flat quartz microcell (Scanlon, Wilmad, or Varian) is appropriate. When the controller is not necessary but maximum signal is, the larger quartz flat cell (Varian) is best. Normally a standard $0.5 \times 3 \times 250$ mm EPR tube is not satisfactory for aqueous samples because of high dielectric loss. Selection of proper spectrometer conditions is governed to a large extent by the nature of the sample and the true width of the resonance lines. The effects of different spectrometer settings on the nitroxide spectrum are illustrated in Chapter 7 and in a recent review by Jost and Griffith (1972). The use of computer methods to separate narrow- and broad-line spectral components (corresponding to mobile and immobile spin labels) and to subtract the spectral

† Editor's note: See Appendix III for a listing of suppliers.

contribution from unbound labels (Jost *et al.*, 1971) has been mentioned earlier. Motion parameters (e.g., $2T_{\parallel}$, $2T_{\perp}$, S, and τ_c) can be rapidly obtained by computer methods (McFarland, 1974).

A spin-labeled enzyme may be obtained in the crystalline form by either labeling the enzyme in solution then crystallizing the labeled protein, or by labeling the enzyme in the crystalline form. For experiments involving crystals of spin-labeled protein (McConnell and Boeyens, 1967; Berliner and McConnell, 1971), reproducible orientation of the sample can be achieved by use of a crystal goniometer (see specifications in Berliner, 1967; such goniometers are commercially available from Varian). In order that the Q of the cavity be retained, the sample is not rotated in the cavity; rather, the laboratory magnetic field is rotated about the cavity. Thus, the larger spectrometers with rotating magnets are most suitable for this type of work.

IV. INFORMATION OBTAINABLE FROM SPIN-LABELED ENZYMES

Nitroxyl radicals exhibit a number of chemical and physical properties that make them extremely useful molecules for studying biochemical systems. These properties render the EPR spectrum of a nitroxide sensitive to various environmental conditions. For example, the spectrum can be affected by the motional constraint and orientation of the spin label in the system, by the polarity of the system, and by the presence of other species in the system such as reducing agents and paramagnetic ions. Conversely, a spin label can affect the properties of the system. It can bring about changes in the relaxation times of nuclei and the lifetimes of fluorophores in excited states.

The diverse properties of these molecules is unquestionably one of the major reasons for their tremendous popularity and wide use in solving biochemical and biophysical problems. The various types of information about the structure and function of enzymes that spin labels can reveal is described in this section.

A. Rates of Catalysis

The first study illustrating that spin labels can be used to study enzyme kinetics was carried out by Berliner and McConnell (1966). In that study, α-chymotrypsin was acylated at pH 4.5 with a *p*-nitrophenyl ester of a spin-labeled carboxylic acid (**IV**, Fig. 2). The EPR spectrum of the strongly immobilized, covalently bound label contained only broad-line components (upward arrows in Fig. 9). However, when the pH of the acyl–enzyme intermediate was raised to 6.8, the spin-labeled acyl group was released, giving a narrow-line spectrum for the freely rotating nitroxide (Fig. 9b). The time-dependent increase in narrow-line amplitude or decrease in broad-line amplitude can be used to measure the rate of deacylation (Fig. 5b). However, the spectral amplitude of a freely tumbling nitroxide is considerably greater than that of a strongly immobilized one. Hence, growth of the narrow-line component is a more sensitive indicator of deacylation. For the deacylation step, a first-order rate constant of $(1.55 \pm 0.1) \times 10^{-3}\ \text{sec}^{-1}$ was calculated.

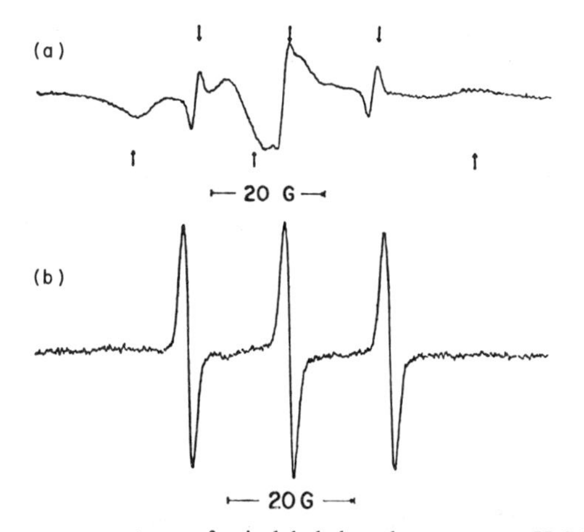

Fig. 9. (a) Resonance spectrum of spin-labeled acyl enzyme at pH 3.5. Broad resonance lines indicated by up arrows (↑) are due to the resonance of the paramagnetic acyl group immobilized at the active site. The three narrow lines—down arrows (↓)—arise from free nitroxide spins in solution due to slow deacylation at this pH. (These narrow lines are actually broadened somewhat by a high modulation amplitude.) (b) Resonance spectrum of the paramagnetic hydrolysis product at pH 4.5. [From Berliner and McConnell (1966).]

The same technique has been used to compare the relative magnitudes of the individual rate parameters for both enantiomers of asymmetric substrates (Flohr and Kaiser, 1972):

$$E + S \underset{K_S}{\overset{k_1}{\rightleftharpoons}} E \cdot S \xrightarrow{k_2} ES' + P_1 \xrightarrow{k_3} E + P_2 \qquad (1)$$

From visible absorption measurements of the rate of release of *p*-nitrophenol and from paramagnetic resonance measurements of the release of spin-labeled carboxylic acid (see Table I), values of K_s, k_2, and k_3 were calculated for both *R* and *S* isomers of **IV** (Fig. 2). Values of k_3 determined by

TABLE I

RATE CONSTANTS FOR HYDROLYSIS OF SPIN-LABELED SUBSTRATES BY α-CHYMOTRYPSIN[a]

	Substrate	$K_s \times 10^4\ M$[b]	$k_2(\text{sec}^{-1})$[b]	$k_3 \times 10^4\ \text{sec}^{-1}$[b]	$k_3 \times 10^4\ \text{sec}^{-1}$[c]
$S(-)$	**IV** (Fig. 2)	5.1 ± 1.2	0.041 ± 0.005	2.5 ± 0.3	2.3 ± 0.2
$R(+)$	**IV** (Fig. 2)	4.1 ± 0.4	0.37 ± 0.02	52 ± 2	45 ± 10

[a] Flohr and Kaiser, 1972.
[b] Constants determined by measurement of *p*-nitrophenol release.
[c] Constants determined by measurement of spin label release.

the two different methods agree to within 8–13% and indicate that the rate of hydrolysis of the dextrorotatory isomer is about twenty times greater than that of the levoratory one (Table I). The absolute configuration of the dextrorotatory molecule resulting from the hydrolysis of **IV** (Fig. 2) has recently been determined by X-ray methods (Ament *et al.*, 1973). Similar kinetic experiments have been carried out on the model enzyme systems cyclohe*x*a-amylose (Flohr *et al.*, 1971) and cyclohe*pt*aamylose (Paton and Kaiser, 1970). A comparison of rate data indicate similar enantiomeric specificities of chymotrypsin and cyclohe*x*aamylose in the catalysis of nonspecific ester substrates such as **IV** (Fig. 2). Cyclohe*pt*aamylose, however, shows no appreciable enantiomeric specificity in the hydrolysis of such substrates.

1-Oxyl-2,2,6,6-tetramethylpiperidinyl-4-phosphate (**X**, Fig. 7) is a substrate for alkaline phosphatase from *Escherichia coli* that gives the corresponding alcohol as the only paramagnetic cleavage product. The pronounced difference in resolution of the proton superhyperfine splitting of the EPR spectra of the substrate and cleavage product serve as the basis for a semiquantitative assay of alkaline phosphatase activity (Mushak *et al.*, 1972). For the assay, the mixture of enzyme and substrate is quickly degassed, and the catalyzed reaction is monitored by loss of the superhyperfine structure of the spectrum and/or the upfield shift of the signal (Fig. 10).

The ability of certain biological molecules such as ascorbate (Kornberg and McConnell, 1971) and glutathione (Morrisett and Drott, 1969) to reduce nitroxides and destroy their paramagnetism is well known. Certain enzyme systems also produce this effect. For example, the reduction kinetics of the spin-labeled phosphate **X** (Fig. 7) and 5-doxylstearic acid (**XXIII**, Fig. 7) by microsomal membranes containing the cytochrome P_{450}–cytochrome P_{450} reductase hydroxylating enzyme system have been studied by Stier and Sackmann (1973). An Arrhenius plot of the reduction

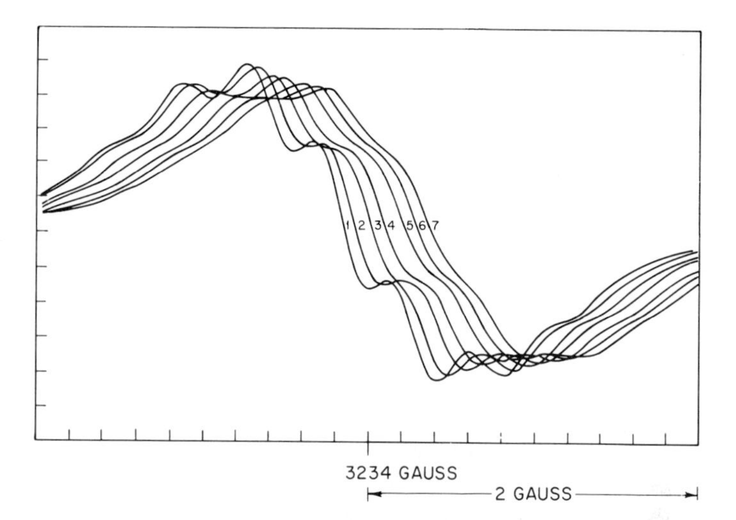

Fig. 10. The effect of enzymatic hydrolysis on the ESR spectrum of **X** (Fig. 5). The curves labeled 1–7 correspond to 2, 11, 21, 36, 61, 91, and 121 min of reaction time, respectively. Temperature 25°C. [Reproduced from Mushak *et al. Biochim. Biophys. Acta* **261**, 337 (1972).]

rate constants revealed a significant difference in the behavior of the water-soluble and lipid-soluble spin labels. The activation energy of the fatty acid reduction decreased abruptly from 30.8 to 8.7 kcal/mole at about 32°C. No such break occurred in the Arrhenius plot of spin-labeled phosphate reduction (activation energy = 13.8 kcal/mole). These results were interpreted to indicate that the reducing enzyme was bounded by a rather rigid phospholipid halo that is in a quasicrystalline structure below 32°C.

The conversion of a nonparamagnetic amine to a nitroxide (just the reverse of the process described in the preceding paragraph) is a novel technique that has been gainfully utilized by Stier and Reitz (1971). 2,2,6,6-Tetramethylpiperidine (**XXI**, Fig. 7) was found to be oxidized by rat and rabbit liver microsomes to the corresponding nitroxide (**XXII**, Fig. 7) in a reaction requiring O_2 and NADPH. An oxidation sequence (Fig. 11) leads from **XXI** (Fig. 7) over the corresponding hydroxylamine (step 1) to the nitroxide (step 2), which is further oxidized to an unknown product (step 3). In addition to these oxidation processes, the nitroxide **XXII** (Fig. 7) is reduced over the *N*-hydroxy derivative to the starting secondary amine (Fig. 11, step 5). The authors deduce this sequence of events by measuring stationary concentrations of radicals formed and by examining the effect of inhibitors, stimulators, and temperature on the overall reaction and single reaction steps that they feel involve three different enzymes.

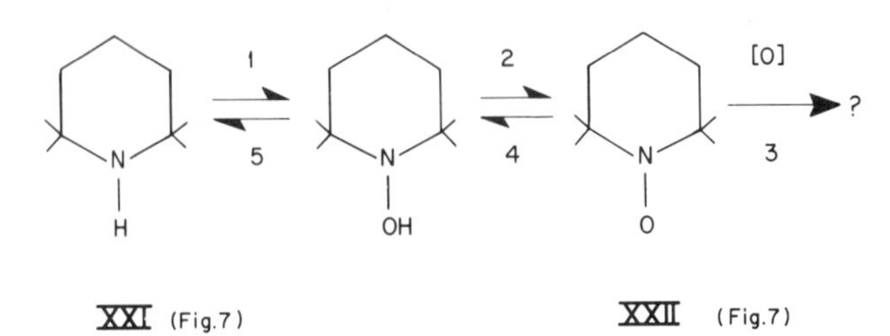

Fig. 11. Pathway proposed by Stier and Reitz (1971) for oxidation of a diamagnetic nitroxide precursor (**XXI**, Fig. 5) to the paramagnetic species by liver microsomes.

B. Mechanisms of Denaturation

Spin labels may be used to study unfolding of the active sites of enzymes. This application is illustrated by the chymotrypsin system, which has been examined by two groups (Morrisett and Broomfield, 1971; Berliner, 1972). In the former study, chymotrypsin was spin-labeled at the active site serine by piperidinyl phospho*n*ate **I** and androstyl phospho*n*ate **II** (Fig. 4). The denaturation of each enzyme derivative as a function of guanidine hydrochloride concentration was then studied by paramagnetic resonance. The EPR spectral changes that were observed are illustrated in Fig. 12. These changes may be expressed as changes in the ratio of resonance line amplitudes. For the experiment illustrated by Fig. 12, the ratio of high field to center field line heights was used. When this ratio was plotted against denaturant concentration, a sigmoidal curve was obtained (Fig. 13). Results obtained in this manner indicated that the concentration of denaturant required to unfold the active site region of α-chymotrypsin labeled with **I** (Fig. 4) is significantly lower than that required when the enzyme is labeled with **II** (Fig. 4). For each of the two derivatives, the denaturant concentration required for unfolding is essentially pH independent. In a similar study employing chymotrypsin labeled with piperidinyl phospho*r*ate **III** (Fig. 4) or pyrrolyl carboxylate **V** (Fig. 2), Berliner (1972) found that when urea was used as the denaturant, the unfolding was pH dependent. These results implied that effects involving charged residues that were involved in active site conformation were swamped out by the ionic denaturant, whereas the neutral denaturant did not obliterate all of these effects.

Mushak and Coleman (1972) have performed a urea denaturation study of human carbonic anhydrase to which the piperidinylsulfonamide **XIII** (Fig. 7) was bound. While 6 *M* urea caused a relatively rapid denaturation of the enzyme as determined by optical rotatory dispersion, circular dichroic, or aromatic difference spectra, only 30% of the sulfonamide was released. At lower urea concentrations, lower levels of spin label were released. To release

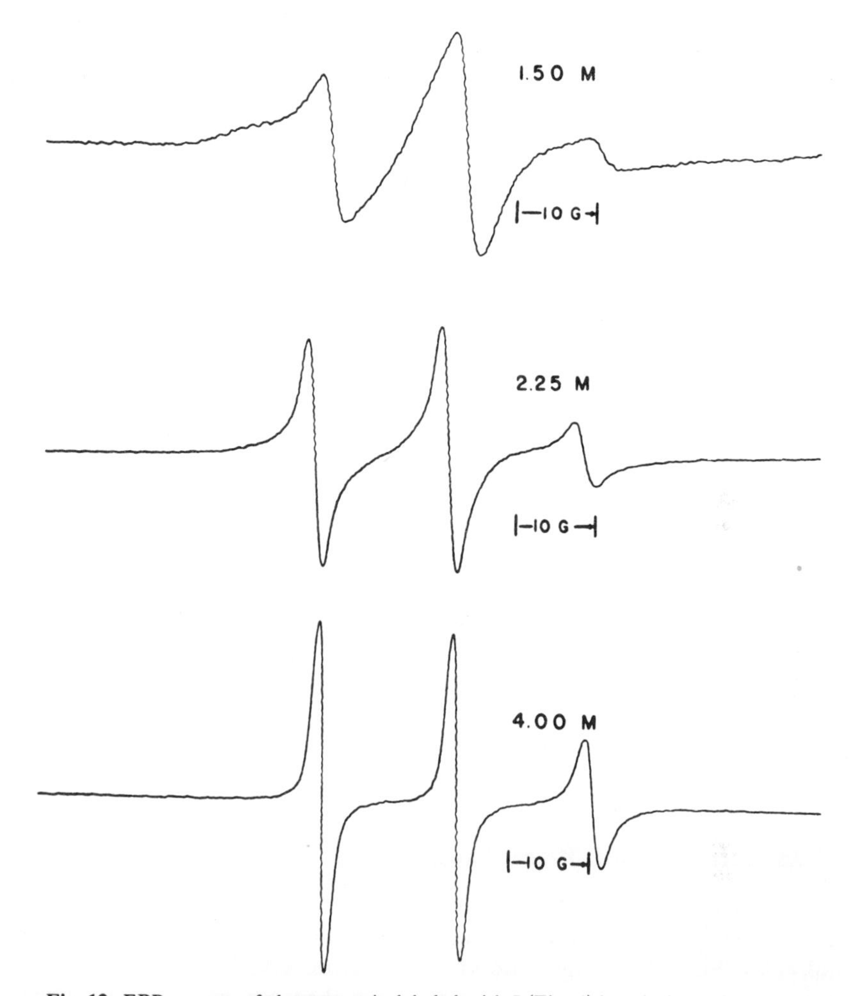

Fig. 12. EPR spectra of chymotrypsin labeled with **I** (Fig. 4) in solutions of progressively increasing guanidine hydrochloride concentration at pH 3.0. The height of the high field resonance line relative to the center line was used as a measure of the amount the mobility of the label was changed by the guanidine hydrochloride. [Reprinted with permission from Morrisett and Broomfield *J. Amer. Chem. Soc.* **93**, 7297 (1971). Copyright by the American Chemical Society.]

all of the spin-labeled sulfonamide, 8 *M* urea was required. Significantly, the stable fraction of the enzyme that was present at urea concentrations below 8 *M* could be observed even when the spin label was added after exposing the native enzyme to urea. These data are consistent with either a stepwise denaturation mechanism or a rapid equilibrium between folded and unfolded conformers at the lower urea levels.

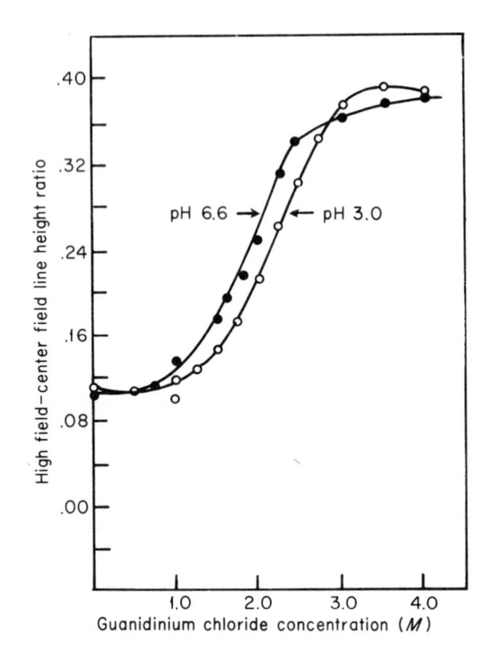

Fig. 13. Plot of change in high field : center field line height ratio of EPR spectra of chymotrypsin spin-labeled with **I** (Fig. 4) as a function of guanidine hydrochloride concentration. [Reprinted with permission from Morrisett and Broomfield, *J. Amer. Chem. Soc.* **93**, 7297 (1971). Copyright by the American Chemical Society.]

C. Distances between Different Groups

The intrinsic paramagnetism of spin-labeled molecules creates perturbations in the magnetic and electronic environment of certain ions and other molecules. Since the magnitude of this effect depends on the distance between the paramagnetic label and the perturbed group, estimates of that distance can be made. Basically, there are four types of experiments that can be used to estimate distances between different groups at the catalytic site (and other sites) on an enzyme. These four types are currently at various stages of development.

1. Distance between a Paramagnetic Spin Label and a Diamagnetic Nucleus

Spin labels, like other paramagnetic species, create fluctuating magnetic fields that enhance the relaxation rate of nuclei within their sphere of influence. (See Chapter 9 for a quantitative treatment of this phenomenon.)

The possibility of gainfully utilizing spin labels to perturb nuclear resonances and thereby measure the intervening distance has been recognized for some time, but only recently has it been exploited. Experiments of this type were first attempted by Sternlicht and Wheeler (1967) and by McConnell *et al.* (1967). In these early studies, a general broadening was observed for all proton resonances of the protein to which a spin label was covalently attached. Because this broadening was so nonspecific, no detailed conclusions could be drawn concerning proximity of particular nuclei to the nitroxyl free electron. In a more fruitful experiment, Roberts *et al.* (1969) found that the noncovalent binding of piperidinyl phosphate **X** (Fig. 7) to ribonuclease was accompanied by a significant broadening of the C-2 proton resonances assigned to His 12 and His 119. The broadening for the His 12 resonance was greater, indicating that the average location of the unpaired electron on the spin label was closer to His 12 than His 119. A numerical value for the intervening distances was not calculated. In a more quantitative study, Mildvan and Weiner (1969a), using a spin-labeled analogue of nicotinamide adenine dinucleotide (**II**, Fig. 7) with the enzyme alcohol dehydrogenase, measured the relaxation rates of bound substrates or pseudosubsubstrates. From these measurements they were able to calculate distances from the protons of bound ethanol, acetaldehyde, or isobutyramide to the paramagnetic label. In a similar study, Krugh (1971) observed changes in the relaxation rates of purine protons of adenosine monophosphate (AMP) when bound to DNA polymerase that simultaneously bound spin labeled adenosine triphosphate (ATP) (**I**, Fig. 7). From these changes, he deduced that the AMP and ATP binding sites were contiguous and that the distance between the nitroxyl moiety on spin-labeled ATP and the C-2 proton of AMP was 7.1 ± 0.6 Å. In a detailed study of internuclear distances on lysozyme, site-directed spin labels were used to perturb the NMR spectra of specific protons on the enzyme and on bound substrates (Wien *et al.*, 1972). Lysozyme covalently tagged with pyrrolylbromoacetamide **I** (Fig. 1) at His 15 broadened the nuclear resonance spectra of *N*-acetyl-α-D-glucosamine and di-*N*-acetyl-D-glucosamine bound at the active site. Using these broadenings, the distance *from* the nitroxide at His 15 *to* the acetamido methyl group of these saccharides was calculated. The measurement was then made in the opposite direction by using spin-labeled saccharides (**IV** and **V**, Fig. 7) bound at the active site to broaden the resonance of the C-2 proton at His 15. In a third type of experiment, the pseudosugar piperidinylacetamide **VI** (Fig. 7), which binds anomalously near Trp 123 (Berliner, 1971), was used to estimate the distance from that residue to the acetamido methyl groups of bound saccharides *and* to the C-2 proton of His 15. The distances calculated from the resonance data were in good agreement with those distances determined from a molecular model of lysozyme built according to X-ray crystallographic coordinates. Recently, Jones *et al.* (1973) have performed distance-measuring experiments on phosphofructokinase which was covalently tagged with piperidinylbromoacetamide **VIII** (Fig. 1) at a single reactive sulfhydryl group. The labeled enzyme retained a high level of catalytic activity and ATP binding capacity. The longitudinal and transverse relaxation times of specific protons in Mg-ATP bound to spin-labeled phosphofructokinase were shorter than those times measured with the native enzyme. This shortening of relaxation times due to dipolar spin–spin interaction was used to calculate distances of 7.2, 7.5, and 7.1 Å from the attached nitroxide to the H-2, H-8, and H-1′ nucleotide protons, respectively. Spin labels enhance the relaxation rate not only of protons on enzymes and bound substrates or inhibitors, but also of protons on water molecules. Studies illustrating this principal have been carried out on liver alcohol dehydrogenase (Mildvan and Weiner, 1969a,b), creatine kinase (Cohn *et al.*, 1971), and citrate synthase (Weidman *et al.*, 1973).

2. Distance between Two Different Spin Labels Attached to the Same Enzyme

When two nitroxyl radicals are attached to the same molecule, their unpaired electrons may undergo magnetic interaction. The magnitude of this interaction depends on the distance between the nitroxyl nitrogens and the motion of the radicals relative to each other. This interaction is large for separations less than 10 Å and small for separations greater than 14 Å (Calvin *et al.*, 1969). When the spin–spin exchange frequency is large compared to the hyperfine frequency, a five-line EPR spectrum with relative line intensities of 1 : 2 : 3 : 2 : 1 is obtained. As the exchange frequency decreases, the intensity ratio becomes 1 : 1 : 1 : 1 : 1, then 2 : 1 : 2 : 1 : 2. When there is no exchange interaction, a simple three-line spectrum (1 : 1 : 1) like that of a single nitroxide is obtained. (A rigorous treatment of spin–spin exchange interactions is presented in Chapter 4.) When a reagent carrying two spin labels (e.g., **IV**, Fig. 4) is attached to the active site of an enzyme, the distance that separates the nitroxyl nitrogens may be governed by the size of that site. Since only small changes in this distance can cause rather large changes in the EPR spectrum, this method has great potential for detecting subtle changes in active site conformation and geometry. This sensitivity is illustrated by changes in the spectrum of biradical **IV** (Fig. 4) attached to Ser 195 of α-chymotrypsin when Met 192 or Met 192 and Met 180 are oxidized (Kosman and Piette, 1972). Even more dramatic differences are observed when the spectra of this label attached to chymotrypsin, trypsin, elastase, and thrombin (Fig. 14) are compared (Hsia *et al.*, 1972). Although double spin labels can exceed single spin labels in sensitivity to changes in their environment, they usually give complicated spectra that

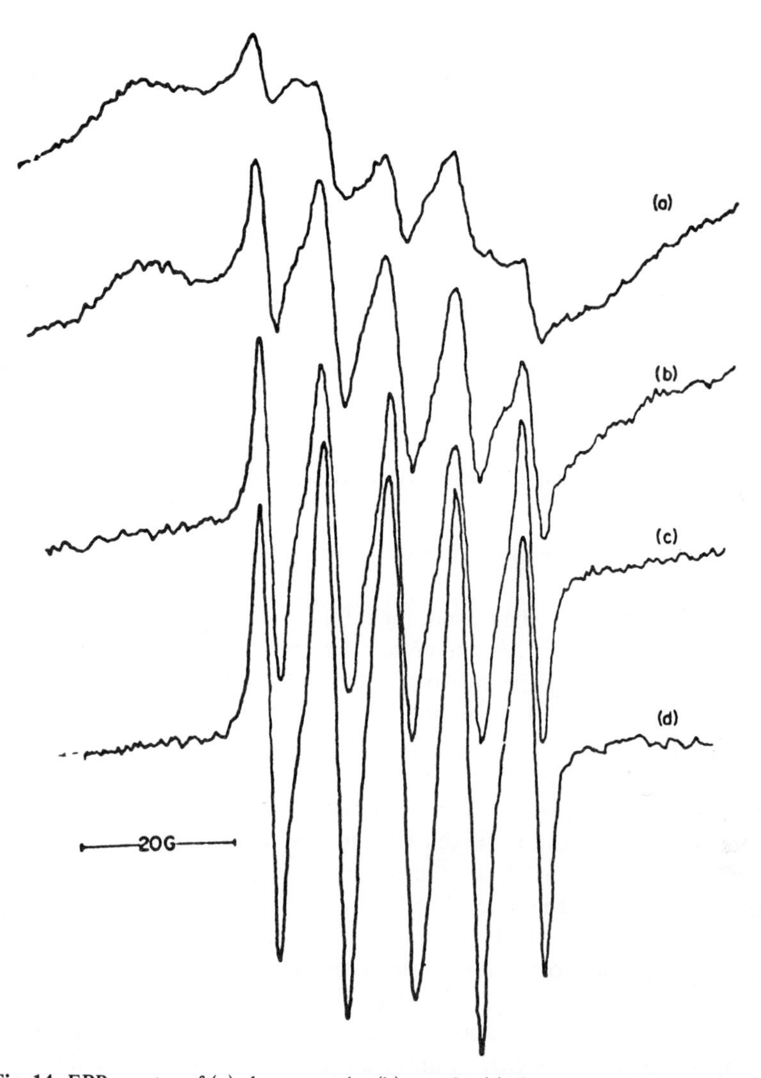

Fig. 14. EPR spectra of (a) chymotrypsin, (b) trypsin, (c) elastase, and (d) thrombin spin labeled at the active-site serine with biradical **IV** (Fig. 4). (Hsia *et al.*, 1972.)

are much more difficult to interpret than monoradical spectra. For this reason, biradical spin labels have not seen the wide use that their monoradical counterparts have. Another application of spin–spin interaction to the study of protein structure involves the covalent attachment or noncovalent binding of two monoradicals at different loci on a protein or peptide. The success of such an experiment requires (a) that two and only two sites are labeled, and (b) that the extent of their occupancy is known. Ovchinnikov (1973) has used this method to estimate the distances between the two ornithine side chains of gramicidin S and six of its structural analogs. In similar fashion, Filatora *et al.* (1973) have employed this technique for measuring the distance between spin labels attached to Arg 1 and Ser 6 of bradykinin.

3. Distance between a Spin Label and a Paramagnetic Ion

When a spin label and a paramagnetic ion are present in the same system, the effect of that ion on the resonance spectrum of the nitroxide is not always a mere broadening. When these groups are rigidly bound to a macromolecule so that their reorientation time is long compared to the electronic relaxation time, the observed line width of the EPR signal is given by

$$\delta H = (g\beta\mu^2\tau/r^6\hbar)(1 - 3\cos^2\theta_R')^2 + \delta H_0 \tag{2}$$

where g, β, and $\hbar$ have their usual meanings, τ is the correlation time for the dipolar interaction, μ is the magnetic moment of the dipole, r is the distance between the ion and the nitroxyl nitrogen, θ_R' is the angle between the applied magnetic field and the line joining the two electronic spins, and δH_0 is the residual line width of the nitroxide in the absence of the paramagnetic ion (Leigh, 1970). Actually, the dipolar effect is observed as a diminution of the signal amplitude rather than a general broadening of the whole spectrum for the reason that Eq. (2) is sensitive to orientation changes. For a certain fraction of orientations, the first term on the right side of the equation will be small compared to the second term on that side, and the signal from this fraction will be governed largely by δH_0 and will yield an unperturbed spectrum. For most other orientations, the first terms are large compared to the second, and the resulting spectra are too broad to observe. Hence, the composite of these two effects is a fraction of the original signal. By determining the extent of signal diminution, one can calculate the distance between the two paramagnetic species using Eq. (2). Taylor *et al.* (1969) have applied this technique to the creatine kinase system. When this enzyme was spin labeled with **I** (Fig. 1) at the essential sulfhydryl group, the estimated distance from the nitroxide to bound Mn^{2+} in the MnADP-spin-labeled kinase was 7–10 Å. Jones *et al.* (1973) have used the same method to map the active site of phosphofructokinase. When this enzyme was labeled with piperidinylbromoacetamide **VIII** (Fig. 1) at a single SH group, the estimated distance from the nitroxyl free electron to manganese in an ATP-containing ternary complex was 12.0 Å.

4. Distance from a Spin Label to a Fluorophore

The ability of species with electron spin multiplicities greater than zero to quench excited electronic states has been known for some time (Birks, 1970). Several groups have used di-*t*-butyl nitroxide as a quencher of excited triplet states (see Green *et al.*, 1973 and references therein). Possible mechanisms for this quenching process include (a) resonance-excitation transfer; (b) collisional or energy-exchange transfer; (c) charge-transfer excited state; (d) electron-exchange-induced intersystem crossing, and (e) vibrational quenching (Green *et al.*, 1973). However, these possibilities can be narrowed down to either (d) or (e). Viscosity studies suggest that this quenching phenomenon is operative over a distance of 4–6 Å. One application of spin-label-induced fluorophore quenching in a biological system involves the tryptophan quenching of albumin by a spin-labeled fatty acid. It has been shown that unlabeled long-chain fatty acids (e.g., oleic acid), upon binding

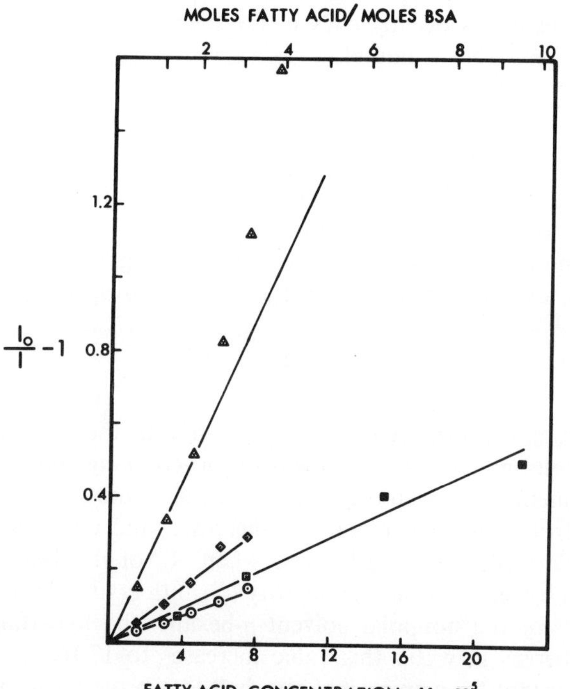

Fig. 15. Stern–Volmer plots of intrinsic tryptophan fluorescence of defatted bovine serum albumin in the presence of fatty acids. The order of quenching efficiency for the bound fatty acid is 5-doxylstearate (12,3-SLFA) (△) ≫ 16-doxylstearate (1,14-SLFA) (◇) > oleate (□) > 12-doxylstearate (5,10-SLFA) (○). (From Morrisett *et al.*, 1975.)

to bovine albumin, diminish the intrinsic tryptophan fluorescence intensity (Spector and John, 1968). The magnitude of this effect varies only slightly when 16-doxylstearate, 12-doxylstearate, or oleate is used (Morrisett *et al.*, 1975). However, when the nitroxyl group is moved up to the C-5 position, which is near one of the tryptophan residues, a striking decrease in intensity (increase in quenching) is observed (Fig. 15). This experiment demonstrates the dependence of nitroxide-induced quenching on the spin label–fluorophore distance and illustrates how the effect may be used to map enzyme active sites.

This phenomenon was also demonstrated by Koblin *et al.* (1973) in a study of erythrocyte membranes. Normally, diamagnetic local anesthetics analogous to **XXIV** (Fig. 7) enhance the fluorescence of 1-anilinonaphthalene-8-sulfonate (ANS) in these membranes. However, when the paramagnetic anesthetic **XXIV** (Fig. 7) was used, it diminished the ANS fluorescence intensity by about 30%. When ascorbate was added to reduce the nitroxide, the intensity increased about 3.5-fold. These results suggested that local anesthetics occupy sites in the membrane which are very near those sites where ANS is bound.

D. Polarity of Binding Sites

The principal values of the hyperfine tensor of a nitroxide are dependent on the polarity of the environment. A hydrophobic environment will shift the equilibrium of Eq. (3) toward the right, yielding a species with much of the unpaired electron density associated with the oxygen atom:

$$^{+}\cdot\overset{|}{\underset{|}{N}}\text{-O:}^{-} \rightleftharpoons \;:\overset{|}{\underset{|}{N}}\text{-O}\cdot \tag{3}$$

In this case, interaction of the free electron with the nitrogen nucleus is minimal resulting in a relatively low isotropic coupling constant, A_0. When the environment becomes more polar, the equilibrium shifts toward the left, electron–nitrogen nuclear interaction increases, and the coupling constant increases. This effect is clearly seen when A_0 of di-*t*-butyl nitroxide is measured in solvents of differing polarity (Griffith *et al.*, 1974). In Table II it is seen that for the nonpolar solvent *n*-hexane, the hyperfine splitting is 15.10 G, whereas in water this value increases to 17.16 G. The isotropic coupling constant for a spin label attached to a protein can be determined only when the label retains a high level of motional freedom. If the attached label is strongly immobilized, A_0 is no longer measurable. However, it has been shown that from spectra of these strongly immobilized nitroxides, the splitting between the low field maxima and high field minima ($2A_{zz}$) can

TABLE II

ISOTROPIC ESR PARAMETERS FOR DI-*t*-BUTYL NITROXIDE[a,b]

No.	Solvent	A_0	g_0
1	Hexane	15.10	2.0061
2	Heptane–pentane (1 : 1)[c]	15.13	2.0061
3	2-Hexene	15.17	2.0061
4	1,5-Hexadiene	15.30	2.0061
5	Di-*n*-propylamine	15.32	2.0061
6	Piperidine	15.40	2.0061
7	*n*-Butylamine	15.41	2.0060
8	Methyl propionate	15.45	2.0061
9	Ethyl acetate	15.45	2.0061
10	Isopropylamine	15.45	2.0060
11	2-Butanone	15.49	2.0060
12	Acetone	15.52	2.0061
13	Ethyl acetate saturated with water	15.59	2.0060
14	*N*,*N*-Dimethylformamide	15.63	2.0060
15	EPA[d] (5 : 5 : 2)[c]	15.63	2.0060
16	Acetonitrile	15.68	2.0060
17	Dimethylsulfoxide	15.74	2.0059
18	*N*-Methylpropionamide	15.76	2.0059
19	2-Methyl-2-butanol	15.78	2.0059
20	EPA[d] (5 : 5 : 10)[c]	15.87	2.0060
21	1-Decanol	15.87	2.0059
22	1-Octanol	15.89	2.0059
23	*N*-methylformamide	15.91	2.0059
24	2-Propanol	15.94	2.0059
25	1-Hexanol	15.97	2.0059
26	1-Propanol	16.05	2.0059
27	Ethanol	16.06	2.0059
28	Methanol	16.21	2.0058
29	Formamide	16.33	2.0058
30	1,2-Ethanediol	16.40	2.0058
31	Ethanol–water (1 : 1)[c]	16.69	2.0057
32	Water	17.16	2.0056
33	10 *M* LiCl aqueous solution	17.52	2.0056

[a] Griffith *et al.*, 1974.

[b] All data measured at room temperature (23°–24°C). Estimated uncertainties are ± 0.02 *G* and ± 0.0001 for A_0 and g_0, respectively, relative to the standard dilute aqueous solution of di-*t*-butyl nitroxide for which $A_0 = 17.16$ and $g_0 = 2.0056$.

[c] By volume.

[d] EPA designates a mixture of ethyl ether (diethyl ether), isopentane (2-methylbutane), and alcohol (ethanol).

be positively correlated with A_0. Using $2A_{ZZ}$ in water and $2A_{ZZ}$ in decalin as extremes, Lassman *et al.* (1973) have developed a hydrophobicity parameter *h*, which is defined in Eq. (4):

$$h = \frac{A_{ZZ}(H_2O) - A_{ZZ}(\text{sample})}{A_{ZZ}(H_2O) - A_{ZZ}(\text{decalin})} \tag{4}$$

This parameter increases from zero for a purely polar environment to unity for an environment with the hydrophobicity of decalin. In cases where the environment is less polar than that of decalin, values of *h* will exceed unity. Leucine aminopeptidase has been tagged with the iodoacetamide analogue of **VIII** (Fig. 1) at one or more sulfhydryl groups and the spectra of this derivative subjected to the analysis described above. A value of 2.7 for *h* was calculated, indicating that the covalently bound label(s) exists in a highly hydrophobic environment (Lassman *et al.*, 1973).

E. Protein Symmetry

Because of the anisotropic nature of the nitrogen nuclear hyperfine tensor and the *g* factor tensor in nitroxides (Griffith *et al.*, 1965), the resonance spectra of these molecules are highly dependent on their orientation in the laboratory magnetic field. For a liquid solution of a nitroxide, there is, of course, an isotropic distribution of orientations, and the resulting spectrum is an average of all these orientations. In a host crystal, however, a nitroxide guest molecule will take up a fixed orientation with respect to the crystallographic axis. In other words, the crystal lattice serves to orient the guest nitroxide in space. Since the crystal can be accurately oriented with respect to the laboratory magnetic field, so can the guest spin label. The dependence of the *g* value and hyperfine splitting on orientation of the label in the magnetic field is illustrated in Fig. 16. When the field is parallel to the N—O bond (*x* axis) or parallel to a line joining the two quaternary carbons (*y* axis), the hyperfine splitting is 6 G (Fig. 16X,Y). When the field is parallel to the nitrogen 2p π orbital that contains the unpaired electron (*z* axis), the coupling constant becomes 32 G (Fig. 16Z). Now suppose that instead of using tetramethyl-1,3-cyclobutadione as the host and di-*t*-butyl nitroxide as the guest (as in Fig. 16), a protein is used as the host and a tightly bound spin-labeled ligand as the guest. In such a system, the resonance spectra of the bound spin label must necessarily show the symmetry of the crystal structure and the symmetry of the protein molecules on which the structure is based (McConnell and Boeyens, 1967). If a crystal symmetry axis is coincident with a molecular symmetry axis, the presence of this molecular axis can often be determined rather easily by use of X-ray diffraction methods. If a crystallographic axis is not coincident with the molecular symmetry axis of

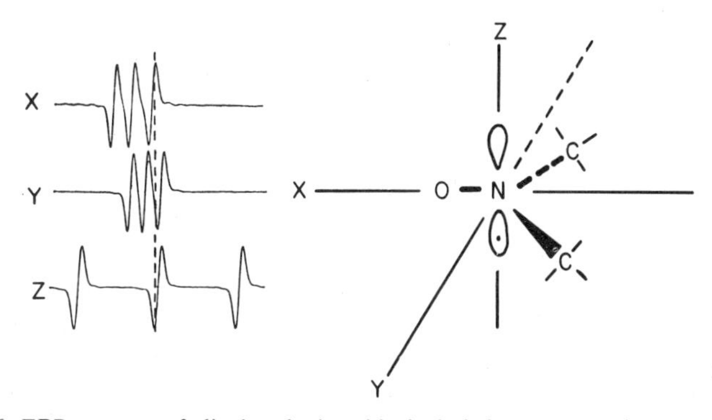

Fig. 16. EPR spectra of di-*t*-butyl nitroxide included as a guest in a crystal host of tetramethyl-1,3-cyclobutanedione. The host crystal orients the guest molecules so that their molecular axes assume fixed orientations with respect to the crystal axes. The crystal has been oriented so that when the X molecular axis of the nitroxide is parallel to the laboratory magnetic field, spectrum X is produced. When the field is parallel to the Y and Z axes, spectra Y and Z are obtained. [Reprinted with permission from Griffith and Waggoner *Accounts Chem. Res.* **2**, 17–24 (1969). Copyright by the American Chemical Society.]

the protein, finding this molecular axis can be very difficult by diffraction techniques. However, this axis should be readily observable by paramagnetic resonance when the protein is spin-labeled. For example, when two spin labels are bound at appropriate positions on a protein and are related by a twofold rotation axis, the resonance spectra of the two labels will usually be different unless the laboratory magnetic field is exactly parallel or perpendicular to the axis. Numerous proteins are composed of identical subunits, and many of these have symmetry axes. Such systems are highly amenable to study by this method.

For suitably spin-labeled symmetrical enzymes, the EPR spectra may be used to detect allosteric conformational changes that accompany the binding of substrates, inhibitors, or activators. In favorable cases, conformational changes brought about by the binding of these ligands will also change the relative orientations of the principal axes of symmetry related spin labels (McConnell and Boeyens, 1967). One obvious use of spin-labeled protein crystals is for comparing their resonance spectra to those obtained for the protein in solution. If the spectrum of an isotropic ensemble of spin-labeled protein crystals (e.g., polycrystalline suspensions) is identical to that obtained for the same protein in solution, this constitutes strong evidence that the conformation in the vicinity of the label is the same in the two states. Such information is of great value to crystallographers who wish to assess the effect of crystal forces on protein conformation during the crystallization process.

Single crystals of monoclinic α-chymotrypsin have been spin labeled with pyrrolidinyl carboxylate **IV** (Fig. 2) at the active site serine (Berliner and McConnell, 1971). The resonance spectra observed for oriented single crystals of the acyl enzyme consist of the superposition of no more than four single molecule spectra. Since the chymotrypsin crystal contains four enzyme molecules per unit cell, this result indicates that each spin label takes up a single orientation relative to the enzyme molecule to which it is attached. The results may be contrasted to those obtained with crystals of hemoglobin that were spin-labeled at the reactive sulfhydryl group on cysteine β-93. In this latter case, the nitroxides assume two isomeric orientations relative to the hemoglobin molecule. In one orientation the label was strongly immobilized and in the other, only partially so (McConnell *et al.*, 1969; Moffat, 1971).

F. Active-Site Geometry

In earlier days, mapping the active site of an enzyme was an extremely tedious process that was attempted by studying the effects on catalytic activity of various chemical modification reagents that were presumably specific for particular amino acid side chains, or by testing an array of inhibitors or substrates for their binding or kinetic constants. Such studies often produced results that by themselves could not be interpreted unambiguously. The advent of spin labels brought new methods for studying the size and shape (geometry) of an enzyme active site and the relative positions of different groups within that site. We have already considered techniques for measuring distances between certain moieties on a protein (Section IV.C). There are two other general types of spin-label experiments that are most often used for this purpose: (a) experiments in which the structure of a covalently or noncovalently binding probe is systematically modified to probe a unique binding site, and (b) experiments in which a single spin label is used to probe the same site on a series of homologous or structurally similar enzymes. These types of experiments are illustrated by the following examples.

The basic method of using spin labels as molecular rulers was first applied by Hsia and Piette (1969) to measure the depth of antibody combining sites. Recently, this method has been used to study the geometry of the active site of bovine erythrocyte carbonic anhydrase B (Chignell *et al.*, 1972). When sulfonamide **XIV** (Fig. 7) was bound to the active site of the enzyme, the motion of the nitroxyl moiety was highly constrained. Experiments with other molecules whose chain length between the aromatic and pyrrolidinyl rings was gradually increased (**XV–XIX** in Fig. 7) indicated that the mobility of the nitroxyl group also increased until, with **XX** (Fig. 7), this group appeared to interact minimally with the active site. These experiments suggested that the active site of this enzyme is a narrow cleft about 14.5 Å deep.

In a subsequent study on human erythrocyte carbonic anhydrase B, Erlich *et al.* (1973) did not observe a completely monotonic increase in the mobility of the nitroxyl group with increasing separation between this group and the sulfonamide moiety. For example, the motion of the nitroxyl group of **XVII** (Fig. 7) bound to the enzyme was significantly greater than that of inhibitor **XVIII** (Fig. 7). The authors interpret this effect to mean that either the active site becomes narrower in a certain region and/or this region is more hydrophobic and hence more strongly binds the pyrrolidine ring. When the same series of labels was repeated on the C isozyme of the same species, a smooth, monotonic increase in mobility of bound sulfonamides **XV–XVIII** (Fig. 7) was observed. The estimate of the active-site depth for this isozyme was 14 Å, which was in excellent agreement with measurements by X-ray diffraction.

Covalently binding spin labels have been used by Kosman *et al.* (1969) and by Berliner and Wong (1974) in studies designed to elucidate the geometry of the active site of chymotrypsin. The former group used piperidinyl derivatives of a succinic ester amide, a succinic diester, and a maleic diester (**I**, **II**, **III**, Fig. 2) to determine the steric relationship between the hydrolytic site containing Ser 195, the aromatic-binding site (which binds the aromatic side chain of natural substrates), and the amide-binding site for the peptide bond which is cleaved by hydrolytic scission. Based on earlier work by Cohen *et al.* (1966; Cohen and Crossley, 1964), Kosman *et al.* (1969) felt that the hydrolytic site was transoid with respect to the aromatic site. The high motional freedom of the maleic diester derivative (**III**, Fig. 2) was considered to be in accord with the prediction that the cis orientation at the double bond would place the nitroxyl moiety out of the active site. However, the similarity of the spectra for the enzyme-bound succinic ester amide and succinic diester led to three possible interpretations for the location of the two nitroxides: (a) they are in the same environment within the active site, (b) they are in different but similar environments within the active site, (c) they are in the same or similar environments that are not related to the catalytic process. In an attempt to determine which of these possibilities was in fact the case, the effect of oxidizing Met 192 and Met 180 was examined. Also, the rates at which the spin-labeled acyl groups were released at pH 7.2 in the presence and absence of indole were studied. Results from these and other experiments led Kosman *et al.* (1969) to propose a mechanistic sequence for the hydrolytic event: (1) the substrate interacts at the aromatic and amide binding sites; (2) the aromatic binding induces a structural change in the enzyme that restricts the orientation of the acyl amide linkage and brings the catalytic centers into correct alignment; (3) acylation of Ser 195 occurs, which results in weakening of interactions in the aromatic and amide binding regions; and finally, (4) deacylation of Ser 195.

Using sulfonylating spin labels (Fig. 3), Berliner and Wong (1974) systematically compared the active site geometry of trypsin and chymotrypsin. The importance of using a variety of different but related spin-labeled reagents becomes apparent in this type of study where it was shown that one label (e.g., *m*-**V**, Fig. 3) attached to each enzyme gives very similar EPR spectra (Fig. 13), while another label (e.g., *m*-**III**, Fig. 3) gives quite different spectra (Fig. 17). The interpretation of the results was considerably aided by

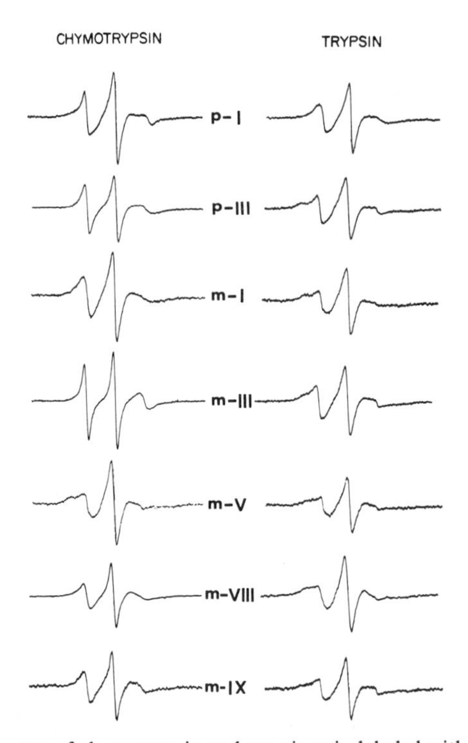

Fig. 17. EPR spectra of chymotrypsin and trypsin spin-labeled with various sulfonylating reagents shown in Fig. 3. In each case, the spectrum indicates equal or higher motional freedom for the spin label attached at the active site of chymotrypsin than for trypsin. [From a chapter by L. J. Berliner entitled Applications of Spin-Labeling to Structure-Confirmation Studies of Enzymes in "Progress in Bioorganic Chemistry, Vol. III," edited by E. T. Kaiser and F. J. Kezdy. Copyright © 1974 by John Wiley & Sons, Inc. Reprinted by permission of the author and publisher.]

a knowledge of the static tertiary structure of these enzymes from X-ray diffraction data (Henderson *et al.*, 1971; Stroud *et al.*, 1971). These X-ray data indicated that the entrance to the side chain binding region of the active site of trypsin was significantly narrower than that of chymotrypsin; i.e., some groups that could be accommodated easily at this locus in chymotrypsin would be excluded from this site in trypsin. Hence, for chymotrypsin, only those spectra of spin labels whose aromatic group could not bind at this locus could be meaningfully compared with spectra of any of the trypsin derivatives. Those derivatives that were considered "comparable" included all of those involving para substituents (*p*-**I** through *p*-**V**, Fig. 3) and some of

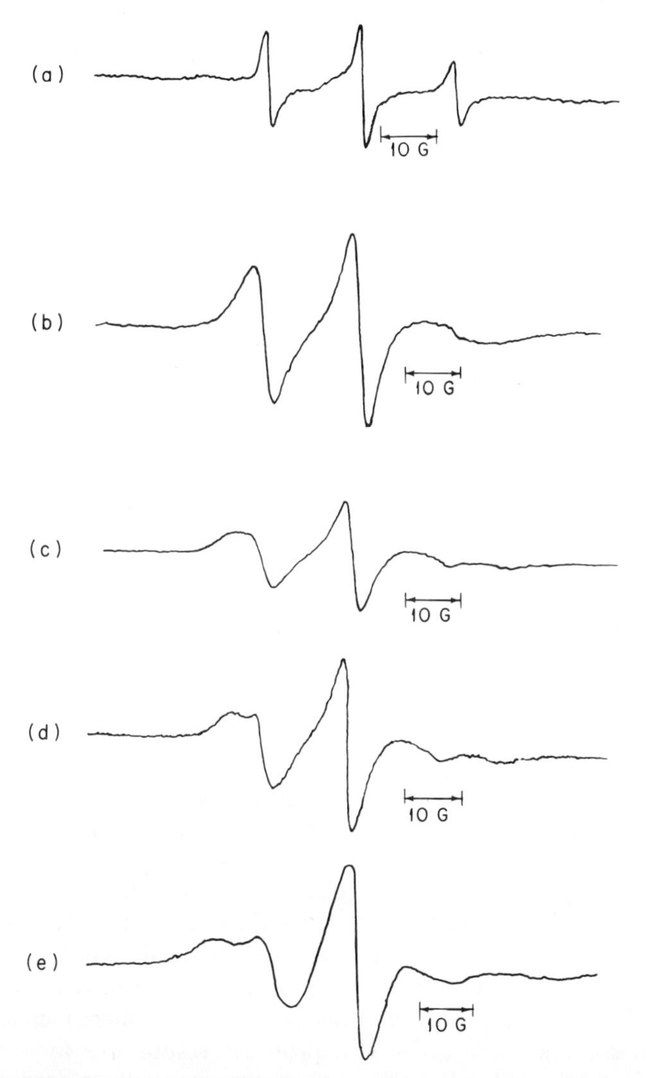

Fig. 18. EPR spectra of (a) acetylcholinesterase, (b) α-chymotrypsin, (c) trypsin, (d) elastase, and (e) subtilisin spin-labeled with **I** (Fig. 4). [From Morrisett and Broomfield (1972).]

those involving meta substituents (*m*-**I**, *m*-**III**, and *m*-**IV** of Fig. 3). These appeared, from model building studies, to give sulfonylated chymotrypsin derivatives in which the aromatic group was excluded from the tosyl hole. In general, comparable spin labels attached to trypsin were more immobilized than when attached to chymotrypsin, implying that the nonspecific site (outside the specific binding site) was more crowded for trypsin than chymotrypsin.

A similar comparison of the active sites of several serine enzymes was carried out by Morrisett and Broomfield (1972) using the phosphonylating reagent **I** (Fig. 4). When this reagent was used to label the active site serine residue of acetylcholinesterase, the nitroxyl group experienced a high degree of rotational mobility. When attached to chymotrypsin, the group was moderately immobilized, and when bound to elastase, trypsin, or subtilisin, the motional freedom of this group was highly constrained (Fig. 18). The methyl substituents on the piperidinyl ring render this group too large to fit into the aromatic binding site of chymotrypsin (Berliner and Wong, 1974). The entrance to this site on trypsin is too narrow (Stroud *et al.*, 1971), and for elastase, it is partially blocked by the side chain of Val 216 (Watson *et al.*, 1970). For subtilisin there is no pronounced cleft or depression at the active site as is present in chymotrypsin (Wright *et al.*, 1969). Since, in all four of these cases the nitroxyl group is not located inside but outside the specificity binding site, the EPR spectra of these four spin-labeled enzymes can be meaningfully compared. The comparison indicates that the nonspecific binding site for the piperidinyl ring is significantly more crowded in elastase, subtilisin, and trypsin than in chymotrypsin. Since no crystallographic data exists for acetylcholinesterase, the relative location of the nitroxyl ring is not known, hence no direct comparison to the other four enzymes can be made.

G. Changes in Active-Site Conformation Caused by Ligand Binding or Proteolysis

The high specificity that so many enzymes exhibit derives, in part, from the very precise spatial relationships that exist between the different functionalities involved in catalysis. Striking changes in catalytic activity often accompany very subtle structural changes in the active-site region. For example, when the active-site serine of subtilisin is chemically converted to cysteine, a new hydrogen bond is formed between the thiol sulfur atom and a nearby imidazole proton. Such an interaction significantly alters the pH dependence of k_{cat}/K_m (Polgar and Bender, 1967). The mere increase in size of the nucleophilic atom ($O \rightarrow S$) is probably responsible for the 33-fold decrease in the K_{cat} value. Spin labels that are strategically located at or near the active sites of enzymes can often detect similar subtle changes.

Such a change was observed with chymotrypsin sulfonylated with *m*-**I** (Fig. 3) at the active site serine (Berliner and Wong, 1974). The enzyme cannot accommodate the aromatic group of this molecule in the tosyl hole. When indole is added, it occupies this tosyl hole and causes steric crowding, which results in an EPR spectrum containing broad-line components that are not observed when indole is absent (Fig. 19, top). When a reagent is used that does allow the aromatic group to bind in the tosyl hole (e.g., *m*-**VII**), indole is blocked out and there is little or no spectral change (Fig. 19, bottom).

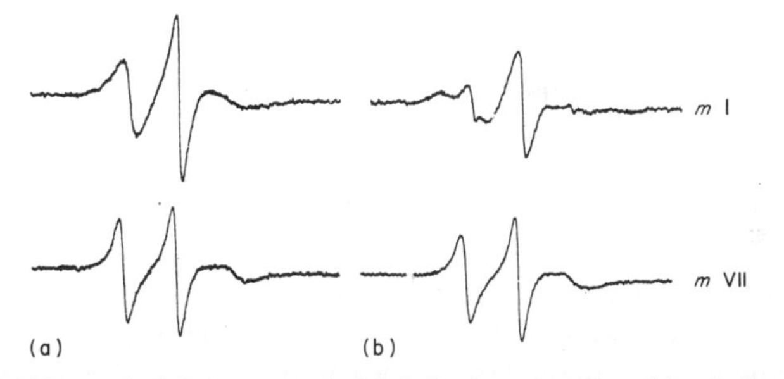

Fig. 19. Effect of indole on spin-labeled chymotrypsin: (a) chymotrypsin; (b) chymotrypsin + indole. When the enzyme is labeled with *m*-**I** (Fig. 3), the aromatic group is thought to reside in a nonspecific binding site outside the tosyl hole. Addition of indole, which binds in the tosyl hole, causes steric crowding of the excluded groups and results in the appearance of a broad line spectral component. In the case of *m*-**VII**, the aromatic group can be accommodated inside the tosyl hole and is least affected by indole. [From a chapter by L. J. Berliner entitled Applications of Spin-Labeling to Structure-Confirmation Studies of Enzymes in "Progress in Bioorganic Chemistry, Vol. III," edited by E. T. Kaiser and F. S. Kezdy. Copyright © 1974 by John Wiley & Sons, Inc. Reprinted by permission of the author and publisher.]

The inactive zymogen chymotrypsinogen is converted to the active enzyme by the tryptic scission of the Arg 15–Ile 16 peptide bond. This cleavage reaction allows the main chain to rotate 180° about an axis defined by the α carbons of residues 19 and 21, resulting in the formation of the important internal ion pair between the α-amino group of Ile 16 and the carboxyl group of Asp 194 (Freer *et al.*, 1970). This ion pair activates the charge relay system responsible for catalytic activity. In the active-site region is located Met 192, which can be exclusively alkylated with iodoacetamide. When a spin-labeled analog of this reagent (**VIII**, Fig. 1) was used to alkylate this residue, the resulting derivative (SL-Met 192-CT) gave a mixed EPR spectrum containing two components, each exhibiting a different *g* value and nitrogen hyperfine coupling constant. This suggested that the nitroxide was residing in both a hydrophobic and hydrophilic environment (Kosman,

1972). If, in fact, this is the case, the rate at which the label shifts from one type of environment to the other must be less than 10^7 sec^{-1}. When SL-Met 192-CT was activated by trypsin, the spectral lines corresponding to nitroxide in a hydrophobic (hydrophilic) environment increased (decreased) and these changes paralleled the rate of zymogen activation as determined by assay of chymotryptic activity.

Citrate synthase catalyzes the condensation of acetyl-coenzyme A with oxaloacetate to form citrate and coenzyme A. When a spin-labeled analogue of acetyl-coenzyme A (**VIII**, Fig. 7) was bound to the enzyme, only narrow-line spectra were observed ($\tau_c \simeq 5 \times 10^{-9}$ sec), even though the analogue exhibited a dissociation constant of 9.3×10^{-5} *M*. In the presence of the substrate, oxaloacetate, a significant decrease in the motional freedom of the nitroxyl ring occurred without any alteration in the concentration of bound spin-labeled coenzyme (Weidman *et al.*, 1973).

H. Rotational Correlation Times of Enzymes

For the large majority of covalently spin-labeled proteins, the bound label experiences motional freedom at or near the bonds that connect the nitroxyl ring to the protein. Hence, the rotational correlation time (τ_c) of the attached label does not accurately reflect that of the whole protein. For a few covalently tagged proteins, there is little or no motional freedom of the label independent of the macromolecule. Chymotrypsin labeled with **IV** (Fig. 2) or **II** (Fig. 4) represents such a case. Using these two derivatives, Shimshick and McConnell (1972) developed a method for accurately measuring τ_c of a protein. The method is based on the theoretical relationship that the resonance positions of the outer hyperfine extrema of a nitroxide EPR spectrum depend on its rotational motion

$$\Delta H(\tau_2) \propto \tau_2^{-2/3} \tag{5}$$

The first step of this method involves theoretical determination of the shift $\Delta H(\tau_2)$ for the high field line as a function of rotational motion by solving a set of Bloch equations coupled by the addition of rotational diffusion terms (McCalley *et al.*, 1972). This calculated shift (relative to the line position at an infinite correlation time) is then plotted as a function of τ_2. In the next step, one measures experimentally the position of the high field hyperfine spectral component (H_h) of the spin-labeled protein relative to an isotropic standard (H_0) as a function of viscosity (η) which is varied by using solutions of different sucrose concentration. A plot of the shift, $H_h - H_0$, versus $(T/\eta)^{2/3}$ is made and extrapolated to infinite viscosity. The value of $H_h - H_0$ at infinite viscosity minus that value in standard aqueous solution is calculated and used to determine τ_2 from the theoretically calculated plot of ln $\Delta H(\tau_2)$ versus ln τ_2. Using this method, a rotational correlation time of 12 nsec was determined for α-chymotrypsin in water at 20°C (Shimshick and McConnell, 1972).

V. DETAILED STUDIES OF SPIN-LABELED ENZYMES

We have now considered which reagents are available for spin-labeling enzymes, how the labeling procedure is carried out, and what information is obtainable from these spin-labeled enzymes. This section describes the application of the spin-labeling method to twenty different enzymes. It does not include studies that have been done on a variety of other noncatalytic proteins such as hemoglobin (which would deserve a chapter all its own) or lipid-binding proteins. This section is not intended to give an in-depth analysis of each study but rather to enumerate the salient features of each study and to direct the reader to the literature relevant to that study.

A. Lysozyme

Hen egg white lysozyme is a well-characterized enzyme that contains 129 amino acids and catalyzes hydrolysis of the β-(1-4) glycosidic bond between *N*-acetylmuramic acid and *N*-acetylglucosamine. A preliminary study of this spin-labeled enzyme was carried out by Sternlicht and Wheeler (1967). The object of their study was to determine the distance of an attached nitroxide (piperidinylmaleimide **IV**, Fig. 1) to other nuclei on the protein by measuring the extent of line broadening due to spin-label-induced relaxation (see Section IV.C.1). Because the site(s) and degree of labeling were unknown. and because the line broadening effect was so nonspecific, no quantitative data on nitroxide–nuclear distances could be obtained. In a later study by Wien *et al.* (1972), successful experiments of this type were carried out. The distance *from* the methyl group of *N*-acetylglucosamine bound to lysozyme in subsite C *to* the nitroxyl nitrogen of pyrrolidinylbromoacetamide **I** (Fig. 1) attached at His 15 was calculated to be 13 Å (α-anomer) or 15 Å (β-anomer). When this distance was measured on a molecular model of the protein built from coordinates determined by X-ray crystallography, it was found to be 18 Å. When this distance was corrected for the length of the attached label (i.e., the distance from the imidazole ring to the nitroxyl nitrogen, about 7 Å) a value of 11 Å was obtained, which compares favorably with the values determined by magnetic resonance. When di-*N*-acetylglucosamine was bound in subsites B and C and the enzyme was covalently tagged with **I** (Fig. 1) at His 15, the distance between the non-reducing acetamidomethyl group (in subsite B) and the nitroxyl nitrogen was calculated to be 18 Å. The corrected X-ray distance (28 Å − 7 Å) was 21 Å. Anticipating that a molecule such as piperidinylacetamide **VI** (Fig. 7) would bind at the active site of lysozyme, Berliner (1971) carried out a 6 Å crystallographic study of a lysozyme–**VI** complex. Surprisingly, the site of highest occupancy by the ligand was not in the catalytic region but in a hydrophobic pocket near Trp 123. This finding was used by Wien *et al.* (1972) to measure the distance from this ligand bound at Trp 123 to the C-2

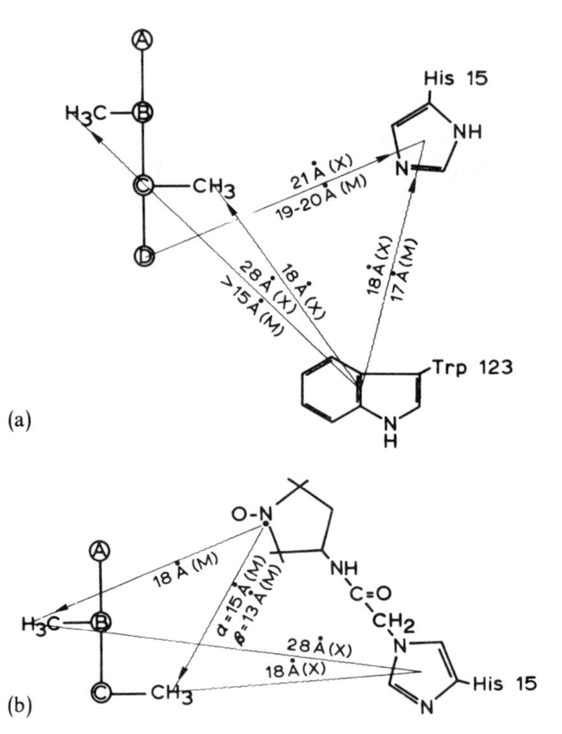

Fig. 20. Diagram comparing distances between groups on lysozyme as determined by measurements on a model based on X-ray diffraction data (X) and determined from proton magnetic resonance line broadening experiments (M). Lines point *from* the site of the spin label *to* the relaxed proton(s) being observed. (a) Noncovalent spin labels. The paramagnetic effect arises from either **VI** (Fig. 5) bound near Trp 123 or the nitroxyl ring of **IV** or **V** (Fig. 5) held in subsite D. The proton resonance lines arise from the C-2 proton of His 15 in one case, and from the acetamidomethyl groups of di-*N*-acetylglucosamine or *N*-acetylglucosamine bound in subsites B and C (or C alone) in the other cases. (b) Covalently spin labeled lysozyme. The paramagnetic effect originates at the nitroxyl moiety attached to His 15. The observed resonances are those of the acetamidomethyl groups of di-*N*-acetylglucosamine or *N*-acetylglucosamine bound in subsites B and C (or C alone). The probable distance between the methylene (attached to N-3 of His 15) and the nitroxyl nitrogen is of the order of 7 Å. [Reprinted with permission from Wien *et al. Biochemistry* **11**, 3707–3716 (1972). Copyright by the American Chemical Society.]

proton of His 15 (17 Å) and the acetamidomethyl group of bound di-*N*-acetylglucosamine (> 15 Å). The corresponding distances measured from the model are 18 Å and 28 Å, respectively. All of these distances are shown in the diagram of Fig. 20.

B. Ribonuclease

Bovine pancreatic ribonuclease A (RNase A) is another well characterized enzyme. It exists as a single chain of 124 amino acids and is cross-linked by four disulfide bonds. Its function is to catalyze the hydrolysis of RNA by cleavage of the 3′-phosphodiester bonds of ribose residues which are linked to pyrimidine bases. Spin-label studies on RNase A were first reported by Smith (1968). Pyrrolidinylbromoacetamide **I**, piperidinylmaleimide **IV**, pyrrolidinylmaleimide **V**, and piperidinyl bromosuccinamide **XI** (Fig. 1) were used to tag this enzyme at pH 5.5 and 8.5. The resulting derivatives were studied in the absence and presence of various homopolymeric nucleic acids. Changes in the EPR spectra were also studied as a function of pH, temperature, and urea concentration. It had been shown earlier that at pH 5.5, RNase was alkylated by bromoacetic acid at either imidazole N-3 of His 12 or N-1 of His 119, each product being totally inactive. Reaction of RNase A with the bromoamide **I** (Fig. 1) at pH 5.5 gave a derivative whose amino acid analysis indicated a gain of 0.5 residue of 3-carboxymethylhistidine and 0.1 residue of ε-carboxymethyllysine, and a loss of 1.1 residue of histidine and 1.1 residue of lysine. The modified enzyme retained 50% of the initial activity against cyclic-CMP. Based on these data and previous experiments of others, Smith (1968, 1971) concluded that RNase had been spin labeled at active site His 12 and to a lesser extent at lysine residues elsewhere. Presumably, enzyme that had not been alkylated was responsible for the residual activity.

Several years later, a careful study of the reaction of RNase A with **I** (Fig. 1) was reported by Daniel *et al.* (1973). The reaction products were separated by ion exchange chromatography and then rigorously characterized. This study clearly demonstrated that His 12 was not alkylated at all, but rather His 105. The labeled enzyme 3-SL-His 105-RNase exhibited 85% of the specific activity of native RNase. Preliminary evidence suggests that the covalently attached spin label is sensitive to conformational changes at the catalytic site. A second derivative, bearing the spin label on a methionine side chain, was also isolated from the reaction mixture.

Since RNase catalyzes the hydrolysis of a cyclic phosphate (cyclic-2′,3′-cytidylic acid), it is not surprising that this enzyme can strongly bind one mole of inorganic phosphate at the active site. Roberts *et al.* (1969) exploited this fact in an experiment in which piperidinyl phosphate (**X**,

Fig. 7) was used to study the active site geometry of RNase. They found that the chemically exchanging nitroxide broadened the nuclear resonance lines assigned to the C-2 protons of His 12 and His 119. The nitroxyl nitrogen was considered closer to His 12, since its C-2 proton resonance was broadened more.

C. DNA Polymerase

DNA polymerase from *E. coli* is a multifunctional enzyme with a molecular weight of about 105,000. It plays an important role in the replication and repair of DNA by catalyzing the formation of a phosphodiester bond between the α-phosphate of a deoxyribonucleoside 5′-triphosphate and the 3′-hydroxyl terminus of a DNA chain. One would suspect that the deoxyribonucleoside triphosphate binding site and the site that binds nucleosides bearing both a 3′-hydroxyl group in the ribose configuration and a 5′-phosphate moiety would be relatively near each other. Krugh (1971) demonstrated that, in fact, these bindings sites were contiguous by loading the former site with spin-labeled ATP (**I**, Fig. 7) and the latter with AMP and observing the nitroxide-enhanced relaxation of the C-2 proton of AMP. The magnitude of this effect indicated that the distance between the nitroxyl nitrogen of **I** and the C-2 proton of AMP was 7.1 $\pm$ 0.6 Å when both ligands were bound to the enzyme.

D. Citrate Synthase

Porcine heart citrate synthase catalyzes the condensation of acetyl-coenzyme A with oxaloacetate to form citrate and coenzyme A. This enzyme binds the pyrrolidinyl-coenzyme A (**VIII**, Fig. 7) with a dissociation constant of 10^{-4} M (Weidman *et al.*, 1973). In this binary complex the bound paramagnetic coenzyme enhanced the longitudinal relaxation rate of water protons by a factor of 3.4. Paramagnetic resonance and proton relaxation rate studies indicated the formation of a ternary complex between citrate synthase, **VIII**, and oxaloacetate. Relaxation rate enhancement for water protons in this ternary complex was about 40% greater than in the binary complex. Citrate, (*R*)-malate, and (*R*, *S*)-tartrate also formed ternary complexes with citrate synthase and **VIII** (Fig. 7). The dissociation constants of these ligands from their complexes was 0.5, 0.4, and $\sim 2 \times 10^{-3}$ M, respectively.

E. Phosphofructokinase

Rabbit skeletal muscle phosphofructokinase is a regulatory enzyme with a protomeric molecular weight of 90,000. However, the smallest fully active oligomer exhibits a molecular weight of 340,000. The enzyme binds nucleotides and divalent cations, and possesses an extremely reactive sulfhydryl group. Jones *et al.* (1972, 1973) have labeled this group covalently with the piperidinyl iodoacetamide **VIII** (Fig. 1) and used three different magnetic resonance techniques to determine the steric relationships between the attached nitroxide, bound manganous ion, and the protons of bound ATP. Bound manganous ion was shown to undergo spin–spin interaction with the nitroxide. The magnitude of this effect was determined from the diminution of the EPR spectral amplitude of the spin label (Taylor *et al.*, 1969) and was used to calculate a nitroxyl nitrogen–Mn^{2+} distance of 12.0 Å. The attached nitroxide enhanced the nuclear relaxation rate of protons on bound ATP.

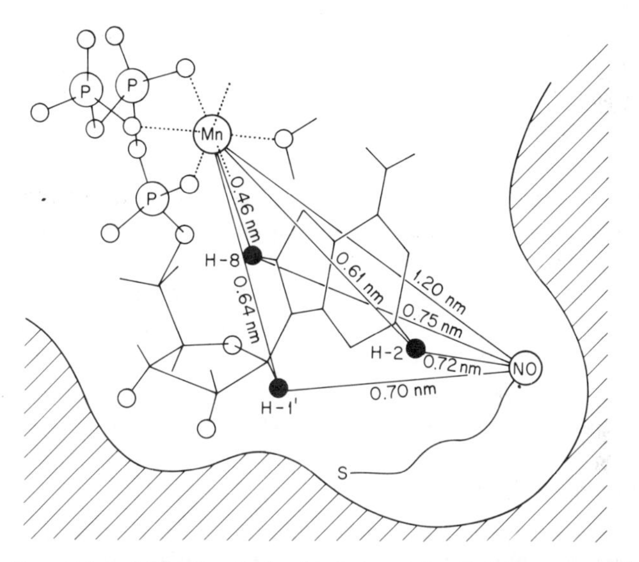

Fig. 21. Proposed model for the relationship between the nitroxide group of the spin label and the neighboring metal–ATP complex bound to phosphofructokinase. [From Jones *et al.* (1973).]

The magnitude of this effect was estimated from an increase in resonance line widths and was used to calculate values of 7.2, 7.5, and 7.0 Å for the distances from the H-2, H-8, and H-1′ ring protons of ATP, respectively, to the nitroxyl nitrogen. Finally, the relaxation rate of solvent water protons was found to be the same for the native and spin-labeled enzymes in either the absence or presence of saturating concentrations of Mg^{2+}-ATP. These results were used to construct a model (shown in Figure 21), for the spatial relationship between the attached nitroxide and bound Mn^{2+}-ATP.

F. Creatine Kinase

ATP-creatine phosphotransferase (creatine kinase) catalyzes the reversible transfer of the terminal phosphoryl group of ATP to an acceptor molecule (such as creatine) to give ADP and a phosphorylated acceptor. The enzyme exists as a dimer of two subunits, each weighing 41,000 daltons. Each subunit contains a reactive sulfhydryl group at its active site. Taylor *et al.* (1969) have labeled both of these SH groups on the rabbit muscle enzyme with the pyrrolidinyl iodoacetamide analogue of **I** (Fig. 1). The EPR spectrum indicated the covalently bound nitroxide was highly immobilized. Titration of the spin-labeled enzyme in the presence of Mg^{2+}, Ca^{2+}, Sr^{2+}, or Ba^{2+} with ADP gave spectra that were all of the same type and that indicated even further immobilization of the nitroxide. With Zn^{2+}, however, the EPR spectrum of the ternary complex was significantly different, suggesting that Zn^{2+}-ADP creatine kinase forms a complex that is structurally different from that formed with the alkaline earth metal ions. A somewhat unexpected observation was that when the ternary complex involved the paramagnetic ions Mn^{2+}, Ni^{2+}, and Co^{2+}, the nitroxide spectrum was not detectably broadened. Rather, an apparent decrease in amplitude of the resonance spectrum was observed. This phenomenon, due to spin–spin interaction between the attached nitroxide and bound paramagnetic ions, was used to calculate a value for the manganese–nitroxyl nitrogen distance. For the ternary complex involving ADP, the distance was estimated to be 7.5 ± 1.5 Å. For the ATP complex, this distance was 11.5 ± 0.6 Å (Cohn *et al.*, 1971). The spin-labeled enzyme was also observed to enhance the water proton relaxation rate about tenfold (Taylor *et al.*, 1971).

G. Alcohol Dehydrogenase

Alcohol dehydrogenase from horse liver (LADH) is an oxidative enzyme that contains 28 sulfhydryl groups. Two of these can be selectively alkylated with iodoacetate, resulting in complete loss of catalytic activity. The enzyme binds 4 moles of Zn^{2+}, two of which are related to catalysis, the other two being of structural importance. The enzyme binds the coenzyme NAD, which is reduced to NADH when substrate ethanol is oxidized. Spin-label studies on this enzyme were first reported by Mildvan and Weiner (1969a,b) and involved the use of an analogue of the coenzyme NAD (piperidinyl-ADP **II**, Fig. 7). When the enzyme was titrated with this analogue, diminution of the EPR spectrum revealed two classes of binding sites for this ligand: two sites bound the analogue with a $K_{dis} = (1.7 \pm 0.8) \times 10^{-5}$ M and five to six sites bound the analogue with a $K_{dis} = (7.5 \pm 0.9) \times 10^{-5}$ M. NADH could displace the bound spin-labeled ligand from the strong binding sites only. The zinc-free enzyme exhibited two strong binding sites for the spin-labeled analogue with $K_{dis} = (2.7 \pm 0.6) \times 10^{-5}$ M and no weak binding sites. The analogue could be completely displaced by NAD (Weiner, 1969). The paramagnetic ligand enhanced the water proton relaxation rate by a factor of 81 (Mildvan and Weiner, 1969a). This factor decreased to a value of 13 as the occupancy of the tight binding site increased from 0 to 2, indicating site–site interaction that is not reflected in the dissociation constants. Similar site–site interaction was detected by decrease of the enhancement factor from 32 to 4.5 in the zinc-free apoenzyme where only two tight binding sites exist. The magnitude of the change in transverse relaxation rates of the protons of bound ethanol, acetaldehyde, and isobutyramide caused by bound **II** (Fig. 7) were used to calculate distances from the nitroxyl nitrogen to different nuclei on these substrates or inhibitors. These distances were: ethanol (methyl), 3.6 Å; ethanol (methylene), 4.1 Å; acetaldehyde (formyl), 3.1 Å; and isobutyramide (methyl), $\leqq$ 3.9 Å (Mildvan and Weiner, 1969b).

This enzyme has been probed in a different fashion by Spallholz and Piette (1972). These workers used pyrrolinylphenanthroline, a spin-labeled chelating agent (**XI**, Fig. 7) that competed with NAD for the coenzyme binding site(s). When bound to the enzyme, **XI** gave an EPR spectrum with a maximum splitting of 66 G, indicating that the nitroxyl moiety was located in a sterically restricted environment. That the spin-labeled ligand was coordinated to the two catalytically essential zinc ions was indicated by the observation that NADH and 1,10-phenanthroline prevented its binding to the enzyme. Two cysteine side chains were alkylated with either the iodo analog of piperidinyl bromoacetamide **VIII** (Fig. 1) or the pyrrolidinyl iodoacetamide **XIX** (Fig. 1). This alkylation was accompanied by a decrease in catalytic activity and binding of the chelating spin label (**XI**, Fig. 7) as indicated in Table III. These results suggested that the zinc ions and sulfhydryl groups which are located within the enzyme's active site(s) are close to each other.

TABLE III

PROPERTIES OF MODIFIED LIVER ALCOHOL DEHYDROGENASE (LADH)[a]

Enzyme	Moles covalently bound reagent/mole LADH	Remaining activity (%)	Moles **XI** (Fig. 7) bound/mole LADH
LADH	0	100	2
LADH-**VIII**[b]	1.5	< 5	≪ 1
LADH-**XIX**[c]	1.0	50	1
LADH-**CM**[d]	2.0	0	< 2

[a] Spallholz and Piette, 1972.
[b] LADH alkylated with the iodo analog of VIII (Fig. 1).
[c] LADH alkylated with XIX (Fig. 1).
[d] LADH carboxymethylated with iodoacetic acid.

H. D-Glyceraldehyde-3-phosphate Dehydrogenase (GPDH)

This enzyme catalyzes the oxidative phosphorylation of D-glyceraldehyde 3-phosphate to 1,3-diphosphoglycerate with concomitant reduction of the coenzyme NAD to NADH. The reaction proceeds via an isolable acyl–enzyme intermediate. The apoenzyme still retains the ability to catalyze the hydrolysis of acyl phosphates and *p*-nitrophenyl esters, both reactions involving the same nucleophilic sulfhydryl groups. GPDH has been alkylated with doxyl aldehyde **XVII** and doxyl iodoacetate **XVIII** (Fig. 1) and acylated with doxyl carboxylate **IX** (Fig. 2) (Balthasar, 1971). The sight of labeling with these reagents was presumed to be the reactive SH group involved in catalysis; however, this was not rigorously demonstrated. The mixed spectra obtained from the enzyme tagged with these spin labels suggested that there might be other sites of labeling also. These experimental, mixed spectra were analyzed by comparing them to spectra generated by computer addition of various amounts of "fluid," "hindered," and "immobilized" single-component spectra. When the apoenzyme was labeled with doxyl iodoacetate **XVIII** (Fig. 1) where $n = 3$, and the resulting EPR spectrum analyzed, the relative concentration of nitroxide in the "fluid," "hindered," or "immobilized" states was estimated to be $< 1\%$, 50%, and 50%, respectively. With increasing values of n, the fraction of spin label in the immobilized state gradually decreased, suggesting a steric gradient for the nitroxyl moiety in the series of bound labels that was studied. Spectra of the enzyme labeled with doxyl aldehyde **XVII** (Fig. 1) or doxyl carboxylate **IX** (Fig. 2) were dominated by the "fluid" component that was due to spin label released during deacylation, and hence were of little value. Addition of the spin-labeled NAD analogue (piperidinyl-ADP **II**, Fig. 7) to GPDH labeled with **XVII**, **XVIII** (Fig. 1), or **IX** (Fig. 2) produced no detectable line changes in spectra of the latter, indicating no spin-spin interaction and suggesting that the covalently bound nitroxides were > 14 Å distant from the noncovalently bound nitroxide (Calvin *et al.*, 1969).

I. Chymotrypsin

No soluble enzyme has been studied by spin-label methods as extensively as chymotrypsin. Since these studies have been critically reviewed elsewhere (Berliner, 1974), they will not be detailed here. Rather, a brief overview of this work is given.

The kinetics of deacylation have been studied for chymotrypsin labeled with **IV** (Fig. 2) by Berliner and McConnell (1966). The function and operation of structural units within the active site of the enzyme were first examined by Kosman *et al.* (1969). Some time later, a study of the conformational integrity and transitions in the modified enzyme were reported

(Kosman and Piette, 1972). Denaturation by guanidine (Morrisett and Broomfield, 1971) and urea (Berliner, 1972) of the active site spin-labeled enzyme has been studied as a function of pH. The orientation of a spin label at the active site of chymotrypsin in the form of single crystals has been examined by magnetic resonance (Berliner and McConnell, 1971). The rotational correlation times of two different spin-labeled chymotrypsin derivatives have been determined by a novel method (Shimshick and McConnell, 1972). Structure–function relationships have been compared for spin-labeled chymotrypsinogen, α-chymotrypsin, and anhydrochymotrypsin (Kosman, 1972). Brief reports by Grigoryan *et al.* (1973) and Hoff *et al.* (1971) of spin-label studies on this enzyme have appeared. Finally, the active site geometry of this enzyme has been carefully studied by use of fifteen different aromatic sulfonyl fluorides (Berliner and Wong, 1974).

J. Trypsin

Trypsin is a proteolytic enzyme that cleaves peptide bonds on the carboxyl side of arginine and lysine side chains. It has a molecular weight of ~ 25,000 and its sequence is highly homologous to that of chymotrypsin. A rigorous comparison of the active-site geometries of trypsin and chymotrypsin has been carried out by Berliner and Wong (1974) using an array of spin-labeled sulfonyl fluorides (Fig. 3). The spectral results at pH 3.5 clearly indicated structural differences at the active site of these two enzymes. The majority of those labels that, upon attachment to chymotrypsin, were shown by model building studies to be excluded from the aromatic binding pocket (*p*-**I**–*p*-**V**, *m*-**I**, *m*-**III**, *m*-**IV**), upon addition of indole, were found to give EPR spectra closely resembling those exhibited by trypsin. Hence, the authors felt that for some of the chymotrypsin and all of the trypsin derivatives, the aromatic portion of the attached spin label was excluded from the "substrate" binding site and was localized chiefly in a general region of the catalytic center (see Section IV.G).

In a comparative study, Morrisett and Broomfield (1972) observed that trypsin labeled with **I** (Fig. 4) at pH 5.5 and 7.7 gave narrow- and broad-line EPR spectra, respectively, both being recorded at pH 3.3. The reason for this spectral difference was attributed to either (a) different modes of binding the nitroxyl group during inhibition at the two different pHs, or (b) very limited proteolysis during the course of the pH 5.5 labeling reaction. Berliner and Wong (1973) showed that the latter alternative was, in fact, the correct one.

Trypsin spin-labeled with **I** (Fig. 4) gives a spectrum very similar to that obtained from elastase and subtilisin labeled with the same reagent (Fig. 18). Since none of these three enzymes can accommodate the nitroxyl ring in its specific substrate binding site, the spectra reflect steric crowding in a general binding site outside the specific site (Morrisett and Broomfield, 1972).

Hsia *et al.* (1972) spin-labeled trypsin with di- and monopiperidinyl phosph*o*rates (**IV** and **VIII**, Fig. 4) and found that the spectra were pH dependent but were not reversible. These results suggest that the labeled enzyme was undergoing autolysis, dephosphorylation, aging, or a combination of these possibilities, none of which were rigorously excluded by the authors.

Berliner *et al.* (1973) have tagged the active site serine of trypsin with piperidinyl phosph*o*nate **I** (Fig. 4) and insolubilized (immobilized) this spin-labeled enzyme by coupling to porous glass beads. The resulting broad-line spectrum contains hyperfine extrema separated by 54–55 G, the same splitting observed when the enzyme was insolubilized first, *then* spin labeled. Spin-labeled trypsin in the presence of saturated sucrose exhibited a splitting (56–57 G) similar to that shown by the same species when coupled to glass beads (Berliner, 1974).

K. Subtilisin

This bacterial protease exhibits a somewhat broader specificity than trypsin or chymotrypsin. When spin labeled at pH 5.5 with piperidinyl phosph*o*nate **I** (Fig. 4), gel filtered at pH 4.3, and measured immediately, the Nagarse enzyme gave a broad-line spectrum (Fig. 18) indicative of considerable motional constraint for the nitroxide (Morrisett and Broomfield, 1972). At pH 3.3, the spectrum contains only narrow-line components, suggesting that at this pH the active-site region of the enzyme is unfolded, a result consistent with the fact that the enzyme has no disulfide bonds to stabilize its tertiary structure. Hsia *et al.* (1972) have labeled the BPN′ and Carlsberg enzymes with the monopiperidinyl phosph*o*rate **VIII** (Fig. 4) and observed no differences in their spectra. However, when these enzymes were labeled with the dipiperidinyl phosph*o*rate nitroxide **IV** (Fig. 4), quite striking spectral differences were observed, illustrating the sensitivity of these bi-radical reagents. Subtilisin Novo has also been spin-labeled with piperidinyl phosph*o*nate **I** (Fig. 4) and examined briefly (Hoff *et al.*, 1971).

L. Thrombin

Thrombin is another of the proteolytic enzymes that is inhibited by diisopropylphosphorofluoridate. It is intimately involved in the cascade of events that leads to formation of the blood clot. Thrombin and trypsin have similar amino acid sequences and exhibit similar specificities toward small synthetic substrates, but they show different specificities toward larger protein substrates. This difference has been detected by a spin label. Thrombin spin labeled with the dipiperidinyl phosph*o*rate **IV** (Fig. 4) gave a spectrum indicating considerably more mobility than was indicated by the spectrum of trypsin labeled with the same reagent (Hsia *et al.*, 1972).

M. Elastase

This enzyme is one of the few agents capable of breaking down elastin, a principal structural component of elastic fibers. It has the same general conformation as α-chymotrypsin and possesses extensive amino acid sequence homologies with the B and C chains of that enzyme. The cavity present at the substrate binding site of chymotrypsin is also present in the elastase molecule, but in the latter, the entrance is partially blocked. Hence, when elastase is inhibited with piperidinyl phospho*n*ate **I** (Fig. 4), the nitroxyl ring must be excluded from the specific binding site. Morrisett and Broomfield (1972) observed a broad line EPR spectrum for this spin-labeled enzyme, indicating steric crowding in the general binding site outside the specific binding-site region. However, Hsia *et al.* (1972) observed a relatively high degree of mobility for the nitroxyl groups of dipiperidinyl phospho*r*ate **IV** (Fig. 4) when attached to this enzyme.

N. Staphylococcal Protease

Dugas (1973) has used piperidinyl phospho*r*ate **III** (Fig. 4) to label the active-site serine of several proteases. One of these was a protease from *Staphylococcal aureus*. This enzyme is a single polypeptide chain with a molecular weight of about 12,000 daltons. When labeled at pH 5.5 then dialyzed at pH 3.5, the enzyme gave a spectrum containing only narrow-line components, indicating high label mobility. When the spectrum was recorded at pH 5.5, it contained broad-line components and appeared similar to that obtained by Morrisett and Broomfield (1972) for trypsin spin-labeled with piperidinyl phospho*n*ate **I**, Fig. 4 (Fig. 18). When the pH was raised to 7.5, the spectrum indicated increased label motion and was similar to that obtained from chymotrypsin labeled with either piperidinyl phospho*r*ate **III**, Fig. 4 (Dugas, 1973) or piperidinyl phospho*n*ate **I**, Fig. 4 (Morrisett and Broomfield, 1972). The apparent stability of this spin-labeled enzyme and the marked pH-dependent changes of its spectra suggest that this will be an exciting system for further study.

O. Cholinesterase

The active-site geometry of cholinesterase from various sources has been studied by two groups. Preliminary reports indicated that the active site(s) of horse serum cholinesterase (Hsia *et al.*, 1969) and electric eel acetylcholinesterase (Morrisett *et al.*, 1969) were more capacious than that of α-chymotrypsin. In a more definitive comparison, Hsia *et al.* (1972) observed that cholinesterases from bovine erythrocytes, electric eel, and horse serum labeled with dipiperidinyl phosphorate **IV** (Fig. 4) have similar spectra, indicating the active sites of these enzymes were quite similar. Membrane-bound and membrane-free electric eel acetylcholinesterase have been active-site labeled with piperidinyl phosphonate **I** (Fig. 4) (Morrisett *et al.*, 1970a,b). Both preparations gave an EPR spectrum which contained narrow line components, suggesting that the active site region is quite large and/or located on the surface of the enzyme in either form. The paramagnetism of the nitroxide was quenched when the spin-labeled membranes were incubated at 25°C for 24 hr. The resonance signal reappeared after addition of nonionic detergent. This effect was attributed to the reducing ability of a nearby sulfhydryl group (Morrisett and Drott, 1969). The quenching effect was not observed with the spin-labeled, membrane-free enzyme. Incubation of the labeled, membrane-free enzyme with pyridine-2-aldoxime resulted in dephosphonylation of the enzyme as determined by the time-dependent increase in spectral amplitude, decrease in spectral line width, and recovery of enzymatic activity. Inhibition of the soluble enzyme with the large, steroidal phosphonylating reagent **II** (Fig. 4) gave a derivative whose spectrum contained broad lines with a total hyperfine splitting of 63 G. Cholinesterase on ox erythrocyte membranes, human erythrocyte membranes, rat brain microsomes, and rat brain nerve ending particles selectively labeled with piperidinyl phosphonate **I** (Fig. 4) gave spectra that were very similar to that obtained with the purified electric eel enzyme (Morrisett *et al.*, 1970b).

P. Adenosine Triphosphatase

Membrane-bound ATPase has been labeled with a site-specific carbodiimide reagent (**III**, Fig. 1). The spin-labeled enzyme gave an EPR spectrum that indicated the motion of the bound nitroxide was highly restricted. The bound label could be reduced to the diamagnetic derivative by succinate, indicating that electrons could be transferred from the respiratory chain to the ATPase system. Formation of the Mn^{2+}–ATP–enzyme ternary complex was attended by a 30% diminution of the resonance signal due to interaction between the two paramagnetic species. Assuming that the correlation time for this interaction was between 10^{-9} and 10^{-8} sec, Azzi *et al.* (1973) calculated a value of 18–25 Å for the distance between the nitroxyl nitrogen and the Mn^{2+} ion (see Section IV.C.3).

Q. Leucine Aminopeptidase

This enzyme cleaves L-amino acids from the amino terminus of a peptide or protein and has been used routinely to determine the optical purity of synthetic polypeptides. This protein has been nonspecifically alkylated on one or more sulfhydryl groups with an analogue of iodoacetamide (**VIII**, Fig. 1). The resulting derivative has been used recently by Lassman *et al.* (1973) to develop a method for scaling the hydrophobicity of spin-labeled regions of a protein.

R. Aspartate Transcarbamylase

This regulatory enzyme exhibits all of the general features of allosterism, including cooperativity in substrate binding and inhibition, and activation through binding at loci that are remote from the catalytic site. The enzyme is involved in pyrimidine biosynthesis and is subject to feedback inhibition by the end product cytidine triphosphate (CTP), which provides a mechanism for turning off synthesis. The enzyme is activated by adenosine triphosphate (ATP). Binding of the substrates aspartate and carbamyl phosphate is cooperative through interactions between catalytic sites on different subunits. X-ray diffraction data suggest the enzyme is composed of six identical regulatory chains and six identical catalytic chains (Wiley *et al.*, 1971). This enzyme has been alkylated with piperidinyl bromoacetamide **VIII** (Fig. 1) (Buckman, 1970). The EPR spectrum of this labeled enzyme was broadened upon addition of CTP and ATP. Succinate and carbamyl phosphate had no effect on the spectrum. Increasing the pH of the enzyme from 6.0 to 10.4 or dissociating it into its subunits by treatment with *p*-chloromercuribenzoate resulted in narrowing of the spectral lines. While these results indicate the attached label(s) is sensitive to important structural changes in this enzyme complex, they would have been much more meaningful if the number and location of attached labels had been accurately determined.

S. Carbonic Anhydrase

Erythrocyte carbonic anhydrase is a zinc-containing metalloenzyme that reversibly catalyzes the hydration of CO_2 and is specifically inhibited by aromatic sulfonamides. The protein is a single polypeptide chain of 260 amino acids and is nearly spherical in shape. Chignell and co-workers have synthesized a variety of spin-labeled sulfonamides (**XIV–XX**, Fig. 7) and used them to measure the depth of the inhibitor binding site (Chignell *et al.*, 1972; Erlich *et al.*, 1973). A detailed description of these experiments appears in Section IV.F and in a recent review by Berliner (1974). Taylor *et al.* (1970) have used a piperidinylhydrazide sulfonamide (**XIII**, Fig. 7) to probe carbonic anhydrase. The similarity of the broad line EPR spectra of spin-labeled Zn^{2+} and Co^{2+} carbonic anhydrase suggests that the conformation at the active site is similar in these two different metalloenzymes. The same spin-labeled ligand was used by Mushak and Coleman (1972) to compare the active-site geometries of several carbonic anhydrase isozymes from various species and to study their denaturation by urea. Hower *et al.* (1971) have reported that piperidinyl estersulfonamide (**XII**, Fig. 7) forms a 1 : 1 complex with an unfractioned mixture of bovine isozymes. Rapid anisotropic motion of the nitroxyl ring occurs about one of the bonds connecting it to the aromatic moiety. The isotropic coupling constant indicates the piperidinyl group resides in a highly polar region.

T. Aspartate Aminotransferase

Two of the four sulfhydryl groups of L-aspartate : 2-oxoglutarate aminotransferase are exposed and have been alkylated with 2,2,6,6-tetramethylpiperidine-1-oxyl 4-iodoacetate, the iodoamide analogue of **VIII** (Fig. 1). The modified enzyme is about 90% as active as the native protein indicating that these groups are not intimately involved in catalysis. The EPR spectrum of the doubly spin-labeled enzyme contained three distinguishable components, some of which became broader when substrates and quasi-substrates were added. Removal of these ligands by gel filtration reversed the effect and restored the original spectrum. When one of the accessible thiol groups (1) was selectively alkylated by maleic acid and the other group (2) was tagged with the spin label, addition of substrate did not induce the spectral changes observed when both groups were labeled. The authors (Polyanovsky *et al.*, 1973) conclude that only the spin label bound to group 1 is sensitive to the interaction of a ligand with the active site. After this group was labeled with [^{14}C]maleic acid and the modified protein was digested with chymotrypsin, an octapeptide spanning sequence positions 41–48 was found to contain the radioactive marker attached to Cys 45. The active site has been estimated to be > 17 Å from this residue.

VI. FUTURE POSSIBILITIES

The foregoing studies aptly demonstrate the successful application of spin labels and magnetic resonance techniques to problems concerning enzyme structure and function. One exciting possibility that has not been discussed is the synthesis of peptides and even small proteins that contain a spin-labeled amino acid at a critical locus where important interactions are known to occur. Weinkam and Jorgensen (1971a) have already prepared a nitronyl nitroxide amino acid (**I**, Fig. 22) whose structure resembles that of histidine, and incorporated it into angiotensin **II** analogues (Weinkam and Jorgensen, 1971b). Unfortunately, the nitronyl nitroxides are less stable and give more complex spectra than the simple nitroxides. A spin-labeled amino acid containing the piperidinyl group (**II**, Fig. 22), whose structure and dimensions closely approximate those of phenylalanine and tyrosine, is

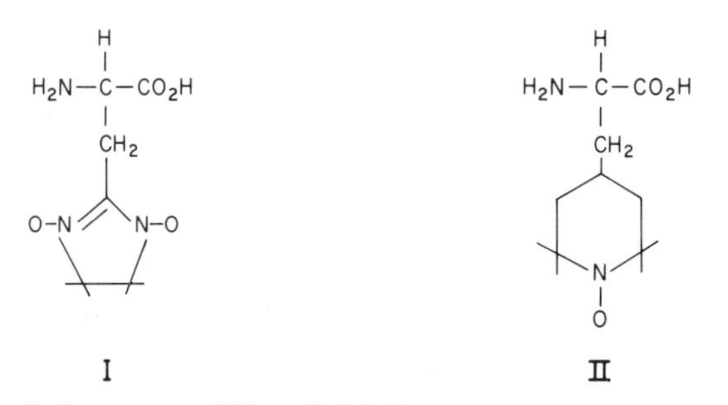

Fig. 22. Chemical structures of (**I**) a spin-labeled histidine analog synthesized by Weinkam and Jorgensen (1971a) and (**II**) a spin-labeled tyrosine analogue. (Sparrow *et al.*, 1975.)

another possibility. This spin-labeled amino acid has now been synthesized and may be used under most of the conditions employed in solid-phase peptide synthesis (Sparrow *et al.*, 1975). It represents a virtually ideal protein reporter group that should be no more perturbing than an intrinsic fluorophore and yet equally or more useful.

Acknowledgments

The author gratefully acknowledges support by HEW grants HL-14194 and HL-05435/34 and by a grant from the John A. Hartford Foundation, Inc. Stimulating discussions and collaborative interactions with a number of colleagues have helped make this chapter possible. These include Drs. R. G. Hiskey, J. H. Harrison, H. H. Dearman, H. R. Drott, K. Akasaka, C. A. Broomfield, L. J. Berliner, R. L. Jackson, J. T. Sparrow, A. M. Gotto, and H. M. McConnell. Special thanks are due to Dr. H. J. Pownall and Miss Debbie Mason for help in preparing and proofing the manuscript. (TBTGA). The author is an established investigator of the American Heart Association.

References

Adler, A. J., Greenfield, N. J., and Fasman, G. D. (1973). Circular dichroism and optical rotatory dispersion of proteins and polypeptides, *Methods Enzymol.* **27**, 675–735.

Ament, S. S., Wetherington, J. B., Moncrief, J. W., Flohr, K., Mochizuki, M., and Kaiser, E. T. (1973). Determination of the absolute configuration of (+)-3-carboxy-2,2,5,5-tetramethyl-1-pyrrolidinyloxy, *J. Amer. Chem. Soc.* **95**, 7896–7897.

Azzi, A., Bragadin, M. A., Tamburro, A. M., and Santato, M. (1973). Site-directed spin labeling of the mitochondrial membrane, *J. Biol. Chem.* **248**, 5520–5526.

Baker, B. R. (1967). "Design of Active-Site Directed Irreversible Enzyme Inhibitors." Wiley, New York.

Balthasar, W. (1971). Spin labeling studies of D-glyceraldehyde-3-phosphate dehydrogenase, *Eur. J. Biochem.* **22**, 158–165.

Barratt, M. D., Dodd, G. H., and Chapman, D. (1969). A spin label for tyrosine residues, *Biochim. Biophys. Acta* **194**, 602–605.

Barratt, M. D., Davies, A. P., and Evans, M. T. A. (1971). Maleimide and isomaleimide pyrrolidine-nitroxide spin labels, *Eur. J. Biochem.* **24**, 280–283.

Beck, W., Schmidtner, K., and Keller, H. J. (1967). Kobalt(II)-halogenid Komplexe mit dem Di-*tert*-butylstickstoffoxid-Radikal, *Chem. Ber.* **100**, 503–511.

Berliner, L. J. (1967). Ph.D. Thesis, Stanford Univ., Stanford, California.

Berliner, L. J. (1971). A 6 Å crystallographic study of a spin-labeled inhibitor complex with lysozyme, *J. Mol. Biol.* **61**, 189–194.

Berliner, L. J. (1972). Urea denaturation of active-site spin-labeled α-chymotrypsin, *Biochemistry* **11**, 2921–2924.

Berliner, L. J. (1974). Applications of spin labeling to structure-conformation studies of enzymes, *Prog. Bioorg. Chem.* **3**, 1–80.

Berliner, L. J., and McConnell, H. M. (1966). A spin-labeled substrate for α-chymotrypsin, *Proc. Nat. Acad. Sci. U.S.* **55**, 708–712.

Berliner, L. J., and McConnell, H. M. (1971). Spin label orientation at the active site of α-chymotrypsin in single crystals, *Biochem. Biophys. Res. Commun.* **43**, 651–657.

Berliner, L. J., and Wong, S. S. (1973). Evidence against two "pH locked" conformations of phosphorylated trypsin, *J. Biol. Chem.* **248**, 1118–1120.

Berliner, L. J., and Wong, S. S. (1974). Spin labeled sulfonyl fluorides as active site probes of protease structure, I. Comparison of the active site environments in α-chymotrypsin and trypsin, *J. Biol. Chem.* **249**, 1668–1677.

Berliner, L. J., Miller, S. T., Uy, R., and Royer, G. P. (1973). An ESR study of the active-site conformations of free and immobilized trypsin, *Biochim. Biophys. Acta* **315**, 195–199.

Birks, J. B. (1970). "Photophysics of Aromatic Molecules," Chapter 10. Wiley (Interscience), New York.

Brand, L., and Witholt, B. (1967). Fluorescence measurements, *Methods Enzymol.* **11**, 776–856.

Buckman, T. (1970). Spin-labeling studies of aspartate transcarbamylase. I. Effects of nucleotide binding and subunit separation, *Biochemistry* **9**, 3255–3265.

Buckman, T., Kennedy, F. S., and Wood, J. M. (1969). Spin labeling of vitamin B_{12}, *Biochemistry* **8**, 4437–4442.

Calvin, M., Wang, H. H., Entine, G., Gill, D., Ferruti, P., Harpold, M. A., and Klein, M. P. (1969). Biradical spin labeling of nerve membranes, *Proc. Nat. Acad. Sci. U.S.* **63**, 1–8.

Canfield, R. E., and Anfinsen, C. B. (1963). Concepts and experimental approaches in the determination of the primary structure of proteins, *In* "The Proteins" (H. Neurath, ed.), Vol. I, pp. 311–378. Academic Press, New York.

Chaiken, I. M., and Smith, E. L. (1969). Reaction of a specific tyrosine residue of papain with diisopropylfluorophosphate, *J. Biol. Chem.* **244**, 4247–4250.

Chignell, C. F., Starkweather, D. K., and Erlich, R. H. (1972). The interaction of some spin-labeled sulfonamides with bovine erythrocyte carbonic anhydrase B, *Biochim. Biophys. Acta* **271**, 6–15.

Cohen, S. G., and Crossley, J. (1964). Kinetics of hydrolysis of dicarboxylic esters and their α-acetamido derivatives by α-chymotrypsin, *J. Amer. Chem. Soc.* **86**, 4999–5003.

Cohen, S. G., Klee, L. H., and Weinstein, S. Y. (1966). Action of α-chymotrypsin on the diethyl esters of fumaric, maleic, and acetylene dicarboxylic acids, *J. Amer. Chem. Soc.* **88**, 5302–5305.

Cohn, M., Diefenbach, H., and Taylor, J. S. (1971). Magnetic resonance studies of the interaction of spin-labeled creatine kinase with paramagnetic manganese–substrate complexes, *J. Biol. Chem.* **246**, 6037–6042.

Craig, L. C., and Chen, H.-C. (1969). On rapid laboratory dialysis, *Anal. Chem.* **41**, 590–596.

Daniel, W. E., Morrisett, J. D., Harrison, J. H., Dearman, H. H., and Hiskey, R. G. (1973).

Spin-labeled ribonuclease A. Selective incorporation of a nitroxide spin label sensitive to active center geometry, *Biochemistry* **12**, 4918–4923.

Drenth, J., Jansonius, J. N., Koekoek, R., Sluyterman, L. A. A., and Wolthers, B. G. (1970). The structure of the papain molecule, *Phil. Trans. Roy. Soc. London* **B257**, 231–236.

Dugas, H. (1973). Introduction à la résonance paramagnétique électronique et à la méthode du marqueur de spin—un article revue, *Can. J. Spectrosc.* **18**, 110–118.

Erlich, R. H., Starkweather, D. K., and Chignell, C. F. (1973). A spin label study of human erythrocyte carbonic anhydrase B and C, *Mol. Pharm.* **9**, 61–73.

Fahrney, D. E., and Gold, A. M. (1963). Sulfonyl fluorides as inhibitors of esterases. I. Rates of reaction with acetylcholinesterase, α-chymotrypsin and trypsin, *J. Amer. Chem. Soc.* **85**, 997–1000.

Filatova, M. P., Reissmann, S., Ravdel, G. A., Ivanov, V. T., Grigoryan, G. L., and Shapiro, A. B. (1973). Konformation suntersuchungen in Bradykinin und Bradykinin-Analoga Hilfe von CD-, NMR- und ESR-messunger, *In* "Peptides 1972" (H. Hanson and H.-D. Jakubke, eds.), pp. 334–340. North-Holland Publ., Amsterdam and American Elsevier, New York.

Flohr, K., and Kaiser, E. T. (1972). Enantiomeric specificity in the chymotrypsin-catalyzed hydrolysis of 3-carboxy-2,2,5,5-tetramethylpyrrolidin-1-oxy *p*-nitrophenyl ester, *J. Amer. Chem. Soc.* **94**, 3675–3676.

Flohr, K., Paton, R. M., and Kaiser, E. T. (1971). Enantiomeric specificity in the cyclohexa-amylose-catalyzed hydrolysis of 3-carboxy-2,2,5,5-tetramethylpyrrolidin-1-oxy *m*-nitrophenyl ester, *Chem. Commun.* **13**, 1621–1622.

Forrester, A. R., Hay, J. M., and Thompson, R. H. (1968). "Organic Chemistry of Stable Free Radicals." Academic Press, New York.

Freer, S. T., Kraut, J., Robertus, J. D., Wright, H. T., and Xuong, N. H. (1970). Chymotrypsinogen: 2.5 Å crystal structure, comparison with α-chymotrypsin, and implications for zymogen activation, *Biochemistry* **9**, 1997–2009.

Green, J. A., Singer, L. A., and Parks, J. H. (1973). Fluorescence quenching by the stable free radical di-*t*-butylnitroxide, *J. Chem. Phys.* **58**, 2690–2695.

Griffith, O. H., and McConnell, H. M. (1966). A nitroxide-maleimide spin label, *Proc. Nat. Acad. Sci. U.S.A.* **55**, 8–11.

Griffith, O. H., and Waggoner, A. S. (1969). Nitroxide free radicals: Spin labels for probing biomolecular structure, *Accounts Chem. Res.* **2**, 17–24.

Griffith, O. H., Cornell, D. W., and McConnell, H. M. (1965). Nitrogen hyperfine tensor and **g** tensor of nitroxide radicals, *J. Chem. Phys.* **43**, 2909–2910.

Griffith, O. H., Keana, J. F. W., Noall, D. L., and Ivey, J. L. (1967). Nitroxide mixed carboxylic–carbonic acid anhydrides. A new class of versatile spin labels, *Biochim. Biophys. Acta* **148**, 583–585.

Griffith, O. H., Dehlinger, P. J., and Van, S. P. (1974). Shape of the hydrophobic barrier of phospholipid bilayers. Evidence for water penetration in biological membranes, *J. Mem. Biol.* **15**, 159–192.

Grigoryan, G. D., Rozantsev, E. G., Teplov, N. E., Ivanov, V. P., and Korkhmazyan, M. M. (1973). Investigation of α-chymotrypsin by the "spin label method" at serine-195, *Izv. Akad. Nauk. Ser. Khim.* 1654–1656.

Gross, E., and Witkop, B. (1962). Nonenzymatic cleavage of peptide bonds: The methionine residues in bovine pancreatic ribonuclease, *J. Biol. Chem.* **237**, 1856–1860.

Gundlach, H. G., Stein, W. H., and Moore, S. (1956). The nature of the amino acid residues involved in the inactivation of ribonuclease by iodoacetate, *J. Biol. Chem.* **234**, 1754–1760.

Hawley, D. M., Roberts, J. S., Ferguson, G., and Porte, A. L. (1967). The structure and absolute stereochemistry of caryophyllene "iodo-nitrosite": A stable nitroxide radical, *Chem. Commun.* 942–943.

Henderson, R., Wright, C. S., Hess, G. P., and Blow, D. M. (1971). α-chymotrypsin: What can we learn about catalysis from X-ray diffraction? *Cold Spring Harbor Symp. Quant. Biol.* **36**, 63–70.

Himmelhock, S. R. (1971). Chromatography of proteins on ion-exchange adsorbents, *Methods Enzymol.* **22**, 273–286.

Hoff, A. J., Oösterbaan, R. A., and Deen, R. (1971). Specific and non-specific binding of an organophosphate spin label to some serine-esterases, *FEBS Lett.* **14**, 17–21.

Hoffman, B. M., Schofield, P., and Rich, A. (1969). Spin-labeled Transfer RNA, *Proc. Nat. Acad. Sci. U.S.* **62**, 1195–1202.

Hower, J. F., Henkens, R. W., and Chesnut, D. B. (1971). A spin label investigation of the active site of an enzyme. Bovine carbonic anhydrase, *J. Amer. Chem. Soc.* **93**, 6665–6671.

Hsia, J. C., and Piette, L. H. (1969). Spin-labeling as a general method in studying antibody active site, *Arch. Biochem. Biophys.* **129**, 296–307.

Hsia, J. C., Kosman, D. J., and Piette, L. H. (1969). Organophosphate spin-label studies of inhibited esterases, α-chymotrypsin and cholinesterase, *Biochem. Biophys. Res. Commun.* **36**, 75–78.

Hsia, J. C., Kosman, D. J., and Piette, L. H. (1972). ESR probing of macromolecules: Spin-labeling of the active sites of the proteolytic serine enzymes, *Arch. Biochem. Biophys.* **149**, 441–451.

Jackson, R. L., Morrisett, J. D., Pownall, H. J., and Gotto, A. M. (1973). The immunochemical and lipid-binding properties of Apolipoprotein-glutamine II, *J. Biol. Chem.* **248**, 5218–5224.

Jones, R., Dwek, R. A., and Walker, I. O. (1972). Conformational states of rabbit muscle phosphofructokinase investigated by a spin label probe, *FEBS Lett.* **26**, 92–96.

Jones, R., Dwek, R. A., and Walker, I. O. (1973). Spin-labelled phosphofructokinase and its interactions with ATP and metal–ATP complexes as studied by magnetic resonance methods, *Eur. J. Biochem.* **34**, 28–40.

Jost, P., and Griffith, O. H. (1972). Electron spin resonance and the spin labeling method, *In* "Methods in Pharmacology" (C. Chignell, ed.), Vol. II, pp. 223–276. Appleton, New York.

Jost, P., Waggoner, A. S., and Griffith, O. H. (1971). Spin labeling and membrane structure, *In* "Structure and Function of Biological Membranes" (L. Rothfield, ed.), pp. 83–144. Academic Press, New York.

Kato, K., and Murachi, T. (1971). Chemical modification of tyrosyl residues of hen egg-white lysozyme by diisopropylphosphorofluoridate, *J. Biochem.* **69**, 725–737.

Keijer, J. H., Wolring, G. Z., and De Jong, L. P. A. (1974). Effect of pH, temperature, and ionic strength on the aging of phosphonylated cholinesterases, *Biochim. Biophys. Acta* **334**, 146–155.

Koblin, D. D., Kaufmann, S. A., and Wang, H. H. (1973). Quenching of 1-anilinonaphthalene-8-sulfonate fluorescence by a spin-labeled local anesthetic, *Biochem. Biophys. Res. Commun.* **53**, 1077–1083.

Kornberg, R. D., and McConnell, H. M. (1971). Inside–outside transitions of phospholipids in vesicle membranes, *Biochemistry* **10**, 1111–1120.

Kosman, D. J. (1972). Electron paramagnetic resonance probing of macromolecules: A comparison of structure/function relationships in chymotrypsinogen, α-chymotrypsin, and anhydrochymotrypsin, *J. Mol. Biol.* **67**, 247–264.

Kosman, D. J., and Piette, L. H. (1972). ESR probing of macromolecules: Spin label study of the conformational integrity and transitions in modified α-chymotrypsin, *Arch. Biochem. Biophys.* **149**, 452–460.

Kosman, D. J., Hsia, J. C., and Piette, L. H. (1969). ESR probing of macromolecules: Function and operation of structural units within the active site of α-chymotrypsin, *Arch. Biochem. Biophys.* **133**, 29–37.

Krugh, T. R. (1971). Proximity of the nucleoside monophosphate and triphosphate binding sites on deoxyribonucleic acid polymerase, *Biochemistry* **10**, 2594–2599.

Lassman, G., Ebert, B., Kuznetsov, A. N., and Damerau, W. (1973). Characterization of hydrophobic regions in proteins by spin labeling technique, *Biochim. Biophys. Acta* **310**, 298–304.

Law, P. Y., Brown, D. G., Lien, E. L., Babior, B. M., and Wood, J. M. (1971). Synthesis and catalytic activity of spin-labeled cobinamide complexes, *Biochemistry* **10**, 3428–3435.

Leigh, J. S. (1970). ESR rigid-lattice line shape in a system of two interacting spins. *J. Chem. Phys.* **52**, 2608–2612.

Likhtenshtein, G. I., and Bobodzhanov, P. K. (1969). New specific paramagnetic label based on trichlorotriazine, *Biofizika* **14**, 741–743.

Lipscomb, W. N., Reeke, G. N., Hartsuck, J. A., Quiocho, F. A., and Bethge, P. H. (1970). The structure of carboxypeptidase A. VIII. Atomic interpretation at 0.2 nm resolution, a new study of the complex of glycyl-L-tyrosine with CPA, and mechanistic deductions, *Phil. Trans. Roy. Soc. London* **B257**, 177–214.

Matthews, B. W., Sigler, P. B., Henderson, R., and Blow, D. M. (1967). Three-dimensional structure of tosyl-α-chymotrypsin, *Nature* (*London*) **214**, 652–656.

McConnell, H. M., and Hamilton, C. L. (1968). Spin-labeled hemoglobin derivatives in solution and in single crystals, *Proc. Nat. Acad. Sci. U.S.* **60**, 776–781.

McConnell, H. M., and McFarland, B. G. (1970). Physics and chemistry of spin labels, *Quart. Rev. Biophys.* **3**, 91–136.

McConnell, H. M., Deal, W., and Ogata, R. T. (1969). Spin-labeled hemoglobin derivatives in solution, polycrystalline suspensions, and single crystals, *Biochemistry* **8**, 2580–2585.

McConnell, H. M., Smith, I. C. P., and Bhacca, N. (1967). Comment on paper by H. Sternlicht and E. Wheeler, *In* "Magnetic Resonance in Biological Systems" (A. Ehrenberg, B. G. Malmstrom, and T. Vanngard, eds.), p. 335. Pergamon, Oxford.

McFarland, B. G. (1974). Spin-label measurements in membranes, *Methods Enzymol.* **33**, 161–197.

McCalley, R. C., Shimshick, E. J., and McConnell, H. M. (1972). The effect of slow rotational motion on paramagnetic resonance spectra, *Chem. Phys. Lett.* **13**, 115–119.

McConnell, H. M. (1967). Spin-labeled protein crystals, *In* "Magnetic Resonance in Biological Systems" (A. Ehrenberg, B. G. Malmstrom, and T. Vanngard, eds.), pp. 313–323. Pergamon, Oxford.

McConnell, H. M., and Boeyens, J. C. A. (1967). Spin-label determination of enzyme symmetry, *J. Phys. Chem.* **71**, 12–14.

Means, G. E., and Feeney, R. E. (1971). "Chemical Modifications of Proteins." Holden-Day, San Francisco, California.

Metz, D. H., and Brown, G. L. (1969). The investigation of nucleic acid secondary structure by means of chemical modification with a cardodiimide reagent, *Biochemistry* **8**, 2312–2328.

Mildvan, A. S., and Weiner, H. (1969a). Interaction of a spin-labeled analogue of nicotinamide-adenine dinucleotide with alcohol dehydrogenase, *J. Biol. Chem.* **244**, 2465–2475.

Mildvan, A. S., and Weiner, H. (1969b). Interaction of a spin-labeled analogue of nicotinamide-adenine dinucleotide with alcohol dehydrogenase. II. Proton relaxation rate and electron paramagnetic resonance studies of binary and ternary complexes, *Biochemistry* **8**, 552–561.

Moffat, J. K. (1971). Spin labelled hemoglobins: A structural interpretation of EPR spectra based on X-ray analysis, *J. Mol. Biol.* **55**, 135–147.

Morrisett, J. D. (1969). Spin-labeled ribonuclease, Ph.D. Thesis, Univ. of North Carolina, Chapel Hill, North Carolina.

Morrisett, J. D., and Broomfield, C. A. (1971). Active site spin-labeled α-chymotrypsin. Guanidine hydrochloride denaturation studies using electron paramagnetic resonance and circular dichroism, *J. Amer. Chem. Soc.* **93**, 7297–7304.

Morrisett, J. D., and Broomfield, C. A. (1972). A comparative study of spin-labeled serine enzymes: Acetylcholinesterase, trypsin, α-chymotrypsin, elastase, and subtilisin, *J. Biol. Chem.* **247**, 7224–7231.

Morrisett, J. D., and Drott, H. R. (1969). Oxidation of the sulfhydryl and disulfide groups by the nitroxyl radical, *J. Biol. Chem.* **244**, 5083–5084.

Morrisett, J. D., Broomfield, C. A., and Hackley, B. E. (1969). A new spin label specific for the active site of serine enzymes, *J. Biol. Chem.* **244**, 5758–5761.

Morrisett, J. D., Broomfield, C. A., and Hackley, B. A. (1970a). Conformational studies on the active site of acetylcholinesterase by electron paramagnetic resonance, *U.S. Army Sci. Conf. Proc.* **3**, 31–36.

Morrisett, J. D., Broomfield, C. A., and Hackley, B. E. (1970b). Conformational studies on the active site of acetylcholinesterase by electron paramagnetic resonance, *Fed. Proc.* **29**, Abs. #2010.

Morrisett, J. D., Wien, R. W., and McConnell, H. M. (1973). The use of spin labels for measuring distances in biological systems, *Ann. N.Y. Acad. Sci.* **222**, 149–162.

Morrisett, J. D., Pownall, H. J., and Gotto, A. M. (1975). Bovine serum albumin: Study of the fatty acid and steroid binding sites using spin-labeled lipids, *J. Biol. Chem.* **250**, 2487–2494.

Motherwell, W. B., and Roberts, J. S. (1972). A convenient nitroxide radical synthesis, *Chem. Commun.* 328–329.

Mushak, P., and Coleman, J. E. (1972). Electron spin resonance studies of spin-labeled carbonic anhydrase, *J. Biol. Chem.* **247**, 373–380.

Mushak, P., Taylor, J. S., and Coleman, J. E. (1972). Proton hyperfine splitting in the ESR spectra of a stable hydroxynitroxide and its esters, *Biochim. Biophys. Acta* **261**, 332–338.

Ogata, R. T., and McConnell, H. M. (1972). Mechanism of cooperative oxygen binding to hemoglobin, *Proc. Nat. Acad. Sci. U.S.* **69**, 335–339.

Ohnishi, S.-i., and McConnell, H. M. (1965). Interaction of the radical ion of chlorpromazine with deoxyribonucleic acid, *J. Amer. Chem. Soc.* **87**, 2293.

Ovchinnikov, Y. A. (1973). The conformational states of peptides as a key to their biological activity, *In* "Peptides 1972" (H. Hanson and H. -D. Jakubke, eds.), pp. 3–77. North-Holland Publ., Amsterdam and American Elsevier, New York.

Paton, R. M., and Kaiser, E. T. (1970). Detection of a "Michaelis" complex by spin labeling in a model enzyme system, *J. Amer. Chem. Soc.* **92**, 4723–4725.

Phillips, D. C. (1967). The hen egg-white lysozyme molecule, *Proc. Nat. Acad. Sci. U.S.* **57**, 484–495.

Polgar, L., and Bender, M. L. (1967). The reactivity of thiol-subtilisin, an enzyme containing a synthetic functional group, *Biochemistry* **6**, 610–620.

Polyanovsky, O. L., Timofeev, V. P., Shaduri, M. I., Misharin, A. Y., and Volkenstein, M. V. (1973). Conformational changes of aspartate aminotransferases in the region of Cys-45 residue observed by means of spin label, *Biochim. Biophys. Acta* **327**, 57–65.

Reiland, J. (1971). Gel filtration, *Methods Enzymol.* **22**, 287–321.

Roberts, G. C. K., Hannah, J., and Jardetzky, O. (1969). Noncovalent binding of a spin-labeled inhibitor to ribonuclease, *Science* **165**, 504–506.

Rosnati, V., and Palazzo, G. (1945). Synthesis of 5-carboline by the Fischer reaction, *Gazz. Chim. Ital.* **84**, 644–646.

Rozantsev, E. G. (1970). "Free Nitroxyl Radicals." Plenum Press, New York.

Rozantsev, E. G., and Shapiro, A. B. (1964). A new stable radical of the indole class, *Izv. Akad. Nauk. SSSR, Ser. Khim.* **6**, 1123–1125.

Schachman, H. K., and Edelstein, S. J. (1973). Ultracentrifugal studies with absorption optics and a split-beam photoelectric scanner, *Methods Enzymol.* **27**, 3–58.

Shimshick, E. J., and McConnell, H. M. (1972). Rotational correlation time of spin-labeled α-chymotrypsin, *Biochem. Biophys. Res. Commun.* **46**, 321–326.

Shuster, L. (1971). Preparative acrylamide gel electrophoresis: Continuous and disc techniques, *Methods Enzymol.* **22**, 412–433.

Singer, S. J. (1967). Covalent labeling of active sites, *Advan. Protein Chem.* **22**, 1–55.

Smith, I. C. P. (1968). A study of the conformational properties of bovine pancreatic ribonuclease A by electron paramagnetic resonance, *Biochemistry* **7**, 745–757.

Smith, I. C. P. (1971). The spin label method, *In* "Biological Applications of Electron Spin Resonance Spectroscopy" (H. Swartz, J. R. Bolton, and D. Borg, eds.), pp. 483–539. Wiley, New York.

Spackman, D. H., Stein, W. H., and Moore, S. (1958). Automatic recording apparatus for use in the chromatography of amino acids, *Anal. Chem.* **30**, 1190–1206.

Spallholz, J. E., and Piette, L. H. (1972). Interaction of spin-labeled analogs of 1,10-phenanthroline and iodoacetamides with horse liver alcohol dehydrogenase, *Arch. Biochem. Biophys.* **148**, 596–606.

Sparrow, J. T. (1975). An improved polystyrene support for solid phase peptide synthesis. *J. Org. Chem.*, submitted.

Sparrow, J. T., Gotto, A. M., and Morrisett, J. D. (1975). Synthesis of a spin-labeled amino acid which resembles tyrosine and phenylalanine. *J. Org. Chem.*, submitted.

Spector, A. A., and John, K. M. (1968). Effects of free fatty acid on the fluorescence of bovine serum albumin, *Arch. Biochem. Biophys.* **127**, 65–71.

Sperling, R., Burstein, Y., and Steinberg, I. Z. (1969). Selective reduction and mercuration of cystine IV–V in bovine pancreatic ribonuclease, *Biochemistry* **8**, 3810–3820.

Steitz, T. A., Henderson, R., and Blow, D. M. (1969). Structure of cystalline α-chymotrypsin. III. Crystallographic studies of substrates and inhibitors bound to the active site of α-chymotrypsin, *J. Mol. Biol.* **46**, 337–348.

Sternlicht, H., and Wheeler, E. (1967). Preliminary magnetic resonance studies of spin-labeled macromolecules, *In* "Magnetic Resonance in Biological Systems" (A. Ehrenberg, B. G. Malmstrom and T. Vanngard, eds.), pp. 325–334. Pergamon, Oxford.

Stier, A., and Reitz, I. (1971). Radical production in amine oxidation by liver microsomes, *Xenobiotica* **1**, 499–500.

Stier, A., and Sackman, E. (1973). Spin labels as enzyme substrates, *Biochim. Biophys. Acta* **311**, 400–408.

Stone, T. J., Buckman, T., Nordio, P. L., and McConnell, H. M. (1965). Spin-labeled biomolecules, *Proc. Nat. Acad. Sci. U.S.* **54**, 1010–1017.

Stroud, R. M., Kay, L. M., and Dickerson, R. E. (1971). The crystal and molecular structure of DIP-inhibited bovine trypsin at 2.7 Å resolution, *Cold Spring Harbor Symp. Quant. Biol.* **36**, 125–140.

Struve, W. G., and Goldstein, D. J. (1971). A new spin label for serine esterases in synaptic membranes, *Fed. Proc.* **30**, Abs. #1332.

Taylor, J. S., Leigh, J. S., and Cohn, M. (1969). Magnetic resonance studies of spin-labeled creatine kinase system and interaction of two paramagnetic probes, *Proc. Nat. Acad. Sci. U.S.* **64**, 219–226.

Taylor, J. S., Mushak, P., and Coleman, J. E. (1970). Electron spin resonance studies of carbonic anhydrase: Transition metal ions and spin-labeled sulfonamides, *Proc. Nat. Acad. Sci. U.S.* **67**, 1410–1416.

Taylor, J. S., McLaughlin, A., and Cohn, M. (1971). Electron paramagnetic resonance and proton relaxation rate studies of spin-labeled creatine kinase and its complexes, *J. Biol. Chem.* **246**, 6029–6036.

Vesterberg, O. (1971). Isoelectric focusing of proteins, *Methods Enzymol.* **22**, 389–412.

Wagner, T. E., and Hsu, C. -J. (1970). A new method for labeling proteins with a stable paramagnetic nitroxide radical, *Anal. Biochem.* **36**, 1–5.

Watson, H. C., Shotten, D. M., Cox, J. M., and Muirhead, H. (1970). Three dimensional Fourier synthesis of tosyl-elastase at 3.5 Å resolution, *Nature (London)* **225**, 806–816.

Weidman, S. W., Drysdale, G. R., and Mildvan, A. S. (1973). Interaction of a spin-labeled analog of acetyl coenzyme A with citrate synthase. Paramagnetic resonance and proton relaxation rate studies of binary and ternary complexes, *Biochemistry* **12**, 1874–1883.

Weiner, H. (1969). Interaction of a spin-labeled analog of nicotinamide-adenine dinucleotide with alcohol dehydrogenase. I. Synthesis, kinetics, and electron paramagnetic resonance studies, *Biochemistry* **8**, 527–533.

Weinkam, R. J., and Jorgensen, E. C. (1971a). Free radical analogs of histidine, *J. Amer. Chem. Soc.* **93**, 7028–7033.

Weinkam, R. J., and Jorgensen, E. C. (1971b). Angiotensin II analogs. VIII. The use of free radical containing peptides to indicate the conformation of the carboxyl terminal region of angiotensin II, *J. Amer. Chem. Soc.* **93**, 7033–7038.

Wien, R. W., Morrisett, J. D., and McConnell, H. M. (1972). Spin-label-induced nuclear relaxation. Distances between bound saccharides, histidine-15, and tryptophan-123 on lysozyme in solution, *Biochemistry* **11**, 3707–3716.

Wiley, D. C., Evans, D. R., Warren, S. G., McMurray, C. H., Edwards, B. F. P., Franks, W. A., and Lipscomb, W. N. (1971). The 5.5 Å resolution structure of the regulatory enzyme, aspartate transcarbamylase, *Cold Spring Harbor Symp. Quant. Biol.* **36**, 285–290.

Wong, S. S., Quiggle, K., Triplett, C., and Berliner, L. J. (1974). Spin labeled sulfonyl fluorides as active site probes of protease structure. II. Spin label syntheses and enzyme inhibition, *J. Biol. Chem.* **249**, 1678–1682.

Wright, C. S., Alden, R. A., and Kraut, J. (1969). Structure of subtilisin BPN′ at 2.5 Å resolution, *Nature (London)* **221**, 235–242.

Wyckoff, H. W., Tsernoglou, D., Hanson, A. W., Knox, J. R., Lee, B., and Richards, F. M. (1970). The 3-dimensional structure of ribonuclease-S. Interpretation of an electron density map at a nominal resolution of 2 Å, *J. Biol. Chem.* **245**, 305–315.

9

Spin-Label-Induced Nuclear Magnetic Resonance Relaxation Studies of Enzymes

THOMAS R. KRUGH

DEPARTMENT OF CHEMISTRY
UNIVERSITY OF ROCHESTER
ROCHESTER, NEW YORK

I. INTRODUCTION

Nuclear magnetic resonance (NMR) spectroscopy has evolved over the last twenty or so years into an exceedingly powerful and important technique in the area of chemistry. In the last dozen years the application of NMR techniques to biological and biochemical problems has been rapidly expanding. There are a number of excellent texts and reviews covering the theory and practice of nuclear magnetic resonance (e.g., Abragam, 1961; Emsley *et al.*, 1965; Bovey, 1969) and its application to biochemical problems (e.g., Dwek, 1973; Bovey, 1972; Sheard and Bradbury, 1970; Phillips, 1973; Jardetzky and Wade-Jardetzky, 1971; Gurd and Keim, 1973).

The use of NMR to extract structural, kinetic, and electronic information from complex molecular systems has now been well established. In systems of biochemical importance there are usually so many nuclei that the unresolved NMR spectra provide little direct information. In addition, the sensitivity of the instruments has been a limiting factor in studying polymers with high molecular weights (MW $\gtrsim$ 15,000). The advent of Fourier transformation techniques and the development of superconducting magnets have greatly increased the sensitivity and the resolution of NMR instruments (Farrar and Becker, 1971; Phillips, 1973), but even with these developments the direct extraction of structural information from the NMR spectrum of high molecular weight systems will be very difficult.

This chapter will discuss the use of spin-label-induced nuclear relaxation to overcome these difficulties and to provide structural and kinetic information. The introduction of a paramagnetic spin-label molecule induces a change in the longitudinal and transverse relaxation times of the nuclei in its environment. The magnitude of the spin-label-induced relaxation will depend upon the particular molecular dynamics and the structural properties of the system under investigation. The changes in the relaxation rates when a small diamagnetic probe molecule binds to a macromolecule have been successfully utilized for a number of years (e.g., see the review by Jardetzky and Wade-Jardetzky, 1971). The use of spin-label-induced nuclear relaxation is more recent; the first reports appeared in 1967 (Sternlicht and Wheeler, 1967; McConnell, 1967) although it was not until 1969 that useful quantitative information was obtained from these studies (Mildvan and Weiner, 1969a,b; Roberts *et al.*, 1969; Taylor *et al.*, 1969). Paramagnetic metal ions may also induce nuclear spin relaxation and the enhancement in the NMR relaxation rates observed when these metal ions bind to enzymes has been effectively utilized in the study of metalloenzymes. In 1970, Mildvan and Cohn provided an excellent review of the use of paramagnetic probes to study aspects of enzyme mechanisms. More recently, Dwek (1972) has provided a detailed review of the theory of proton relaxation enhancement probes and discussed the applications and limitations to systems containing macromolecules.

The use of spin labels or paramagnetic ions to induce nuclear relaxation provides independent but complementary means of probing macromolecules, especially the active sites of enzymes. This chapter will discuss the theory and methodology of the combined use of NMR and spin-labeling techniques to probe the active sites of enzymes. The reader is referred to Chapter 8 by Morrisett for a discussion of the use of spin labels in the study of the active sites of enzymes using ESR techniques.

II. NUCLEAR SPIN RELAXATION

A. Physical Description

In the presence of a static magnetic field H_0, nuclei such as protons with a spin $I = \frac{1}{2}$ may exist in two different energy levels. The populations of these two energy levels are determined by the Boltzmann distribution, where the energy difference is $\Delta E = h\nu = 2\mu H_0$, where μ is the magnetic moment of the proton. In the absence of an external field ($H_0 = 0$), the spin levels are degenerate and thus have equal populations. Turning on the magnetic field disturbs the equilibrium population, and the spin system responds by approaching the new equilibrium population of the spin levels in an exponential manner with a time constant T_1. The net result is a growth in magnetization aligned along the magnetic field direction (denoted as M_z). The total energy of the spin system decreases, with the net change in energy being transferred to the surrounding spins (called the lattice). Thus T_1 is called the spin–lattice relaxation time. The NMR experiment takes advantage of the fact that a radio-frequency field (of frequency $\nu_0 = 2\mu H_{\text{eff}}/h$) will induce transitions between the nuclear energy levels. If the level of the radio-frequency field H_1 is sufficiently large, then the Boltzmann distribution of the spins is disturbed. This condition is termed "saturation" and will be important in our discussion of the techniques used to measure the relaxation times. The superposition of a radio-frequency field along either the x or y direction for a short time will cause the magnetization to nutate into the xy plane. This magnetization will decay to its equilibrium value ($M_x = M_y = 0$) with a time constant T_2, called the spin–spin relaxation time or the transverse relaxation time.

T_1 relaxation processes are generally dominated by the magnetic dipole–dipole interaction of a nucleus with the fluctuating magnetic dipoles of the surrounding nuclei (the lattice). The magnetic dipoles of the surrounding nuclei fluctuate as a result of thermal motion, and only the component of this motion equal to the Larmor precession frequency will be effective in producing T_1 relaxation. Small molecules generally tumble much faster than the Larmor precession frequency, and thus T_1 relaxation times are usually in the range of seconds (e.g., T_1 of H_2O is $\sim$ 3 sec.). If the small molecule binds to a macromolecule, the rate of rotation decreases, and thus the T_1 relaxation processes become more effective and T_1 is reduced.

There are additional relaxation processes that contribute to T_2 relaxation but not T_1 relaxation. These arise when the motion of the molecule is not sufficiently rapid to average out the effects of neighboring spins. Static magnetic field inhomogeneities also contribute to the dephasing of the magnetization in the xy plane and these must be considered in the measurement of T_2.

B. Theory of Induced Relaxation

The magnitude of dipole–dipole relaxation depends, in part, upon the magnitude of the magnetic moments of the interacting spins. The electron magnetic moment is ~700 times larger than the proton magnetic moment, and thus the relaxation of a proton in the vicinity of an unpaired electron is generally dominated by the electron–nuclear dipole–dipole interaction. The relaxation rates for the direct interaction of a nuclear spin with a paramagnetic ion are well represented by the Solomon (1955)–Bloembergen (1957) equations given in Eqs. (1) and (2) (see also Reuben *et al.*, 1970; Krugh, 1971; Dwek, 1972). ω_I and ω_S are the Larmor precession frequencies of the proton and electron, respectively. Assuming that $\omega_S \gg \omega_I$, we have

$$\frac{1}{T_{1M}} = \frac{2}{15} S(S+1) \frac{\gamma_I^2 g^2 \beta^2}{r^6} \left\{ \frac{3\tau_{C1}}{1 + \omega_I^2 \tau_{C1}^2} + \frac{7\tau_{C2}}{1 + \omega_S^2 \tau_{C2}^2} \right\} + \frac{2}{3} S(S+1) \left(\frac{A}{\hbar} \right)^2 \left\{ \frac{\tau_{C2}}{1 + \omega_S^2 \tau_{C2}^2} \right\} \tag{1}$$

$$\frac{1}{T_{2M}} = \frac{1}{15} S(S+1) \frac{\gamma_I^2 g^2 \beta^2}{r^6} \left\{ 4\tau_{C1} + \frac{3\tau_{C1}}{1 + \omega_I^2 \tau_{C1}^2} + \frac{13\tau_{C2}}{1 + \omega_S^2 \tau_{C2}^2} \right\} + \frac{1}{3} \left(\frac{A}{\hbar} \right)^2 S(S+1) \left\{ \tau_{C1} + \frac{\tau_{C2}}{1 + \omega_S^2 \tau_{C2}^2} \right\} \tag{2}$$

where γ_I is the magnetogyric ratio, β is the Bohr magneton, g is the electronic g factor, r is the separation between the electron spin and the nuclear spin, $A/\hbar$ is the electron nuclear hyperfine coupling constant in hertz, and the τ terms are correlation times. The first terms in both equations arise from dipole–dipole interaction between the electron spin and the nuclear spin, while the second terms arise from scalar coupling of the two spins. In these experiments we will generally be able to neglect the scalar terms, since the unpaired electron and the nuclear spin are on separate molecules. This is not true for proton relaxation enhancements of water molecules where the scalar contribution to T_{2M} may be large. We note that these equations were derived with the assumptions that the unpaired electron spin is a fixed distance from the nuclear spin and that the vector connecting these spins undergoes isotropic rotational diffusion. The correlation times are given by

$$\frac{1}{\tau_{C1}} = \frac{1}{\tau_R} + \frac{1}{\tau_{e1}}; \qquad \frac{1}{\tau_{e1}} = \frac{1}{\tau_{1S}} + \frac{1}{\tau_M} \tag{3}$$

$$\frac{1}{\tau_{C2}} = \frac{1}{\tau_R} + \frac{1}{\tau_{e2}}; \qquad \frac{1}{\tau_{e2}} = \frac{1}{\tau_{2S}} + \frac{1}{\tau_M} \tag{4}$$

where τ_M is the lifetime of a nucleus in the immediate vicinity of the paramagnetic ion, τ_R is the rotational correlation time of the vector between the unpaired electron and the nuclear spin, τ_{1S} is the longitudinal relaxation time of the electron and τ_{2S} is the transverse relaxation time of the electron.

We can investigate the dependence of $1/T_{1M}$ and $1/T_{2M}$ as a function of the correlation time if we assume that either the longitudinal and transverse relaxation times of the electron are equal or that the rotational correlation time $\tau_R \ll \tau_{1e}$ and τ_{2e}. The latter condition will generally be true with the spin-label studies reported in this review. Inspection of Eqs. (1) and (2) shows that the terms that contain τ_{C2} are important when $\omega_S \tau_{C2} \leqq 1$. Since $\omega_S \cong 700\omega_I$, the terms in τ_{C2} are generally important only at short correlation times. It is possible that $\tau_{C2} \ll \tau_{C1}$ if $\tau_{2S} \ll \tau_{1S}$ and if both τ_{1S} and $\tau_{2S} \ll \tau_R$. This situation could occur for a spin-label substrate bound to a large macromolecule that had a very long rotational correlation time ($\tau_R \sim 10^{-7}$ sec). In a later section of this chapter we will discuss the determination of the correlation times. For the present illustration we will assume that there is only one correlation time.† The graph in Fig. 1 shows the frequency dependence of the dipolar term in the Solomon–Bloembergen equation for T_{1M} relaxation (Eq. 1). The correlation time function goes through a maximum when $\omega_I \tau_C = 1$. A comparison between the correlation functions for the dipolar portions of T_{1M} and T_{2M} relaxation is shown in Fig. 2 for experiments at 100 MHz. The $f'(\tau_C)$ has been divided by a factor of two before plotting so that both Eqs. (1) and (2) have the same leading coefficients. This allows a direct comparison between the spin-label-induced relaxation for T_{1M} and T_{2M}. The major difference between the T_{1M} and T_{2M} relaxation is that the linear term in $f'(\tau_C)$ results in an ever increasing value

† Terms of the form $\tau/(1 + \omega^2\tau^2)$ have a maximum value when $\omega\tau = 1$. When the frequency of the NMR experiment is 60 MHz, the maximum contributions of the τ_{C2} terms to $1/T_{1M}$ and $1/T_{2M}$ are 1.4×10^{-11} and 2.6×10^{-11}, respectively. Considering the form of Eqs. (1) and (2), we see that the τ_{C2} terms will be negligible unless $\tau_{C1} \gtrsim 10^{-10}$ sec. Thus, the differentiation between the τ_{C1} and τ_{C2} terms will be important only when the transverse relaxation time of the electron is very short. This situation can be easily detected in the ESR spectrum of the bound radical.

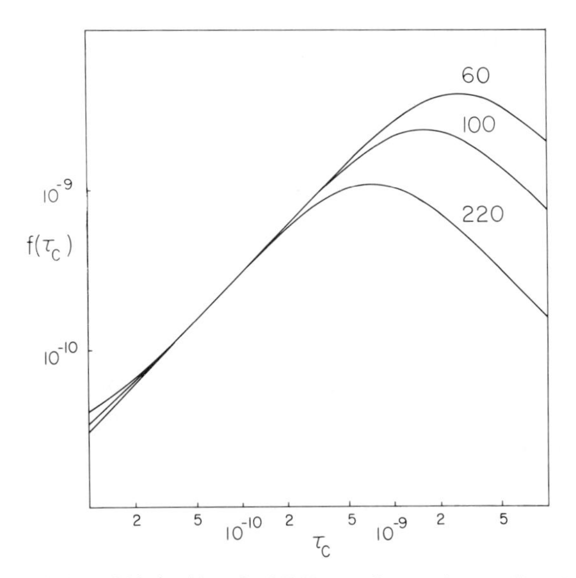

Fig. 1. The variation of $f(\tau_C)$ with τ_C for NMR experiments done at 60, 100, and 220 MHz. The function $f(\tau_C)$ is defined in the legend to Fig. 2.

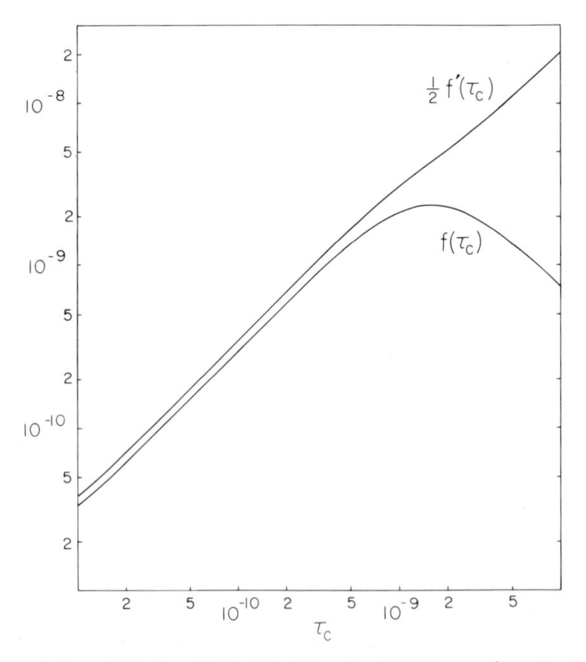

Fig. 2. The variation of $f(\tau_C)$ and $f'(\tau_C)/2$ with τ_C for NMR experiments at 100 MHz.

$$f(\tau_C) = \left\{\frac{3\tau_C}{1 + \omega_I^2\tau_C^2} + \frac{7\tau_C}{1 + \omega_S^2\tau_C^2}\right\}, \qquad f'(\tau_C) = \left\{4\tau_C + \frac{3\tau_C}{1 + \omega_I^2\tau_C^2} + \frac{13\tau_C}{1 + \omega_S^2\tau_C^2}\right\}$$

for $f'(\tau_C)$ as the correlation time increases, while $f(\tau_C)$ goes through a maximum value. The value of $f'(\tau_C)$ is always larger than $f(\tau_C)$ except when $\omega_S^2 \tau_C^2 \ll 1$. In this region, T_{1M} is equal to T_{2M}. In Fig. 2 we note that in the region where $\omega_I \tau_C < 1 < \omega_S \tau_C$ the ratio of T_{1M}/T_{2M} is relatively constant and approximately equal to $\frac{7}{6}$.

Small molecules in aqueous solution generally have rotational correlation times of the order of 10^{-11} sec. When the small molecule binds to a macromolecule, the rotational correlation time is increased and the graphs in Figs. 1 and 2 show that the increased correlation time will lead to an enhancement in the relaxation rates. Note that the correlation time discussed here is for the vector between the unpaired electron spin on the nitroxide radical and the nuclear spin. The basis for the structural measurements lies in the observation that if one can experimentally measure $1/T_{1M}$ and, or, $1/T_{2M}$ and either determine or estimate $f(\tau_C)$, then Eqs. (1) and (2) can be used to calculate internuclear separations and thus map out the active sites of enzymes. The measurement or the determination of these parameters is not always straightforward. If the value of r (the electron–nuclear separation) is time dependent due to rotational motion or the presence of different conformers, then an average value of r will be determined. This value will be heavily weighted toward the value of closest approach due to the r^{-6} dependence in the relaxation rate equations. This same consideration would hold if a spin label were interacting with identical substrates bound in two distinct binding sites or if two or more spin-label molecules were interacting with the same substrate molecule. For these reasons, it is important to determine the stoichiometry of substrate binding as well as the number of sites and the location of the sites on the enzyme that have been spin labeled.

C. Types of Experiments

Spin-label-induced nuclear relaxation experiments can be divided into two general areas: (1) those in which the nitroxide radical molecule is covalently attached to the macromolecule; and (2) those in which the nitroxide radical is noncovalently bound to a particular site (or sites) on the macromolecule.† The two types of experiments may be further divided into groups as follows:

† In an effort to provide a consistent nomenclature in this book, we will hereafter refer to a covalently attached nitroxide molecule as a *spin label* and a noncovalently attached nitroxide radical as a *spin probe*. However, the reader will find that the term spin label is commonly used in the literature for both covalently and noncovalently bound nitroxides. We will also hereafter refer to the binding of spin labels (or spin probes) to enzymes instead of the more general term macromolecule, since most of the spin-label-induced nuclear relaxation experiments to date have involved enzymes.

1. Covalently Attached Spin Labels

(a) Interaction with substrates. The nitroxide radical may affect the relaxation of substrates bound to the enzyme.

(b) Interaction with enzyme groups. The nitroxide radical will enhance the relaxation rate of enzyme groups in the vicinity of the bound radical. These experiments will, of course, be limited to the small group of enzymes for which NMR signals may be observed and assigned to particular groups on the enzyme.

(c) Interaction with water protons. The paramagnetic contribution to the water relaxation may be monitored as substrates, coenzymes, inhibitors, etc. are added to the solution.

2. Noncovalently Bound Spin Probes

(a) Interaction with substrates. The spin probe and the substrate (or other noncovalently bound molecules) interact when both are bound to the enzyme. This is probably the most useful form of the spin-labeling experiments in that the nitroxide radical is incorporated into a substrate analog which then binds to a *particular* site on the enzyme.

(b) Interaction with enzyme groups. The spin probe will enhance the relaxation of the enzyme nuclei in the vicinity of the binding site. As discussed above these experiments are somewhat limited.

(c) Interaction with water protons. The enhancement in the water relaxation rate when the spin probe binds to the enzyme may be monitored as a function of the concentration of a second molecule that binds to the enzyme. A change in the proton relaxation enhancement indicates that the second substrate is affecting the local environment of the spin probe molecule.

The schematic illustration of these interactions is shown in Figs. 3 and 4. The multiple equilibria involved in the use of spin probes (Fig. 4) complicate the interpretation of the data. For simplicity in the illustrations we have indicated only the direct interactions between the electron spin and the nuclear spin when both are bound to the enzyme, since these direct interactions are used with Eqs. (1) and (2) to calculate structural parameters. The experimentally measured relaxation rates $1/T_1$ and $1/T_2$ contain both diamagnetic and paramagnetic contributions from which $1/T_{1M}$ and $1/T_{2M}$ must be extracted. This will be the subject of the following section.

D. Determination of $1/T_{1M}$ and $1/T_{2M}$

The experimental details for measuring T_1 and T_2 will be discussed in Section III of this chapter. For the present we wish to illustrate the extraction of $1/T_{1M}$ and $1/T_{2M}$ from the experimentally observed relaxation rates, $(1/T_1)_{obs}$ and $(1/T_2)_{obs}$. In the absence of chemical exchange, one would

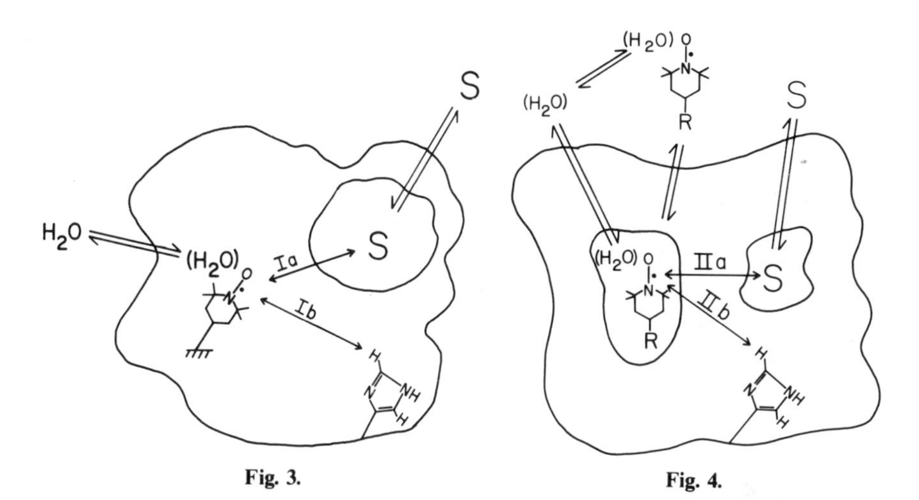

Fig. 3. Fig. 4.

Fig. 3. Schematic illustration of the possible experiments using a spin-labeled enzyme. In this illustration, the six-membered nitroxide ring has been covalently attached to the enzyme and induces nuclear relaxation in the bound substrate, the C-2 proton of histidine, and the water molecules.

Fig. 4. Schematic illustration of spin-probe experiments. In this illustration, the spin probe binds at one enzyme binding site and induces nuclear relaxation in the bound substrate, the C-2 proton of histidine, and the water molecules.

expect to observe separate relaxation rates for molecules that exist in different chemical environments. The NMR signals from these two molecules may exhibit individual resonances and in this case the measurement of the relaxation rates for the nuclei in the two environments would be straightforward. If the molecule undergoes chemical exchange between the two environments, the relaxation rates may be dramatically affected. In fact, the exchange process itself becomes a relaxation mechanism.

It is possible to modify the Bloch equations for nuclear magnetic resonance to include the effect of chemical exchange of nuclei between paramagnetic and diamagnetic environments (McConnell, 1958; Carrington and McLachlan, 1967; Wien *et al.*, 1972). The practical use of spin-label-induced relaxation to determine structural parameters will be limited to those cases in which the nuclei of interest are in the fast-exchange limit. Wien *et al.* (1972) have shown that the term "fast exchange" refers to conditions for which

$$1/\tau \gg (f_A/T_{2A}), (f_M/T_{2M}) \tag{5}$$

$$(f_A/T_{2A}) + (f_M/T_{2M}) \gg \tau f_A^2 f_M^2 (\omega_A - \omega_M)^2 \tag{6}$$

where $\tau = \tau_A + \tau_M$ (τ_M is the lifetime of the nucleus in the paramagnetic environment and τ_A is the lifetime of the nucleus in the diamagnetic environment; f_A is the fraction of the time the nuclear spin is in the paramagnetic environment; $f_M = 1 - f_A$; ω_A is the angular resonance frequency of the nuclear spin in the diamagnetic environment; ω_M is the angular resonance frequency in the paramagnetic environment. Physically, the fast exchange conditions require that the average lifetime of a nucleus in the diamagnetic environment τ_A be much shorter than T_{2A} and that the average lifetime of a nucleus in the paramagnetic environment τ_M be much shorter than T_{2M}. In addition, the conditions in Eq. (6) insure that the exchange rate does not contribute to the linewidth. If these conditions are satisfied for transverse relaxation, then the spin–lattice relaxation T_1 will also be in the fast-exchange limit, and the observed relaxation rates will be

$$\left(\frac{1}{T_1}\right)_{\text{obs}} = \frac{f_A}{T_{1A}} + \frac{f_M}{T_{1M}} \tag{7}$$

$$\left(\frac{1}{T_2}\right)_{\text{obs}} = \frac{f_A}{T_{2A}} + \frac{f_M}{T_{2M}} \tag{8}$$

The stringent requirements for fast-exchange conditions given in Eqs. (5) and (6) must be met if the populations of the A and M sites are about equal. In the limit where $f_A \gg f_M$, the assumptions of Swift and Connick (1962) and Luz and Meiboom (1964) that $\tau_A \gg \tau_M$ leads to the condition of fast exchange when $\tau_M \ll T_{2M}$, T_{1M}. This definition of fast exchange is usually appropriate for paramagnetic metal ion or spin-label-induced relaxation of water protons or substrate nuclei if the substrate concentration is much larger than the spin label or paramagnetic metal ion concentration. Most of the experiments described in this chapter fall into this category. Eq. (5) can be rewritten to show that not only must $\tau_M \ll T_{2M}$, T_{1M}, but the inequality $\tau_A \ll \mathrm{T}_{2A}$, T_{1A} must also hold for fast exchange in those experiments when the inequality $f_A \gg f_M$ is not appropriate.

Each of the relaxation rates T_{1A}, T_{2A}, T_{1M}, and T_{2M} given in Eqs. (7) and (8) can be thought of as having both paramagnetic and diamagnetic contributions. Thus we write that

$$\left(\frac{1}{T_1}\right)_{\text{obs}} = \left(\frac{f_M}{T_{1M}}\right)_{\text{E\dot{R}S}} + \left(\frac{f_A}{T_{1A}}\right)_{\text{E\dot{R}}} + \left(\frac{f_M}{T_{1M}}\right)_{\text{ES}} + \left(\frac{f_A}{T_{1A}}\right)_{\text{S}} \tag{9}$$

$$\left(\frac{1}{T_2}\right)_{\text{obs}} = \left(\frac{f_M}{T_{2M}}\right)_{\text{E\dot{R}S}} + \left(\frac{f_A}{T_{2A}}\right)_{\text{E\dot{R}}} + \left(\frac{f_M}{T_{2M}}\right)_{\text{ES}} + \left(\frac{f_A}{T_{2A}}\right)_{\text{S}} \tag{10}$$

where the first term on the right hand side of Eqs. (9) and (10) is the contribution of the spin-label- (or spin-probe-) induced relaxation of the nuclei in the enzyme–spin label–substrate complex (EṘS). These are the values that we wish to use in Eqs. (1) and (2) to determine structural parameters. The second term on the right-hand side of these equations represents the paramagnetic contribution to the nuclear relaxation rates when the nuclei of interest are not in the enzyme–spin label complex ($E\dot{R}$). This is the outer-sphere contribution to the relaxation rate referred to by Mildvan and Cohn (1970) ($1/T_{1(os)}$) and by Dwek (1972) ($1/T''_{1A}$). The last two terms on the right hand side of Eqs. (9) and (10) represent all the other diamagnetic contributions to the nuclear relaxation rates. The binding of the substrate S to the enzyme E to form the enzyme–substrate complex ES may provide a significant contribution to the diamagnetic relaxation [given as $(1/T_{1M})_{ES}$ and $(1/T_{2M})_{ES}$ in Eqs. (9) and (10)], which has been used extensively to probe the active sites of enzymes and to measure enzyme substrate binding constants and exchange rates (e.g., see Jardetzky, 1964; Sykes, 1969; Bovey, 1972).

The form of Eqs. (9) and (10) is different from that used by Mildvan and Cohn (1970) and Dwek (1972) in their review articles that dealt primarily with paramagnetic-metal-ion-induced relaxation and my work on DNA polymerase (Krugh, 1971) since in these cases $f_A \gg f_M$ and the approximations of Swift and Connick (1962) and Luz and Meiboom (1964) were appropriate. The correspondence among the different notations is illustrated below for T_1 relaxation. Equation (9) has been rearranged to Eq. (11) for easy comparison with Eq. (12).

$$\left(\frac{f_M}{T_{1M}}\right)_{E\dot{R}S} = \left(\frac{1}{T_1}\right)_{obs} - \underbrace{\left(\frac{f_M}{T_{1M}}\right)_{ES} - \left(\frac{f_A}{T_{1A}}\right)_{S}} - \left(\frac{f_A}{T_{1A}}\right)_{E\dot{R}} \tag{11}$$

$$\downarrow\downarrow \qquad \downarrow\downarrow \qquad \qquad \qquad \downarrow\downarrow$$

$$\frac{1}{T_{1p}} = \frac{pq}{T_{1M} + \tau_M} = \left(\frac{1}{T_1}\right)_{obs} - \frac{1}{T_{1(o)}} - \frac{1}{T_{1(os)}} \tag{12}$$

Equation (12) is the notation used in the review article by Mildvan and Cohn (1970). Similar notation was used by Dwek (1972) with the exceptions that $1/T_{1(os)}$ was changed to $1/T''_{1A}$ and $1/T_{1(o)}$ became $1/T_{1A(o)}$. In the paper by Krugh (1971), $1/T_{1(obs)}$ was changed to $1/T_{1\,bound}$ and $1/T_{1(o)}$ became $1/T_{1\,free}$. In the above equations, $p = f_M$ and q is the coordination number of the paramagnetic molecule. Under the conditions used to derive Eq. (12), $f_A = 1 - p \cong 1$. We note that Eq. (11) was derived under the assumptions of fast chemical exchange, while Eq. (12) was derived under the assumption that $f_A \gg f_M$. Thus Eq. (12) contains a τ_M term, while Eq. (11) has already assumed that $\tau_M \ll T_{1M}$. One can thus see a subtle difference in the two types of experiments, since it is theoretically possible to determine τ_M from experiments in which Eq. (12) is appropriate by simply varying the temperature to influence the relative magnitudes of T_{1M} (or T_{2M}) and τ_M. In those experiments where f_A is not much larger than f_M, the possible extraction of kinetic parameters is not as straightforward [e.g., see the discussion by Wien *et al.* (1972)]. Analogous expressions can be written for the transverse relaxation processes, providing that the system is in the fast-exchange limit as given by Eqs. (5) and (6). The measurement of each of the terms in Eqs. (9) and (10) and the necessary control experiments will be discussed for each type of experiment outlined in Section II.C. Before completing this section, we would like to add that there are two other terms that should be added to Eqs. (9) and (10) for spin-probe experiments. These terms arise because the spin-probe molecule is noncovalently bound to the enzyme and thus some fraction of $\dot{R}$ will be free in solution. This fraction of $\dot{R}$ will contribute to both the relaxation of free and bound substrate, and thus Eq. (9) should contain a term of the form $(1/T_{1A})_{\dot{R}\cdots S}$ for the paramagnetic contribution of the free nitroxide to the relaxation rate of the free substrate as well as a term for the paramagnetic contribution of the free nitroxide to the relaxation rate of the enzyme-bound substrate, $(1/T_{1M})_{ES\cdots\dot{R}}$. Similar terms should appear in Eq. (10) for transverse relaxation. These terms may be important in some experiments, especially when high concentrations of radicals are used.

E. Frequency and Temperature Dependence of $1/T_{1M}$ and $1/T_{2M}$

We have already seen in Figs. 1 and 2 that $f(\tau_C)$ is a function of the Larmor frequency, and thus $1/T_{1M}$ is a function of the frequency of the

TABLE I

TEMPERATURE AND FREQUENCY DEPENDENCE OF $1/T_{1M}$ AND $1/T_{2M}$[a]

	$\omega_I^2\tau_C^2 < 1$ τ_C dominated by			$\omega_I^2\tau_C^2 > 1$ τ_C dominated by		
	τ_R	τ_S	τ_M	τ_R	τ_S	τ_M
$d(1/T_{1M})/dT$	$-$	$\pm$	$-$	$+$	$\mp$	$+$
$d(1/T_{1M})/d\omega$	0[c]	0, $\pm$[b]	0[c]	$-$	$-$[b]	$-$
$d(1/T_{2M})/dT$	$-$	$\pm$	$-$	$-$	$\pm$	$-$
$d(1/T_{2M})/d\omega$	0[c]	0, $\pm$[b]	0[c]	0[d]	0, $\pm$[b]	0[d]
T_{1M}/T_{2M}		7/6[c]			>1	

[a] Assuming fast chemical exchange. A positive (+) sign means that the given parameter (e.g., $1/T_{1M}$) will increase in magnitude as the temperature or frequency increases.

[b] Will depend upon the frequency dependence and magnitude of τ_{1S} and τ_{2S}.

[c] Provided $\omega_S^2\tau_C^2 > 1$.

[d] There will be a small frequency dependence when $\omega_I\tau_C \approx 1$.

experiment. If $\omega_I \tau_C \geqq 1$ the value of $f(\tau_C)$ will be strongly dependent upon the Larmor frequency of the experiments, while in the region where $\omega_I \tau_C < 1 < \omega_S \tau_C$ the value of $f(\tau_C)$ is frequency independent. Thus the frequency dependence of $1/T_{1M}$ can provide useful information about the value of τ_C. The form of $f'(\tau_C)$ shows that the transverse relaxation is relatively independent of the Larmor frequency except in the regions where either $\omega_I \tau_C \cong 1$ or $\omega_S \tau_C \cong 1$.

The temperature dependence of the spin-label-induced nuclear relaxation can provide especially important information. A change in temperature is equivalent to a change in τ_C, and thus the temperature dependence of $1/T_{1M}$ and $1/T_{2M}$ will depend upon whether τ_R, τ_M, or τ_S determines the value of τ_C [Eqs. (3) and (4)]. For spin-label- and spin-probe-induced relaxation, the τ_{1S} and τ_{2S} terms will generally be much longer than τ_R except under unusual circumstances such as solutions with very large concentrations of spin labels ($\geqq 1$ mM).† The value of τ_S may be determined from the linewidth of the EPR spectra of the enzyme-bound nitroxide radical (e.g., see Chapters 2, 3, and 8, or Poole and Farach, 1971). The temperature dependence of τ_S could also be determined from the EPR spectra. The temperature dependence of τ_R is given by

$$\tau_R = \tau_R^0 \exp(E_R/RT)$$

in which it can be seen that the tumbling rate $(1/\tau_R)$ increases as the temperature is raised. The exchange rate, $1/\tau_M$, always increases as the temperature is raised. With this information and the use of Figs. 1 and 2, we may calculate the temperature dependence and the frequency dependence of $1/T_{1M}$ and $1/T_{2M}$ assuming fast chemical exchange (Table I). The temperature dependence of the spin-label-induced relaxation $1/T_{1p}$ can help to determine if the system is in the fast-exchange region. If $1/T_{1p}$ decreases with increasing temperature, then the system must be in the fast-exchange region for T_1 relaxation ($\tau_M < T_{1M}$), since we would expect to observe the opposite behavior if

† The relaxation times of the unpaired electron in a typical nitroxide radical bound to an enzyme are probably of the order of 10^{-7} sec in dilute solutions. For example, an inorganic nitroxide known as Fremy's salt has an electronic relaxation time of 3.4×10^{-7} sec (Kooser *et al.*, 1969). A power saturation experiment on the EPR spectra of a covalently bound spin label on lysozyme showed that τ_S was approximately 3.5×10^{-7} sec (Wien *et al.*, 1972). However, when the concentration of the nitroxide is $\geqq 1 \times 10^{-3}$ M, spin exchange can effectively shorten the value of τ_S. At very high concentrations of nitroxide radicals, τ_S can be short enough to produce exchange-narrowing of the lines. A dramatic illustration of this effect comes from the work of Kreilick and coworkers (Kreilick, 1973), who have used nitroxide radicals as solvents and have been able to record the NMR spectra of the protons and carbon nuclei of nitroxide radicals. They have also recently determined the electronic relaxation times of nitroxide radicals as a function of concentration, temperature, and magnetic field strength (R. W. Kreilick and M. S. Davis, personal communication).

the value of $1/T_{1p}$ was limited by the exchange rate ($\tau_M \geqq T_{1M}$). Similarly, the temperature dependence of the transverse relaxation rates can help to determine if the fast-exchange approximations are appropriate for T_2 relaxation. However, it is a little more difficult to be certain that the transverse relaxation is in the fast exchange limit without a knowledge of the "off" rate constants for binding, because the chemical exchange may contribute to $(1/T_2)_{obs}$ [Eq. (6)]. It is possible for the spin–lattice relaxation to be in the fast-exchange limit while the transverse relaxation is not in the fast-exchange limit (i.e., $T_{2M} \leqq \tau_M \ll T_{1M}$). This situation could easily occur if (a) $\omega_I \tau_C \geqq 1$ or (b) if the term $(\omega_A - \omega_M)^2$ were unusually large due to a large diamagnetic shift upon binding or a pseudocontact interaction with the unpaired electron. If both $1/T_{1p}$ and $1/T_{2p}$ increase as the temperature is raised, then the system is certainly not in the fast-exchange limit. It is possible that $1/T_{1p}$ (or $1/T_{1M}$) would increase as the temperature is raised provided that $\omega_I^2 \tau_C^2 > 1$. However, $1/T_{2p}$ (or $1/T_{2M}$) would decrease at the same time. Under these same conditions, the value of $1/T_{1p}$ would decrease as the frequency of the experiment was increased (e.g., from 60 to 100 MHz) while $1/T_{2p}$ would be frequency independent. If the value of τ_C is determined by τ_S (i.e., $\tau_S \ll \tau_R$ and τ_M), then the exact nature of the frequency and temperature dependence of the relaxation rates is less certain. However, the frequency and temperature dependence of τ_S could be determined by EPR spectroscopy and compared with the NMR results. If the system is not in the fast chemical exchange limit, then the temperature dependence of $1/T_{1p}$ and $1/T_{2p}$ can be used to extract the value of τ_M^{-1}, the "off" rate constant, and to investigate the thermodynamics of binding.

F. Determination of the Correlation Time τ_C

The determination of structural parameters from the spin-label-induced nuclear relaxation measurements requires a knowledge of the value of $f(\tau_C)$ and or $f'(\tau_C)$. These correlation functions can be easily calculated if the values of τ_R, τ_S, and τ_M are known. The relaxation times of the unpaired electron, τ_{1S} and τ_{2S}, and the rotational correlation time of the electron spin may be determined from the EPR spectra (see the discussion in Chapter 8 by Morrisett). However, the value of τ_R that is appropriate to the present experiments is the rotational correlation time of the vector between the unpaired electron spin and the nucleus of interest. This value may be significantly different from the rotational correlation time of the electron spin or the rotational correlation time estimated from the molecular weight of the enzyme. The best way to determine τ_C is to measure both $1/T_{1M}$ and $1/T_{2M}$ at several different frequencies. The ratio between $1/T_{1M}$ and $1/T_{2M}$ is given by

$$\frac{\dfrac{1}{T_{1M}}}{\dfrac{1}{T_{2M}}} = \frac{2\left\{\dfrac{3\tau_{C1}}{1+\omega_I^2\tau_{C1}^2} + \dfrac{7\tau_{C2}}{1+\omega_S^2\tau_{C2}^2}\right\}}{\left\{4\tau_{C1} + \dfrac{3\tau_{C1}}{1+\omega_I^2\tau_{C1}^2} + \dfrac{13\tau_{C2}}{1+\omega_S^2\tau_{C2}^2}\right\}} \tag{13}$$

assuming there is no contribution to the relaxation from scalar coupling. Once again we are formally faced with the question of two correlation times, τ_{C1} and τ_{C2}, but EPR spectroscopy could be used to determine both τ_{e1} and τ_{e2} in those unusual circumstances where this may present a problem. The value of τ_{C2} will always be less than or equal to τ_{C1} and thus these terms will only be important at very short correlation times (when $\omega_S\tau_{C2} \leqq 1$). From the form of Eq. (13), it is clear that the ratio method of determining τ_C is only useful in the region where $\omega_I\tau_C \geqq 1$ since the ratio of $(1/T_{1M})/(1/T_{2M})$ is constant in the region where $\omega_I\tau_C < 1 < \omega_S\tau_C$. To determine the correlation time τ_C, one calculates the ratio of the experimentally observed relaxation rates $(1/T_{1M})/(1/T_{2M})$ and then solves Eq. (13) for the value of τ_{C1}. A graphical solution is probably easiest, and in Fig. 5 we have plotted the ratio of the spin-label-induced nuclear relaxation rates against the value of the correlation time at three different frequencies. These frequencies correspond to the Larmor precession frequencies of protons (100 MHz), carbon-13 (25 MHz) and phosphorus-31 (40 MHz) nuclei in a 23.5 kG magnetic field. The correlation time τ_C can be read off the graph in Fig. 5 for a given value of T_{1M}/T_{2M}, provided the system is in the fast chemical exchange limit for both T_1 and T_2 relaxation. For example, if the ratio of T_{1M}/T_{2M} at 100 MHz were 18, then from Fig. 5 the value of τ_C would be 8×10^{-9} sec. If the

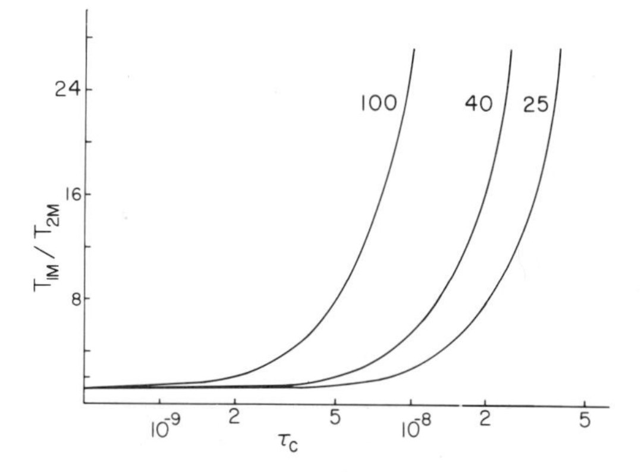

Fig. 5. Plot of the ratio of T_{1M}/T_{2M} versus τ_C for NMR experiments done at 100, 40, and 25 MHz.

experiments had been performed at 40 MHz or 25 MHz, the ratio of T_{1M}/T_{2M} would have been 3.7 and 2.1, respectively.

The frequency dependence of $1/T_{1M}$ is also a convenient means of determining the value of τ_C. The ratio $T_1(\omega_1)/T_1(\omega_2)$ is plotted against the correlation time for three separate experiments in Fig. 6. Thus, the value of τ_C can be determined by measuring T_{1M} at two or more frequencies and then reading the value of τ_C that corresponds to the experimental value(s) of $T_{1M}(\omega_1)/T_{1M}(\omega_2)$ from the graph in Fig. 6. For example, if the ratio of T_{1M} (220 MHz)/T_{1M} (100 MHz) was found to be 4.5, then the correlation time τ_C is 5×10^{-9} sec. For the same system we would expect to observe T_{1M} (100 MHz)/T_{1M} (60 MHz) equal to 2.4 and T_{1M} (220 MHz)/T_{1M} (60 MHz) = 10.6. From Fig. 6 we note that this method of determining τ_C is most sensitive when ω_1 and ω_2 are very different. Sensitivity or the availability of instrumentation will often determine the limits on the range of ω_1 and ω_2.

In summary, the most certain way to determine the value of $f(\tau_C)$ and $f'(\tau_C)$ is to measure the temperature dependence and the frequency dependence of both $1/T_{1M}$ and $1/T_{2M}$. In some cases it may be possible to determine τ_C by a knowledge of τ_R, τ_S, and τ_M obtained from other techniques. For example, if τ_R were much smaller than either τ_S or τ_M, then τ_C would equal τ_R. There is a necessary caveat here because the value of τ_R is for the vector between the nuclear spin and the electron spin. This vector could have a different correlation time than the overall rotational correlation time of the enzyme due to internal motion or rotation of either the molecule containing the nucleus of interest or the spin-label molecule when they are bound to the enzyme. In those cases where the detailed relaxation measurements are not possible, the estimation of τ_C can be very useful.

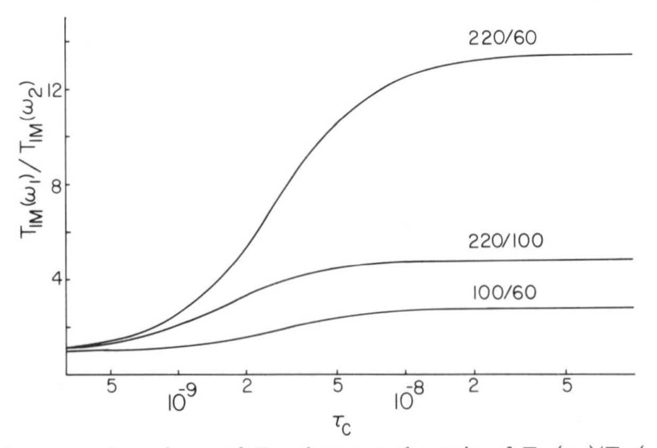

Fig. 6. Frequency dependence of T_{1M} shown as the ratio of $T_{1M}(\omega_1)/T_{1M}(\omega_2)$ for three different frequency ranges.

G. Distance Measurements

Once the values of T_{1M} or T_{2M} and τ_C are known, the calculation of the separation between the unpaired electron and the nuclei of interest is straightforward with the use of Eqs. (1) and (2) and the appropriate constants. These equations are of the form

$$\frac{1}{T_{1M}} = \frac{2C}{r^6} f(\tau_C) + \frac{1}{2}\left(\frac{A}{\hbar}\right)^2 \left\{\frac{\tau_{C2}}{1 + \omega_S^2 \tau_{C2}^2}\right\} \tag{14}$$

$$\frac{1}{T_{2M}} = \frac{C}{r^6} f'(\tau_C) + \frac{1}{4}\left(\frac{A}{\hbar}\right)^2 \left\{\tau_{C1} + \frac{\tau_{C2}}{1 + \omega_S^2 \tau_{C2}^2}\right\} \tag{15}$$

where the second term in each equation arises from the scalar interaction. Since the distance calculations imply that the molecules are separated, the scalar terms will be either negligible or zero. The only exception to this rule might be proton relaxation enhancements of water molecules in the "coordination sphere" of nitroxide radicals, but in this case the value of r is not particularly interesting. Rearranging Eqs. (14) and (15) with the assumption that the hyperfine coupling is negligible, we get

$$r = C_{T_1}(f(\tau_C)T_{1M})^{1/6} \tag{16}$$

$$r = C_{T_2}(f'(\tau_C)T_{2M})^{1/6} \tag{17}$$

where $f(\tau_C)$ and $f'(\tau_C)$ are defined in the caption to Fig. 2. The values of the constants C_{T_1} and C_{T_2} (for T_{1M} and T_{2M} in seconds and r in angstroms) are listed in Table II for a variety of experiments. We note in the above equations that the value of r depends upon the sixth root of the product of $f(\tau_C)$ and T_{1M} or $f'(\tau_C)$ and T_{2M}, and thus rather large uncertainties in the values of $f(\tau_C)$ and T_{1M} or $f'(\tau_C)$ and T_{2M} will not appreciably affect the value of r. In fact, if the product $f(\tau_C)T_{1M}$ or $f'(\tau_C)T_{2M}$ is in error by a factor of two, the calculated value of r would be in error by $2^{1/6} = 1.12$ or 12%. Even an error

TABLE II

CONSTANTS TO BE USED WITH EQS. (16) AND (17)

Nuclei observed	Frequency (MHz)	C_{T_1}	C_{T_2}	$\omega_I(\times 10^{-8})$ (rad sec^{-1})	$\omega_S(\times 10^{-11})$ (rad sec^{-1})
Protons	220	540	480	13.8	9.12
Protons	100	540	480	6.28	4.14
Protons	60	540	480	3.77	2.49
Fluorine-19	94.08	529	471	5.91	4.14
Phosphorus-31	40.5	400	356	2.54	4.14
Carbon-13	25.15	341	304	1.58	4.14

of a factor of ten in the product of $f(\tau_C)T_{1M}$ or $f'(\tau_C)T_{2M}$ would produce only a 47% error in the value of r. Thus, the sixth root dependence allows the determination of rather precise separations. In the absence of any knowledge of the correlation time, it is still possible to determine relative distances between an unpaired electron and two or more nuclei. Conversely, the relative distances will be free of any error in the estimation of τ_C, providing that both (or all) nuclei have the same value of τ_C.

If the system is not in the fast-exchange limit it is still possible to determine a lower limit on the value of T_{1M}, which provides an upper limit on the value of r. The same holds true for T_{2M}, except that a little more caution is required because of the possibility of exchange broadening. Thus, even in the absence of fast exchange it may be possible to obtain geometric information. It is interesting to ask what limits there are on the value of r that may be determined? Theoretically, of course, there is no limit, provided that we can determine the value of T_{1M}. However, as r decreases the magnitude of T_{1M} and T_{2M} become so small that eventually T_{1M} and T_{2M} are less than τ_M and the system is no longer in the fast exchange limit necessary for the precise distance calculations. The shortest distance measurable is thus limited by the exchange rate of the system under study. The exchange rates of substrates vary by several orders of magnitude, but for the sake of argument, we will consider an enzyme–substrate system with $\tau_M = 1 \times 10^{-6}$ sec. This would limit the measurement of T_{1M} to $T_{1M} > 1 \times 10^{-5}$ sec. Assuming that the experiments were done at 100 MHz and that $\tau_C = 1.5 \times 10^{-9}$ sec, then from Eq. (16) we calculate that the minimum distance measurable is 3 Å. Under the same conditions, the minimum value of r that could be determined by T_2 relaxation measurements is 3.2 Å. Three angstroms is probably the smallest practical value of r that could be determined, although there is no absolute minimum for a favorable system. Since we are discussing spin-label-induced relaxation, it is not likely that one would be interested in precise measurements from induced relaxation at separations much less than 3 Å, since the geometry of any molecule that close to the spin-label group is most likely going to be distorted from its normal geometry by the presence of the spin-label moiety. At large values of r (> 20 Å) the value of T_{1M} and T_{2M} become so long that they become experimentally difficult to determine. To illustrate this point, let $r = 20$ Å, $\tau_C = 1.5 \times 10^{-9}$ sec, and $\tau_M = 1 \times 10^{-6}$ sec as above. From Eqs. (16) and (17), we calculate that T_{1M} and T_{2M} at 100 MHz would be 1 and 0.6 seconds, respectively. From Eqs. (7) and (8), we see that these values of T_{1M} and T_{2M} would significantly affect the observed values of T_1 and T_2 only if f_A and f_M are of the same order of magnitude. Under very favorable conditions (e.g., if $f_M > f_A$ or if $f_A \cong 0$), the upper limit on r may be 30 Å, but certainly not much more than this. Spin-label-induced nuclear relaxation experiments can thus provide an important window to look at the geometry of enzymes and enzyme–substrate complexes within 20–30 Å of the location of the nitroxide. Placing the window (the spin label) in a position that affords the best view and produces the smallest environmental impact is an important aspect of spin labeling.

III. MEASUREMENT OF RELAXATION TIMES

The measurement of the spin–lattice and transverse relaxation times may be performed by using either pulsed NMR techniques or continuous wave (CW) NMR techniques. Until the recent development of pulsed Fourier transformation NMR (FT-NMR), the use of pulsed techniques was generally limited to very concentrated solutions. Solvent water protons at 111 *M* satisfy this requirement, and as mentioned previously, water relaxation measurements have been extensively used in conjunction with paramagnetic metal ions to investigate enzyme mechanisms (e.g., see Mildvan and Cohn, 1970). Relaxation measurements on systems containing many nuclei have generally utilized continuous-wave techniques because of the selectivity and sensitivity of the experiments. FT-NMR has drastically changed this situation, and as a result of recent instrumental developments, relaxation measurements will become a common and valuable diagnostic tool. The reader is referred to the book by Farrar and Becker (1971) for an introduction to the theory and methods of pulse and Fourier transform NMR. The application of FT-NMR to the important area of carbon-13 NMR is thoroughly covered in texts by Stothers (1972) and Levy and Nelson (1972). The present section will discuss the relaxation measurements as they apply to spin-label-induced nuclear relaxation.

A. Pulsed Techniques

The use of pulsed, or transient, techniques for measuring relaxation times dates back to the early days in the development of NMR (Bloembergen *et al.*, 1948; Torrey, 1949; and Hahn, 1950). The experimental approach is generally straightforward. The nuclei are allowed to come to equilibrium in the magnetic field and then a short pulse of radio-frequency energy is applied to rotate or nutate the magnetization by the desired angle. The magnitude and the duration of the rf pulse determine the angle through which the magnetization vector is rotated. In presently available commercial FT spectrometers, the time required to nutate the magnetization by 90° is of the order of 20–50 μsec, and thus relaxation times greater than 1 msec may be easily measured. After the initial pulse (usually 180° for T_1 measurements and 90° for T_2 measurements), the return of the magnetization to its equilibrium value is monitored as a function of time by means of appropriate sampling pulses. The various types of pulse sequences generally used are discussed below.

1. Pulsed Methods for Measuring T_1

The inversion–recovery pulse sequence is probably the most versatile and has been widely used for measuring T_1. The initiation pulse is a 180° pulse that reverses the direction of the magnetization (M_z goes from M_0 to $-M_0$). The system is allowed to relax for a period of time τ, and then a 90° sampling pulse is applied as illustrated in Fig. 7. During the waiting period τ, longitudinal relaxation occurs, which causes M_z to go from $-M_0$ through zero to its equilibrium value of M_0, as illustrated in Fig. 7. The amount of relaxation depends upon the relationship between τ and T_1, and thus the decay rate of M_z can be established. The equation governing the relaxation† is

$$M_z = M_0(1 - 2\exp(-\tau/T_1)) \tag{18a}$$

which shows that $M_z \approx -M_0$ when $\tau \ll T_1$ and $M_z = M_0$ when $\tau \gg T_1$. A plot of $\ln(M_0 - M_z)$ versus τ will yield a straight line with a slope of $-1/T_1$. The value of T_1 can also be determined by measuring the value of τ that produces a null in the signal after the 90° pulse. Equation (18a) can be rearranged to show that $\tau_{\text{null}} = T_1 \ln 2$. For samples with only a single resonance line, the measurement of τ_{null} is both easy and reliable. Fourier transformation techniques and the graphical method of determining T_1 are required for samples with multiple lines. Freeman and Hill (1971) have developed a variation of the inversion–recovery pulse sequence to reduce the danger that a slow drift in spectrometer gain might introduce an error in the measurement of T_1. This can be particularly important when extensive time averaging is required to complete the experiments. It is also possible to measure T_1 by using a series of three 90° pulses (Hahn, 1950), but this technique is less commonly used. Freeman and Hill (1971) have described the use of a repetitive series of 90° pulses in conjunction with FT-NMR to measure T_1. This approach is similar to the progressive saturation method described in the section on continuous wave methods. We also note that McDonald and Leigh (1973) have developed a convenient pulse sequence for measuring long T_1's.

† Campbell and Freeman (1973) have discussed the influence of cross-relaxation on NMR spin–lattice relaxation times and have shown that cross-relaxation effects may introduce errors in the measurement of T_1 of the order of 10% (or more for the case of a strongly coupled spin system). Werbelow and Marshall (1973) have recently considered the possibility of observing nonexponential methyl nuclear relaxation for macromolecules and concluded that even though an internally rotating methyl group on a large macromolecule should show nonexponential relaxation, it will probably be difficult to observe experimentally. Since there are presently no experimental observations of nonexponential nuclear relaxation of macromolecules in solution the nonexponential contributions to the relaxation rates are generally assumed to be negligible.

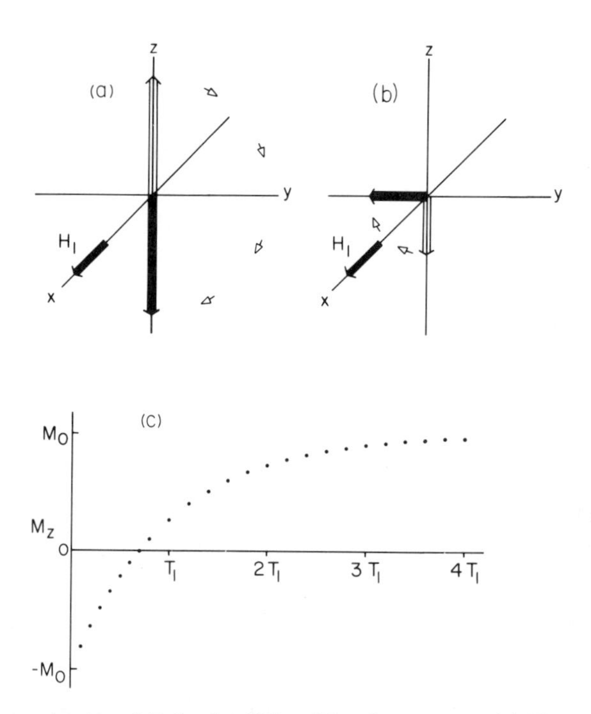

Fig. 7. Determination of T_1 by the $180°$–τ–$90°$ pulse sequence. (a) The magnetization is inverted by a $180°$ pulse ($M_0 \rightarrow -M_0$). (b) After a waiting period τ (equal to 0.3 T_1), a $90°$ sampling pulse is applied. (c) Plot of the amplitude of the magnetization detected immediately after the $90°$ sampling pulse versus the delay time τ in units of T_1. Note that each point represents a separate $180°$–τ–$90°$ sequence.

2. Pulsed Methods for Measuring T_2

The basic problem with measuring T_2 is that magnetic field inhomogeneities contribute to the dephasing of the magnetization in the *xy* plane (i.e., T_2 relaxation). Hahn (1950) was the first to propose a method to overcome this difficulty. This technique, called the *spin-echo method*, consists of a $90°$–τ–$180°$ pulse sequence and the observation of a spin echo at a time 2τ. The $180°$ pulse reverses the dephasing effect of the static magnetic field inhomogeneities, which results in the formation of a spin echo. The Carr–Purcell (1954) pulsed method for T_2 measurements uses a series of $180°$ pulses after the initial $90°$ pulse with a time interval of 2τ after each $180°$ pulse. The Carr–Purcell pulse sequence is thus

$$90°\text{–}\tau\text{–}180°\text{–}2\tau\text{–}180°\text{–}2\tau\text{–}180° \cdots$$

where the amplitude of the envelope of the train of echoes gives a measure of the exponential decay of the magnetization due to T_2 relaxation of the sample. The Meiboom and Gill (1958) modification of the Carr–Purcell method helps to minimize the problems of diffusion and pulse adjustments and is commonly used in commercial pulse spectrometers. Vold *et al.* (1973) have recently discussed the errors in the measurements of transverse relaxation rates by wideband multiple pulse techniques.

B. Continuous Wave Methods

1. Determination of T_1

a. Direct Method To measure T_1 of a resonance line by the direct method (Bloembergen *et al.*, 1948), one sits on the center of the resonance line and increases the value of H_1 until saturation occurs. The value of H_1 is then quickly decreased to a level well below saturation ($\sim$ 10 dB), and the response of the spin system is recorded as a function of time by means of an external recorder. The return to equilibrium is governed by spin–lattice relaxation processes, and thus the time constant for the exponential return to equilibrium is T_1. This method is relatively straightforward and may be used to measure the T_1 values of each resonance in a relatively complex spectrum; it is not suitable for very narrow resonance lines or when T_1 is much less than 0.5 sec.

b. Progressive Saturation The progressive saturation method was also outlined by Bloembergen *et al.* (1948) and complements the direct method because it is applicable to smaller values of T_1 for which the direct method cannot be employed. However, Mildvan *et al.* (1967) have reported that the errors in the measurement of T_1 by the progressive saturation method can be as large as $\pm 40\%$. The progressive saturation method is based upon the measurement of the NMR absorption line, under slow passage conditions, as a function of the level of H_1. The signal amplitude at the center of resonance is given by

$$\text{signal amplitude} \sim \frac{\gamma H_1 M_0 T_2}{1 + (\gamma H_1)^2 T_1 T_2} \tag{18b}$$

where we see that the amplitude is linearly proportional to H_1 for values of H_1 below the onset of saturation [i.e., $(\gamma H_1)^2 T_1 T_2 \ll 1$]. The signal amplitude reaches a maximum value when $(\gamma H_1)^2 T_1 T_2 = 1$ and then decreases with increasing levels of H_1 as the resonance line is progressively saturated. The value of T_1 may be calculated from the relation at maximal signal amplitude, $(\gamma H_1)^2 T_1 T_2 = 1$, if the absolute value of H_1 and T_2 are independently known. The value of T_2 may be determined from the methods discussed below. As a result of the rather large uncertainties associated with the progressive saturation method, it is generally useful as a means of determining the order of magnitude of T_1.

c. Adiabatic Half-Passage Sykes (1969) demonstrated that the adiabatic half-passage technique could be used to measure both T_1 and T_2 for signals with low signal-to-noise ratio. This technique had long been used to measure $T_{1\rho}$, the spin–lattice relaxation time in the rotating frame, and T_2 (Meiboom, 1961). As in all other experiments to measure T_1, one must perturb the equilibrium population of the spin energy levels. The adiabatic passage technique takes advantage of the fact that the magnetization M_0 will follow the "effective field" during a sweep into or through resonance, provided that the sweep rate satisfies the adiabatic passage conditions (Abragam, 1961, pp. 35, 66). The net result of a sweep onto the center of resonance is that M_0 is aligned along the field H_1. This magnetization then decays with a time constant $T_{1\rho}$. Sykes (1969) observed that the value of T_1 could be determined by first setting the field to the center of resonance (using a saturating value of H_1) and then pulsing the field far off resonance for a period of time t_1, followed by an adiabatic sweep back to the center of resonance. During the off-resonance time t_1, spin–lattice relaxation processes begin to restore M_z to its equilibrium value M_0. T_1 can thus be determined by an appropriate plot of the value of the amplitude of the magnetization returned to the center of resonance as a function of the time spent off resonance. An example of these experiments is shown in Fig. 8. An important consideration of these experiments is that the spectrometer

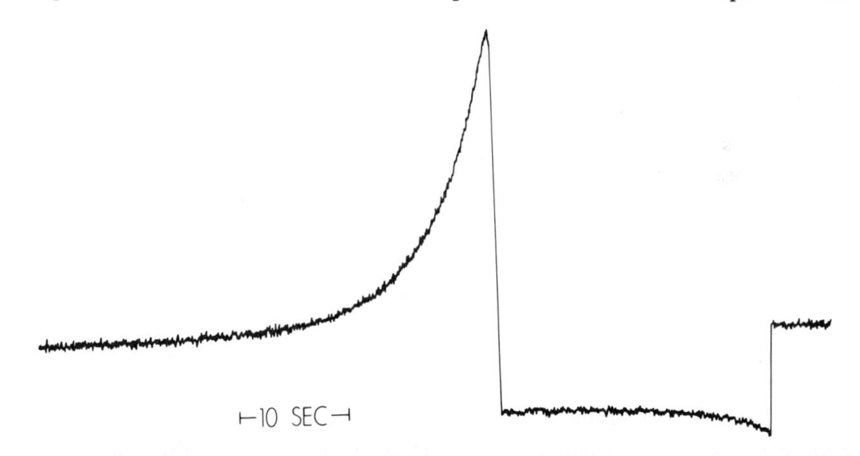

Fig. 8. Experimental dispersion mode magnetization (M_y) of acetate protons as a function of time for 0.067 *M* sodium acetate and 0.10 *M* adenosine-5′-monophosphate in D_2O. The spectrum was recorded from right to left. The discontinuity at the right corresponds to the jump-off resonance. The return to resonance is begun at the sharp rise near the center of the figure. The acetate protons are 286 Hz upfield from the HDO resonance. The off-resonance position was 100 Hz upfield and the H_1 field was ~6 Hz. The T_{1p} from this spectrum is 5.7 ± 0.2 sec, and from a series of experiments, $T_1 = 6.5 \pm 0.3$ sec. [From Neely and Krugh (1973)].

remain exactly at the center of resonance. Slightly different modifications have been reported that utilize the field/frequency lock of the spectrometer to maintain stability (Sykes and Wright, 1970; Krugh, 1971; Neely and Krugh, 1973); this presents the advantage that weak signals can be time averaged, which is an important consideration in the study of high molecular weight enzymes (Krugh, 1971). The adiabatic half-passage technique may be utilized on samples containing several resonance lines; the effect of neighboring resonances and large solvent absorptions on the relaxation times determined by the adiabatic half-passage technique is discussed by Neely and Krugh (1973). The adiabatic half-passage technique is suitable for measuring T_1 values longer than ~ 0.3 sec.

2. Determination of T_2

a. Linewidth Measurements The transverse relaxation time T_2 determines the natural linewidth of a resonance, and thus T_2 may be easily determined from the high-resolution spectrum. The relationship between T_2 and the full-width at half-maximum height for a Lorenztian line is $\Delta\nu_{1/2} = (\pi T_2)^{-1}$ where $\Delta\nu_{1/2}$ is given in hertz and T_2 is in seconds. This expression is valid only when the spectrum is recorded under slow-passage conditions with the value of H_1 set well below saturation. Any inhomogeneities in the static magnetic field will also contribute to the observed linewidth

$$(1/T_2)_{\text{obs}} = 1/T_2 + 1/T_2^* \tag{19}$$

where $1/T_2^*$ is the contribution due to magnetic field inhomogeneities. A well-tuned 100 MHz high-resolution NMR will have a value of T_2^* in the range of 1–3 seconds. The value of $1/T_2^*$ may be easily determined by including in the sample a reference compound known to have a very narrow intrinsic linewidth ($T_2^* \ll T_2$). The value of $1/T_2^*$ is then subtracted from all the other lines in the spectra. The determination of T_2 from linewidth measurements is essentially limited to those situations in which $T_2 < T_2^*$.

b. Adiabatic Half-Passage The adiabatic half-passage technique was originally devised as a means of measuring T_2 (Meiboom, 1961; Solomon, 1959); we have already discussed the use of this technique for measuring T_1. $T_{1\rho}$ is measured as the time constant for the exponential decay of the magnetization after the sweep to the center of resonance (Fig. 8). The value of T_2 may be obtained as the limiting value of $T_{1\rho}$ as $H_1 \to 0$ [i.e., $T_{1\rho} \to T_2$ as $H_1 \to 0$ (Sykes, 1969)]. The adiabatic half-passage technique is suitable for measuring T_2 values greater than 0.1 sec and thus complements the linewidth measurements, which are suitable when T_2 is less than ~ 1 sec.

IV. ENZYME STUDIES

This section will discuss the application of spin-label-induced nuclear relaxation to study the structure and function of enzymes. This discussion will highlight the methodology of the experiments and will be presented in terms of the types of experiments discussed in Section II.C of this chapter. It is not always possible to classify the reported studies into one of the subgroups presented in Section II.C, because more than one type of experiment may be performed on a particular enzyme. An example of this is the recent work of Wien *et al.* (1972) in which both spin-label and spin-probe molecules were utilized. These studies will be presented separately in the appropriate sections. Spin-label studies designed to elucidate the structure of the active sites of enzymes generally utilize both NMR and ESR techniques. The reader is referred to Chapter 8 by Morrisett, which included a review of the studies reported in the literature. To avoid unnecessary duplication, we will only present a few studies in detail. The examples were chosen to illustrate the methodology, the promise, and the problems associated with the spin-label-induced nuclear relaxation experiments.

A. Covalently Attached Spin Labels

The attachment of a spin label at a particular site on an enzyme provides a unique means of probing the environment of the enzyme. If the probe molecule is placed at or near the active site of the enzyme, then the magnetic resonance experiments may be used to determine the geometry of the active site. Covalently attaching the molecule to more distant regions of the enzyme presents equally intriguing possibilities, such as the detection of allosteric conformational changes.

The covalently bound spin label will enhance the relaxation rates of substrate protons, enzyme groups, and the water protons of the "hydration sphere" of the nitroxide (Type 1a, 1b, and 1c experiments). All three types of experiments have been performed. However, the NMR spectra of spin-labeled enzymes (Type 1b experiments) have not yet provided any quantitative data concerning distances between specific enzyme protons and the covalently linked spin label, because the NMR spectra exhibit a rather nonspecific broadening of all the protein resonances (Sternlicht and Wheeler, 1967; McConnell, 1967; and Wien *et al.*, 1972). The interactions of the covalently bound spin label with both substrate protons and water protons have provided useful quantitative results as illustrated below.

1. Relaxation Enhancements of Substrate Protons by a Covalently Bound Spin Label

a. Yeast Alcohol Dehydrogenase Sloan and Mildvan (1974) covalently attached a spin label to the sulfhydryl group of Cys 43 of yeast alcohol dehydrogenase. This enzyme was chosen because a one-site stoichiometric modification (of cysteine residue 43) had been established by Harris (1964). This specificity is an important requirement in the spin-labeling procedure, since it defines a unique location of the spin label. A second attractive feature of the yeast alcohol dehydrogenase is that the rate of dissociation of the enzyme–NADH complex is fast enough ($> 491\ sec^{-1}$) to satisfy the fast chemical exchange conditions between free and bound coenzyme. A drawback of this approach is the observation that the introduction of the iodoacetamide spin label inactivates the yeast alcohol dehydrogenase, even though the dissociation constants of the substrates and coenzymes agreed with their respective K_M and K_I values for the active enzyme. The dissociation constants for the spin-labeled enzyme were determined by measuring the effects of the substrates, inhibitors, and coenzymes on the spin–lattice relaxation times of the water protons. These experiments are analogous to the enhancement of the water relaxation rates observed for the binding of paramagnetic ions to macromolecules (Mildvan and Cohn, 1970; Dwek, 1972). The paramagnetic contribution to the relaxation rates of the protons and pyrophosphate phosphorous of NADH (the reduced form of nicotinamide adenine dinucleotide) and the methylene protons of isobutyramide (IBA) in binary and ternary complexes were measured by pulsed NMR techniques and used to calculate the distances between the enzyme-bound spin label and these nuclei. Sloan and Mildvan (1974) determined the diamagnetic contributions to the relaxation rates by measuring the relaxation times of NADH and isobutyramide bound to two analogous diamagnetic enzyme systems. These systems were the yeast alcohol dehydrogenase enzyme inactivated with iodoacetamide and the diamagnetic amine derived by reducing the spin-labeled enzyme with excess ascorbate. The contribution of the spin-labeled enzyme to the relaxation of free substrate [the "outer sphere" contribution, $1/T_{1(os)}$ in Eq. 12] was shown to be negligible by measuring the change in $1/T_{1p}$ as the NADH was displaced with excess NAD. The correlation time τ_C was determined from the measurements of $1/T_{1M}$ and $1/T_{2M}$ at both 100 and 220 MHz as discussed in Section II.F of this chapter. The calculated distances between the unpaired electron and the various nuclei are shown in Fig. 9 for both binary ($E\dot{R}$–NADH) and ternary ($E\dot{R}$–NADH–IBA) complexes. The reader is referred to Sloan and Mildvan (1974) for a detailed discussion of these results, and a comparison with X-ray crystallographic results.

b. Other Enzymes The enhancements in the nuclear relaxation rates of substrates bound to several other spin-labeled enzymes have also been studied, including lysozyme (Wien *et al.*, 1972), creatine kinase (Cohn and Reuben, 1971), and phosphofructokinase (Jones *et al.*, 1972, 1973).

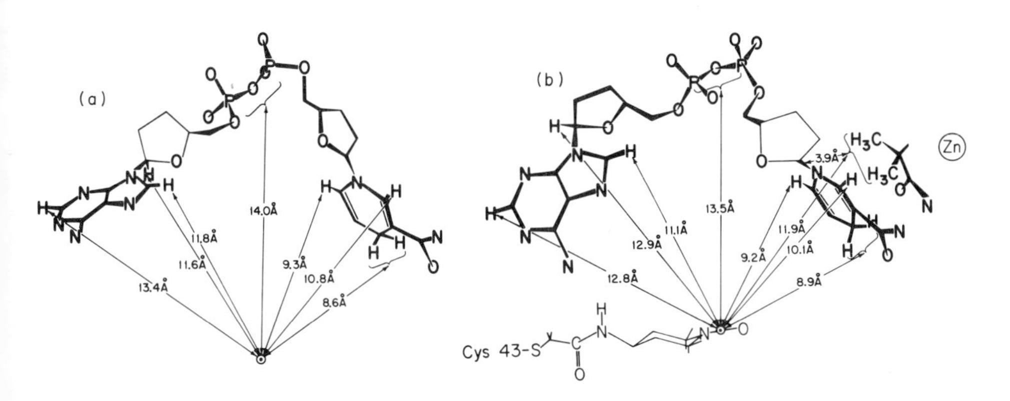

Fig. 9. Geometry of the (a) binary and (b) ternary EṘ–NADH–isobutyramide complexes consistent with the distances between the unpaired electron of the spin label and the protons and phosphorous of bound NADH and isobutyramide. [Reprinted with permission from Sloan and Mildvan, *Biochemistry* **13**, 1711–1718 (1974). Copyright by the American Chemical Society.]

2. Relaxation Enhancements of Water Protons by a Covalently Bound Spin Label

Spin labels will enhance the relaxation rates of water protons in an analogous manner to the proton relaxation enhancement (PRE) of paramagnetic metal ions (Mildvan and Cohn, 1970; Dwek, 1972). These studies provide a convenient means of determining the binding constants and the number of binding sites by using PRE as a measure of substrate binding. For example, Sloan and Mildvan (1974) used PRE measurements on spin-labeled yeast alcohol dehydrogenase to determine the dissociation constants of the substrates. These experiments are, of course, limited to those cases in which the binding of the substrate causes a change in the environment or mobility of the spin label, since it is these effects that produce a change in the PRE. Taylor *et al.* (1969) found that the PRE of spin-labeled creatine kinase was a useful monitor of the effects of substrate binding, while Jones *et al.* (1973) observed that the PRE of spin-labeled rabbit skeletal muscle phosphofructokinase did not detect conformational changes in the enzyme when the substrate MgATP was added.

B. Noncovalently Bound Spin Probes

The use of spin-label substrate analogues (spin probes) provides an easy method of directing the unpaired electron to a unique site or a limited number of sites on the enzyme. On the other hand, the experimentalist must now face the problem of synthesizing a substrate analogue (containing a nitroxide moiety) that will bind to the enzyme without significantly distorting the binding site. The spin-probe-induced nuclear relaxation of substrate molecules, enzyme groups, and water protons have been reported for a number of systems.

1. Relaxation Enhancements of Substrate Protons by a Noncovalently Bound Spin Probe

a. DNA Polymerase DNA polymerase is a multifunctional enzyme that plays a role in the reproduction and repair of DNA. The physiological properties and the functions of this enzyme have been reviewed by Kornberg (1969). In the presence of magnesium, the enzyme has two nucleotide binding sites, one that binds deoxyribonucleoside triphosphates and a separate site that binds nucleotides that have both a 3′-hydroxyl group in the ribose configuration and a 5′-phosphate linkage. The specificity of the 3′-hydroxyribonucleotide binding site suggested that this site is related to the site that binds the primer terminus of a DNA chain. If this site is the primer terminus binding site, then the two nucleotide binding sites should be adjacent to one another. The proximity of the two binding sites was determined by binding a paramagnetic analogue of adenosine triphosphate (ATP) in the triphosphate binding site and measuring the spin-probe-induced nuclear relaxation of the C-2 proton of adenosine monophosphate (AMP) bound in the monophosphate binding site (Krugh, 1971). The chemical structures of TEMPO-ATP [*N*-6(2,2,6,6 tetramethylpiperidine-1-oxyl)adenosine 5′-triphosphate] and AMP are shown in Fig. 10. These experiments illustrate the techniques and a few of the subtleties that may be involved in measuring the value of T_{1M} used in the distance calculations. The use of a spin-probe molecule means that there will always be some fraction of the spin-probe molecules bound to the enzyme with the remaining fraction free in solution. The relative proportions of these two species will depend upon the equilibrium constant for binding of the spin probe. Both the bound and the free spin probes will contribute to the relaxation of the nuclei of both the bound and the free AMP molecules. For the binding of TEMPO-ATP to DNA polymerase, the contribution of the free TEMPO-ATP to the relaxation rates of the bound AMP is negligible [this is the $(1/T_1)_{ES \cdots \dot{R}}$ term mentioned earlier]. However, the contribution of the bound TEMPO-ATP to the relaxation rates of the free AMP molecules $(1/T_1)_{E\dot{R} \cdots S)}$ is not negligible. The magnitude of this contribution must be measured and used to correct the observed paramagnetic induced relaxation (Krugh, 1971). A convenient means of determining this nonspecific contribution to the relaxation rates of molecules free in solution is to include in the

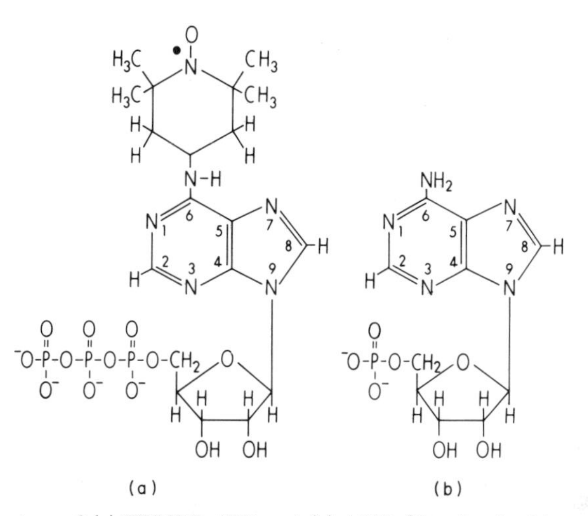

Fig. 10. Structure of (a) TEMPO-ATP and (b) AMP. [Reprinted with permission from Krugh, *Biochemistry*, **10**, 2594 (1971). Copyright by the American Chemical Society.]

solutions a convenient ion, such as acetate ion, that does not bind to the enzyme. The diamagnetic contributions to the relaxation rates may be estimated by measuring the relaxation rates of the system in the absence of the spin-probe molecule or in the presence of the amide derivative of the spin probe. An alternative approach is to take advantage of the equilibria involved in binding the spin probe by adding an excess of diamagnetic ligand to compete with the spin probe for the enzyme binding site. An approximate 100-fold excess of thymidine 5′-triphosphate (TTP) was used in the DNA polymerase experiments. This approach has the advantage that the competing ligand is added to the same solution used for the measurement of $1/T_{1p}$. It is also important to note that when the spin probe is "chased off" the enzyme by the addition of a competing ligand, the magnitude of $(1/T_1)_{E\dot{R}\cdots S}$ is diminished while the contribution of the free spin probe to the relaxation rates of the AMP molecules $(1/T_1)_{\dot{R}\cdots S}$ and $(1/T_1)_{ES\cdots\dot{R}}$ is increased due to the population changes in the two states. These effects must be considered in evaluating the magnitude of T_{1M} and T_{2M} to be used in the distance calculations. The temperature and frequency dependence of the relaxation rates can be used to determine the value of τ_C and to insure that the system is in the fast chemical exchange limit, as discussed in Sections II.E and F. Sensitivity limitations and the thermal instability of the DNA polymerase prevented the execution of these detailed experiments. However, the transverse relaxation rate $1/T_{2M}$ was found to be twenty-two times larger than the spin–lattice relaxation rate $1/T_{1M}$, thus providing assurance that the fast chemical exchange approximation is valid for the spin–lattice relaxation time (i.e.,

$T_{1M} \gg \tau_M$). This is often a useful approach if the temperature dependence of the relaxation rates cannot be measured. The correlation time for the experiments with DNA polymerase was $\tau_C \cong 9 \times 10^{-9}$ sec and the separation between the unpaired electron on TEMPO-ATP and the C-2 proton of AMP was calculated to equal 7.1 ± 0.6 Å (estimated error). This separation is about the value expected from a model of a DNA chain if the TEMPO-ATP is placed in the position of the next nucleotide residue to be added to the primer terminus. Although this is not positive proof that the monophosphate binding site is the primer terminus binding site, when the NMR results are combined with the biochemical evidence, it does offer rather convincing evidence that the monophosphate binding site is the primer terminus binding site.

b. Other Enzymes Mildvan and Weiner (1969a,b) have studied liver alcohol dehydrogenase using a paramagnetic analogue of the coenzyme NAD, and Wien *et al.* (1972) have studied lysozyme using a variety of spin-labeling techniques.

2. Relaxation Enhancements of Enzyme Nuclei by a Noncovalently Bound Spin Probe

Enzymes exhibit complicated NMR spectra, and only in favorable cases is it possible to assign a resonance line to a particular amino acid group of the enzyme. For example, the C-2 protons of the histidine residues of ribonuclease and lysozyme have been assigned and serve as a useful probe of enzyme structure. These types of experiments are presently limited to the study of low molecular weight enzymes. The development of high frequency (>300 MHz) NMR spectrometers, Fourier transformation techniques, and the chemical or biochemical synthesis of partially deuterated or ^{13}C-enriched enzymes should greatly expand the potential of using spin probes to enhance the relaxation rates of enzyme groups.

Roberts *et al.* (1969) synthesized a spin-probe analogue of phosphate ion and found that when it binds to ribonuclease the broadening of the C-2 protons of the histidine residues were in the order His 12 > His 48 ≫ His 105, His 119, suggesting that the separation between these residues and the spin-probe molecule were also in this order. Wien *et al.* (1972) have monitored the broadening of the C-2 proton His-15 of lysozyme when three different spin probes were bound to the enzyme. Spin-probe saccharides, which bound to the active site, provided a measure of the distance from subsite D to His 15. In another experiment, a spin-probe molecule that bound near Trp 123 was used to estimate the distance from that residue to the acetamido methyl groups of bound saccharides, and to the C-2 proton of His 15. The NMR results for both ribonuclease and lysozyme were consistent with the structures of the enzymes determined by X-ray crystallography.

3. Proton Relaxation Enhancement (PRE) of Water Molecules by a Noncovalently Bound Spin Probe

These experiments are analogous to the PRE experiments using paramagnetic metal ions (Mildvan and Cohn, 1970; Dwek, 1972). The binding of the spin probe to a macromolecule will generally enhance the relaxation of the water protons associated with the nitroxide moiety of the spin probe. If the addition of other substrates changes either the accessibility of the solvent water molecules to the hydration sphere or the rotational motion of the water molecule–nitroxide complex, then the proton relaxation enhancement will be influenced. The substrate may either enlarge, diminish, or not affect the PRE of the spin probe. In fact, the binding of the spin probe to the macromolecule may also either enlarge, diminish, or not change the PRE of the solvent water. For example, if the spin probe bound to the enzyme in an orientation that prevented the exchange of water molecules between the hydration sphere of the nitroxide and the bulk solvent water, then the PRE would be diminished. The same situation would occur if the addition of a substrate inhibited the exchange of water molecules either directly for steric reasons or indirectly as a result of a conformational change of the enzyme. Consequently, the changes in the PRE of water molecules that result from either the binding of the spin probe to the enzyme or the addition of substrates are especially useful for determining the binding constants and the number of binding sites available (as well as the associated thermodynamic and kinetic information). The utility of this approach was demonstrated by the work of Mildvan and Weiner (1969a) in their study of the binding of a spin probe analogue of nicotinamide–adenine dinucleotide with liver alcohol dehydrogenase. Weidman *et al.* (1973) have recently studied the interaction of a spin probe analogue of acetylcoenzyme A with citrate synthase using both relaxation rate studies of solvent water protons and paramagnetic resonance studies.

V. CONCLUSIONS

In this chapter we have discussed the use of spin-label-induced relaxation of nuclear magnetic resonance signals to determine the geometrical, equilibrium, thermodynamic, and kinetic properties of enzymes and enzyme–substrate complexes. The potential of these experiments seems only to be limited by the necessity of designing appropriate spin-probe substrate analogues or the covalent spin labeling of functional groups of the enzyme. The variety of experiments performed to date indicates that these considerations are not major limitations. The use of solid phase synthesis or biochemical synthesis to incorporate enriched (e.g., ^{13}C or ^{2}H) amino acids or spin-label analogues of amino acids (such as TEMPO-alanine, Chapter 8,

Section II, Fig. 22) into various regions of the enzyme offers a great deal of promise. Although this chapter has dealt primarily with applications to enzyme systems, the techniques are applicable to a wide variety of other systems. For example, Kornberg and McConnell (1971) incorporated a phosphatidylcholine spin label into phosphatidylcholine vessicles and used the broadening of the *N*-methyl proton signal to show that the system undergoes rapid lateral diffusion.

REFERENCES

Abragam, A. (1961). "The Principles of Nuclear Magnetism." Oxford Univ. Press, London and New York.

Bloembergen, N. (1957). Proton relaxation times in paramagnetic solutions, *J. Chem. Phys.* **27**, 572–573.

Bloembergen, N., Purcell, E. M., and Pound, R. V. (1948). Relaxation effects in nuclear magnetic resonance absorption, *Phys. Rev.* **73**, 679–712.

Bovey, F. A. (1969). "Nuclear Magnetic Resonance Spectroscopy." Academic Press, New York.

Bovey, F. A. (1972). "High Resolution NMR of Macromolecules." Academic Press, New York.

Campbell, I. D., and Freeman, R. (1973). Influence of cross-relaxation on NMR spin-lattice relaxation times, *J. Magn. Res.* **11**, 143–162.

Carr, H. Y., and Purcell, E. M. (1954). Effects of diffusion on free precession in nuclear magnetic resonance experiments, *Phys. Rev.* **94**, 630–638.

Carrington, A., and McLachlan, A. D. (1967). "Introduction to Magnetic Resonance," pp. 208–210. Harper, New York.

Cohn, M., and Reuben, J. (1971). Paramagnetic probes in magnetic resonance studies of phosphoryl transfer enzymes, *Accounts Chem. Res.* **4**, 214–222.

Dwek, R. A. (1972). Proton relaxation enhancement probes, *Advan. Mol. Relaxation Processes* **4**, 1–53.

Dwek, R. A. (1973). "Nuclear Magnetic Resonance (N.M.R.) in Biochemistry." Oxford Univ. Press, London and New York.

Emsley, J. W., Feeney, J., and Sutcliffe, L. H. (1965). "High Resolution Nuclear Magnetic Resonance Spectroscopy," Vol. I and II. Pergamon, Oxford.

Farrar, T. C., and Becker, E. D. (1971). "Pulse and Fourier Transform NMR." Academic Press, New York.

Freeman, R., and Hill, H. D. W. (1971). Fourier transform study of NMR spin-lattice relaxation by "progressive saturation," *J. Chem. Phys.* **54**, 3367–3377.

Gurd, F. R. N., and Keim, P. (1973). The prospects for carbon-13 nuclear magnetic resonance studies in enzymology, *Methods Enzymol.* **27**, 836–911.

Hahn, E. L. (1950). Spin echoes, *Phys. Rev.* **80**, 580–594.

Harris, J. I. (1964). The structure and catalytic activity of thiol dehydrogenases, *In* "Structure and Activity of Enzymes" (T. W. Goodwin, J. I. Harris, and B. S. Hartley, eds.), pp. 97–109. Academic Press, New York.

Jardetzky, O. (1964). The study of specific molecular interactions by nuclear magnetic relaxation measurements, *Advan. Chem. Phys.* **7**, 499–531.

Jardetzky, O., and Wade-Jardetzky, N. G. (1971). Applications of nuclear magnetic resonance spectroscopy to the study of macromolecules, *Ann. Rev. Biochem.* **40**, 605–634.

Jones, R., Dwek, R. A., and Walker, I. O. (1972). Conformational states of rabbit muscle phosphofructokinase investigated by a spin label probe, *FEBS Lett.* **26**, 92–96.

Jones, R., Dwek, R. A., and Walker, I. O. (1973). Spin labeled phosphofructokinase and its interactions with ATP and metal-ATP complexes as studied by magnetic resonance methods, *Eur. J. Biochem.* **34**, 28–40.

Kooser, R. G., Volland, W. V., and Freed, J. H. (1969). ESR relaxation studies on orbitally degenerate free radicals. I. Benzene anion and tropenyl, *J. Chem. Phys.* **50**, 5243–5257.

Kornberg, A. (1969). Active center of DNA polymerase, *Science* **163**, 1410–1418.

Kornberg, R. D., and McConnell, H. M. (1971). Lateral diffusion of phospholipids in a vesicle membrane, *Proc. Nat. Acad. Sci. U.S.* **68**, 2564–2568.

Kreilick, R. W. (1973). NMR studies of organic radicals, *In* "NMR of Paramagnetic Molecules: Principles and Applications" (G. N. La Mar, W. DeW. Horrocks, Jr., and R. H. Holm, eds.), pp. 595–626. Academic Press, New York.

Krugh, T. R. (1971). Proximity of the nucleoside monophosphate and triphosphate binding sites on deoxyribonucleic acid polymerase, *Biochemistry* **10**, 2594–2599.

Levy, G. C., and Nelson, G. L. (1972). "Carbon-13 Nuclear Magnetic Resonance for Organic Chemists." Wiley, New York.

Luz, Z., and Meiboom, S. (1964). Proton relaxation in dilute solutions of cobalt(II) and nickel(II) ions in methanol and the rate of methanol exchange of the solvation sphere, *J. Chem. Phys.* **40**, 2686–2692.

McCalley, R. C., Shimshick, E. J., and McConnell, H. M. (1972). The effect of slow rotational motion on paramagnetic resonance spectra, *Chem. Phys. Lett.* **13**, 115–119.

McConnell, H. M. (1958). Reaction rates by nuclear magnetic resonance, *J. Chem. Phys.* **28**, 430–431.

McConnell, H. M. (1967). Comment on article by H. Sternlicht and E. Wheeler, *In* "Magnetic Resonance in Biological Systems" (A. Ehrenberg, B. G. Malmstrom, and T. Vanngard, eds.), p. 335. Pergamon, Oxford.

McDonald, G. G., and Leigh, J. S. (1973). A new method for measuring longitudinal relaxation times, *J. Magn. Res.* **9**, 358–362.

Meiboom, S. (1961). Nuclear magnetic resonance study of the proton transfer in water, *J. Chem. Phys.* **34**, 375–388.

Meiboom, S., and Gill, D. (1958). Modified spin-echo method for measuring nuclear relaxation times, *Rev. Sci. Instrum.* **29**, 688–691.

Mildvan, A. S., and Cohn, M. (1970). Aspects of enzyme mechanisms studied by nuclear spin relaxation induced by paramagnetic probes. *Adv. Magn. Resonance* **33**, 1.

Mildvan, A. S., Leigh, J. S., and Cohn, M. (1967). Kinetic and magnetic resonance studies of pyruvate kinase. III. The enzyme–metal–phosphoryl bridge complex in the fluorokinase reaction, *Biochemistry* **6**, 1805–1818.

Mildvan, A. S., and Weiner, H. (1969a). Interaction of a spin-labeled analog of nicotinamide adenine dinucleotide with alcohol dehydrogenase. II. Proton relaxation rate and electron paramagnetic resonance studies of binary and ternary complexes, *Biochemistry* **8**, 552–562.

Mildvan, A. S., and Weiner, H. (1969b). Interaction of a spin-labeled analogue of nicotinamide adenine dinucleotide with alcohol dehydrogenase. III. Thermodynamic, kinetic, and structural properties of ternary complexes as determined by nuclear magnetic resonance. *J. Biol. Chem.* **244**, 2465–2475.

Neely, J. W., and Krugh, T. R. (1973). Measurement of relaxation times by the adiabatic half passage ($T_{1\rho}$) technique, *J. Magn. Res.* **9**, 84–91.

Phillips, W. D. (1973). Nuclear magnetic resonance spectroscopy of proteins, *Methods Enzymol.* **27**, 825–836.

Poole, C. P., Jr., and Farach, H. A. (1971). "Relaxation in Magnetic Resonance." Academic Press, New York.

Reuben, J., Reed, G. H., and Cohn, M. (1970). Note on the distinction between transverse and longitudinal electron relaxation times obtained from nuclear relaxation studies, *J. Chem. Phys.* **52**, 1617.

Roberts, G. C. K., Hannah, J., and Jardetzky, O. (1969). Noncovalent binding of a spin labeled inhibitor to ribonuclease, *Science* **165**, 504–506.

Sheard, B., and Bradbury, E. M. (1970). Nuclear magnetic resonance in the study of biopolymers and their interaction with ions and small molecules, *Progr. Biophys. Mol. Biol.* **20**, 187–246.

Sloan, D. L., and Mildvan, A. S. (1974). Magnetic resonance studies of the geometry of bound nicotinamide adenine dinucleotide and isobutyramide on spin labeled alcohol dehydrogenase, *Biochemistry* **13**, 1711–1718.

Solomon, I. (1955). Relaxation processes in a system of two spins, *Phys. Rev.* **99**, 559–565.

Sternlicht, H., and Wheeler, E. (1967). Preliminary magnetic resonance studies of spin labeled macromolecules, *In* "Magnetic Resonance in Biological Systems" (A. Ehrenberg, B. G. Malmstrom, and T. Vanngard, eds.), pp. 325–334. Pergamon, Oxford.

Stothers, J. B. (1972). "Carbon-13 NMR Spectroscopy." Academic Press, New York.

Swift, T. J., and Connick, R. E. (1962). NMR-relaxation mechanisms of ^{17}O in aqueous solutions of paramagnetic cations and the lifetime of water molecules in the first coordination sphere, *J. Chem. Phys.* **37**, 307–320.

Sykes, B. D. (1969). An application of transient nuclear magnetic resonance methods to the measurement of biological exchange rates. The interaction of trifluoroacetyl-D-phenylalanine with the chymotrypsins, *J. Amer. Chem. Soc.* **91**, 949–955.

Sykes, B. D., and Wright, J. M. (1970). Measurement of nuclear spin relaxation times on an HA-100 spectrometer, *Rev. Sci. Instrum.* **41**, 876–877.

Taylor, J. S., Leigh, J. S., Jr., and Cohn, M. (1969). Magnetic resonance studies of spin-labeled creatine kinase system and interaction of two paramagnetic probes, *Proc. Nat. Acad. Sci. U.S.* **64**, 219–226.

Torrey, H. C. (1949). Transient nutations in nuclear magnetic resonance, *Phys. Rev.* **76**, 1059–1068.

Vold, R. L., Vold, R. R., and Simon, H. E. (1973). Errors in measurements of transverse relaxation rates, *J. Magn. Res.* **11**, 283–298.

Weidman, S. W., Drysdale, G. R., and Mildvan, A. S. (1973). Interaction of a spin labeled analog of acetyl coenzyme A with citrate synthase. Paramagnetic resonance and proton relaxation rate studies of binary and ternary complexes, *Biochemistry* **12**, 1874–83.

Werbelow, L. G., and Marshall, A. G. (1973). Internal rotation and nonexponential methyl nuclear relaxation for macromolecules. *J. Magn. Res.* **11**, 299–313.

Wien, R. W., Morrisett, J. D., and McConnell, H. M. (1972). Spin-label-induced nuclear relaxation. Distances between bound saccharides, histidine-15, and tryptophan-123 on lysozyme in solution, *Biochemistry* **11**, 3707–3716.

10

Anisotropic Motion in Liquid Crystalline Structures

JOACHIM SEELIG

DEPARTMENT OF BIOPHYSICAL CHEMISTRY
BIOCENTER OF THE UNIVERSITY OF BASEL
BASEL, SWITZERLAND

I. CLASSIFICATION OF LIQUID CRYSTALS

Liquid crystals are substances which share some properties with both liquids and solids. On the one hand they may be as fluid as water, and on the other hand the constituent molecules are ordered in a very regular pattern similar to a crystal lattice. Depending on the chemical composition, two classes of liquid crystals can be distinguished. *Thermotropic* liquid crystals consist of *one* component. The liquid crystalline phase is formed upon heating the substance to a well-defined transition point. *Lyotropic* liquid crystals are prepared by treating certain amphiphilic molecules such as fatty acids or lipids with a controlled amount of water or other solvent. Lyotropic liquid crystals may consist of two or more components. Both classes of liquid crystal can occur in various structures (mesophases), which can be characterized by their optical properties and also by means of X-ray diffraction. (For reviews about liquid crystals see Brown *et al.*, 1971; Brown and Labes, 1972.)

Thermotropic liquid crystals are conventionally divided into smectic, nematic, and cholesteric mesophases. *Smectic* liquid crystals consist of a series of layers in which the long axes of the rodlike molecules are perpendicular to the plane of the layer and more or less parallel to each other (Fig. 1a). The molecules rotate around their long molecular axes and diffuse

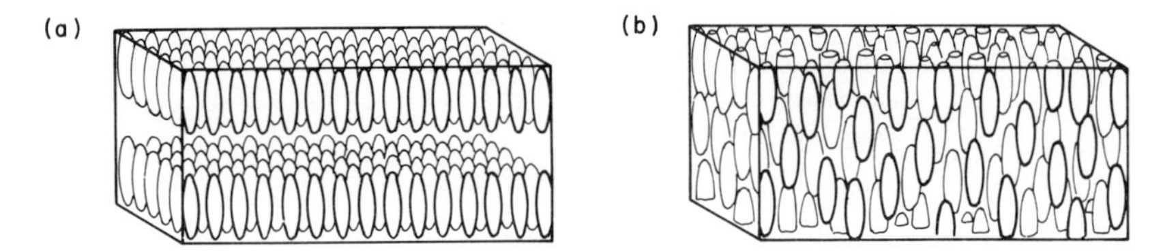

Fig. 1. Thermotropic liquid crystals. (a) Smectic phase, two-dimensional order: The molecules are aligned parallel to each other and separated into layers; (b) nematic phase, one-dimensional order: The molecules are aligned parallel to each other, but are not separated into strata.

freely within the plane (lateral diffusion) of the layer, but the rotation perpendicular to the long molecular axis is severely hindered. In *nematic* liquid crystals the long molecular axes are also parallel to each other, but the molecules are not separated into layers (Fig. 1b). *Cholesteric* liquid crystals can be constructed schematically by superimposing layers of nematic liquid crystals with a uniform twist, as is shown in Fig. 2. This helical structure gives rise to a large optical activity. Cholesteric liquid crystals are the most optically active systems known. Nematic and smectic phases have no optical rotating power, unless the constituting molecules themselves show optical activity.

Lyotropic mesophases are made up from amphiphilic molecules. These molecules are characterized by containing two groups of different solubility, namely a polar group and a nonpolar hydrophobic group. Depending on the type of amphiphilic substance and the amount of water added, lyotropic phases may also assume a variety of structures. This polymorphism is even broader than that for thermotropic liquid crystals. A systematic classification may be found in Brown *et al.* (1971). In the following we consider only two main classes, namely the so-called *lamellar* and the *hexagonal* mesophases.

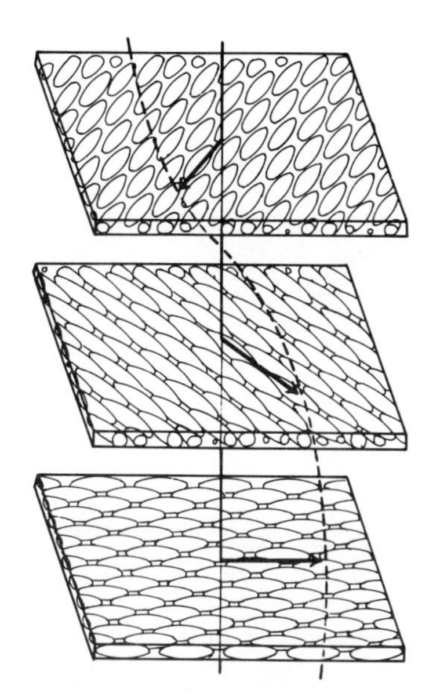

Fig. 2. Cholesteric liquid crystals.

In the *lamellar* phase the amphiphilic molecules are arranged in coherent double layers of molecules separated from each other by layers of water. The polar groups are in contact with water, while the nonpolar tail forms the hydrocarbon core of the double layer (Fig. 3a). In analogy to thermotropic liquid crystals, the *lamellar* phase is also called a *lyotropic smectic* liquid crystal.

The *hexagonal* phase consists of rodlike cylindrical particles with a lipophilic core in an aqueous environment (Fig. 3b). The polar groups are lying again on the outside of the cylinders. The name for this phase originates from the fact that the cylinders are grouped together in a hexagonal array, so that each rod is surrounded by six neighbors.

Micelles, finally, may be regarded as primitive precursors of liquid crystals. Certain classes of amphiphilic molecules, such as fatty acids or soaps, can be dissolved in water just as a normal salt. If the concentration of the amphiphilic compound is then increased above a certain critical limit, the so-called critical micelle concentration, the lipophilic portions of the molecule will associate in order to minimize the contact with the water molecules. Micelles are rapidly fluctuating systems. The lifetime of an individual micelle is rather short and it has neither a well-defined geometrical shape nor a long-range order of the constituting molecules.

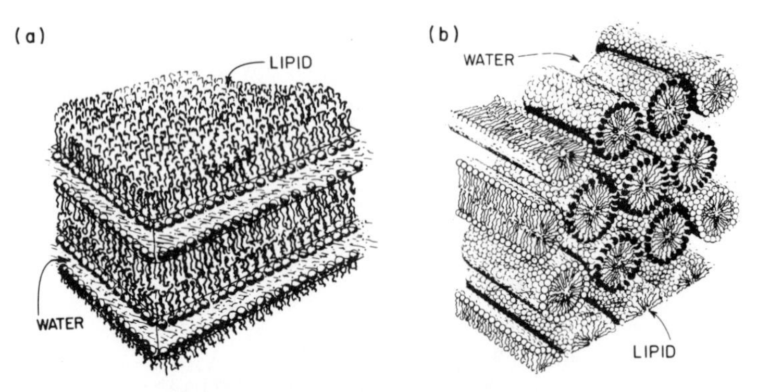

Fig. 3. Lyotropic liquid crystals. (a) Bilayer (lyotropic smectic phase); (b) cylinders (lyotropic hexagonal phase). (From "A Review of the Structure and Physical Properties of Liquid Crystals," G. H. Brown *et al.* © The Chemical Rubber Co., 1971. Used by permission of the Chemical Rubber Co.)

Micelles should not be confused with *vesicles* (Huang, 1969). Vesicles are spherical structures with a single bilayer of lipid molecules as an outer membrane. The interior of such vesicles is tightly sealed from the outside. Vesicles can thus be regarded as a lamellar mesophase bent into a globular structure.

Liquid crystals play an important role in living systems (Fergason and Brown, 1968). This is due to several reasons. (1) They are sensitive to external forces, i.e., they respond to heat, light, mechanical pressure, and chemical environment. (2) They are fluid enough to allow a fast lateral diffusion and distribution of matter and energy. (3) Finally, liquid crystals easily adjust themselves to a multiplicity of shapes required by the geometry of the system, while, at the same time, they maintain a very high degree of order on a molecular level.

The lipid bilayer, i.e., the lyotropic smectic mesophase, is the most prominent liquid crystalline structure associated with the living state, since it is the basic structure of many biological membranes. Therefore much of this and the two following chapters will be devoted to lipid bilayers. Some of the other mesophases described above also occur in living systems. For example, bile from human beings and animals under normal conditions is a *micellar* solution of lecithin, cholesterol, and sodium cholate. An increase of the cholesterol content, which occurs under pathological conditions, leads to the formation of *cholesteric* liquid crystals and finally to crystalline gall stones (Olszewski *et al.*, 1973). However, the importance of mesophases other than the bilayer for the living state has only begun to be realized.

Thermotropic liquid crystals have received much attention in recent years for completely different reasons. Since their ordering can easily be altered by electric fields or heat, the interest focuses on the technical application of such systems in electronic displays and in thermal mapping (Brown and Labes, 1972).

II. THE RELEVANCE OF SPIN PROBES TO THE INVESTIGATION OF LIQUID CRYSTALS

The systematic classification of liquid crystalline mesophases has shown that we have a rather detailed knowledge about the overall geometry of such systems. However, we have little reliable knowledge about the molecular structure and the dynamical properties within these aggregates. This is especially true for lyotropic mesophases, which are composed of rather flexible molecules. The problem is best illustrated for the case of a lipid bilayer. Two questions can be asked considering a cross section of such a bilayer: (1) How are the hydrocarbon chains arranged at various distances from the lipid–water interface, and (2) what is the rate of motion of the chain segments? The first question describes a *structural* problem since it investigates the orientation of the hydrocarbon chains relative to each other and to the bilayer surface. The second question asks for the *dynamical* properties of the various chain segments in the bilayer. Ultimately both questions merge into the general problem of how the rates of molecular movements are connected with the specific structure of the liquid crystalline phase.

Spin probes are ideally suited to investigate these structural and dynamical aspects. One important reason is the high fluidity of most liquid crystals, leading to EPR spectra with sharp resonance lines (" fast tumbling region "). The *ordering* of the molecules is then related in a simple manner to the *position* and the *separation* (hfs) of the resonance lines, while the *dynamical properties* are reflected in the *width* of the resonance lines. With some computational effort quantitative details about the rates of motion can be extracted from the linewidth. Spin probes thus provide information about the molecular architecture of liquid crystals that so far has not been available by other spectroscopic techniques.

A spin probe study of liquid crystals should proceed in two steps. The first is an analysis of the EPR spectra, yielding physical parameters about the orientation and the motion of the spin probe in the host system. The second step is then the interpretation of these parameters in terms of relevant molecular models for the liquid crystalline phase. In the following section we therefore develop most of the theory for the spectra interpretation, while the remainder of this chapter will focus on the results obtained for various phases.

III. ANISOTROPIC MOTION OF SPIN PROBES IN LIQUID CRYSTALS†

The EPR spectra of spin probes in liquid crystals are determined primarily by the *anisotropic* motion of the molecules, i.e., by the preferential rotation of the molecules around the long molecular axis (cf. Figs. 1 and 3). Rotations perpendicular to this axis are severely hindered, but angular fluctuations of small amplitude do occur. Depending on the rate of these fluctuations, two different pictures must be used to describe the EPR spectra. If the fluctuations are *fast*, the measuring device cannot follow the molecular movements but detects the *average* position of the molecules. It is then convenient to define the average orientation by means of order parameters (Saupe, 1968). If the fluctuations are very *slow*, the instrument sees a *pseudostatic* distribution of molecules, with the molecular axes pointing at various directions of space. The EPR spectrum is then a superposition of signals, the overall line shape being determined by the geometry of the distribution.

The terms "fast" and "slow" depend, of course, on the time resolution of the measuring device and also, in part, on the molecular parameters. For a nitroxide radical studied with a conventional 9-GHz EPR spectrometer, a motion can be considered fast if the motional frequencies are of the order of 10^9 Hz or larger. Such frequencies are sufficient for an effective averaging of the **g** and **A** tensors. Under the same experimental conditions a pseudostatic distribution of spin probes is obtained for rotational frequencies less than 10^6–10^7 Hz. In the following we confine ourselves to the fast tumbling region, mainly because the fluctuations in liquid crystals clearly belong to this frequency range and also because simple rules can be given for the interpretation of the spectra. The slow tumbling region, which comprises the frequency window between the pseudostatic distribution and the fast tumbling range, is of less importance for most liquid crystals and requires a completely different formalism, which has already been discussed in Chapter 3.

Limiting the discussion to frequencies greater than 10^8–10^9 Hz, i.e., to correlation times shorter than approximately 3×10^{-9} sec, has certain practical advantages. First, starting from simple geometric considerations it is possible to predict the *positions* of the resonance lines for a given anisotropy of the motion and a given geometry of the sample without using much mathematics. This allows a straightforward evaluation of order parameters and tilt angles and is thus extremely valuable in the everyday use of spin

† Editor's note: More discussion on models for anisotropic motion is found in Chapter 12.

labels. This approach will be discussed in Section III.A. Second, if a line shape simulation of the EPR spectrum is required, this can be performed by means of perturbation theory. Such an analysis yields additional information about the rates of motion and the method will be treated briefly in Section III.B.

A. Position of the Resonance Lines

1. Time-Averaged Hyperfine Tensor and g Tensor

The long axes of the molecules in a liquid crystalline phase tend to orient parallel to each other. The average orientation of an ensemble of molecules, grouped together in a rather small region of the sample, can then be characterized by a director z', the motion of the molecules having cylindrical symmetry about z'. The directions of the z' axes of the microregions will not necessarily be uniform over a macroscopic sample, but we neglect this effect for the moment and consider just one homogeneous microregion.

We ascribe a Cartesian coordinate system x, y, z to the nitroxide moiety (Hubbell and McConnell, 1969; Seelig, 1970). The x axis is extended in the direction of the N–O bond and the z axis in the direction of the nitrogen 2p π orbital. The y axis follows by defining a right-handed coordinate system. This choice of the coordinate system has the advantage that the hyperfine tensor **A** and the **g** tensor become diagonal in this molecular frame (at least to a good approximation; cf. Chapter 6):

$$\mathbf{A} = \begin{pmatrix} A_{xx} & 0 & 0 \\ 0 & A_{yy} & 0 \\ 0 & 0 & A_{zz} \end{pmatrix}, \qquad \mathbf{g} = \begin{pmatrix} g_{xx} & 0 & 0 \\ 0 & g_{yy} & 0 \\ 0 & 0 & g_{zz} \end{pmatrix} \tag{1}$$

Next the **A** tensor and the **g** tensor must be transformed from the molecule-fixed coordinates x, y, z to space-fixed coordinates x', y', z'. This is achieved by using the (orthogonal) transformation matrix **D**

$$\begin{pmatrix} x' \\ y' \\ z' \end{pmatrix} = \mathbf{D} \begin{pmatrix} x \\ y \\ z \end{pmatrix} \tag{2}$$

with

$$\mathbf{D} = \begin{pmatrix} \cos\eta_1 & \cos\eta_2 & \cos\eta_3 \\ \cos\xi_1 & \cos\xi_2 & \cos\xi_3 \\ \cos\theta_1 & \cos\theta_2 & \cos\theta_3 \end{pmatrix} \tag{3}$$

Actually, only the direction cosines between the molecular axes x, y, z and the space-fixed director axis z' are of importance. We therefore write them down explicitly:

$$\mathbf{x} \cdot \mathbf{z}' = \cos\theta_1, \qquad \mathbf{y} \cdot \mathbf{z}' = \cos\theta_2, \qquad \mathbf{z} \cdot \mathbf{z}' = \cos\theta_3 \tag{4}$$

(**x**, **y**, **z**, and **z**′ are unit vectors along the corresponding axes).

The other direction cosines will be eliminated very soon. Applying the transformation matrix **D** leads to the following new tensors:

$$\mathbf{A}' = \mathbf{DAD}^{-1}, \qquad \mathbf{g}' = \mathbf{DgD}^{-1} \tag{5}$$

This is just a simple geometric coordinate transformation. However, the angles η_i, ξ_i, and θ_i are not static, but fluctuate rapidly in time. The essential step is then to take into account the specific physical properties of the system, that is, the rotational symmetry of the motion around **z**′. The *time-averaged* **A**′ tensor and **g**′ tensor must be *invariant* against rotations around z' and must therefore have axial symmetry:

$$\langle \mathbf{A}' \rangle = \begin{pmatrix} A_\perp & 0 & 0 \\ 0 & A_\perp & 0 \\ 0 & 0 & A_\parallel \end{pmatrix}, \qquad \langle \mathbf{g}' \rangle = \begin{pmatrix} g_\perp & 0 & 0 \\ 0 & g_\perp & 0 \\ 0 & 0 & g_\parallel \end{pmatrix} \tag{6}$$

Using the orthogonality properties of the **D** matrix, the time-averaged components of the **g**′ tensor are found to be

$$g_\perp = \tfrac{1}{2}(1 - \langle \cos^2\theta_1 \rangle)(g_{xx} - g_{yy}) + \tfrac{1}{2}(1 - \langle \cos^2\theta_3 \rangle)(g_{zz} - g_{yy}) + g_{yy} \tag{7}$$

$$g_\parallel = \langle \cos^2\theta_1 \rangle (g_{xx} - g_{yy}) + \langle \cos^2\theta_3 \rangle (g_{zz} - g_{yy}) + g_{yy} \tag{8}$$

Analogous expressions can be written for $A_\perp$ and $A_\parallel$, but before doing this, we introduce the following simplification. From single-crystal studies it is known that $A_{xx} \approx A_{yy}$ and using this approximation we obtain the simpler formulas

$$A_\perp = \tfrac{1}{2}(1 - \langle \cos^2\theta_3 \rangle)(A_{zz} - A_{xx}) + A_{xx} \tag{9}$$

$$A_\parallel = \langle \cos^2\theta_3 \rangle (A_{zz} - A_{xx}) + A_{xx} \tag{10}$$

The meaning of $A_\parallel$, $A_\perp$, $g_\parallel$, and $g_\perp$ can best be illustrated by recalling for a moment the situation in single-crystal experiments (Fig. 4). In single crystals A_{xx} and g_{xx}, for example, are the experimentally observed hyperfine splitting and g factor if the magnetic field is applied along the x axis of the NO radical. Analogously, in liquid crystals $A_\parallel$ and $g_\parallel$ are the experimentally observed hyperfine splitting and g factor if the magnetic field is applied along the director axis z'. On the other hand, with the external field perpen-

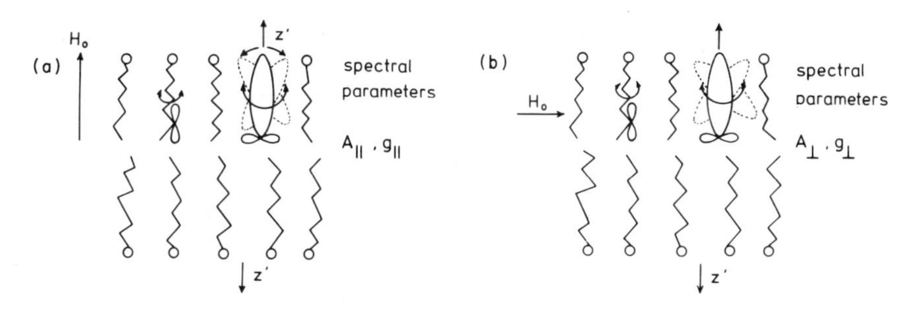

Fig. 4. Spectral parameters for oriented liquid crystals. A fatty acid spin probe and a steroid spin probe are incorporated into a lipid bilayer. The axis of motional averaging z' is the normal on the bilayer surface. In (a) the magnetic field is applied parallel to z'. The experimental spectra correspond to the spectra shown in Figs. 6 and 7 (solid lines) and the experimental parameters are $A_{\parallel}$ and $g_{\parallel}$. In (b) the magnetic field is aligned perpendicular to z'. The experimental spectra correspond to the dashed spectra in Figs. 6 and 7. The experimental parameters are $A_{\perp}$ and $g_{\perp}$.

dicular to z', the measured values are $A_{\perp}$ and $g_{\perp}$. In the general case, where the magnetic field makes an angle φ with the axis of motional averaging, the following relations hold:

$$A_{\text{exp}} = (A_{\parallel}^2 \cos^2 \varphi + A_{\perp}^2 \sin^2 \varphi)^{1/2} \tag{11}$$

$$g_{\text{exp}} = (g_{\parallel}^2 \cos^2 \varphi + g_{\perp}^2 \sin^2 \varphi)^{1/2} \tag{12}$$

With these equations we have already derived the basic results for the interpretation of the spin probe spectra in liquid crystals.

2. Order Parameters

For a given set of experimental parameters A_{exp}, g_{exp}, and φ, the above equations can be used to determine the mean angular fluctuations $\langle \cos^2 \theta_i \rangle$ in liquid crystals. This is the most interesting application of spin probes in such mesophases. It has been found convenient to represent the fluctuations in terms of order parameters S_{ii}, which are defined as follows:

$$S_{ii} = \tfrac{1}{2}(3\langle \cos^2 \theta_i \rangle - 1), \qquad i = 1, 2, 3 \tag{13}$$

The order parameters are part of a 3×3 order matrix which has been introduced for the interpretation of NMR spectra in liquid crystalline solvents (Saupe, 1964). In the NMR experiments the principal axes of the molecule-fixed interaction tensors are generally unknown and the order matrix may then contain up to five independent elements. In the case of the NO radical the principal axes of the **g** and **A** tensors are known, however, and there is no reason or practical advantage to elaborate on the matrix formalism. To be in accordance with the NMR nomenclature we retain the double indexing of S, but actually there are only three order parameters left, corresponding to the fluctuations of the x, y, and z axes of the NO group. Due to the orthogonality relation of the cosines

$$\cos^2 \theta_1 + \cos^2 \theta_2 + \cos^2 \theta_3 = 1 \tag{14}$$

only two order parameters are independent. The third parameter is then fixed by the relation

$$S_{11} + S_{22} + S_{33} = 0 \tag{15}$$

Combining Eqs. (7)–(10) with Eq. (13) yields the following convenient expressions:

$$S_{33} = (A_{\parallel} - A_{\perp})/(A_{zz} - A_{xx}) \tag{16}$$

$$S_{11} = [3g_{\parallel} - (g_{xx} + g_{yy} + g_{zz}) - 2S_{33}(g_{zz} - g_{yy})]/2(g_{xx} - g_{yy}) \tag{17}$$

The order parameter S_{33}, i.e., the order parameter of the nitrogen 2p π orbital, can thus be determined from a measurement of the hyperfine splittings alone (with the assumption $A_{xx} = A_{yy}$), while the determination of S_{11} requires additional g-factor measurements.

The vocable "order parameter" for S_{ii} is perhaps misleading, since S_{ii} is not only related to the statistical order of the system but also to the geometry of the spin probe.

Let us illustrate the influence of the molecular geometry with three different spin probes, where the nitrogen 2p π orbital (z axis) makes the angles $\xi = 0$, 54.7, and 90° with the long molecular axis (cf. Fig. 5). We consider the hypothetical situation where the long molecular axis remains exactly parallel to the z' axis, that is, angular fluctuations do not occur. The nitrogen 2p π orbital is then rotating around z' at angles $\theta_3 = \xi = 0$, 54.5, and 90°, respectively. Inserting these values into Eq. (13) leads to order parameters

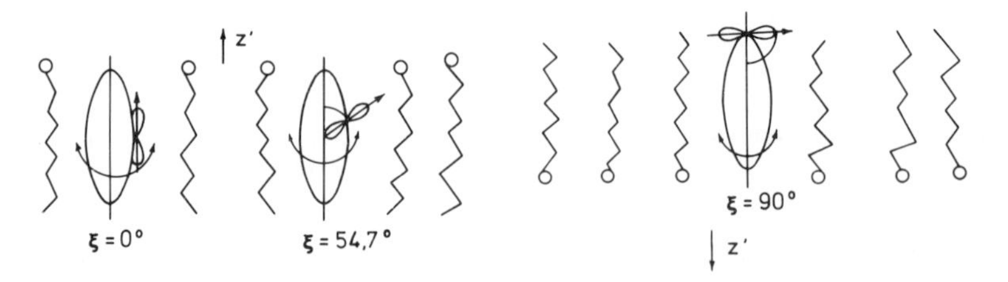

Fig. 5. Order parameter and molecular geometry. A rigid molecule is aligned parallel to z' and rotating around its long molecular axis. The order parameter S_{mol} is thus $S_{\mathrm{mol}} = 1$. The nitrogen 2p π orbital is attached at three different angles $\xi = 0$, 54.7, and 90°. The order parameter S_{33} is then given by $S_{33} = 1$, 0, and $-1/2$, respectively.

$S_{33} = 1$, 0, and -0.5, respectively. Thus even if the molecular order is the same, different values S_{33} are observed depending on the orientation of the NO group with respect to the rest of the molecule.

For a given spin probe with a rigid molecular frame the order parameters are, of course, a measure of the anisotropy or randomness of the motion. A totally random motion leads to $A_{\parallel} = A_{\perp} = a_N$ and $g_{\parallel} = g_{\perp} = g$, with

$$a_N = \tfrac{1}{3}(A_{xx} + A_{yy} + A_{zz}) \tag{18}$$

and

$$g = \tfrac{1}{3}(g_{xx} + g_{yy} + g_{zz}) \tag{19}$$

a_N and g are the isotropic hyperfine splitting constant and the isotropic g factor, respectively. From Eqs. (7)–(10) and (13) it then follows that the isotropic motion is characterized by $\langle \cos^2 \theta_i \rangle = \frac{1}{3}$ and $S_{ii} = 0$. (This corresponds to an apparent angular fluctuation of 54.7°, which is half the tetrahedral angle. $\theta = 54.7°$ is an *apparent* fluctuation angle because the measured quantity is the average of $\cos^2 \theta_i$.) Any positive or negative deviations from $S_{ii} = 0$ indicate an anisotropic motion of the molecule. However, it should be borne in mind that an experimental result $S_{ii} = 0$ can also be caused by a combination of an anisotropic motion with a specific orientation of the probing group.

In general we are most interested in the order parameter S_{mol} of the long molecular axis. S_{mol} can be determined from the experimental S_{33} by the following relation:

$$S_{mol} = S_{33}[(3\cos^2 \xi - 1)/2]^{-1} \tag{20}$$

where ξ denotes the angle between the long molecular axis and the corresponding NO axis. Equation (20) is valid only under the condition $S_{11} = S_{22}$, which is approximately fulfilled in many spin label experiments.†

3. Line Positions and Sample Geometry

So far we have considered the EPR spectrum of a homogeneous microregion. We have seen that the position of the resonance lines depends on the molecular fluctuations (characterized by the order parameters) as well as on the orientation of the director axis $\mathbf{z}'$ with respect to the external magnetic field $\mathbf{H}_0$ (characterized by φ). We now proceed to discuss macroscopic samples of liquid crystals where the $\mathbf{z}'$ axes of the domains are not necessarily aligned parallel to each other, but may also be oriented in a cylindrical fashion or distributed at random. The appearance of the EPR spectrum is strongly influenced by the specific type of distribution.

† Editor's note: See also Chapter 12 and Appendix IV, p. 567, for more discussion on order parameters.

a. Planar Oriented Samples The simplest situation is a parallel alignment of the $\mathbf{z}'$ axes, where all microregions have the same orientation with respect to $\mathbf{H}_0$. In *nematic* liquid crystals such a parallel alignment is easily induced by applying a strong magnetic field. The rodlike molecules orient themselves generally parallel to $\mathbf{H}_0$. For this reason nematic liquid crystals have become very popular as anisotropic solvents in NMR spectroscopy (Saupe, 1968; Diehl and Khetrapal, 1969; Meiboom and Snyder, 1971).

Similarly, thermotropic *smectic* phases can also be aligned in a magnetic field when cooled down slowly from the nematic or isotropic phase. Once aligned, the order persists for a long time even in the absence of the magnetic field. Ordered smectic phases offer the possibility of arbitrary control of the direction of orientation (Luz and Meiboom, 1973).

Lyotropic liquid crystals are more difficult to orient. One example has been reported where a *lyotropic nematic* phase was aligned in a magnetic field (Lawson and Flautt, 1967). This specific system shows the rather unusual property that the preferred orientation of the molecules is perpendicular to the magnetic field.

Ordered bilayers (*lyotropic smectic* mesophase) can be prepared by pressing the phase between two closely spaced, parallel glass surfaces (de Vries and Berendsen, 1969; Seelig, 1970). Shearing forces and the contact with the glass surfaces arrange the microcrystalline regions in parallel layers on the supporting plates. The system has then a unique axis of orientation, which is identical with the normal of the glass plates. This axis becomes the optical axis of the liquid crystalline system. It is therefore possible to observe the ordering process by means of a polarizing microscope. A well-ordered sample shows pseudoisotropy, that is, it appears homogeneously black if viewed under crossed polarizers. Ordered bilayers can also be oriented with respect to the magnetic field, the direction of the molecular alignment being independent of the magnetic field.

The formal treatment developed in Section A.1 can be applied to planar oriented samples without any further modification. Let us illustrate this for two examples of practical importance. Figures 6 and 7 show the EPR spectra of two spin probes having the same rotational frequency ($\tau_{20} = \tau_{22} = 2.3 \times 10^{-10}$ sec) and the same order parameter of the long molecular axis ($S_{\text{mol}} = 0.6$), but having different orientations of the nitrogen 2p π orbital. In Fig. 6 the nitrogen 2p π orbital is *parallel* to the long molecular axis. This corresponds to the schematic diagram in Fig. 5a and is realized in practice by fatty acid spin labels **I**(m, n) (the $n + 2$ doxyl derivatives of an

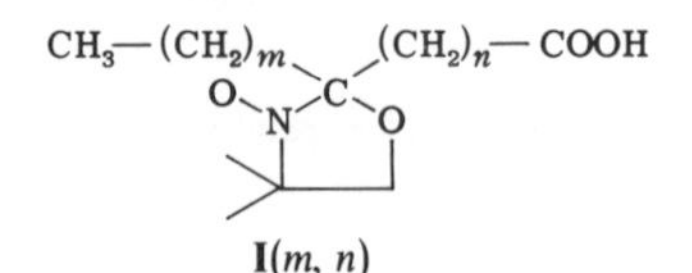

I(m, n)

$(m + n + 3)$-carbon fatty acid). In Fig. 7 the nitrogen 2p π orbital is *perpendicular* to the axis of rotation. This geometry is represented schematically by Fig. 5c. Typical practical examples are the steroid spin probes **II**.

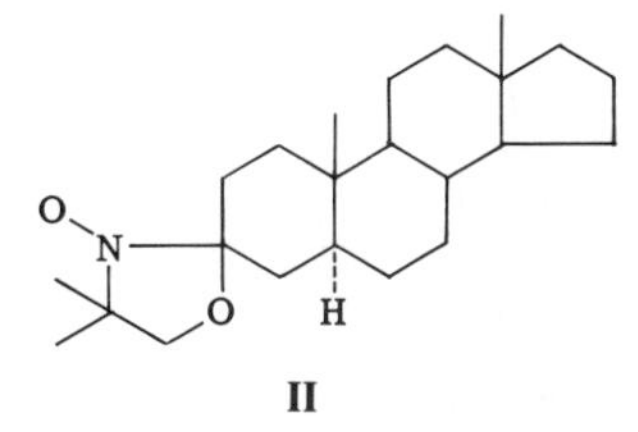

II

Before proceeding, a brief comment on the influence of the solvent polarity is necessary (Seelig, 1970; Hubbell and McConnell, 1971). It has been found that the hyperfine splitting constant a_N is sensitive to changes in the solvent polarity, increasing polarity leading to increasing values of a_N. (The solvent dependence of the g factor is small and can be neglected here.) If both orientations $\mathbf{H}_0 \parallel \mathbf{z}'$ and $\mathbf{H}_0 \perp \mathbf{z}'$ have been measured, as has been done in Figs. 6 and 7, the polarity of the spin label environment can easily be determined according to

$$a'_N = \tfrac{1}{3}(A_{\parallel} + 2A_{\perp}) \tag{21}$$

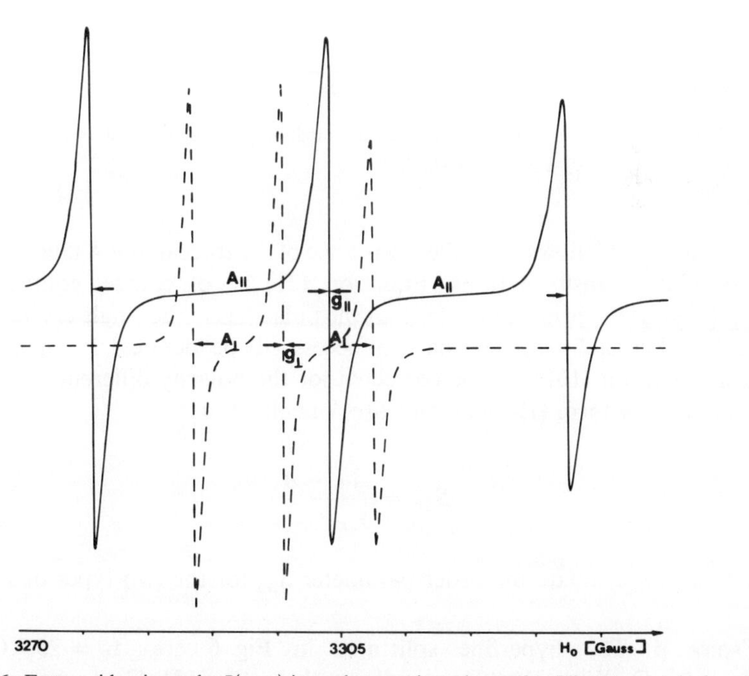

Fig. 6. Fatty acid spin probe **I**(m, n) in a planar oriented sample of liquid crystals. (——) Magnetic field $H_0 \parallel z'$; (---) magnetic field $H_0 \perp z'$.

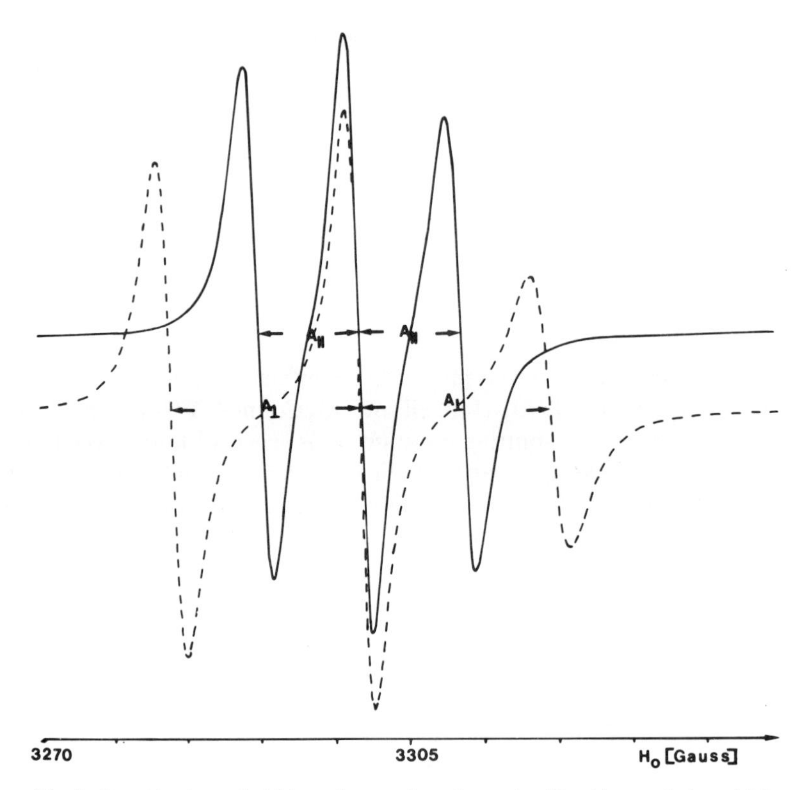

Fig. 7. Steroid spin probe **II** in a planar oriented sample of liquid crystals. (——) Magnetic field $H_0 \parallel z'$; (---) magnetic field $H_0 \perp z'$.

Equation (21) follows from the invariance of the trace of the **A** tensor against coordinate transformations. Equation (21) is, of course, equivalent to Eq. (18) but the prime indicates that the polarities in the single crystal and in the liquid crystalline phase may not necessarily be identical. In this case, our basic equation (16) must be corrected for the polarity difference and takes the following form (Hubbell and McConnell, 1971):

$$S_{33} = \frac{A_{\parallel} - A_{\perp}}{A_{zz} - A_{xx}} \frac{a_{\mathrm{N}}}{a'_{\mathrm{N}}} \tag{22}$$

Let us now evaluate the order parameter S_{33} for the two types of spectra.

Figure 6 The hyperfine splittings in Fig. 6 are $A_{\parallel} = 25.7$ G and $A_{\perp} = 9.7$ G. With Eq. (21) it follows that $a'_{\mathrm{N}} = 15.0$ G. As approximate molecular parameters we use the following single-crystal data: $A_{xx} = A_{yy} = 5.8$ G, $A_{zz} = 30.8$ G, $a_N = 14.1$ G. This yields

$$S_{33} = \frac{25.7 - 9.7}{30.8 - 5.8} \frac{14.1}{15.0} = 0.60$$

At this point we should admit that Figs. 6 and 7 are actually computer simulations of experimental spectra (Schindler and Seelig, 1973), for which we know the molecular data quite accurately, namely

$$A_{xx} = 6.95 \text{ G} \qquad A_{yy} = 5.35 \text{ G} \qquad A_{zz} = 33 \text{ G}$$

$$g_{xx} = 2.00872 \qquad g_{yy} = 2.00616 \qquad g_{zz} = 2.0027$$

$$S_{33} = 0.60 \qquad S_{11} = -0.35 \qquad S_{22} = -0.25$$

Thus our simple evaluation of S_{33} is identical with the exact value for S_{33}.

Figure 7 The experimental data are $A_{\parallel} = 9.4$ G and $A_{\perp} = 17.8$ G. This yields $a_N' = 15.0$ G and

$$S_{33} = \frac{9.4 - 17.8}{30.8 - 5.8} \frac{14.1}{15.0} = -0.31$$

Applying Eq. (20), the order parameter of the molecular axis is found to be

$$S_{mol} = -0.31 \times [(3 \cos^2 90° - 1)/2]^{-1} = 0.62$$

The computer simulation of Fig. 7, which actually represents an experimental spectrum for a steroid spin probe in a bilayer, was based on the following set of molecular parameters:

$$A_{zz} = 6.6 \text{ G} \qquad A_{yy} = 5.0 \text{ G} \qquad A_{zz} = 30.8 \text{ G}$$

$$g_{xx} = 2.0080 \qquad g_{yy} = 2.0050 \qquad g_{zz} = 2.0035$$

Furthermore, in the calculation the nitrogen 2p π orbital was not oriented exactly perpendicular to the axis of rotation, but the z (x, y) axis of the NO group makes the angle 85° (110°, 21°) with this axis. The exact order parameters are therefore

$$S_{33} = -0.30, \qquad S_{22} = -0.17, \qquad S_{11} = 0.47, \qquad S_{mol} = 0.60$$

From these two examples we may therefore conclude that even if the assumptions $A_{xx} = A_{yy}$ and $S_{11} = S_{22}$ ($S_{33} = S_{22}$ in the case of the steroid) are not valid, the use of Eqs. (22) and (20) still leads to rather precise values for S_{33} and S_{mol}.

The spectral differences between Figs. 6 and 7 can easily be understood by means of a purely qualitative argument. In Fig. 6 the z axis of the NO group is extended parallel to the axis of rotation. The separation $A_{\parallel}$ therefore corresponds to the almost pure $A_{zz} \approx 30$ G component, while $A_{\perp}$ is determined by the average $\frac{1}{2}(A_{xx} + A_{yy}) \approx 6$ G. In Fig. 7 the y axis of the NO group is parallel to the axis of rotation. Then $A_{\parallel}$ corresponds to the almost pure $A_{yy} \approx 6$ G component, while $A_{\perp}$ is determined by the average $\frac{1}{2}(A_{zz} + A_{xx}) \approx 18$ G. However, regardless of the different values for $A_{\parallel}$, $A_{\perp}$, and S_{33}, both spin probes yield the same result with respect to the ordering of the long molecular axes.

b. Random Samples A random sample is characterized by a completely isotropic distribution of the z' axes. This means either a chaotic collection of microregions or a spherical ordering of the domains. The first case is realized experimentally by filling the mesophase in an EPR sample tube of large diameter, while a spherical ordering is typical for lipid vesicles. Both arrangements yield the same type of spectra, which arise from a superposition of signals analogous to those shown in Fig. 6 or Fig. 7, but summed over all directions of space. The results are shown in Figs. 8a and 8b and correspond to labels **I**(m, n) and **II**. The spectra were calculated with the same set of parameters as used for Figs. 6 and 7. The spectral differences between Fig. 6 (Fig. 7) and Fig. 8a (Fig. 8b) are caused exclusively by the different geometries.

A random distribution of spin probes combined with an anisotropic rotation of the molecules is often encountered in biological systems, for example, in membranes. Since a homogeneous planar orientation of such systems is difficult to prepare, we are left with the problem of how to evaluate the order parameters. Fortunately, this is easy enough for S_{33} if the inner and outer hyperfine peaks are well resolved from each other, as, for example, in Fig. 8a. The separation of the outer peaks corresponds to twice the largest splitting in the oriented system, i.e., to $2A_{\parallel}$ in Fig. 6, while the separation of the inner peaks corresponds to twice the smallest splitting in the oriented phase, i.e., to $2A_{\perp}$ in Fig. 6 (Hubbell and McConnell, 1971). All further steps are then the same as illustrated earlier in the evaluation of S_{33} for the spectra of Figs. 6 and 7.

This type of analysis has several pitfalls, which shall be discussed briefly. The first problem is encountered in evaluating the spectra of random distributions of spin probes **II** (Fig. 8b). The separation of the outer peaks again corresponds to twice the largest splitting in the oriented case, but this separation must now be identified with $2A_{\perp}$, as can be seen from Fig. 7. Furthermore, the inner hyperfine extrema are not well resolved, and a judicial guess for a_N' must be made before S_{33} can be determined. This leads us to the second difficulty encountered with random distributions. If the motion is too slow or not sufficiently anisotropic, the inner and outer extrema are no

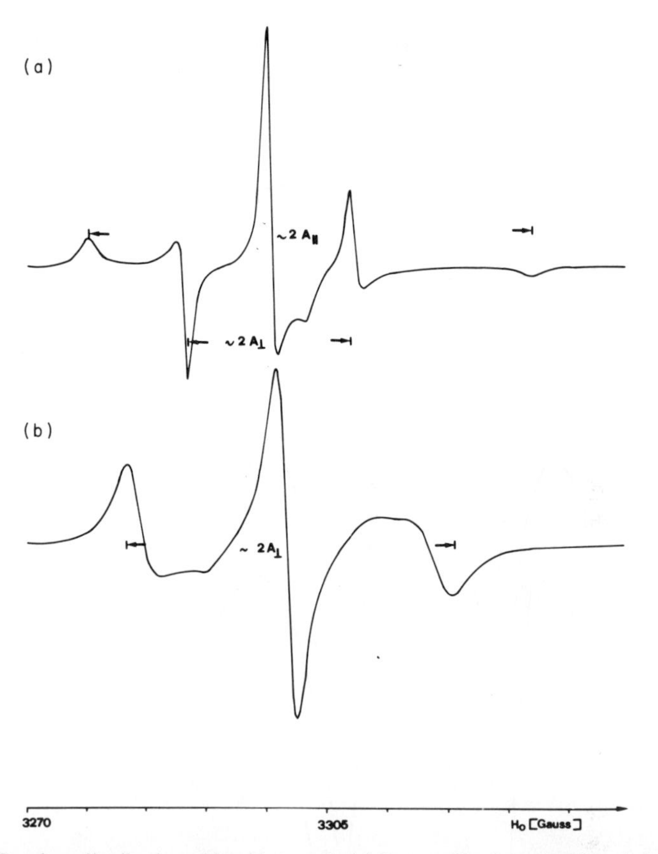

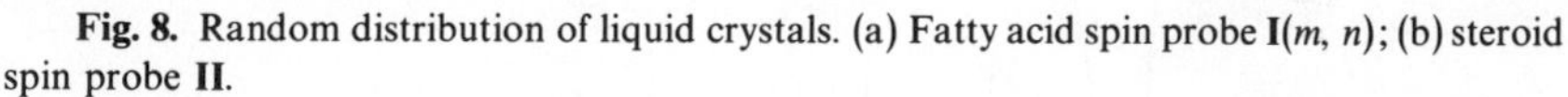

Fig. 8. Random distribution of liquid crystals. (a) Fatty acid spin probe **I**(m, n); (b) steroid spin probe **II**.

longer resolved. This is illustrated for labels **I**(m, n) in Fig. 9. In Fig. 9a the motion is slower by a factor of ten ($\tau_{20} = 3.0 \times 10^{-9}$ sec) as compared to Fig. 8a, but otherwise the molecular parameters are the same. As the order parameter decreases, the inner and outer hyperfine extrema move closer together and eventually collapse. The above method of evaluating S_{33} then breaks down and for order parameters $S_{33} < 0.4$ only the computer simulation of the spectrum yields the correct answer.

When working with biological membranes one should also be aware that the spin probes might be incorporated into domains with slightly different order parameters. This introduces an additional difficulty in evaluating S_{33}.

c. Cylindrically Oriented Samples A cylindrical orientation is obtained by sucking smectic liquid crystals into glass capillaries of very small diameter

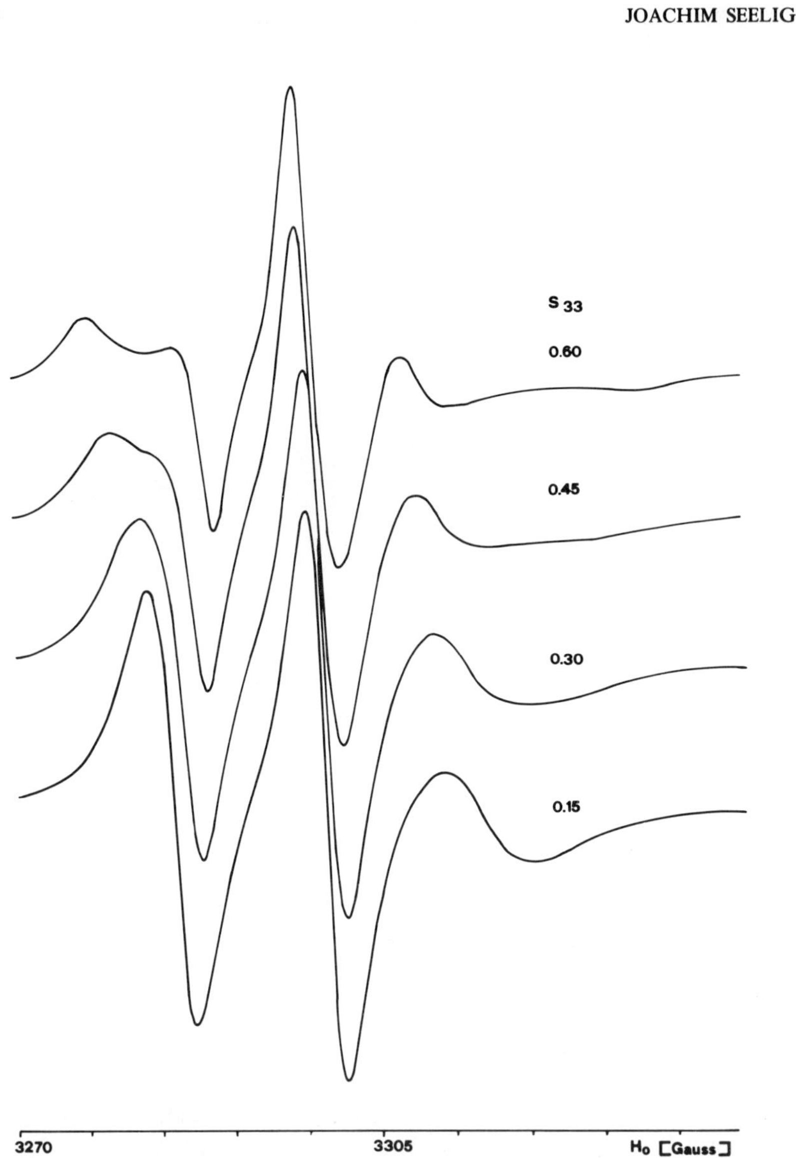

Fig. 9. Random distribution of liquid crystals (nematic or smectic phase). Influence of the order parameter on the spectral shape.

(Schindler and Seelig, 1973; Seelig and Limacher, 1974). The lyotropic phase is oriented on the glass surface and forms concentric cylinders around the axis of the capillary. The normal of the smectic plane is thus oriented perpendicular to the cylinder axis. The EPR spectra are again dependent on the orientation of the magnetic field with respect to the cylinder axis. If the magnetic field is applied parallel to the capillary axis, the spectrum is identical with that of an *oriented* sample with $\mathbf{H}_0 \perp \mathbf{z}'$ (dashed spectra in Figs. 6 and 7). However, if $\mathbf{H}_0$ is perpendicular to this axis, the spectrum resembles that of a *random* distribution. This is shown in Fig. 10. The intensity of the outer hyperfine extrema has increased because more spin probes

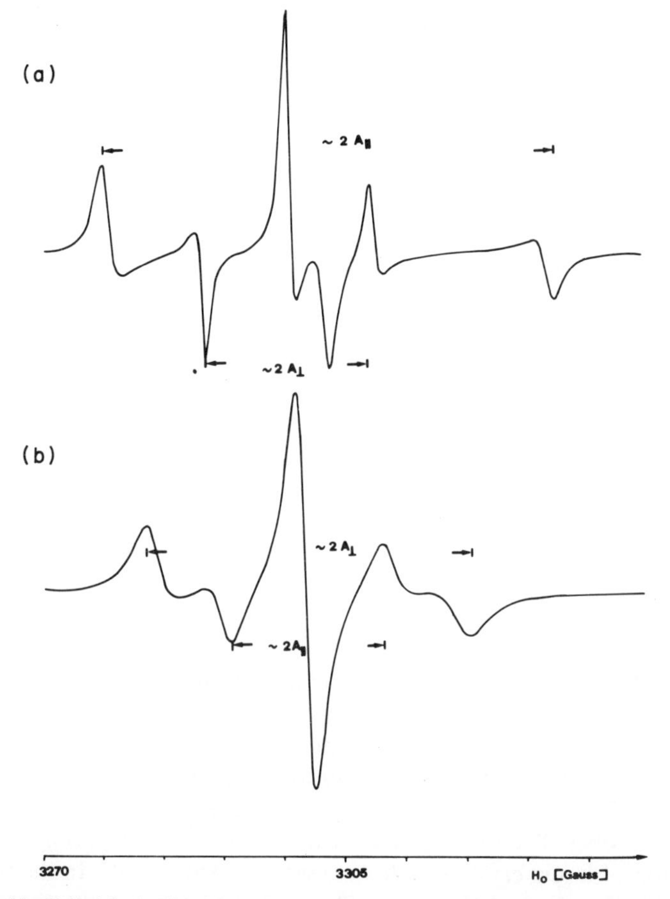

Fig. 10. Cylindrical orientation of liquid crystals (nematic or smectic phase). The magnetic field H_0 is aligned perpendicular to the cylinder axis. (a) Fatty acid spin probe **I**(m, n); (b) Steroid spin probe **II**.

are aligned parallel to $\mathbf{H}_0$ in the cylindrical than in the random distribution. Also, the resolution of inner and outer hyperfine extrema in the case of spin probe **II** has improved. These peculiar properties of the cylindrical geometry have not received much attention as yet, but it should be noted that such an ordering has a close resemblance to the orientation of the lipid bilayers in a nerve fiber.

4. Tilted Liquid Crystals

Tilted mesophases (Brown *et al.*, 1971; Luz and Meiboom, 1973) were first detected by means of X-ray diffraction. It was observed that in a number of smectic phases the thickness of the strata did not correspond to the length of the molecules. As a possible explanation it was suggested that the molecules within a stratum might be inclined. Thermotropic smectic phases have been divided therefore into several subgroups, the most important being the smectic A and the smectic C phase. In the smectic A modification the long axes of the molecules are perpendicular to the plane, while in the smectic C phase they are tilted away from the normal of the planes. The molecules in the smectic C phase execute two types of motion. The first is a rapid rotation of the molecules around their long axes, while the second corresponds to a slow reorientation about the normal of the smectic plane. Since the tilt angle $\bar{\vartheta}$ remains constant, the latter motion can be described as a precession of the molecules on a cone of vertex angle $2\bar{\vartheta}$, with the normal on the smectic plane as its axis. On the time scale of the EPR experiment the precession frequency is so slow that this motion may be considered as pseudostatic. Tilted liquid crystalline phases have a lower symmetry than the uniaxial systems discussed in the previous sections, and the exact description of magnetic interactions becomes complicated (Luz and Meiboom, 1973). An *approximate* picture may be obtained by using the following distribution function $p(\vartheta)$, where ϑ is the angle between the normal of the smectic plane and the molecular axis (McFarland and McConnell, 1971; McConnell and McFarland, 1972):

$$p(\vartheta) \sim \exp[-(\vartheta - \bar{\vartheta})^2/2\vartheta_0^2] \tag{23}$$

Equation (23) assumes that the molecules are distributed around the normal of the smectic plane according to a Gaussian distribution. The sharpness of this distribution is determined by the spread angle ϑ_0. The smaller is ϑ_0, the better is the alignment of the molecules along the tilt direction $\bar{\vartheta}$.

The specific geometry of a tilted smectic phase is reflected in the spectra of the spin probes. The tilt angle can be determined by comparing a random distribution of liquid crystals with a planar oriented sample. In the random distribution the effect of the tilt angle is averaged out and the spectrum is identical to Fig. 8. The order parameter S_{33} can be evaluated as described above from the separation of the inner and outer hyperfine extrema. In the planar oriented sample the position of the lines is shifted, however. If the magnetic field is applied perpendicular (parallel) to the smectic plane, the hyperfine splitting is smaller (larger) for tilted liquid crystals than for a nontilted phase with the same order parameter. This is shown in Fig. 11. The observed changes in the hyperfine splitting can be understood at least qualitatively by means of Eq. (11), where φ should be replaced by the tilt angle $\bar{\vartheta}$.

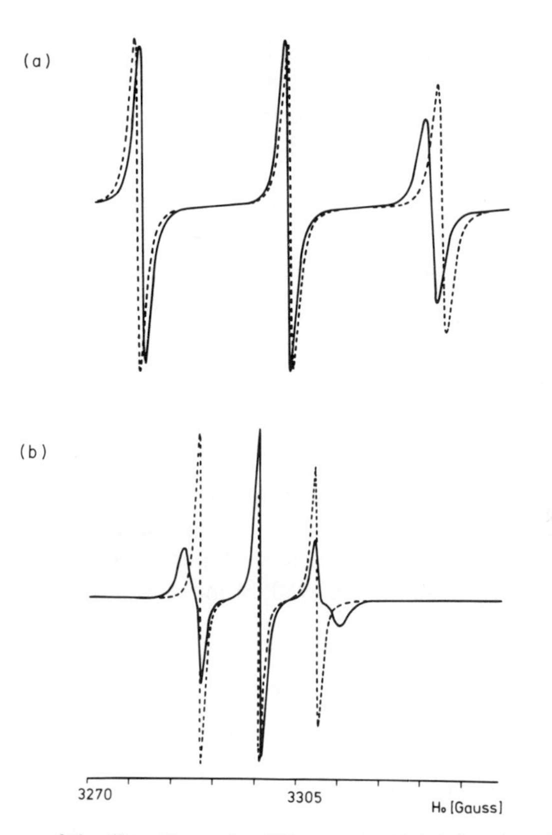

Fig. 11. Influence of the tilt angle on the EPR spectra of label **I**(*m*, *n*) in planar oriented liquid crystals: (a) magnetic field $H_0 \parallel z'$; (b) magnetic field $H_0 \perp z'$; (---) no tilt, spread $\vartheta_0 = 3.4°$; (——) tilt $\bar{\vartheta} = 20°$, spread $\vartheta_0 = 3.4°$.

For a well-oriented sample Eq. (11) can also be used to estimate the magnitude of the tilt angle, but in general a very careful line shape analysis is required to obtain precise values for $\bar{\vartheta}$. This is due to the asymmetric broadening of lines, as shown in Fig. 11, and also to the fact that a large spread angle (poorly oriented sample) causes similar spectral changes. Only a computer simulation of the spectrum provides sufficient accuracy to distinguish between the two parameters.

B. Linewidth and Spectral Simulations

The dynamical properties of a spin probe are reflected in the transversal relaxation times $T_2(M)$. M is the nuclear spin quantum number (^{14}N: $M = +1, 0, -1$). Each of the three resonance lines in a nitroxide EPR spectrum is therefore characterized by a specific relaxation time. $T_2(M)$ is easily evaluated from the *width* of the resonance lines. Assuming Lorentzian line shapes and recording the spectrum in the derivative mode, $T_2(M)$ is related to the linewidth $\Delta\nu(M)$ as follows:

$$\Delta\nu(M) = 1/\sqrt{3}\,\pi T_2(M) \tag{24}$$

[$\Delta\nu(M)$ is the separation of the maximum and minimum of the resonance line, measured in frequency units.] In the case of the NO radical the dominant contributions to $T_2(M)$ are the *motional modulations* of the hyperfine-tensor and **g**-tensor interactions. The quantitative analysis of these interactions in terms of a T_2 formula combined with a precise measurement of the linewidth provides information concerning the *rate* of *molecular reorientation.*

If the motion is isotropic, that is, if the reorientations around the three molecular axes occur with equal probability and equal rate, $T_2(M)$ in a liquid is easily evaluated by means of perturbation theory (Freed and Fraenkel, 1963, 1964; Kivelson, 1960; Stone *et al.*, 1965; Poggi and Johnson, 1970):

$$\begin{aligned} T_2(M)^{-1} = {} & [\tfrac{1}{20}b^2(3 + 7u) + \tfrac{1}{15}(\Delta\gamma H_0)^2(\tfrac{4}{3} + u) \\ & + \tfrac{1}{8}b^2M^2(1 - \tfrac{1}{5}u) + \tfrac{1}{5}b\,\Delta\gamma H_0 M(\tfrac{4}{3} + u)]\tau_r \end{aligned} \tag{25}$$

In Eq. (25) H_0 is the applied magnetic field, τ_r is the correlation time, and

$$\begin{aligned} b &= \tfrac{2}{3}[A_{zz} - \tfrac{1}{2}(A_{xx} + A_{yy})] \\ \Delta\gamma &= (|\beta|/\hbar)[g_{zz} - \tfrac{1}{2}(g_{xx} + g_{yy})] \\ u &= 1/(1 + \omega_0^2\tau_r^2) \end{aligned}$$

The nonsecular terms containing u are important for very short correlation times, but are neglected in some of the T_2 formulas found in the literature. In addition to the **A**- and **g**-tensor interactions there are also other contributions to $T_2(M)$, which are not included in Eq. (25), such as spin rotational interactions, unresolved proton hyperfine splitting, and instrumental modulation broadening. These broadening mechanisms are summarized as residual linewidth. The residual linewidth does not depend on M and is added as a constant to the $T_2(M)$ formula. A critical evaluation of the factors involved in the determination of $T_2(M)$ for spin probes has been given by Poggi and Johnson (1970).

Equation (25) shows that the isotropic reorientation of the spin probe is characterized by a *single* correlation time. The correlation time τ_r depends on the effective radius r and the local viscosity η of the solvent, in the following way:

$$\tau_r = 4\pi\eta r^3/3kT \tag{26}$$

The measurement of the linewidth thus not only allows a determination of the correlation time but also yields an estimate of the local viscosity of the solvent as experienced by the spin probe during rotation.

If we consider the *anisotropic* motion in a liquid crystalline phase, the linewidth analysis becomes rather tedious. No simple rules analogous to those derived for the line positions have been established as yet. This is due to several reasons. First, the spin probe is partially aligned and the rotational reorientations around the three molecular axes can proceed at different rates. In the case of liquid crystals it is reasonable to assume an *axially symmetric* diffusion tensor and the reorientation is characterized by at least *two* correlation times. Second, the linewidth depends not only on the *rate* of motion but also on the *orientation* of the spin probe with respect to the magnetic field. The influence of the orientation can be seen by comparing the linewidth for the two orientations $\mathbf{H}_0 \parallel \mathbf{z}'$ and $\mathbf{H}_0 \perp \mathbf{z}'$ in Fig. 5 or Fig. 6. Third, the linewidth formula depends strongly on the model chosen to describe the reorientation (cf. Chapters 2 and 3). Finally, the above-mentioned problem of the residual linewidth is also encountered in the case of anisotropic rotations. In liquid crystals and bilayer membranes the observed peak-to-peak linewidth is of the order of 1.5–3 G. Computer simulations of the spectra indicate that up to 50% of the total linewidth must be attributed to the residual linewidth.

In the earliest simulations of anisotropic spin probe spectra these difficulties were circumvented by assuming Lorentzian or Gaussian lines of *constant* linewidth. The averaging over all orientations was performed by means of Monte Carlo methods (Itzkowitz, 1967; Hubbell and McConnell, 1971; Jost *et al.*, 1971). From the parameters of the Monte Carlo calculation a crude estimate of the correlation time for the rotation around the long molecular axis was obtained. The agreement between the experimental and calculated spectra was then improved by a semiempirical application of perturbation theory (McFarland and McConnell, 1971; McConnell and McFarland, 1972). The relaxation time was assumed to be of the form

$$1/T_2(M) = A_M + B_M M + C_M M^2 \tag{27}$$

The parameters A_M, B_M, and C_M were dependent on the orientation of the sample and had to be optimized for each M. Since a rather large set of empirical parameters had to be used in Eq. (27), the next consequent step was the rigorous application of perturbation theory, calculating the linewidth as a function of the correlation times and the orientation with respect to the magnetic field (Luckhurst and Sanson, 1972; Schindler and Seelig, 1973).

Let us briefly outline the essential aspects of the theory (Schindler and Seelig, 1973). The spin Hamiltonian in the molecular reference frame is given by

$$\hbar\mathscr{H}(t) = \beta_e \mathbf{S}(t)\mathbf{g}\mathbf{H}_0 + \mathbf{S}(t)\mathbf{A}\mathbf{I}(t) \tag{28}$$

The nuclear Zeeman term has been neglected, and $\mathbf{S}(t)$ and $\mathbf{I}(t)$ are the time-dependent electronic and nuclear spin operators, respectively. The magnetic field $\mathbf{H}_0$ is known only in the space-fixed system of the laboratory. The molecule-fixed components of the angular momenta and magnetic field must therefore be expressed in terms of their space-fixed components. In principle, this could be done by means of similar trigonometric operations as applied in Section III.A. A more elegant approach is to rewrite Eq. (28) in the terms of spherical coordinates (cf. Chapter 2):

$$\hbar\mathscr{H}(t) = \sum_{l,\, m,\, \lambda} F_\lambda^{(l,\, -m)} A_\lambda''^{(l,\, m)}(t) \tag{29}$$

The spatial factors $F_\lambda^{(l,\, -m)}$ and the irreducible tensors $A_\lambda^{(l,\, m)}$ are listed in Table I. The coordinate transformation can then be performed by means of

TABLE I

IRREDUCIBLE TENSOR OPERATORS $A_\lambda^{(l,\, m)}$ AND SPATIAL FACTORS $F_\lambda^{(l,\, -m)}$

$A_A^{(0,\,0)} = [S_z I_z + \frac{1}{2}(S^+ I^- + S^- I^+)]$	$F_A^{(0,\,0)} = ha$
$A_A^{(2,\,0)} = (\frac{8}{3})^{1/2}[S_z I_z - \frac{1}{4}(S^+ I^- + S^- I^+)]$	$F_A^{(2,\,0)} = (\frac{3}{2})^{1/2} h\, \Delta a$
$A_A^{(2,\,\pm 1)} = \mp(S_z I^\pm + S^\pm I_z)$	$F_A^{(2,\,\pm 1)} = 0$
$A_A^{(2,\,\pm 2)} = S^\pm I^\pm$	$F_A^{(2,\,\pm 2)} = \frac{1}{2} h\, \delta a$
$A_G^{(0,\,0)} = S_z$	$F_G^{(0,\,0)} = H_0 \beta_e g$
$A_G^{(2,\,0)} = (\frac{8}{3})^{1/2} S_z$	$F_G^{(2,\,0)} = (\frac{3}{2})^{1/2} H_0 \beta_e\, \Delta g$
$A_G^{(2,\,\pm 1)} = \mp S^\pm$	$F_G^{(2,\,\pm 1)} = 0$
$A_G^{(2,\,\pm 2)} = 0$	$F_G^{(2,\,\pm 2)} = \frac{1}{2} H_0 \beta_e\, \delta g$
$a = \frac{1}{3}(A_{xx} + A_{yy} + A_{zz})$	$g = \frac{1}{3}(g_{xx} + g_{yy} + g_{zz})$
$\Delta a = \frac{1}{6}(2A_{zz} - A_{xx} - A_{yy})$	$\Delta g = \frac{1}{6}(2g_{zz} - g_{yy} - g_{xx})$
$\delta a = \frac{1}{2}(A_{xx} - A_{yy})$	$\delta g = \frac{1}{2}(g_{xx} - g_{yy})$

Wigner rotation matrices $D_{m',\, m}^{(l)}$. We consider two rotations. The first is a transformation from the molecular coordinate system to the coordinate system of the liquid crystalline phase (defined by the director axis $\mathbf{z}'$), the second is a transformation from the liquid crystal to the coordinate system of the magnetic field:

$$\hbar\mathscr{H}(t) = \sum_{l,\, m,\, \lambda} F_\lambda^{(l,\, -m)} \sum_p D_{p,\, m}^{(l)}(\alpha\beta\gamma)(t) \sum_q D_{q,\, p}^{(l)}(\alpha'\beta'\gamma') A_\lambda^{(l,\, q)} \tag{30}$$

The time dependence is now contained in the rotation matrices $D^{(l)}_{p,m}(\alpha\beta\gamma)(t)$. The total spin Hamiltonian is then divided into two parts, which are evaluated separately:

$$\hbar\mathscr{H}(t) = \hbar\overline{\mathscr{H}} + \hbar[\mathscr{H}(t) - \overline{\mathscr{H}}] \tag{31}$$

The time-averaged part $\hbar\overline{\mathscr{H}}$ determines the line positions. Its calculation starting from Eq. (30) is greatly simplified by invoking the symmetry of the liquid crystalline phase. Evaluating the time-averaged Hamiltonian leads to identical results as were derived previously.

The advantage of irreducible tensor operators lies in calculating the dynamical part of the spin Hamiltonian, $\hbar[\mathscr{H}(t) - \overline{\mathscr{H}}]$, which determines the linewidth. In the fast tumbling region $\hbar[\mathscr{H}(t) - \overline{\mathscr{H}}]$ can be evaluated by means of perturbation theory. Contrary to the case for time-averaged Hamiltonian, *no exact* solution can be given for the dynamical part, but various approximations are possible depending on the model chosen for the molecular reorientation in the liquid crystals and on the number of terms included in the calculation. A detailed discussion is beyond the scope of this chapter and the reader is referred to the original literature cited earlier and to Chapters 2 and 3. Our own experience in simulating spectra of spin probes **I**(m, n) and **II** in liquid crystalline bilayers can be summarized as follows: (1) The molecular reorientation can be described sufficiently well using an anisotropic reorientation model, characterized by two correlation times τ_{20} and τ_{22}. τ_{20} is associated with the rate of fluctuations perpendicular to the long molecular axis, and τ_{22} measures the rate of rotation around this axis. For stiff, rodlike molecules dissolved in a liquid crystalline phase τ_{22} is distinctly shorter than τ_{20}, since the rotation around the long molecular axis experiences less friction than the rotation perpendicular to this axis. The ratio τ_{22}/τ_{20} depends not only on the geometry of the spin probe, but also is influenced by the structure of the phase. (2) In simpler linewidth calculations only the secular terms are considered. We find that pseudosecular and nonsecular terms may contribute up to 20% of the total linewidth and should therefore always be included in the linewidth calculation. (3) An extremely important problem is the choice of the correct molecular parameters (g and A tensors). Single-crystal data generally refer to an environment which is different from that in a liquid crystalline phase. Furthermore, spin probes used in single-crystal studies are structurally different from those used in mesophases. Notwithstanding the usefulness of single-crystal data in evaluating line *positions*, they are too crude an approximation for *linewidth* calculations in liquid crystals. The problem is further complicated by the fact that the A and g tensors must be transformed into the coordinate system of the anisotropic diffusion tensor.

The principal axes of this diffusion tensor are not known a priori and inspection of molecular models can only provide an approximate solution to this problem.

Linewidth calculations are therefore rather time-consuming, since a consistent set of molecular parameters must be established for each phase and each spin label. But if these parameters have been found, all spectra—regardless of the sample geometry or the numerical value of the correlation time—can be simulated with the same perfect accuracy. This yields precise information about the mobility of the molecules in the mesophase and some of the results will be discussed in more detail in the following section.

IV. NITROXIDE SPIN PROBE INVESTIGATIONS OF LIQUID CRYSTALLINE MESOPHASES

Although spin probes may resemble the host molecules in many respects (dimensions, stiffness, chemical structure), they do not form liquid crystalline phases by themselves. Spin probes only indirectly reflect the ordering of the host system. The interpretation of the following results in terms of the structure of the pure liquid crystalline phase should therefore be regarded as an approximation.

It is not intended to give a complete coverage of all nitroxide spin probe studies performed with liquid crystals. Instead this section is restricted to just a few applications in order to illustrate the potential of the technique.

A. Thermotropic Liquid Crystals

The most important aspect of a liquid crystalline phase is the orientational order of the constituent molecules. This can be measured approximately by spin probes. The first attempt to determine the order parameter of a *nematic* phase with nitroxides has been made by Ferrutti *et al.* (1969). It was observed that the probe molecules were generally slightly less ordered than the host molecules. This indicated that the spin probes introduced a small perturbation into the system. Nevertheless, nitroxide spin probes can be useful in several respects.

1. They can provide information about anisotropic solute–solvent interaction (Luckhurst, 1973). A knowledge of this interaction is important since it provides another test for the various statistical theories that have been advanced to describe nematic liquid crystals.
2. NO spin probes can easily detect phase transitions. For example, 4-*n*-butyloxybenzilidene-4′-acetoaniline forms a nematic phase between 110 and 98°C when there is a transition to a smectic A mesophase. Using steroid

nitroxides, a discontinuity in the order parameter has been detected at the nematic–smectic A transition, which demonstrates the first-order character of this transition (Luckhurst and Seteka, 1972). A second example is bis(4′-*n*-octyloxybenzol)-2-chloro-1,4-phenylenediamine. This molecule displays an extremely long nematic range, from 59 to 179°C. Studies with vanadyl acetylacetonate (VAAC) as a paramagnetic probe showed an increase of the order parameter with decreasing temperature. At 110°C a discontinuity of the slope of this curve was observed, which was interpreted as a second-order phase change. Subsequent experiments with nitroxide spin probes revealed, however, that the discontinuity is not present in the ordering curves of the nitroxides (Shutt *et al.*, 1973). The difference between the two measurements is probably due to an incorrect evaluation of the VAAC order parameter. At 110°C the molecular motions become too slow to effectively average the hyperfine anisotropies of VAAC, causing an abrupt contraction of the VAAC splittings. This difficulty does not arise with nitroxide probes because here the hyperfine splitting is much less than that of VAAC: 15 G for NO compared to 107 G for VAAC. Thus the dynamic range of NO probes is extended by a factor ten toward lower rates compared to VAAC.

3. An analysis of the linewidth yields the correlation times and the diffusion coefficient of the anisotropic motion (Luckhurst and Sanson, 1972). Such an analysis has been attempted for the smectic A mesophase of 3-*N*-(4′-ethoxybenzilideneamino)-6-*n*-butylpyridine (Luckhurst *et al.*, 1974). This compound forms a smectic A liquid crystal in the temperature range 29–49°C. At 49°C there is a phase transition to a nematic phase. This phase was doped with a steroid spin label and a homogeneous orientation of the smectic phase was obtained in the following way. The sample was heated until the nematic phase was formed and a strong magnetic field was applied. The molecules were thus aligned more or less parallel to the magnetic field. Next, the temperature was lowered slowly below the nematic–smectic A transition point. By this procedure the macroscopic ordering of the molecules was also maintained in the smectic A phase. The oriented phase could be rotated perpendicular to the direction of the magnetic field without impairing the homogeneous alignment of the microregions. The angular variation of the linewidth of the steroid spin probe was measured and interpreted with two different models for the molecular reorientation. Using the "strong collision" model the correlation time τ_{20} was found to be $\tau_{20} \sim 3 \times 10^{-8}$ sec rad^{-1}, while the correlation time τ_{22} was determined to be $\tau_{22} \sim 3 \times 10^{-9}$ sec rad^{-1}. On the other hand, applying the so-called "diffusion model" for the molecular reorientation yielded the principal components of the diffusion tensor of the steroid probe, namely $R_{\parallel} = 1.3 \times 10^8$ rad sec^{-1} and $R_{\perp} = 2.6 \times 10^6$ rad sec^{-1}. Both models explain the experimental data equally well and both approaches show that the motion of

the steroid spin probe is essentially a fast rotation of the molecule around its long axis combined with distinctly slower fluctuations perpendicular to the director axis of the liquid crystalline phase.

As a final example of spin probe investigations of thermotropic liquid crystals, let us consider a cholesteric mesophase (Luckhurst, 1973; Luckhurst and Smith, 1973). The cholesteric phase is induced by adding 1% cholesteryl chloride to the nematic phase of 4,4'-dimethoxyazoxybenzene. If a steroid nitroxide spin probe is dissolved in this cholesteric liquid crystal, the experimental EPR spectrum looks very similar to that shown in Fig. 10b. The explanation is the following: As is shown in Fig. 2, the cholesteric phase consists of a helical arrangement of nematic layers. In the EPR experiment the applied magnetic field aligns the helix axis perpendicular to the field direction. As long as the magnetic field is small, the distribution of the molecules around this helix axis has cylindrical symmetry, inducing a cylindrical distribution of steroid spin probes. The precise simulation of the EPR spectrum provides interesting information about the physical characteristics of the cholesteric phase.

B. Lipid Bilayers

Lipid bilayers belong to the class of *lyotropic* liquid crystals. Their preparation is rather simple. For example, mixing of potassium oleate (70–80 wt %) with a controlled amount of water (30–20 wt %) results in the formation of a bilayer phase which is stable at least in the range 20–80°C (Ekwall *et al.*, 1969; Axel and Seelig, 1973). The lamellar structure is easily verified by X-ray diffraction. Another lyotropic system with bilayer structure is a ternary mixture of sodium decanoate (~ 30 wt %), decanol (~ 40 wt %), and water (~ 30 wt %) (Seelig, 1970). Actually, there exists a wide variety of fatty acids, long-chain alcohols, and other amphiphilic substances that alone or in combination with each other form lyotropic liquid crystals of lamellar structure (Porter and Johnson, 1967; Brown and Labes, 1972). The above examples are therefore in no way unique, but they are mentioned here because their molecular characteristics have been investigated extensively by means of spin labels.

Physopholipid bilayers constitute a special class of lyotropic lipid bilayers. The chemical structure and the functional properties of phospholipids are more complicated than those of the simple soaplike molecules described here. Nevertheless, phospholipid bilayers have many *physical properties* in common with soaplike bilayers. The latter can therefore be regarded as suitable and simple model systems for investigating structural properties of lipid membranes. Compared to biological membranes or phospholipids, such systems have the following advantages: (a) The chemical composition is well defined and can be changed arbitrarily; (b) quite often the bilayer is composed of just one type of molecule, so that it is easy to study the influence of chain length, double bonds, water content, or variations in the polar group; (c) the overall geometry and the dimensions can easily be determined by X-ray diffraction; (d) it is possible to obtain a homogeneous orientation of the bilayer system, by pressing the phase between closely spaced quartz plates.

1. Segmental Order Parameters in Lipid Bilayers

The experimental EPR spectra of spin probes **I**(m, n) and **II** in oriented or randomly distributed bilayers closely resemble the computed spectra shown in Figs. 6–10. The evaluation of the order parameters is thus straightforward. Which information about the molecular architecture of the bilayer can be deduced from such experiments? As a representative example, we choose a bilayer composed of sodium decanoate (28 wt %), decanol (42 wt %), and water (30 wt %) (Seelig, 1970; Seelig *et al.*, 1972). In Fig. 12 the experimental order parameters S_{33}, determined with stearic acid spin labels **I**(m, n), are plotted as a function of the position of the NO group. The order parameter decreases as the nitroxide group is moved down the chain. Close to the water–lipid interface the motion of the spin probe is highly anisotropic, but becomes more and more isotropic toward the bilayer interior. From this spin probe behavior a preliminary picture of the bilayer can

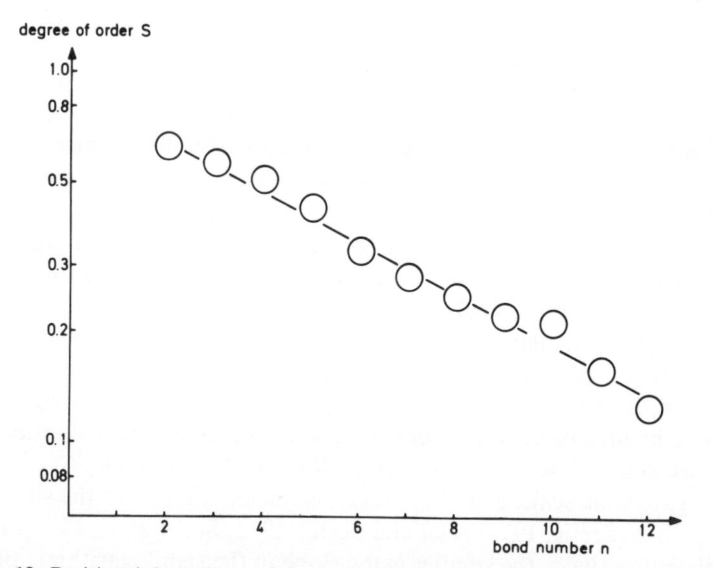

Fig. 12. Positional dependence of the order parameter in a lipid bilayer as detected with fatty acid spin probes **I**(m, n). [Reprinted with permission from Seelig *et al. J. Amer. Chem. Soc.* **94**, 6364–6371 (1972). Copyright by the American Chemical Society.]

be derived. The spin probes **I**(m, n) detect well-ordered lipid chains close to the polar surface with the hydrocarbon chains rotating rapidly around their long molecular axes. In the central part of the bilayer the lipid chains are disordered. The transition from an ordered to a disordered configuration proceeds gradually. This behavior has been observed for virtually all lipid membranes, which so far have been investigated with spin labels **I**(m, n), including phospholipid bilayers and biological membranes. Sometimes the terms "crystalline" and "fluid" are invoked to characterize the ordered and disordered regions of the bilayer. This is, however, misleading, since it implies different rates of motion. Actually, the rate of motion close to the polar region is equal to or even faster than that in the bilayer interior (cf. below).

A simple statistical model for the decrease of the order parameter is the following. We picture the polymethylene chain as a chain of freely jointed segments. We assume that each chain segment undergoes angular fluctuations with respect to the adjacent segments. These fluctuations endow the hydrocarbon chain with a certain flexibility. In the bilayer the hydrocarbon chain is anchored rigidly in the polar region. One end of the chain is thus fixed, while the other end is free to move. The overall motion experienced by a given chain segment is the larger the more segments contribute to it, that is, the larger the number of methylene units between the segment under investigation and the polar group. This "wormlike" model yields at least a qualitative explanation for the experimental results (Seelig, 1970).

The molecular basis for the chain flexibility are rapid interconversions between trans and gauche conformations of the individual carbon–carbon bonds. This rotational isomeric model is widely used to describe the configurational properties of polymers in solution (cf. Flory, 1969). It is essentially a linear Ising model and in the case of polymethylene chains in solution the chain properties are exclusively determined by the energy difference between trans and gauche conformations E_{rot} ($\sim$ 500–800 cal mole^{-1}). This model can be adapted to lipid bilayers in the following way (Seelig, 1971): The essential difference between an isotropic solution and a lipid bilayer is the intermolecular interaction between adjacent hydrocarbon chains in the lipid bilayer. This energy can be accounted for by an apparent increase in the energy difference between trans and gauche conformations. In the linear Ising model the pure intramolecular bond rotation energy is replaced by an empirical parameter E_σ which comprises contributions from intra- as well as intermolecular forces. If the order parameter S_{33} is plotted on a logarithmic scale, as in Fig. 12, E_σ is related directly to the slope of the straight line (Seelig, 1971; Axel and Seelig, 1973; Seelig *et al.*, 1974). The less flexible is the chain, the smaller is the slope in the semilogarithmic plot and the larger is E_σ. For the sodium decanoate–decanol bilayer (Fig. 12) E_σ is found to be 1400–1500 cal mole^{-1}. Subtracting the bond rotation potential leads to an interaction energy of 700–800 cal mole^{-1} per CH_2 unit for this bilayer. This result agrees with calorimetric and monolayer studies on similar systems.

The conceptual difficulty inherent to this statistical model is the following (Seelig and Niederberger, 1974b): The model describes the properties of individual chains. The interactions between adjacent chains are taken into account by an empirical parameter, but otherwise the chains are considered as completely independent of each other. The model does not explain how the lipid molecules are packed in the bilayer. This shortcoming is avoided by a two-dimensional cooperative model for the chain packing (Seelig and Niederberger, 1974b). The experimental basis for this new model is provided by recent deuterium label experiments, which provide another means for measuring the segmental order parameters. The deuterium data clearly deviate from the corresponding spin label results (Seelig and Seelig, 1974; Seelig and Niederberger, 1974a,b), and a certain revision of previous spin label interpretations is necessary. The spin probes **I**(m, n) introduce a small perturbation into the bilayer. Thus the spin probes do not detect the structure of the pure lipid bilayer, but reflect the response of the bilayer under a small physical stress. This is not meant to discriminate against spin labels. Proteins are known to penetrate the lipid phase of membranes, and they may deform the neighboring lipid chains in a similar way as the nitroxide group. The spin probes indicate the ease of hydrocarbon chain distortions at various depths in the bilayer.

After this brief aside into statistical mechanics we describe a few spin probe experiments that involve a bending or folding of the hydrocarbon chains. If the stearic acid probes **I**(m, n) are incorporated into a sodium octanoate–octanoic acid bilayer, the order parameter decreases faster than in the bilayer shown in Fig. 12. Approximately in the middle of the bilayer S_{33} approaches zero, and a further shift of the NO group down the chain leads to negative values for S_{33} (Seelig *et al.*, 1972). In this bilayer the lipid region is a few angstroms shorter than the stearic acid molecule. Therefore, in order to keep its tail in a hydrophobic environment, the molecule is forced into a bent configuration.

Negative order parameters are also observed for a bilayer composed of oleic acid molecules (Axel and Seelig, 1973; Seelig *et al.*, 1974). Here the decisive factor is the cis double bond, which introduces a bent configuration in the otherwise straight hydrocarbon chain. This bent configuration is maintained in the bilayer and is detected by the spin probes **I**(m, n). Since negative order is meaningless in a statistical sense, both examples illustrate the fact that the order parameters S_{ii} are determined by both the *configuration* of the spin probe and the *randomness* of the motion.

A folding of the hydrocarbon chain back on itself can be observed for dicarboxylic acid labels **III**(m, m) incorporated into lipid bilayers. If label **III**(6, 6) is dissolved in a sodium octanoate–octanoic acid bilayer, EPR

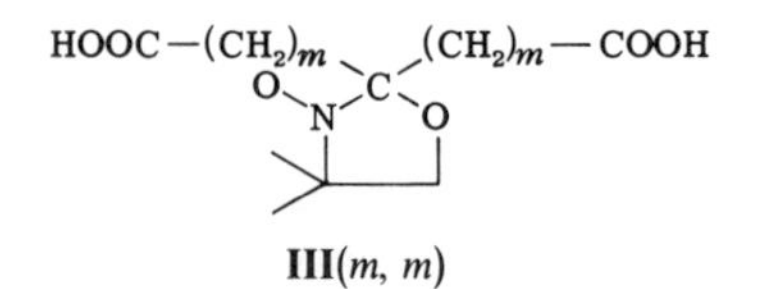

III(m, m)

spectra as shown in Fig. 13 are obtained (Seelig *et al.*, 1972). These spectra can be interpreted as a superposition of two different configurations of the spin label **III**(6, 6). Signal A comes from the linear, extended form of the molecule. The two carboxyl groups are anchored in *opposite* polar regions of the bilayer. Signal B is due to a folded configuration of the label, where both carboxyl groups lie in the same polar region. Increasing temperature shifts the equilibrium toward the folded form. The temperature dependence of the equilibrium constant of this folding process also allows a determination of the E_σ parameter (Limacher and Seelig, 1972).

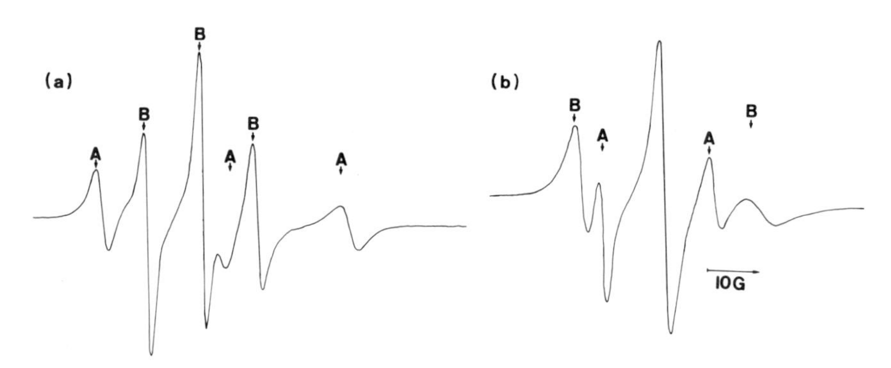

Fig. 13. Folding of fatty acid chains. Spin label **III**(6, 6) in an oriented lipid bilayer. (a) Applied magnetic field parallel to the bilayer normal; (b) applied magnetic field perpendicular to the bilayer normal. [Reprinted with permission from Seelig *et al. J. Amer. Chem. Soc.* **94**, 6364–6391 (1972). Copyright by the American Chemical Society.]

2. Rates of Molecular Motion

A detailed line shape analysis has been performed for spin probe **I**(13, 2) intercalated in a sodium decanoate–decanol bilayer (Schindler and Seelig, 1973). For this spin probe the correlation times τ_{22} and τ_{20}, referring to the rotation around and perpendicular to the long molecular axis, are found to be identical. At 22°C their value is $\tau_{20} = \tau_{22} = 2.3 \times 10^{-10}$ sec. The same type of analysis applied to the steroid spin probe **II** yields the following parameters (Schindler and Seelig, 1974): $\tau_{20} = 1.8 \times 10^{-9}$ sec and

$\tau_{22} = 3.2 \times 10^{-10}$ sec. Both spin probes indicate almost the same rate for the rotation around the long molecular axis (τ_{22}): The rotational diffusion constant is approximately $R_{\parallel} = 7 \times 10^{8}$ sec^{-1} for both spin probes. The two spin probes yield different results with respect to the molecular reorientation perpendicular to the long molecular axis (τ_{20}). This difference is characteristic of the different rigidity of the two labels. Label **I**(13, 2) has a flexible hydrocarbon chain and the equality $\tau_{20} = \tau_{22}$ is probably due to the fast rotational isomerizations between trans and gauche conformations. On the other hand, the steroid molecule **II** has a completely rigid molecular frame and the reorientation perpendicular to the long molecular axis is hindered by the neighboring lipid chains. Thus the correlation time τ_{20} increases and the rotational diffusion coefficient decreases ($R_{\perp} = 9 \times 10^{7}$ sec^{-1}) compared to the rotation around the molecular axis. The line shape analysis of spin probes **I**(m, n) with the label group further down the chain has not reached the same state of perfection. Our preliminary data indicate, however, a small decrease of the rates of motion toward the central part of the bilayer.

The temperature dependence of the correlation times in the sodium decanoate–decanol bilayer has also been measured. In the range 8–44°C label **I**(13, 2) and steroid label **II** indicate an exponential temperature dependence of τ_{20} and τ_{22}. The activation energy is approximately 7 kcal mole^{-1}. Below 8°C the correlation times increase in an irregular fashion. Both types of label indicate a structural change in the bilayer, characterized by a tilting of the fatty acid chains with respect to the bilayer normal. The tilt angle increases gradually with decreasing temperature.

The correlation times can also be interpreted in terms of a microviscosity η_{local} of the bilayer phase. Taking into account the different molecular geometries, both spin probes **I**(13, 2) and **II** yield the same result for this local viscosity, namely $\eta_{\text{local}} \approx 4 \times 10^{-2}$ P at 22°C.

C. Lyotropic Liquid Crystals with Cylindrical Structure

In a hexagonal phase the amphiphilic molecules are arranged in rods of cylindrical cross section (Fig. 3b). Such a liquid crystalline phase is formed, for example, by a ternary mixture of the following chemical composition: sodium octanoate (43 wt %), decanol (8 wt %), and water (49 wt %) (Ekwall *et al.*, 1968). This phase has been investigated with spin labels **I**(m, n) (Seelig and Limacher, 1974). Figure 14a shows the experimental EPR spectrum of label **I**(4, 2) for a random distribution of the hexagonal phase. This spectrum differs appreciably from that of a bilayer (Fig. 8a). It shows less fine structure and the outer extrema are smeared out to some extent. On the other hand, the linewidth of the central peak is practically the same in both

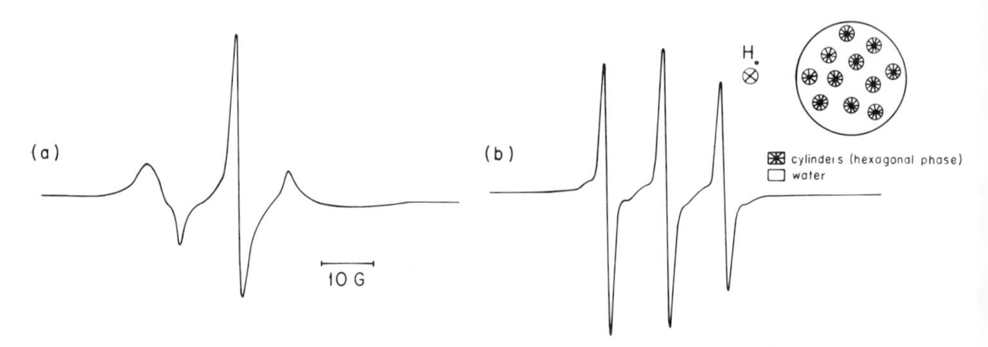

Fig. 14. Lyotropic hexagonal phase (cylindrical rods). Spin probes **I**(m, n). (a) Random distribution of cylindrical rods; (b) cylinders, oriented in glass capillaries. Magnetic field aligned parallel to the cylinder axis. [From Seelig and Limacher (1974) with permission of Gordon and Breach.]

cases. The rotation of the hydrocarbon chains in the cylinders is therefore as rapid as that in the bilayer.

Sucking the hexagonal phase into glass capillaries produces a perfect ordering of the amphiphilic cylinders, such that the rods are aligned parallel to the long axis of the capillary. If the magnetic field is then applied along the capillary axis, an EPR spectrum consisting of just three sharp lines is obtained (Fig. 14b). This spectrum is now very similar to that obtained for oriented lamellar liquid crystals (Fig. 6, dashed spectrum).

The spectral differences between the hexagonal and the bilayer phases can be explained by the rapid translational diffusion of the lipid molecules. The line shape analysis yields a microviscosity of approximately 0.04 P for both phases, which corresponds to a lateral diffusion of 17 Å within 10^{-8} sec. Since the diameter of a cylindrical rod is only 26 Å, a diffusion along a distance of 17 Å may change the direction of the nitrogen 2p π orbital by as much as 90°. This rotation around the cylinder axis thus leads to a partial averaging of the A-tensor anisotropies $A_{\parallel}$ and $A_{\perp}$. The separation of the outer extrema in Fig. 14a corresponds to approximately $\frac{1}{2}(A_{\parallel} + A_{\perp})$. The diffusion around the cylinder axis has *no* effect on the EPR spectra as long as the magnetic field is *perpendicular* to the plane of the rotation (Fig. 14b). In this case the observed splitting corresponds to $A_{\perp}$. The translational diffusion has also no effect in lipid bilayers, since the curvature of these bilayers is too small to produce an appreciable change in the direction of the 2p π orbital within a correlation time of 10^{-8} sec. On the other hand, if the translational diffusion becomes small, as may be the case in hexagonal phases composed of phospholipid molecules, the spectrum of the nonoriented hexagonal phase may become identical with that of the corresponding bilayer phase.

ACKNOWLEDGMENTS

I wish to thank Dipl.-Phys. Hansgeorg Schindler for computing the spectra. This work was supported by the Swiss National Science Foundation under grant no. 3.8620.72.

REFERENCES

Axel, F., and Seelig, J. (1973). Cis double bonds in liquid crystalline bilayers, *J. Amer. Chem. Soc.* **95**, 7972–7978.

Brown, G. H., and Labes, M. M. (eds.) (1972). "Liquid Crystals 3," Gordon and Breach, New York.

Brown, G. H., Doane, J. W., and Neff, V. D. (1971). "A Review of the Structure and Physical Properties of Liquid Crystals," Butterworths, London and Washington, D.C.

de Vries, J. J., and Berendsen, H. J. C. (1969). Nuclear magnetic resonance measurements on a macroscopically ordered smectic liquid crystalline phase, *Nature (London)* **221**, 1139–1140.

Diehl, P., and Khetrapal, C. L. (1969). *In* "NMR—Basic Principles and Progress," Vol. 1, Chapter 1. Springer Verlag, New York.

Ekwall, P., Mandell, L., and Fontell, K. (1968). The three component system sodium caprylate–decanol–water, *Acta Polytechn. Scand. Chem. Incl. Mat. Ser.* **74**, III, 1–116.

Ekwall, P., Mandell, L., and Fontell, K. (1969). Ternary systems of potassium soap, alcohol, and water, *J. Colloid Interface Sc.* **31**, 508–539.

Fergason, J. L., and Brown, G. H. (1968). Liquid crystals and living systems, *J. Amer. Oil Chem. Soc.* **45**, 120–127.

Ferrutti, P., Gill, D., Harpold, M. A., and Klein, M. P. (1969). ESR of spin-labeled nematogen like probes dissolved in nematic liquid crystals, *J. Chem. Phys.* **50**, 4545–4550.

Flory, P. J. (1969). "Statistical Mechanics of Chain Molecules." Wiley (Interscience), New York.

Freed, J. H., and Fraenkel, G. K. (1963). Theory of linewidth in electron spin resonance spectra, *J. Chem. Phys.* **39**, 326–348.

Freed, J. H., and Fraenkel, G. K. (1964). Linewidth study in epr spectra: the para and other dinitrobenzene anions, *J. Chem. Phys.* **40**, 1815–1829.

Huang, C. (1969). Studies on phosphatidylcholine vesicles. Formation and physical characteristics, *Biochemistry* **8**, 344–351.

Hubbell, W. L., and McConnell, H. M. (1969). Motion of steroid spin labels in membranes, *Proc. Nat. Acad. Sci. U.S.* **63**, 16–22.

Hubbell, W. L., and McConnell, H. M. (1971). Molecular motion in spin-labeled phospholipids and membranes, *J. Amer. Chem. Soc.* **93**, 314–326.

Itzkowitz, M. S. (1967). Monte Carlo simulation of the effects of molecular motion on the epr spectrum of nitroxide free radicals, *J. Chem. Phys.* **46**, 3048–3056.

Jost, P., Libertini, L. J., Hebert, V. C., and Griffith, O. H. (1971). Lipid spin labels in lecithin multilayers. A study of motion along fatty acid chains, *J. Mol. Biol.* **59**, 77–98.

Kivelson, D. (1960). Theory of esr linewidths of free radicals, *J. Chem. Phys.* **33**, 1094–1106.

Lawson, K. D., and Flautt, T. J. (1967). Magnetically oriented lyotropic liquid crystalline phases, *J. Amer. Chem. Soc.* **89**, 5489–5491.

Limacher, H., and Seelig, J. (1972). Folding of fatty acid chains in liquid crystalline bilayers, *Angew. Chem. Int. Ed.* **11**, 920–921.

Luckhurst, G. R. (1973). Magnetic resonance studies of thermotropic liquid crystals, *Mol. Cryst. Liq. Cryst.* **21**, 125–159.

Luckhurst, G. R., and Sanson, A. (1972). Angular dependent linewidths for a spin probe dissolved in a liquid crystal, *Mol. Phys.* **24**, 1297–1311.

Luckhurst, G. R., and Seteka, M. (1972). Molecular Organization in the nematic and smectic A mesophases of 4-*n*-butyloxybenzylidene-4′-acetoaniline, *Mol. Cryst. Liq. Cryst.* **19**, 179–187.

Luckhurst, G. R., and Smith, H. J. (1973). The structure of a cholesteric mesophase perturbed by a magnetic field, *Mol. Cryst. Liq. Cryst.* **20**, 319–341.

Luckhurst, G. R., Seteka, M., and Zannoni, Z. (1974). An electron resonance investigation of molecular motion in the smectic A mesophase of a liquid crystal, *Mol. Phys.* **28**, 49–68.

Luz, Z., and Meiboom, S. (1973). Nuclear magnetic resonance studies of smectic liquid crystals, *J. Chem. Phys.* **59**, 275–295.

McConnell, H. M., and McFarland, B. G. (1972). The flexibility gradient in biological membranes, *Ann. N.Y. Acad. Sci.* **195**, 207–217.

McFarland, B. G., and McConnell, H. M. (1971). Bent fatty acid chains in lecithin bilayers, *Proc. Nat. Acad. Sci. U.S.* **68**, 1274–1278.

Meiboom, S., and Snyder, L. C. (1971). Nuclear magnetic resonance spectra in liquid crystals and molecular structure, *Accounts Chem. Res.* **4**, 81–87.

Olszewski, M. F., Holzbach, R. T., Saupe, A., and Brown, G. H. (1973). Liquid crystals in human bile, *Nature* (*London*) **242**, 336–337.

Poggi, G., and Johnson, C. S. (1970). Factors involved in the determination of rotational correlation times for spin labels, *J. Magn. Res.* **3**, 436–445.

Porter, R. S., and Johnson, J. F. (1967). "Ordered Fluids and Liquid Crystals." Advan. Chem. Ser. 63, Amer. Chem. Soc., Washington, D.C.

Saupe, A. (1964). Kernresonanzen in kristallinen Flüssigkeiten und kristallin flüssigen Lösungen, *Z. Naturforsch.* **19a**, 161–171.

Saupe, A. (1968). Neuere Ergebnisse auf dem Gebiet der flüssigen Kristalle, *Angew. Chem.* **80**, 99–115.

Schindler, H., and Seelig, J. (1973). EPR spectra of spin labels in lipid bilayers, *J. Chem. Phys.* **59**, 1841–1850.

Schindler, H., and Seelig, J. (1974). EPR spectra of spin labels in lipid bilayers II. Rotation of steroid spin probes. *J. Chem. Phys.* **61**, 2946–2949.

Seelig, J. (1970). Spin label studies of oriented liquid crystals (a model system for bilayer membranes), *J. Amer. Chem. Soc.* **92**, 3881–3887.

Seelig, J. (1971). On the flexibility of hydrocarbon chains in lipid bilayers, *J. Amer. Chem. Soc.* **93**, 5017–5022.

Seelig, J., and Limacher, H. (1974). Lipid molecules in lyotropic liquid crystals with cylindrical structure (a spin label study), *Mol. Cryst. Liq. Cryst.* **25**, 105–112.

Seelig, J., and Niederberger, W. (1974a). Deuterium labeled lipids as structural probes in liquid crystalline bilayers, *J. Amer. Chem. Soc.* **96**, 2069–2072.

Seelig, J., and Niederberger, W. (1974b). Two pictures of a lipid bilayer. A comparison between deuterium label and spin label experiments, *Biochemistry* **13**, 1585–1588.

Seelig, J., and Seelig, A. (1974). Deuterium magnetic resonance studies of phospholipid bilayers, *Biochem. Biophys. Res. Comm.* **57**, 406–411.

Seelig, J., Limacher, H., and Bader, P. (1972). Molecular architecture of liquid crystalline bilayers, *J. Amer. Chem. Soc.* **94**, 6364–6371.

Seelig, J., Axel, F., and Limacher, H. (1974). Molecular architecture of bilayer membranes. *Ann. N.Y. Acad. Sci.* **222**, 588–596.

Shutt, W. E., Gelerinter, E., Fryburg, G., and Sheley, C. F. (1973). Electron paramagnetic resonance studies of a viscous nematic liquid crystal. II. Evidence counter to a second order phase change, *J. Chem. Phys.* **59**, 143–147.

Stone, T. J., Buckman, T., Nordio, P. L., and McConnell, H. M. (1965). Spin-labeled biomolecules. *Proc. Nat. Acad. Sci. U.S.* **54**, 1010–1017.

11

Oriented Lipid Systems as Model Membranes†

IAN C. P. SMITH and KEITH W. BUTLER

DIVISION OF BIOLOGICAL SCIENCES
NATIONAL RESEARCH COUNCIL OF CANADA
OTTAWA, CANADA

I. TYPES OF SYSTEM

A. Membranes and the Need for Model Systems

Biological membranes are involved either directly or indirectly in almost all biochemical processes. As well as forming barriers with very specific permeability properties, they are involved in complex chemical transformations such as the chain of reactions of oxidative phosphorylation in mito-

† Issued as NRCC Publication Number 13953.

chondria. They represent a principal area of research in biophysics today, since an understanding of their structure–function relationships will lead to a better understanding of such processes as nerve transmission and brain function.

The composition and morphology of biological membranes are complex. The probability of a uniform character throughout the membrane structure is low. The principal components are phospholipids such as phosphatidylcholine (Fig. 1), neutral lipids such as cholesterol (Fig. 1), and proteins. A

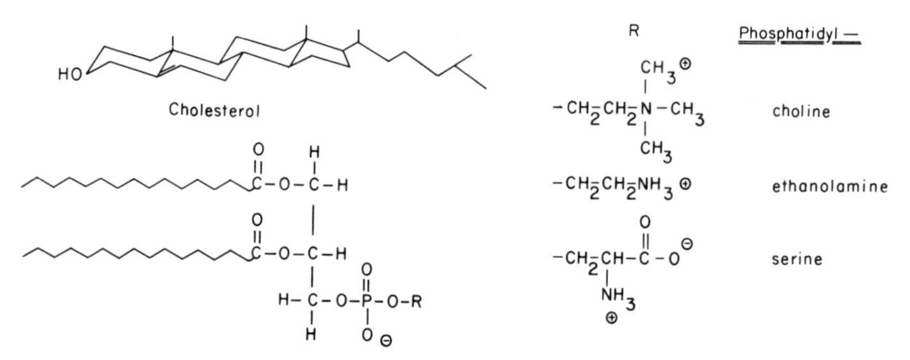

Fig. 1. Some of the common components of real and model membrane systems. A wide variety of fatty acids may be attached to the 1 and 2 positions of the glycerol, leading to a series of phospholipids with the same head group. In phosphatidic acid, R = H, and in phosphatidylinositol, R = inositol.

large body of evidence points to the presence of phospholipid bilayers in membranes, but it does not discount the possible presence of a wide variety of other structures as well. Because of the proposed ubiquity of phospholipid bilayers, a knowledge of their properties in the absence of a large number of other complicating factors is essential. The two principal model bilayer systems that have been used in ESR investigations are liposomes and oriented multibilayers. Much of the theory and formalism required in these systems has already been described in Chapter 10. This chapter will concentrate more on applications of the oriented multibilayer system to problems of biological interest. Liposome studies will be described in Chapter 12. A comprehensive review (Schreier-Muccillo and Smith, 1976) and a concise summary (Keith *et al.*, 1973) on spin-probe studies of real and model membranes are available.

B. Oriented Lipid Multibilayers

In early studies of lipids it was found that successive layers of lipid could be deposited on glass plates by repeated dipping into monomolecular films deposited at an air–water interface (Blodgett, 1935; Blodgett and Langmuir,

1937). Subsequent X-ray diffraction studies have demonstrated that the organization of the system is similar to that shown in Fig. 2 (Levine *et al.*, 1968; Levine and Wilkins, 1971). This system is excellent for study by ESR because it can be oriented on quartz plates and studied as a function of the angle between the magnetic field and the plane of the multibilayers. Analysis of a large number of spectra taken at different angles gives a much higher probability that the model for spin-probe behavior is unambiguous. A final

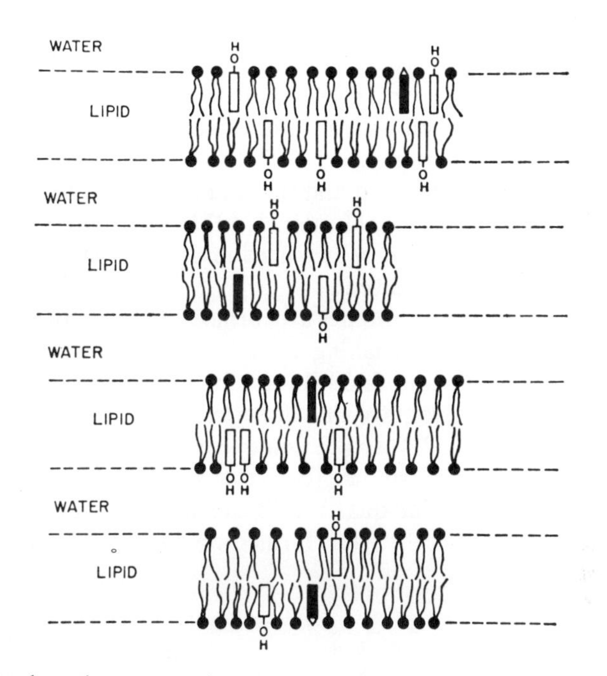

Fig. 2. A schematic representation of a portion of a multibilayer system. The light rectangles represent cholesterol, the circles with two chains, phospholipids, and the black rectangles, spin probes. The concentration of spin probes in the diagram is five to ten times that used in a typical experiment. The plane of the bilayers is parallel to the plane of the supporting quartz plates. [From Smith (1971).]

check on the model comes from studies at different ESR frequencies (Mailer *et al.*, 1974). The presence of bilayers can be confirmed by low-angle X-ray diffraction measurements on the sample used for the ESR study (Butler *et al.*, 1973b). A variety of membrane-active agents can be introduced into the system via the bathing solution, or by codeposition with the lipids (Smith, 1971). Thus, a high degree of control can be exercised over this system and interactions studied systematically to develop an overall view of the more complex mixtures found in biological membranes.

Multibilayers also permit experiments to be performed that require physical manipulation of the films. Devaux and McConnell (1972) were able to prepare lecithin films in which the spin probes were initially confined to discrete areas by depositing the spin probe on one quartz plate and lecithin on another. After the spin probe spot was cut to dots whose size was measured microscopically, the two plates were placed together and the rate of diffusion of the spin probes through the lecithin was measured (see Section III.B).

Multibilayers are formed by evaporating chloroform solutions of the lipid and the desired spin probe. In the one-stage method, films are formed by filling the flat area of quartz ESR cells with the chloroform solution and drawing air through the cell with a water aspirator (Hsia *et al.*, 1970) or by blowing wet nitrogen through the cell (Butler *et al.*, 1970b). Alternatively, chloroform solutions of the lipid and the spin probe may be evaporated on pieces of microscope cover glass (Jost *et al.*, 1971; Butler *et al.*, 1973b; Mailer *et al.*, 1974). The sample is generally placed under high vacuum for a period of time to remove residual solvent.

In a second approach, liposomes are formed by drying the lipid to a film in a flask, adding water, and shaking. A portion of this suspension is placed on a piece of cover glass and dried (Libertini *et al.*, 1969). Films formed on cover glasses have the advantage that imperfections may be removed with a razor blade (Jost *et al.*, 1971), and the sample can be used for both ESR and low-angle X-ray diffraction studies (Butler *et al.*, 1973b).

Films formed in the flat quartz cells are usually hydrated by adding aqueous solutions to the cell. Films formed on cover glass can be hydrated

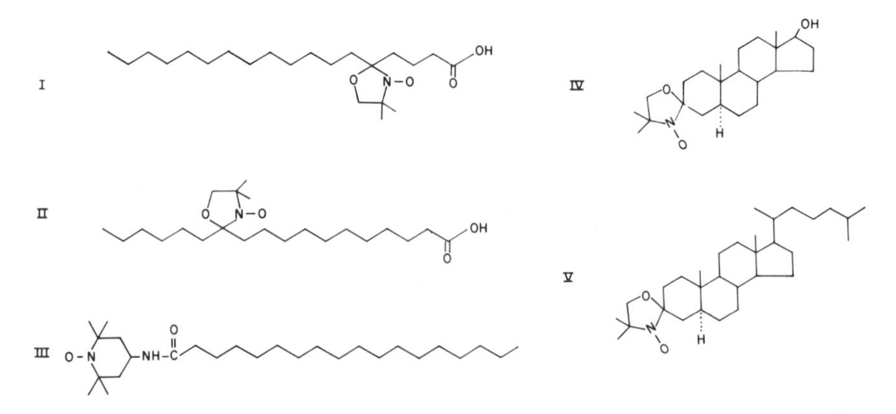

Fig. 3. Spin probes commonly used in studies of oriented multibilayers. **I** and **II** are the 5- and 12-doxyl derivatives of stearic acid, respectively. The doxyl group may be at other positions, such as carbon 16. **III** is the stearamide spin probe. **IV** and **V** are 3-doxylandrostan-17β-ol and 3-doxylcholestane, respectively.

by immersion in aqueous solution or by mounting the films in chambers held at known relative humidities.†

In still another method of producing ordered multilayers, dried lipid and a known percentage water are mixed together and placed between glass plates. Motion of one plate relative to the other imparts a shearing force to the lipid, serving to orient it into multilayers parallel to the glass plates (de Vries and Berendsen, 1969; Seelig, 1970; Hemminga and Berendsen, 1972). This type of system was described in detail in Chapter 10.

Throughout this chapter we shall use the symbols $A'_{\parallel}$ and $A'_{\perp}$ for the hyperfine splittings observed when the applied magnetic field is parallel or perpendicular, respectively, to the plane of the multibilayers. The formulas of the principal spin probes employed are given in Fig. 3.

II. NATURE OF THE SPECTRA

A. Isolated Spin Probes

1. Cholestane and Androstane Spin Probes

The rigid steroid spin probes such as the 3-doxyl derivatives of cholestane-3-one (Fig. 3) and androstane-3-one-17-ol provide a good starting point for a discussion of the ESR spectra of lipid spin probes in oriented multibilayers. Figure 4 shows the directions and magnitudes of the principal hyperfine splittings, as well as the orientation of the spin probe in a highly ordered lipid system. Even for the ordered system at least two limiting cases of spectral behavior can be expected, depending upon whether or not the steroid probe can undergo rapid rotation about its long axis.

Figures 4a and b refer to the case where no such rotation occurs. With the magnetic field perpendicular to the bilayer plane, it would also lie along the y principal axis of the nitroxide group and one would expect an ESR spectrum of three narrow lines separated by 6 G.‡ With the magnetic field parallel to the multibilayer plane the situation is more complex. There is no reason why any orientation of the plane of the steroid nucleus with respect to rotation about the steroid long axis should be more or less probable. Thus, all angles of rotation will be present in the sample. In one such orientation (Fig. 4b), the external magnetic field will lie along the z axis of

† Editor's note: See also Chapter 7.

‡ The hyperfine splitting and g values used in this argument are those measured for the 3-doxylcholestane spin probe in oriented multibilayers of highly ordered dipalmitoyllecithin–cholesterol (Hsia *et al.*, 1971). Because both quantities are known to vary with their environment, these values are most representative for this probe in a lipid system. For spectral simulation it is usually necessary to vary these parameters slightly to account for variations in environment.

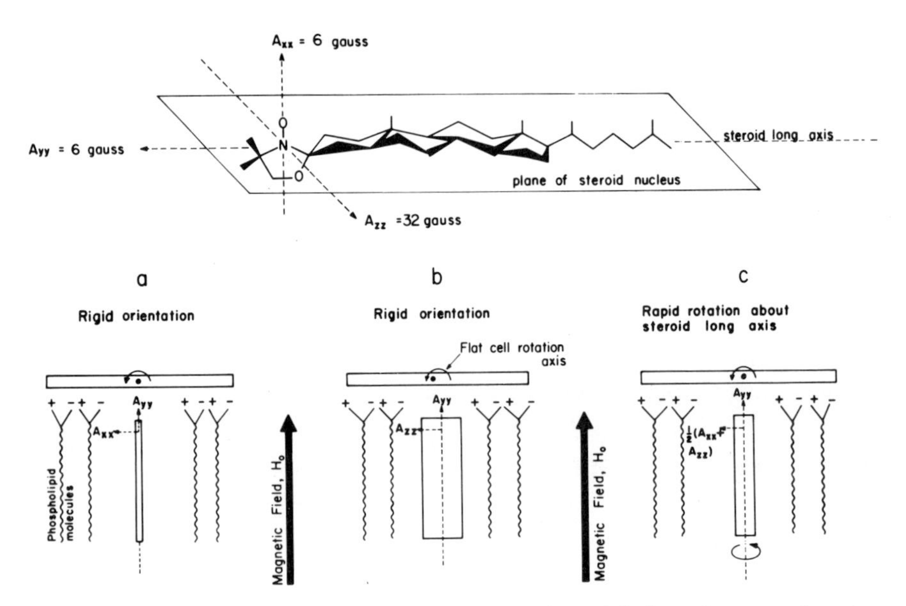

Fig. 4. The principal axes of the cholestane spin probe and the hyperfine separations that would be observed under different conditions (see text). (a) $H_0 \parallel A_{yy}$: observe 6 G, rotate flat cell 90°; $H_0 \parallel A_{xx}$: observe 6 G; (b) $H_0 \parallel A_{yy}$: observe 6 G, rotate flat cell 90°; $H_0 \parallel A_{zz}$: observe 32 G; (c) $H_0 \parallel A_{yy}$: observe 6 G, rotate flat cell 90°; $H_0 \parallel (A_{xx} + A_{zz})/2$: observe 19 G. [From Smith (1971).]

the nitroxide moiety and an ESR spectrum of three narrow lines separated by 32 G, centered at a field corresponding to $g_{zz} \cong 2.002$ would be observed. In the other extreme case (Fig. 4a), the field lies along the x direction, and an ESR spectrum of three narrow lines separated by 6 G, centered on $g_{xx} = 2.009$, would be observed. At any intermediate orientation between those represented in Figs. 4a and b the hyperfine splittings and the field positions of the centers of the spectra would be intermediate to those of the two extreme cases. Thus, since all orientations are equally probable, the observed spectrum would be an envelope of all the spectra described above; it would be a powder spectrum characterized by principal hyperfine splittings of 6 and 32 G and g values of approximately 2.009 and 2.002. Such a spectrum is shown in Fig. 5.

On the other hand, if rapid rotation about the long axis does occur, averaging of A_{zz} and A_{xx} (and g_{zz} and g_{xx}) occurs and one observes an ESR spectrum of three narrow lines separated by $(A_{zz} + A_{xx})/2 \cong 19$ G and centered on $(g_{zz} + g_{xx})/2 \cong 2.006$ when the field is parallel to the bilayer plane (Fig. 4c). If the rate of rotation is intermediate rather than rapid, i.e., comparable to $(A_{zz} - A_{xx})$ or $|g_{xx} - g_{zz}|\beta \mathbf{H} h^{-1}$ expressed in frequency

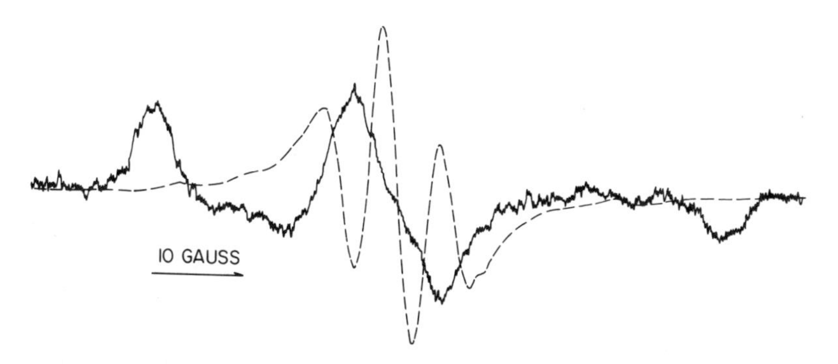

Fig. 5. ESR spectra of 3-doxylcholestane in a dry film of cholesterol–lecithin 2 : 1 mole ratio. The solid and dashed spectra were obtained with the film parallel and perpendicular to the magnetic field of the spectrometer, respectively. Note the 64 G separation between the hyperfine extrema in the parallel spectrum (corresponding to a hyperfine splitting of 32 G), which is indicative of lack of rotation of the probe about any axis. [From Hsia *et al.* (1970).]

units, then line broadening of the type described in Chapters 2 and 3 occurs, and a spectrum such as the dashed one in Fig. 6b is obtained. With the field perpendicular to the bilayer plane the motion will have little effect on the ESR spectrum because the field lies along the axis about which the motion is taking place. Thus, an ESR spectrum of three narrow lines separated by approximately 6 G (A_{yy}), centered on $g_{yy} = 2.006$, is obtained.

In both the cases described above, if the ordering is less than perfect, the spectra will be composites of spectra characterized by hyperfine splittings greater than 6 G (field perpendicular to the bilayer plane) and less than 32 (or 19) G (field parallel to the bilayer plane). Quantitative descriptions of such spectra are best realized by spectral simulation using a variety of models for the probe distribution (Jost *et al.*, 1971; Lapper *et al.*, 1972).

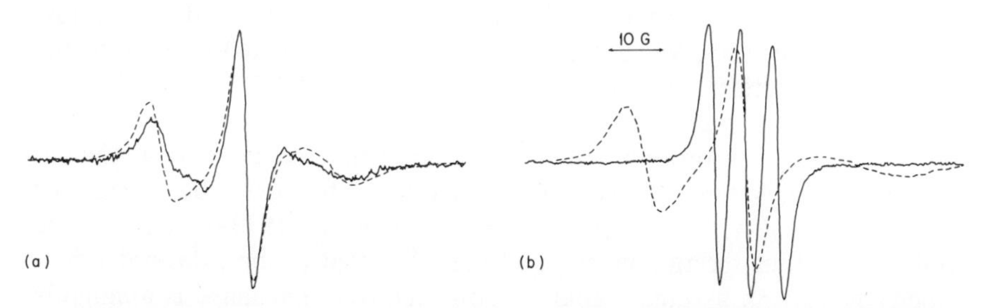

Fig. 6. Spectra of the cholestane spin probe in hydrated films of lipid from the white matter of bovine brain. The solid and dashed spectra were observed with the films perpendicular and parallel to the magnetic field of the spectrometer, respectively. (a) A low degree of angular dependence is observed with the cholesterol-free lipid. (b) A high degree of angular dependence is observed with brain lipid containing 33 mole% cholesterol. In both the probe is rotating at an intermediate rate about the steroid long axis. [From Butler *et al.* (1970b).]

One common more complex case occurs when this probe can undergo a rapid conical motion about a normal to the bilayer plane (Lapper *et al.*, 1972). This rapid motion causes an averaging of the hyperfine splittings observed in the parallel and perpendicular orientations; the wider the amplitude of the motion, the more similar these splittings will be. A characteristic of this system is that three relatively narrow ESR lines are seen in each orientation (Fig. 7). Addition of cholesterol in this case results in a decrease in the *amplitude* of the rapid motion, to give a greater difference between the hyperfine *splittings* observed in the complimentary directions, and a decrease in the *rate* of the motion, to give *broader* resonances in the spectra obtained with the applied field parallel to the bilayer plane. These effects will be discussed in greater detail in Section III.B.

Another common complex case observed for 3-doxylcholestane is where the probe can undergo rapid motion only about the steroid long axis, and a wide distribution of orientations of the steroid long axis is possible. In the limit of a very wide (random) distribution, the resultant ESR spectrum is an envelope of spectra with hyperfine splittings varying from 19 G $[(A_{zz} + A_{xx})/2]$ to 6 G (A_{yy}). Such is the case for multibilayers of the lipids from ox brain white matter from which the cholesterol has been removed (Butler *et al.*, 1970b). The ESR spectra of Fig. 6a are almost identical, as would be expected for observation of a random array of probe orientations from directions either perpendicular or parallel to the multibilayer plane. Addition of cholesterol results in a significant narrowing of the distribution width, and the parallel and perpendicular spectra correspond to those of a well-oriented probe rotating at an intermediate rate about the steroid long axis. For degrees of order intermediate to the two extremes described above, spectra such as those in Fig. 8 are obtained. Increasing distribution widths for the probe long axis are demonstrated by the decrease in the amplitude of the peak marked B and the increase in amplitude of the broad peak at low field. It has been shown that the ratio B/C can be related directly to the distribution width (Lapper *et al.*, 1972; Neal *et al.*, 1975).

When narrow resonances are obtained for the 3-doxylcholestane probe in oriented systems, order parameters characterizing the degree of alignment of the long steroid axis (see Chapter 10) can be obtained: $S = (A'_{\parallel} - A'_{\perp})/(A_{\parallel} - A_{\perp})$. (Schreier-Muccillo *et al.*, 1973a). It should be emphasized that for this probe, and the fatty acid probes, the order parameter formalism assumes rapid motion such that line shape is minimally influenced by the rate of motion. Failure to demonstrate that this assumption is valid can lead to inaccurate order parameters and deceptive discontinuities in plots of order parameter versus temperature (Cannon *et al.*, 1975).

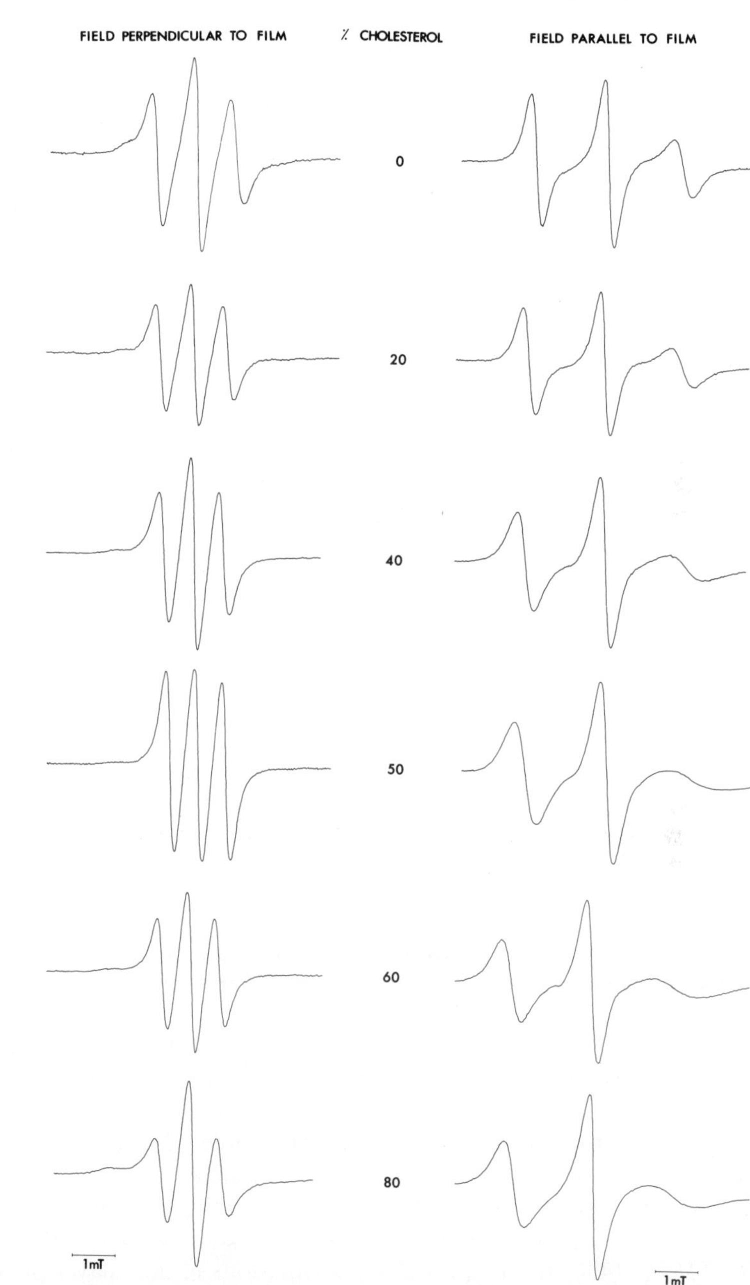

Fig. 7. The effect of cholesterol on the ESR spectra of 3-doxylcholestane in hydrated multibilayers of egg lecithin. Note the decrease in linewidth in the perpendicular spectra up to 50 mole% cholesterol and the increase, especially in the high field resonance, in the linewidth in the parallel spectra. [From Lapper *et al.* Reproduced by permission of the National Research Council of Canada from *Can. J. Biochem.* **50**, 969–981 (1972).]

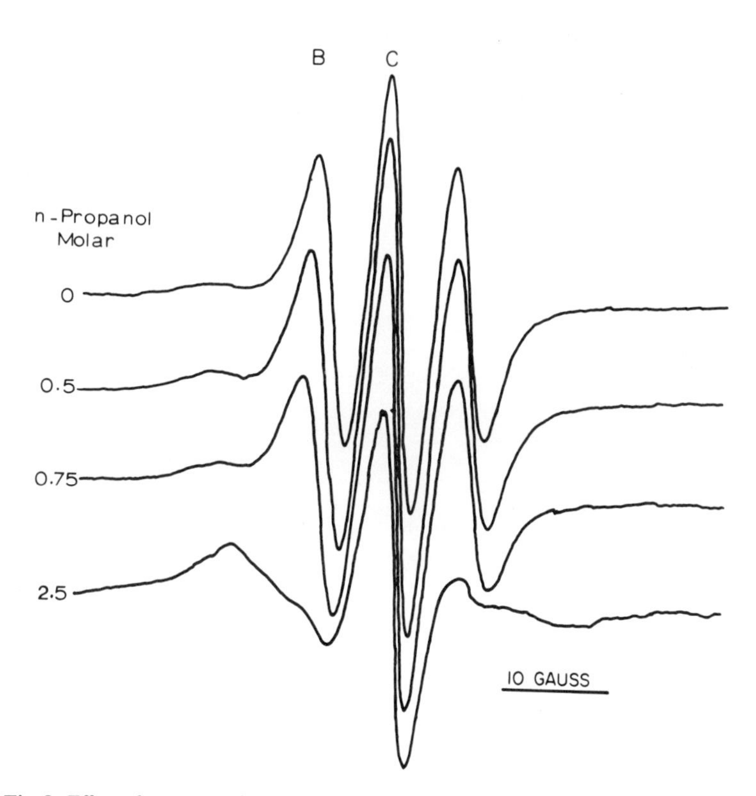

Fig. 8. Effect of *n*-propanol concentration in 0.9% NaCl, 10 m*M* phosphate buffer, pH 7.0 on films formed of the lipids from human red cell ghosts and 3-doxylcholestane. The plane of the film is perpendicular to the magnetic field of the spectrometer. [From Paterson *et al.* (1972).]

2. Fatty Acid Spin Probes

When a doxyl group is attached to a fatty acid chain, the arrangement of nitroxide principal axes is complementary to that of the common cholestane and androstane spin probes. Figure 9 shows that the largest hyperfine splitting (32 G) occurs when the magnetic field is parallel to the fatty acid long axis (hence perpendicular to the plane of a well-ordered multibilayer system). Another difference in behavior relative to the steroid probes is the lack of rigidity of the group to which the doxyl moiety is attached. Thus, rapid segmental motion due to a large number of gauche–trans interconversions of the fatty acid chain will result in partial averaging of g and hyperfine tensor anisotropy. If this motion is rapid on the ESR time scale, the ESR spectra in the parallel and perpendicular orientations of the bilayer planes relative to the magnetic field will consist of three narrow lines with separations greater than 6 G, and less than 32 G, respectively. An order parameter

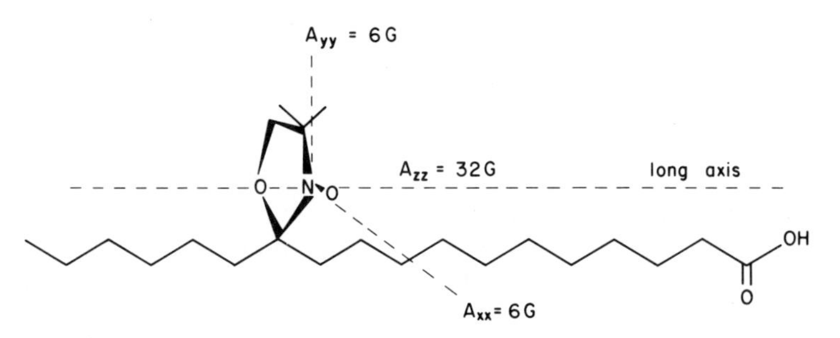

Fig. 9. The principal axes and hyperfine splittings of 12-doxylstearic acid. The axes and splittings of the other doxylstearic acids are analogous.

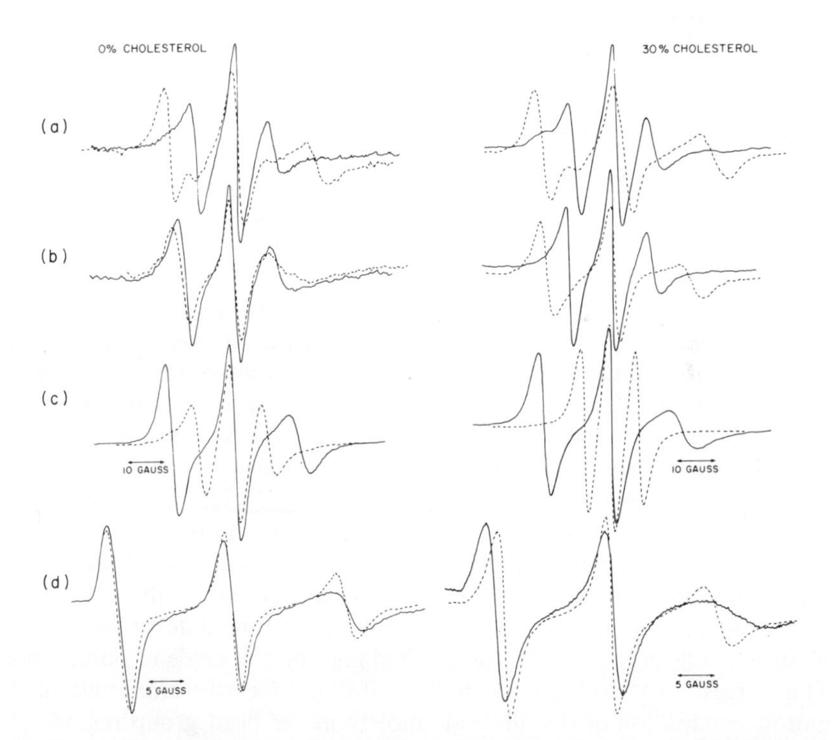

Fig. 10. ESR spectra of films of egg lecithin containing zero and 30% cholesterol. Solid and dashed spectra were obtained with the multibilayers parallel and perpendicular to the applied magnetic field, respectively. The spin probes are (a) **I**, 5-doxylstearic acid; (b) **II**, 12-doxylstearic acid; (c) **V**, 3-doxylcholestane; and (d) **III**, the stearamide spin probe (see Fig. 3). [From Schreier-Muccillo *et al.* (1973a).]

similar to that described in Chapter 10 for the lyotropic mesophases can be calculated and related to the average amplitude of the rapid segmental motion (Schreier-Muccillo *et al.*, 1973a). A comparison of the ESR spectra of the 5-doxyl stearate (**I**), 12-doxyl stearate (**II**), 3-doxylcholestane (**V**), and *N*-oxyltetramethylpiperidinylstearamide (**III**, see Section II.A.3) probes in egg lecithin with and without cholesterol is shown in Figure 10. When the rate of segmental motion becomes intermediate on the ESR time scale, the spectra become more complex and can only be quantitatively interpreted by computer simulations considering both correlation times and order parameters explicitly (Polnaszek *et al.*, 1973).† Failure to observe three well-defined resonances for the fatty acid probes in multibilayer films is indicative of (i) poorly prepared (inhomogeneous) film, (ii) intermediate rate of motion of the probe, (iii) a wide distribution of possible amplitudes of the rapid motion.

This discussion of the fatty acid spin probes applies equally well to the spectra of phospholipid spin probes in which one of the acyl chains is spin labeled (Hubbell and McConnell, 1971).

3. The Stearamide Spin Probe

The stearamide spin probe (Fig. 3) has been shown by ascorbate reduction studies to intercalate in lipid bilayers so as to have the nitroxide moiety in the head group region exposed to solvent (Schreier-Muccillo *et al.*, 1974). Even in highly ordered bilayers only a slight angular dependence of the ESR spectra is evident. This is due to rapid motion of wide amplitude in this region of the bilayer. This preferred location for the nitroxide moiety makes the stearamide spin probe of value for exploring lipid–protein interfaces in complex systems (Verma *et al.*, 1973b). Typical spectra of this probe in ordered bilayers are shown in Fig. 10 (spin label III). A common occurrence in ESR spectra of the stearamide probe is a low field resonance narrower than the center field resonance, as shown in Fig. 10 for both parallel and perpendicular orientations of the cholesterol-free films. Addition of 30 mole% cholesterol, which improves the degree of order and decreases the rate of motion of the fatty acid moieties of the lecithin, results in a greater degree of anisotropy seen by the head group probe and a dependence of the relative widths of the low and center field lines on the angle of observation. This suggests that addition of cholesterol diminishes the amplitude of the anisotropic motion of the nitroxide moiety in the head group region. The influence of the rate of anisotropic motion on the ESR line shape depends upon the direction of observation because the direction of the field deter-

† Editor's note: See Chapter 3 for a complete discussion of this problem.

mines the maximum and minimum values of the g and hyperfine splitting values between which the nitroxide interconverts. A more complete discussion of such linewidth effects is given in Section III.B.

B. Interacting Spin Probes

When free radicals get sufficiently close to one another their unpaired electrons can interact via one or both of two mechanisms. These interactions have been put to use in elucidating lateral diffusion, order, and lipid–lipid distances in oriented lipid multibilayers.

1. Exchange Interaction

This is the most commonly encountered of the two mechanisms. It is due to the overlap of the orbitals containing the two unpaired electrons. Depending on the magnitude of this overlap, which is determined by the separation between the orbitals, several types of ESR spectra can result. When the overlap is large, i.e., when the so-called exchange integral is greater than the hyperfine splitting, the overall ESR spectrum will reflect electron–nuclear interactions with the magnetic nuclei of both interacting molecules. In the simplest case of interaction between two identical nitroxide radicals, the ESR spectrum will consist of five lines of intensity 1 : 2 : 3 : 2 : 1 with hyperfine separation $A_N/2$, where A_N is the corresponding hyperfine splitting for an isolated nitroxide. Such spectra are usually observed for nitroxide biradicals in which the nitroxide moieties are part of the same molecular framework (Dupeyre *et al.*, 1965). A biradical spin probe has been employed in oriented lipid bilayers (Keana and Dinerstein, 1971), focussing on the anisotropy of the electron–electron dipole–dipole interaction (see Section II.B.2) as a measure of order. When the overlap between the nitroxide orbitals is weak, or is strongly modulated by rapid motion, the resonances formerly of intensity 2 in the spectra described above become broad and in some cases unobservable (Luckhurst and Pedulli, 1970).

Spin exchange can also occur in solution of monoradicals at high concentration. At the moment of collision, a biradical state is formed. The greater the frequency of collision, the greater the contribution biradical states can make to the ESR line shape. At an intermediate rate of collision, relative to the frequency equivalent of the energy separation between the biradical and monoradical states, severe line broadening occurs. At even higher rates of collision a single resonance will appear—this is the so-called exchange-narrowing limit. These cases are illustrated in Fig. 11.

Exchange broadening is often observed in lipid films when the spin probes are excluded to form aggregates insoluble in water or are highly

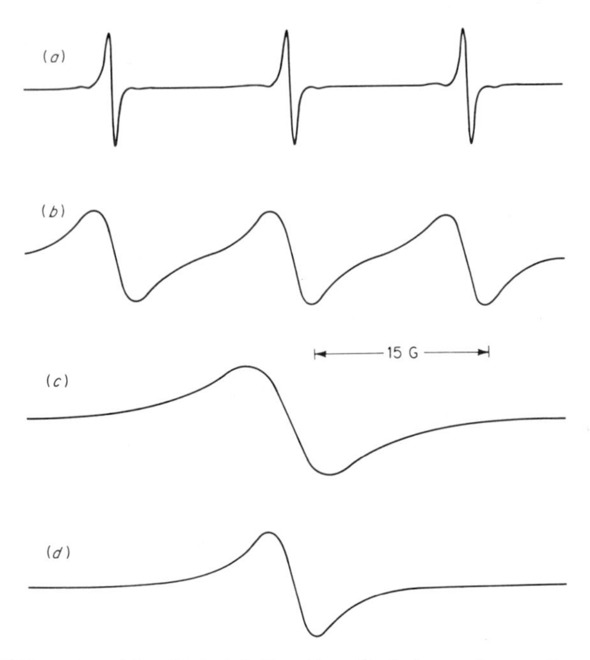

Fig. 11. ESR spectra of the di-*t*-butyl nitroxide radical at room temperature at (a) 10^{-4} *M*, (b) 10^{-2} *M*, and (c) 10^{-1} *M* in ethanol. (d) is the pure liquid nitroxide. Spectra (b) and (c) illustrate the exchange-broadening phenomenon, while (d) shows the exchange narrowing that takes place at very high radical concentrations. [From Wertz and Bolton (1972).]

localized due to preferential solubility in one lipid component (Schreier-Muccillo *et al.*, 1973b; Butler *et al.*, 1974). It can be used to determine the rate of lateral diffusion in lipid films by depositing a highly concentrated spot of spin probe in a lipid film and following its diffusion and concomitant dilution (Devaux and McConnell, 1972) (see Section III.B and Fig. 14). Because it is dependent on the frequency of molecular collisions, the extent of exchange broadening in lipid systems usually increases with increasing temperature, in a fashion opposite to that of electron–electron dipole–dipole interaction (see Sections II.B.2 and III.A).

2. Electron–Electron Dipole–Dipole Interaction

The interaction between the dipole moments of two unpaired electrons is determined by their relative orientations and mutual separation. If both reorient rapidly in the applied magnetic field, this interaction is averaged to zero regardless of concentration. If they reorient at an intermediate rate relative to the frequency corresponding to the energy difference between the electron–electron energy states, line broadening will be observed. In this case the magnitude of the line broadening will decrease with increasing temperature as the rate of molecular reorientation in the magnetic field increases. In lipid systems with high spin-probe concentrations both exchange broadening and dipole–dipole broadening may be present. The principal contributor can be determined by a study of the temperature dependence (Scandella *et al.*, 1972).

In highly ordered systems, spin probes have well-defined orientations with respect to one another. Discrete electron–electron dipole–dipole energy levels are thus present at high spin-probe concentrations and may be manifest in the ESR spectra as extra splittings of the resonances. The magnitude of the splittings in these systems depend on the separation between the electrons, and on the direction of the magnetic field relative to the line joining the electrons (Marsh and Smith, 1972, 1973). Thus, the dipole–dipole splittings may be used to estimate the separation between spin probes in a given system (see Sections III.A and B and Fig. 15).

C. Reduction of Nitroxide Probes by Sodium Ascorbate

It has been known for some time that sodium ascorbate reduces nitroxide moieties to diamagnetic species, presumably their hydroxylamine analogues. In dilute aqueous solution at room temperature the reduction is instantaneous. However, if the spin probe is in a protected environment, it may be reduced more slowly or not at all by the hydrophilic ascorbate ion. This was used in the studies of the diffusion of phospholipids from the inner to the outer monolayer of a vesicular bilayer system (Kornberg and McConnell, 1971). In like fashion, the rate of nitroxide reduction can be used to study the rate of penetration of the ascorbate ion into oriented phospholipid bilayers containing spin probes, and hence the relative depths of the nitroxide moieties of a series of spin probes. For a given spin probe, the relative reduction rates are indicative of the permeability of the bilayers under a variety of conditions (Schreier-Muccillo *et al.*, 1975; Butler, 1975). Some applications of this technique to the multibilayer system are given in Section III.I.

III. APPLICATIONS

A. Lipid Composition and the Ability to Form Ordered Bilayers

Bilayers are generally accepted to form at least part of the lipid architecture of biological membranes. Because these membranes contain a large variety of lipids, it is valuable to know the chemical requirements for bilayer formation and the properties of bilayers formed from different lipids. If films containing spin probes show significant differences between the ESR spectra taken with the film perpendicular and parallel to the magnetic field of the spectrometer, the existence of well-ordered bilayers can be assumed. In addition, from the shape of the spectra and the splittings, $A_\perp$ and $A_\parallel$, the fluidity of the film and the nature and amplitude of the motion of the spin probes can be calculated (see Section II and Chapters 10 and 12).

Schreier-Muccillo *et al.* (1973b) showed that the ability to form well-ordered bilayers depended upon the size and charge of the headgroup; the number, length, and degree of saturation of the acyl chains; the presence of cholesterol, the temperature; and the pH, type, and concentration of ions in the aqueous phase. Well-ordered films were observed with phosphatidylcholine, phosphatidylserine, and phosphatidylethanolamine over the pH range where they have no net charge and with monolaurin and monopalmitin above their transition temperatures. Below their chain-melting transition temperatures, exchange-broadened spectra (Section II.B.1) were observed for these saturated monoglycerides. This indicates that the lipid is too tightly packed to dissolve the spin probe and a separate phase containing a high concentration of spin probe is formed. Strong exchange interactions broaden the ESR lines, and no information can be drawn from the spectra about the existence of bilayers or about molecular orientation or motion. Unsaturated monoglycerides and saturated and unsaturated diglycerides had angular-independent spectra at all temperatures with and without cholesterol. Boggs and Hsia (1973b) showed that a mixed monoglyceride produced angular-dependent spectra at room temperature only if cholesterol was added, and that phosphatidylinositol, cardiolipin, ceramides, and sphingomyelin with either fatty acid or cholestane spin probes produced spectra characteristic of ordered bilayers.

With films formed of phosphatidic acid, Schreier-Muccillo *et al.* (1973b) observed no angular dependence of the ESR spectra, even in the presence of high concentrations of calcium (see Section D). However, X-ray diffraction measurements clearly showed the presence of bilayers, leading to the conclusion that the hydrocarbon chains were randomly oriented within the bilayers rather than being fully extended normal to the plane of the bilayers. This illustrates how these methods can complement each other. X-ray diffraction can demonstrate the existence of bilayers and measure their thickness and spacing. If the lipids are crystalline, X-ray diffraction can measure the spacing of the acyl chains. Above the liquid crystal transition temperature, the chains are simply seen as fluid (Tardieu *et al.*, 1973). It is in this area that spin probes have the advantage of measuring the degree of orientation and fluidity.

X-ray diffraction has also been used to demonstrate lipid phases other than lamellar. Rand and Sengupta (1972) showed that calcium ions converted cardiolipin from a lamellar to a hexagonal phase in which the lipid forms cylinders instead of sheets. With a series of fatty acid spin probes Boggs and Hsia (1973a) demonstrated that the ESR spectra of an aqueous suspension of cardiolipin in a calcium-containing solution were indistinguishable from the spectra of the corresponding liposomes. Orientational S parameters (Section II.A, and Chapters 10 and 12) have been derived from spin probes in cardiolipin-calcium complexes (Hegner *et al.*, 1973). The S parameter is a measure of the molecular distribution of the probe. A large value of S does not necessarily imply the existence of bilayers in an unknown system. When these cylinders were oriented on a quartz plate, some spectral anisotropy resulted, but not that characteristic of multibilayers.

B. Cholesterol–Lipid Interaction

Cholesterol is a common component of vertebrate membranes, and similar sterols are found in vascular plants, algae, fungi, and microorganisms. It has been shown in monolayer studies (Leathes, 1925; de Bernard, 1958) that the total area of a film formed from egg lecithin and cholesterol is less than those of films formed from the same number of molecules of lecithin and cholesterol separately. This condensing effect of cholesterol was observed with films formed of synthetic lecithins: 1-stearoyl-2-lauroyl, ditetradecanoyl, dioleoyl, and 1-stearoyl-2-oleoyl, but not distearoyl or didecanoyl. Condensation was also noted with 1-stearoyl-2-oleoylphosphatidylethanolamine (Demel *et al.*, 1967).

It has been observed that the addition of cholesterol increases the angular anisotropy of the spectra of the 3-doxylcholestane spin probe in hydrated egg lecithin multilayers (Hsia *et al.*, 1970, 1971; Smith, 1971; Lapper *et al.*, 1972; Schreier-Muccillo *et al.*, 1973b). Lapper *et al.* (1972) used several theoretical models to simulate the ESR spectra of the 3-doxylcholestane spin probe in hydrated films of egg lecithin containing a series of increasing cholesterol concentrations. Experimentally, as the cholesterol concentration increased from zero to 55%, the $A'_{\perp}$ decreased from 9.4 to 6.7 G and $A'_{\parallel}$ increased from 17 to 18.9 G. In addition, the linewidths in the perpendicular orientation decreased, and those in the parallel orientation increased over this same cholesterol concentration range. The model that best fitted these observations was a random walk model in which the spin probe could move rapidly within a cone of limited volume. As the concentration of cholesterol increased, the half angle of the cone characterizing the random walk decreased from 46° to 17° at 55 mole% cholesterol and the rate of motion decreased.

The model of Lapper *et al.* (1972) proposes that the changes in shape of the resonances in the parallel spectra are due to changes in the rate of the conical random walk. The sensitivity of the line shapes to this rate depends on the frequency of the ESR experiment. Mailer *et al.* (1974) observed spectra of the 3-doxylcholestane spin probe in hydrated lecithin and lecithin–cholesterol films parallel and perpendicular to the magnetic field of spectrometers at 9.5, 24, and 35 GHz (Fig. 12). As in the previous experiments, cholesterol broadened the lines in the parallel spectra (solid lines, Fig. 12), but whereas the linewidths were in the order $W_0 < W_1 < W_{-1}$ at 9.5 GHz, they were in the order $W_1 < W_0 < W_{-1}$ at 24 and 35 GHz (where the subscripts 1, 0, and −1 refer to the hyperfine lines at low, intermediate, and high field, respectively). This reversal of the widths of the resonances at low and intermediate fields can be understood in terms of the data in Fig. 13. The ΔH values indicate the change in field position for a given ESR line due to rotation about the *y* axis (the principal modulation caused by the conical motion of the spin probe). Thus, at 9 GHz, this rotation results in changes in

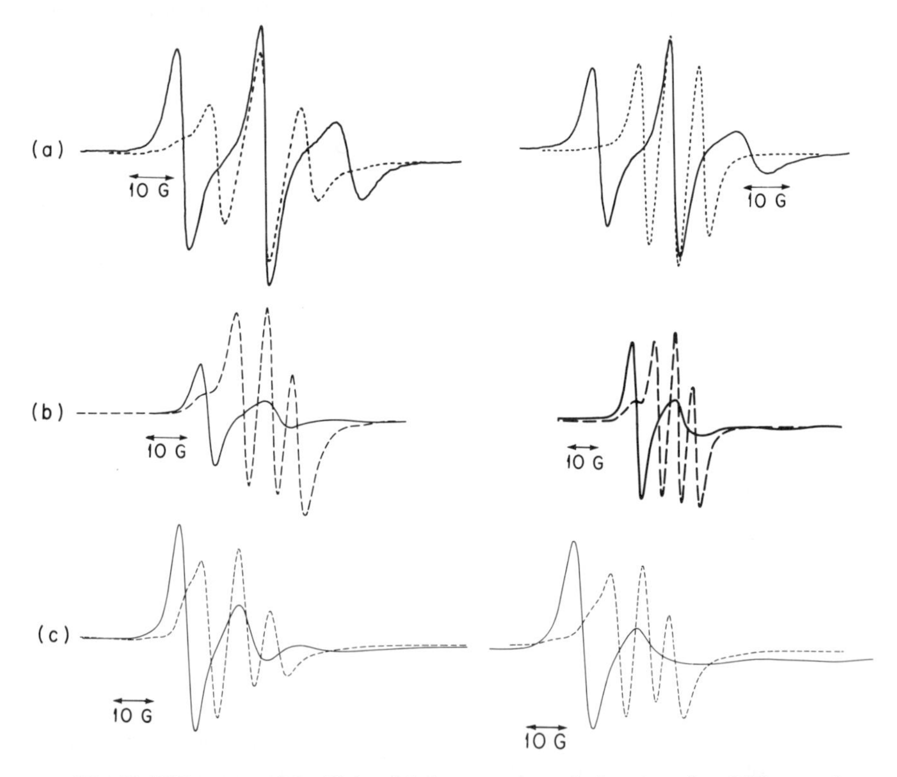

Fig. 12. ESR spectra of the 12-doxylcholestane spin probe in oriented multibilayers of egg lecithin: (*left*) no cholesterol, (*right*) 30 mole% cholesterol; (a) 9.5 GHz, (b) 24 GHz, (c) 35 GHz. The solid and dashed spectra were obtained with the applied magnetic field parallel and perpendicular to the bilayer surface, respectively. [From Mailer *et al.* (1974).]

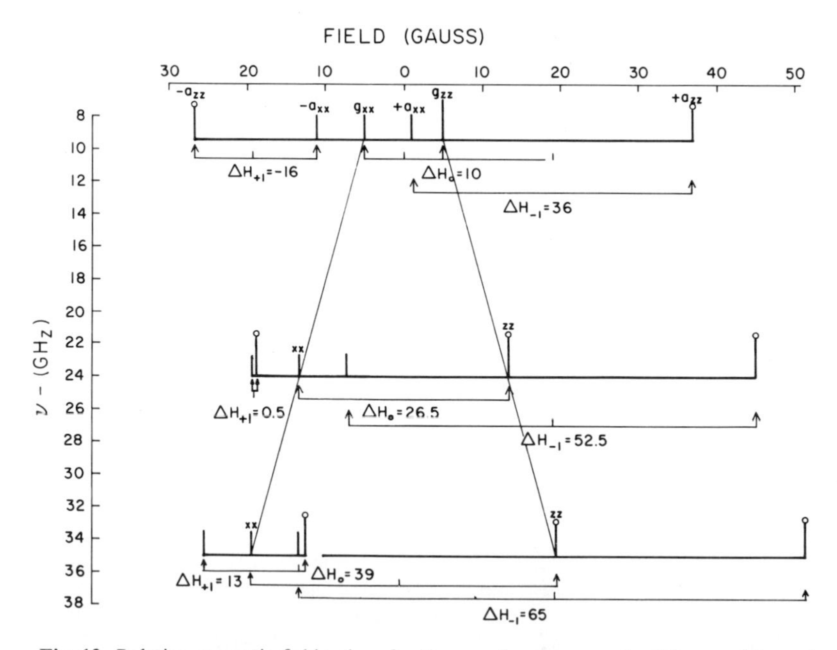

Fig. 13. Relative magnetic field values for the x and z components of the g and hyperfine splitting tensors of the 3-doxylcholestane spin probe as a function of the observing spectrometer frequency. The ΔH values indicate the change in field position experienced by a given hyperfine line due to rotation about the steroid long axis. [From Mailer *et al.* (1974).]

the field positions of the low field (1) and center field (0) lines of -16 and $+10$ G, respectively. However, at 24 GHz, the low field line moves by only 0.5 G, whereas the central line moves by 26.5 G due to rotation relative to the magnetic field. The larger the change in position of the resonance, the greater is the influence of the rate of the tumbling on the width of the resonance. From the relative widths of the resonances the correlation time for the motion was estimated; the values obtained from the spectra at three microwave frequencies were in good agreement. Thus, the variable-frequency ESR served to confirm and quantitate the earlier model of Lapper *et al.* (1972). The addition of 30 mole% cholesterol increased the correlation time characterizing the probe motion from 1.8×10^{-9} to 3.8×10^{-9} sec, corresponding to a significant decrease in fluidity accompanying the increase in the degree of order.

Devaux and McConnell (1972) measured the rate of lateral diffusion in lecithin multilayers by observing the rate of decrease in exchange broadening of the spectra as spin-labeled lecithin diffused out from an applied dot into the bulk of the unlabeled lecithin (Fig. 14). They found that the time taken for the spectral change was 1.5 to 2 times as long in an equimolar

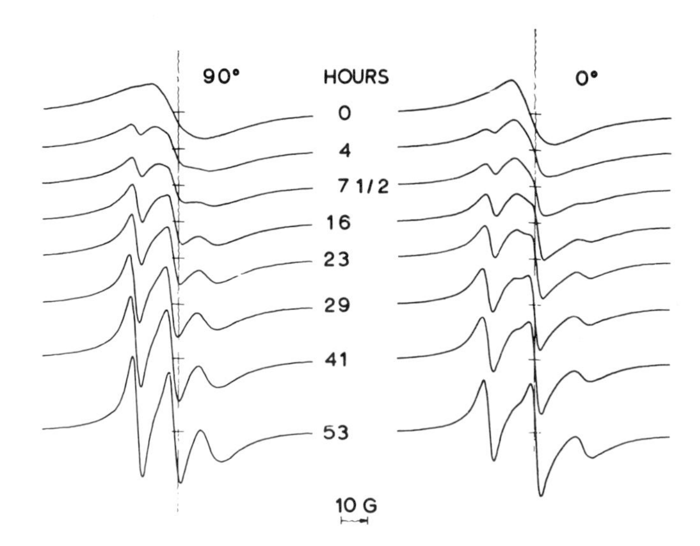

Fig. 14. ESR spectra of clusters of spin-labeled lecithin in bilayers of dipalmitoyllecithin. Spectra were taken at the indicated times after deposition of a concentrated patch of spin probe, with the applied magnetic field perpendicular (90°) and parallel (0°) to the film. [Reprinted with permission from Devaux and McConnell *J. Amer. Chem. Soc.* **94**, 4475–4481 (1972). Copyright by the American Chemical Society.]

mixture of lecithin and cholesterol as it was in lecithin alone, indicating a decrease in the fluidity and in the lateral diffusion coefficient.

In a study employing a series of fatty acid probes as well as the 3-doxylcholestane and 3-doxylandrostane spin probes, Schreier-Muccillo *et al.* (1973a) found that all probes used showed an increase in ESR spectral anisotropy as cholesterol was added to egg lecithin films. The spectra of the androstane spin probe were in accord with the model proposed for the cholestane spin probe by Lapper *et al.* (1972) (see above). The data for the fatty acid spin probes indicated that cholesterol increases the degree of extension of the fatty acid chains (decreases the probability of gauche conformations in the hydrocarbon chains) and decreases the amplitude of motion of the chains as a whole. With dipalmitoyllecithin, the degree of order of the fatty acid spin probes as measured by the S parameter decreased on increasing the cholesterol content from 20 to 50 mole%. The probability of gauche conformations was relatively unaffected but the amplitude of motion was found to increase with increasing cholesterol concentration. These results corroborate the hypothesis that a biological function of cholesterol is to stabilize membrane fluidity. It decreases the fluidity of fluid lecithins and fluidizes lecithins that are below their transition temperatures.

Marsh and Smith (1973) intercalated high concentrations (about 8 mole %) of cholestane spin probe into films to measure the lateral separation of the spin probes by the magnitude of the electron–electron dipolar interaction. Figure 15 shows the extra splittings observed in ESR spectra taken with the magnetic field perpendicular to the plane of multibilayers of dipalmitoyllecithin of varying cholesterol content. The larger the splitting, the closer together are the spin probes. It was found that the cholestane spin-probe separation decreased with increasing cholesterol concentration in hydrated egg and dioleoyllecithin bilayers, and increased with increasing cholesterol concentration in hydrated dipalmitoyllecithin bilayers. These changes in packing of the molecules correspond respectively to the condensing and fluidizing effects of cholesterol previously mentioned. The condensing effect

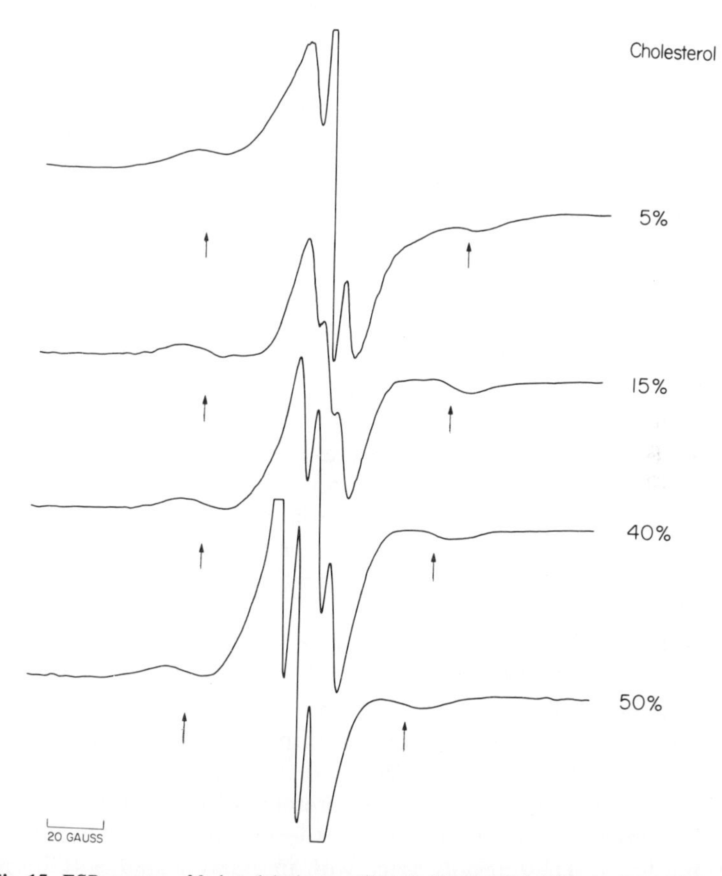

Fig. 15. ESR spectra of 3-doxylcholestane (8 mole % of total lipid) in hydrated films formed of dipalmitoyllecithin plus the indicated amount of cholesterol. Spectra were observed at room temperature with the magnetic field perpendicular to the multibilayer plane. The arrows indicate the resonances arising from interacting spin probe pairs. [From Marsh and Smith (1973).]

of cholesterol could be due to the cholesterol fitting into cavities left in the lecithin or to specific lecithin–cholesterol interactions that change the conformation of the lecithin. From the magnitude of the observed change in spin probe spacing and from values of lecithin and cholesterol areas and condensing effects measured in monolayer studies, the authors concluded that each explanation accounts for approximately half of the condensing effect. Lecithin is not the only lipid whose molecular orientation is influenced by cholesterol. Cholesterol has been found to promote order with saturated monoglycerides (Schreier-Muccillo *et al.*, 1973b) with mixed lipids from beef brain, human red cell ghosts, and morning glory cells (Butler *et al.*, 1970b) and with phosphatidylethanolamine, phosphatidylserine, sphingosine, cerebrosides, and ceramides (Boggs and Hsia, 1973b).

Multibilayers with intercalated spin probes have also been used to study which structural characteristics of cholesterol are important in influencing the molecular arrangement of lipids. Butler *et al.* (1970b) studied films formed of lipids isolated either from beef brain white matter or from human red cell ghosts. Cholesterol was removed from these lipids by silica gel chromatography and films were formed of the decholesterolized lipid and the cholestane spin probe plus a series of concentrations of the test steroids. With no added steroid, no angular dependence of the ESR spectra was observed. Both cholesterol and the plant sterol β-sitosterol, the 24β-ethyl derivative of cholesterol were highly effective at producing order. However, 5-androsten-3β-ol, which lacks the tail portion of cholesterol, was much less effective in promoting order. The 3β-hydroxyl group was found to be vital in producing order; cholestanone produced only a minor degree of order, and cholestane was totally ineffective. The presence of double bonds in the steroid nucleus was unimportant, as cholesterol and cholestanol were equally effective at all concentrations, and ergosterol, another plant steroid with two double bonds in the steroid nucleus and one in the side chain, was highly effective in promoting order at low concentrations. Its solubility in phospholipids is limited (Demel *et al.*, 1972), obviating its incorporation at high concentrations. The shape of the steroid nucleus is vitally important; 5β compounds have a bent steroid nucleus and are totally ineffective in promoting order, whereas all the steroids that promote order have a planar nucleus. It was concluded that the presence of a planar nucleus and a 3β-hydroxyl group are required to promote order, and the presence of a hydrocarbon chain at C-17 enhances this ability. The presence of a polar group at C-17 or on the hydrocarbon chain abolishes the ordering ability for mammalian cell membrane lipids. These results agree with monolayer studies (Ghosh and Tinoco, 1972) that indicated that cholesterol and β-sitosterol were about equally effective in condensing a wide variety of synthetic lecithins. 5α-Androstane-3β-ol also exerted a condensing effect, but to a much lesser extent than cholesterol or β-sitosterol.

In a later study using fatty acid and cholestane spin probes in egg lecithin, Hsia *et al.* (1972) found essentially the same dependence of ordering ability on sterol structure. In contrast to the studies using mammalian lipids and the previously mentioned monolayer study, β-sitosterol was found to be much less effective than cholesterol as a condensing agent.

A dramatic influence of cholesterol on the degree of order has been found in multibilayers of the lipids from the white matter of bovine brain (Neal *et al.*, 1975). Figure 16 shows the ESR spectra of the 3-doxylcholestane spin probe in this system with the magnetic field perpendicular to the plane of the

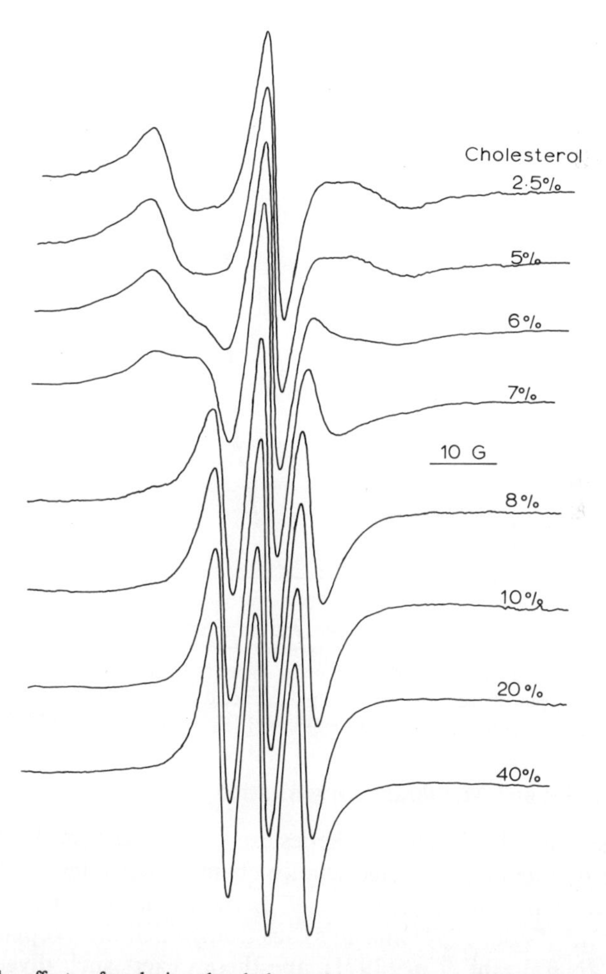

Fig. 16. The effects of replacing the cholesterol in decholesterolized beef brain lipid. The ESR spectra are of 3-doxylcholestane in hydrated multibilayers oriented perpendicular to the magnetic field of the spectrometer at 23°C. [From Neal *et al.* (1975).]

bilayers. As described in Section II.A.1, the spectra at low cholesterol concentrations are due to a wide distribution of probe long axes with rapid rotation occurring about the long axes. The ratio of amplitudes of the resonance approximately 6 G to low field of the spectral center to that of the central resonance (Fig. 16, 7 and 8% cholesterol) provides a convenient measure of the degree of order (Butler *et al.*, 1970b). This amplitude ratio has been shown to depend upon the width of the distribution of spin probe long axes (Lapper *et al.*, 1972; Neal *et al.*, 1975); it approaches 1 for a high degree of order and a narrow distribution width, and 0 for a low degree of order and a wide distribution. It is represented as D/E in Fig. 17, where a significant ordering effect of cholesterol occurs over a very narrow concentration range.

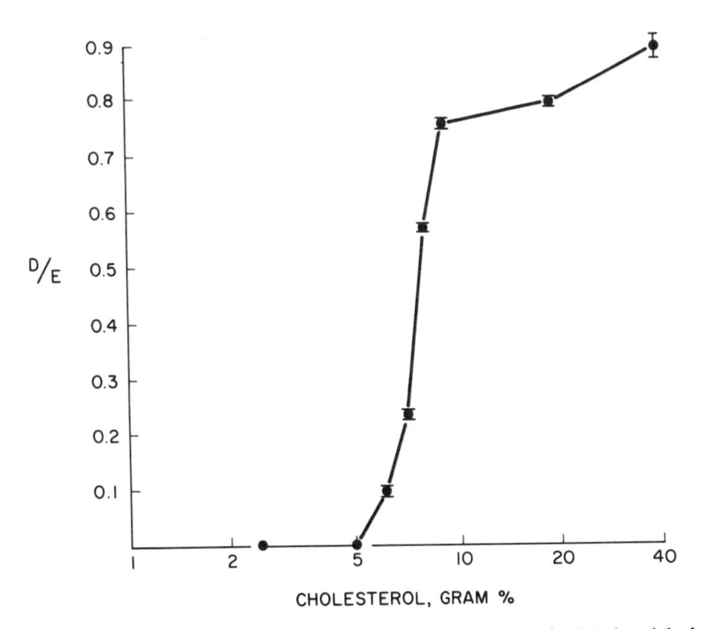

Fig. 17. Dependence of the arbitrary order index D/E (see text) of 3-doxylcholestane in oriented multibilayers of decholesterolized lipid extracted from the white matter of bovine brain as a function of concentration of added cholesterol [From Neal *et al.* (1975).]

C. Anesthetics and Membrane Organization

It is generally held that anesthetics exert their effect on the membrane of the nerve by blocking the conductance changes necessary for the transmission of action potentials (Seeman, 1972). As well as blocking conduction in whole axons (Skou, 1954) and in axons from which the axoplasm had been removed (Narahashi *et al.*, 1971), anesthetics exert such diverse effects as protecting red cells against osmotic lysis (Roth and Seeman, 1971) and increasing the surface pressure of monolayers of nerve tissue lipids spread on Ringer solutions (Skou, 1961). In each system, the order of potency of the anesthetic agents is the same as their order of potency at blocking transmission of action potentials in nerves. In the monolayer studies, each anesthetic, when applied at the minimum concentration causing nerve blockage, caused the same size of increase in the surface pressure of the film. In addition, pH changes modified the nerve-conduction blocking and the film-expanding actions of anesthetics in the same manner.

Spin probes can be used to measure the effects of general and local anesthetics on molecular order and motion in lipid films. It was found that flowing gas mixtures containing chloroform or butane rapidly decreased the spectral anisotropy of cholestane spin probes in lecithin or brain lipid films (Butler, 1975). From free energy considerations, Schneider (1968) had concluded that alcohols interacted with hydrophobic binding regions in membranes. Paterson *et al.* (1972) investigated the effect of a series of alcohols of varying chain length on the spectra of the 3-doxylcholestane spin probe in films formed from brain or erythrocyte lipids. The films were hydrated with a phosphate buffer at pH 7 and the ESR spectra measured. They were then exposed to the same buffer containing a series of increasing alcohol concentrations, and ESR spectra were taken after equilibration at each alcohol concentration. As the concentration of alcohol increased, a point was reached where the degree of spectral anisotropy began to decrease rapidly (Fig. 8). In the spectra observed with the films perpendicular to the magnetic field of the spectrometer, the hyperfine separation increased and the B/C ratio (see Fig. 8 and Section II.A.1), decreased. As the chain length of the alcohols increased, the concentration required to bring about a given decrease in the B/C ratio decreased (Fig. 18). The dependence of the concentration required for the discontinuous decrease in order on the chain length of the alcohol paralleled that for protection of red cells against osmotic hemolysis, alteration of the conductivity of black lipid membranes, and inhibition of tadpole reflexes, suggesting a common mechanism for these effects—disordering of membrane lipids.

Butler *et al.* (1973a) performed similar studies with local anesthetics using phosphate buffers over a range of pH. As in the case of the alcohols, order decreased with increasing anesthetic concentration. In addition, at a fixed concentration, order decreased with increasing pH. Generally, there was a correlation between the pK of the anesthetic and the pH required to bring about disorder.

Bovine brain lipid in which the cholesterol content has been reduced to 5% produces films with only a slight degree of spectral anisotropy. Hydrating these films with solutions containing low concentrations of local anesthetic leads to a massive increase in anisotropy. This ordering effect of local

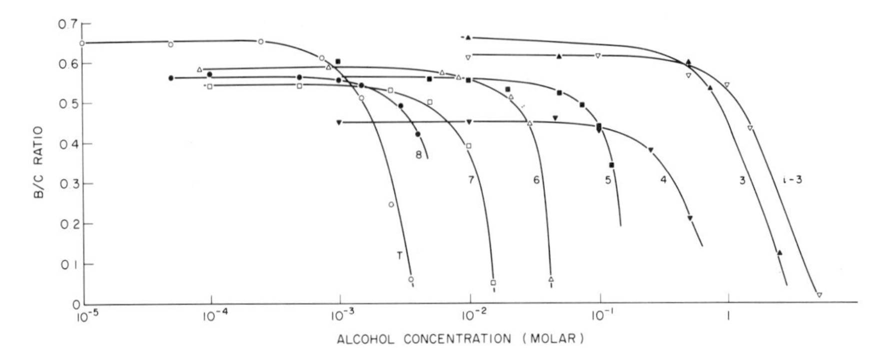

Fig. 18. The disordering effect of alcohols on human red cell ghost lipid films at pH 7, 23°C. The arbitrary order index *B/C* (equivalent to *D/E* in Fig. 17 and *B/C* in Fig. 8) was measured using 3-doxylcholestane with the films perpendicular to the applied magnetic field. T indicates the data for thymol, *i*-3 for isopropanol, and the other numbers for long chain alcohols containing the indicated number of carbons atoms. [From Paterson *et al.* (1972).]

anesthetics responded differently to pH than did the disordering effect described above. For each local anesthetic there was a pH range where ordering was maximal. This range was the same as that where higher concentrations of the same local anesthetics first began to disorder films containing their normal concentration of cholesterol. It was concluded that whereas the disordering effects of some local anesthetics, such as procaine, are mainly or entirely due to the uncharged form of the drug, the ordering effect requires a balance of the charged and uncharged forms of the drug.

This study has been extended to a series of other local anesthetics and to tranquilizers such as promazine (Neal *et al.*, 1975). All compounds tested produced first an increase in spectral anisotropy for films of brain lipid containing 5% cholesterol, then a decrease, as their concentration in the hydrating buffer increased. The concentrations of the drugs required to give maximum ordering was related to their pharmacological potency. In addition, interactions between salts and drugs were observed. It was found that the presence of calcium in the hydrating fluid decreased the concentration of tetracaine required to produce ordering. Calcium alone had no effect on order and did not potentiate the ordering action of pentobarbitone.

The effects of calcium and tetracaine on the permeability into lipid films has been measured by observation of the rate of reduction of spin probes by sodium ascorbate (see also Sections II.C and III.I). With films formed from lipids of the white matter of bovine brain, both calcium and tetracaine increased the permeability of the lipids to sodium ascorbate, and their effects were additive. The influence of calcium and tetracaine on permeability varied with the lipid used to form the film. Egg lecithin films given a net negative charge by incorporation of either phosphatidic acid or dicetyl phosphate responded in the same manner as brain white matter lipids. However, the permeability of phosphatidylserine or the lipids from bovine brain gray matter, was less in the presence of tetracaine and calcium together than in the presence of calcium alone. In the absence of calcium, tetracaine caused an increase in the permeability into bilayers of these lipids (Butler, 1975).

D. pH and Salt Effects

The pH of the aqueous phase, and the nature and concentration of the salt present have profound influences on the structure, stability, and conductance of membranes. Changes in pH can alter the charge and structure of proteins and hence influence their interactions with lipids (see Section III.G; also Chapter 12). Drug effects on membranes and model systems are also dependent on pH and the presence of salts (see Sections III.C and E).

The influence of pH and salts can be exerted directly on lipids, changing or shielding their charges and hence altering the manner in which they pack together. There is also the possibility of the formation of salt cross-links between lipid molecules. Films formed from the lipids extracted from the white matter of bovine brain and the 3-doxylcholestane spin probe exhibit very little anisotropy when hydrated with distilled water (Butler *et al.*, 1970a). Addition of nonelectrolytes such as sucrose has no effect, but the addition of salt results in a large increase in ESR spectral anisotropy. This effect is thus one of charge, not of osmotic or vapor pressure. It was shown to be due to the cation and to depend on the cation valence. The order of effectiveness in promoting spectral anisotropy was $NaCl = KCl = LiCl < MgCl_2 = CaCl_2 < LaCl_3 < ThCl_4$, whereas the chloride, thiocyanate, sulfate, and chromate salts of the same cation were equal in effectiveness. This influence of cations depended upon the lipid having a net negative charge. Addition of ions to the hydrating solution had no effect on the degree of order in films of egg lecithin (which has no net charge at pH 7) plus cholesterol. When the films contained dicetyl phosphate (which has a net negative charge at pH 7), the influence of cations was the same as with brain lipids. It was concluded that cations act by reducing the surface charge density of the bilayers. With the reduction in repulsion between headgroups the films can contract and achieve a higher degree of order.

The effects of cations are not the same with all negatively charged lipids. Calcium ions cause cardiolipin to form a hexagonal, rather than a lamellar phase (Rand and Sengupta, 1972), and cause films of phosphatidylserine to disintegrate (Schreier-Muccillo *et al.*, 1973b). No order was found in films of phosphatidic acid, with or without cholesterol, no matter what calcium concentration was present in the hydrating medium (Schreier-Muccillo *et al.*, 1973b). X-ray diffraction studies proved that a bilayer structure was present at 1 *M* calcium, indicating that the disorder in phosphatidic acid films was intralamellar. If the phosphatidic acid is diluted with lecithin to the point where phosphatidic acid constitutes 5–10% of the film, calcium again acts to increase spectral anisotropy.

All the charged groups on phospholipids may be titrated; hence the net charge on phospholipid films can be manipulated by changing the pH of the buffer used to hydrate the films. In an experiment where this was done with films containing spin probes, it was found that the maximum degree of spectral anisotropy occurred at pH values where the molecules had no net charge (Smith, 1971; Schreier-Muccillo *et al.*, 1973b). This supports the hypothesis that a net charge of the headgroups leads to molecular repulsion, an expansion of the film, and a decrease in ESR spectral anisotropy. Effects of ions and surface charge on permeability into bilayers and liposomes have also been demonstrated using the spin-probe method (see Section III.I).

E. The Influence of Uncouplers of Oxidative Phosphorylation

Uncouplers of oxidative phosphorylation are compounds that permit the metabolism of intermediates of the citric acid cycle to proceed with little or no concomitant production of ATP. The manner in which these compounds uncouple ATP production from oxidative phosphorylation is not at all clear. It has been suggested that uncouplers increase the permeability of the mitochondrial membranes. While it has been demonstrated that uncouplers can increase the conductivity of black lipid membranes, there has been no correlation between the ability to increase conductivity and uncoupling activity.

In an ESR study (Smith, 1971; Verma *et al.*, 1973a) films were formed of brain lipid or lecithin, with or without cholesterol, and the 3-doxylcholestane spin probe. When any of the films was first hydrated with a buffer, and then exposed to the same buffer containing 2×10^{-5} *M* uncoupler (dinitrophenol, pentachlorophenol, or dicumarol), a decrease in the B/C ratio of the ESR spectra in the perpendicular orientation (see Section II.A.1) was noted (Fig. 19). With films formed of either brain lipid or of lecithin plus cholesterol, the disordering effect was maximal at the pK of the uncoupler and approximately 10 min after the addition of the uncoupler. After 60–90 min, the disordering effect had vanished. With films of cholesterol-free lecithin, the effects were greater in magnitude than in the presence of cholesterol. The maximum effect occurred at pH 4.0 with all uncoupling agents, and the effect increased with time up to a constant value.

The results with cholesterol-containing films are compatible with a model in which a dimer of the protonated and unprotonated forms of the

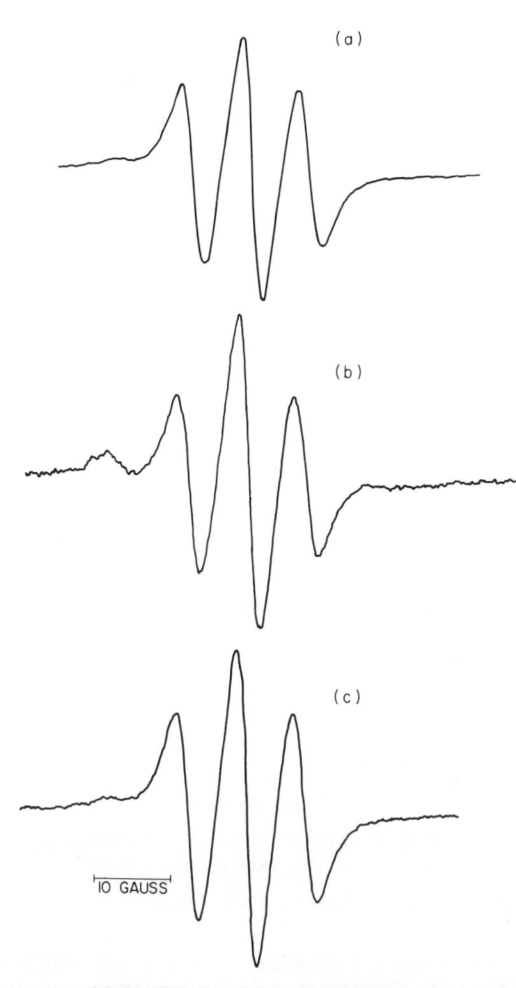

Fig. 19. ESR spectra of 3-doxylcholestane in multibilayers of egg lecithin and cholesterol (2 : 1 mole ratio) oriented perpendicular to the applied magnetic field. The films were hydrated with 0.1 *M* citrate–phosphate buffer, pH 4.0 with (a) no dinitrophenol, (b) 2×10^{-5} *M* dinitrophenol (10 min after addition), and (c) 2×10^{-5} *M* dinitrophenol (30 min after addition). [From Verma *et al.* (1973a).]

uncoupler is required for action; the results with cholesterol-free lecithin are not. The effects of uncouplers are clearly dependent on the presence or absence of cholesterol, the pH of the medium, and the time elapsed after addition of the uncoupler. The disordering effect of the uncouplers could cause conductivity changes or they could bring about changes in the lipid–protein interactions by altering the packing of the lipids (Paterson *et al.*, 1972).

F. Visual Pigments and Lipid Order

Photosynthesis and vision are associated with the lamellar structures found in chloroplasts and retinal rods, respectively. Illumination probably brings about changes in the potentials across or permeability through these structures. Investigations have been made on the effect of photosensitive pigments on the molecular structure of bilayers, and the effect of light, by observing the ESR spectra of the 3-doxylcholestane spin probe intercalated in lipid multibilayers.

Verma *et al.* (1974b) observed that visual pigments can influence the architecture of lipid films in the absence of light. The effect depended upon the lipid used to form the film as well as upon the pigment. Both all-trans retinal and retinol in concentrations up to 20 gm% increased the spectral anisotropy of egg lecithin films in a manner similar to cholesterol. The ratio of intensities of the low field and central resonances observed in the perpendicular orientation (B/C in Section II.A.1) increased and the width of the distribution of the probe long axis orientations decreased with increasing concentrations of the visual pigments. Concentrations of visual pigments greater than 20 gm% disordered the films. Films formed from a mixture of phosphatidylethanolamine and lecithin were less sensitive to this disordering effect than were films formed of pure phosphatidylethanolamine; formation of a Schiff base was postulated to be responsible in part for disordering by retinal. Disordering effects of visual pigments were also observed with films formed of phosphatidylserine, sphingomyelin plus cholesterol, or brain lipid. Part of the reason for the decrease in spectral anisotropy with concentrations of retinal above a certain level is formation of a second phase consisting mainly of retinal and the spin probe. Films formed of only retinal and the probe produced ESR spectra characteristic of a rigid random distribution of probes.

Verma *et al.* (1972) studied the effect of light on egg lecithin films containing the 3-doxylcholestane spin probe and various pigments. In this investigation, the concentration of the pigments was kept below the level required to induce disorder in the dark. It was found that illumination caused an increase in the width of the distribution of the probe long axis (disorder) for films containing chlorophyll or all-trans retinal or retinol. No effect of illumination was noted for pigment-free films; if they were then bathed in aqueous solutions of methylene blue, they became photosensitive. All light-induced disordering was maximal when the films were illuminated within the pigment absorption bands in the visible wavelength region. The effect of illumination on films containing chlorophyll *a* depended upon the nature of the anions present in the bathing solution (Verma *et al.*, 1972). The light-induced swelling of chloroplasts is also anion-dependent.

Gendel *et al.* (1972) have observed an increase in the correlation time of mono- and biradical spin probes intercalated in frog retinal rods upon illumination. From the nature of these changes the authors concluded that there is a change in the conformational state of the membrane. Verma *et al.* (1973b) found similar effects in rod outer segment membranes from bovine retina. The spin probe studies of oriented lipid systems show that photosensitive pigments can bring about changes in the orientational distribution of the probe in bilayers, and suggest an explanation for some of the effects observed in the more complex organelles. These changes in membrane lipid order could be involved in the biological process.

G. Protein–Lipid Interactions

The major components of biological membranes are proteins and lipids, which may assume different structures in different membranes (Urry, 1972). Membrane proteins can alter the structural and permeability properties of phospholipids (Juliano *et al.*, 1971; Kaplan, 1973), and membrane enzymes require phospholipids with specific headgroups and fatty acid composition for activity (Kimelberg and Papahadjopoulos, 1972; Jain, 1973; Eletr *et al.*, 1973). Interactions between lipids and proteins have been extensively studied using a variety of approaches. ESR studies have been made of the effects of liposomes on labeled protein and of the effects of protein on labeled liposomes, as well as analogous studies using lipid films. Experiments and spectral interpretation for the liposome systems will be discussed extensively in Chapter 12.

In the oriented multibilayer experiments (Smith, 1971; Butler *et al.*, 1973b; Smith *et al.*, 1974), films formed of lipid isolated from the white matter of ox brain and 3-doxylcholestane were hydrated with aqueous solutions of proteins or peptides of known amino acid composition. ESR spectral characteristics of low, intermediate, or high degrees of order were obtained, depending upon the protein used and the pH of the medium. The limiting cases, highly angular-dependent and angular-independent spectra were observed only when the pH of the hydrating solution was below the p*K* of the protein. Apparently then, pronounced effects of proteins on negatively charged brain lipid films are observed only when the proteins have a positive charge. Proteins such as lysozyme and ribonuclease, which produce a high degree of order in the lipid multilayers, act like multivalent cations (see Section III.D). This effect is independent of the protein conformation in solution; oxidized ribonuclease A, in which the four disulfide bonds are ruptured, is as effective in promoting order as the native compound, and 8 *M* urea had no influence on the effectiveness of lysozyme or ribonuclease in promoting order. X-ray diffraction studies showed that planar periodic bilayer structures were present in brain lipid samples hydrated with proteins that promoted either high or low degrees of order (Butler *et al.*, 1973b). The disordered films were thus shown to have wide distributions of probe orientations within the oriented bilayers. No strong correlations could be found between amino acid composition or hydrophobicity and the effects of proteins on lipid films (Butler *et al.*, 1973b; Davis *et al.*, 1973). The heme proteins, catalase and hemoglobin were both strong disordering agents while cytochrome *c* was much less effective (Butler *et al.*, 1973b).

Smith *et al.* (1974) found that polylysines ranging in molecular weight from 2800 to 280,000 all caused brain-lipid films to show highly angular dependent ESR spectra. Copolymers of lysine and alanine and of lysine and phenylalanine in various ratios produced films whose spectra had little or no angular dependence. Addition of these copolymers to the hydrating salt solution reduced the B/C ratio, their effectiveness depending upon the absolute concentration of alanine or phenylalanine residues.

In contrast with the results with copolymers of lysine with alanine or phenylalanine, the disordering effects of copolymers of lysine and leucine were not simply dependent on the content of leucine, an amino acid abundant in brain white matter protein (Folch-Pi and Stoffyn, 1972). Lys–Leu (17:1) simply disordered as the concentration increased, but Lys–Leu (2.7:1) first increased the characteristic amplitude ratio obtained in the perpendicular orientation (Section II.A.1) (Fig. 20) and then decreased it as the concentration increased. Pumping Lys–Leu 2.7 : 1 slowly past the film showed that it was the amount of polypeptide to which the film was exposed, not just the concentration, that is important in bringing about a given effect. The initially high amplitude ratio gradually dropped as the solution passed through the cell.

An interesting observation by Smith *et al.* (1974) is related to the nature of the multibilayer system. Films were first allowed to reach an intermediate degree of order by hydration with a buffer, and then were exposed to the same buffer containing an order-promoting or a disordering polypeptide. Disordering polypeptides caused a massive decrease in the B/C ratio of the films; ordering polypeptides had little or no effect. The ordering peptides could not penetrate the first several partially ordered bilayers and thus could not exert their effect on the bulk of the sample, whereas the order-reducing peptides could penetrate and interact with the entire sample.

No protein or polypeptide tested influenced the degree of order in films of egg lecithin. If 5% of either phosphatidic acid or phosphatidylserine was included in the egg lecithin films, proteins had much the same effect as they did on films of brain lipids. Charge interaction is obviously important for the influence of proteins on lipid structure.

All the preceding experiments were performed with the cholestane spin probe. By using other spin probes, it is possible to examine various depths

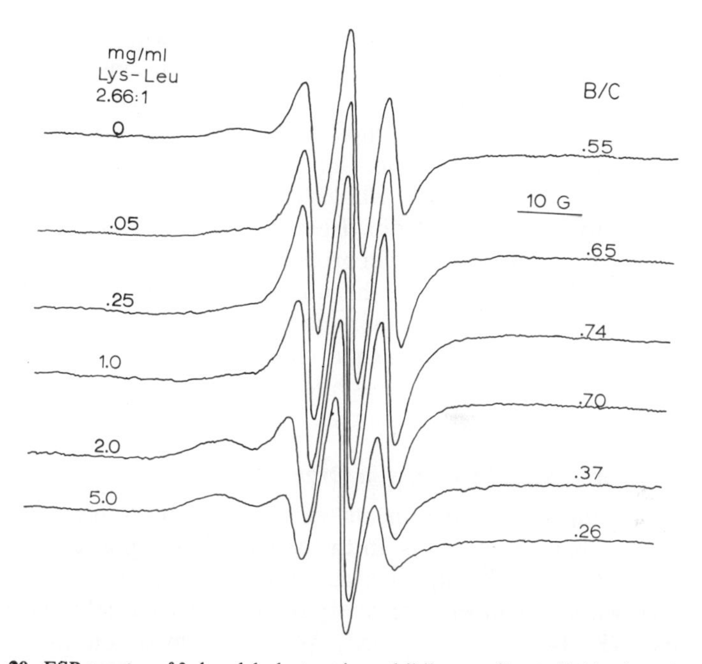

Fig. 20. ESR spectra of 3-doxylcholestane in multibilayers of brain lipid oriented perpendicular to the applied magnetic field. The films were hydrated with 0.14 *M* NaCl, 0.01 *M* tris, pH 7.4, plus the given concentrations of lysine–leucine copolymer. [Reprinted with permission from Smith *et al.* "Protides of the Biological Fluids, Vol. 21" (H. Peeters, ed.), 1974, Pergamon Press Ltd.]

within the lipid bilayer (see Chapters 10 and 12). The degree of order was measured as the difference between the hyperfine separations obtained in the parallel and perpendicular orientations for spin probes consisting of stearic acid with a doxyl group at the 5, 12, or 16 position (Fig. 3). The effect of each protein on brain lipid films containing the stearic acid spin probes paralleled the results with the cholestane spin probe. The order or disorder extends throughout the bilayers; it is not confined to the headgroup region. These experiments show that the oriented multilayer system is useful for examining the effects of proteins on lipid structure. The effect of proteins and polypeptides depends upon the amino acid composition, concentration, and pH.

Melittin, an oligopeptide toxin from bee venom, is asymmetric in distribution of amino acids. The first twenty residues are principally apolar and the last six at the carboxyl end of the molecule are polar. Melittin increases the glucose and ion permeability of liposomes at concentrations greater than 10^{-6} *M* and the capacity of lipids to absorb 12-doxylstearic acid at higher concentrations (Williams and Bell, 1972). These effects were independent of the nature of the headgroup of the phospholipid.

The effect of melittin on order in planar multilayers of various lipids was measured by Verma *et al.* (1974a). Increasing concentrations of melittin lead to an increase in the width of the orientational distribution of the 3-doxylcholestane probe long axis. Phosphatidylethanolamine was more sensitive to melittin than was egg lecithin; melittin at 10^{-5} M randomized completely the probe orientation. Cholesterol reduced further the sensitivity of egg lecithin to melittin. The 5-doxylstearic acid probe was less sensitive to melittin than was the cholestane spin probe, but increasing concentrations of melittin increased the amplitude of the motion of the spin probe and reduced the rate of that motion. The authors concluded that melittin reorganizes both the headgroup and hydrocarbon chain regions in the phospholipid films, illustrating that both electrostatic and hydrophobic forces are involved in lipid–protein interactions.

H. Temperature-Induced Changes in Organization

Variable temperature experiments are more difficult to perform on multibilayer films than on solutions. In the case of films deposited on conventional quartz flat cells for aqueous samples, the cell is too wide to fit the commercial variable temperature cavity insert, and a special insert must be constructed (K. Butler, unpublished). With films made on cover glasses, solvent evaporation becomes a problem at elevated temperature. Nonetheless, with proper attention to these difficulties, variable-temperature measurements can be made.

Jost *et al.* (1971) studied 5, 12, and 16-doxylstearate spin probes in egg lecithin multibilayers over the temperature range $-40°$ to 33°C at a low degree of hydration. At -40°C, the spectra taken with the magnetic field parallel and perpendicular to the plane of the film were virtually identical. As the temperature increased, the angular dependence of the ESR spectra increased. At all temperatures the degrees of anisotropy were in the order 5-doxylstearate > 12-doxylstearate > 16-doxylstearate.

Marsh (1974) using fully hydrated egg lecithin multibilayers found the same dependence of anisotropy on nitroxide position. It decreased with increasing temperature over the range 5°–65°C. This was attributed to an increasing amplitude of motion of the spin probe. From this dependence of anisotropy, measured as the S parameter, on nitroxide position he calculated P_g and P_t, the probabilities of any carbon–carbon single bond being present as the gauche or trans conformer, respectively. From the variation in these terms with temperature, he calculated the activation energy for gauche–trans interconversion, and showed that the addition of 30% cholesterol increased this activation energy by 1.5 kcal/mole of CH_2. Thus, studies performed at different temperatures can be used to explore molecular interactions and derive activation energies.

Garrigou-Lagrange and Smith (1975) investigated the influence of temperature on hydrated films of distearoyllecithin containing the doxylstearic acid probes. Below the chain-melting temperature, the ESR spectra were composed of broad lines due to a wide distribution of relatively immobile probes; above this temperature (58°C), the spectra were composed of narrow lines with an angular-dependent hyperfine separation. Further increase in temperature increased the rate of anisotropic motion without a significant decrease in the order parameter.

Below the temperature for melting of the fatty acid chains, hydrated films of monoglycerides excluded the 3-doxylcholestane probe into regions of high probe concentration to produce exchange-broadened ESR spectra (Schreier-Muccillo *et al.*, 1973b) (see Sections II.B.1 and III.A). By raising the temperature or adding cholesterol, the fluidity of the system became sufficiently high to permit entry of the probes and their study by ESR. The temperature range over which a high degree of order could be maintained has been found to be a useful indicator of the organization-inducing properties of local anesthetics (Butler *et al.*, 1973a) (see Section III.C).

Schreier-Muccillo *et al.* (1975) have measured the temperature and time dependence of the rate of reduction of spin probes in lipid bilayers by ascorbate as an indicator of the permeability to ascorbate (Section III.I). They used the 3-doxylcholestane, 3-doxylandrostane-17-ol, and three doxylstearic acid spin probes to measure reduction rates and activation energies as a function of depth within the bilayer. The reduction rate was increased by a net positive charge and decreased by a net negative charge added to the bilayer. As the nitroxide group was buried deeper within the bilayer, the rate of reduction decreased. However, the activation energy was not influenced by any of these alterations. On the other hand, addition of cholesterol caused the activation energy to increase from 6 to 15 kcal/mole.

I. Diffusion into Bilayers

Permeability is an important characteristic of biological membranes involved in excitability and the regulation of cell volume and composition. Its measurement usually involves the net movement of molecules or ions or the flux of radioactive tracers through the membrane from the aqueous compartment on one side to the aqueous compartment on the other side. Recently, spin probes have been used to study part of this process, namely, the penetration into rather than transport across bilayers.

Kornberg and McConnell (1971) formed liposomes containing a lecithin spin probe with the nitroxide attached to the choline nitrogen. This nitroxide group is exposed to the aqueous phase. By treating these liposomes with sodium ascorbate at 0°C, at which temperature the liposomes were shown to be impermeable to the ascorbate ion, they rapidly reduced all the nitroxide groups in the outward-facing half of the bilayer. They were able to show a difference in ESR spectra between the probes on the two sides of the bilayer and to measure the rate at which molecules moved from one side of the bilayer to the other (transverse diffusion).

If lipid films are hydrated with an ascorbate-containing solution, there is no protected half of the bilayers, and all probes with the nitroxide group exposed to the aqueous phase are rapidly reduced. Spin probes whose nitroxide groups are within the lipid are protected and are destroyed more slowly: Figs. 21 and 22. This rate of reduction depends upon the lipid

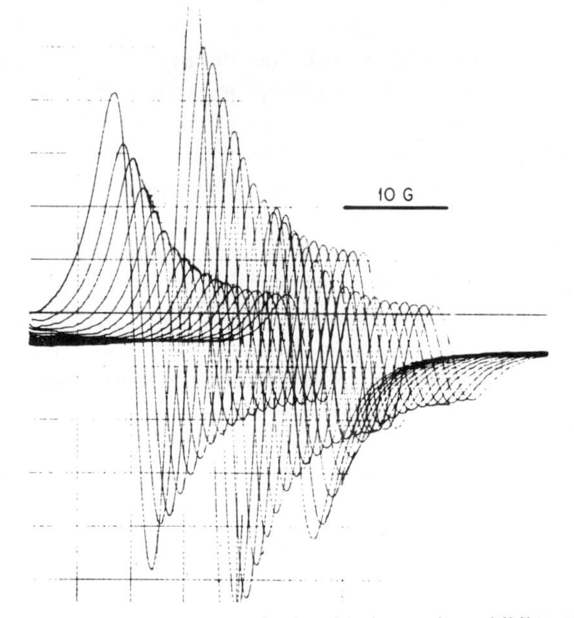

Fig. 21. Decay of the ESR spectrum of 3-doxylcholestane in multibilayers formed of 30% phosphatidylserine, 35% phosphatidylcholine, and 35% cholesterol in the presence of 10 m*M* ascorbate at pH 6.6 and 8.3°C. Spectra were taken every minute after hydration, and the magnetic field was offset for each spectrum for clarity. [From Schreier-Muccillo *et al.* (1975).]

composition and the spin probe used. Schreier-Mucillo *et al.* (1975) found that as the doxyl group was located deeper within the bilayer, the half-time of reduction increased. This result is compatible with increased distance for diffusion of the ascorbate into the bilayer. Adding a positive charge to the film, which would be expected to attract the negatively charged ascorbate ion, increased the rate of reduction; adding a negative charge to the film reduced the rate of reduction. The activation energy of the process was found to be independent of film charge. Cholesterol, on the other hand,

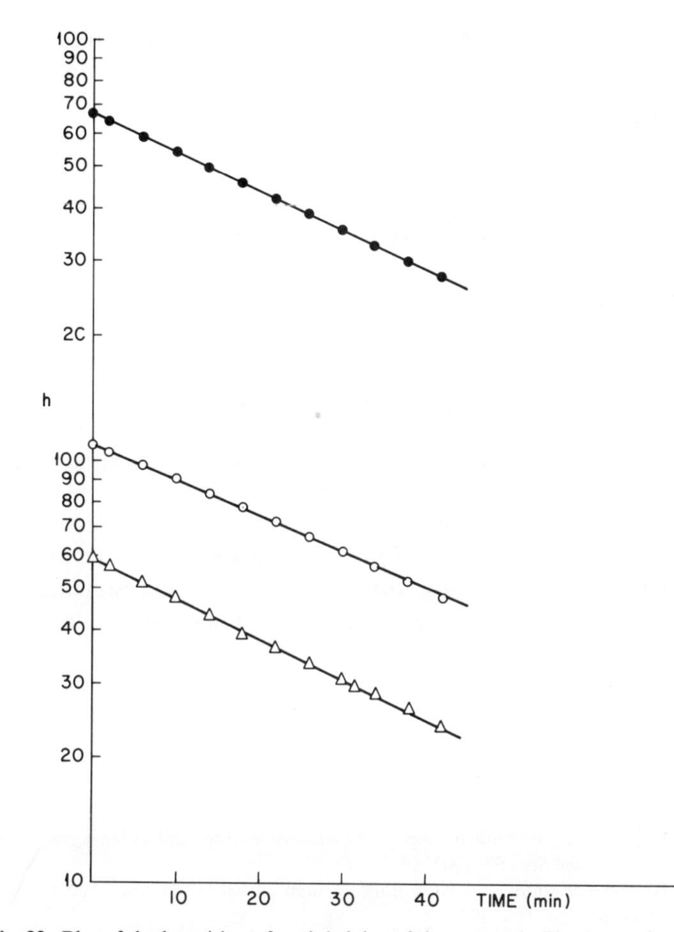

Fig. 22. Plot of the logarithm of peak heights of the spectra in Fig. 21 as a function of time (● = h_0, ○ = h_1, Δ = h_{-1}). The amplitudes of the three resonances decrease exponentially at the same rate. [From Schreier-Muccillo *et al.* (1975).]

increased the activation energy from 6 to 15 kcal. It was concluded that permeation takes place in two steps: partition into and diffusion through the membrane.

Papahadjopoulos (1970) demonstrated that calcium, and local anesthetics in the absence of calcium, increased the permeability of phosphatidylserine vesicles for $^{22}Na^+$. In the presence of calcium, local anesthetics brought about a decrease in permeability at concentrations lower than those required to increase permeability in its absence. Butler (1975) used the rate of reduction of spin probes by sodium ascorbate to study the effects of local anesthetics and calcium on permeability into the bilayer. Calcium and local anesthetics were found to influence the packing of the lipid molecules in the films; the effects depended on the nature of the lipids as well as on the charge of the film.

IV. CONCLUSION

The ESR studies of spin probes in oriented multibilayer systems have provided a high level of understanding of the spectra and the physical processes responsible for them. Both were essential before the extremely complex spectra observed in vesicles or biological membranes could be approached with any sort of confidence. The future for the spin-probe technique in membrane systems involves more rigorous interpretation of the spectra in terms of both time- and order-dependent processes. This is a costly process in time and computing funds, but it is essential if reliable parameters related to lipid motion and order are to be extracted from the ESR spectra. The frequency of the ESR spectrometer provides a valuable additional experimental variable. The ESR data should also be interpreted in terms of the models evolving from nuclear magnetic resonance studies (^{1}H, ^{13}C, ^{31}P, ^{2}H) on similar systems, keeping in mind the qualitative differences that may exist due to the perturbative influence of the nitroxide moiety.

REFERENCES

Blodgett, K. B. (1935). Films built by depositing successive monomolecular layers on a solid surface, *J. Amer. Chem. Soc.* **57**, 1007–1022.

Blodgett, K. B., and Langmuir, I. (1937). Built-up films of barium stearate and their optical properties, *Phys. Rev.* **51**, 964–982.

Boggs, J. M., and Hsia, J. C. (1973a). Orientation and motion of amphiphilic spin labels in hexagonal lipid phases, *Proc. Nat. Acad. Sci. U.S.* **70**, 1406–1409.

Boggs, J. M., and Hsia, J. C. (1973b). Structural characteristics of hydrated glycerol and sphingo-lipids. A spin label study, *Can. J. Biochem.* **51**, 1451–1459.

Butler, K. W. (1975). Drug-Biomolecule interactions: Spin-probe study of effects of anesthetics on membrane lipids, *J. Pharm. Sci.* **64**, 497–501.

Butler, K. W., Dugas, H., Smith, I. C. P., and Schneider, H. (1970a). Cation-induced organization changes in a lipid bilayer model membrane, *Biochem. Biophys. Res. Commun.* **40**, 770–776.

Butler, K. W., Smith, I. C. P., and Schneider, H. (1970b). Sterol structure and ordering effects in spin-labeled phospholipid multibilayer structures. *Biochim. Biophys. Acta* **219**, 514–517.

Butler, K. W., Schneider, H., and Smith, I. C. P. (1973a). The effects of local anesthetics on lipid multilayers. A spin probe study, *Arch. Biochem. Biophys.* **154**, 548–554.

Butler, K. W., Hanson, A. W., Smith, I. C. P., and Schneider, H. (1973b). The effects of protein on lipid organization in a model membrane system. A spin probe and X-ray study, *Can. J. Biochem.* **51**, 980–989.

Butler, K. W., Tattrie, N. H., and Smith, I. C. P. (1974). The location of spin probes in two phase mixed lipid systems, *Biochim. Biophys. Acta* **363**, 351–360.

Cannon, B., Polnaszek, C. F., Butler, K. W., Eriksson, L. G., and Smith, I. C. P. (1975). The fluidity and organization of mitochondrial membrane lipids of the brown adipose tissue of cold-adapted rats and hamsters as determined by nitroxide spin probes, *Arch. Biochem. Biophys.* **167**, 505–518.

Davis, M. A. F., Hauser, H., Leslie, R. B., and Phillips, M. C. (1973). Protein hydrophobicity and lipid–protein interaction, *Biochim. Biophys. Acta* **317**, 214–218.

de Bernard, L. (1958). Associations moléculaires entre les Lipides, *Bull. Soc. Chim. Biol.* **40**, 161–170.

Demel, R. A., Van Deenen, L. L. M., and Pethica, B. A. (1967). Monolayer interactions of phospholipids and cholesterol, *Biochim. Biophys. Acta* **135**, 11–19.

Demel, R. A., Bruckdorfer, K. R., and Van Deenen, L. L. M. (1972). The effect of sterol structure on the permeability of liposomes to glucose, glycerol, and Rb^+, *Biochim. Biophys. Acta* **255**, 321–330.

Devaux, P., and McConnell, H. M. (1972). Lateral diffusion in spin-labeled phosphatidylcholine multilayers, *J. Amer. Chem. Soc.* **94**, 4475–4481.

de Vries, J. J., and Berendesen, H. J. C. (1969). Nuclear magnetic resonance measurements on a macroscopically ordered smectic liquid crystalline phase, *Nature* (*London*) **221**, 1139–1140.

Dupeyre, R. M., Lemaire, H., and Rassat, A. (1965). A stable biradical in the nitroxide series, *J. Amer. Chem. Soc.* **87**, 3771–3772.

Eletr, S., Zakim, D., and Vessey, D. A. (1973). A spin-label study of the role of phospholipids in the regulation of membrane-bound microsomal enzymes, *J. Mol. Biol.* **78**, 351–362.

Folch-Pi, J., and Stoffyn, P. J. (1972). Proteolipids from membrane systems, *Ann. N.Y. Acad. Sci.* **195**, 86–107.

Garrigou-Lagrange, C., and Smith, I. C. P. (1975). A spin-probe study of motion and order in multibilayer films of lecithins containing saturated fatty acids (in preparation).

Gendel, L. Y., Fedorovich, I. B., Ostrovskii, M. A., Shapiro, A. B., and Kruglyakova, K. E. (1972). Photoinduced conformational transitions in photoreceptive membranes of rod external segments as investigated by the method of spin probes, *Dokl. Akad. Nauk. USSR* **205**, 1469–1472.

Ghosh, D., and Tinoco, J. (1972). Monolayer interactions of individual lecithins with natural sterols, *Biochim. Biophys. Acta* **266**, 41–49.

Hegner, D., Schummer, U., and Schnepel, G. H. (1973). The effect of calcium on the temperature-induced phase changes in liquid-crystalline cardiolipin structure, *Biochim. Biophys. Acta* **307**, 452–458.

Hemminga, M. A., and Berendsen, H. J. C. (1972). Magnetic resonance in ordered lecithin cholesterol multilayers, *J. Magn. Resonance* **8**, 133–143.

Hsia, J. C., Schneider, H., and Smith, I. C. P. (1970). Spin-label studies of oriented phospholipids: Egg lecithin, *Biochim. Biophys. Acta* **202**, 399–402.

Hsia, J. C., Schneider, H., and Smith, I. C. P. (1971). A spin label study of the influence of cholesterol on phospholipid multibilayer structures, *Can. J. Biochem.* **49**, 614–622.

Hsia, J. C., Long, R. A., Hruska, F. E., and Gesser, H. D. (1972). Steroid–phosphatidylcholine interactions in oriented multibilayers, *Biochim. Biophys. Acta* **290**, 22–31.

Hubbell, W. L., and McConnell, H. M. (1971). Molecular motion in spin-labeled phospholipids and membranes, *J. Amer. Chem. Soc.* **93**, 314–320.

Jain, M. K. (1972). "The Bimolecular Lipid Membrane." Van Nostrand-Reinhold, Princeton, New Jersey.

Jost, P., Libertini, L. J., Herbert, V. C., and Griffith, O. H. (1971). Lipid spin labels in lecithin multilayers, *J. Mol. Biol.* **59**, 77–98.

Juliano, R. L., Kimelberg, H. K., and Papahadjopoulos, D. (1971). Synergistic effects of a membrane protein (spectrin) and Ca^{++} on the Na^{+} permeability of phospholipid vesicles, *Biochim. Biophys. Acta* **241**, 894–905.

Kaplan, J. H. (1973). The effect of lysozyme on anion and cation diffusion from phospholipid liposomes, *Biochim. Biophys. Acta* **311**, 1–5.

Keana, J. F. W., and Dinerstein, R. J. (1971). A new highly anisotropic dinitroxide ketone spin label. A sensitive probe for membrane structure, *J. Amer. Chem. Soc.* **93**, 2808–2810.

Keith, A. D., Sharnoff, M., and Cohn, G. E. (1973). A summary and evaluation of spin labels as probes for biological membrane structure, *Biochim. Biophys. Acta* **300**, 379–419.

Kimelberg, H. K., and Papahadjopoulos, D. (1972). Phospholipid requirements for (Na^{+} + K^{+})-ATPase activity, *Biochim. Biophys. Acta* **282**, 277–292.

Kornberg, R. D., and McConnell, H. M. (1971). Inside–outside transitions of phospholipids in vesicle membranes, *Biochemistry* **10**, 1111–1120.

Lapper, R. D., Paterson, S. J., and Smith, I. C. P. (1972). A spin label study of the influence of cholesterol on egg lecithin multibilayers, *Can. J. Biochem.* **50**, 969–981.

Leathes, J. B. (1925). Croonian lectures on the role of fats in vital phenomena, *Lancet* **208**, 853–856.

Levine, Y. K., and Wilkins, M. H. F. (1971). Structure of oriented lipid bilayers, *Nature (London) New Biol.* **230**, 69–72.

Levine, Y. K., Bailey, A. I., and Wilkins, M. H. F. (1968). Multibilayers of phospholipid bimolecular leaflets, *Nature (London)* **220**, 577–578.

Libertini, L. J., Waggoner, A. S., Jost, P. C., and Griffith, O. H. (1969). Orientation of lipid spin labels in lecithin multilayers, *Proc. Nat. Acad. Sci. U.S.* **64**, 13–19.

Luckhurst, G. R., and Pedulli, G. F. (1970). Interpretation of biradical electron resonance spectra, *J. Amer. Chem. Soc.* **92**, 4738–4739.

Mailer, C., Taylor, C. P. S., Schreier-Muccillo, S., and Smith, I. C. P. (1974). The influence of cholesterol on molecular motion in egg lecithin bilayers—A variable frequency electron spin resonance study of a cholestane spin probe. *Arch. Biochem. Biophys.* **163**, 671–678.

Marsh, D. (1974). Intermolecular steric interactions and lipid chain fluidity in phospholipid multibilayers—A spin label study, *Can. J. Biochem.* **52**, 631–636.

Marsh, D., and Smith, I. C. P. (1972). Interacting spin labels as probes of molecular separation within phospholipid bilayers, *Biochem. Biophys. Res. Commun.* **49**, 916–922.

Marsh, D., and Smith, I. C. P. (1973). An interacting spin label study of the fluidizing and condensing effects of cholesterol on lecithin bilayers, *Biochim. Biophys. Acta* **298**, 133–144.

Narahashi, T., Frazier, D. T., Deguchi, T., Cleaves, C. A., and Ernau, M. C. (1971). The active form of pentobarbital in squid giant axons. *J. Pharmocol. Exp. Ther.* **177**, 25–33.

Neal, M. J., Butler, K. W., Polnaszek, C. F., and Smith, I. C. P. (1975). A spin-label study of the ordering effect of cholesterol and anesthetics on lipid bilayers, *Molec. Pharmacol.* (in press).

Papahadjopoulos, D. (1970). Phospholipid model membranes III. Antagonistic effects of Ca^{++} and local anesthetics on the permeability of phosphatidylserine vesicles, *Biochem. Biophys. Acta* **211**, 467–477.

Paterson, S. J., Butler, K. W., Hwang, P., Labelle, J., Smith, I. C. P., and Schneider, H. (1972). The effects of alcohols on lipid bilayers. *Biochim. Biophys. Acta* **266**, 597–602.

Polnaszek, C. F., Bruno, G. V., and Freed, J. H. (1973). ESR line shapes in the slow-motional region: Anisotropic liquids, *J. Chem. Phys.* **58**, 3185–3199.

Rand, R. P., and Sengupta, S. (1972). Cardiolipin forms hexagonal structures with divalent cations, *Biochim. Biophys. Acta* **255**, 484–492.

Roth, S., and Seeman, P. (1971). All lipid-soluble anaesthetics protect red cells, *Nature (London) New Biol.* **231**, 284–285.

Scandella, C. J., Devaux, P., and McConnell, H. M. (1972). Rapid lateral diffusion of phospholipids in rabbit sarcoplasmic reticulum, *Proc. Nat. Acad. Sci. U.S.* **69**, 2056–2060.

Schneider, H. (1968). The intramembrane location of alcohol anesthetics, *Biochim. Biophys. Acta* **163**, 451–458.

Schreier-Muccillo, S., and Smith, I. C. P. (1976). "Spin Label Probes of Biological Membranes," to be published by Academic Press, New York.

Schreier-Muccillo, S., Marsh, D., Dugas, H., Schneider, H., and Smith, I. C. P. (1973a). A spin probe study of the influence of cholesterol on motion and orientation of phospholipids in oriented multibilayers and vesicles, *Chem. Phys. Lipids* **10**, 11–27.

Schreier-Muccillo, S., Butler, K. W., and Smith, I. C. P. (1973b). Structural requirements for the formation of ordered lipid multibilayers, *Arch. Biochem. Biophys.* **159**, 297–311.

Schreier-Muccillo, S., Marsh, D., and Smith, I. C. P. (1975). A spin probe study of permeability properties of phospholipid membranes, *Arch. Biochem. Biophys.* (in press).

Seelig, J. (1970). Spin label studies of oriented smectic liquid crystals, *J. Amer. Chem. Soc.* **92**, 3381–3387.

Seeman, P. (1972). The membrane actions of anesthetics and tranquilizers, *Pharmacol. Rev.* **24**, 583–655.

Skou, J. C. (1954). Local anesthetics. I. The blocking potencies of some local anaesthetics and of butyl alcohol determined on peripheral nerves, *Acta Pharmacol. Toxicol.* **10**, 281–291.

Skou, J. C. (1961). The effect of drugs on cell membranes with special reference to local anaesthetics, *J. Pharm. Pharmacol.* **13**, 204–217.

Smith, I. C. P. (1971). A spin label study of the organization and fluidity of hydrated phospholipid multibilayers, *Chimia* **25**, 349–360.

Smith, I. C. P., Williams, R. E., and Butler, K. W. (1974). Spin label studies of the influence of peptides on the organization of lipids in bilayers, *In* "Protides of the Biological Fluids" (H. Peeters, ed.), Vol. 21, pp. 215–222. Pergamon, Oxford.

Tardieu, A., Luzzati, V., and Reman, F. C. (1973). Structure and polymorphism of the hydrocarbon chains of lipids, *J. Mol. Biol.* **75**, 711–733.

Urry, D. W. (1972). Conformation of protein in biological membranes and a model transmembrane channel, *Ann. N.Y. Acad. Sci.* **195**, 108–125.

Verma, S. P., Schneider, H., and Smith, I. C. P. (1972). Spin probe studies of photo-induced structural changes in phospholipid multibilayers containing light-sensitive pigments, *FEBS Lett.* **25**, 197–200.

Verma, S. P., Schneider, H., and Smith, I. C. P. (1973a). Organizational changes in phospholipid multibilayers induced by uncouplers of oxidative phosphorylation, *Arch. Biochem. Biophys.* **154**, 400–406.

Verma, S. P., Berliner, L. J., and Smith, I. C. P. (1973b). Cation-dependent light-induced structural changes in visual receptor membranes, *Biochem. Biophys. Res. Commun.* **55**, 704–709.

Verma, S. P., Wallach, D. F. H., and Smith, I. C. P. (1974a). The action of melittin on phosphatide multibilayers as studied by infrared dichroism and spin labeling. A model approach to lipid–protein interaction, *Biochim. Biophys. Acta* **345**, 129–140.

Verma, S. P., Schneider, H., and Smith, I. C. P. (1974b). Effects of visual pigments on the organization of phospholipid multibilayer model membranes—A spin probe study, *Arch. Biochem. Biophys.* **162**, 48–55.

Wertz, J. E., and Bolton, J. R. (1972). "Electron Spin Resonance: Elementary Theory and Practical Applications." McGraw-Hill, New York.

Williams, J. C., and Bell, R. M. (1972). Membrane matrix disruption by melittin, *Biochim. Biophys. Acta*, **288**, 255–262.

12

Lipid Spin Labels in Biological Membranes

O. HAYES GRIFFITH and PATRICIA C. JOST

INSTITUTE OF MOLECULAR BIOLOGY
AND
DEPARTMENT OF CHEMISTRY
UNIVERSITY OF OREGON
EUGENE, OREGON

I. INTRODUCTION

The basic unit of life is the cell. Each living cell is bounded by a thin flexible envelope, the plasma membrane. The plasma membrane is only a small percentage of the total membrane in cells of higher organisms. Membranes also surround the cell nucleus and represent the basic structural component of various organelles such as mitochondria and chloroplasts. In addition, the cytoplasm is filled with an extensive membrane system, the endoplasmic reticulum. These various membranes play a crucial role in many cellular processes. For example, the plasma membrane regulates the transport of molecules into and out of the cell and is intimately involved in normal and abnormal cell–cell interactions. These interactions, in turn, are important in differentiation, mobility, cell division, and cell recognition. Organelle membranes are the site of important cellular functions. For example, mitochondrial and chloroplast membranes are the sites for critical energy conversions (e.g., the generation of adenosine triphosphate). Protein biosynthesis occurs on the polyribosomes bound to the endoplasmic reticulum membranes, and many enzymatic activities are linked to these cytoplasmic membranes. Clearly, there is a diversity in function and composition among these many membranes. What has been apparent for many years is that there seem to be some useful generalizations that apply to all cellular membranes. They are all composed of lipids (predominantly phospholipids) and protein, and all have the same general appearance and dimensions in transmission electron micrographs (although at high resolution significant differences are often observed).

Within the past decade, there has been gradual acceptance of a general model for the molecular structure of membranes. One version of this model is sketched in Fig. 1. Much of the phospholipid is organized into bilayers, with the polar head groups at the aqueous interfaces and the hydrocarbon tails forming a central lipoid region. The membrane proteins are predominantly globular with hydrophobic surfaces buried in the bilayer and hydrophilic surfaces exposed to the aqueous environments (Singer, 1971, and references contained therein).

As a first approximation, proteins can be classified into two categories according to their solubility properties. Some membrane proteins require detergents or other harsh treatments for solubilization. In Fig. 1 these intrinsic, or integral, proteins are shown embedded in the phospholipid bilayer. Another class of membrane proteins is removed from the membranes by mild aqueous treatments (e.g., changes in salt concentration). These extrinsic or peripheral proteins are shown adhering to the polar surface of the phospholipid bilayer or interacting with other proteins. Examples of intrinsic (integral) membrane proteins include the major glycoprotein of the red cell

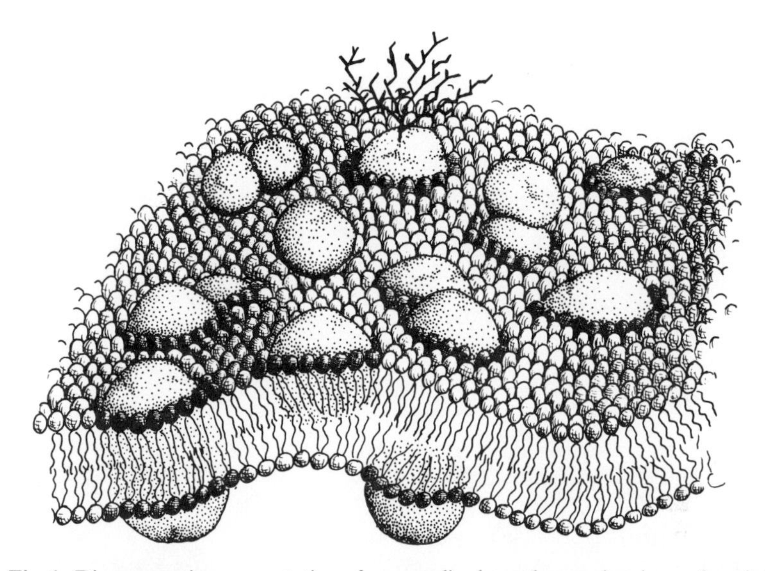

Fig. 1. Diagrammatic representation of a generalized membrane, showing a phospholipid bilayer continuum in which proteins (intrinsic or integral) are imbedded. Other proteins (extrinsic or peripheral) are shown associated with the surface of the membrane. The bushlike structure on one protein represents the carbohydrate of a glycoprotein. The phospholipid in contact with the imbedded proteins is shaded darker to indicate that it may differ in properties from the rest of the bilayer. (Sketch by Denita Widders.)

plasma membrane, the cytochrome oxidase complex of the inner mitochondrial membrane, and cytochrome b_5 of the liver endoplasmic reticulum membranes. Examples of extrinsic (peripheral) membrane proteins are spectrin of the red cell plasma membrane and cytochrome c of the inner mitochondrial membrane.

The general model in Fig. 1 should be viewed as a crude working hypothesis that fits many experimental observations and is only the latest in a long line of constantly evolving models. Many techniques are currently being used to test certain aspects of this model. And equally important, the question of the molecular basis of the *differences* in various membranes that serve a wide variety of functions can now be approached. Current membrane research is focused on testing the soundness of these generalizations about membrane structure, and proceeding beyond this point to understand the molecular basis of the structure and function of specific specialized membranes. Our aim here is to discuss how one spectroscopic technique, spin labeling, is contributing to these research goals. Rather than to provide an inclusive review of the literature, we have chosen instead to select a few illustrations of the various ways in which spin labeling can be used to obtain new information about membrane structure and function. The common

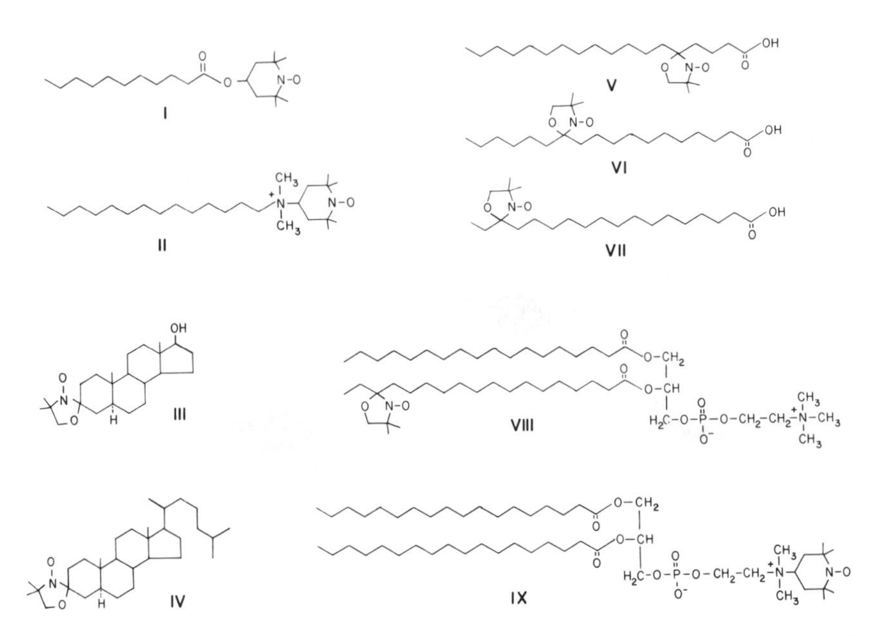

Fig. 2. Representative lipid spin labels. **I**, Waggoner *et al.* (1968); Rozantsev (1970); **II, III**, Hubbell and McConnell (1969a); **IV**, Keana *et al.* (1967); **V–VIII**, Keith *et al.* (1968); Waggoner *et al.* (1969); Hubbell and McConnell (1971); Jost *et al.* (1971); **IX**, Kornberg and McConnell (1971).

theme is the use of lipid spin labels, some of which are given in Fig. 2. In this chapter we show how the ESR data from these spin labels are analyzed to obtain information regarding molecular motion in membranes, lipid order in membranes, the polarity profile across the membrane, the identification and extent of the bilayer, and preliminary approaches to lipid–protein interactions in the plane of the membrane. Other important applications of these lipid spin labels are given in Chapter 10 (liquid crystals), Chapter 11 (oriented samples and influence of perturbants on membrane model systems), and Chapter 13 (lateral diffusion and phase separations).

II. MOLECULAR MOTION IN MEMBRANES

A. Models for Anisotropic Motion

1. Anisotropic Motion in the Intermediate Field Approximation

Simple models can be used to calculate line shapes that are a reasonably good approximation of the experimental spectra and can serve to reduce complicated line shapes to a set of physically meaningful parameters for comparison of related systems.

As discussed in Chapter 2, the general spin Hamiltonian that describes the behavior of a collection of nitroxide spin labels is $\hat{\mathscr{H}} = \hat{\mathscr{H}}_{\text{Zeeman}} + \hat{\mathscr{H}}_{\text{hyperfine}} + \hat{\mathscr{H}}_{\text{J}} + \hat{\mathscr{H}}_{\text{D}}$. The $\hat{\mathscr{H}}_{\text{J}}$ and $\hat{\mathscr{H}}_{\text{D}}$ are the electron–electron exchange Hamiltonian and the electron–electron dipole Hamiltonian. These terms may be neglected except in experiments such as lateral diffusion where high local concentrations of the spin label are used (see Chapters 11 and 13) or when using dinitroxide spin labels (see Chapter 4). The spin Hamiltonian $\hat{\mathscr{H}} = \hat{\mathscr{H}}_{\text{Zeeman}} + \hat{\mathscr{H}}_{\text{hyperfine}}$ is often written in the nitroxide coordinate system (x, y, z) diagrammed in Fig. 3, because the **g** and **A** matrices are diagonal in this system, with principal values g_{xx}, g_{yy}, g_{zz}, and A_{xx}, A_{yy}, A_{zz}. The x axis is along the N—O bond, z is parallel to the nitrogen and oxygen 2p orbitals containing the unpaired electron, and y is perpendicular to the x, z plane.

In discussing anisotropic motion it is convenient to express $\hat{\mathscr{H}}$ in the laboratory coordinate system XYZ, where Z is defined to be the direction of the laboratory magnetic field. In the X, Y, Z system, **g** and **A** become the nondiagonal matrices **g**′ and **A**′ calculated from the similarity transformations $\mathbf{g}' = \mathbf{L} \cdot \mathbf{g} \cdot \mathbf{L}^{\tau}$ and $\mathbf{A}' = \mathbf{L} \cdot \mathbf{A} \cdot \mathbf{L}^{\tau}$, where primes indicate quantities in the laboratory coordinate system, τ denotes transpose and **L** is the transformation matrix, which rotates the nitroxide x, y, z axes into the laboratory X, Y, Z axes.

$$\mathbf{A}' = \begin{pmatrix} A_{XX} & A_{XY} & A_{XZ} \\ A_{YX} & A_{YY} & A_{YZ} \\ A_{ZX} & A_{ZY} & A_{ZZ} \end{pmatrix}$$

$$= \begin{pmatrix} l_{Xx} & l_{Xy} & l_{Xz} \\ l_{Yx} & l_{Yy} & l_{Yz} \\ l_{Zx} & l_{Zy} & l_{Zz} \end{pmatrix} \begin{pmatrix} A_{xx} & 0 & 0 \\ 0 & A_{yy} & 0 \\ 0 & 0 & A_{zz} \end{pmatrix} \begin{pmatrix} l_{Xx} & l_{Yx} & l_{Zx} \\ l_{Xy} & l_{Yy} & l_{Zy} \\ l_{Xz} & l_{Yz} & l_{Zz} \end{pmatrix} \tag{1}$$

The l's are the direction cosines between the nitroxide x, y, z axes and the laboratory X, Y, Z axes. For example l_{Zx} is the cosine of the angle between

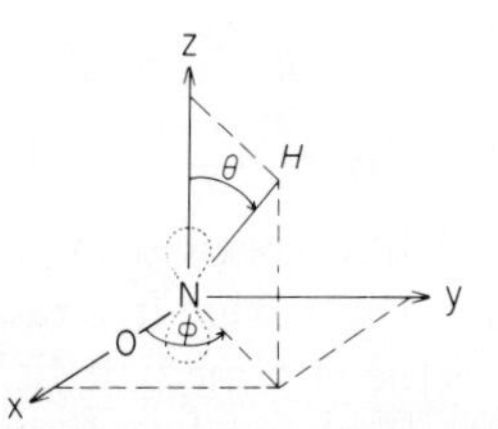

Fig. 3. The nitroxide coordinate system, where H indicates the magnetic field and θ is the angle between the external magnetic field and the nitroxide z axis.

the N—O bond and the laboratory Z axis. Expansion of Eq. (1) and of the corresponding $\mathbf{g}'$ matrix gives the useful relations

$$g_{ZZ} = g_{xx} l_{Zx}^2 + g_{yy} l_{Zy}^2 + g_{zz} l_{Zz}^2 \tag{2}$$

$$A_{ZZ} = A_{xx} l_{Zx}^2 + A_{yy} l_{Zy}^2 + A_{zz} l_{Zz}^2 \tag{3}$$

$$A_{XZ} = A_{xx} l_{Xx} l_{Zx} + A_{yy} l_{Xy} l_{Zy} + A_{zz} l_{Xz} l_{Zz} \tag{4}$$

$$A_{YZ} = A_{xx} l_{Yx} l_{Zx} + A_{yy} l_{Yy} l_{Zy} + A_{zz} l_{Yz} l_{Zz} \tag{5}$$

The spin Hamiltonian becomes in the laboratory coordinate system

$$\hat{\mathscr{H}} = g_{ZZ} \beta_e H \hat{S}_Z + A_{ZZ} \hat{S}_Z \hat{I}_Z + A_{XZ} \hat{S}_Z \hat{I}_X + A_{YZ} \hat{S}_Z \hat{I}_Y \tag{6}$$

The g_{ZZ}, A_{ZZ}, A_{XZ}, A_{YZ} terms are related to the nitroxide principal values g_{xx}, g_{yy}, g_{zz} and A_{xx}, A_{yy}, A_{zz} by Eqs. (2)–(5). The less important terms (i.e., $\hat{S}_X$, $\hat{S}_Y$), leading to small off-diagonal matrix elements, have been omitted. With this simplification, the 6×6 secular determinant for nuclear spin $I = 1$ reduces to two 3×3 determinants, and each of these factor to give one eigenvalue directly and the other two as roots of a quadratic equation (see, for example, Libertini and Griffith, 1970). The resulting six eigenvalues of Eq. (6) are

$$E_{(\pm 1/2, \pm 1)} = \pm \tfrac{1}{2} g_{ZZ} \beta_e H \pm \tfrac{1}{2} [A_{ZZ}^2 + A_{XZ}^2 + A_{YZ}^2]^{1/2} \tag{7}$$

$$E_{(\pm 1/2, 0)} = \pm \tfrac{1}{2} g_{ZZ} \beta_e H \tag{8}$$

The allowed transitions are $\Delta M_S = \pm 1, \Delta M_I = 0$, so that for all orientations of the spin label in the magnetic field, the ESR spectrum of a nitroxide consists of $2I + 1 = 3$ lines of equal intensity, the center line position characterized by g and the distance between adjacent lines given by the splitting constant A, where

$$g = g_{ZZ} \tag{9}$$

$$A = (A_{ZZ}^2 + A_{XZ}^2 + A_{YZ}^2)^{1/2} \tag{10}$$

Substituting the quantities of Eqs. (2)–(5) reduces Eqs. (9) and (10) to the more familiar forms

$$g = g_{xx} l_{Zx}^2 + g_{yy} l_{Zy}^2 + g_{zz} l_{Zz}^2 \tag{11}$$

$$A = (A_{xx}^2 l_{Zx}^2 + A_{yy}^2 l_{Zy}^2 + A_{zz}^2 l_{Zz}^2)^{1/2} \tag{12}$$

Equations (11) and (12) have been shown to give a very good fit to accurate single crystal data for nitroxide free radicals. This level of approximation is often referred to as the intermediate field treatment, since retention of the pseudosecular terms [$\hat{S}_Z \hat{I}_x$ and $\hat{S}_Z \hat{I}_y$ in Eq. (6)] is tantamount to allowing $H_{hf} > H$, i.e., the hyperfine field can exceed the laboratory field.

This is appropriate for 9.5 GHz or 35 GHz ESR spectra of nitroxide spin labels. Another convenient, although much less accurate approximation is the high field approximation where $H \gg H_{hf}$. In this limit, the pseudosecular terms do not contribute significantly, and the g and A parameters are given by $g = g_{ZZ}$ and $A = A_{ZZ}$ or the equivalent expression $A = (A_{xx} l_{Zx} + A_{yy} l_{Zy} + A_{zz} l_{Zz})$. The expressions for g are the same in the two treatments, whereas the equations for A are not.

Introducing rapid molecular motion, Eq. (6) becomes

$$\hat{\mathscr{H}} = \langle g_{ZZ} \rangle \beta_e H \hat{S}_Z + \langle A_{ZZ} \rangle \hat{S}_Z \hat{I}_Z + \langle A_{XZ} \rangle \hat{S}_Z \hat{I}_X + \langle A_{YZ} \rangle \hat{S}_Z \hat{I}_Y \tag{13}$$

and therefore the g and A value Eqs. (9) and (10) are

$$g = \langle g_{ZZ} \rangle \tag{14}$$

$$A = (\langle A_{ZZ} \rangle^2 + \langle A_{XZ} \rangle^2 + \langle A_{YZ} \rangle^2)^{1/2} \tag{15}$$

The angular brackets in these equations indicate spatial averaging resulting from the rapid fluctuations of the angles between the nitroxide axes and laboratory magnetic field direction. The nature of the motion is not specified other than it is rapid on the ESR time scale; i.e., $\tau \ll 10^{-8}$ sec since for nitroxides $\{|A_{zz} - A_{xx}|\}^{-1} \simeq 1 \times 10^{-8}$ sec and $h\{|g_{xx} - g_{zz}|\beta H\}^{-1} \simeq 3 \times 10^{-8}$ sec (at 9.5 GHz), where τ is the rotational correlation time. Thus, instead of a general line-shape theory, we deal here with a relatively simple treatment that is widely used to describe anisotropic motion in membrane model systems and biological membranes. It is clear from Eqs. (2)–(5) that the evaluation of the terms $\langle g_{ZZ} \rangle$, $\langle A_{ZZ} \rangle$, $\langle A_{XZ} \rangle$, $\langle A_{YZ} \rangle$ involves only integrals over products of two direction cosines (e.g., $\langle l_{Zx}^2 \rangle$, $\langle l_{Xx} l_{Zx} \rangle$, etc.). The rapid motion limit is popular because it yields realistic line shapes and because existing computer programs that do not contain time explicitly (e.g., powder ESR spectral programs) can be used to calculate the spectra simply by substituting motion-averaged parameters for the customary input parameters g_{xx}, g_{yy}, g_{zz} and A_{xx}, A_{yy}, A_{zz}.

The limiting behavior of these equations is illustrated in Figs. 4 and 5. Of special interest are the spectra obtained along the three nitroxide axes when the molecule is rigid and oriented (Fig. 4). For example, when $H \parallel x$, $l_{Zx}^2 = 1$, all other products of direction cosines are zero, and Eqs. (2)–(5) reduce to $g_{ZZ} = g_{xx}$, $A_{ZZ} = A_{xx}$, and $A_{XZ} = A_{YZ} = 0$. Therefore, Eqs. (14) and (15) become simply $g = g_{xx}$ and $A = A_{xx}$. Similarly, for $H \parallel y$, $g = g_{yy}$, $A = A_{yy}$ and for $H \parallel z$, $g = g_{zz}$, $A = A_{zz}$. The g values g_{xx}, g_{yy}, and g_{zz} of spectra (a)–(c) of Fig. 4 differ, as can be seen by the distance each center line is shifted from the dashed reference line. The splitting constants A_{xx} and A_{yy} of spectra (a) and (b) are very nearly equal and are much smaller than the coupling constant A_{zz} of the spectrum (c). The actual measured values for

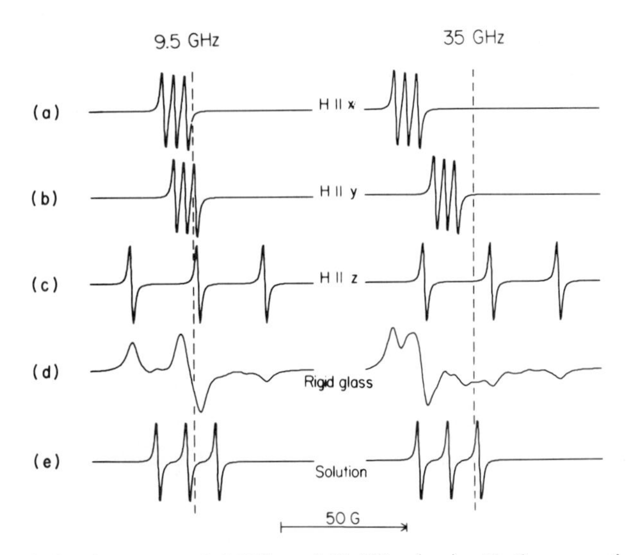

Fig. 4. Calculated spectra at 9.5 GHz and 35 GHz: (a–c) with the magnetic field along each axis of a rigidly oriented nitroxide; (d) a collection of randomly oriented nitroxides in the absence of motion, and (e) the same collection undergoing rapid isotropic motion. The principal values are those of 2-doxylpropane, $A_{xx} = 5.9$, $A_{yy} = 5.4$, $A_{zz} = 32.9$, $g_{xx} = 2.0088$, $g_{yy} = 2.0058$, $g_{zz} = 2.0022$. (Jost *et al.*, 1971.)

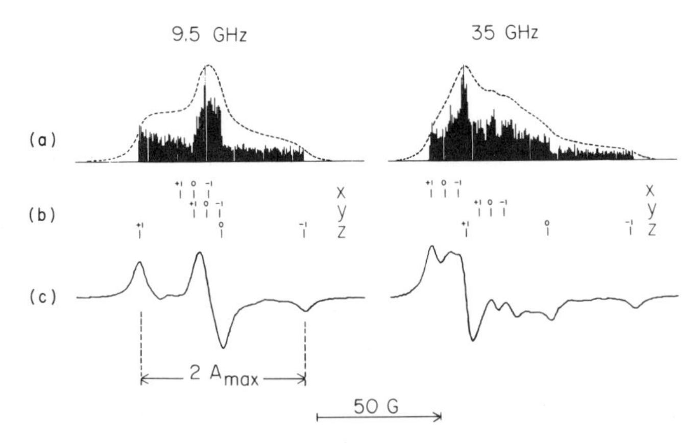

Fig. 5. The rigid glass (highly immobilized) line shape at microwave frequencies of 9.5 GHz and 35 GHz: (a) the computer-generated stick spectra and the absorption envelopes (arbitrary vertical scale); (b) line positions with the magnetic field along the nitroxide *x*, *y*, and *z* axes; (c) first-derivative line shapes. The input parameters are the same as in Fig. 4, with peak-to-peak first derivative line width 4.0 G; Lorentzian line shape, and calculations are to 0.5 G resolution.

various nitroxides are tabulated in Appendix II. The g and A values are of course independent of the laboratory magnetic fields selected. The splittings of each individual spectrum are the same at 9.5 GHz (3.4 kG) and 35 GHz (12.5 kG), but spectra are shifted in proportion to the magnitude of the laboratory field because of the Zeeman equation, $h\nu = g\beta H$. As a result, the superposition of spectra for a collection of nitroxides gives a field-dependent line shape. This is illustrated by the 9.5 GHz and 35 GHz rigid glass spectra (d) in Fig. 4.

The corresponding computed spectra of rapidly tumbling nitroxides in solution at 9.5 and 35 GHz are given as (e) in Fig. 4. In this limit of rapid isotropic motion the nitroxide axes point with equal probability through all surface elements of a unit sphere, thus $\langle l_{Zx}^2 \rangle = \frac{1}{3}$, $\langle l_{Zy}^2 \rangle = \frac{1}{3}$, $\langle l_{Zz}^2 \rangle = \frac{1}{3}$ and the other six integrals ($\langle l_{Xx} l_{Zx} \rangle$, etc.) are zero.† Equations (14) and (15) become $g = g_0 = \frac{1}{3}(g_{xx} + g_{yy} + g_{zz})$ and $A = A_0 = \frac{1}{3}(A_{xx} + A_{yy} + A_{zz})$ so that the solution spectra of Fig. 4e lie between the extremes of the $H \parallel x$, $H \parallel y$, $H \parallel z$ spectra of the oriented spin labels. We include this example to emphasize that Eq. (15) and not Eq. (12) must be used to calculate the motion-averaged A parameters, even though the final computer program utilizes Eq. (12). This is also true in Monte Carlo calculations, where computer-generated random walks replace the integrals over products of direction cosines. This problem does not occur in the g value expressions or the high field approximation since these equations are linear.

The superposition of the spectra of nitroxides at all possible orientations of the magnetic field to give the rigid glass or powder spectrum is shown in more detail in Fig. 5. The principal orientations are given in (b), where +1, 0, −1 refer to the quantum number of the z component of the ^{14}N nuclear spin. The stick spectra (Fig. 5a) are a sum of these three principal spectra plus all intermediate line positions resulting from other orientations of the spin label in the magnetic field calculated using Eqs. (11) and (12). When the intrinsic linewidth is applied to the individual stick spectra, the corresponding absorption line shape (dashed line) and first derivative spectra (c) are generated. At 9.5 GHz, the overlap is extensive and the only parameter readily measured is operationally defined as the distance between the outermost peaks ($2A_{max}$) and corresponds approximately to $2A_{zz}$. At 35 GHz the distance between outermost peaks does not correspond to $2A_{zz}$. However, many more peaks are observed, and it is possible in favorable cases to

† For example with the coordinates discussed in the next section, $l_{Zz} = \cos\theta$ and

$$\langle l_{Zz}^2 \rangle = \frac{\int_{\phi=0}^{2\pi} \int_{\theta=0}^{\pi} \cos^2\theta \sin\theta \, d\theta \, d\phi}{\int_{\phi=0}^{2\pi} \int_0^{\pi} \sin\theta \, d\theta \, d\phi} = \frac{4\pi/3}{4\pi} = \frac{1}{3}$$

The quantity $\sin\theta \, d\theta \, d\phi$ is the area element in spherical coordinates.

estimate all six parameters, g_{xx}, g_{yy}, g_{zz}, A_{xx}, A_{yy}, A_{zz}, as can be seen in the figure.

The rigid glass spectrum and solution spectrum (Fig. 4d,e) represent the extreme cases of no molecular motion (on the ESR time scale) and rapid isotropic motion. The spectra observed with lipid spin labels in biological membranes lie somewhere between these extremes. To make the problem more interesting, there is almost always preferential motion about the long axes of these asymmetric lipid spin labels.

2. Rapid Rotation about a Molecular Axis

In order to describe motion and orientation of spin labels in membranes it is convenient to introduce two additional coordinate systems, X_1, Y_1, Z_1 and x_2, y_2, z_2. The X_1, Y_1, Z_1 axes characterize the sample or specimen; for example, Z_1 is chosen to indicate the normal to the bilayer plane in phospholipid multilayers. The molecular axes x_2, y_2, z_2 refer to the geometry of the lipid spin labels. For spin labels having a well-defined long axis, this axis is defined to be z_2 or R. The notations we use for the four coordinate systems and the transformations connecting any two coordinate systems are summarized in Fig. 6. As a helpful mnemonic device, the macroscopic coordinate systems (laboratory and sample) are shown as capital letters. The microscopic molecular and nitroxide axes are indicated by lower case letters.

Specifying all four coordinate systems of Fig. 6 is only necessary in the general case. In practice, there is usually considerable simplification. The special cases where one of the principal nitroxide axes (x, y, or z) coincides with the long axis of the lipid spin label are shown in Fig. 7. Notice that in the case of the doxylfatty acid spin label (c) the nitroxide z axis coincides with $R(z_2)$, and the largest splitting occurs when H is parallel to the long axis of the molecule.

Several different choices of angles relating any two coordinate systems have been used in the spin-labeling literature. In order to avoid confusion, the basic equations (1)–(15) are written in the direction cosine formalism,

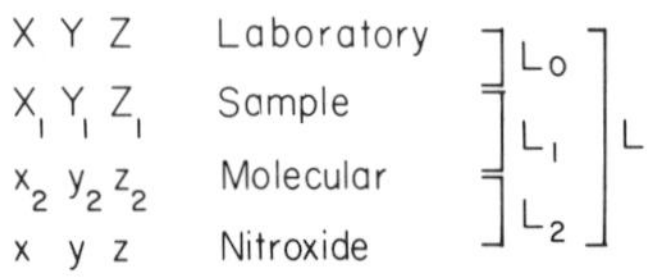

$X\ Y\ Z$	Laboratory	L_0 (Laboratory–Sample)	L (Laboratory–Nitroxide)
$X_1\ Y_1\ Z_1$	Sample	L_1 (Sample–Molecular)	
$x_2\ y_2\ z_2$	Molecular	L_2 (Molecular–Nitroxide)	
$x\ y\ z$	Nitroxide		

Fig. 6. The four coordinate systems and designations of the transformation matrices relating the laboratory, sample, molecular, and nitroxide axis systems. Capital letters indicate macroscopic systems, while lower case letters refer to molecular systems. [From Van *et al.* (1974).]

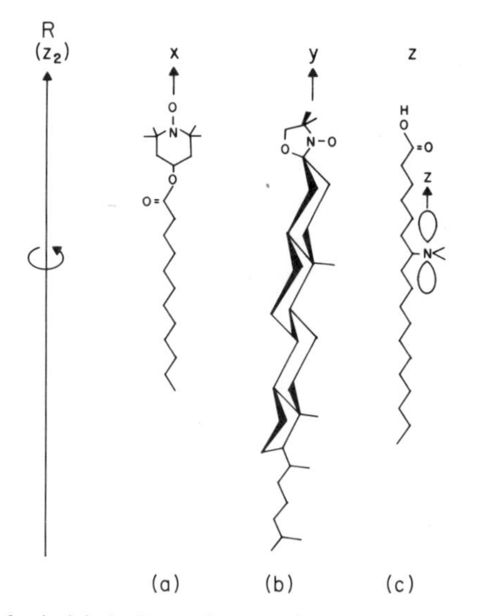

Fig. 7. Examples of spin labels that undergo rapid rotation about the nitroxide *x*, *y*, or *z* axes. The molecular axis *R* is defined to be parallel to the long axis of these asymmetric molecules. The spin labels are (a) 2,2,6,6-tetramethyl-4-piperidinol-1-oxyl dodecanoate, (b) 3-doxyl-5α-cholestane, and (c) 7-doxylstearic acid. [Adapted from Van *et al.* (1974).]

which is valid for any of the coordinate systems. The set of angles that is most widely used in applying these equations is shown in Fig. 8. This choice of Eulerian angles has the advantage that θ and ϕ correspond to the familiar angles in spherical coordinates. Hence, Fig. 8 reduces to Fig. 3 for the case where the nitroxide axes are to be related directly to the laboratory axes ($\psi = 0$ because the Hamiltonian is invariant to rotation about the field axis c). With this choice, $l_{Zx}^2 = \sin^2\theta\cos^2\phi$, $l_{Zy}^2 = \sin^2\theta\sin^2\phi$, and $l_{Zz}^2 = \cos^2\theta$ in Eqs. (11) and (12).

Anisotropic motion combined with isotropic tumbling at low or intermediate frequencies requires a complex line-shape analysis. However, there are limiting situations that are sufficiently simple to be of general interest and provide considerable insight into some of the general line shapes observed in membrane spin-labeling experiments. One example is rapid rotation about the long axis R in the absence of any other motion (i.e., a "molecular spinning top"), shown in Fig. 7.

In this model, the relationship between the molecular axes and the nitroxide axes is fixed by the geometry of the spin label, and both axis systems rotate rapidly in the sample X_1, Y_1, Z_1 coordinate system. The position of the sample in the magnetic field fixes the relationship between X_1, Y_1, Z_1 and X, Y, Z. The motion averaging involves only the angles θ_1,

ϕ_1, ψ_1 of the $\mathbf{L}_1$ matrix relating the x_2, y_2, z_2 and the X_1, Y_1, Z_1 axes. If the overall transformation matrix $\mathbf{L}$ is written as a product of three terms,† $\mathbf{L} = \mathbf{L}_0 \mathbf{L}_1^\tau \mathbf{L}_2^\tau$, then $\mathbf{A}' = \mathbf{L}_0 \mathbf{L}_1^\tau \mathbf{L}_2^\tau \mathbf{A} \mathbf{L}_2 \mathbf{L}_1 \mathbf{L}_0^\tau$. Hence with motion averaging, $\langle \mathbf{A}' \rangle = \mathbf{L}_0 \langle \mathbf{L}_1^\tau \mathbf{A}_{x_2 y_2 z_2} \mathbf{L}_1 \rangle \mathbf{L}_0^\tau$ where $\mathbf{A}_{x_2 y_2 z_2} = \mathbf{L}_2^\tau \mathbf{A} \mathbf{L}_2$. This stepwise procedure simplifies the calculation by confining the motion averaging to one simple matrix, $\langle \mathbf{L}_1^\tau \mathbf{A}_{x_2 y_2 z_2} \mathbf{L}_1 \rangle$. For rotation about a molecular axis (taking $z_2 \parallel Z_1$), this matrix contains the simple normalized integrals $\langle \cos^2 \psi_1 \rangle = \int_0^{2\pi} \cos^2 \psi_1 \, d\psi / \int_0^{2\pi} d\psi_1 = \frac{1}{2}$ and

$$\langle \cos \psi_1 \sin \psi_1 \rangle = \langle \cos \psi_1 \rangle = \langle \sin \psi_1 \rangle = 0$$

This procedure yields the following useful equations relating the motion-averaged principal values to the single-crystal parameters:

$$\bar{A}_\parallel^\mathrm{r} = A_{zz} \cos^2 \theta_2 + A_{yy}(1 - \cos^2 \theta_2) + (A_{xx} - A_{yy}) \sin^2 \theta_2 \cos^2 \psi_2 \quad (16)$$

$$\bar{A}_\perp^\mathrm{r} = A_{zz} \tfrac{1}{2}(1 - \cos^2 \theta_2) + A_{yy} \tfrac{1}{2}(1 + \cos^2 \theta_2) + (A_{xx} - A_{yy}) \tfrac{1}{2}(1 - \sin^2 \theta_2 \cos^2 \psi_2) \quad (17)$$

The corresponding expressions for $\bar{g}_\parallel^\mathrm{r}$ and $\bar{g}_\perp^\mathrm{r}$ are exactly the same as Eqs. (16) and (17) with g_{xx}, g_{yy}, g_{zz} replacing A_{xx}, A_{yy}, A_{zz}.

Transforming with the $\mathbf{L}_0$ matrix yields the motion average matrix elements of $\mathbf{g}'$ and $\mathbf{A}'$

$$\langle g_{ZZ} \rangle = \bar{g}_\parallel^\mathrm{r} \cos^2 \theta_0 + \bar{g}_\perp^\mathrm{r} \sin^2 \theta_0 \quad (18)$$

$$\langle A_{ZZ} \rangle = \bar{A}_\parallel^\mathrm{r} \cos^2 \theta_0 + \bar{A}_\perp^\mathrm{r} \sin^2 \theta_0 \quad (19)$$

$$\langle A_{XZ} \rangle = (\bar{A}_\perp^\mathrm{r} - \bar{A}_\parallel^\mathrm{r}) \sin \theta_0 \cos \theta_0 \quad (20)$$

$$\langle A_{YZ} \rangle = 0 \quad (21)$$

and substituting these quantities into Eqs. (14) and (15) gives

$$g = \bar{g}_\parallel \cos^2 \theta + \bar{g}_\perp \sin^2 \theta \quad (22)$$

and

$$A = \{[\bar{A}_\parallel \cos^2 \theta + \bar{A}_\perp \sin^2 \theta]^2 + [(\bar{A}_\parallel - \bar{A}_\perp) \sin \theta \cos \theta]^2\}^{1/2} \quad (23)$$

which can be written as

$$A = (\bar{A}_\parallel^2 \cos^2 \theta + \bar{A}_\perp^2 \sin^2 \theta)^{1/2} \quad (24)$$

† The product can logically take three forms $\mathbf{L} = \mathbf{L}_0 \mathbf{L}_1 \mathbf{L}_2^\tau$, $\mathbf{L} = \mathbf{L}_0 \mathbf{L}_1^\tau \mathbf{L}_2^\tau$, or $\mathbf{L} = \mathbf{L}_0^\tau \mathbf{L}_1^\tau \mathbf{L}_2^\tau$, depending on whether one selects a molecular reference, specimen reference, or laboratory frame of reference. In the specimen reference system used here, the specimen is fixed in space and the spin labels tumble or rotate rapidly in the specimen. The magnetic field is then rotated to various positions about the specimen. The final answer is of course the same in all three cases (for details, see Van *et al.*, 1974).

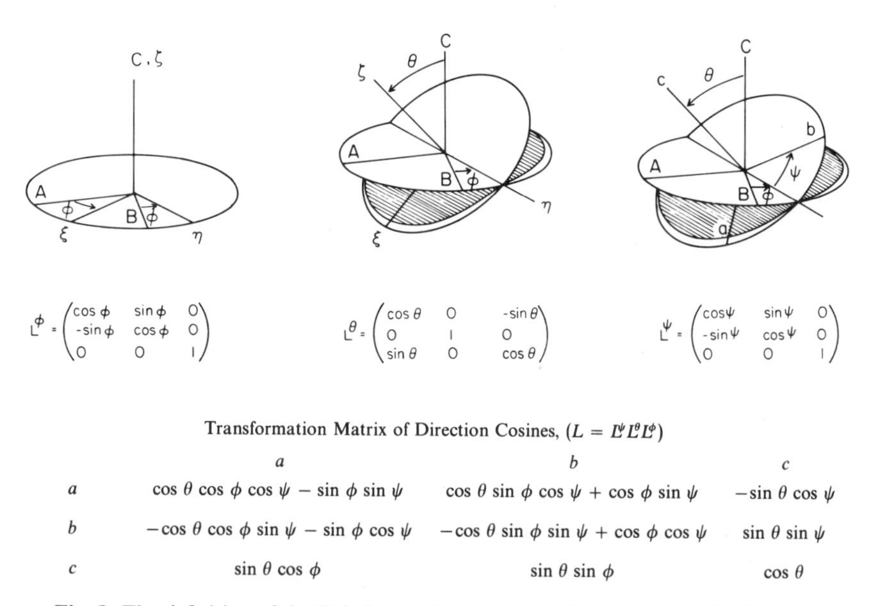

Transformation Matrix of Direction Cosines, ($L = L^\psi L^\theta L^\phi$)

	a	*b*	*c*
a	$\cos\theta\cos\phi\cos\psi - \sin\phi\sin\psi$	$\cos\theta\sin\phi\cos\psi + \cos\phi\sin\psi$	$-\sin\theta\cos\psi$
b	$-\cos\theta\cos\phi\sin\psi - \sin\phi\cos\psi$	$-\cos\theta\sin\phi\sin\psi + \cos\phi\cos\psi$	$\sin\theta\sin\psi$
c	$\sin\theta\cos\phi$	$\sin\theta\sin\phi$	$\cos\theta$

Fig. 8. The definition of the Eulerian angles chosen to relate two generalized coordinate systems (*A*, *B*, *C* and *a*, *b*, *c*). In solving a specific problem, these two general coordinate systems can represent any two of the four specific coordinate systems of Fig. 6. [Reprinted from Van *et al.* (1974).]

where, to emphasize the generality of these equations, we have omitted the superscript r (and the subscript on θ_0). In summary, rapid rotation of the spin labels reduces the Hamiltonian to an axially symmetric form [i.e., Eq. (13) combined with Eqs. (18)–(21)], and thus rapid rotation yields axially symmetric g and A equations [Eqs. (22)–(24)], even though the original g and A matrices were not assumed to have axial symmetry. Equations similar to (16) and (17) were originally derived using the high field approximation (Seelig, 1970), and it is interesting to note that these equations are valid in the intermediate field treatment, as shown by the derivation outlined above (Van *et al.*, 1974).

In most of the lipid spin labels shown in Fig. 2, the molecular axes coincide with one of the nitroxide axes. In these cases, when the only motion present is rapid rotation about one principal axis, the motion-averaging of the A and g tensors is greatly simplified. For example, when the nitroxide x axis is the axis of rotation, $\theta_2 = 90°$, $\psi_2 = 0°$ and Eqs. (16) and (17) reduce to $\bar{A}^r_\parallel = A_{xx}$ and $\bar{A}^r_\perp = \frac{1}{2}(A_{yy} + A_{zz})$. Similarly, for rapid rotation about the y axis $\theta_2 = 90°$, $\psi_0 = 90°$, so that $\bar{A}^r_\parallel = A_{yy}$ and $\bar{A}^r_\perp = \frac{1}{2}(A_{xx} + A_{zz})$. With the rotation about the nitroxide z axis, $\theta_2 = 0$ (ψ_2 need not be specified), so $\bar{A}^r_\parallel = A_{zz}$ and $\bar{A}^r_\perp = \frac{1}{2}(A_{xx} + A_{yy})$. The g value expressions are equivalent. To calculate the spectra, $\bar{A}^r_\parallel$, $\bar{A}^r_\perp$ and $\bar{g}^r_\parallel$, $\bar{g}^r_\perp$ are sometimes supplied as input values for $\bar{A}_{\parallel}$, $\bar{A}_{\perp}$ and $\bar{g}_{\parallel}$, $\bar{g}_{\perp}$ in a computer program based on Eqs. (22)–(24). More commonly, a computer program based on Eqs. (11) and (12) is used, in which case A_{xx}, A_{yy}, A_{zz} become $\bar{A}^{r}_{\perp}$, $\bar{A}^{r}_{\perp}$, $\bar{A}^{r}_{\parallel}$ and g_{xx}, g_{yy}, g_{zz} become $\bar{g}^{r}_{\perp}$, $\bar{g}^{r}_{\perp}$, $\bar{g}^{r}_{\parallel}$.

The results of line-shape calculations using Eqs. (11) and (12) for each of the simple cases of *rapid rotation* about a nitroxide axis are shown in Fig. 9. In (a) are shown the stick spectra and the corresponding absorption line shape. Just below the stick spectra are indicated the line positions corresponding to the motion-averaged principal values (b). It is easily seen from this type of diagram why rotation about the x axis yields the three peaks of descending line heights, why the ESR spectrum for rotation about the y axis is remarkably symmetrical about the center line, and why the spectrum for rotation about the z axis looks very much like the 9.5 GHz rigid glass spectrum of Fig. 5 (z axis rotation and the rigid glass are more easily distinguished at 35 GHz).

For comparison, experimental spectra obtained with an appropriate spin label rotating rapidly (or undergoing large amplitude motion) in the linear cavity of an inclusion crystal are given in (c). The general line-shape features of the calculated and companion experimental spectra correspond sur-

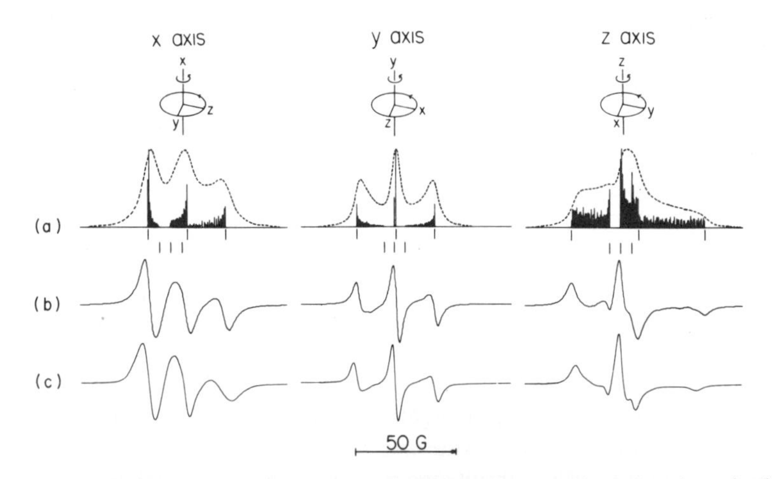

Fig. 9. Theoretical line shapes and experimental examples of rapid rotation about the three principal nitroxide axes ($\alpha = 90°$, $\gamma = 0°$). The 9.5 GHz stick spectra and absorption (a) and the corresponding first derivative spectra (b) were calculated using the principal values for 2-doxylpropane (see legend for Fig. 4); the peak-to-peak first derivative linewidths are 5.0 G, 3.0 G, and 3.8 G for x-axis, y-axis, and z-axis rotation, respectively. (c) For x-axis rotation the experimental spectrum is of 2,2,6,6-tetramethyl-4-piperidinol-1-oxyl dodecanoate trapped in β-cyclodextrin (Birrell *et al.*, 1973a); the example of y-axis rotation is di-t-butyl nitroxide trapped in thiourea (Birrell *et al.*, 1973a). The example of z-axis motion is 7-doxyl-stearate trapped in γ-cyclodextrin (Birrell *et al.*, 1973b). All three samples were at room temperature.

prisingly well. However, as we extend the models to more complex types of motion, it is possible to show that the spin labels in the inclusion hosts are also undergoing motion that does not perfectly conform to the simple rotation model. The motion of the spin label about the x axis at room temperature is not completely averaging the y and z components (restricted oscillation), while the spin label with predominantly z-axis motion is allowed some small additional motion in the other planes (wobble of the long axis). But, for the most part, the dominant motion occurring can be described as rapid rotation about a single axis, and Fig. 9 serves to highlight the characteristic line shapes associated with this kind of anisotropic motion. Given the asymmetry of the lipid spin labels and a microscopically ordered domain such as the membrane bilayer regions, many of the spectral features seen in Fig. 9 are also seen in spectra from biological samples.

3. OSCILLATION

Another useful model involves rapid restricted oscillations in a plane. This would result, for example, if the motion about the long axes of the spin labels of Fig. 7 were restricted. In this case, the motion-averaged matrix elements of the spin Hamiltonian lack axial symmetry. The resulting general equations for an arbitrary orientation of the nitroxide x, y, z axes with respect to the molecular x_2, y_2, z_2 axes can be derived but are complex. However, when the molecular axis is parallel to a nitroxide axis, the equations reduce to a simple form. A specific example of interest observed in membranes occurs with limited oscillation of the steroid spin labels (Fig. 7b), since these spin labels resemble the cholesterol found in most plasma membranes. The oscillations are described by a half-amplitude α in the nitroxide xz plane. Evaluating the corresponding integrals over the products of direction cosines in the Hamiltonian matrix is straightforward. For example $\langle \cos^2 \psi_1 \rangle = \int_{-\alpha}^{+\alpha} \cos^2 \psi_1 \, d\psi_1 / \int_{-\alpha}^{+\alpha} d\psi_1 = \frac{1}{2}(1 + P)$ where

$$P = (\sin \alpha \cos \alpha)/\alpha$$

The resulting motion-averaged ESR parameters are related to α by the equations

$$\bar{A}^p_{xx} = A_{xx} + (A_{zz} - A_{xx})\tfrac{1}{2}(1 + P) \tag{25}$$

$$\bar{A}^p_{yy} = A_{zz} - (A_{zz} - A_{xx})\tfrac{1}{2}(1 + P) \tag{26}$$

$$\bar{A}^p_{zz} = A_{yy} \tag{27}$$

The g-value expressions are equivalent and all six equations are valid in the intermediate field treatment as well as the high field approximation. The resulting ESR line shapes for restricted oscillation about the y axis of the steroid spin labels are shown in the center spectra of Fig. 10.

The $\bar{A}^p_{xx}$, $\bar{A}^p_{yy}$, $\bar{A}^p_{zz}$ and $\bar{g}^p_{xx}$, $\bar{g}^p_{yy}$, $\bar{g}^p_{zz}$ are substituted for A_{xx}, A_{yy}, A_{zz} and g_{xx}, g_{yy}, g_{zz} in Eqs. (11) and (12) to calculate the line positions. Equations (22)–(24) cannot be used because the principal values lack axial symmetry. The equations for oscillations about the x and z axes are obtained from Eqs. (25)–(27) by cyclic permutation. The resulting ESR line shapes for randomly oriented samples are given as a function of α in Fig. 10. This model has the correct limiting behavior, i.e., as $\alpha \to 0$, $P \to 1$, and all spectra reach the rigid glass limit shown for each series. In the other extreme, as the amplitude of oscillations approach complete rotation $\alpha \to 90°$ and $P \to 0$, and the spectra on the bottom row of each series represent rapid rotation about the x, y, or z nitroxide axes. For intermediate ranges of oscillations about the x or y axes, α may be estimated from the distance between the outermost lines of the 9.5 GHz spectra. For example, the experimental spectrum (c) of x-axis motion in Fig. 9 is better simulated with $\alpha = 75°$ (instead of $\alpha = 90°$), indicating some slight restriction of rotation in the inclusion crystal.

Note that for small values of α it is difficult to distinguish the type of

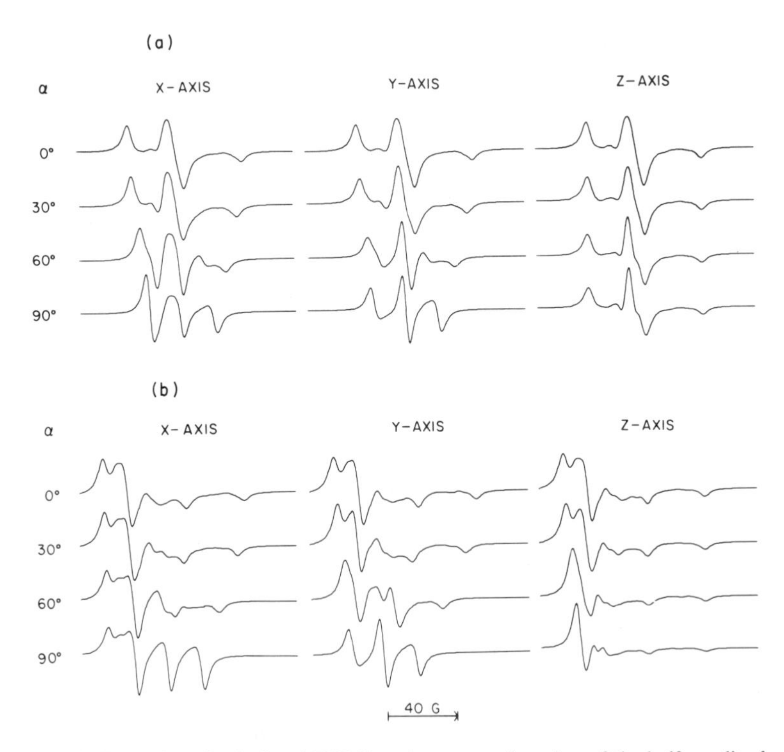

Fig. 10. Examples of calculated ESR line shapes as a function of the half amplitude α for oscillations about the x, y, and z nitroxide axes. The principal A- and g-value parameters are the same as in Fig. 4. [Reprinted from Van *et al.* (1974).]

molecular motion, whereas for $\alpha > 30°$, the spectral features are distinctive. Figure 10 also illustrates the difficulties in detecting motion about the z axis at 9.5 GHz. In contrast, at 35 GHz, marked changes occur in the spectral line shape as α increases from 0° to 90°.

4. Wobbling of the Long Molecular Axis

An important type of molecular motion occurs when the long axis of the rotating spin label is rapidly "wobbling" in the sample. This anisotropic motion has been variously described as a restricted random walk of the R axis on the surface of a sphere, as rapid motion within a cone, and as rapid fluctuations of the angle θ_1 between the R axis and the sample axis (Z_1). The simplest mathematical approximation to this kind of molecular motion is based on Fig. 11. The angle θ_1 fluctuates randomly between the limits of $\theta_1 = 0$ and a maximum half amplitude of $\theta_1 = \gamma$. To simplify the mathematical expressions, the spin label is assumed to rotate rapidly about the axis R (i.e., $\alpha = 90°$). The effects of molecular motion are taken into account by spatial averages over the appropriate angles appearing in the matrix elements of the transformed spin Hamiltonian. Wobbling of the long axis is a symmetric motion, and the resulting spin Hamiltonian has axial symmetry. The most important normalized integral in this treatment turns out to be

$$\langle \cos^2 \phi_1 \cos^2 \theta_1 \rangle = \frac{\int_{\theta_1=0}^{2\pi} \int_{\theta_1=0}^{\gamma} \cos^2 \phi_1 \cos^2 \theta_1 \sin \theta_1 \, d\theta_1 \, d\phi_1}{\int_{\phi_1=0}^{2\pi} \int_{\theta_1=0}^{\gamma} \sin \theta_1 \, d\theta_1 \, d\phi_1}$$

$$= \frac{1}{3} \frac{(1 - \cos^3 \gamma)}{(1 - \cos \gamma)} = \frac{1}{3}(1 + \cos \gamma + \cos^2 \gamma) \equiv W \quad (28)$$

where the angles are defined in Figs. 6 and 8.

The resulting motion-averaged parameters $\bar{A}_{\parallel}^{\mathrm{w}}$ and $\bar{A}_{\perp}^{\mathrm{w}}$ are

$$\bar{A}_{\parallel}^{\mathrm{w}} = \bar{A}_{\perp}^{\mathrm{r}} + (\bar{A}_{\parallel}^{\mathrm{r}} - \bar{A}_{\perp}^{\mathrm{r}})W \quad (29)$$

$$\bar{A}_{\perp}^{\mathrm{w}} = \bar{A}_{\perp}^{\mathrm{r}} + (\bar{A}_{\parallel}^{\mathrm{r}} - \bar{A}_{\perp}^{\mathrm{r}})\tfrac{1}{2}(1 - W) \quad (30)$$

where $\bar{A}_{\parallel}^{\mathrm{r}}$ and $\bar{A}_{\perp}^{\mathrm{r}}$ are given by Eqs. (16) and (17). The g-value expressions are equivalent. The matrix elements $\langle g_{ZZ} \rangle$, $\langle A_{ZZ} \rangle$, $\langle A_{XZ} \rangle$, and $\langle A_{YZ} \rangle$ are the

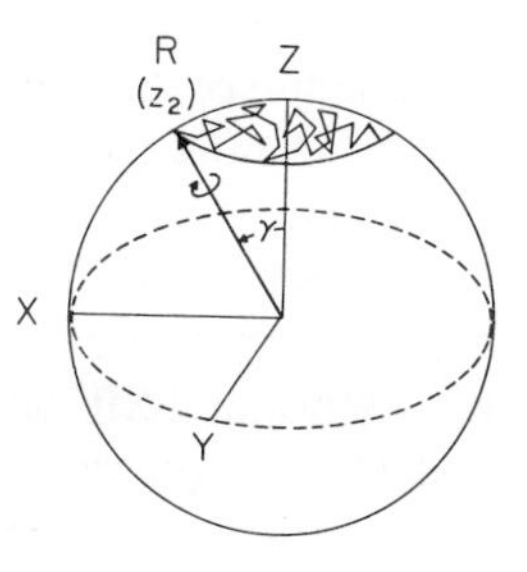

Fig. 11. Diagrammatic representation of the wobble (restricted random walk) model. The long axis R of the spin label is confined to a cone of half amplitude γ about the sample Z_1 axis. Z_1 is usually the normal to the bilayer plane.

same as in Eqs. (18)–(21) with the index w replacing r and therefore Eqs. (14) and (15) reduce to the axially symmetric forms given by Eqs. (22)–(24). The spectral line positions are calculated by substituting $\bar{g}_{\parallel}^{w}$, $\bar{g}_{\perp}^{w}$, $\bar{A}_{\parallel}^{w}$, $\bar{A}_{\perp}^{w}$ for $\bar{g}_{\parallel}$, $\bar{g}_{\perp}$, $\bar{A}_{\parallel}$, $\bar{A}_{\perp}$ in Eqs. (22) and (23), or by substituting $\bar{g}_{\perp}^{w}$, $\bar{g}_{\perp}^{w}$, $\bar{g}_{\parallel}^{w}$ and $\bar{A}_{\perp}^{w}$, $\bar{A}_{\perp}^{w}$, $\bar{A}_{\parallel}^{w}$ for g_{xx}, g_{yy}, g_{zz} and A_{xx}, A_{yy}, A_{zz} in Eqs. (11)–(12). Equations (29) and (30) were first derived for wobble of the nitroxide *z* axis in the high field approximation (Jost *et al.*, 1971) and were later verified as correct in the intermediate field treatment (Israelachvili *et al.*, 1974; Van *et al.*, 1974).

It is worth noting that for rapid rotation the following equation applies:

$$A_0 = \tfrac{1}{3}(\bar{A}_{\parallel} + 2\bar{A}_{\perp}) \tag{31}$$

This can be seen by direct substitution of Eqs. (16) and (17) or of Eqs. (29) and (30) into Eq. (31). Equation (31) has been verified experimentally for free radicals rapidly rotating in the tubular channels of inclusion crystals (Griffith, 1964; Birrell *et al.*, 1973a,b). The importance of this equation is that it provides a simple relationship between the motion averaged $\bar{A}_{\parallel}$ and $\bar{A}_{\perp}$ parameters, and the motion-independent A_0 parameter, so that if any two of these quantities are measured, the third can be estimated.

Motion in the wobble model can be described by the single parameter γ, which gives the maximum half-amplitude of the motion (see Fig. 11). When $\gamma \to 0$, $W \to 1$, and the wobble model reduces to rotation about the molecular axis. In the opposite limit of large amplitude motion $\gamma \to \pi/2$, $W \to \frac{1}{3}$, so that from Eqs. (29) and (30), $\bar{A}_{\perp}^{w} = \bar{A}_{\parallel}^{w} = \frac{1}{3}(\bar{A}_{\parallel}^{r} + 2\bar{A}_{\perp}^{r}) = A_0$. This removes all anisotropy, and the spectrum is that of the nitroxide freely tumbling in solution.

The equations presented here for the wobble model may be used for any spin-label geometry. They are most frequently used, however, to describe the line shapes of the doxyl fatty acids and doxyl phospholipids. This corresponds to wobble of the nitroxide *z* axis, and the motion-averaged parameters from Eqs. (29) and (30) become

$$\bar{A}_{\parallel}^{w} = \tfrac{1}{2}(A_{xx} + A_{yy}) + [A_{zz} - \tfrac{1}{2}(A_{xx} + A_{yy})]W \tag{32}$$

and

$$\bar{A}_{\perp}^{w} = \tfrac{1}{2}(A_{xx} + A_{yy}) + [A_{zz} - \tfrac{1}{2}(A_{xx} + A_{yy})]\tfrac{1}{2}(1 - W) \tag{33}$$

where W is defined by Eq. (28).

The use of the wobble model allows a systematic variation of the motion-averaged values as a function of a single parameter γ. An example of this approach is given in Table I and Fig. 12. The motion-averaged values obtained from Eqs. (32) and (33) and the equivalent expressions for $\bar{g}_{\parallel}$ and $\bar{g}_{\perp}$ are listed in Table I. These values were then used as input parameters to a conventional ESR powder spectra computer program based on Eqs. (11) and (12) to generate Fig. 12. The specific program used here is EULOR (Libertini *et al.*, 1974).

The resulting stick spectra and corresponding absorption and first-derivative line shapes calculated are shown in Fig. 12. Below each stick spectrum is indicated the position of the parallel and perpendicular spectral components defined by the motion-averaged parameters of Table I. This figure shows very clearly the full range of motion from $\gamma = 0°$ to the isotropic spectra at $\gamma = 90°$. This model can never give the rigid glass spectrum, because even at $\gamma = 0°$, the spin label is still rotating about the long axis. However, for all practical purposes, at 9.5 GHz the spectrum for $\gamma = 0°$ and the rigid glass spectrum are indistinguishable. Thus, Fig. 12 can be used to represent the full range of motion from the rigid glass to the isotropic limit.

The rapid motion treatment has limitations. One limitation is that only motion that is rapid on the ESR time scale is formally treated. A second problem with this treatment, and with other line shape treatments currently in use, is the inability to differentiate between frequency and amplitude of motion. For example, decreasing amplitudes of an oscillation at a fixed frequency and decreasing frequencies at a fixed amplitude are difficult to distinguish. The wobble model illustrates this point. Figure 12 gives the line shape changes when the maximum *amplitude* of a rapid random walk is increased from zero to motion over the full sphere ($0° > \gamma > 90°$), an equally valid picture of the molecular events described by the line shapes of Fig. 12 is that, for all spectra shown, the movement of the rapidly rotating spin label extends over the full sphere, but that the *frequency* of wobble increases from the top to the bottom of the figure. (In this interpretation, γ becomes a frequency parameter instead of an amplitude parameter.) Physically speaking, there is a real difference between amplitude and frequency. Experimentally, however, distinguishing amplitude from frequency is troublesome because of the time window used in the experiment. The problem can be likened to photographing the side of a pair of disks, one rotating and the other rocking back and forth. Both disks are labeled with a red dot on the

TABLE I

A SET OF MOTION-AVERAGED PARAMETERS CALCULATED FROM THE WOBBLE MODEL[a, b]

γ	$\bar{A}_{\parallel}$	$\bar{A}_{\perp}$	$\bar{g}_{\parallel}$	$\bar{g}_{\perp}$	γ	$\bar{A}_{\parallel}$	$\bar{A}_{\perp}$	$\bar{g}_{\parallel}$	$\bar{g}_{\perp}$
0	32.9	5.6	2.0022	2.0073	60	21.5	11.3	2.0043	2.0062
15	32.0	6.1	2.0024	2.0072	75	17.7	13.2	2.0050	2.0059
30	29.4	7.4	2.0029	2.0070	90	14.7	14.7	2.0056	2.0056
45	25.7	9.3	2.0035	2.0066					

[a] Jost *et al.* (1971).

[b] Single-crystal values (2-doxylpropane) used are $A_{xx} = 5.9$, $A_{yy} = 5.4$, $A_{zz} = 32.9$, $g_{xx} = 2.0088$, $g_{yy} = 2.0058$, $g_{zz} = 2.0022$.

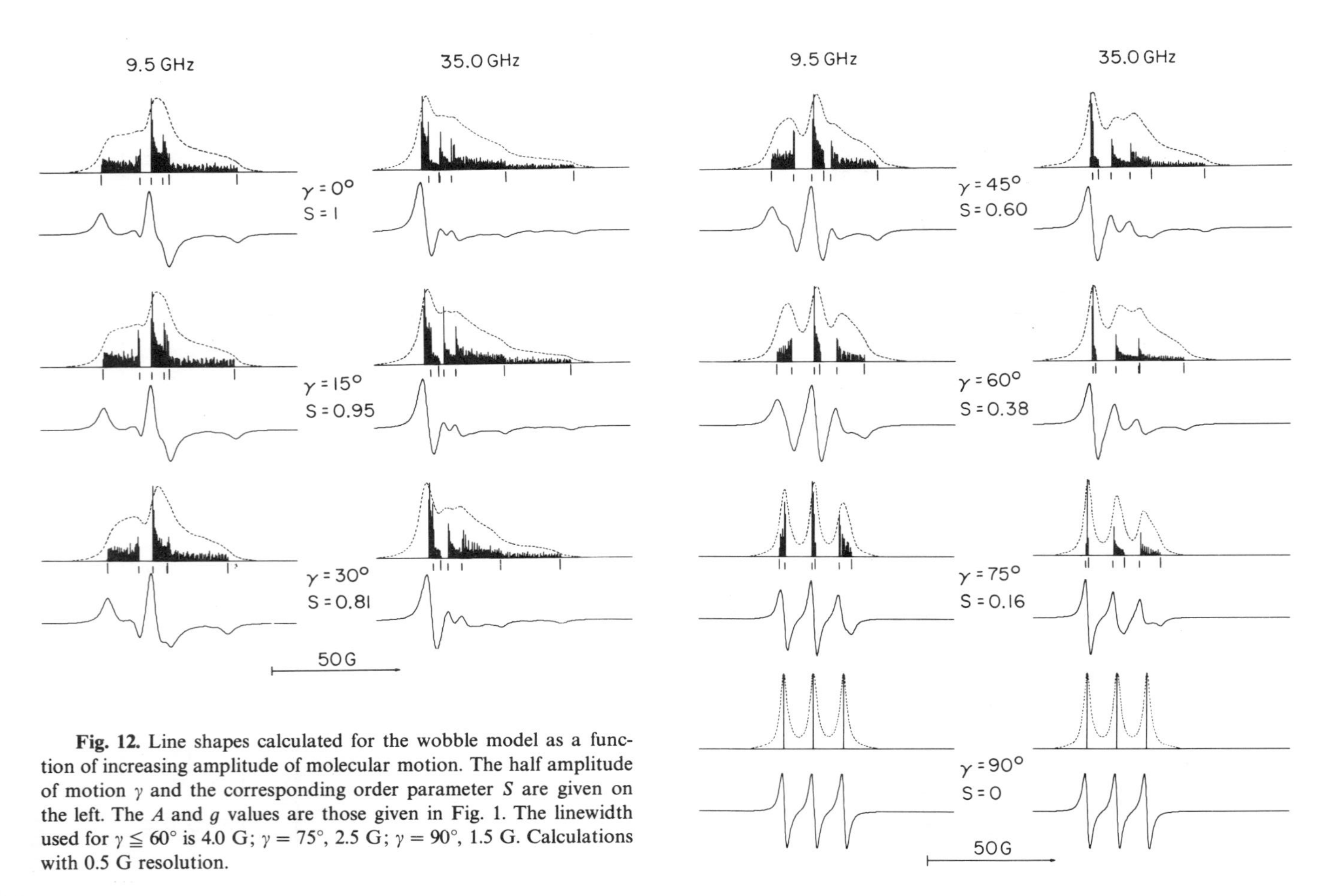

Fig. 12. Line shapes calculated for the wobble model as a function of increasing amplitude of molecular motion. The half amplitude of motion γ and the corresponding order parameter S are given on the left. The A and g values are those given in Fig. 1. The linewidth used for $\gamma \leqq 60°$ is 4.0 G; $\gamma = 75°$, 2.5 G; $\gamma = 90°$, 1.5 G. Calculations with 0.5 G resolution.

rim. At a fast camera shutter speed both disks appear stationary. At an intermediate shutter speed, the blurred arc of the red dot will begin to appear to be much the same for the two disks. At a slow shutter speed, the differences in amplitude of motion are clearly distinguished. The rotating disk will produce a red circular blur, while the rocking disk will still give a short red arc. In this crude analogy, the ESR experiment corresponds to the intermediate shutter speed, and this shutter speed is fixed for the conventional ESR spectrometer.

In biological membranes, the problem is complex, since undoubtedly many frequencies and amplitudes are involved. Unlike some spin-labeled proteins in solutions, lipid spin labels are highly asymmetric and when present in membranes are not free to tumble in all directions. (This is confirmed by the experiments with oriented samples; see Section III.) Therefore, in biological membranes both frequency and amplitude must be considered. This is taken into account in the above models by treating the motion as rapid and characterizing the maximum amplitude by a single parameter (e.g., α or γ).

5. Order Parameters

The simple rotation, oscillation, and wobble models are presented above to provide some physical insight into the origin of various line shapes. Another useful approach is based on the 3 × 3 order matrix introduced by Saupe to characterize molecular order in liquid crystals (Saupe, 1964). The matrix elements are given by $S_{ij} = \frac{1}{2}\langle 3l_{Ni}\, l_{Nj} - \delta_{ij}\rangle$, where N is the fixed reference axis of the sample (e.g., the optic axis or orientation axis Z_1) and i, $j = 1, 2, 3$ define a rectangular axis system fixed in the molecule. The S matrix resembles the dipolar part of the **A** matrix in that it is a symmetrical 3 × 3 matrix whose elements depend on the choice of the coordinate system and whose trace is zero. When diagonalized there are only two linearly independent matrix elements, and for the axially symmetric case, this reduces to only one independent matrix element, designated S_{33} or simply S, where

$$S = \tfrac{1}{2}(3\langle l_{N3}^2\rangle - 1) \tag{34}$$

The early applications of order parameters in spin labeling involved liquid crystals (Ferruti *et al.*, 1969; Corvaja *et al.*, 1970; Seelig, 1970). A discussion of these applications is given in Chapter 10.

The order parameter is dependent on the frequency and amplitude of the molecular motion. Thus, to characterize a complex system completely, an infinite number of order parameters would be required, covering the range of no motion to high-frequency molecular motion. As a consequence, an order parameter measured from nuclear magnetic resonance data need not equal the order parameter measured from a spin-labeling experiment or a fluorescent-labeling experiment. As a practical matter, since we sample only one frequency in the conventional ESR experiment, it is convenient to use $\delta = (A_{zz} - A_{xx}) \simeq 10^8\ \text{sec}^{-1}$ as a dividing line to reduce the problem to two-order parameters. When $\tau < 1/\delta$ an order parameter measures molecular motion, and when $\tau > 1/\delta$, a spatial order parameter results. So even in this simple view there are two order parameters, one dealing with molecular motion (S_{motion}) and the other describing the orientation (S_{space}) of the spin label with respect to the sample axis. The use of an order parameter, while mathematically correct, does not necessarily imply either orientation or anisotropic motion. We discuss molecular motion here and the orientation aspects in Section III. In the spin-labeling literature, the order parameter is usually estimated from the ESR line positions of disordered samples and is usually used as a parameter of molecular motion; therefore, in this section we denote S_{motion} by S.

The ESR experiment detects only the motion and orientation of the N—O group, so the order parameter is initially related to the x, y, z coordinates. However, it is the behavior of the lipid to which the N—O group is attached that is important, so the order matrix is usually transformed to the molecular coordinate system, with z_2 as the long axis (R) of the spin label. It is in the molecular coordinate system that the S matrix often has axial symmetry, hence one independent parameter S. When the molecular z_2 and the nitroxide z axes coincide, the problem is straightforward and the (motion) order parameter is†

$$S = (\bar{A}_{\parallel} - \bar{A}_{\perp})/(A_{zz} - \tfrac{1}{2}(A_{xx} + A_{yy})) \tag{35}$$

where $\bar{A}_{\parallel}$ and $\bar{A}_{\perp}$ are measured from the experimental spectra and A_{xx}, A_{yy}, and A_{zz} are the principal values measured in the absence of molecular motion and in an environment of similar polarity.

† At first glance it may not be obvious that Eq. (35) is equivalent to the definition of S given by Eq. (34). To show that it is equivalent note that $\langle\cos^2\phi_1\rangle = 1$, so that $\langle\cos^2\phi_1\cos^2\theta_1\rangle = \langle\cos^2\theta_1\rangle$. Instead of evaluating this integral, as was done in the wobble model, we express Eqs. (32) and (33) in the more general forms

$$\bar{A}_{\parallel} = \tfrac{1}{2}(A_{xx} + A_{yy}) + [A_{zz} - \tfrac{1}{2}(A_{xx} + A_{yy})]\langle\cos^2\theta_1\rangle \tag{32a}$$

$$\bar{A}_{\perp} = \tfrac{1}{2}(A_{xx} + A_{yy}) + [A_{zz} - \tfrac{1}{2}(A_{xx} + A_{yy})]\tfrac{1}{2}[1 - \langle\cos^2\theta_1\rangle] \tag{33a}$$

Subtracting Eq. (33a) from Eq. (32a) and solving for $\langle\cos^2\theta_1\rangle$ yields

$$\langle\cos^2\theta_1\rangle = \frac{2}{3}\frac{(\bar{A}_{\parallel} - \bar{A}_{\perp})}{[A_{zz} - \tfrac{1}{2}(A_{xx} + A_{yy})]} + \frac{1}{3}$$

Substituting this expression for $\langle l_{N3}^2\rangle$ in Eq. (34) yields Eq. (35). However, when the molecular z_2 axis and the nitroxide z axis do not coincide, Eqs. (32) and (33) are replaced by Eqs. (29) and (30) (again with $\langle\cos^2\phi_1\rangle$ replacing W). Combining equations as above, the result is $S = (\bar{A}_{\parallel} - \bar{A}_{\perp})/(\bar{A}_{\parallel}^{\text{r}} - \bar{A}_{\perp}^{\text{r}})$, which may be written in the specific form $S = (\bar{A}_{\parallel}^{\text{w}} - \bar{A}_{\perp}^{\text{w}})/(\bar{A}_{\parallel}^{\text{w}} - \bar{A}_{\perp}^{\text{w}})_{\gamma\to 0}$. [See discussion following Eq. (36).]

There is a relationship between S and the γ parameter of the wobble model. Substituting Eq. (32) and (33) for $\bar{A}_{\parallel}$ and $\bar{A}_{\perp}$ in Eq. (35) gives

$$S = \tfrac{1}{2}(3W - 1) = \tfrac{1}{2}(\cos\gamma + \cos^2\gamma) \tag{36}$$

The S values derived from Eq. (36) are listed below the corresponding values of γ in Fig. 12. These S values can also be obtained from Eq. (35) and Table I. Figure 13 is a plot of S versus γ, which is useful in comparing values of the parameters in the literature. Note that S does not vary linearly with γ. A change in S from 1 to 0.9 corresponds to an increase in the half amplitude of 20°. Whereas, a change in S from 0.6 to 0.5 corresponds to an increase of only 7.5° in the half amplitude of the motion. In the general case where the molecular and nitroxide axes do not necessarily coincide, $S = (\bar{A}_{\parallel}^{w} - \bar{A}_{\perp}^{w})/(\bar{A}_{\parallel}^{w} - \bar{A}_{\perp}^{w})_{\gamma\to 0}$, where $\bar{A}_{\parallel}^{w}$ and $\bar{A}_{\perp}^{w}$ are obtained from Eqs. (29) and (30), and $(\bar{A}_{\parallel}^{w} - \bar{A}_{\perp}^{w})_{\gamma\to 0}$ is the limit of $(\bar{A}_{\parallel}^{w} - \bar{A}_{\perp}^{w})$ as γ approaches zero (S corresponds to S_{mol} of Chapter 10). With these substitutions and letting $A_{xx} = A_{yy}$,

$$S = (\bar{A}_{\parallel}^{w} - \bar{A}_{\perp}^{w})/[(A_{zz} - A_{xx})\tfrac{1}{2}(3\cos^2\theta_2 - 1)] = \tfrac{1}{2}(\cos\gamma + \cos^2\gamma)$$

Thus, the plot of Fig. 13 is independent of the angle θ_2 between the nitroxide z axis and the molecular axis z_2 (or R). In practice, $\bar{A}_{\parallel}$ and $\bar{A}_{\perp}$ are measured from the experimental spectra (the line positions are marked below the stick spectra of Figs. 9 and 12), then $(\bar{A}_{\parallel} - \bar{A}_{\perp})$ is divided by $(A_{zz} - A_{xx})$ and again by $\tfrac{1}{2}(3\cos^2\theta_2 - 1)$ to obtain the order parameter S in the molecular coordinate system (Brown *et al.*, 1971). The parameter γ is obtained from $\bar{A}_{\parallel}$, $\bar{A}_{\perp}$ and Eqs. (29) and (30) or from the graph of Fig. 13. Note that it is often preferable to emphasize the quantities most closely related to the experiment (i.e., $\bar{A}_{\parallel}$ and $\bar{A}_{\perp}$) and then to estimate S or γ.

Neither S nor γ is sufficient to completely describe the system. For example S is equal to unity for all the very different line shapes of Figs. 9 and

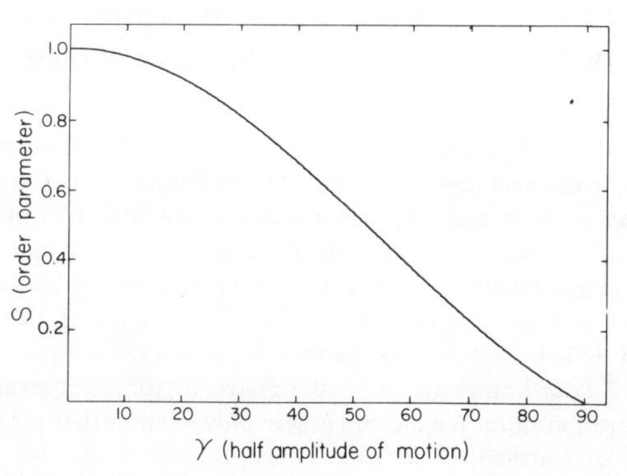

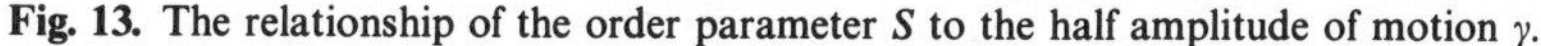

Fig. 13. The relationship of the order parameter S to the half amplitude of motion γ.

10, since this parameter does not take into account the amplitude or frequency of motion about the long molecular axis R.† When the molecular R axis and the nitroxide z axis coincide, oscillation has little effect on the ESR spectra because $A_{xx} \simeq A_{yy}$. The R and z axes approximately coincide in the fatty acid and related phospholipid spin labels (Fig. 7). Thus, the S parameter or γ are usually used to describe the effects of motion on the 9.5 GHz ESR spectra of fatty acids and phospholipid spin labels in biological membranes (Fig. 12). The advantage of S is that it is well known in the liquid crystal literature, and to a good approximation it varies linearly with the position of the outermost extrema of the ESR spectrum. The advantage of the wobble model is that it provides a useful physical description, and the parameter γ is linear with the amplitude of the motion.

There are other possible ways to reduce complex line shapes to a set of parameters. Some of these approaches are described in Chapters 2, 3, and 10. A common method is simply to report $2A_{max}$, the distance between the outermost ESR lines, when this quantity is measurable. The advantage of $2A_{max}$ is that there is little ambiguity about the measurement. Combining Eqs. (31) and (35) gives

$$S = \tfrac{3}{2}(\bar{A}_{\parallel} - A_0)/[A_{zz} - \tfrac{1}{2}(A_{xx} + A_{yy})] \tag{37}$$

where A_0, A_{xx}, A_{yy}, and A_{zz} are motion-independent constants, so that if polarity effects are ignored $2\bar{A}_{\parallel}$ ($\simeq 2A_{max}$) is linearly related to S. All of these treatments are approximations, but they are useful in studies of membranes where interpretations are usually based on *relative* changes in the ESR line shapes.

B. Flexibility Profile. Segmental Motion along the Lipid Chains

1. Experimental Examples

In order to determine the motional degree of freedom of membrane lipids, the N—O moiety is attached directly to the hydrocarbon chain as in spin labels I–V of Fig. 2 (Keith *et al.*, 1968; Waggoner *et al.*, 1969; Hubbell and McConnell, 1971; Jost *et al.*, 1971). Using a set of these doxyl fatty acids

† At least one additional parameter is required. One example is the oscillation parameter α. Another approach is to define a new order matrix (S_{ij}^{x}) relating one of the short axes (x_2) of the molecule with respect to the sample X_1 axis (recall that the first-order matrix relates the long molecular axis z_2 with respect to the sample Z_1 axis). This order matrix can of course be diagonalized, but it is not axially symmetric. For the specific case of limited amplitude oscillations in the nitroxide x, z plane discussed above in connection with the steroid spin label of Fig. 2b, $S_{11}^{x} = \frac{1}{4}(1 + 3P)$, $S_{33}^{x} = -\frac{1}{2}$ and $S_{22}^{x} = -(S_{11}^{x} + S_{33}^{x}) = \frac{1}{4}(1 - 3P)$, where $P = (\sin\alpha\cos\alpha)/\alpha$. To our knowledge, the order matrix S_{ij}^{x} (or the equivalent S_{ij}^{y}) has not been used in the membrane literature. We include it here only to show that both parameters α and γ can be related to order matrices.

or phospholipids, it is then possible to plot the extent of the local methylene chain motion along the lipid chains in either isotropic (Hubbell and McConnell, 1971) or oriented preparations (Jost *et al.*, 1971). This method has also been used to study a model liquid crystal system of sodium decanoate and *n*-decyl alcohol (see Chapter 10). Typical results for membrane phospholipids are shown in Fig. 14. In (a) is the isotropic sample. In (b), the *same preparation* is oriented on a thin glass slide with the magnetic field parallel (solid spectra) or perpendicular (dashed spectra) to the normal to the slide. In the random sample, the outermost lines move in and the linewidths become narrower as the doxyl group is translated from the C-5 to the C-7, the C-12 and finally the C-16 position. In the corresponding oriented samples, the linewidths also become narrower and the line positions of the parallel and perpendicular spectra gradually approach each other as the doxyl group is moved toward the hydrocarbon end. This marked effect is clearly the result of increased motional freedom along the hydrocarbon chains, culminating in fluidlike behavior at the hydrocarbon end.

An example of the flexibility profile in a biological membrane (*Acholeplasma laidlawii* membranes) is shown in Fig. 15. The parameter $2A_{max}$ $(2T_m)$

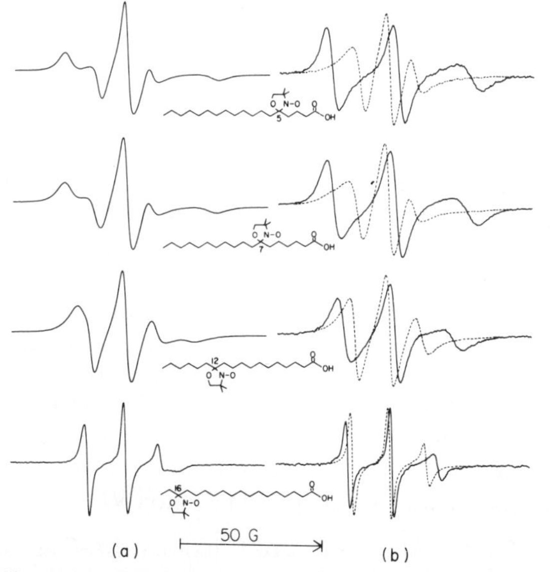

Fig. 14. An illustration of the flexibility profile of phospholipid bilayers. The sample is egg phosphatidylcholine (a) randomly oriented on small glass beads and (b) oriented on a glass support slide. In the oriented samples, the solid line is recorded with the magnetic field along the normal to the slide, and the dashed line with the sample rotated 90°. The samples contained the doxylstearic acids (phospholipid : spin label molar ratio 150 : 1) and were maintained at 81% relative humidity at room temperature.

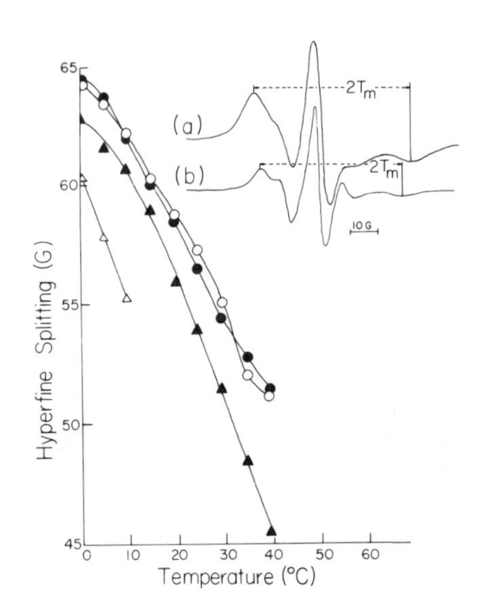

Fig. 15. The temperature dependence of the large splitting $2A_{max}(2T_m)$ of doxylfatty acids diffused into *Acholeplasma laidlawii* membranes. The spin labels are 5-doxylstearic (○), 5-doxylpalmitic (●), 8-doxylstearic (▲), and 12-doxylstearic (△) acids. Spectra (a) and (b) are of 5-doxylpalmitic acid in the membranes and vesicles of lipids extracted from the membranes, respectively. [Adapted from Rottem *et al.* (1971).]

is plotted for 5-, 8-, and 12-doxyl fatty acids diffused into the membranes. Note that for any temperature investigated $2A_{max}$ decreases monotonically as the doxyl group is moved toward the center of the bilayer. These data also illustrate how the outermost splittings change as a function of temperature. In the insert, the spectrum of the 5-doxylpalmitic acid in the membranes [spectrum (a)] is compared to the same spin label in an aqueous dispersion of the lipids extracted from the membranes [spectrum (b)], showing general similarity in the line shape. Despite small quantitative differences in measured values, the flexibility profile of these biological membranes closely resembles that seen in phospholipid model membranes. In membrane samples with a high protein content, an additional more immobilized phase is frequently seen, and this is discussed in Section VI.

2. Analysis: A Practical Example of Calculating Order Parameters

An analysis of the oriented samples of Fig. 14b is given in Section III of this chapter. Here we discuss measurements on the randomly oriented samples of Fig. 14a. Our aim is to present the type of analysis used in the membrane literature in sufficient detail that readers new to this technique can proceed directly with the interpretation of their own data.

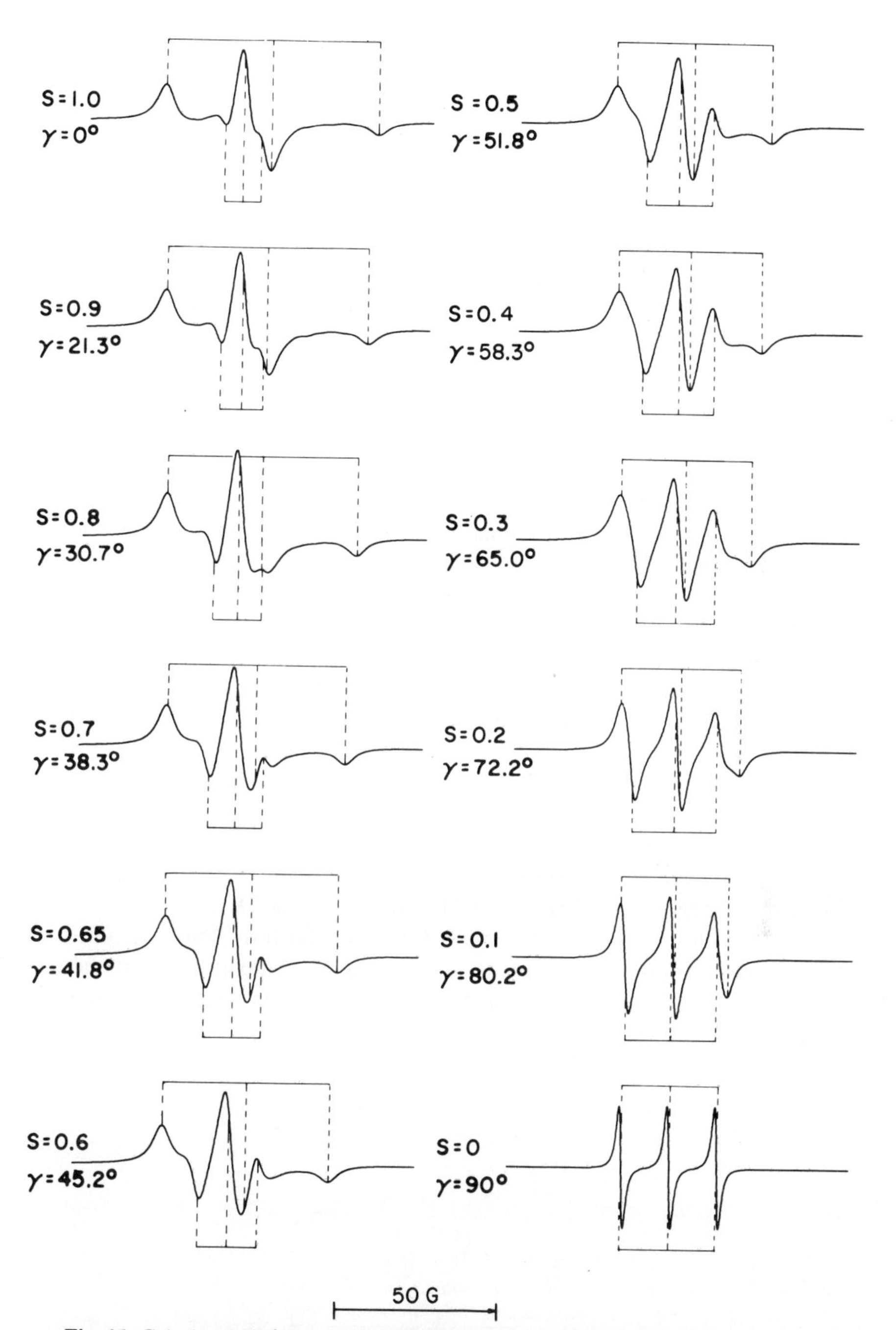

Fig. 16. Calculated ESR line shapes for randomly oriented samples as a function of the order parameter S or γ. The dashed lines indicate the precise positions of the three ESR lines defined by $\bar{A}_{\parallel}$ and $\bar{g}_{\parallel}$ (large splittings) and $\bar{A}_{\perp}$ and $\bar{g}_{\perp}$ (small splittings). Quantitative values are given in Table II. Calculated with 0.1 G resolution.

It is straightforward to measure the maximum splitting $2A_{max}$ (also called $2A_m$ or $2T_m$) from the spectrum. For Fig. 14a, A_{max} equals 27.3, 27.1, 21.2, and 17.0 G from top to bottom, respectively. A number of studies report this parameter alone. To proceed further and determine S or γ, one additional measurement is needed. This second parameter can be either $\bar{A}_\perp$ or A_{zz}. We now illustrate several variations in the calculations of S (and γ).

For the moment we set aside the question of polarity effects and consider only the measurement of $\bar{A}_\perp$. It is useful to look at computer simulations in order to see how to make this measurement. One set of line shapes is given in Fig. 12. Another set generated in the same way but having a different emphasis is shown in Fig. 16. The dashed vertical lines show the exact positions defined by $2\bar{A}_\parallel$ and $2\bar{A}_\perp$. It is clear from Fig. 16 that the maximum splitting $2A_{max}$ is a very good measure of $2\bar{A}_\parallel$. Furthermore, the minimum splitting $2A_{min}$ (first minimum to last maximum) is close but not exactly equal to $2\bar{A}_\perp$. The measurement of A_{max} and A_{min} is illustrated in more detail in Fig. 17. This spectrum is an enlargement of the spectrum for $S = 0.6$ of Fig. 16.

Table II presents the relevant quantitative data associated with Fig. 16. The first two columns give the independent variable as S or γ. The input parameters for each value of S are calculated from Eqs. (31) and (35), and the set of principal values for 2-doxylpropane. Note that $A_0 = \frac{1}{3}(A_{xx} + A_{yy} + A_{zz}) = 14.73$ G with this choice of principal values. The spin label is assumed to be rapidly rotating so that $\bar{g}_\parallel$ and $\bar{g}_\perp$ are calculated from similar expressions. The spectral line positions with $H \parallel R$ and $H \perp R$, determined by the sets of parameters $\bar{A}_\parallel$, $\bar{g}_\parallel$ and $\bar{A}_\perp$, $\bar{g}_\perp$ are listed in columns 5 to 10 (marked "Input") of Table II.

The next four columns of Table II (marked "Output") are the positions of the first maximum, the first minimum, the last maximum, and the last minimum of the computed line shapes of Fig. 16. For $S = 0.8$, 0.9, and 1.0,

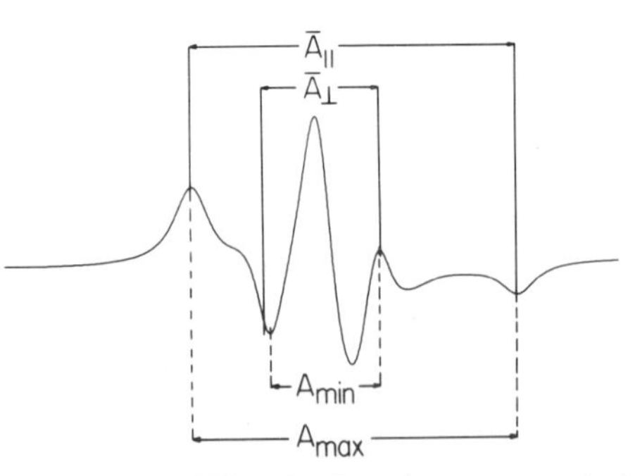

Fig. 17. The $S = 0.6$ spectrum of Fig. 16 enlarged to compare the input values of $\bar{A}_\parallel$ and $\bar{A}_\perp$ (indicated at the top) with the measured values of A_{max} and A_{min} (indicated at the bottom). Values are given in Table II.

TABLE II

RELATIONSHIP BETWEEN FIRST-DERIVATIVE PEAK POSITIONS AND THE FUNDAMENTAL QUANTITIES $\bar{A}_{\parallel}$, $\bar{A}_{\perp}$ [a]

Motion parameters		Input parameters		Computer listing of relevant line positions[b]										Measured values of splitting, S^{app}		
				Input						Output						
				$\bar{A}_{\parallel}$			$\bar{A}_{\perp}$									
S	γ	$\bar{A}_{\parallel}$	$\bar{A}_{\perp}$	+1	0	−1	+1	0	−1	1	2	3	4	A_{max}	A_{min}	S^{app} [c]
1.0	0.0	32.90	5.65	29.7	62.5	95.3	48.2	53.8	59.4	29.5	39.4	—	95.3	32.90	—	—
0.9	21.3	31.00	6.56	30.9	61.9	92.9	47.7	54.2	60.7	30.8	40.4	—	92.9	31.05	—	—
0.8	30.7	29.26	7.47	32.1	61.3	90.5	46.9	54.3	61.7	32.0	40.6	—	90.6	29.30	—	—
0.7	38.3	27.46	8.37	33.4	60.8	88.2	46.4	54.7	63.0	33.3	47.1	63.2	88.2	27.45	8.05	0.71
0.65	41.8	26.54	8.83	33.9	60.4	86.9	46.0	54.8	63.6	33.9	46.8	63.6	87.0	26.55	8.40	0.67
0.6	45.2	25.64	9.28	34.5	60.1	85.7	45.8	55.0	64.2	34.6	46.6	64.1	85.8	25.6	8.75	0.62
0.5	51.8	23.82	10.19	35.8	59.6	83.4	45.1	55.2	65.3	35.8	46.0	64.9	83.3	23.75	9.45	0.52
0.4	58.3	22.01	11.09	37.1	59.1	81.1	44.5	55.5	66.5	37.1	45.5	65.9	80.8	21.85	10.20	0.43
0.3	65.0	20.19	12.00	38.3	58.4	78.5	44.0	55.9	67.8	38.4	45.0	67.0	78.4	20.00	11.00	0.33
0.2	72.2	18.36	12.92	39.6	57.9	76.2	43.3	56.2	69.1	39.7	44.2	68.3	76.0	18.15	12.05	0.22
0.1	80.2	16.54	13.83	40.7	57.2	73.7	42.6	56.4	70.2	40.6	43.6	69.6	73.7	16.55	13.00	0.13
0.0	90.0	14.73	14.73	42.0	56.7	71.4	42.0	56.7	71.4	41.3	42.9	70.7	72.3	15.50	13.90	0.06

[a] All values are in gauss. Linewidths used were 4.0 G, $S = 1$ to $S = 0.3$; 3.0 G, $S = 0.2$; 2.5 G, $S = 0.1$; 1.5 G, $S = 0$.

[b] Output lines 1, 2, 3, 4 refer to the position of first maximum, first minimum, last maximum, and last minimum of the first-derivative spectrum (see Fig. 16). Relative line positions are on an arbitrary scale of 0 to 100 G, calculated to 0.1 G. The magnetic field increases from left to right. The microwave frequency was taken to be 9.5 GHz; +1, 0, −1 refer to the z component of the ^{14}N nuclear spin.

[c] The approximate order parameter (S^{app}) defined as $S^{app} = (A_{max} - A_{min})/[A_{zz}^{c} - \frac{1}{2}(A_{xx}^{c} + A_{yy}^{c})]$ where A_{xx}^{c}, A_{yy}^{c}, and A_{zz}^{c} are the single-crystal values used throughout this table.

the relevant last maximum is not defined, and therefore A_{min} cannot be measured for this range of S values. The A_{max} and A_{min} values listed are the differences between "Output" columns 1 and 4 and between 2 and 3, respectively.

To calculate S, the simplest approximation is to assume $A_{max} = \bar{A}_{\parallel}$ (a very good assumption) and $A_{min} \simeq \bar{A}_{\perp}$ (a fair assumption, see Table II). With these assumptions and ignoring differences in polarity, the approximate order parameter is

$$S^{app} = (A_{max} - A_{min})/[A_{zz}^{c} - \tfrac{1}{2}(A_{xx}^{c} + A_{yy}^{c})] \tag{38}$$

where A_{xx}^{c}, A_{yy}^{c}, and A_{zz}^{c} are single-crystal values. The S^{app} values are given in the last column of Table II. Note that the agreement between S^{app} and S is very good. This agreement is somewhat misleading, since in these computed spectra, no polarity correction of the single-crystal values is needed. In practice, the polarity of the single crystal usually differs from the polarity of the experimental sample. In order to avoid confusion between the single-crystal parameters and the parameters in the environment of the sample, we temporarily label the single crystal values A_{xx}^{c}, A_{yy}^{c}, A_{zz}^{c}, and A_{0}^{c}. In the absence of additional information, we assume that all values change proportionally with polarity. Thus, $A_{xx} = kA_{xx}^{c}$, $A_{yy} = kA_{yy}^{c}$, $A_{zz} = kA_{zz}^{c}$, and $A_0 = kA_0^{c}$, where $A_0^{c} = \frac{1}{3}(A_{xx}^{c} + A_{yy}^{c} + A_{zz}^{c})$. Substituting these quantities in Eq. (35) yields

$$S = (\bar{A}_{\parallel} - \bar{A}_{\perp})/[A_{zz}^{c} - \tfrac{1}{2}(A_{xx}^{c} + A_{yy}^{c})]k \tag{39}$$

In order to obtain k and at the same time provide a better estimate of $\bar{A}_{\perp}$, the data of Table II can be used to provide a correction term for $A_{\min}$. A good discussion of this approach (Seelig, 1970; Hubbell and McConnell, 1971) is given in Appendix IV by Gaffney. In her method

$$\bar{A}_{\perp} = A_{\min} + 1.4(1 - S^{\text{app}}) \tag{40}$$

where all values are in gauss. From Table II, the same constant 1.4 ± 0.08 is obtained. This is remarkable agreement considering that different principal values, different linewidths, and different computer programs were used. A somewhat better approximation to the data of Table II is given by the expression (valid for $S^{\text{app}} < 0.8$)

$$\bar{A}_{\perp} = A_{\min} + 1.32 + 1.86 \log(1 - S^{\text{app}}) \tag{41}$$

As S approaches zero, the correction depends strongly on the linewidth of the computed spectra. It can be seen from Fig. 16 that when $S = 0$, $2A_{\max} = 2A_0 + w$, and $2A_{\min} = 2A_0 - w$, where w is the peak-to-peak first-derivative linewidth. Subtracting these two quantities gives $A_{\max} - A_{\min} = w$. In other words, for small values of S, the calculated linewidth can make a large contribution to the correction factor. In practice, this problem is taken care of, because as S decreases, progressively narrower linewidths are required to match the experimental spectra. Once $\bar{A}_{\parallel}$ and $\bar{A}_{\perp}$ are available, then A_0 is calculated from Eq. (31) and k is given by $k = A_0/A_0^{c}$.

Another variation in determining S and γ relies on the measurement of A_{zz} rather than $\bar{A}_{\perp}$. The experimental values of A_{zz} are obtained by measuring $A_{\max}$ with the sample at $-196°$C ($A_{\max}^{-196}$). The polarity correction in this case is given by $k = A_{\max}^{-196}/A_{zz}^{c}$.† Therefore, from Eq. (37) or Eq. (39)

$$S = \tfrac{3}{2}[\bar{A}_{\parallel} - kA_0^{c}]/[A_{zz}^{c} - \tfrac{1}{2}(A_{xx}^{c} + A_{yy}^{c})]k \tag{42}$$

† Even in single crystals there is some molecular motion at room temperature. For example, $A_{zz}^{c}(23°\text{C}) = 32.9$ G and $A_{zz}^{c}(-196°\text{C}) = 33.6°$ G for 2-doxylpropane (L. J. Libertini, unpublished data).

which can be rewritten as

$$S = \frac{3}{2}\left\{\frac{(\bar{A}_{\parallel}/k) - \frac{1}{2}(A^{c}_{xx} + A^{c}_{yy})}{A^{c}_{zz} - \frac{1}{2}(A^{c}_{xx} + A^{c}_{yy})}\right\} - \frac{1}{2} \tag{43}$$

Thus, the need to estimate $\bar{A}_{\perp}$ is replaced by the necessity to obtain a spectrum at liquid nitrogen temperature. Both methods have shortcomings. Measurements of A^{-196}_{max} assume that the integrity of the sample is unaltered by freezing. Estimations of $\bar{A}_{\perp}$ using a correction factor derived from calculated spectra suffer from the fact that calculated line shapes at best differ in some details from the experimental spectra. These small line-shape differences most frequently occur in the regions used in measuring A_{min} and produce changes that are not accounted for by Eqs. (40) and (41).

For the convenience of the reader we summarize these variations on methods of calculating an order parameter for randomly oriented samples containing the doxyl fatty acid and phospholipid labels. These methods have in common the measurement of A_{max} from the experimental sample and a set of single-crystal parameters, A^{c}_{xx}, A^{c}_{yy}, and A^{c}_{zz}. All quantities are in gauss.

Method I. Given: Measured values of A_{max} and A_{min}

1. Assume $A_{max} = \bar{A}_{\parallel}$, $A_{min} \simeq \bar{A}_{\perp}$ and the polarity of the single crystal approximates that of the experimental sample.
2. Calculate an approximate-order parameter S^{app} from Eq. (38).

Method II. Given: Measured values of A_{max} and A_{min}

1. Assume $A_{max} = \bar{A}_{\parallel}$.
2. Calculate $\bar{A}_{\perp}$ from Eq. (40) or Eq. (41).
3. Calculate polarity correction k for the single-crystal values; $k = A_0/A^{c}_0$, where $A_0 = \frac{1}{3}(\bar{A}_{\parallel} + 2\bar{A}_{\perp})$, and $A^{c}_0 = \frac{1}{3}(A^{c}_{xx} + A^{c}_{yy} + A^{c}_{zz})$.
4. Calculate S from Eq. (39).

Method III. Given: Measured values of A_{max} and A^{-196}_{max}

1. Assume $A_{max} = \bar{A}_{\parallel}$.
2. Assume $A^{-196}_{max} = A_{zz}$.
3. Calculate the polarity correction k for the single crystal values; $k = A^{-196}_{max}/A^{c}_{zz}$.
4. Calculate S from Eq. (43). Note that, alternatively, $\bar{A}_{\perp}$ can be calculated using Eq. (31) (with $A_0 = kA^{c}_0$), and then S can be calculated from Eq. (39).

Methods II and III yield better approximations than Method I. The values of γ can be obtained using these same approximations, from Eq. (32) or from S using Eq. (36) (or Fig. 13).

A worked example using these three methods is given in Table III. The raw data, A_{max} and A_{min}, are measured from the experimental spectra of hydrated egg lecithin (left column, Fig. 14). Parallel samples, sealed in quartz ESR tubes were cooled in a liquid nitrogen Dewar to measure A_{max}^{-196}. The values of S and γ were then calculated using Methods I–III. The maximum variation in S values for a given spin label is 10% or less in this example.

Molecular motion along fatty acid chains has also been measured using proton magnetic resonance (Chan *et al.*, 1972; Horwitz *et al.*, 1972; Kohler *et al.*, 1972; Lee *et al.*, 1972), ^{13}C magnetic resonance (Metcalfe *et al.*, 1971; Godici and Landsberger, 1974), ^{19}F magnetic resonance (Birdsall *et al.*, 1971), and deuterium magnetic resonance (Seelig and Neiderberger, 1974). All of these techniques support the idea that there is increased segmental motion toward the center of the bilayer. The differences observed are attributed to two factors: first, the perturbation introduced by the spin label (see Chapter 10); second, the time windows of the experiments are different (Gaffney and McConnell, 1974) (see Section II.A.4). The main point is that the spin-labeling data provides an operationally defined parameter S (or γ) and that it is the relative change in this parameter at some depth in the bilayer and not the absolute value that provides a handle on the experimental variable.

TABLE III

DATA[a] AND CALCULATED MOTION PARAMETERS FOR FIG. 14

Spin-label position	A_{max} [a]	A_{min} [a]	A_{max}^{-196} [b]	Method I[c] S^{app}	Method I[c] γ^{app}	Method II[c] $\bar{A}_{\perp}$	Method II[c] S	Method II[c] γ	Method II[c] A_0	Method III[c] S	Method III[c] γ	Method III[c] A_0
5	27.3	8.4	33.3	0.69	39°	8.8	0.67	40°	14.9	0.67	40°	14.9
7	27.1	8.6	33.1	0.68	40°	9.0	0.65	42°	15.0	0.67	40°	14.8
12	21.2	9.9	32.6	0.41	58°	10.8	0.39	59°	14.3	0.37	60°	14.6
16	17.0	11.8	32.4	0.19	73°	13.0	0.15	76°	14.3	0.14	77°	14.5

[a] Estimated uncertainty ±0.1 G.
[b] Estimated uncertainty ±0.3 G.
[c] Single-crystal values same as for Table I; Methods I, II, and III are described in the text.

III. CHARACTERIZING MACROSCOPIC ORDER

Some of the most compelling evidence for the organization of the lipids into bilayers in biological membranes comes from studies on macroscopically oriented samples. Orientation of samples permits direct characterization of the ordered regions. The direction of preferential alignment and the degree of order are important properties that can be determined from the ESR spectra of lipid spin labels in oriented membrane preparations.

A. Gaussian Distribution of Orientations

An idealized cross section of a phospholipid bilayer is represented in Fig. 18. The fatty acid spin label (A) intercalates with the long hydrophobic chain normal to the plane of the bilayer. The rigid spirane structure of the doxyl group places the nitroxide z axis (arrow) parallel to the extended lipid chain. Thus, the direction of maximum splitting will be observed when the magnetic field is normal to the plane of the bilayer (i.e., along the arrow A). The splitting becomes progressively smaller as the bilayer is rotated in the magnetic field, and the minimum splitting occurs when the bilayer is parallel to the magnetic field. The opposite effect is seen with the steroid spin label (B), where the z axis is perpendicular to the long axis of the steroid molecule.

In oriented samples of real bilayers and membranes, the order is not nearly as perfect as shown in Fig. 18. A realistic physical description is that a Gaussian distribution of orientations exists about the alignment direction Z_1 (Z_1 is the sample axis, and is usually normal to the support surface). The orientation of the spin label is characterized by the angle θ_1 between the sample Z_1 axis and the molecular z_2 axis. For a collection of spin labels, we require an orientation distribution function $P(\theta_1)$ such that $P(\theta_1)\ d\theta_1$ gives

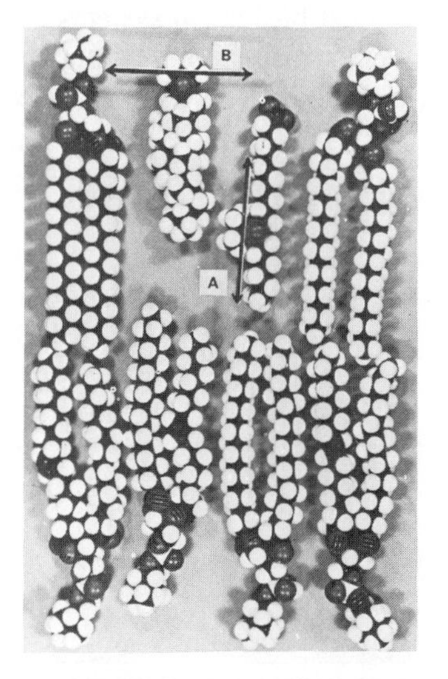

Fig. 18. Space-filling (Corey–Pauling–Kolton) molecular models arranged to approximate the organization of a phospholipid bilayer region, showing the orientation of two lipid spin labels, 12-doxylstearic acid (A) and 3-doxyl-5α-cholestane (B). The arrows indicate the direction of the nitroxide z axis. [Reproduced from Jost *et al.* (1971).]

the fraction of spin labels with z_2 axes lying within a zone bounded by θ_1 and $\theta_1 + d\theta_1$ on the surface of a unit sphere. In the Gaussian distribution model,

$$P(\theta_1) = N^{-1} \sin \theta_1 \exp[-(\theta_1 - \delta)^2/(2\sigma^2)] \tag{44}$$

where N is a normalizing constant, σ is the orientation distribution parameter and δ is the angle of tilt. The parameter σ is the standard deviation from the mean of the Gaussian curve, but does not have this precise meaning in Eq. (44) because of the function $\sin \theta_1$ that results from the area element in spherical coordinates. The angle of tilt parameter δ is the angle between the sample axis Z_1 and the maximum of the distribution function. Figure 19 shows the shape of the distribution function as a function of σ when $\delta = 0$. At small values of σ, the distribution approaches the familiar Gaussian shape. As σ is increased, the distribution becomes progressively less Gaussian, and as $\sigma \to \infty$, it becomes a sine function. The maximum of the curve shifts from 0° to 90° as σ is increased from 0 to ∞. This is a result of the increasing area element from the pole to the equator and not of any preferential alignment at 90°. Introducing an angle of tilt also shifts the maximum of the curve from 0°. The angle of tilt parameter can range from 0° to 90°, and the plots corresponding to Fig. 19 for $\delta = 90°$ are given elsewhere (Libertini *et al.*, 1974).

1. Planar Samples: The Wheat Field Model

Macroscopic order in most biological systems is examined by preparing flat layered samples. Aqueous dispersions of phospholipids spontaneously form stacked bilayers (multilayers) on a flat surface when excess water is

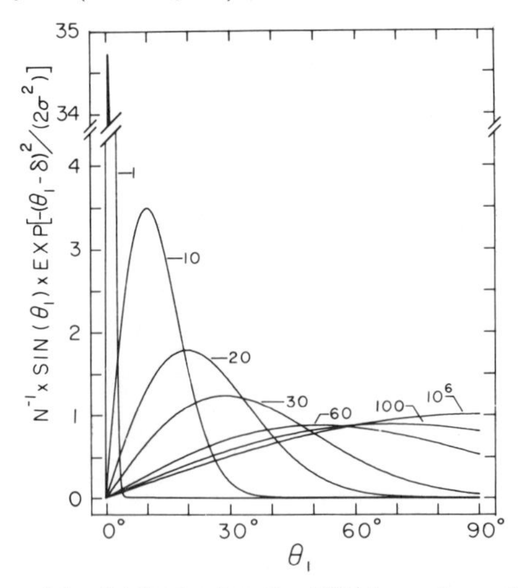

Fig. 19. The shape of the distribution function $P(\theta_1)$ for various values of the distribution parameter σ (angle of tilt (δ) is zero). [Reproduced from Libertini *et al.* (1974).]

removed (Libertini *et al.*, 1969). Variations of this approach include depositing the sample on the sides of a flat ESR cell (Hsia *et al.*, 1970), or on a glass slide in a hydration chamber (Jost *et al.*, 1971), or shearing the sample between two flat plates (McFarland and McConnell, 1971). In fact, it is very difficult to obtain randomly oriented samples, and it has been necessary to resort to coating fine glass beads (Jost and Griffith, 1973) or compressing the phospholipids between two nesting hemispheres (McFarland and McConnell, 1971) to obtain randomly oriented multilayers. Biological membranes can be oriented by settling on a glass surface, by centrifugation, or in the case of some specialized membranes by flow techniques.

The most important parameter in characterizing the order is the Gaussian distribution parameter σ. Figure 20 provides physical insight into the meaning of this parameter and can be thought of as analogous to a windswept wheat field. Each stick in Fig. 20 gives the direction of the molecular z_2 axis of one spin label in the sample. For small values of σ, all of these axes are pointed very nearly normal to the plane of the bilayer, since the angle of tilt in this example is zero. As σ is increased, the degree of orientation decreases progressively, and as $\sigma \rightarrow \infty$ ($\sigma = 10^6$ in Fig. 20), there is a random distribution of the vectors. For all practical purposes, randomness is achieved at much smaller values of σ, and very little orientation effect is seen in the ESR spectra for $\sigma \geqq 300°$. Figure 20 is not intended to represent the cross-sectional view of a bilayer. It represents the cross-sectional view of the spin label z_2 axes, which in the case of doxyl fatty acids and phospholipids is the direction of the nitroxide $2p_z$ orbital. These spin labels are flexible structures and it is well established that the distribution of the $2p_z$ orbitals is narrow near the polar head groups and becomes progressively more random

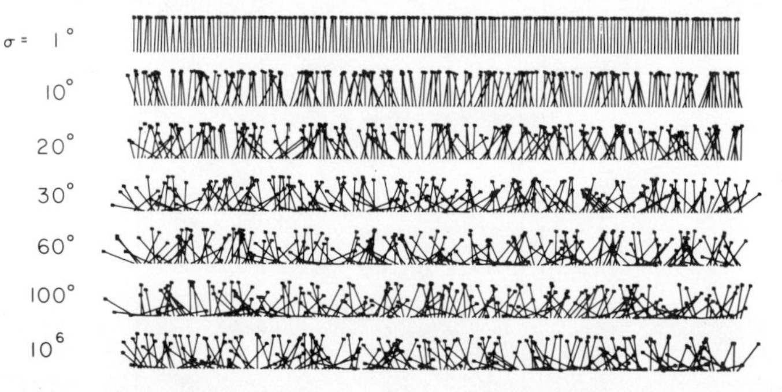

Fig. 20. The directions of the spin label z_2 axes as a function of the orientation distribution parameter σ, i.e., the wheat field model. The orientation of each spin label with respect to a plane is indicated by projecting vectors for angles of θ_1, each randomly assigned an angle ϕ_1 (into and out of the paper). Note the progressive decrease in order as the orientation distribution parameter σ is increased, so that for large values ($\sigma = 10^6$) the vectors are randomized. Angle of tilt $\delta = 0°$. [Reproduced from Libertini *et al.* (1974).]

toward the center of the bilayer, reflecting some increased disorder toward the center of the bilayer.

The parameters σ and δ are determined for a sample by matching the experimental spectra with spectra calculated applying Eq. (44). A computer program (EULOR) incorporating this Gaussian distribution function is described by Libertini *et al.* (1974). Values of σ can then be interpreted quantitatively by referring to Fig. 21. This figure gives the fraction of spin labels whose average positions fall within the cone of half-angle θ_1. For a cone size of $\theta_1 = 90°$, all z_2 axes lie within this cone, regardless of the value of σ, because of the normalization procedure. Conversely, when $\theta_1 = 0$, the percent of spin labels is zero because the area element at the pole approaches zero. Any intermediate value of θ_1 can be used as a reference cone angle, and a convenient value is $\theta_1 = 30°$. According to Fig. 21, if the experimentally determined value is $\sigma = 20°$ (and $\delta = 0$), then approximately 70% of the spin labels orient within this reference cone. Whereas, if the sample is randomly oriented, only 13% are found within this cone.

In some systems, the angle of tilt parameter is nonzero. As a practical matter, there is an interplay between σ and δ in the line-shape simulations. In some cases only an upper limit for δ has been established (Libertini *et al.*, 1969; Birrell *et al.*, 1973a), and in other experiments values of δ have been reported (McFarland and McConnell, 1971; Birrell and Griffith, 1974). Molecular motion and, in particular, lateral diffusion tend to obscure the effects of an angle of tilt, and it is not always possible to determine this parameter. In the phospholipid bilayer systems examined thus far, the angle of tilt reflected is largest near the polar head group and becomes smaller, and

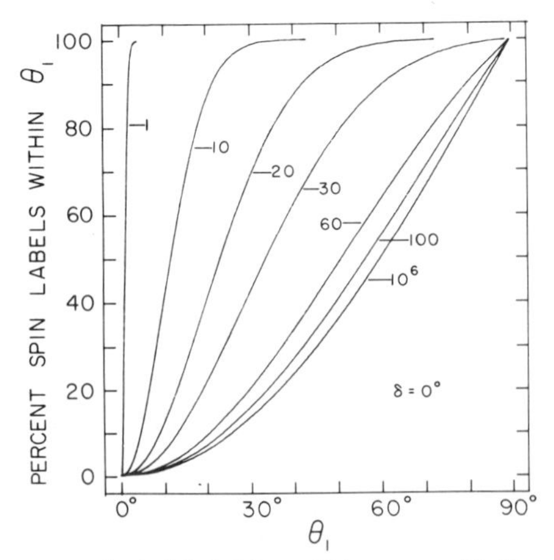

Fig. 21. The percentage of spin labels lying within a cone of half angle θ_1, as a function of the orientation distribution parameter when the angle of tilt is zero. [Reproduced from Libertini *et al.* (1974).]

possibly zero, toward the center of the bilayer. Thus, in a simplified picture, the fatty acid chains may be viewed as bent somewhere in the middle of the chain (McFarland and McConnell, 1971). Another interesting observation is that the magnetic field apparently can induce an orientation of phospholipids (Gaffney and McConnell, 1974). This does not alter θ_1, so the above treatment of orientation distributions remains essentially correct. There is no effect on the spectrum parallel to Z_1. With the magnetic field perpendicular to Z_1, the magnetic field tends to create a nonrandom distribution of the angle ϕ_1. This effect can be taken into account by replacing $P(\theta_1)$ by a distribution function $P(\theta_1, \phi_1)$ which depends upon both the polar and azimuthal angles. However, this introduces a new parameter and the effects on line-shape simulations are small, so for the purposes of this chapter this magnetic field effect is neglected.

It should be kept in mind that all of the effects discussed here concern the degree of orientation and angle of tilt of the doxyl group and are then interpreted as reflecting the average orientation of the lipid backbone. This doxyl group is a perturbation, and, although it is not possible to measure the magnitude of this perturbation, the observation of fairly high degrees of orientation in some multilayers and membranes suggests that the spin labels give a reasonable representation of the behavior of the unlabeled phospholipids.

2. Cylindrical Samples

Some biological membrane samples exhibit cylindrical symmetry. An example is an oriented nerve. In such systems, the orientation distribution is described by setting $\delta = 90°$ in Eq. (44). This is equivalent to rolling up the planar sample. A high degree of orientation (small σ) corresponds to all spin labels very nearly perpendicular to the cylindrical axis, and as σ increases, the orientation becomes progressively more random, as in the case of the planar sample. An experimental application to biological samples is discussed below, and an example of liquid crystals exhibiting cylindrical symmetry is discussed in Chapter 10.

B. Molecular Motion in Oriented Samples

Biological samples invariably exhibit some degree of molecular motion. The effects of this motion are combined with the effects of orientation in the final ESR line shapes. In principle, the Gaussian distribution model can be combined with general line shape theories of molecular motion. However, molecular motion is often treated simply using the models for rapid anisotropic motion discussed above. In these limiting cases, the principal A and g values are simply replaced by their motion-averaged counterparts in the computer program, and time does not explicitly enter into the calculations.

To characterize an oriented sample, at least three parameters must be determined, one describing the molecular motion present and the others specifying the degree of orientation and the angle of tilt. A reasonable approach is to separate the problem into two parts by examining both randomly oriented samples and the corresponding oriented specimen. The degree of motion (i.e., γ or S) is determined on the random sample. Then the line shapes of the oriented samples are generated with the motion parameter fixed, and the orientation parameters σ and δ are varied.

As an example we consider the experimental spectra of phospholipid bilayers of Fig. 14. The values of the motion parameter γ (or S) are given in Table III. With these values of γ and the single-crystal parameters corrected for polarity, using A_{zz}^{-196}, the parameters σ and δ were varied and the resulting spectra, generated by the computer program EULOR, were compared for visual fit of the experimental spectra in (b) of Fig. 14. Reasonably good fits are obtained at the 5 position for $\sigma = 20°$ and $\delta = 25°$. Similarly, for 7-doxylstearic acid $\sigma \simeq 20°$, $\delta \simeq 25°$; for 12-doxylstearic acid $\sigma \simeq 30°$, $\delta \leqq 15°$; for 16-doxylstearic acid $\sigma \simeq 35°$, $\delta < 15°$. These numbers should not be taken too seriously. The determination of these two parameters is more ambiguous than the determination of the motion parameter γ. For example, lower values of δ combined with higher values of σ also give reasonable splittings and general line shapes (see discussion in Section III.C). However, if complete anisotropy data from the oriented sample are available, it is possible to get a better estimate of δ (Birrell and Griffith, 1974), and this narrows the range of acceptable σ values.

An experimental example of a sample with cylindrical symmetry is used in Fig. 22. The sample is an invertebrate nerve bundle (walking leg nerve from the lobster *Homarus americanus*), which has been the subject of several ESR studies (Hubbell and McConnell, 1969b; Simpkins *et al.*, 1971; Jost *et al.*, 1973d). The spin label 5-doxylstearic acid was diffused into the nerve bundles. As was the case with the planar samples it is useful to examine both the randomly oriented and ordered specimen. For the randomly oriented sample, the molecular motion parameter is $\gamma = 42°$ (or $\gamma = 35°$, uncorrected for polarity), determined by matching the experimental and calculated line shapes shown at the top of Fig. 22. To orient the sample, the spin-labeled nerve bundle was mounted in polyethylene tubing with excess buffer. The ESR spectra with the magnetic field along the cylindrical axis ($H_{90°}$) and normal to this axis ($H_{0°}$) are shown in the center of Fig. 22. For the calculated spectra $\gamma = 42°$, $\delta = 90°$, and σ is varied to match the experimental spectrum. Setting $\delta = 90°$ effectively "rolls up" the analogous planar distribution, so that the minimum splitting is observed at $H_{90°}$, and the orientation is essentially random when the magnetic field is perpendicular to the axis of the nerve bundle. In this case, $\sigma = 30°$ provided a relatively good

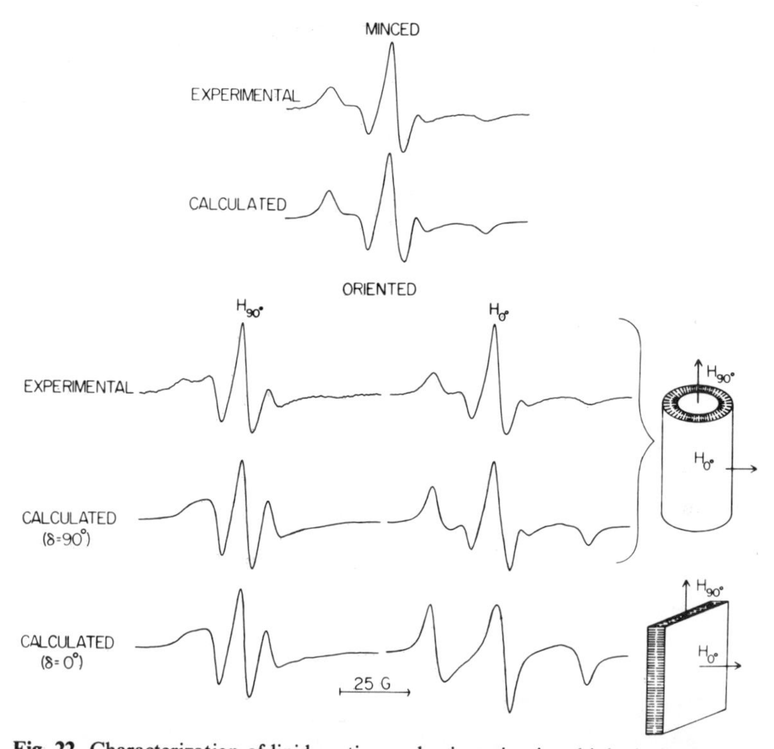

Fig. 22. Characterization of lipid motion and orientation in a biological membrane with cylindrical symmetry. The sample is a lobster nerve bundle labeled by diffusion with 5-doxylstearic acid. In the calculated spectra $\gamma = 42°$ ($k = 1.054$), $\sigma = 30°$, and the cylindrical symmetry is introduced by making δ (angle of tilt) $= 90°$. The nerve bundle can be "unrolled" into the corresponding planar sample by making $\delta = 0°$ (bottom row). By introducing different intrinsic widths for each line and supplying an orientation dependent portion of the linewidth, the calculated line shapes can be improved (Libertini *et al.*, 1974; Gaffney and McConnell, 1974) however these additional parameters were omitted here for simplicity. [Adapted from Libertini *et al.* (1974).]

visual fit. In the bottom row (Fig. 22) are the corresponding spectra calculated with $\delta = 0°$; this corresponds to "unrolling" the nerve bundle into a planar sample so that it can then be visually compared with other flat samples. The value of carrying out this last step is to obtain a better feeling for the degree of orientation in the nerve bundle as compared to bilayers in the conventional flat multilayer model systems. In this case simulations at the bottom of Fig. 22 can be compared to the spectra at the top of Fig. 14. The molecular motion in the nerve bundle ($\gamma = 42°$) is about the same as in the phospholipid model system ($\gamma = 40°$). The orientation in the nerve bundle is remarkably high ($\sigma = 30°$) considering the nature of the sample, but still less ordered than the model system ($\sigma = 20°$). Referring back to

Fig. 21, approximately 40% of the spin labels lie within a cone of half angle of 30° for $\sigma = 30°$, whereas this number increases to 70% within the same 30° cone when $\sigma = 20°$.

The degree of order in biological samples measured by the spin-labeling technique always represents a lower limit because of macroscopic disorder within the sample. This is illustrated in Fig. 23. The longitudinal section (Fig. 23b) clearly shows that the alignment of axons is imperfect, and therefore the microscopic order may be considerably greater than implied by the reported value of σ. Perhaps the most important aspect of these orientation experiments is that they establish the preferred direction of the alignment of the lipid spin labels. In the samples examined thus far, this alignment direction has always been consistent with that expected for the classical bilayer regions, and this provides strong evidence for the existence of bilayer regions in biological membranes.

C. Ordering the Order Parameters

One way of characterizing order is to specify the two parameters σ and δ as discussed above. Another way to characterize these oriented samples is to define operationally an anisotropy parameter S_{aniso} by the equation

$$S_{\text{aniso}} = \Delta A / \Delta A_{\text{max}} \tag{45}$$

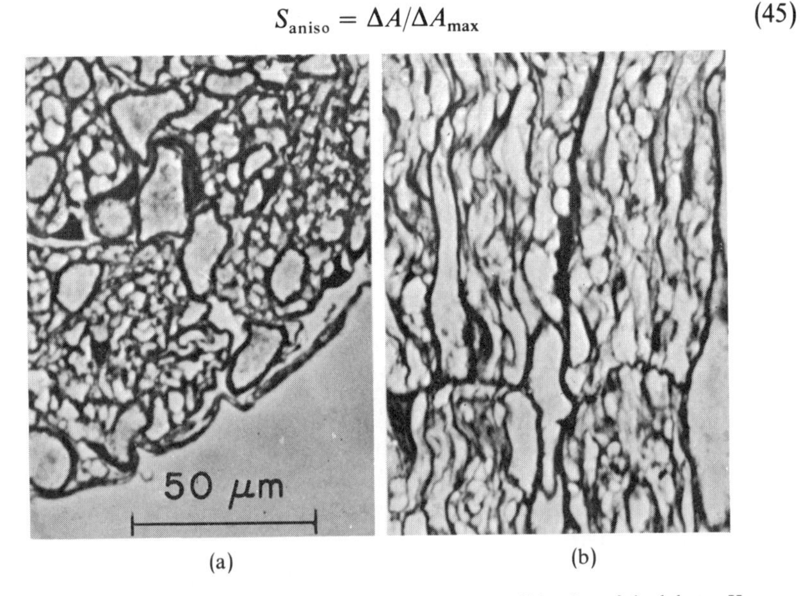

Fig. 23. Optical micrographs of nerve bundle from the walking leg of the lobster *Homarus americanus* (a) cross section, (b) longitudinal section. The nerve bundle was fixed with osmium tetroxide, dehydrated, imbedded in Epon and sectioned. The section was exposed to *p*-phenylenediamine to enhance the contrast (Estable-Puig *et al.*, 1965). [Reproduced from Libertini *et al.* (1974).]

where ΔA is the difference between the splittings observed when the magnetic field is parallel and perpendicular to the plane of the sample, and ΔA_{max} is the maximum observable anisotropy given by the familiar relation $\Delta A_{max} = A_{zz} - \frac{1}{2}(A_{xx} + A_{yy})$. Returning to our example in Fig. 14, the splittings are measured along the baseline as the distance between the low field and center field lines of the pairs of spectra in Fig. 14b. The difference in splittings for each pair of spectra is designated as ΔA. The measured values of ΔA are 12.3, 10.1, 5.4, and 2.4 G for the 5-, 7-, 12-, and 16-doxylstearic acids, respectively. A set of single-crystal parameters are chosen (in this case, the values for 2-doxylpropane) and corrected for polarity as described in Section II. The resulting ΔA_{max} values are 27.3, 27.4, 27.0, and 26.8 G for this set of samples. Therefore, S_{aniso} is 0.45, 0.37, 0.20, and 0.09, respectively, for the 5-, 7-, 12-, and 16-doxylstearic acids oriented in egg lecithin multilayers.

In this chapter we have mentioned a number of parameters—S_{motion}, S_{space}, S_{aniso}, γ, σ, and δ—used to characterize the ESR data. In Section II, the relation between the parameters γ and S_{motion} is discussed. Our purpose here is to explore the relationships between these and the remaining parameters. From the basic definition of order parameters given in Eq. (34), S_{space}, expressed in the coordinate system of Fig. 6,

$$S_{space} = \tfrac{1}{2}(3\langle \cos^2 \theta_1 \rangle - 1) \tag{46}$$

In the case of spatial order the expectation value $\langle \cos^2 \theta_1 \rangle$ is given by

$$\langle \cos^2 \theta_1 \rangle = \frac{\int P(\theta_1) \cos^2 \theta_1 \sin \theta_1 \, d\theta_1}{\int P(\theta_1) \sin \theta_1 \, d\theta_1} \tag{47}$$

where $P(\theta_1)$ is the distribution function defined by Eq. (44). Combining Eqs. (46) and (47)

$$S_{space} = \frac{3}{2} \frac{\int P(\theta_1) \cos^2 \theta_1 \sin \theta_1 \, d\theta_1}{\int P(\theta_1) \sin \theta_1 \, d\theta_1} - \frac{1}{2} \tag{48}$$

Eq. (48) relates the parameters σ and δ to the spatial order parameter S_{space}. This equation has been numerically integrated as a function of σ and δ, and a summary of the results is shown in Fig. 24. This figure can be used to calculate a value of S_{space} given one set of parameters σ and δ. It also points out several interesting features of Eq. (48). The interplay between σ and δ discussed earlier is illustrated quantitatively. For example, a spatial-order parameter of 0.5 corresponds to $\sigma = 29°$, $\delta = 0°$; or $\sigma = 25°$, $\delta = 10°$; or $\sigma = 20°$, $\delta = 20°$; or $\sigma = 13°$, $\delta = 30°$. This does not mean that the line shapes are constant over this range of σ and δ, but it does show that increasing values of σ compensated by decreasing values of δ can lead to similar splittings. Figure 24 also shows the asymptotic behavior of S_{space}. As $\sigma \to \infty$,

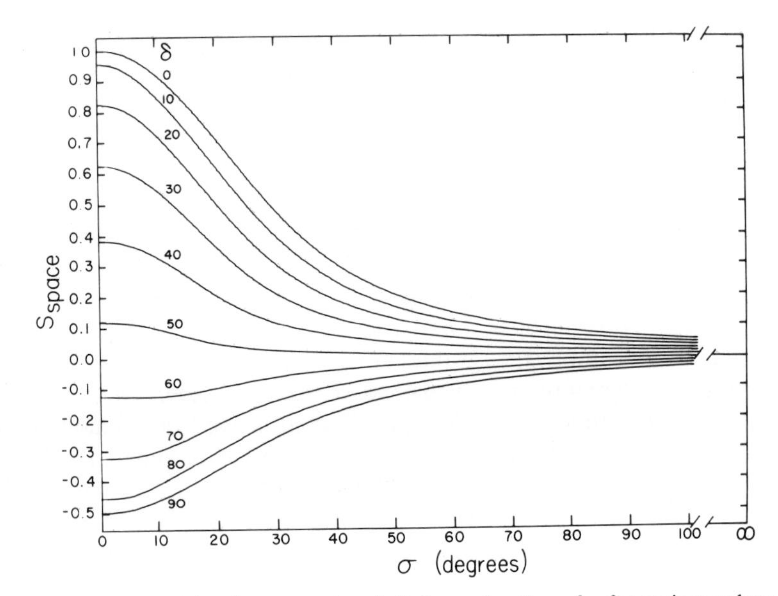

Fig. 24. The spatial order parameter plotted as a function of σ for various values of δ. $S_{\text{space}} = 0$ corresponds to a randomly oriented sample.

$S_{\text{space}} \to 0$, and the larger the value of σ, the smaller is the effect of the angle of tilt parameter δ.

The overall order in the sample is a function of all the order parameters. A useful approximate relationship is

$$S_{\text{aniso}} \simeq S_{\text{motion}} \cdot S_{\text{space}} \tag{49}$$

This equation serves to relate the order parameters measured from isotropic and ordered samples. The range of all three parameters is from 0 to 1. The two parameters S_{space} and S_{aniso} are zero for randomly oriented samples. Note that if the sample is very well ordered (small σ) and there is no angle of tilt, then $S_{\text{space}} \simeq 1$, and $S_{\text{motion}} \simeq S_{\text{aniso}}$. In other words, the motion order parameter can be estimated from the splittings of the oriented sample. This occasionally occurs in some synthetic liquid crystalline samples (see Chapter 10), but rarely in phospholipid model systems or biological membranes. Equation (49) is intended only to suggest a general relationship. It is not sufficiently rigorous to calculate, for example, S_{space}, given S_{aniso} and S_{motion}. In biological samples, the best procedure appears to be first to determine S_{motion} (or γ) from the randomly oriented samples, and then to characterize the ordered samples by σ and δ or by S_{aniso}.

IV. POLARITY PROFILE

In 1895, Overton postulated from experiments on plant cells that there must exist a lipoid barrier to diffusion between the cytoplasm of the cell and the external environment. These classical experiments of Overton and later Danielli and Davson established that the rates of diffusion of most substances across the plasma membrane are many orders of magnitude lower than would occur in the absence of the membrane and, furthermore, that the relative diffusion rates parallel the oil–water partition coefficients (with the exception of certain ions and sugars) (Davson and Danielli, 1943). The early work, combined with extensive recent studies, provides convincing evidence that fluid phospholipid bilayers are responsible for this hydrophobic barrier in biological membranes. It is of interest to measure the shape of this hydrophobic barrier in the direction normal to the bilayer surface. We outline below how spin labels can be used as probes of the local electric fields within membrane model systems and biological membranes. The shape of the hydrophobic barrier is obtained by attaching the spin label to various positions along the lipid chain and then measuring the small solvent shifts from the ESR spectra. The solvent shifts are then plotted against a function of distance across the membrane. The resulting graph is called a polarity profile, by analogy with the electron density profiles obtained from X-ray and neutron diffraction studies of biological membranes.

The existence of a small solvent dependence of the A and g parameters of nitroxides is well known (Briere *et al.*, 1965, 1966; Kawamura, 1967). In general, A_0 decreases by 1 to 2 G and g_0 increases slightly (~ 0.0005) as the nitroxide is transferred from water to hydrocarbon solvents. Solvent effects of spin labels in micelles, vesicles, and membranes have been used empirically in a number of studies (Waggoner *et al.*, 1967, 1968; Hubbell and McConnell, 1968, 1971; Griffith *et al.*, 1971; Seelig *et al.*, 1972).

The change in A_0 can be viewed as resulting from changes in the relative contributions of the two canonical structures of the neutral free radical

The unpaired electron of the nitroxide free radical is localized on the oxygen atom in structure **I** and on the nitrogen atom in structure **II**. Solvents that tend to stabilize the ionic structure **II** will increase the unpaired spin density on nitrogen and, therefore, increase the ^{14}N splitting constant A_0 or A_{zz}. This simple description ignores problems associated with changes in free-radical geometry and the formation of charge transfer complexes in certain solvents that could result in withdrawing spin density from both the nitrogen and oxygen atoms, rather than redistributing spin density as implied in structures **I** and **II**.

The isotropic splitting constant (A_0) is linearly related to the π electron spin densities on nitrogen (ρ_N) and oxygen (ρ_O) through the familiar relation

$$A_0 = Q_N \rho_N + Q_{NO} \rho_O \tag{50}$$

where Q_N and Q_{NO} are constants (Wertz and Bolton, 1972). The value of Q_N lies between 18 G and 24 G and $Q_{NO} \ll Q_N$, so that the small dependence on ρ_O may be neglected for the purposes of this discussion. Thus, Eq. (50) reduces to

$$A_0 = Q_N \rho_N \tag{51}$$

or

$$\Delta A_0 = Q_N \, \Delta\rho_N \tag{52}$$

where $\Delta\rho_N$ and ΔA_0 are the changes in ρ_N and A_0 caused by solvent perturbations or external electric fields. With the simplifying assumption that all perturbations about the spin label contribute to an average electric field **E**, $\Delta\rho_N$ increases with the component of the electric field along the N—O bond (the x axis) according to the relation

$$\Delta\rho_N = C_1 \mathbf{E}_x + C_2 \mathbf{E}_x^2 \tag{53}$$

The quadratic term can be neglected and from first-order perturbation theory, using reasonable molecular orbital parameters, $C_1 \simeq 1.5 \times 10^{-9}$ (V/cm)$^{-1}$ (Griffith *et al.*, 1974). Therefore, Eqs. (52) and (53) predict that an electric field of 10^7 V cm^{-1} could produce a change in the A_0 component of 0.4 G on a favorably oriented sample. This is only a rough estimate, but it is useful in developing a feeling for the order of magnitude of the electric field effects.

A. Solvent Effects: Van der Waals Interactions and Hydrogen Bonding

The average electric field in Eq. (53) is a vector sum $\mathbf{E} = \sum_i \mathbf{E}_i$ where the $\mathbf{E}_i$ values represent all electric fields from nearby molecules and ions. Neglecting quadratic terms, it follows from Eqs. (52) and (53) that ΔA_0, the total change in A_0, can be represented by a sum

$$\Delta A_0 = \Delta A_v + \Delta A_h + \Delta A_{dl} + \Delta A_\phi \tag{54}$$

where ΔA_v is associated with van der Waals interactions, ΔA_h is the hydrogen bonding term, ΔA_{dl} results from the electric fields of the charged double layer of the membrane and ΔA_ϕ takes into account the membrane potential. The total splitting constant (A_0) is given by

$$A_0 = A_0^{\varepsilon=1} + \Delta A_0 \tag{55}$$

where $A_0^{\varepsilon=1}$ is the splitting constant in the absence of local electric fields (i.e., unit dielectric constant in a field-free environment). In principle $A_0^{\varepsilon=1}$ could be measured from gas phase data, but this is usually impractical because of low vapor pressure of the spin labels and because of line broadening due to spin–rotation interactions. In order to estimate $A_0^{\varepsilon=1}$ and to separate the effects of ΔA_v and ΔA_h, it is convenient to examine the solvent effect in homogeneous solutions where ΔA_{dl} and ΔA_ϕ are of course zero.

A simple plot of A_0 versus g_0 does not separate ΔA_v and ΔA_h, because both parameters are affected by van der Waals and hydrogen-bonding interactions. However, a plot of A_0 against a function of the dielectric constant can provide insight into the relative magnitudes of ΔA_v and ΔA_h. A useful functional relationship is provided by the Onsager model, which is a reasonable approximation when there are no specific directional interactions between solvent and solute molecules (i.e., $\Delta A_h = 0$). In the Onsager model, the solvent is treated as a continuous medium of dielectric constant ε, and the solute molecule is represented by a sphere of radius r, with a point electric dipole of moment μ at the center (Onsager, 1936). When the polar molecule (e.g., nitroxide) is immersed in the isotropic medium, it polarizes the surrounding solvent molecules, and this polarization gives rise to a "reaction" field, E_R. For the model spin label, di-*t*-butyl nitroxide, $\mu \simeq 3$ debye, $n_0^2 \simeq 2$, and $r \simeq 4$ Å (derived from a density of 0.86), and with these numbers the general expression for E_R reduces to

$$E_R \simeq [(\varepsilon - 1)/(\varepsilon + 1)](1.8 \times 10^7) \quad \text{V cm}^{-1} \tag{56}$$

Equation (56) indicates that a plot of A_0 (or ΔA_0) versus the function $(\varepsilon - 1)/(\varepsilon + 1)$ should yield a straight line for solvents in which $\Delta A_h \ll \Delta A_v$. Di-*t*-butyl nitroxide has been examined in over thirty solvents to test this relationship (see Table II, Chapter 8). Solvents such as hexane, methyl propionate, ethyl acetate, 2-butanone, acetonitrile, and dimethylsulfoxide exhibit a linear relationship between A_0 and $(\varepsilon - 1)/(\varepsilon + 1)$. An extrapolation to $(\varepsilon - 1)/(\varepsilon + 1) \rightarrow 1$ yields an intercept of $A_0^{\varepsilon=1} = 14.85 \pm 0.05$ G (Griffith *et al.*, 1974). The slope of the curve for di-*t*-butyl nitroxide is found experimentally to be

$$\Delta A_0 = 0.8(\varepsilon - 1)/(\varepsilon + 1) \tag{57}$$

Combining Eqs. (56) and (57) yields $\Delta A_0 = (0.8\ E_R)/(1.8 \times 10^7)$ V cm^{-1}, so that an electric field of 10^7 V cm^{-1} corresponds to a ΔA_0 of about 0.4 G. This agrees with the estimate obtained independently from the molecular orbital perturbation calculation. It is worth noting that instead of plotting the A_0 versus the reaction field, it is equally valid to plot A_0 versus the dipole moment (μ^*) induced by the reaction field (Seelig *et al.*, 1972). Either way, an absolute value for E_R should not be taken too seriously, since a number of assumptions are involved and the Onsager equations contain a $1/r^3$ term, so the final number is rather sensitive to the choice of the molecular radius.

The most interesting result of this treatment is that alcohols and water are shown to have values of ΔA_0 that lie well above the line drawn from Eq. (56) and the nonhydrogen bonding solvent data. For example, di-*t*-butyl nitroxide in ethanol gives $A_0 = 16.06$ G so that $\Delta A_0 = 16.06$ G $-$ 14.85 G $= 1.21$ G. Equation (57) predicts $\Delta A_0 = 0.74$ so evidently $\Delta A_v \simeq 0.8$ G and $\Delta A_h \simeq 0.4$ G. Water is the most important example, and here $A_0 = 17.16$ G, $\Delta A_v = 0.78$, so that $\Delta A_h = 1.5$ and $\Delta A_h > \Delta A_v$. Thus, in Eq. (54) the hydrogen-bonding term can be large, and it is this fact that makes possible the study of water penetration in biological membranes using lipid spin labels.

The solvent effect of the doxyl lipid spin labels is similar to that of di-*t*-butyl nitroxide, although the splittings have not been determined with the same degree of accuracy. The 5-, 12-, and 16-doxyl fatty acids exhibit approximately the same solvent shifts. The approximate relationship between the doxyl lipid and di-*t*-butyl nitroxide (DTBN) data is $A_0(\text{doxyl}) = 0.80\, A_0(\text{DTBN}) + 2.0$, hence $\Delta A_0(\text{doxyl}) \simeq 0.80\, \Delta A_0(\text{DTBN})$. From extrapolations of homogeneous solvent data, $A_0^{\varepsilon=1} = 13.85 \pm 0.09$ G and $\Delta A_0 \simeq 0.64(\varepsilon - 1)/(\varepsilon + 1)$ for the various positional isomers of the doxyl fatty acids.

The accuracy of the present measurements does not permit the use of Eq. (56) to measure the local dielectric constant within biological membranes. Small changes in ΔA_0 correspond to large changes in dielectric constant. Furthermore, any water molecules present would lead to erroneously high estimates of the dielectric constant.

B. Electric Field Effects of the Charged Double Layer and Membrane Potential

The ionic head groups of the phospholipids form a charged double layer on either side of the hydrophobic bilayer. Once some assumptions are made regarding the geometry of the polar head groups, the term ΔA_{dl} can be estimated from Eq. (53) for any given charge distribution. The simplest model approximates each double layer by a pair of infinite plane sheets with uniform charge per unit area of $+6$ and -6, separated by a distance *a*. For this model, the electric potential ϕ at points beyond the pair of infinite plane sheets is constant; hence the electric field is zero and $\Delta A_{\text{dl}} = 0$. However, this model is an oversimplification, since the distances between phosphate groups *b* is not negligible compared to the distance from the phosphate to the N—O group within the bilayer. The spin label clearly does not see a uniform charged plane.

A better approximation is sketched in Fig. 25. In Fig. 25, the double layers are represented by hexagonal grids of positive charges and negative charges. Since the orientations of the charged groups are unknown, two limiting cases have been tested. In one extreme, the polar groups are in the fully extended conformation, and in the other limit they are bent 90° so that all charges are coplanar. The electrostatic potential ϕ at point Y caused by one double layer is the sum of potentials ϕ_{ij} due to the charged pairs. Thus $\phi = \sum \phi_{ij}$ where $\phi_{ij} = q(4\pi\varepsilon)^{-1}(1/r_1^{ij} - 1/r_2^{ij})$ and r_1 and r_2 are the distances from Y to the negative and positive groups of the ijth charged pairs (see Fig. 25). The effect of the double layers on either side of the membrane are summed and ϕ is differentiated to obtain the electric field at point Y. The N—O groups of the 5-, 12-, and 16-doxyl lipids in the all trans configuration are estimated from molecular models to be at 9, 18, and 21 Å, respectively, from the negative charge. With $a = 5$ Å, $b = 10$ Å, $d = 52$ Å, a grid size of 10^4 charged pairs, and assuming an average dielectric constant of $\varepsilon = 2$ in the bilayer, the component of the electric field normal to the bilayer plane at 9 Å is 5×10^5 V cm^{-1} in the extended double layer model and about 7×10^5 V cm^{-1} when the polar head groups are bent by 90°. In both models, the electric field is less than 5×10^{-3} V cm^{-1} at 18 Å and 21 Å. Combining Eqs. (52) and (53) for an electric field of 7×10^5 V cm^{-1}, and assuming for the moment that the N—O bonds are pointed normal to the bilayer,

$$\Delta A_{\text{dl}} = Q_{\text{N}} C_1 E_x \simeq 20 \times (1.5 \times 10^{-9}\ \text{V}^{-1}\ \text{cm})(7 \times 10^5\ \text{V cm}^{-1}) \simeq 0.02$$

The corresponding values of ΔA_{dl} for the C-12 and C-16 positions are vanishingly small.

This order of magnitude electric field calculation indicates that the ΔA_{dl} term at the C-5 position is very small and ΔA_{dl} is negligible as the N—O

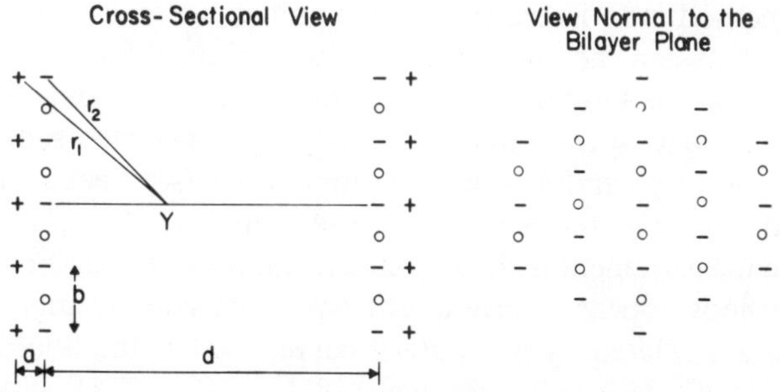

Fig. 25. The configuration of charges upon which the electric fields of the double layer are calculated. Negative signs represent phosphate groups and positive signs indicate the protonated amines in the fully extended configuration. The circles are positions of the protonated amines in the collapsed double layer limit. [From Griffith *et al.* (1974).]

group is moved to positions further into the bilayer. Introducing a net negative charge on the bilayer by displacing some of the positive charges in the above calculation (i.e., the Helmholtz approximation) increases $\Delta A_{\rm dl}$, but not by a significant amount. Furthermore, these estimates represent upper limits to $\Delta A_{\rm dl}$ because the N—O bond is assumed to be parallel to the normal to the bilayer plane, whereas experimentally it is known that a Gaussian distribution of orientations exists in which the N—O groups are more nearly perpendicular to the normal. In view of these considerations, we conclude that the $\Delta A_{\rm dl}$ term is very small outside the immediate vicinity of the double layer and is negligible at the C-5, C-12, and C-16 positions.

The last term ΔA_{ϕ} is easily estimated. For example, if an electrostatic potential of 40 mV exists across a membrane of effective thickness 40 Å, the electric field perpendicular to the bilayer plane is 40×10^{-3} V/40×10^{-8} cm $= 10^5$ V cm^{-1}. The 5-, 12-, and 16-doxyl lipids experience the same electric field, but the upper limit is only $\Delta A_{\phi} = Q_{\rm N} C_1 E_x = 20 \times (1.5 \times 10^{-9}\ \text{V}^{-1}\ \text{cm}^{-1})(10^5\ \text{V cm}^{-1}) \simeq 0.003$ G. This is of course only an order-of-magnitude estimate, but it is again an upper limit because of the orientation factor, and we conclude that the ΔA_{ϕ} term is negligible.

C. Evidence for Water Penetration into Lipid Bilayers

In principle, the polarity profile could be determined by measuring one of the four splitting constants A_0, A_{xx}, A_{yy}, A_{zz} for the N—O group at various positions along the lipid chain. In practice, none of these parameters can be measured directly because of the partial motion averaging of the electron–nuclear dipolar interaction. There are three indirect methods of obtaining the polarity profile. One method is to estimate $\bar{A}_{\parallel}$ and $\bar{A}_{\perp}$ with the spin labels in randomly oriented samples and then to obtain A_0 from Eq. (31). The second method also involves Eq. (31), but in this case the maximum and minimum splittings are estimated from oriented samples. The third method is to freeze the sample and measure $2A_{\rm max}^{-196}$ ($\simeq 2A_{zz}$) as the distance between the outermost splittings of the 9.5 GHz ESR spectra. Each method has its limitations. In the first method it is sometimes difficult to measure $\bar{A}_{\perp}$. The second method requires well-oriented samples and no angle of tilt or very careful computer simulations (see Section III,C). Both the first and second methods require the assumption of Eq. (31). In the third method, the measurement is direct, but the sample is frozen. Fortunately, all of these methods appear to give qualitatively the same results. The A_0 and $A_{\rm max}^{-196}$ data are related by a standard curve, so that the final plot can be given in terms of relative changes in either A_0 or A_{zz}. The first two methods were discussed in the flexibility profile and orientation sections of this chapter. Here we use the third method to construct a polarity profile and to obtain information regarding water penetration into biological membranes.

A standard curve of A_0 versus A_{max}^{-196} for homogeneous solvents is given in Fig. 26. The A_0 values were measured from the sharp three-line spectra of the lipid spin labels in dilute solution. The solution was then cooled to liquid nitrogen temperature for the measurement of A_{max}^{-196}. In each *homogeneous* solvent, the 5-, 12-, and 16-doxylstearic acid derivatives give very nearly the same A_0 and A_{max}^{-196} values, and these three data points are represented by one point in Fig. 26.

Plotted on the standard curve are representative A_{max}^{-196} data needed for the polarity profile. The samples are spin labels diffused into membranes of the microsomal fraction from calf liver and into the lipids extracted from this fraction (note that the corresponding A_0 values are not determined in this approach, and the circles represent the intersection of the A_{max}^{-196} values with the standard curve). The most striking feature of these data is that the three points of each preparation spread out along the standard curve, instead of superimposing as in the case of the homogeneous solvents. Figure 26 also

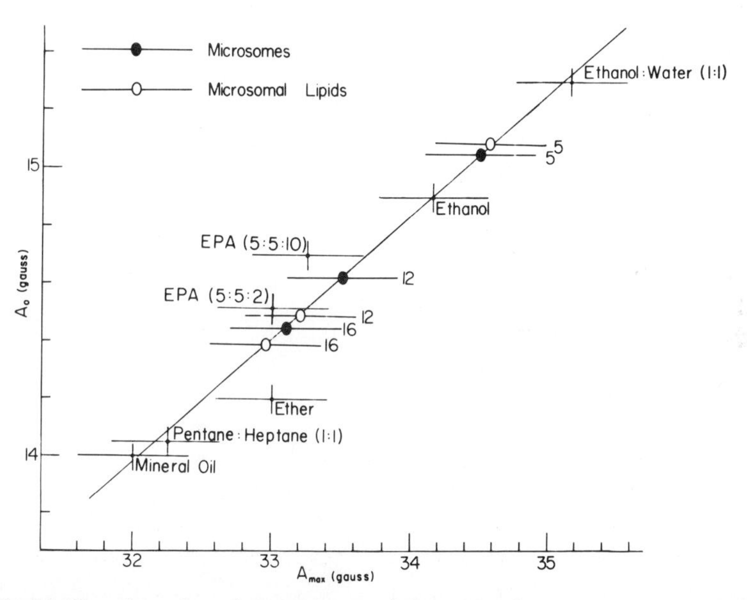

Fig. 26. The solvent dependence of the two nitrogen hyperfine parameters, A_0 and A_{max}^{-196}. Hyperfine splittings were measured for 5-, 12-, and 16-doxylstearic acids in ethanol–water (1 : 1) and ethanol. To improve the solubility in hydrocarbon solvents at low temperature, the methyl esters of 5-, 12-, 16-doxylstearic acids were used to obtain A_0 and A_{max} values in diethyl ether–2-methylbutane (1 : 1), and mineral oil. In ethanol, the fatty acid spin labels and the methyl esters yield the same A_0 and A_{max} values within experimental error. No other solvents were examined using both the fatty acids and methyl esters. A_{max} values for the solvents, microsomal lipid vesicles, and membranes of the microsomal fraction were determined at $-196°C$. [From Griffith *et al.* (1974).]

shows that the polarity profile of the microsomal lipids is essentially the same as that observed in the membranes of the microsomal fraction, thus providing additional evidence that these membranes (largely endoplasmic reticulum) contain phospholipid bilayers. The horizontal error bars are large because the lines are broad in the rigid glass ESR spectra.

The data of Fig. 26 clearly indicate that the N—O group at the C-5 position in the lipid bilayers is reporting an environment more polar than ethanol, whereas labels at the C-12 and C-16 positions are in a more hydrocarbonlike environment. These data alone do not specify the origin of the polarity effect. However, the calculations suggest that it is the ΔA_h term. This hypothesis is testable by replacing the aqueous dispersion of vesicles by lipid films supported on glass wool in a hydration chamber. The samples are then equilibrated at 100% relative humidity or dehydrated over P_2O_5. Figure 27 gives the results of one set of experiments. The upper curve shows that the polarity profile of the hydrated film is essentially the same as for the vesicles in solution. In contrast, the lower curve of the dehydrated sample is almost flat. Removal of water effectively abolishes the polarity profile, confirming the importance of the ΔA_h term. Both the fatty acid and phospholipid spin

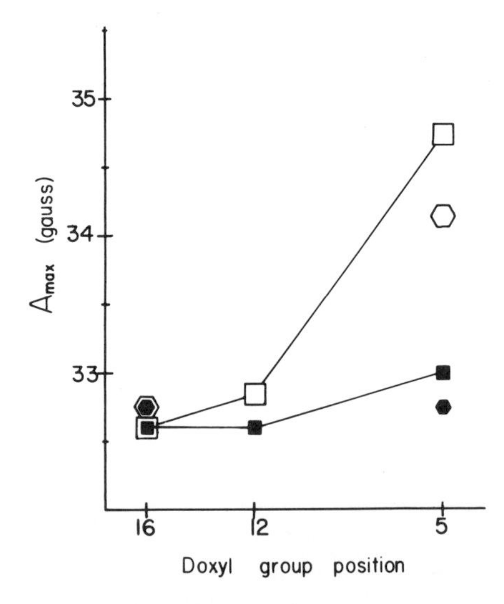

Fig. 27. Effect of dehydration on microsomal lipid vesicles as a function of the doxyl group position along the hydrocarbon chains. The open points (□, ⬡) were recorded for samples that had been equilibrated at 100% relative humidity at room temperature, and the solid points (■, ⬢) were recorded on dehydrated samples. Squares and hexagons represent data obtained using the fatty acid spin labels (**V–VI**) and the L-α-lecithin spin labels, respectively. All samples were frozen at −196°C before recording the ESR spectra. Lines connect the data points for the fatty acid spin labels. [From Griffith *et al.* (1974).]

labels show this marked effect of hydration when the N—O group is at the C-5 position near the polar head group. At the C-12 position the two points in Fig. 26 are within the experimental uncertainty of ± 0.4 G and, as expected, there is no observable hydration effect at the C-16 position.

These observations are summarized in the composite polarity profile of Fig. 28. Distances along the horizontal axes were obtained from Corey–Pauling–Koltun molecular models of the spin labels and from known dimensions of lipid bilayers and membranes. On the y axis, the point at 40 Å is simply the ΔA_0 value for the doxyl fatty acids in aqueous solution and the point at 30 Å is an indirect measurement using spin labels in a concentrated

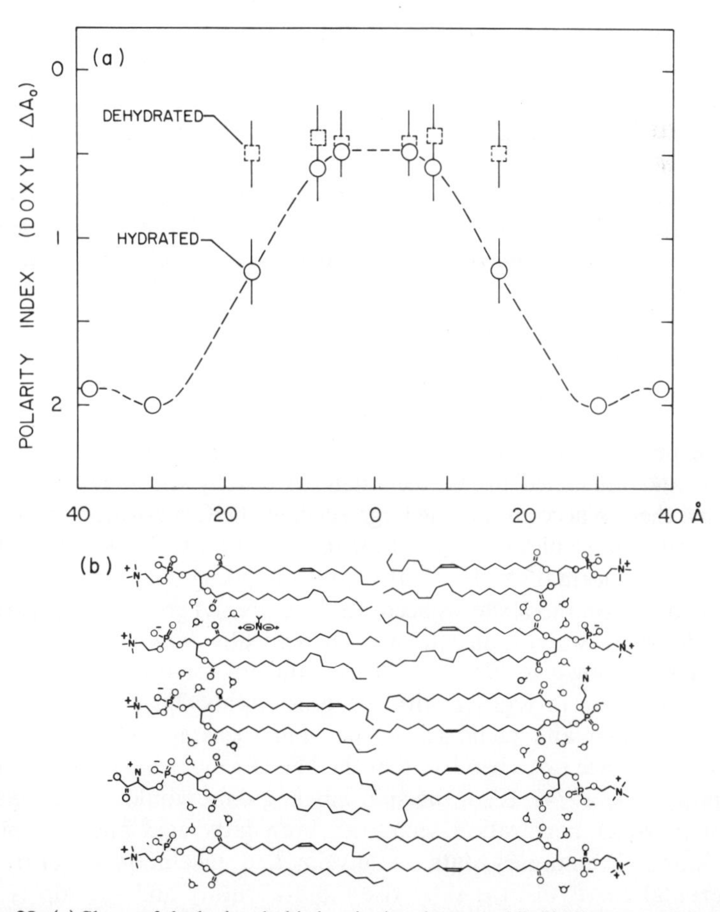

Fig. 28. (a) Shape of the hydrophobic barrier in microsomal lipid bilayers as derived from spin-labeling data. (b) Cross-sectional sketch of a lipid bilayer including a few water molecules drawn to scale with the normal O—H···O hydrogen bonding distance of 2.8 Å. [From Griffith *et al.* (1974).]

$LiCl_3$ solution to approximate the environment of the charged double layer. The important points within the bilayer are derived from the A_{max}^{-196} values by first converting to A_0 using Fig. 26, and then subtracting 13.85 G to give the final ΔA_0 plotted in Fig. 28. This polarity profile can be improved by using spin labels bound in the polar head group region, by filling in all points from C-3 to C-16, and by performing X-ray diffraction experiments on the same samples to obtain a more accurate estimate of the overall bilayer thickness. Nevertheless, the essential features are clear. The curve can be approximated by a trapezoid, and the shape is sensitive to the amount of water present.

Another method of obtaining barrier shapes involves current–voltage curves from ion transport across black lipid films (Hall *et al.*, 1973). These barrier shapes are for positively charged complexes in the bilayer, whereas the spin-labeling experiments involve the stabilization of polar structures in the bilayer, so the results are not strictly comparable. Hall *et al.* (1973) concluded that the barrier in phosphatidylethanolamine films to the transport of potassium ion by nonactin is trapezoidal. Defining d to be the width of the base and $(1 - 2n)d$ to be the width at the top of the trapezoid, the parameter n ranges from zero (rectangular shape) to 0.5 (triangular shape). On this scale, Hall *et al.* (1973) report $n \simeq 0.27$ for the phosphatidylethanolamine bilayer. The polarity profiles for the hydrated and dehydrated microsomal lipids of Fig. 28 are approximately trapezoidal, having $n \simeq 0.36$ and $n \simeq 0.19$, respectively, so the polarity profile resembles the barrier shape derived from the transport measurements.

The penetration of water molecules reduces n. If we assume that the labeled and unlabeled lipids intercalate to exactly the same depth in the bilayers, then to account for the large effect at the C-5 position, the oxygens of the water molecules must penetrate to at least the C-2 position in order to establish the normal O—H···O hydrogen bonding distance of 2.8 Å. This would correspond to a site adjacent to the carbonyl group on the hydrocarbon side. A few water molecules are schematically drawn at this position in the diagram shown in Fig. 28b. These data do not prove that these spin labels intercalate in register with the other lipids. The N—O groups are somewhat polar and could cause the lipid to extend further out at the aqueous interface or to bend so that the N—O groups are exposed to water (Griffith *et al.*, 1974). If so, the spin labeling data would overestimate the extent of water penetration seen with both fatty acid and phospholipid spin labels. Preliminary data on myelin and myelin lipids also give a trapezoidal curve, but the sides rise more abruptly and the top is higher on the scale, suggesting a more saturated hydrocarbonlike center. It is evident that the shape of the hydrophobic barrier will vary both with the composition of the polar head groups and with the lipid side chains.

V. MEASURING THE FRACTION OF MEMBRANE LIPIDS IN THE FLUID BILAYER

Not all membrane lipid is in a fluid bilayer. Membranes are heterogeneous. Some lipid is in contact with membrane proteins, and this phase exhibits a much lower degree of mobility than the fluid bilayer (see Section VI). A second source of heterogeneity can occur within the lipid phase itself, where more than one lipid is present. Depending on the temperature and lipid composition, phase separations into fluid and paracrystalline regions may occur (see Chapter 13). Our purpose in this section is to discuss two methods of estimating the fraction of the total membrane lipid that exists as fluid bilayer and to discuss briefly the limitations and advantages of these methods.

A. Determining the Fluid Bilayer by Spectral Subtraction

When two spectral components are present, the composite line shape Σ can be represented as the sum

$$\Sigma = \chi_f F + \chi_b B \tag{58}$$

where

$$\chi_f + \chi_b = 1 \tag{59}$$

and χ_f is the fraction of the fluid component, χ_b is the fraction of the bound (more immobilized) component, and F and B are the ESR line shapes of the pure fluid and bound components, respectively.

An example of a composite spectrum is shown in Fig. 29. This is the spectrum of 16-doxylstearic acid diffused into partially lipid-depleted cytochrome oxidase. This system is discussed in detail in Section VI. The details of spectral subtraction are given in Chapter 7 and in Klopfenstein *et al.* (1972). (See Figs. 14 and 15 of Chapter 7.) Here we use the spectrum of Fig. 29 as an example of the use of spectral subtraction to estimate the fraction of lipid in the fluid bilayer χ_f. Figure 29 exhibits at least two spectral components. The outermost peaks of the bound component (b) are indicated by solid arrows, and the outermost peaks of the fluid component (f) are indicated by dashed arrows. In order to perform the spectral subtraction, one component must be obtained separately or computer simulated. The sample used for Fig. 29 contains 0.24 mg phospholipid/mg protein. Cytochrome oxidase samples that are lipid-depleted to below 0.2 mg phospholipid/mg protein exhibit only the broad component. The splittings and overall appearance of the broad component are similar in the two

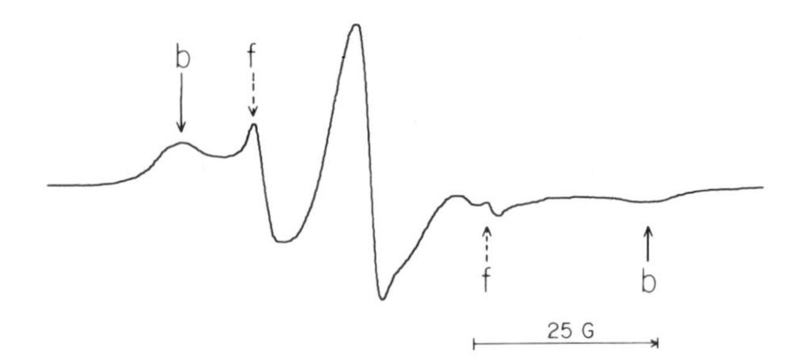

Fig. 29. ESR spectrum of 16-doxylstearic acid diffused into a buffered aqueous dispersion of cytochrome oxidase (0.24 mg phospholipid/mg protein) at 24°C. Arrows b and f identify the outermost lines of the bound and fluid components, respectively. [Adapted from Jost *et al.* (1973a).]

samples, and we use the sample of lower lipid content to represent the immobilized line shape B. The spectral subtraction is then accomplished in five steps:

1. The initial spectra, Σ (composite) and B (bound), are adjusted for slanting baseline and to set the baseline at zero.
2. Both are integrated twice to obtain the absorption of each in arbitrary units. Let I_Σ and I_b represent the relative absorption of the composite and bound spectra.
3. Spectral titration is performed by subtracting I_b from I_Σ, the endpoint being recognized by negative values of the baseline or reversal of peaks. From this spectral titration is obtained f, the fraction of the unnormalized bound spectrum required to reach the endpoint.
4. The absorption contributed by the fluid component I_f is calculated as $I_f = I_\Sigma - f(I_b)$ in arbitrary units.
5. The fraction of spin label in the fluid lipids is $\chi_f = I_f / I_\Sigma$.

As an example, for Fig. 29 and the corresponding immobilized spectrum, $I_\Sigma = 3.512 \times 10^7$ and $I_b = 5.095 \times 10^7$. From the spectral subtraction, $f = 0.65$. Therefore, $I_f = 3.51 \times 10^7 - 0.65(5.09 \times 10^7) = 0.20 \times 10^7$, and $\chi_f = (0.20 \times 10^7)/(3.51 \times 10^7) = 0.06$. Thus, in Fig. 29 the sharp peaks represent only 6 % of the total absorption. In this example the remaining 94% of the spin label is immobilized by protein complexes.

The subtraction procedure requires that the two spectra Σ and B (or F) be shifted into register (see Klopfenstein *et al.*, 1972) and assumes that there is a detectable endpoint. It is evident that F may be somewhat heterogeneous (more than one fluid environment) without invalidating the approach, providing the bound component can be reasonably well approximated.

B. Determining the Fraction of Fluid Lipid by Lipid–Water Partitioning of a Small Spin Label

Another method of determining the fraction of lipid in the fluid state is based on the partitioning of a rapidly tumbling spin label between the lipid and aqueous phases (Hubbell and McConnell, 1968; McConnell *et al.*, 1972). In both environments, the spectrum consists of three sharp lines characterized by the two parameters A_0 and g_0. Because of the solvent effects, A_0 is slightly smaller and g_0 is slightly larger in the lipid environment (see Section IV.B). The interplay of these A- and g-value shifts cause the spectra to overlap asymmetrically, so that at 9.5 GHz only the high field lines are resolved (at 35 GHz all of the lines are usually resolved) (Griffith *et al.*, 1971).

Figure 30 illustrates the spectrum of the small spin label TEMPO in water (a) and partitioning between water and lipid bilayers (b). The ratio A/B depends on the partition coefficient of the spin label between the lipid and aqueous phases, the relative amounts of lipid and water, and the line-widths in the two environments. The lipid–water partition coefficient of TEMPO increases with increased fluidity of the lipid phase. TEMPO tends to be excluded from paracrystalline regions of lipid bilayers.

The first step in this procedure (McConnell *et al.*, 1972) is to measure the TEMPO parameter $A/B \times 10^2$ for a series of lipid samples known to be in a fluid state (above the transition temperature). An example is shown in Fig. 31.

To remove the dependence in concentration, these measurements are converted to the "specific TEMPO parameter"

$$\rho = ((A/B) \times 10^2)/[\text{Fluid lipid}] \tag{60}$$

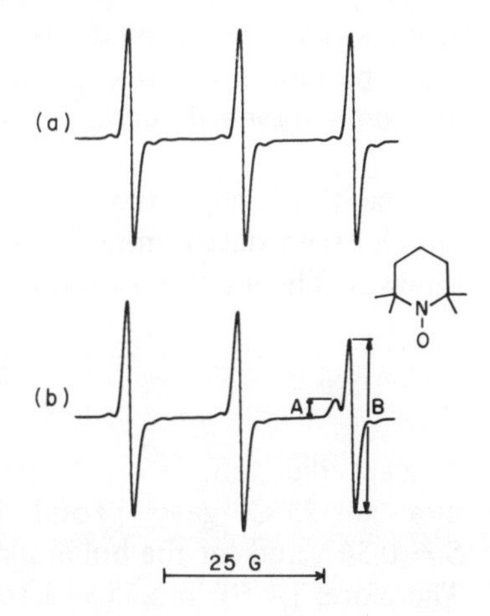

Fig. 30. ESR spectrum of the spin label TEMPO (5×10^{-5} *M*) in (a) water and (b) egg lecithin (phosphatidylcholine) in water (25 mg/ml) at 23°C. The small spin label TEMPO is shown at the right.

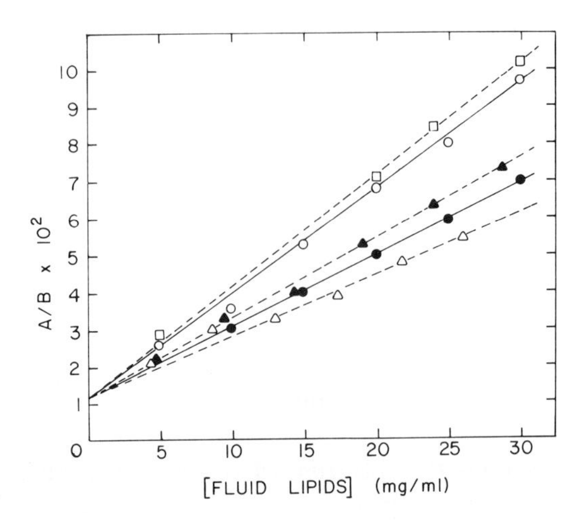

Fig. 31. Plots of the TEMPO parameter versus the concentration of total lipid for various lipid mixtures in the fluid state at 25°C. The solid lines correspond to mixtures of cholesterol and egg lecithin (w/w) of 20% (●) and 16.4% (○) cholesterol. The dashed lines are for mixtures of cholesterol and lipids extracted from rabbit sarcoplasmic reticulum (w/w), 20.9% (△), 10.5% (▲), and 3.0% (□) cholesterol. The order parameters (from 12-doxylstearic acid) are, from top to bottom: 0.347, 0.358, 0.385, 0.408, and 0.448. [Adapted from McConnell *et al.* (1972).]

where [Fluid lipid] is fluid lipid concentration (in this case the total lipid) expressed as milligrams of lipid per milliliter, and $(A/B)' = (A/B) - (A/B)_0$, $(A/B)_0$ being the ratio of the two peak heights in buffer and would be zero except for the small ^{13}C satellites bracketing the three main ESR lines (see Fig. 30).

The second step is to obtain a measure of the mobility in the lipid bilayers. This is done by diffusing in 12-doxylstearic acid and measuring the order parameter by one of the methods discussed in Section II.B.2. To provide a standard curve, ρ is plotted against the order parameter S as in Fig. 32.

The third step involves measuring $(A/B)'$ and S for the membrane sample, then determining ρ for the sample from the standard curve of ρ versus S. The fraction of fluid lipid is

$$\chi_f = \frac{[\text{Fluid lipid}]}{[\text{Total lipid}]} = \frac{(A/B)' \times 10^2}{\rho[\text{Total lipid}]} \tag{61}$$

As a specific example, one preparation of membranes of sarcoplasmic reticulum at 25°C gave [Total lipid] = 15.74 mg/ml, $(A/B) \times 10^2 = 5.35$, $S = 0.347$, and for the buffer alone, $(A/B)_0 = 1.16$ (McConnell *et al.*, 1972). Therefore, $(A/B)' = 5.35 - 1.16 = 4.19$. From Fig. 32 (for $S = 0.347$) ρ is

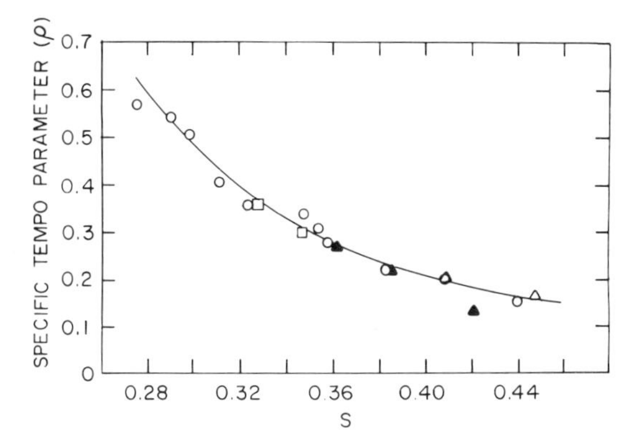

Fig. 32. Plot of the specific TEMPO parameter ρ versus the order parameter S (for 12-doxylstearic acid) in mixtures of cholesterol with egg lecithin (○) and lipids extracted from sarcoplasmic reticulum (△, ▲, and □) at 25°C. [From McConnell *et al.* (1972).]

0.310. The fraction of fluid lipid is, from Eq. (61), $\chi_f = (4.19)/(0.310) \times (15.74) = 0.86$. Most of the lipid in this membrane sample at 25°C, therefore, is in the fluid bilayer state.

C. Advantages and Limitations

Each method for determining the fraction of fluid lipid bilayer has its advantages and limitations. The principal limitations of the spectral titration approach is that it requires digitizing the spectra and necessitates isolation or simulation of one component of the multicomponent system. The main advantages of this method are that it is conceptually simple, does not require concentration determinations with respect to the aqueous volume (recall that the lipid spin label partitions only between the fluid and nonfluid regions of the membrane), and it is not critically dependent on temperature. The advantage of the TEMPO method is that no computer is required. The limitations lie in the sensitivity of this method to errors in determination of concentration, careful regulation of the temperature, and construction of the standard curve.

The two techniques for measuring the fraction of membrane lipid that is in fluid bilayers are complementary. Spectral subtraction is most reliable in the lower ranges of fluid bilayer where the endpoints are clearly defined. The TEMPO method is best suited for samples with high fractions of fluid bilayer, where the spectrum is dominated by the fluid bilayer component, so that the measurement of the order parameter is well defined (i.e., not ambiguous due to overlapping of spectra from both the bound and fluid components).

VI. LIPID-PROTEIN INTERACTIONS IN MEMBRANES

A. Peripheral Membrane Proteins

The peripheral proteins form a major class of membrane proteins. These proteins are removed from the membrane by mild aqueous treatments and are evidently bound electrostatically to the surfaces of the membrane. A number of peripheral proteins has been reported, but the most extensively studied protein of this class is cytochrome *c*. Cytochrome *c* is a globular basic protein of the mitochondrial electron transport chain. It has a remarkably asymmetric charge distribution and a net positive charge of +8 at neutral pH (Dickerson, 1972). Cytochrome *c* binds electrostatically to lipid bilayers containing acidic and neutral phospholipids. The amount of protein bound to bilayers increases with the amount of negatively charged phospholipid (e.g., cardiolipin or phosphatidylinositol) in the lipid mixture. Classically, the cytochrome *c*-lipid complex was detected when the red color of the protein was found associated with the lipid pellet on centrifugation (Green and Fleischer, 1963). These complexes have been examined by X-ray diffraction and optical spectroscopy (Blaurock, 1973; Gulick-Krzywicki *et al.*, 1969; Steineman and Läuger, 1971). Covalent spin labeling of the protein has been used to probe the orientation of cytochrome *c* on lipid bilayers (Vanderkooi *et al.*, 1973).

Lipid spin labels can be used in this cytochrome *c*-phospholipid system to assess any effects of the adsorbed peripheral protein on the bilayer structure. Among the criteria that can be examined are the fluidity gradient, the spectral anisotropy of oriented samples, the response to controlled hydra-

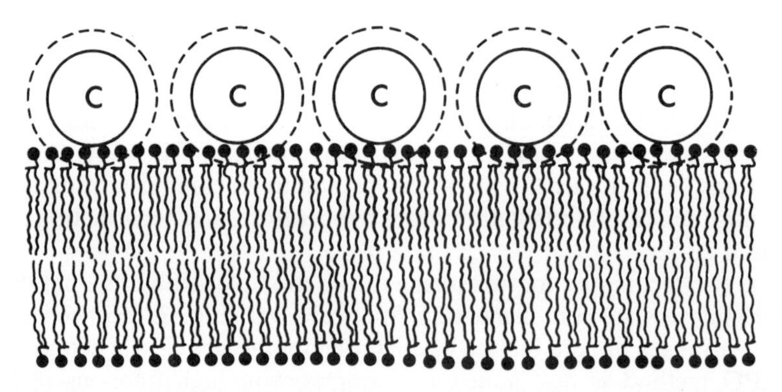

Fig. 33. Diagrammatic sketch of the association of the globular cytochrome *c* proteins with the phospholipid bilayer. This is a cross-sectional view with aqueous phases above and below the plane of the bilayer. The dashed lines are included to suggest the interaction of ionic groups of the protein with ionic groups of the phospholipids. [Reproduced from Van and Griffith (1975).]

tion of disordered samples, and the polarity profile. The experiments are performed by incorporating the spin label into lipid bilayers and examining these properties in the lipid mixture alone and in the cytochrome *c*–phospholipid complex. The essential result of these kinds of experiments (using 5-, 12-, and 16-doxylstearic acids and 3-doxyl-5α-cholestane) is that cytochrome *c* has only small effects on the properties of the bilayer (Van and Griffith, 1974) (see also Chapter 11). These spin-labeling results are consistent with the sketch of Fig. 33, where cytochrome *c* is depicted as sitting on the bilayer without affecting the major features of the lipid bilayer. These lipid spin labels may be insensitive to small changes in bilayer structure associated with the polar head groups. This general picture does not preclude disturbances of the bilayer very near the polar head groups or the possibility that cytochrome *c* induces lateral phase separation.

B. Integral Membrane Proteins

The integral membrane proteins form a second major class of membrane proteins. These proteins require detergent treatment for isolation and frequently cannot be entirely freed of phospholipid without irreversible inactivation. In order to study lipid–protein interactions of these integral membrane proteins, it is desirable to reduce the heterogeneity as much as possible by isolating the protein complex and studying its lipid-binding properties. One such system is cytochrome oxidase isolated from the beef heart inner mitochondrial membrane. Cytochrome oxidase (cytochrome aa_3) is the terminal member of the mitochondrial electron transport chain (overall reaction $4H^+ + O_2 + 4e^- \rightarrow 2H_2O$). As it is usually isolated, the protein complex and its associated phospholipid spontaneously form closed membranous vesicles (Chuang *et al.*, 1970). These cytochrome oxidase membranes are not patches of the intact membrane; rather, they are composed of just one of the major protein complexes together with some of the lipids from the inner mitochondrial membrane. Cytochrome oxidase is classed as an integral protein complex, since the use of detergents is necessary in its isolation and it is not water soluble in its monomeric form. The complex is only partially characterized, but when it is prepared by the method of Sun *et al.* (1968), it consists of approximately six different polypeptide chains of molecular weight ranging from 10,000 to 36,000 (Capaldi and Hayashi, 1972; Kuboyama *et al.*, 1972). Cytochrome oxidase membranes form a convenient model system with an enzymatic assay (oxidation of cytochrome *c*). An additional useful feature of this system is that the phospholipid–protein ratio can be varied by lipid depletion without irreversible inactivation.

Lipid spin labels have been diffused into aqueous dispersions of cytochrome oxidase, varying the lipid content of this model system. Representative data utilizing the 16-doxylstearic acid are shown in the left column of

Fig. 34 (Jost *et al.*, 1973a). Several observations can be made from these spectra. First, the top spectrum (a) is characteristic of a highly immobilized spin label. This spectrum is observed whenever the lipid content is below ~2 mg phospholipid/mg protein. At the other extreme, when no protein is present [spectrum (e)] the sharper three-line spectrum characteristic of this spin label in fluid bilayers is observed. At intermediate levels, the spectra (b, c, d) are composite.

In order to verify that these composite spectra represent sums of the *same* two spectral components and to determine the ratios of the two components

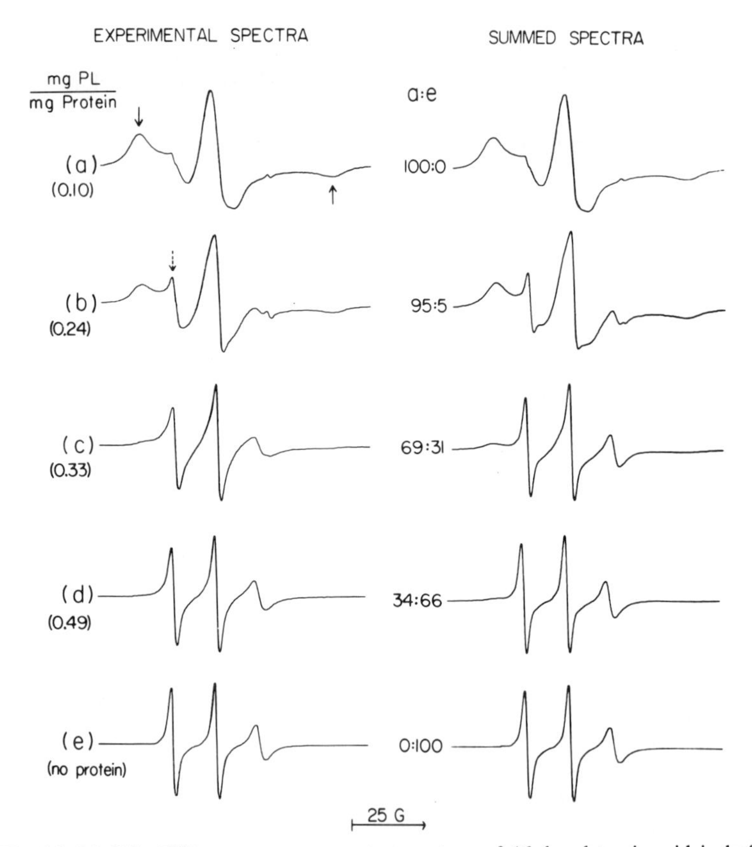

Fig. 34. 9.5 GHz ESR spectra at room temperature of 16-doxylstearic acid in buffered aqueous suspensions of cytochrome oxidase (left) compared with synthesized spectra obtained by summing spectra (a) and (e). The experimental samples were obtained by lipid depletion of the membranous cytochrome oxidase vesicles (d). Spectrum (e) is of lipids isolated from sample (d). On the right, the ratios (a : e) are the fraction of the total absorption contributed by spectrum (a) and (e). Note the general similarity of the synthesized spectra to the corresponding experimental spectra (b, c, d). [Adapted from Jost *et al.* (1973a).]

in the final line shape, the spectra can be synthesized by summing the two putative spectral components—bound lipid (a) and fluid bilayer (e). The results of this procedure are shown in the right column of Fig. 34. For example, spectrum (b) was reasonably well approximated by summing 95 parts of spectrum (a) and 5 parts of spectrum (e). Spectra (c) and (d) can be approximated by the same procedure, varying the proportions used. As a result of this approach, we conclude that the fluid bilayer in spectra (a)–(e) is 0, 5, 31, 66, and 100% of the total lipid present. The remainder of the lipid is immobilized. A slightly different and somewhat more precise method, based on spectral subtraction of spectrum (a) from each of the composite spectra, gives 6, 35, and 57% of the lipid as the fluid bilayer components of spectra (b), (c), and (d), respectively. Subtraction avoids the problem of whether or not there are several related mobilities in the fluid bilayer. Since the bound component is not observed in the absence of protein, it is reasonable to ascribe the immobilization to interaction of the lipid with the protein complex.

These same data can provide an estimate of the actual amount of phospholipid immobilized by the protein, providing it is assumed that the distribution of the lipid spin label faithfully reflects the distribution of the phospholipid between the two environments, i.e., that the binding constant of the spin label acyl chain is the same as that of a phospholipid acyl chain. With this assumption, the amount of bound lipid C_b in units of milligrams phospholipid per milligram of protein is simply $C_b = X_b C_t$, where C_t is the total lipid (in phospholipid per milligram of protein ratio as determined by phosphate and protein analysis) and X_b is the fraction of immobilized component. For example, for spectrum (b), $C_t = 0.24$ mg phospholipid/mg protein, $X_b = 0.95$, and $C_b = (0.95)(0.24) = 0.23$ mg phospholipid/mg protein. Similarly, for spectra (b) and (c) (see Fig. 33), the bound component represents 0.23 and 0.17 mg phospholipid/mg protein. Using spectral subtraction, these amounts are 0.22, 0.21, and 0.21 mg phospholipid/mg protein for (b), (c), and (d), respectively (Griffith *et al.*, 1973). A number of other samples has also been analyzed, and the remarkable observation is that there is about 0.2 mg bound phospholipid/mg protein, regardless of the amount of fluid lipid present.

Given that the amount of immobilized phospholipid is ~ 0.2 mg phospholipid/mg protein, it is interesting to try to picture this component in molecular terms. The simplest model is to visualize this lipid as coating the hydrophobic surfaces of the protein complex. An estimation of the amount of phospholipid required to surround cytochrome oxidase is summarized in Fig. 35, where dimensions of the protein are taken from Vanderkooi *et al.* (1972) based on electron microscopy. The result is that about 0.2 mg phospholipid/mg protein would be needed to coat the hydrophobic surface, which is the same amount as that calculated from the spin-labeling data. Of course, irregularities in the perimeter, protein–protein contacts or association of some phospholipid molecules by only one acyl chain would alter this number somewhat.

In view of the assumptions used in the treatment of the spin-labeling data, it is important to correlate these data with other measurements. Thus far, four other pieces of relevant data are available on cytochrome oxidase. First, electron micrographs of preparations at various lipid levels indicate that closed vesicles form only at higher lipid levels (> 0.3 mg phospholipid/mg protein), suggesting that there is a minimum extent of fluid bilayer necessary for vesicle formation (Chuang *et al.*, 1970; Jost *et al.*, 1973c). Second, successive lipid extractions (cold aqueous acetone) readily reduce lipid–protein ratios to nearly 0.2, and the remaining lipid is more difficult to remove (Griffith *et al.*, 1973). The third piece of data involves enzyme activity. Enzyme activity, measured by oxidation of cytochrome *c*, is essentially absent when lipid levels are very low. Addition of phospholipid to these inactive lipid-depleted complexes restores activity, with maximal activity reached at about 0.2 mg phospholipid/mg protein. The fourth item involves parallel spin-labeling data using a steroid spin label 3-doxyl-5α-androstan-17β-ol (**III**, Fig. 2). This rigid spin label executes an oscillatory motion instead of a wobbling motion, so the line shapes are very different (see Fig. 10), providing a good test for the subtraction method, as well as

Assumptions

MW protein complex = 210,000
MW phospholipid = 775
Diameter of one aliphatic chain = 4.8 Å (X-ray)
Shape of complex, approx. rectangular, 52 × 60 Å (em)

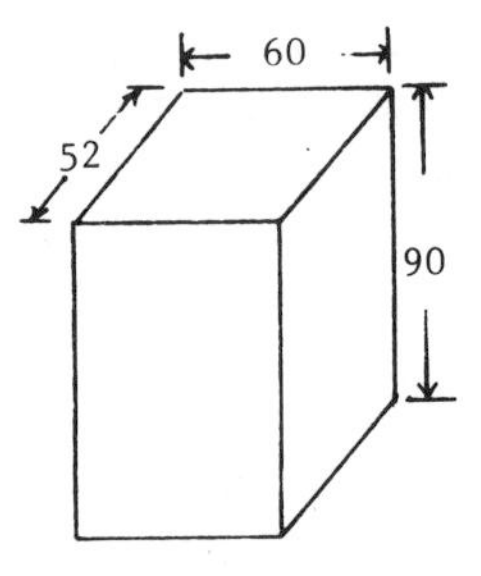

therefore

Number of phospholipid molecules per protein complex

$$= \frac{\text{protein perimeter}}{\text{diameter of aliphatic chain}} \times \frac{2}{2 \text{ chains/pl}}$$

$$= \frac{2(52 + 60 \text{ Å})}{4.8 \text{ Å}} = 47 \text{ molecules phospholipid/protein complex}$$

Mg pl/mg protein in boundary layer = 47(775/210,000) ~ 0.2 mg pl/mg protein

Fig. 35. The assumptions and calculations used for estimating the number of lipid molecules that can be accommodated in a single lipid layer around the cytochrome oxidase complex. In this estimate, the protein is approximated by a rectangular solid shape (the assumption of a cylindrical shape does not significantly alter the result). Note that this calculation assumes that both lipid chains in a single phospholipid molecule associate with the protein surface, that the chains are tightly packed, and that there are no protein–protein contacts; hence, it represents only a crude estimate of the amount of boundary lipid that can be accommodated.

introducing a different polar head group. The result is essentially the same; that is, there is about 0.2 mg of immobilized phospholipid/mg protein (Jost *et al.*, 1973b).

To summarize these experiments on membranous cytochrome oxidase, the spin-labeling data indicate, without any doubt, that lipid in contact with integral protein is immobilized. The data also indicate that the number of lipid-binding sites is independent of the extent of the fluid bilayer regions. A third point is that only two major environments are observed rather than a continuous change in one lipid environment. This argues that the protein complex has a well-defined three-dimensional structure and does not have polypeptide chains extending randomly out into the bilayer. Consistent with this view is that even with very limited fluid bilayer regions present, the fluid regions are not greatly perturbed by the presence of the protein and exhibit the classical behavior of extended fluid bilayers in pure lipid systems.

These experiments, as analyzed, give little information regarding binding constants, so that immobilization should not be equated with tight binding. Assuming the lipid spin label behaves like the other lipids in the system, the amount of lipid immobilized is 0.2 mg phospholipid/mg protein, which can be accounted for in terms of a boundary layer surrounding the protein, as shown in Fig. 36. This model, while not unique, is the simplest model consistent with all of the data. Speculating as to the biological significance, this boundary layer may act as a solvation shell and prevent indiscriminate protein aggregation (i.e., precipitation) in the plane of the bilayer.

Independent evidence for a boundary layer or a halo surrounding intrinsic proteins has been developed using another interesting spin-labeling technique (Stier and Sackman, 1973). This technique utilizes measurement of the rate of decay of the ESR signal as a function of temperature. In this experiment, the spin label 5-doxylstearic acid was diffused into membranes of the microsomal fraction isolated from rabbit liver. The center peak height was then monitored as a function of time for each of several temperatures.

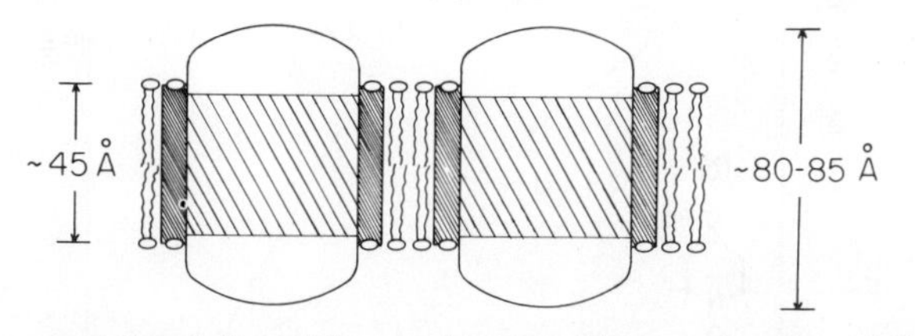

Fig. 36. A cross-sectional view of the cytochrome oxidase membranes as deduced from the available data. The membrane is composed of limited bilayer regions into which the protein complex is intercalated with hydrophobic regions (lightly cross-hatched) surrounded by an immobilized boundary layer of lipid (densely cross-hatched). The boundary layer is evidently not ordered, and the side chains have been deliberately obscured to indicate this. [Reprinted from Jost *et al.* (1973c).]

The rate of signal loss is greatly increased by the addition of NADPH, and the authors attribute this enzymatic reduction to the cytochrome P_{450}–cytochrome P_{450} reductase hydroxylating enzyme system. The initial reaction proceeds at a constant rate with zero-order kinetics, and the essential results are shown in the upper curve of the Arrhenius plot of Fig. 37. The rate of reduction of the lipid spin label below 32°C proceeds slowly, whereas above 32°C, a marked increase in the rate of signal destruction is seen. Three controls were used. First, the signal destruction of the watersoluble spin label TEMPO phosphate (4-hydroxy-2,2,6,6-tetramethylpiperidine-1-oxyl dihydrogen phosphate) (see Fig. 35) was studied over the same temperature range. There is no break in the Arrhenius plot, and this provides evidence that the break in the upper curve is not due to changes in enzyme activity at 32°C. As a second control, the order parameter of 5-doxylstearic acid was measured over the same temperature range and no abrupt changes (which might indicate a possible phase transition) were observed. The third control involved the measurement of lateral diffusion (see Chapter 13) of another lipid spin label, 16-doxylstearic acid, at several temperatures. The results confirmed that the rates of diffusion in the bulk phase were characteristic of

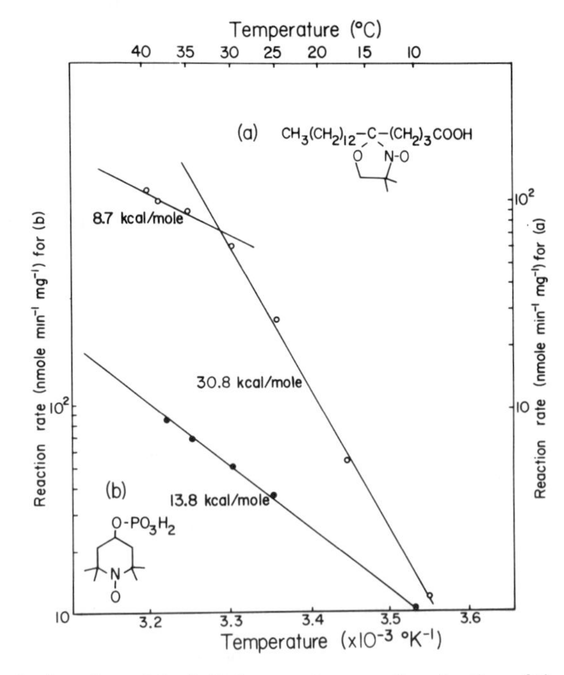

Fig. 37. Arrhenius plots of the initial rates of enzymatic reduction of the two spin labels shown: (a) 5-doxylstearic acid (open circles, upper curve); and (b) TEMPO phosphate (closed circles, lower curve). [Modified from Stier and Sackmann (1973).]

fluid lipid bilayers over the range of 20°–40°C. The most logical model derived from these data is that the reducing protein is surrounded by a halo of phospholipid that is in quasicrystalline state below 32°C and undergoes a phase transition to a fluid state at 32°C, allowing the lipid spin label to reach the protein and consequently to be reduced. This halo presumably includes immobilized lipid, as in the case of cytochrome oxidase, but also probably includes a more extended lipid domain. There are also fluorescence measurements that indicate a certain amount of lipid in a bacterial membrane is not in the fluid bilayer state (Träuble and Overath, 1973).

Membrane proteins differ in the relative proportions exposed to the aqueous phase and buried in the lipid bilayer. An interesting example of a membrane protein that is fairly well characterized is cytochrome b_5, isolated from liver endoplasmic reticulum. Cytochrome b_5 contains one heme group and consists of a single polypeptide chain. When isolated with detergents, the intact cytochrome b_5 can be freed of phospholipid and has a molecular weight of 16,700 (Spatz and Strittmatter, 1971). When the membrane is treated with proteolytic enzymes, the heme-bearing portion of cytochrome b_5 is recovered, and this hydrophilic fragment has now been characterized by X-ray crystallography (Mathews *et al.*, 1971).

Cytochrome b_5 is a good system for studying the lipid-binding properties of the hydrophilic and hydrophobic portions of an amphipathic membrane protein. Figure 38 compares the ESR spectra of fatty acid spin labels in aqueous dispersions of the intact (detergent-isolated) cytochrome b_5, the trypsin-released hydrophilic fragment, and as a control, the lipids isolated

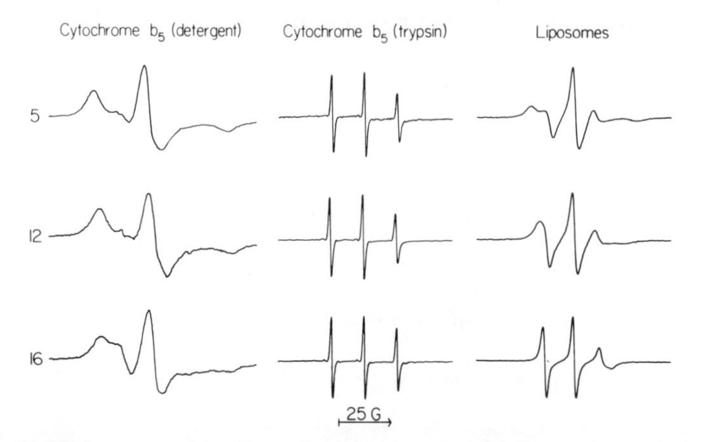

Fig. 38. ESR spectra of 5-, 12-, and 16-doxylstearic acids in solution with detergent-extracted cytochrome b_5, trypsin-extracted cytochrome b_5, and aqueous dispersions of liver microsomal lipids at room temperature. [Reproduced from Dehlinger *et al.* (1974).]

from the microsomal fraction (predominantly endoplasmic reticulum) (Dehlinger *et al.*, 1974). Note that the 5-, 12-, and 16-doxylstearic acids are all immobilized by the intact cytochrome b_5, whereas no immobilization is seen with the trypsin-released fragment. The spectra of the fatty acid spin labels in the liposomes simply illustrates the classical behavior of these lipid spin labels in fluid bilayers (cf. Fig. 14). No comparable flexibility profile is seen when these spin labels interact with the protein. Intact cytochrome b_5, as might be expected, also immobilizes the corresponding phospholipid spin labels.

The polarity of the spin label environment in this system can be determined by the method outlined in Section IV. With intact cytochrome b_5, the values of A_{max}^{-196} for 5-, 12-, and 16-doxylstearic acid are very nearly the same and represent a polarity between ethanol and ethanol–water, 1 : 1 (see Fig. 26). In other words, the usual polarity gradient present in the phospholipid bilayers is not observed when the spin labels are in contact with the protein. The lipid-binding regions on the hydrophobic segment of the protein are more polar than the interior of the bilayer (this effect probably is due to hydrogen bonding between the polypeptide and the N—O moiety).

These spin-labeling data and other biochemical experiments (Strittmatter *et al.*, 1972; Rogers and Strittmatter, 1973; Enomoto and Sato, 1973) are consistent with the model illustrated in Fig. 39. The sites of immobilized lipid are confined to one region of the protein, the peptide segment not present in the trypsin-released fragment. The hydrophobic tail (cross-hatched) is responsible for anchoring the molecule in the bilayer. The lipid in contact with this hydrophobic tail is strongly immobilized (but perhaps not strongly bound, since the binding constant is unknown). These data are

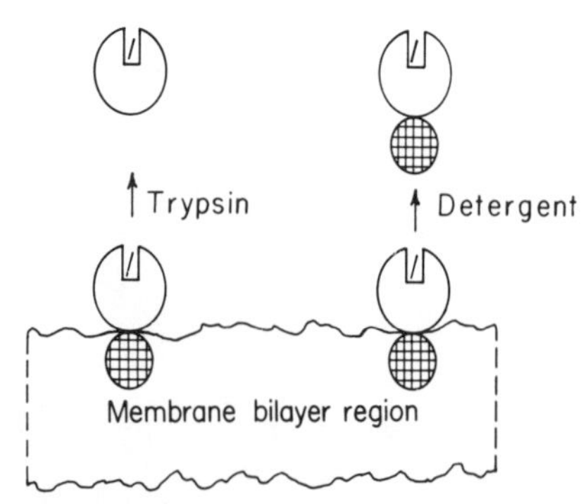

Fig. 39. A pictorial representation of the relationships between detergent-extracted cytochrome b_5 and trypsin-extracted cytochrome b_5, and their relationship to the membrane. [Reproduced from Dehlinger *et al.* (1974).]

consistent with those on the cytochrome oxidase complex from the mitochondrial membrane.

In summary, lipid spin labels in contact with the hydrophobic surfaces of integral membrane proteins exhibit very different spectral properties from the same spin labels in lipid bilayers. First, in contact with protein the nitroxide groups of all the lipid spin labels tested are immobilized, regardless of their position along the hydrocarbon chain, whereas in the bilayer the usual flexibility profile is observed. Second, there is some evidence (although sketchy) that the lipid in contact with protein surface is disordered, in contrast to the order observed in the bilayer (Jost *et al.*, 1973c). Third, the polarity of the environment on the protein surface is independent of the nitroxide position, while the lipid spin labels report a definite polarity profile in the bilayer. Finally, dehydration has very little effect on the spectra of the protein-bound spin labels but has a marked effect on motion, orientation, and polarity in the lipid bilayer.

The selected experiments discussed in this chapter were chosen only to illustrate some possible approaches, and this chapter is not intended to be a review of the field of spin labeling of membranes. Spin labeling is a relatively new technique, and although there is a rapidly developing literature, it is clear that the field is in an early stage of development. It is hoped that the approaches outlined here, combined with those of other chapters, will help others planning to use spin labeling in their studies of biological systems.

ACKNOWLEDGMENTS

This investigation was supported by Public Health Service Research grants CA10337 and COA grant I KO4 CA23359 from the National Cancer Institute. We gratefully acknowledge the technical assistance of Miss Dee Brightman and informal criticism and discussions with various members of the "Eugene Membrane Group." We are especially indebted to Dr. Charles A. Burke for his valuable discussions on the problem of order in lipid bilayers and for the computer generation of Fig. 24.

REFERENCES

Birdsall, N. J. M., Lee, A. G., Levine, Y. K., and Metcalfe, J. C. (1971). Fluorine-19 NMR of monofluorostearic acids in lecithin vesicles, *Biochim. Biophys. Acta* **241**, 693–696.

Birrell, G. B., and Griffith, O. H. (1974). Determination of angle of tilt in phospholipid multilayers, *Fed. Proc.* **33**, 1342.

Birrell, G. B., Van, S. P., and Griffith, O. H. (1973a). Electron spin resonance of spin labels in organic inclusion crystals. Models for anisotropic motion in biological membranes, *J. Amer. Chem. Soc.* **95**, 2451–2458.

Birrell, G. B., Griffith, O. H., and French, D. (1973b). Observation of *z*-axis anisotropic motion of a nitroxide spin label, *J. Amer. Chem. Soc.* **95**, 8171–8172.

Blaurock, A. E. (1973). Structure of a lipid–cytochrome *c* membrane, *Biophys. J.* **13**, 290–298.

Briere, R., Lemaire, H., and Rassat, A. (1965). Nitroxydes. XV. Synthèse et étude de radicaux libres stables piperidiniques et pyrrolidinique, *Bull. Soc. Chim. Fr.* **32**, 3273–3283.

Briere, R., Rassat, A., Rey, P., and Tchoubar, B. (1966). Nitroxydes. XXI. Mise en evidence par resonance paramagnetique electronique de l'effet specifique des sels sur la solvation des radicaux nitroxydes, *J. Chim. Phys.* **63**, 1575–1576.

Brown, G. H., Doane, J. W., and Neff, V. D. (1971). "A Review of the Structure and Physical Properties of Liquid Crystals." CRC Press, Cleveland, Ohio.

Capaldi, R. A., and Hayashi, H. (1972). The polypeptide composition of cytochrome oxidase from beef heart mitochondria, *FEBS Lett.* **26**, 261–263.

Chan, S. I., Seiter, C. H. A., and Feigenson, G. W. (1972). Anisotropic and restricted molecular motion in lecithin bilayers, *Biochim. Biophys. Res. Commun.* **46**, 1488–1492.

Chuang, T. F., Awasthi, Y. C., and Crane, F. L. (1970). Model mosaic membrane: cytochrome oxidase, *Proc. Indiana Acad. Sci. 1969* **79**, 110–120.

Corvaja, C., Giacometti, G., Kopple, K. D., and Zianddin. (1970). Electron spin resonance studies of nitroxide radicals and biradicals in nematic solvents. *J. Amer. Chem. Soc.* **92**, 3919–3924.

Davson, H., and Danielli, J. F. (1943). "The Permeability of Natural Membranes." Cambridge Univ. Press, London and New York.

Dehlinger, P. J., Jost, P. C., and Griffith, O. H. (1974). Lipid binding to the amphipathic membrane protein cytochrome b_5, *Proc. Nat. Acad. Sci. U.S.* **71**, 2280–2284.

Dickerson, R. (1972). The structure and history of an ancient protein, *Sci. Amer.* **226**, April, 58–72.

Enomoto, K., and Sato, R. (1973). Incorporation *in vitro* of purified cytochrome b_5 into liver microsomal membranes, *Biochem. Biophys. Res. Commun.* **51**, 1–7.

Estable-Puig, J. F., Bauer, W. C., and Blumberg, J. M. (1965). Paraphenylenediamine staining of osmium-fixed plastic-embedded tissue for light phase microscopy, *J. Neuropathol. Exp. Neurol.* **24**, 531–535.

Ferruti, P., Gill, D., Harpold, M. A., and Klein, M. P. (1969). ESR of spin-labeled nematogen-like probes dissolved in nematic liquid crystals, *J. Chem. Phys.* **50**, 4545–4550.

Gaffney, B. J., and McConnell, H. M. (1974). The paramagnetic resonance of spin labels in phospholipid membranes. *J. Magn. Res.* **16**, 1–28.

Godici, P. E., and Landsburger, F. R. (1974). The dynamic structure of lipid membranes. A ^{13}C nuclear magnetic resonance study using spin labels, *Biochemistry* **13**, 362–368.

Green, D. E., and Fleischer, S. (1963). The role of lipids in mitochondrial electron transfer and oxidative phosphorylation, *Biochim. Biophys. Acta* **70**, 554–582.

Griffith, O. H. (1964). ESR and molecular motion of the RCH_2CHOOR' radicals in X-irradiated ester-urea inclusion compounds, *J. Chem. Phys.* **41**, 1093–1105.

Griffith, O. H., Libertini, L. J., and Birrell, G. B. (1971). The role of lipid spin labels in membrane biophysics, *J. Phys. Chem.* **75**, 3417–3425.

Griffith, O. H., Jost, P. C., Capaldi, R. A., and Vanderkooi, G. (1973). Boundary lipid and fluid bilayer regions in cytochrome oxidase model membranes, *Ann. N.Y. Acad. Sci.* **222**, 561–573.

Griffith, O. H., Dehlinger, P. J., Van, S. P. (1974). Shape of the hydrophobic barrier of phospholipid bilayers. Evidence for water penetration in biological membrane, *J. Membrane Biol.* **15**, 159–192.

Gulik-Krzywicki, T., Schecter, E., Luzzati, V., and Faure, M. (1969). Interactions of proteins and lipids: Structure and polymorphism of protein–lipid–water phases, *Nature* (*London*) **223**, 1116–1121.

Hall, J. E., Meade, C. A., and Szabo, G. (1973). A barrier model for current flow in lipid bilayer membranes, *J. Membrane Biol.* **11**, 75–97.

Horwitz, A. F., Horsley, W. J., and Klein, M. P. (1972). Magnetic resonance studies on membrane and model membrane systems: Proton magnetic relaxation rates in sonicated lecithin dispersions, *Proc. Nat. Acad. Sci. U.S.* **69**, 590–593.

Hsia, J. C., Schneider, H., and Smith, I. C. P. (1970). Spin label studies of oriented phospholipids: Egg lecithin, *Biochim. Biophys. Acta* **202**, 399–402.

Hubbell, W. L., and McConnell, H. M. (1968). Spin-label studies of the excitable membranes of nerve and muscle, *Proc. Nat. Acad. Sci. U.S.* **61**, 12–16.

Hubbell, W. L., and McConnell, H. M. (1969a). Motion of steroid spin labels in membranes, *Proc. Nat. Acad. Sci. U.S.* **63**, 16–22.

Hubbell, W. L., and McConnell, H. M. (1969b). Orientation and motion of amphiphilic spin labels in membranes, *Proc. Nat. Acad. Sci. U.S.* **64**, 20–27.

Hubbell, W. L., and McConnell, H. M. (1971). Molecular motion in spin-labeled phospholipids and membranes, *J. Amer. Chem. Soc.* **93**, 314–326.

Israelachvili, J., Sjösten, J., Göran Erickson, L. E., Ehrström, M., Gräslund, A., and Ehrenberg, A. (1974). Theoretical analysis of the molecular motion of spin labels in membranes. ESR spectra of labeled *Bacillus subtilis* membranes, *Biochim. Biophys. Acta* **339**, 164–172.

Jost, P. C., and Griffith, O. H. (1973). The molecular reorganization of lipid bilayers by osmium tetroxide. A spin-label study of orientation and restricted *y*-axis anisotropic motion in model membrane systems, *Arch. Biochem. Biophys.* **159**, 70–81.

Jost, P., Libertini, L. J., Hebert, V. C., and Griffith, O. H. (1971). Lipid spin labels in lecithin multilayers. A study of motion along fatty acid chains, *J. Mol. Biol.* **59**, 77–98.

Jost, P. C., Griffith, O. H., Capaldi, R. A., and Vanderkooi, G. (1973a). Evidence for boundary lipid in membranes, *Proc. Nat. Acad. Sci. U.S.* **70**, 480–484.

Jost, P. C., Capaldi, R. A., Vanderkooi, G., and Griffith, O. H. (1973b). Lipid-protein and lipid–lipid interactions in cytochrome oxidase model membranes, *J. Supramol. Struct.* **1**, 269–280.

Jost, P. C., Griffith, O. H., Capaldi, R. A., and Vanderkooi, G. (1973c). Identification and extent of fluid bilayer regions in membranous cytochrome oxidase, *Biochim. Biophys. Acta* **311**, 141–152.

Jost, P., Brooks, U. J., and Griffith, O. H. (1973d). Fluidity of phospholipid bilayers and membranes after exposure to osmium tetroxide and glutaraldehyde, *J. Mol. Biol.* **76**, 313–318.

Kawamura, T., Matsunami, S., and Yonezawa, T. (1967). Solvent effects on the *g*-value of di-*t*-butyl nitric oxide, *Bull. Chem. Soc. Japan* **40**, 1111–1115.

Keana, J. F. W., Keana, S. B., and Beetham, D. (1967). A new versatile ketone spin label, *J. Amer. Chem. Soc.* **89**, 3055–3056.

Keith, A. D., Waggoner, A. S., and Griffith, O. H. (1968). Spin-labeled mitochondrial lipids in *Neurospora crassa*, *Proc. Nat. Acad. Sci. U.S.* **61**, 819–826.

Klopfenstein, C., Jost, P., and Griffith, O. H. (1972). The dedicated computer in electron spin resonance spectroscopy, *In* "Computers in Chemicals and Biochemical Research" (C. Klopfenstein, and C. Wilkins, eds.), pp. 175–221. Academic Press, New York.

Kohler, S. J., Horwitz, A. F., and Klein, M. P. (1972). Magnetic resonance studies on membranes and model membrane systems. IV. Comparison of yeast and egg lecithin dispersions, *Biochem. Biophys. Res. Commun.* **49**, 1414–1421.

Kornberg, R. D., and McConnell, H. M. (1971). Inside–outside transitions of phospholipids in vesicle membranes, *Biochemistry* **10**, 1111–1120.

Kuboyama, M., Yong, F. C., and King, S. E. (1972). Studies on cytochrome oxidase. VIII. Preparation and some properties of cardiac cytochrome oxidase, *J. Biol. Chem.* **247**, 6375–6383.

Lee, A. G., Birdsall, N. J. M., Levine, Y. K., and Metcalfe, J. C. (1972). High resolution proton relaxation studies of lecithins, *Biochim. Biophys. Acta* **255**, 43–56.

Libertini, L. J., and Griffith, O. H. (1970). Orientation dependence of the electron spin resonance spectrum of di-*t*-butyl nitroxide, *J. Chem. Phys.* **53**, 1359–1367.

Libertini, L. J., Waggoner, A. S., Jost, P. C., and Griffith, O. H. (1969). Orientation of lipid spin labels in lecithin multilayers, *Proc. Nat. Acad. Sci. U.S.* **64**, 13–19.

Libertini, L. J., Burke, C. A., Jost, P. C., and Griffith, O. H. (1974). An orientation distribution model for interpreting ESR line shapes of ordered spin labels, *J. Magn. Res.* **15**, 460–476.

Mathews, F. S., Argos, P., and Levine, M. (1971). The structure of cytochrome b_5 at 2.0 Å resolution, *Cold Spring Harbor Symp. Quant. Biol.* **36**, 387–395.

McConnell, H. M., Wright, K. L., and McFarland, B. G. (1972). The fraction of the lipid in a biological membrane that is in a fluid state: A spin label assay, *Biochem. Biophys. Res. Commun.* **47**, 273–281.

McFarland, B. G., and McConnell, H. M. (1971). Bent fatty acid chains in lecithin bilayers, *Proc. Nat. Acad. Sci. U.S.* **68**, 1274–1278.

Metcalfe, J. C., Birdsall, N. J. M., Feeney, J., Lee, A. G., Levine, Y. K., and Partington, P. (1971). Carbon-13 NMR spectra of lecithin vesicles and erythrocyte membranes, *Nature* (*London*) **233**, 199–210.

Onsager, L. (1936). Electric moments of molecules in liquids, *J. Amer. Chem. Soc.* **58**, 1486–1493.

Rogers, M. J., and Strittmatter, P. (1973). Lipid–protein interactions in the reconstitution of the microsomal reduced nicotinamide adenine dinucleotide–cytochrome b_5 reductase system, *J. Biol. Chem.* **248**, 800–806.

Rottem, S., Hubbell, W. L., Hayflick, L., and McConnell, H. M. (1970). Motion of fatty acid spin labels in the plasma membrane of *Mycoplasma*, *Biochim. Biophys. Acta* **219**, 104–113.

Rozantsev, E. G. (1970). "Free Nitroxyl Radicals." Plenum Press, New York.

Saupe, A. (1964). Kernresonanzen in kristallinen Flüssigkeiten und kristallinflüssigen Lösungen, Teil I, *Z. Naturforsch.* **19A**, 161–171.

Seelig, J. (1970). Spin label studies of oriented liquid crystals (a model system for bilayer membranes), *J. Amer. Chem. Soc.* **92**, 3881–3887.

Seelig, J., and Niederberger, W. (1974). Two pictures of a lipid bilayer. A comparison between deuterium label and spin-label experiments, *Biochemistry* **13**, 1585–1588.

Seelig, J., Limacher, H., and Bader, P. (1972). Molecular architecture of liquid crystalline bilayers. *J. Amer. Chem. Soc.* **94**, 6364–6371.

Simpkins, H., Jay, S., and Panko, E. (1971). Changes in the molecular structure of axonal and red blood cell membranes following treatment with phospholipase A_2, *Biochemistry* **10**, 3579–3585.

Singer, S. J. (1971). The molecular organization of biological membranes, *In* "Structure and Function of Biological Membranes" (L. I. Rothfield, ed.), pp. 145–222. Academic Press, New York.

Spatz, L., and Strittmatter, P. (1971). A form of cytochrome b_5 that contains an additional hydrophobic sequence of 40 amino acid residues, *Proc. Nat. Acad. Sci. U.S.* **68**, 1042–1046.

Steineman, A., and Läuger, P. (1971). Interaction of cytochrome *c* with phospholipid monolayers and bilayer membranes, *J. Membrane Biol.* **4**, 74–85.

Stier, A., and Sackmann, E. (1973). Spin labels and enzyme substrates. Heterogeneous lipid distribution in liver microsomal membranes, *Biochim. Biophys. Acta* **311**, 400–408.

Strittmatter, P., Rogers, M. J., and Spatz, L. (1972). The binding of cytochrome b_5 to liver microsomes, *J. Biol. Chem.* **247**, 7188–7194.

Sun, F. F., Prezbindowski, K. S., Crane, F. L., and Jacobs, E. E. (1968). Physical state of cytochrome oxidase relationship between membrane formation and ionic strength, *Biochim. Biophys. Acta* **153**, 804–818.

Träuble, H., and Overath, P. (1973). The structure of *Escherichia coli* membranes studied by fluorescence measurements of lipid phase transitions, *Biochim. Biophys. Acta* **307**, 491–512.

Van, S. P., and Griffith, O. H. (1975.) Bilayer structure in phospholipid–cytochrome *c* model membranes, *J. Membrane Biol.* **20**, 155–170.

Van, S. P., Birrell, G. B., and Griffith, O. H. (1974). Rapid anisotropic motion of spin labels. Models for motion averaging of the ESR parameters. *J. Magn. Res.* **15**, 444–459.

Vanderkooi, G., Senior, A. E., Capaldi, R. A., and Hayachi, H. (1972). Biological membrane structure. III. The lattice structure of membranous cytochrome oxidase, *Biochim. Biophys. Acta* **274**, 38–48.

Vanderkooi, J., Erecinska, M., and Chance, B. (1973). Cytochrome *c* interactions with membranes. II. Comparative study of the interaction of *c* cytochromes with the mitochondrial membrane, *Arch. Biochem. Biophys.* **157**, 531–540.

Waggoner, A. S., Griffith, O. H., and Christensen, C. R. (1967). Magnetic resonance of nitroxide probes in micelle-containing solutions, *Proc. Nat. Acad. Sci. U.S.* **57**, 1198–1205.

Waggoner, A. S., Keith, A. D., and Griffith, O. H. (1968). Electron spin resonance of solubilized long-chain nitroxides, *J. Phys. Chem.* **72**, 4129–4132.

Waggoner, A. S., Kingzett, T. J., Rottschaeffer, S., and Griffith, O. H. (1969). A spin-labeled lipid for probing biological membranes, *Chem. Phys. Lipids* **3**, 245–253.

Wertz, J. E., and Bolton, J. R. (1972). "Electron Spin Resonance: Elementary Theory and Practical Applications" (W. P. Orr and A. Stryker-Rodda, eds.), p. 125. McGraw-Hill, New York.

13

Molecular Motion in Biological Membranes

HARDEN M. McCONNELL

DEPARTMENT OF CHEMISTRY, STANFORD UNIVERSITY
STANFORD, CALIFORNIA

I. INTRODUCTION

In this closing chapter, we shall discuss a number of applications of the spin label technique to the study and solution of a few selected problems in membrane biophysics. The problems to be discussed have been limited to those that we feel are of rather general interest, that is, of interest to the physicist, the physical chemist, and the molecular biologist. The topic of membrane biophysics was chosen because subtle dynamical aspects of membrane molecular structure are likely to play significant roles in biological function. Magnetic resonance methods can be used to study some of the dynamical properties of biological membranes and model membranes (phospholipid bilayers).

It must be emphasized that this chapter is *not* a review of the spin label technique or of many applications of spin labels to membranes, nor does it claim in any way to deal with the most significant work. It has been limited in scope by the immediate interests of its author, by the availability of figures from earlier publications, and by the knowledge that many other theoretical and experimental aspects of the technique are dealt with comprehensively by other authors elsewhere in this book. For a guide to much original and interesting work by Russian scientists in this field, the reader is referred to the recent monograph by Likhtenstein (1974).

To begin our discussion, let us consider thermal phase transitions and the "fluidity" of phospholipid bilayers.

II. "FLUID" MEMBRANES

Figure 1 shows the paramagnetic resonance spectrum of the simple nitroxide spin label Tempo (2,2,6,6-tetramethylpiperidine-1-oxyl) (**I**) in an

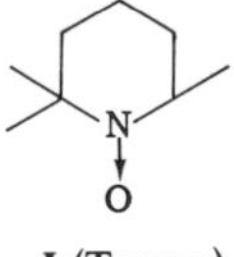

I (Tempo)

aqueous solution. Except for superhyperfine structure due to isotopes of low abundance, the resonance spectrum consists of three lines, corresponding to the three components of the nitrogen nuclear spin quantum number, 1, 0, −1 (low-field line, center line, high-field line).

Figure 2 shows paramagnetic resonance spectra of Tempo in an aqueous solution containing a dispersion of the phospholipid, dielaidoylphosphatidylcholine at three temperatures. The high-field line clearly has two components, *H* and *P*. The signal *P* arises from Tempo dissolved in water and the signal *H* arises from Tempo dissolved in a fluid hydrophobic environment that is provided by the hydrocarbon interior of the phospholipid

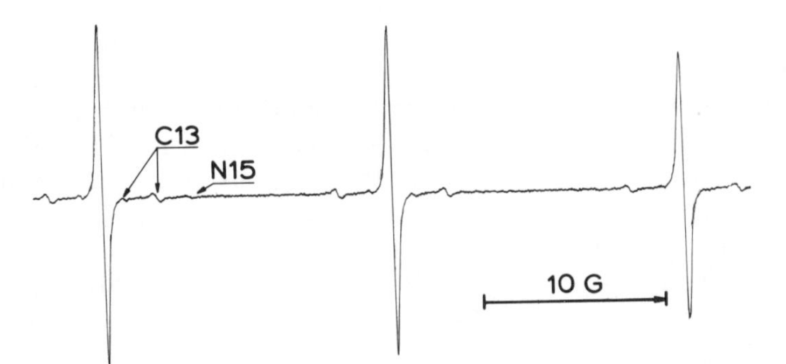

Fig. 1. The paramagnetic resonance spectrum of the nitroxide spin label Tempo (2,2,6,6-tetramethylpiperidine-1-oxyl) in aqueous solution.

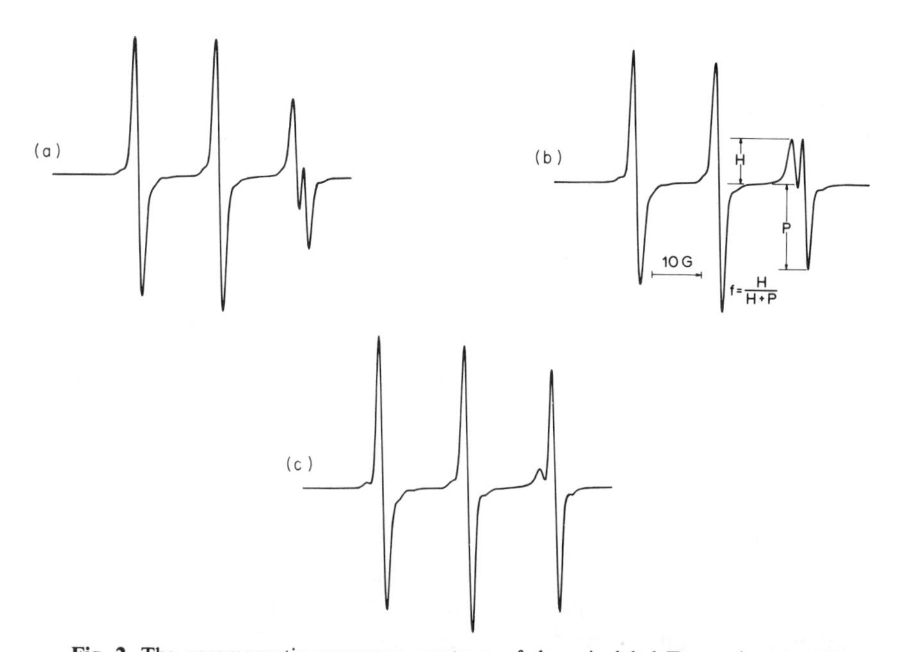

Fig. 2. The paramagnetic resonance spectrum of the spin label Tempo in an aqueous dispersion of the phospholipid, dielaidoylphosphatidylcholine at three temperatures; (a) 25.2°C is above the hydrocarbon chain melting transition temperature, (b) 12.6°C is nearly equal to the transition temperature, and (c) 9.8°C is below the transition temperature. The parameter $f = H(H + P)$ is an approximate measure of the fraction of Tempo dissolved in the hydrophobic region of the bilayer, since signal H arises from Tempo in this region and signal P arises from Tempo in the polar aqueous environment. [Reprinted with permission from Wu and McConnell *Biochemistry* **14**, 847–854 (1974). Copyright by the American Chemical Society.]

bilayer. Signals that overlap with signal H can be observed when Tempo is dissolved in a fluid hydrocarbon such as decane. Each component of the hyperfine triplet in Fig. 2 is a doublet, but due to small, compensating effects of solvent polarity on the g factor and hyperfine splitting, only the high-field resonance signals are resolved by a spectrometer operating at 9.5 GHz and 3300 G. [At 35 GHz and 10,000 G both the high-field signal and the center signal are clearly resolved into H and P doublets (Gaffney and McNamee, 1974).] If we let H and P represent the amplitudes (peak heights) of signals such as H and P in Fig. 2, then the plots of $f = H/(H + P)$ versus temperature shown in Fig. 3 exhibit the well-known endothermic phase transitions in a number of phospholipids. The largest increase in solubility of Tempo in these bilayer membranes arises from the phase transition in which there is a "melting" of the hydrocarbon chains.

Paramagnetic resonance spectra such as those shown in Fig. 2 were observed originally in aqueous solutions containing a number of biological

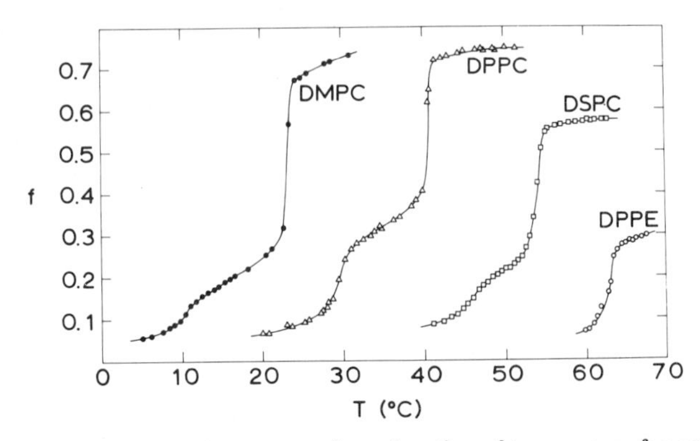

Fig. 3. The Tempo spectral parameter *f* as a function of temperature for aqueous dispersions of dimyristoylphosphatidylcholine (●), dipalmitoylphosphatidylcholine (△), distearoylphosphatidylcholine (□), and dipalmitoylphosphatidylethanolamine (○). [Reprinted with permission from Shimshick and McConnell *Biochemistry* **12**, 2351–2360 (1973). Copyright by the American Chemical Society.]

membranes; from such spectra it was concluded that at least some biological membranes contain hydrophobic regions of low viscosity, due to the phospholipids (Hubbell and McConnell, 1968).

III. LATERAL PHASE SEPARATIONS

As discussed later, phospholipid molecules in the "fluid" state (above the hydrocarbon chain melting temperature) undergo rapid lateral diffusion. In contrast, the rate of transverse "flip-flop" of phospholipids from one side of a membrane to the other is much slower. Kornberg and McConnell (1971a) measured the rate of transmembrane flip-flop of a head-group spin label in egg lecithin phosphatidylcholine vesicles, and found it to be of the order of 6.5 hr^{-1} at 30°C. The rate of flip-flop of a head-group phospholipid spin label in excitable membrane vesicles prepared from the electroplax of *Electrophorus electricus* was found to be much higher (~ 4.7 min^{-1} at 15°C), but still many orders of magnitude larger than the step rate of lateral diffusion (McNamee and McConnell, 1973). In experiments with biological membranes it is always possible that denatured protein can serve as a catalyst for flip-flop. The rate of fusion and fission of phospholipid vesicles is also normally quite slow, as discussed later. Thus, in considering the effects of temperature on mixtures of phospholipids, one is led to the idea that these systems should exhibit *lateral phase separations within the plane of the membrane*, as illustrated schematically in Fig. 4.

Studies of the binding of Tempo to a number of binary mixtures of phospholipids do in fact show that the binding data can be interpreted in terms of lateral phase separations of the phospholipids. For example, the data in Fig. 5 show the binding of Tempo to binary mixtures of two phosphatidylcholines, dielaidoyl- and dipalmitoylphosphatidylcholine. As discussed in the figure legends, these data can be accounted for in terms of the phase diagram shown in Fig. 6. According to this phase diagram, any binary mixture of the two phospholipids at a temperature above the fluidus curve is in a totally homogeneous fluid state—that is, the two components are homogeneously mixed with one another, and all the molecules undergo rapid lateral diffusion in the plane of the bilayer membrane. At temperatures

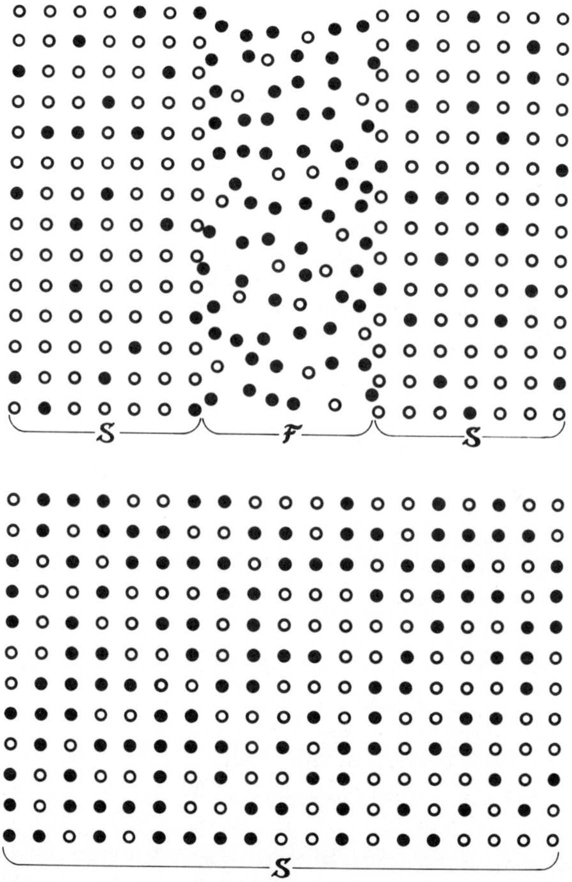

Fig. 4. Schematic illustration of lateral phase separations in the plane of a lipid bilayer membrane, the membrane consisting of two hypothetical lipids, black lipids and white lipids. The bilayer is seen from above; in one case a solid domain (S) is rich in the higher melting white lipid and is in equilibrium with the fluid domain (F), which is rich in the lower melting black lipid. The *cubic* arrangement of lipids in the crystalline solid solution phase is only schematic. [From Shimshick *et al.* (1973).]

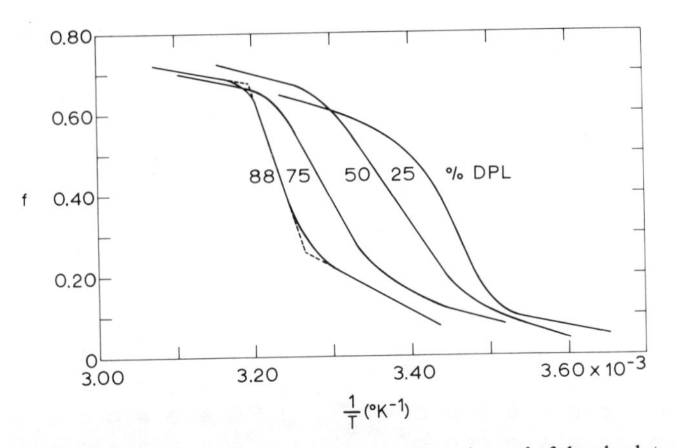

Fig. 5. Plots of the Tempo spectral parameter f vs. the reciprocal of the absolute temperature for mixtures of dielaidoyl- and dipalmitoylphosphatidylcholine in the presence of excess water. Curves for mixtures of composition 25, 50, 75, and 88% dipalmitoylphosphatidylcholine are shown. Each curve exhibits two major changes in slope (break points), which correspond to the beginning and end of the lateral phase separation. Dashed lines on the 88% curve show how the temperatures are defined by extensions of the experimental curves. Data points were taken at less than 1°C intervals and point scatter was no greater than the thickness of the lines used to draw the curves. [From Grant *et al.* (1974).]

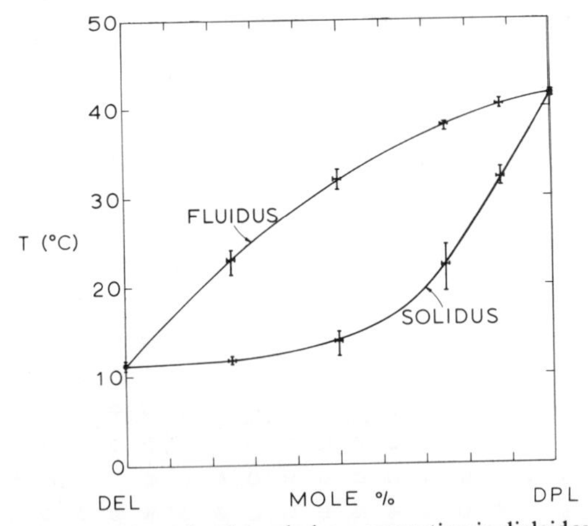

Fig. 6. Phase diagram representing lateral phase separation in dielaidoylphosphatidylcholine (DEL)–dipalmitoylphosphatidylcholine (DPL) mixtures as derived from the paramagnetic resonance data in Fig. 5. The x axis gives the mole percent of DPL in the binary mixture. Conditions corresponding to points above the fluidus line represent totally fluid lipid mixtures and points below the solidus line represent totally solid lipid. For points between these two curves, fluid and solid coexist, the domain compositions being predicted by intersection points with the fluidus and solidus lines, respectively, of a horizontal line through the point. [From Grant *et al.* (1974).]

below the solidus curve, the molecules are homogeneously mixed (solid solution case) but their rate of lateral motion is much less. At temperatures and compositions between the solidus and fluidus curves in Fig. 6 one infers the coexistence of two domains, one composed of lipids in "solid" solution and the other composed of lipids in "fluid," or liquid solution. It has been shown by Grant *et al.* (1974) that the domain structure predicted by the phase diagram in Fig. 6 can be visualized by rapid quenching and freeze-fracture electron microscopy. (See Fig. 7.)

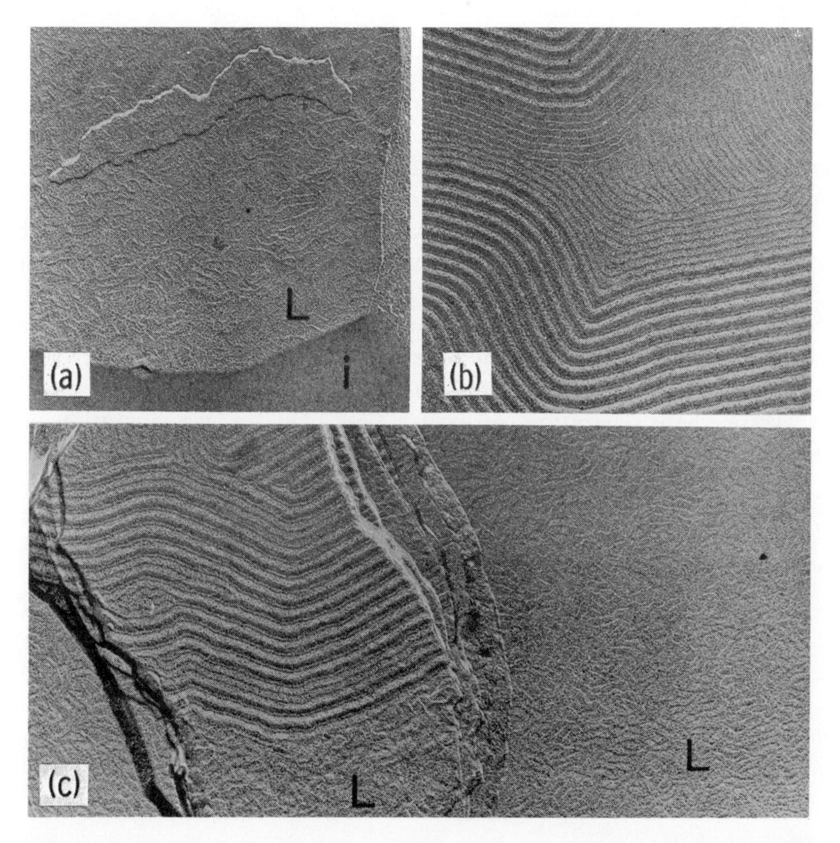

Fig. 7. Electron microphotographs of platinum–carbon replicas of liposome fracture faces for 50–50 mole % mixture of dielaidoyl- and dipalmitoylphosphatidylcholine quenched from various temperatures. Direction of shadow is from bottom to top of page. Magnification is × 69,000. (a) Liposome quenched from 36°, showing "fluid" fracture face; i, ice; L, lipid. (b) Typical liposome fracture face appearance when quenched from 10–11°C. Note the highly ordered pattern characteristic of the "solid"-phase lipid, and the fact that no "fluid" lipid is present. (c) Multilamellar structure quenched from 30°C, showing appreciable areas of highly ordered (solid) lipid in equilibrium with fluid lipids. [From Grant *et al.* (1974).]

For a number of other phase diagrams representing lateral phase separations in the plane of bilayer membranes, see Shimshick and McConnell (1973a,b). In these studies it was generally found that where any two lipids of a pair of lipids differ significantly in their physical properties (e.g., melting temperatures) they often exhibit solid-phase immiscibility, even though the fluid phases are homogeneous. The close approach of a portion of the solidus curve in Fig. 6 to a horizontal line indicates a case of near solid-phase immiscibility. Figure 8 shows an interesting example of solid-phase immiscibility, a binary mixture of cholesterol and dimyristoylphosphatidylcholine. Here the horizontal line between 0 and ~ 20% cholesterol denotes the region of solid-phase immiscibility; in this composition range, below the melting temperature of the phospholipid, this diagram predicts that the lipid bilayer will be found to consist of domains of the pure phospholipid and a solid solution containing cholesterol and dimyristoylphosphatidylcholine. This domain structure for this particular pair of lipids has not yet been visualized by freeze-fracture electron microscopy.

A particularly interesting phase diagram has recently been obtained by Wu and McConnell (1974) and is given in Fig. 9. This diagram shows a zone of fluid-phase immiscibility, denoted $L_1 + L_2$. In this region the diagram predicts a separation of the lipids into two fluid phases, one richer in dielaidoylphosphatidylcholine and the other richer in dipalmitoylphosphatidylethanolamine. This separation may be a pure lateral phase separation, a transverse phase separation, or even a (less likely) phase separation into noncontinuous bilayers. The first two possibilities are illustrated in Fig. 10.

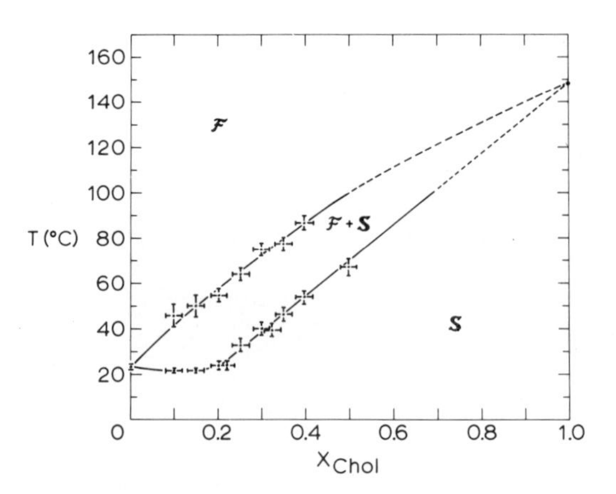

Fig. 8. Phase diagram of a dimyristoylphosphatidylcholine–cholesterol dispersion in excess water. The regions in which the fluid phase (F), the solid phase (S), and an equilibrium mixture of fluid and solid phases (F + S) are stable are indicated. [From Shimshick and McConnell (1973b).]

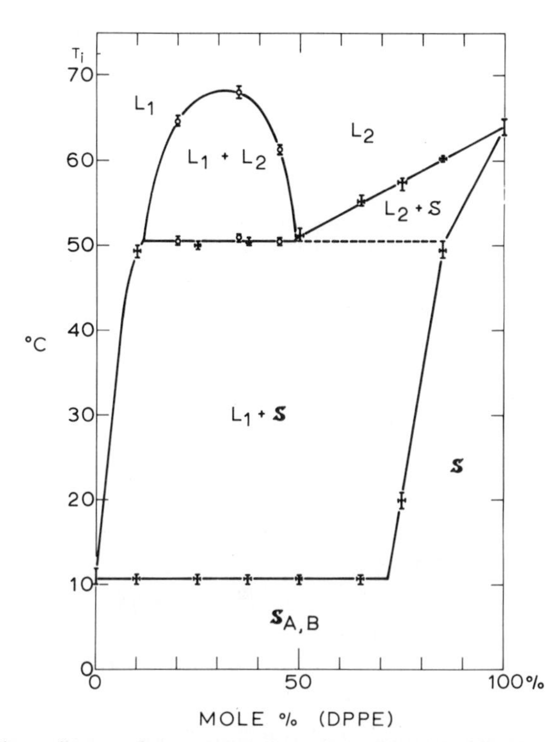

Fig. 9. Phase diagram for aqueous dispersions of the dielaidoylphosphatidylcholine–dipalmitoylphosphatidylethanolamine (DEPC–DPPE) binary system. +, Data taken from temperature breaks in experimental Tempo spectral parameter curves; ○, data taken from temperature breaks in the paramagnetic resonance spectra of **V**(1, 14) present at low concentrations ($\gtrsim 1\%$ mole) in the binary mixture. [Reprinted with permission from Wu and McConnell *Biochemistry* **14**, 847–854 (1974). Copyright by the American Chemical Society.]

An asymmetric distribution of lipids across a bilayer membrane should, on purely theoretical grounds, give rise to a bending of the membrane, as illustrated in Fig. 10d. For a discussion of this subject, see Wu and McConnell (1974).

Lateral phase separations of phospholipids in binary mixtures of phospholipids affect the lateral distributions of membrane proteins. For example, Grant and McConnell (1974) have reconstituted membranes containing glycophorin and a binary mixture of lipids and have shown from freeze-fracture electron microscopy that when both fluid and solid lipids coexist, the glycophorin is excluded from the solid phase. A freeze-fracture microphotograph illustrating this effect is shown in Fig. 11. Glycophorin is the major sialoglycoprotein of the human red blood cell, and was isolated and purified by the method of Marchesi *et al.* (1972). In order to obtain the photograph in Fig. 11, glycophorin was reconstituted into a 50 : 50 mole %

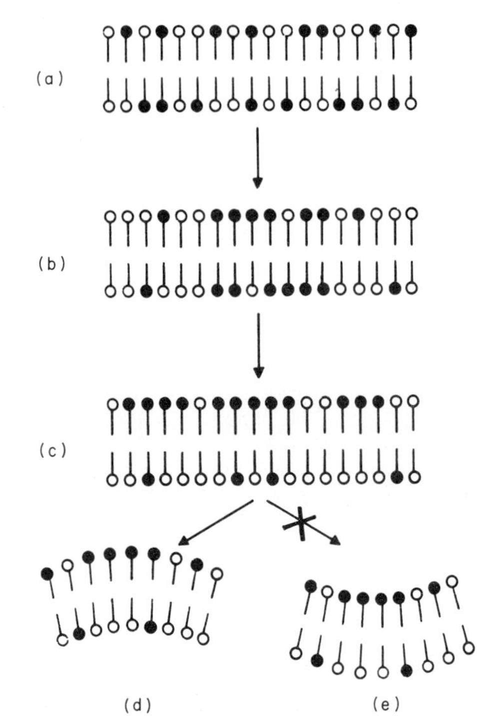

Fig. 10. Schematic representation of possible consequences of limited fluid-fluid miscibility of phospholipids. (a) Homogeneous fluid solution of black and white lipids. (b) Partial fluid–fluid immiscibility arising from lateral phase separations into regions relatively rich in white lipids and regions relatively rich in black lipids. (c) Transverse phase separation so that one half of the bilayer is rich in black lipids and the other half is rich in white lipids. (d, e). Stable and unstable polarized distortions of the membrane from a planar shape due to asymmetric transverse distribution of lipids. [Reprinted with permission from Wu and McConnell *Biochemistry* **14**, 847–854 (1974). Copyright by the American Chemical Society.]

mixture of dielaidoylphosphatidylcholine and dipalmitoylphosphatidylcholine, and this ternary mixture was rapidly quenched from a temperature of approximately 20°C, which corresponds to a point in the middle of the phase diagram in Fig. 6. The "particles" in this photograph are glycophorin molecules (or small polymers of them); the photograph shows clearly that these particles are excluded from the solid-phase lipids, which show a clear banded structure. Similar experiments have been carried out by Kleemann and McConnell (unpublished) in reconstituted binary mixtures containing the Ca^{2+} ATPase from sarcoplasmic reticulum. Again, when both fluid-phase lipids and solid-phase lipids coexist, the membrane proteins (freeze-fracture "particles") are excluded from the solid phase. An analogous behavior of the membrane protein rhodopsin reconstituted into a pure lipid phase has been studied by Chen and Hubbell (1973) and Hong and Hubbell (1972).

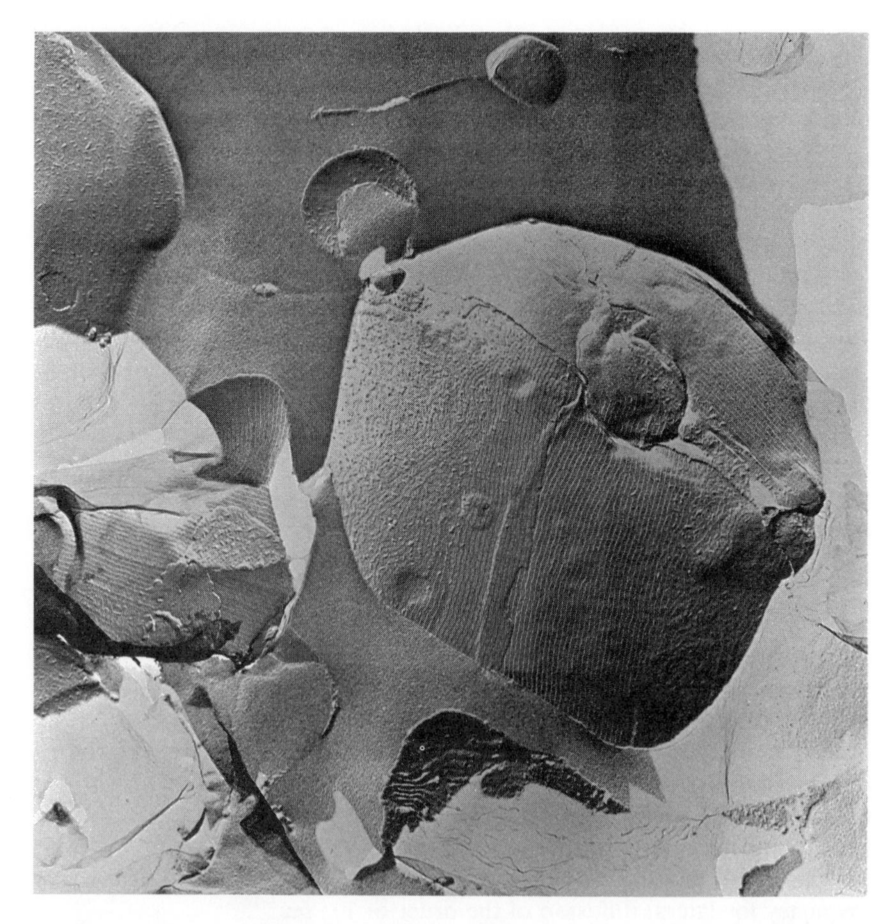

Fig. 11. Freeze-fracture electron microphotograph of a binary mixture of phospholipids containing the major sialoglycoprotein from the human erythrocyte, glycophorin. The lipids are 50 mole % dielaidoylphosphatidylcholine and 50 mole % dipalmitoylphosphatidylcholine; this sample was quenched from an ambient temperature of ~ 22°C, corresponding to ~ 50% solid and ~ 50% fluid lipids (see Fig. 6). The "particles" are due to glycophorin molecules or small polymers of glycophorin molecules and are seen to be confined almost exclusively to the fluid-phase lipids.

IV. LATERAL DIFFUSION

The fact that lateral lipid-phase separations in binary mixtures of phospholipids is rapid and reversible is due to the fact that the rate of lateral diffusion of lipid molecules in the fluid phase is rapid. The first studies of the rates of lateral diffusion of phosphatidylcholine were made by Kornberg and McConnell (1971b), who studied the proton nuclear resonance linewidth of the $N{-}CH_3$ protons in phosphatidylcholine vesicles containing a small concentration of the head-group spin label **II**. These experiments established

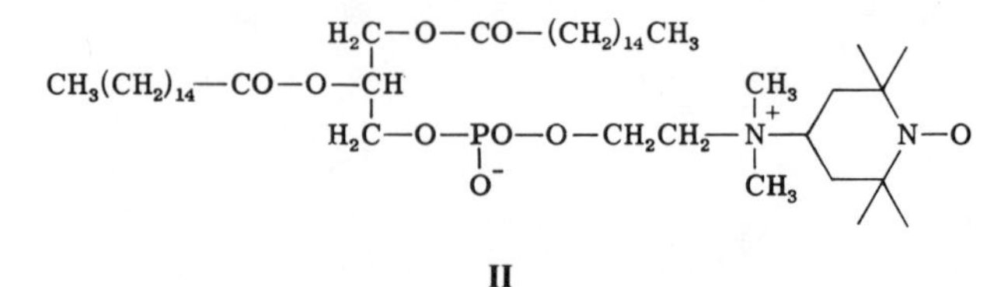

II

that lateral diffusion of the lipids must be rapid, on at least the time scale of the nuclear resonance experiment, in that the proton resonance linewidth was much larger than could be obtained from a static, random distribution of spin labels. In other words, the proton resonance signal was broadened so strongly that each $N{-}CH_3$ group had to undergo close interaction with the spin label, due to lateral diffusion. A detailed analysis of the data showed that the frequency of the translation step for lateral diffusion is large compared to 3×10^3 sec.

Figure 12 gives a schematic illustration of a method used by Devaux and McConnell (1972) to measure the rate of lateral diffusion of phospholipids. A concentrated patch of phospholipid spin labels (of the order of 0.2 mm in diameter) was inserted into a multibilayer system and the time evolution of the paramagnetic resonance spectra was recorded. With increasing time, the concentration of label became smaller, as did the spin exchange broadening of the resonance spectra. The observed spectra were then analyzed in terms of a distribution of concentrations determined by the lateral diffusion and in terms of a set of reference spectra obtained experimentally, corresponding to various concentrations of spin label. This study yielded a diffusion constant $D = 2 \times 10^{-8}$ cm^2 sec^{-1} at room temperature, corresponding to a step frequency for lateral diffusion of the order of 10^7 sec^{-1}.

An entirely different method for determining the rate of lateral diffusion of lipids in bilayer membranes was used by Sackmann and Träuble (1972a,b) and Träuble and Sackmann (1972). These investigators analyzed the paramagnetic resonance spectra of the androstane spin label **III** in dipalmitoyl-

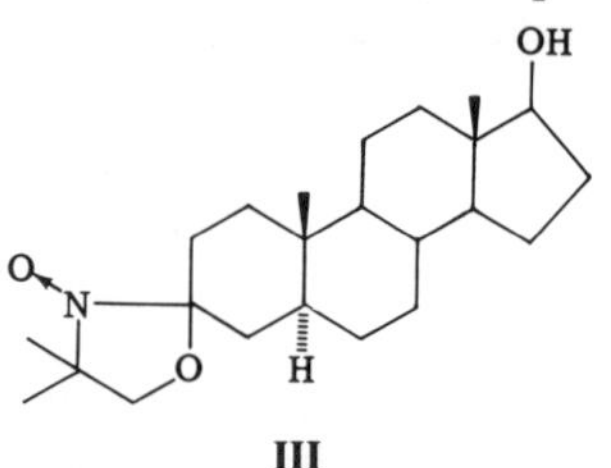

III

phosphatidylcholine membranes in terms of resonance broadening by spin exchange collisions. This theoretical line shape analysis yielded a diffusion constant $D \simeq 10^{-8}$ cm^2 sec^{-1} for the steroid label, of the same order as that measured by the totally independent method used by Devaux and McConnell (1972).

The studies of Sackmann and Träuble (1972a,b) and Devaux and McConnell (1972) were both carried out in "fluid" phospholipid bilayer membranes. The rate of lateral diffusion of phospholipids in the *solid* phase of phospholipid bilayers is not known. The most that can be said at the present time is that the time constant for equilibrium to be reached in certain mixtures of phosphatidylcholines between solid solutions and fluid solutions is relatively short, of the order of a few seconds or less, based on unpublished

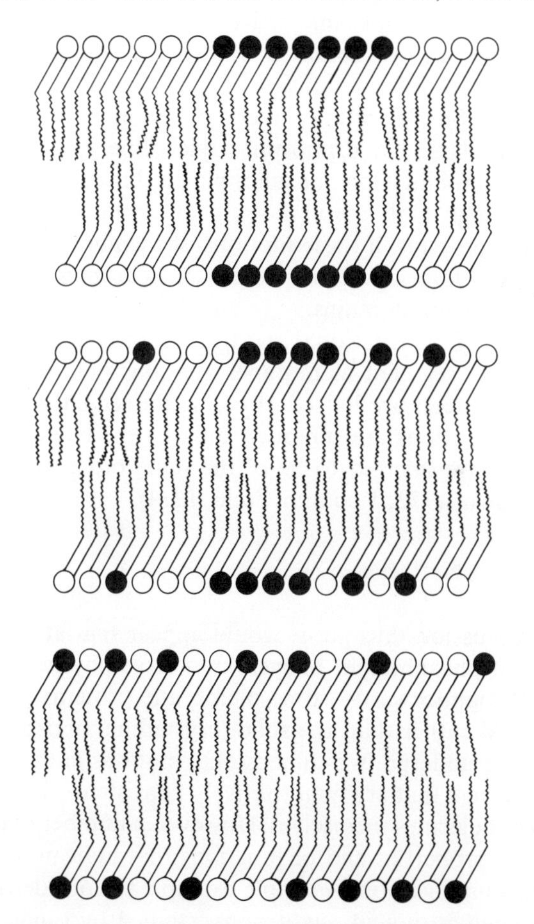

Fig. 12. Schematic model of spin-labeled lipids (black polar head groups) diffusing in bilayer planes of a host phospholipid, didihydroserculoylphosphatidylcholine (white polar head groups). [Reprinted with permission from Devaux and McConnell *J. Amer. Chem. Soc.* **94**, 4475–4481 (1972). Copyright by the American Chemical Society.]

temperature jump experiments by Shimshick and McConnell (unpublished, 1973). However, relatively rapid achievement of equilibrium between solid and fluid domains of lipids may not involve diffusion *in* the solid phase, but may rather involve the time constants for melting and crystallization of the solid domains. An upper bound for the time constant for mixing of fluid and solid domains of lipids may be obtainable from ^{13}C nuclear resonance experiments.

The spin label exchange line broadening effect has been used by McConnell *et al.* (1972), Scandella *et al.* (1972), Devaux *et al.* (1973), and Devaux and McConnell (1974) to estimate the rate of lateral diffusion of spin labeled phospholipids in the sarcoplasmic reticulum membrane of rabbit skeletal muscle. These investigators obtained diffusion constants of the order of 10^{-8} cm^2 sec^{-1} for the lateral diffusion of spin labeled phospholipids in this membrane at 37°C. Similar rates of lateral diffusion were obtained for both phosphatidylethanolamine and phosphatidylcholine.

A high lateral mobility for phospholipids in biological membranes implies a correspondingly high lateral mobility and rotational diffusion mobility for membrane proteins, if these motions of membrane proteins are not inhibited by special forces, such as protein–protein interactions or interactions with lipids in solid domains.

Optical studies of the rotational and translational diffusion rates of rhodopsin in rod outer segment membranes are consistent with the idea that these membrane proteins are indeed "free," in the sense that their rotational and translational motions are inhibited only by the lateral motions of the host lipids, which are inferred to be of the same order as those discussed (Cone, 1972; Brown, 1972; Poo and Cone, 1974).

V. MEMBRANE FUNCTION

From our discussion thus far, it would appear that at least some membrane proteins in phospholipid bilayers may have physical properties that are relatively straightforward—some membrane proteins prefer to be dissolved in the fluid rather than the solid (crystalline) phospholipid regions, and some proteins undergo two-dimensional motions in the fluid regions of phospholipid bilayers that are quantitatively consistent with the lateral motions of the lipid molecules themselves. However, a number of proteins in cell membranes have properties and/or functions that indicate that this picture may be an oversimplification. A simple example to consider involves sugar transport in *E. coli* fatty acid auxotrophs, studied by Linden *et al.* (1973). The particular fatty acid auxotroph designated 30Eβox$^-$ cannot synthesize unsaturated fatty acids and cannot β-oxidize them. Figure 13 shows an Arrhenius plot of the rate of β-galactoside uptake into these cells for cells

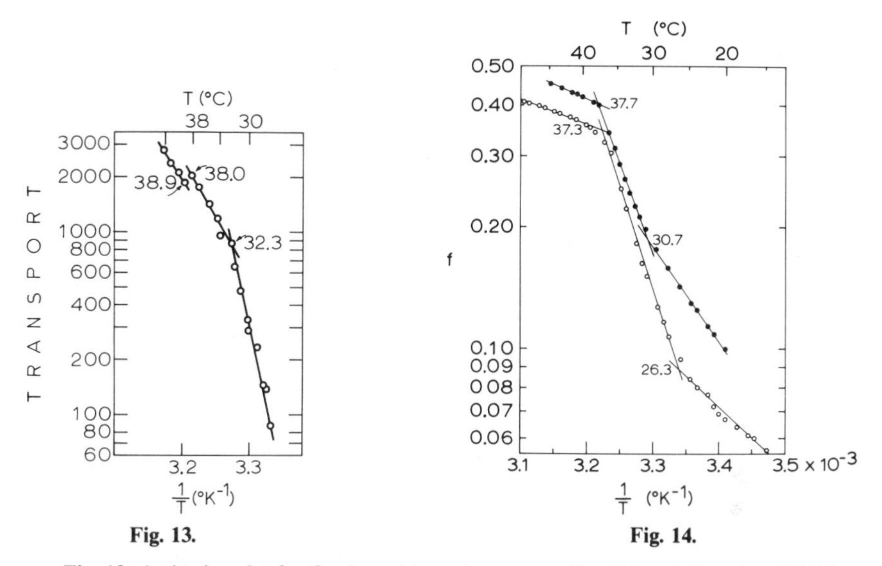

Fig. 13. Arrhenius plot for β-galactoside transport by cells of *E. coli* (βox$^-$) at 37°C in a medium supplemented with elaidic acid. The units for transport of *o*-nitrophenyl-β-galactoside are nmol/20 min per 1×10^9 cells. [From Linden *et al.* (1973).]

Fig. 14. Tempo spectral parameter *f* as a function of the reciprocal of the absolute temperature $1/T$ for *E. coli* (βox$^-$) inner membranes (●) and extracted inner membrane phospholipids (○) when the cells were grown at 37°C on a medium enriched in elaidic acid. [Data taken from Linden *et al.* (1973).]

grown at 37° with elaidic acid as the unsaturated fatty acid supplement. It will be seen that there is a remarkable *increase* in the rate of transport with *decreasing* temperature at temperatures in the range 38.1–38.8°C. It will also be seen that at a lower temperature, ~ 32°C, there is a discontinuous change in slope (but not in the transport rate itself) for the uptake of β galactoside. The transport proteins involved in the active uptake of these sugars are localized in the inner membranes of these cells.

A plot of the binding of the spin label Tempo to these inner membranes is shown in Fig. 14. These binding data are similar to those reported by Shimshick and McConnell (1973a,b) for binary mixtures of phospholipids, the higher temperature corresponding to the onset of the lateral phase separation and the lower temperature corresponding to the completion of the lateral phase separation. In terms of fatty acid composition, these inner membranes are *approximately* a binary mixture of fatty acids. See Linden *et al.* (1973). This interpretation of the spin label data is also supported, at least semiquantitatively, by freeze-fracture studies of the inner membranes of these cells (Kleemann and McConnell (1974)). Again, the membrane proteins (freeze-fracture particles) tend to be excluded from regions of solid lipid when the sample is quenched from ambient temperatures where solid lipids are present.

The substance of these observations is that the onset of the solid lipid phase separation with decreasing temperature has a remarkable and unexpected effect on transport through biological membranes. From a physical point of view, an even more surprising feature of this result is that the enhancement of transport is so pronounced just at the temperature corresponding to the *onset* of the lateral phase separation, where only a small proportion of solid phase lipid is calculated to be present. Our interpretation of this effect involves the consideration of two distinct factors:

(a) At the temperature where both solid and fluid lipids coexist in equilibrium the lipid bilayer should have an enhanced lateral compressibility (and extensibility). This enhanced compressibility, along with its associated fluctuations in lipid density, should facilitate the insertions of protein into and through lipid bilayers and should also facilitate expansion–contraction cycles involving alterations in protein structure within the lipid bilayer.

(b) In order for a small proportion of the solid-phase lipids to be effective in presenting the transport proteins with a kinetically significant enhancement of lateral compressibility and density fluctuations, these solid lipids should be present near the transport proteins at the temperature corresponding to the onset of the lateral phase separation. This could be understood if the membrane proteins serve as nucleation centers for the solidification and melting of membrane lipids.

There is little doubt that lipid molecules in the vicinity of intrinsic membrane proteins are strongly perturbed by these proteins. For example, Hong and Hubbell (1972) have shown that the incorporation of as little as 0.0035 mole % of rhodopsin in phosphatidylcholine bilayers produces a very significant effect on the ordering of the hydrocarbon chains. Also, Jost *et al.* (1973), Dehlinger *et al.* (1974), and Stier and Sackmann (1973) have used spin labels to obtain evidence for a relatively rigid halo of lipids in the immediate vicinity of membrane proteins.

Of great interest in connection with this discussion is the question of the behavior of the lipids surrounding an intrinsic membrane protein when both fluid- and solid-phase lipids coexist. One interesting effect related to this problem has been observed by Grant and McConnell (1974). Spin labeled glycophorin incorporated in pure dimyristoylphosphatidylcholine shows a temperature dependent paramagnetic resonance spectrum indicating a transition temperature of 24.5 ± 0.3°C, slightly above the transition temperature of the pure lipid. Tempo binding to the same lipid–glycophorin membrane indicates a transition temperature of 23.5 ± 0.5°C, slightly below that of the pure phospholipid. This result is consistent with the idea that there is a formation of a highly temperature dependent halo around this membrane protein at the transition temperature of the lipid in spite of the fact that this membrane protein tends to be excluded from pure, solid (crystalline) regions of phospholipids. This is precisely the structural environment that may facilitate the function of membrane transport proteins.

VI. THE FLEXIBILITY GRADIENT

The high lateral mobility of phospholipids in the fluid phase of lipids has its origin in the liquid-like (or liquid-crystal-like) state of the hydrocarbon chains. The motions of these hydrocarbon chains in phospholipid bilayers and in biological membranes has been the subject of numerous experimental and theoretical studies. Many of these have been based on analyses of the paramagnetic resonance spectra of the fatty acid spin labels **IV**(m, n) and the phospholipid spin labels **V**(m, n).

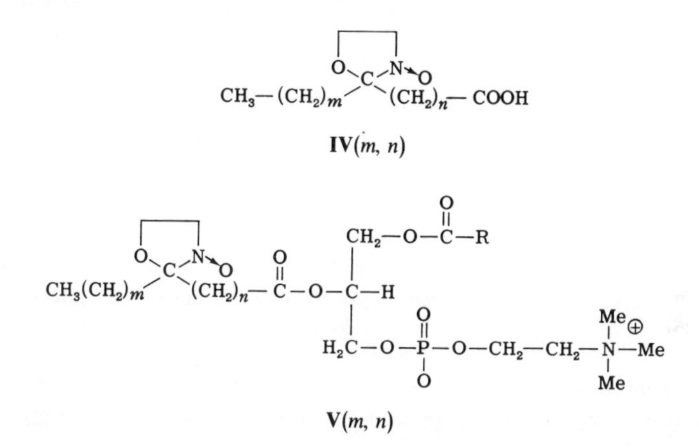

IV(m, n)

V(m, n)

Theoretical analyses of spectra of this type have been discussed extensively in the literature (Hubbell and McConnell, 1971; McFarland and McConnell, 1970; Hubbell and McConnell, 1969a,b; Seelig, 1970; Schindler and Seelig, 1973; Jost *et al.*, 1971; Hemminga and Berendsen, 1972; Gaffney and McConnell, 1974), and are also discussed elsewhere in this book (see Chapters 3 and 12). An essential point is that the resonance spectra of these labels can be accounted for quantitatively by decomposing the exact, time-dependent spin Hamiltonian for the problem $\mathscr{H}(t)$ into a time-averaged, effective Hamiltonian $\mathscr{H}'$ and a time dependent perturbation $\mathscr{H}(t) - \mathscr{H}'$ characterized by short correlation times:

$$\mathscr{H}(t) = \mathscr{H}' + [\mathscr{H}(t) - \mathscr{H}'] \qquad (1)$$

The time dependence of $\mathscr{H}(t)$ arises from rotational motions of the paramagnetic nitroxide group. Hubbell and McConnell (1971) have shown how it is relatively simple to estimate the elements of $\mathscr{H}'$ from observed paramagnetic resonance spectra. Improved approximations have been described by Gaffney and McNamee (1974). Ultimately, of course, the criterion as to whether or not the decomposition in Eq. (1) is accurate depends on comparisons between observed and calculated spectra, particularly at different applied magnetic field strengths (Gaffney and McConnell, 1974a). (See also Appendix IV.) The hyperfine and g tensors of $\mathscr{H}'$ are related to those of $\mathscr{H}$ by time averages of the squares of the direction cosines of the principal axes of $\mathscr{H}$ in the $\mathscr{H}'$ axis system (Hubbell and McConnell, 1969a,b). Seelig (1970) pointed out that these angular averages can be expressed in terms of "order parameters," well known earlier in the study of liquid crystals (Saupe, 1968). The most useful of these order parameters is S, where

$$S = \tfrac{1}{2}\langle 3 \cos^2 \vartheta - 1 \rangle_{\mathrm{T}} \tag{2}$$

Here $\vartheta = \vartheta(t)$ is the time dependent angle between the π-orbital axis (i.e., the z axis, which is the axis of the largest ^{14}N nuclear hyperfine splitting) and the principal axis z', the axis of the largest average hyperfine splitting. The angular brackets in Eq. (2) denote a time average over a characteristic time T. For nitroxide spin labels the time T is determined by the reciprocal of the anisotropy of the hyperfine interaction and Zeeman interactions, in frequency units. This places T in the range 10^{-7}–10^{-9} sec. In order for the decomposition in Eq. (1) to be useful, and in order for the order parameter S in Eq. (2) to have physical significance, it is necessary that the correlation time(s) τ that characterize the molecular motion be short compared to T,

$$\tau \ll T \tag{3}$$

Although order parameters S have been used extensively in discussions of the paramagnetic resonance spectra of spin labels in membranes, their interpretation and their measurement require some care:

(i) As mentioned, S is defined only for high frequency motions, $\tau^{-1} \gg T^{-1}$.

(ii) According to current usage, Eq. (2) applies to the time average over the motions of an individual molecule. Since different molecules in a sample may have different average orientations that persist for times much longer than T, the order parameter need not reflect information about the average ordering of the molecules in a sample. For example, a sample might be totally disordered (isotropic distribution of orientations of x', y', z') and at the same time be characterized by order parameters $S \simeq 1$. In this sense "order parameter" is a poor term. The term "frequency–amplitude order parameter" is more appropriate. That is, the magnitude of S in Eq. (2) is determined by the amplitudes of molecular motions that take place with frequencies larger than $\sim 1/\pi T$.

(iii) Since the order parameter S is a frequency–amplitude parameter, its magnitude can depend on the characteristic frequency $1/\pi T$. In principle, the molecular motion of a given molecule may have two different order parameters if different magnetic interactions are used to study the motion, providing of course $T \gg \tau$ is valid in each case.

(iv) In the interpretation of the paramagnetic resonance spectra of spin labels, one must distinguish carefully between highly anisotropic, high frequency motion and slow, isotropic motion, since these can have qualitatively similar effects on some spectral features (see McCalley *et al.*, 1972 and Chapter 2 by Freed).

(v) As discussed later, the analysis of paramagnetic resonance spectra can lead to a determination of *an ensemble-average* order parameter S_e,

$$S_e = \tfrac{1}{2}\langle 3 \cos^2 \vartheta - 1\rangle_e \tag{4}$$

Here ϑ is the angle between the principal hyperfine axis z and some specified axis fixed relative to the sample, and the average is over all molecules in the sample at a given instant of time. That is, for a large number of molecules labeled $k = 1, 2, 3, \ldots, N$,

$$S_e = (1/N) \sum_{k=1}^{k=N} \tfrac{1}{2}(3 \cos^2 \vartheta_k - 1) \tag{5}$$

This ensemble-average order parameter is more useful for some purposes than is the frequency–amplitude order parameter.

In the fatty acid spin labels **IV**(m, n) and the phospholipid spin labels **V**(m, n) the π-orbital axis of the nitroxide group is parallel to the hydrocarbon chain when this chain has its fully extended, all-trans conformation (see Fig. 15). The first observations made with the labels **IV**(m, n) showed, as expected, that the π-orbital axes are more nearly perpendicular than parallel to the surface of biological membranes and phospholipid bilayers (Hubbell and McConnell, 1969b; Libertini *et al.*, 1969). However, the studies of Hubbell and McConnell (1969b) yielded an unanticipated result. It was found that in both biological membranes (Hubbell and McConnell, 1969b) and in phospholipid bilayers (Hubbell and McConnell, 1971) the order parameters $S = S(n)$ decrease rapidly with increasing n, where n is the number of methylene groups that separate the paramagnetic oxazolidine ring from the carboxyl carbon atoms, in both the fatty acid labels **IV**(m, n) and the phospholipid labels **V**(m, n) (see Fig. 16). This gradient in order parameter was also found in smectic liquid crystals of decanol and decanoic acid with the fatty acid labels **IV**(m, n) (Seelig, 1970). This apparent increase in the hydrocarbon chain motion as one approaches the terminal methyl groups

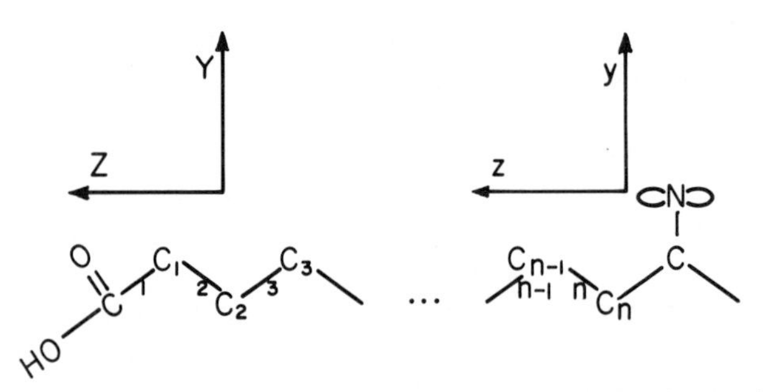

Fig. 15. Spin label **IV**(m, n) in extended, all-trans conformation. The Z axis is parallel to the chain direction. The principal hyperfine axis z is perpendicular to the N-oxyloxazolidine ring, is parallel to the π-orbital axis that holds the odd electron, and is also parallel to Z when the polymethylene chain is extended. [Reprinted with permission from Hubbell and McConnell *J. Amer. Chem. Soc.* **93**, 314–326 (1971). Copyright by the American Chemical Society.]

has been referred to as the "flexibility gradient" and has been the subject of numerous subsequent experimental and theoretical studies. This flexibility gradient is of considerable interest and poses a number of problems. One problem of course is whether the flexibility gradient is a property of hydrocarbon chains in label-free bilayers or whether the apparent flexibility gradient arises partially or entirely from the paramagnetic oxazolidine ring. Assuming that the flexibility gradient is a property of label-free bilayers, what is the physical significance of these frequency–amplitude order parameters in terms of bilayer structure? In particular, do the ensemble-average

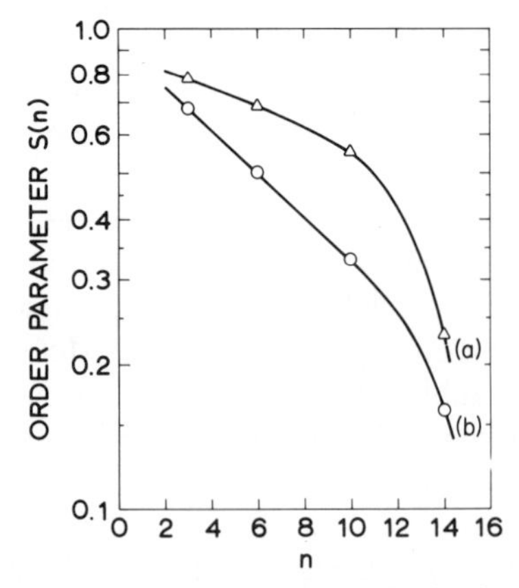

Fig. 16. The order parameter $S(n)$ as a function of n in the spin labels **V**(m, n) in fully hydrated multibilayers of (a) egg lecithin and cholesterol, 2 : 1 mole ratio, and (b) egg lecithin alone. [From McConnell and McFarland (1972), with permission of the New York Academy of Sciences.]

order parameters also show a flexibility gradient, or does this flexibility gradient only exist in a special frequency domain? Finally, if the ensemble-average order parameters also show a flexibility gradient, there is the question of how topologically linear chains can be packed together in a plane where one end of the chain apparently has more motional freedom than does the other.

McFarland and McConnell (1971) have proposed a general model of bilayer structure that seeks to account for the flexibility gradient in terms of a *cooperative* tilting of the hydrocarbon chains of the phospholipid molecules near the polar head groups. The manner in which this allows more motional freedom near the terminal methyl groups is indicated schematically in Fig. 17. We start with a group of hydrocarbon chains near the polar head groups that are collectively tilted relative to the normal to the bilayer surface, whereas near the terminal methyl groups the chains are perpendicular to the bilayer surface. In Fig. 17 there is clearly more space for molecular motions near the terminal methyl groups than near the polar groups. Thus, with this sort of overall packing, larger amplitude high-frequency motions can take place near the terminal methyl groups than near the polar head groups. In the configuration sketched in Fig. 17, with the chains tilted at 30°, the carbon atom density near the polar head groups is higher by about 13% than the carbon atom density near the terminal methyl groups. However, preferential motion (due to gauche$^+$, gauche$^-$, trans isomerizations) near the terminal methyl groups *increases* the carbon atom

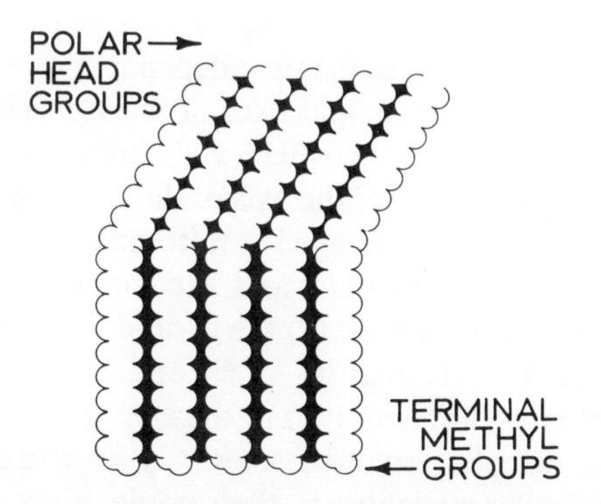

Fig. 17. Schematic, idealized representation of bent fatty acid chains. The purpose of this diagram is to illustrate the elementary geometrical fact that a cooperative bending of chains near the polar head groups reduces the motional freedom of these chains relative to those near the terminal methyl groups. Trans ↔ gauche$^+$ ↔ gauche$^-$ isomerizations of these chains near the terminal methyl groups are therefore more probable than near the polar head groups. This effect *increases* the carbon atom density gradient in the direction of the normal to the bilayer. [See text, McFarland and McConnell (1971), and McConnell and McFarland (1972).]

density in this region (McConnell and McFarland, 1972), so there is no evident conflict between this model and the conclusion by Wilkins *et al.* (1971) that the carbon atom density in the hydrocarbon chain region of bilayers is nearly constant across the bilayer. In support of this proposal, McFarland and McConnell pointed out that a study of the paramagnetic resonance spectra of phospholipid spin labels in lipid bilayers (see following discussion and Appendix B) showed measurable π-orbital tilt ($\sim 30°$) when the oxazolidine ring was near the polar head group, as in **V**(10, 3) (5-doxylpalmitoyl lecithin). The main significance of this result is *not* that the π-orbital axis is tilted, which could arise from packing effects other than the proposed tilt of the hydrocarbon chains, but that the π-orbital tilt *can be measured*. This signifies that the oxazolidine ring together with its molecular environment must have some persistent structure that lasts for a time long compared to the reciprocal of the frequency separations for spectra calculated with and without this persistent tilt. This π-orbital tilt is estimated to last at least 10^{-7} sec. This is in sharp contrast to the motion of the nitroxide group when this group is near the terminal methyl groups, as in **V**(1, 14) (15-doxylpalmitoyl lecithin). Here the amplitude and frequency of the motion are both high; the resonance spectra are sharp and nearly isotropic. The motion is both rapid ($\sim 10^{-9}$ sec) and of large amplitude (order parameter $S \simeq 0.1$).

In order to develop further our discussion of the order parameter gradient, let us describe in somewhat more detail the analysis of the resonance spectra of the phospholipid labels **V**(m, n).

The interpretation of the paramagnetic resonance spectra of the spin labels **V**(m, n) incorporated into coplanar arrays of oriented phospholipid bilayers requires consideration of the time dependent as well as time-average angular distributions of the principal axes of the spin Hamiltonian $\mathscr{H}$. Let the principal axes of $\mathscr{H}$ be x, y, z, where x is along the N $\rightarrow$ O bond direction and z is along the direction of the axis of the π orbital that holds the odd electron. The axes of the time-average Hamiltonian $\mathscr{H}'$ are denoted x', y', z'. In the calculations made by Gaffney and McConnell (1974), the paramagnetic resonance spectra depend on (a) the g'- and T'-tensor elements of $\mathscr{H}'$, (b) the time dependent terms in $\mathscr{H}(t) - \mathscr{H}'$, and thus the tensor elements g', T', g, and T, together with the motion of the axes x', y', z' relative to x, y, z, and (c) the angular distribution of the axes x', y', z' corresponding to different orientations of different labels in the sample. Since calculated and observed spectra are in good agreement when $\mathscr{H}'$ has axial symmetry, the distribution function for the x', y', z' axis system requires the use of only one angle ϑ, the angle between z' and the normal to the bilayer plane. Let $\rho(\vartheta)\, d\vartheta$ be the probability that z' is found between ϑ and $\vartheta + d\vartheta$. In absence of any field-induced lipid orientation (see, however, Gaffney and McConnell,

1974b) this distribution function must be cylindrically symmetric; it has been assumed to have the form

$$\rho(\vartheta) = \sin\vartheta \exp[-(\vartheta - \bar{\vartheta})^2/2\vartheta_0^2] \tag{6}$$

The quantity $\bar{\vartheta}$ is a "tilt parameter" and ϑ_0 is a measure of the spread of the angular distribution of orientations. The parameters ϑ_0 and $\bar{\vartheta}$ are determined by comparisons of observed and computer-calculated spectra.

The rapid anisotropic motion of the instantaneous axis system for any given nitroxide free radical can be conveniently referred to the "average" principal axis system x', y', z' for the same radical. Let Θ be the angle between z and z', the axes of the largest ^{14}N nuclear hyperfine splitting. Gaffney and McConnell (1974a) assumed the following instantaneous distribution of z about z':

$$\rho(\Theta) = \sin\Theta \exp(-\Theta^2/2\Theta_0^2) \tag{7}$$

The rapid molecular motion $\Theta = \Theta(t)$ averages the magnetic interactions in $\mathcal{H}$ to yield those in $\mathcal{H}'$; the order parameter S that is determined by comparison of $\mathcal{H}$ with $\mathcal{H}'$ leads to a determination of the distribution parameter Θ_0. In other words,

$$S = \frac{\int_0^\pi \sin\Theta[\exp(-\Theta^2/2\Theta_0^2)]\frac{1}{2}(3\cos^2\Theta - 1)\, d\Theta}{\int_0^\pi \sin\Theta \exp(-\Theta^2/2\Theta_0^2)\, d\Theta} \tag{8}$$

Finally, we consider the "instantaneous" distribution $\rho_e(\vartheta_i)$. Here ϑ_i is the instantaneous angle between the hyperfine axis z and the bilayer normal, and $\rho_e(\vartheta_i)\, d\vartheta_i$ is the probability that any hyperfine axis z can be found between ϑ_i and $\vartheta_i + d\vartheta_i$. The quantity $\rho_e(\vartheta_i)$ is referred to as an *ensemble distribution function* since it measures a probability distribution function appropriate to the entire sample and not necessarily to a localized region. The ensemble distribution function $\rho_i(\vartheta_i)$ (giving the distribution of z relative to the bilayer normal) is calculated from a convolution of $\rho(\vartheta)$ (giving the distribution of z' relative to the bilayer normal) and $\rho(\Theta)$ (given the distribution of z relative to z') by methods described by Gaffney and McConnell (1974a).

Figure 18 shows plots of ensemble distributions for two phospholipid labels **V**(m, n) in different bilayers. These plots exhibit distinct maxima; thus the resonance data demonstrate a distinct tilt of the label orientations. Plots of $\rho_e(\vartheta_i)/\sin\vartheta_i$ also show distinct maxima for $\vartheta_i > 0$. Note that the distribution function for **V**(1, 14) is more isotropic than the distribution function for **V**(10, 3), but that the distribution function for **V**(1, 14) itself is far from isotropic. For an isotropic distribution $\rho_i(\vartheta_i)$ varies as $\sin\vartheta_i$, which has a maximum at $\vartheta_i = 90°$. Gaffney and McConnell (1974a) have calculated the distribution functions $\rho_i(\vartheta_i)$ for the labels **V**(10, 3), **V**(5, 10), and **V**(1, 14).

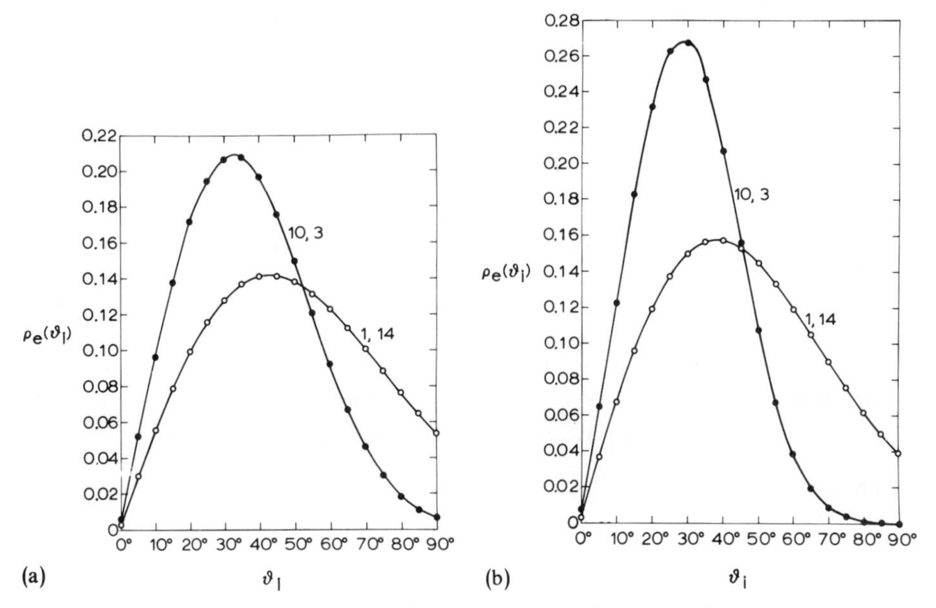

Fig. 18. Plots of the ensemble distribution function $\rho_e(\vartheta_i)$ vs. ϑ_i for (a) hydrated bilayers of egg lecithin and (b) egg lecithin–cholesterol (2 : 1 mole ratio). Plots for labels **V**(10, 3) (●—●) and label **V**(1, 14) (○—○) are included. For the **V**(10, 3) plot: in egg lecithin, $\bar{\vartheta} = 30°$, $\vartheta_0 = 5°$, $\Theta_0 = 21°$, $S' = 0.69$; in egg lecithin–cholesterol, $\bar{\vartheta} = 25°$, $\vartheta_0 = 5°$, $\Theta_0 = 16°$, $S' = 0.79$. For the **V**(1, 14) plot: in egg lecithin, $\bar{\vartheta} = 0°$, $\vartheta_0 = 5°$, $\Theta_0 = 48°$, $S' = 0.16$; in egg lecithin–cholesterol, $\bar{\vartheta} = 0°$, $\vartheta_0 = 5°$, $\Theta_0 = 42°$, $S' = 0.23$. [From Gaffney and McConnell (1974a).]

From these distribution functions one can calculate the ensemble-average order parameters S_e by an equation equivalent to Eqs. (4) and (5):

$$S_e = \int_0^\pi \tfrac{1}{2}(3\cos^2\vartheta_i - 1)\rho_i(\vartheta_i)\,d\vartheta_i \Big/ \int_0^\pi \rho_i(\vartheta_i)\,d\vartheta_i \tag{9}$$

Calculated ensemble-average order parameters for **V**(10, 3), **V**(5, 10), and **V**(1, 14) are $S_e = 0.41$, 0.30, and 0.16, respectively (Gaffney and McConnell, 1974a). These order parameters still show a gradient, but not as large a gradient as the amplitude–frequency order parameters.

The foregoing discussion makes it clear that *the flexibility gradient measured by spin labels involves both a frequency gradient and an amplitude gradient.* The gradient in the ensemble-average order parameter is a pure amplitude of motion effect. On the other hand, there must also be a frequency gradient for certain molecular motions; otherwise the order parameters S would everywhere be equal to S_e.

We now turn to the question of the extent to which the flexibility gradient is a property of phospholipid bilayers in the absence of a "perturbing" spin label. Fortunately, Seelig and Seelig (1974a,b) have synthesized specifically deuterium labeled phospholipids analogous to the spin labeled phospholipids **V**(m, n), where deuterium atoms are located on carbon atom $n + 2$, the same carbon atom to which the oxazolidine ring is attached in spin label **V**(m, n). The nuclear quadrupole splittings in the deuterium nuclear resonance spectra have been analyzed by Seelig and Seelig (1974a,b) in terms of order parameters that can be compared directly with those deduced from the spin label resonance spectra. Order parameters, derived from the deuterium resonance spectra are given in Fig. 19. The "deuterium resonance order parameters" also show a flexibility gradient; *therefore, there can be no doubt that a flexibility gradient exists,* since the deuterium atoms can have no significant structural perturbation on the lipid bilayers. The deuterium resonance derived order parameters are frequency–amplitude order parameters. Here the characteristic time T over which the molecular motion is averaged [cf. Eq. (2)] must only be larger than the inverse of the deuterium nuclear quadrupole splitting in C–D bonds, i.e., of the order of 10^{-5}–10^{-6} sec.

There is clearly a quantitative difference between the deuterium resonance, frequency–amplitude order parameters and those derived from the spin label resonance data. Seelig and Seelig (1974a,b) have attributed this difference to the perturbation effect of the paramagnetic oxazolidine ring in the

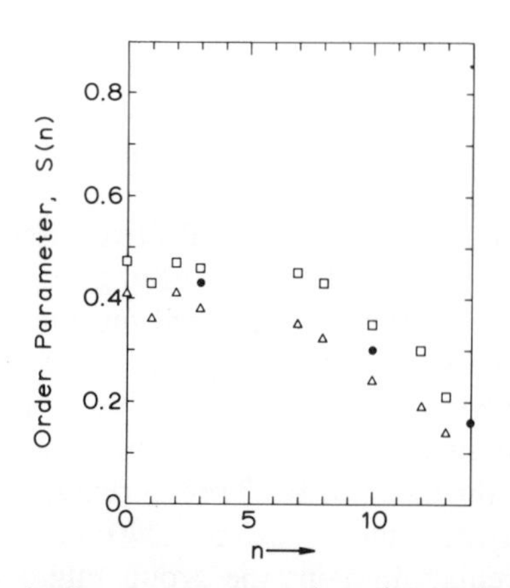

Fig. 19. *Frequency–amplitude* order parameters (△, □) derived from quadrupole splittings in the deuterium nuclear magnetic resonance spectra of selectively labeled phospholipids compared with *ensemble-average* order parameters (●) derived from the paramagnetic resonance spectra of spin label phospholipids. The points △ and □ refer to selectively deuterium labeled L-α-dipalmitoylphosphatidylcholine, where the deuterium atoms are on carbon atoms $n + 2$ and the temperatures are 41 and 57°C, respectively. Data taken from Seelig and Seelig (1974a,b). Spin label derived ensemble-average order parameters refer to the phospholipid spin labels **V**(m, n) in egg lecithin at 25°C. [●, from Gaffney and McConnell (1974a). □, △ from Seelig and Seelig, 1974a,b.]

labels **IV**(m, n). Obviously we can hardly expect the derived order parameters to be identical in the two cases. However, Gaffney and McConnell (1974a) have emphasized that there may be another and more interesting source of this discrepancy, namely, that lipid bilayers have molecular motions in the polar head-group region that are short on the deuterium resonance time scale (10^{-5}–10^{-6} sec) but long on the paramagnetic resonance time scale (10^{-7}–10^{-9} sec). Support for this idea can be seen in Fig. 19, which also gives the ensemble-average order parameters for spin labels **V**(10, 3), **V**(5, 10), and **V**(1, 14). It will be seen that there is no significant difference between these ensemble-average order parameters and the frequency–amplitude parameters from deuterium nuclear resonance spectra. (Unfortunately, the lipids and temperatures employed are not identical in the two cases.)

A simple way to understand this result from a physical point of view is to return to Fig. 17. If the lifetime of the collective tilted state near the polar head groups is between 10^{-7} and 10^{-5} sec, then motion of the chains in this region of the bilayer averages nuclear quadrupole interactions, but not hyperfine interactions. In this case deuterium resonance, frequency–amplitude order parameters should be equal to deuterium resonance, ensemble-average order parameters. The latter are evidently equal to the spin-label, ensemble-average order parameters to within the experimental uncertainties.

If the relatively slow motion of the type discussed here does exist, then it is likely that this motion can be detected using the nonlinear response techniques discussed in Appendix A. *Assuming* for the moment that this slow motion *is* detected, an interesting question will remain. Namely, is the motion slow because there is a high effective viscosity for independent motion of chain segments near the polar head group, or is the motion slow because it requires the collective, concerted motion of many phospholipid molecules, as might be suggested by Fig. 17? With the latter possibility in mind, it can be estimated that a circular group of ten or so phospholipid molecules could exhibit a concerted rotational motion with a rotational diffusion constant of the order of 10^{-6} sec. It remains to be seen whether or not a "collective relaxation mode" exists. (The collective group of molecules need not form a circular domain; the group might be linear and have a wavelike motion.)

At this time it is not known whether the flexibility gradient in lipid bilayers and membranes serves any special biophysical function. It is possible that the flexibility gradient plays no crucial role in membrane function, but is merely a consequence of a restriction of all motions of hydrocarbon chains that would otherwise give rise to energetically unfavorable hydrocarbon chain–water contacts. On the other hand, it is hard to imagine that these highly cooperative interactions have not been exploited for some useful purpose by cellular membranes.

In addition to their use as probes of membrane dynamics and structure, the labels **IV**(m, n) and **V**(m, n) can sometimes be used in an empirical way to detect physiological changes in membrane "fluidity." For example, as few as 1–3 molecules of the prostaglandin PGE_2 reproducibly increases the order parameter of **IV**(10, 3) (5 doxyl-palmitate) bound to the human erythrocyte by the order of 1%. A comparable number of molecules of the prostaglandin PGE_1 reproducibly decrease this order parameter (Kury and McConnell, 1973). The effect of carbamyl choline on the order parameter of **IV**(10, 3) was used to demonstrate the existence of an acetylcholine, muscarinic receptor in the human erythrocyte (Huestis and McConnell, 1974). It appears likely that further studies of this sort will be carried out in the future.

VII. MEMBRANE FUSION

Membrane fusion and/or fission accompanies a number of biological events, including the infection of cells by lipid-enveloped viruses and cell division. In every case it can be safely assumed that the fusion of two biological membranes involves the fusion of two phospholipid bilayers. Spin labels

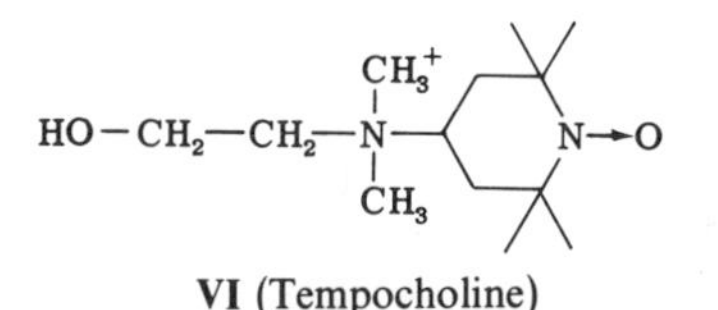

VI (Tempocholine)

II, **V**(m, n), and tempocholine **VI** have proved to be convenient in studies of fusion of lipid bilayers. For example, label **V**(1, 14) can be incorporated at such high concentrations in phospholipid vesicles that the resonance signal is strongly exchange-broadened. Thus the fusion of phospholipid vesicles of pure **V**(1, 14) with sarcoplasmic reticulum membranes can readily be followed from the change in the resonance line shape, since once a vesicle of **V**(1, 14) fuses with this biological membrane, the local high concentration of the label quickly decreases through rapid lateral diffusion (Scandella *et al.*, 1972). This effect has also been used to follow the fusion of exogeneous lipids with intact, viable cells of *Mycoplasma laidlawii* (Grant and McConnell, 1973).

One simple model system for the study of membrane fusion is an aqueous dispersion of small, single-compartment phosphatidylcholine vesicles. These vesicles can be prepared by the sonication of aqueous dispersions of egg lecithin or from pure phosphatidylcholines at temperatures above their chain melting transition temperatures. These vesicles have outer diameters of the order of 250 Å. Two vesicles of this type are illustrated schematically in Fig. 20, along with a possible fusion event. For 250 Å vesicles, elementary geometrical considerations show that if a vesicle fuses with another vesicle in such a way that the total quantity of lipids, the bilayer thickness ($\sim$ 50 Å), and the spherical shape are preserved, the resulting fused vesicle has a contained volume about four times the volume of a single vesicle and a ratio of inner surface area to outer surface area of $\sim$ 0.5 instead of 0.3. In addition to lipid mixing, the fusion of these vesicles with one another leads to four changes that were studied by Taupin and McConnell (1972), using spin labels.

(i) When the label **V**(7, 6) (8-doxylpalmitoyl lecithin) is incorporated at low concentrations in dipalmitoylphosphatidylcholine vesicles, it is found

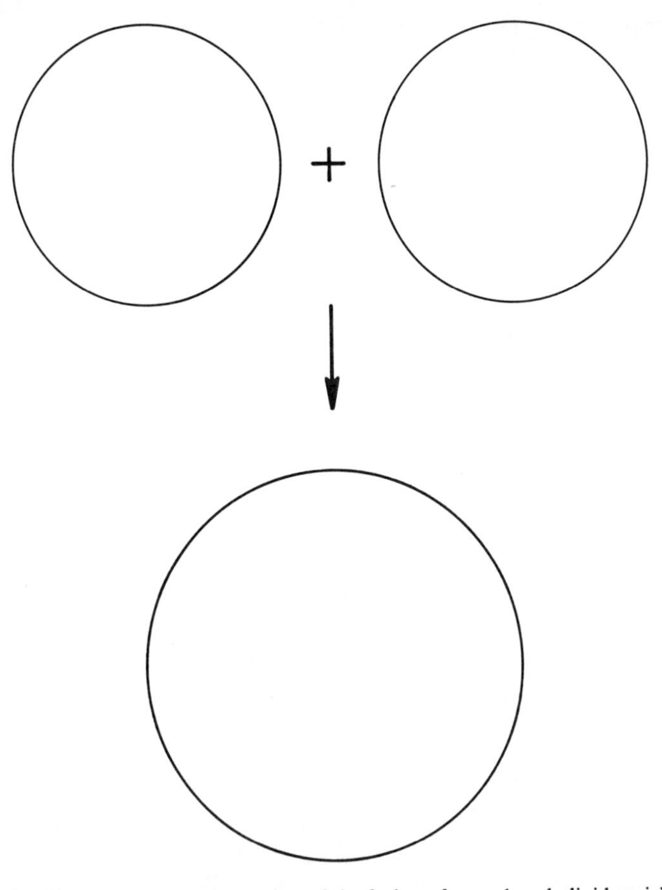

Fig. 20. Schematic representation of the fusion of two phospholipid vesicles.

that, with the passage of time, the chain melting transition temperature of this lipid, as monitored by the resonance spectrum of **V** (7, 6), becomes sharper and sharper. This is the expected result of fusion, since the fused vesicles have less curvature strain, less inside–outside lipid heterogeneity, and a larger number of molecules that can undergo cooperative transitions.

(ii) The increase of "protected tempocholine signal" provides a measure of the increase of internal volume on fusion. Taupin and McConnell prepared vesicles of dipalmitoylphosphatidylcholine in the presence of tempocholine in solution. The "outside" tempocholine signal was destroyed by first reducing the temperature to 0°C and then adding ascorbic acid. It was found that when vesicles were maintained at 48°C and the outside protected signal was measured at various times after vesicle preparation, the protected signal increased by $\sim$ 10–30% during the first hour and then more slowly thereafter.

(iii) During the elementary fusion step depicted in Fig. 20, there must be a transfer of outward-facing phospholipids to inward-facing lipids, since on fusion the ratio of outward-facing lipids to inward-facing lipids decreases. This expected effect was demonstrated experimentally. When these vesicles were "doped" with spin label **II** (the phosphatidylcholine label), two effects were observed that are consistent with the idea of fusion. First, by double integration of the (derivative-curve) spectra, the ratio of the number of inward-facing labels to the total number of labels was found to increase by $\sim$ 51% during the first hour and by $\sim$ 20% after 20 hr. Second, in the small, single-compartment vesicles, inward-facing labels showed spectra that were distinctly more immobilized than those of the outward-facing labels. With the passage of time, the spectra of the inward-facing labels become less strongly immobilized and approach the spectra of the outward-facing labels.

(iv) Mixing of internal contents. The most surprising result of the study by Taupin and McConnell (1972) is the evidence they obtained for the mixing of the internal contents of vesicles. One experiment carried out by Taupin and McConnell (1972) involves a mixing of two populations of vesicles. One population of vesicles contained a high concentration of $MnCl_2$ (0.5 M) on the inside only. The second population contained 3×10^{-3} M tempocholine, inside and outside. Control experiments showed that the sharp resonance triplet of tempocholine was strongly broadened by Mn^{2+} concentrations above 0.02 M. The two populations of vesicles were mixed and maintained at 48°C. The resulting "protected tempocholine signal" was much weaker than that observed in experiments without $MnCl_2$. The results are only explained if there is a preferential mixing of the internal contents of the two populations of vesicles. In contrast to this result, studies of the fusion of small phospholipid vesicles with cells of *Mycoplasma laidlawii* indicate that the fusion takes place *without* the delivery of the internal contents of the vesicles into the cytoplasma of the *laidlawii* cells (Grant and McConnell, 1973).

This discussion has shown how many features of phospholipid membrane fusion can be studied using spin labels. Clearly this is a field of study where only the surface has been scratched, and many more experiments suggest themselves. For more recent work, see Prestegard and Fellmeth (1974).

APPENDIX A. ROTATIONAL CORRELATION TIMES

Many biophysical applications of nitroxide spin labels depend on the fact that the paramagnetic resonance spectra of the labels are affected by molecular motions characterized by correlation times that fall in the range 10^{-10}–3×10^{-7} sec. For example, labels incorporated in proteins in solution often have correlation times in this range, and the resonance spectra are highly sensitive to biophysical and biochemical events that change these correlation times. It has been clear for a long time that the applicability of the spin label technique to biophysical problems might be extended significantly if the resonance spectra were sensitive to molecular motions with correlation times greater than 10^{-7} sec, since many interesting motions must have these longer rotational correlation times. For example, the rotational diffusion time of the visual pigment rhodopsin in the rod outer segment membranes is known from optical studies to be $\sim 20 \times 10^{-6}$ sec (Cone, 1972; Brown, 1972; Poo and Cone, 1974), a period two orders of magnitude above the mentioned upper limit of 3×10^{-7}. This upper limit is of course only an approximate figure and is determined by the intrinsic inhomogeneous resonance linewidths ($\sim 1/\pi T_2$) and the hyperfine and g-factor anisotropy of nitroxide spin labels. This upper limit refers to conventional, nonsaturated, slow-passage paramagnetic resonance spectra, and has been verified in both experimental studies and theoretical calculations. (For example, see Fig. 21.) Detailed theoretical treatments of the resonance spectra are to be found in the literature and elsewhere in this book. Even so, it is instructive to give a rough, order-of-magnitude estimate of this upper value of τ_2, since it will be useful in the later discussion.

In order to estimate the slowest possible isotropic rotational diffusion of a protein that can be detected from the normal paramagnetic resonance absorption spectrum, we select two points A and B on the resonance spectrum that are distinct from one another (resolved) and at the same time are as close in frequency (applied field) as possible. Let the frequency separation of A and B be $\Delta\nu$. (The points A and B need not be distinct maxima or minima.) In order for the points A and B to be resolved, it is necessary that the frequency separation be at least as large as $1/\pi T_2$, where T_2 is the *inhomogeneous linewidth parameter*. Let us now calculate how much angular rotation of the protein $\Delta\vartheta$ is necessary to produce a change in resonance frequency of

$$\Delta\nu \simeq 1/\pi T_2 \tag{A.1}$$

This is

$$\frac{\Delta\vartheta}{\pi/2}(T_\parallel - T_\perp) = \frac{1}{\pi T_2} \tag{A.2}$$

$$\Delta\vartheta = \frac{1}{2T_2}\frac{1}{T_\parallel - T_\perp} \tag{A.3}$$

Here $T_\parallel$ and $T_\perp$ are the hyperfine splittings parallel and perpendicular to the principal hyperfine axis of the largest hyperfine splitting (π-orbital axis).

Let us now calculate the time required for this angular displacement. It is

$$t = \tfrac{3}{2}\tau_c\,\overline{(\Delta\vartheta)^2} \tag{A.4}$$

$$t = \tfrac{3}{2}\tau_c\,\tfrac{1}{4}(1/T_2^2)[1/(T_\parallel - T_\perp)^2] \tag{A.5}$$

From general principles of magnetic resonance theory, we know that the

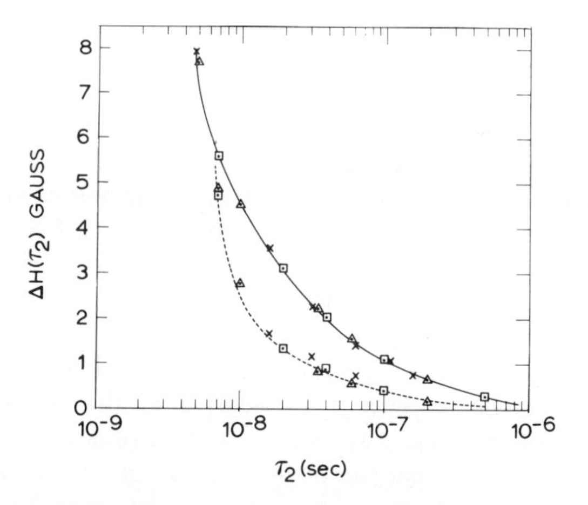

Fig. 21. Inward shifts ΔH in the paramagnetic resonance spectra of nearly strongly immobilized spin labels of the high-field (——) and low-field (- - -) extrema of the derivative curve spectrum as a function of τ_2 calculated for $T_\parallel = 34.3$ G, $T_\perp = 6.5$ G, and $\Delta g = 0.0051$; □, calculated for nonaxial parameters; △, calculated for $T_\parallel = 30.8$ G, $T_\perp = 5.8$ G, and $\Delta g = 0.0053$; (all of the above have $T_2 = 2.4 \times 10^{-8}$ sec) (McCalley *et al*., 1972); ×, Messenger and Gordon calculation (private communication) for $T_\parallel = 30.8$ G, $T_\perp = 5.8$ G, $\Delta g = 0.0053$, and $T_2 = 3.3 \times 10^{-8}$ sec; ○, Goldman *et al.* (1972) for $T_\parallel = 32.0$ G, $T_\perp = 6.0$ G, $\Delta g = 0.048$, and $T_2 = 2.2 \times 10^{-8}$ sec. [From McCalley *et al.* (1972).]

spectra at A and at B and between A and B will be modified when $t \lesssim 1/(\pi \, \Delta\nu)$, that is, when

$$t \lesssim T_2 \tag{A.6}$$

Thus the paramagnetic resonance absorption spectrum is sensitive to motions with correlation times equal to or less than

$$\tau_c = \tfrac{8}{3} T_2^3 (T_{\parallel} - T_{\perp})^2 \tag{A.7}$$

For an order-of-magnitude calculation, $T_{\parallel} - T_{\perp} = 70$ MHz,

$$T_2 = 2.4 \times 10^{-8}$$

(inhomogeneous linewidth for the outer signals of nitroxide radicals), the estimate yields $\tau_c \sim 2 \times 10^{-7}$ sec, in good order-of-magnitude agreement with the theoretical calculations leading to the results summarized in Fig. 21.

Recently certain paramagnetic resonance techniques have been developed that offer promise for the measurement of much longer correlation times for spin labels. These methods may generally be referred to as forms of "*saturation transfer spectroscopy*." The essential idea is that the spin system is partially saturated at one spectral position and the transfer of saturation is monitored at a second spectral position $\Delta\nu$ away. In the ideal case the transfer of spin saturation is brought about exclusively by the rotational diffusion of the molecule. This general method greatly increases the upper limit on correlation times that can be measured. It has been shown both experimentally and by detailed theoretical calculations that times as long as 10^{-4} sec may be determined. As shown below, saturation transfer spectroscopy increases the longest correlation time that can be measured by a factor of the order of $T_1/T_2 \simeq 300$, where $T_2 \simeq 2.4 \times 10^{-8}$ sec is the effective inhomogeneous T_2 and $T_1 \simeq 6.6 \times 10^{-6}$ sec.

As we have noted, two just resolved spectral points A and B separated in frequency by $\Delta\nu$ correspond to molecular orientations that are converted into one another in a time t given in Eq. (A.4). Spin saturation introduced into the system at one spectral point (say A) will thus be detectable at the second point (B) provided t is of the order of or less than T_1. Thus the longest correlation time that can be detected using saturation transfer spectroscopy is

$$\tau_c \approx \tfrac{8}{3} T_1 T_2^2 (T_{\parallel} - T_{\perp})^2$$

In this case the upper value of τ_c that can be determined is ~ 300 times larger, i.e., $\sim 10^{-4}$ sec.

For descriptions of saturation transfer spectroscopy, see Hyde and Dalton (1972), Hyde and Thomas (1973), Thomas and McConnell (1974), and Smigel *et al.* (1974).

APPENDIX B. DIRECT EXPERIMENTAL EVIDENCE FOR A PHOSPHOLIPID SPIN LABEL TILT

Figure 22 gives a set of paramagnetic resonance spectra providing quite direct experimental evidence supporting the idea that the principal π-orbital axis of **IV**(10, 3) is tilted by about 30° relative to the bilayer normal in hydrated egg lecithin multilayers.

Figure 22a shows the paramagnetic resonance spectrum of an isotropic sample of egg lecithin multilayers containing the fatty acid label **IV**(10, 3). As has been shown elsewhere, the indicated separation of the outer hyperfine extrema is equal to $2T'_{\parallel}$, exactly twice the hyperfine splitting to be found in a perfectly oriented sample when the applied field is parallel to the principal axis z'. When the collective tilt in the lipid bilayer is destroyed by the use of tetracaine, z' is in fact perpendicular to the bilayer plane, as can be seen by a comparison of the spectra in Figs. 22a and 22b. In spectrum (b), the applied field is exactly parallel to the principal axes z' of all the spin labels. In going from spectrum (b) to spectrum (c), the applied field direction is moved 30° from the bilayer normal, so the applied field is now 30° from the principal

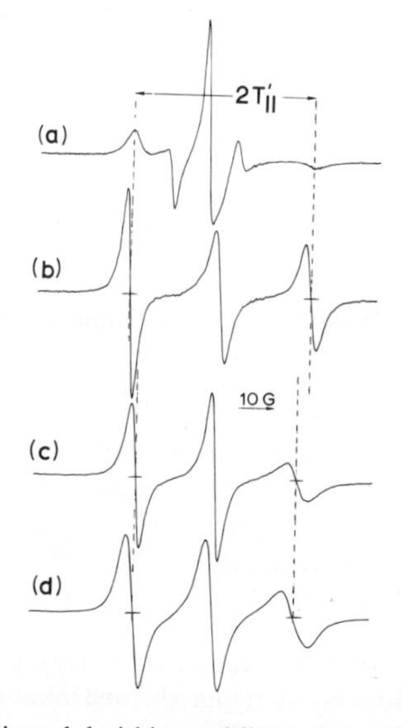

Fig. 22. Spectra for oriented lecithin multilayers containing the fatty acid spin label **IV**(10, 3). (a) Isotropic distribution of hydrated lecithin containing tetracaine (10 : 1 mole ratio). (b) Spectrum of oriented multilayers of lecithin containing tetracaine with the magnetic field perpendicular to the bilayer surface. (c) Same as (b) except that the magnetic field makes an angle of 30° with the normal to the bilayer plane. (d) Spectrum of oriented multilayers of egg lecithin (tetracaine free) taken with the magnetic field perpendicular to bilayer surface. The value of $2T'_{\parallel}$ for **IV**(10, 3) is not changed by tetracaine in these experiments. [From McConnell and McFarland (1972) with permission of the New York Academy of Sciences.]

axis z' of each label. The hyperfine splitting accordingly decreases. The oriented multilayer sample used to obtain the spectrum in (d) contains *no* tetracaine, and the π orbital is thought to be tilted $\sim 30^\circ$ from the bilayer normal. When the applied field direction is parallel to the bilayer normal in this case it should make an angle of 30° with *all* the principal axes z' in this sample. Thus the spectrum, and particularly the hyperfine splittings in (c) and (d), should be the same. They evidently are.

References

Brown, P. K. (1972). Rhodopsin rotates in the visual receptor membrane, *Nature* (*London*) *New Biol.* **236**, 35–38.

Chen, Y. S., and Hubbell, W. L. (1973). Temperature- and light-dependent structural changes in rhodopsin lipid membranes, *Exp. Eye Res.* **17**, 517–532.

Cone, R. A. (1972). Rotational diffusion of rhodopsin in the visual receptor membrane, *Nature* (*London*) *New Biol.* **236**, 39–43.

Dehlinger, P. J., Jost, P. C., and Griffith, O. H. (1974). Lipid binding to the amphipathic membrane protein cytochrome b_5, *Proc. Nat. Acad. Sci. U.S.* **71**, 2280–2284.

Devaux, P., and McConnell, H. M. (1972). Lateral diffusion in spin-labeled phosphatidylcholine multilayers, *J. Amer. Chem. Soc.* **94**, 4475–4481.

Devaux, P., and McConnell, H. M. (1974). Equality of the rates of lateral diffusion of phosphatidylethanolamine and phosphatidylcholine spin labels in rabbit sarcoplasmic reticulum, *Ann. N.Y. Acad. Sci.* **222**, 489–496.

Devaux, P., Scandella, C. J., and McConnell, H. M. (1973). Spin–spin interactions between spin-labeled phospholipids incorporated into membranes, *J. Magn. Res.* **9**, 474–485.

Gaffney, B. J., and McNamee, C. M. (1974). Spin label measurements in membranes, *Methods Enzymol.* **32B**, 161–198.

Gaffney, B. J., and McConnell, H. M. (1974a). The paramagnetic resonance spectra of spin labels in phospholipid membranes, *J. Magn. Res.* **16**, 1–28.

Gaffney, B. J., and McConnell, H. M. (1974b). Effect of a magnetic field on phospholipid membranes, *Chem. Phys. Lett.* **24**, 310–313.

Goldman, S. A., Bruno, G. V., and Freed, J. H. (1972). Estimating slow motional rotational correlation times for nitroxides by electron spin resonance, *J. Phys. Chem.* **76**, 1858–1860.

Grant, C. W. M., and McConnell, H. M. (1973). Fusion of phospholipid vesicles with viable, *Acholeplasma laidlawii*, *Proc. Nat. Acad. Sci. U.S.* **70**, 1238–1240.

Grant, C. W. M., and McConnell, H. M. (1974). Glycophorin in lipid bilayers. *Proc. Nat. Acad. Sci. U.S.* **71**, 4653–4657.

Grant, C. W. M., Wu, S. H. W., and McConnell, H. M. (1974). Lateral phase separations in binary lipid mixtures: Correlation between spin label and freeze-fracture electron microscopic studies, *Biocheim. Biophys. Acta* **363**, 151–158.

Hemminga, M. A., and Berendsen, H. J. C. (1972). Magnetic resonance in ordered lecithin–cholesterol multilayers, *J. Magn. Res.* **8**, 133–143.

Hong, K., and Hubbell, W. L. (1972). Preparation and properties of phospholipid bilayers containing rhodopsin, *Proc. Nat. Acad. Sci.* **69**, 2617–2621.

Hubbell, W. L., and McConnell, H. M. (1968). Spin-label studies of the excitable membranes of nerve and muscle, *Proc. Nat. Acad. Sci. U.S.* **61**, 12–16.

Hubbell, W. L., and McConnell, H. M. (1969a). Motion of steroid spin labels in membranes. *Proc. Nat. Acad. Sci. U.S.* **63**, 16–22.

Hubbell, W. L., and McConnell, H. M. (1969b). Orientation and motion of amphiphilic spin labels in membranes. *Proc. Nat. Acad. Sci. U.S.* **64**, 20–27.

Hubbell, W. L., and McConnell, H. M. (1971). Molecular motions in spin-labeled phospholipids and membranes, *J. Amer. Chem. Soc.* **93**, 314–326.

Huestis, W. H., and McConnell, H. M. (1974). A functional acetylcholine receptor in the human erythrocyte. *Biochem. Biophys. Res. Commun.* **57**, 726–732.

Hyde, J. S., and Dalton, L. R. (1972). Very slowly tumbling spin labels: Adiabatic rapid passage, *Chem. Phys. Lett.* **16**, 568–572.

Hyde, J. S., and Thomas, D. D. (1973). New EPR methods for the study of very slow motion: application to spin-labeled hemoglobin, *Ann. N.Y. Acad. Sci.* **222**, 680–692.

Jost, P. C., Capaldi, R. A., Vanderkooi, G., and Griffith, O. H. (1973). Lipid–protein and lipid–lipid interactions in cytochrome oxidase model membranes, *J. Supra. Struct.* **1**, 269–280.

Jost, P. C., Libertini, L. J., Hebert, U. C., and Griffith, O. H. (1971). Lipid spin labels in lecithin multilayers. A study of motion along fatty acid chains, *J. Mol. Biol.* **59**, 77–98.

Kleemann, W., and McConnell, H. M. (1974). Lateral phase separations in *Escherichia coli* membranes, *Biochim. Biophys. Acta* **345**, 220–230.

Kornberg, R. D., and McConnell, H. M. (1971a). Inside–outside transitions of phospholipids in vesicle membranes. *Biochemistry* **10**, 1111–1120.

Kornberg, R. D., and McConnell, H. M. (1971b). Lateral diffusion of phospholipids in a vesicle membrane, *Proc. Nat. Acad. Sci. U.S.* **68**, 2564–2568.

Kury, P. G., Ramwell, P. W., and McConnell, H. M. (1973). The effect of prostaglandinds E_1 and E_2 on the human erythrocyte as monitored by spin labels, *Biochem. Biophys. Res. Commun.* **56**, 478–483.

Libertini, L. J., Waggoner, A. S., Jost, P., and Griffith, O. H. (1969). Orientation of lipid spin labels in lecithin multilayers, *Proc. Nat. Acad. Sci. U.S.* **64**, 13–19.

Likhtenstein, G. T. (1974). "Spin Labeling." Nauka, Moscow.

Linden, C., Wright, K., McConnell, H. M., and Fox, C. F. (1973). Phase separations and glucoside uptake in *E. coli* fatty acid auxotrophs, *Proc. Nat. Acad. Sci. U.S.* **70**, 2271–2275.

Marchesi, V. T., Tillack, T. W., Jackson, R. L., Segrest, J. P., and Scott, R. E. (1972). Chemical characterization and surface orientation of the major glycoprotein of the human erythrocyte membrane, *Proc. Nat. Acad. Sci. U.S.* **69**, 1445–1449.

McCalley, R. C., Shimshick, E. J., and McConnell, H. M. (1972). The effect of slow rotational motion on paramagnetic resonance spectra, *Chem. Phys. Lett.* **13**, 115–119.

McConnell, H. M., and McFarland, B. G. (1970). Physics and chemistry of spin labels, *Quart. Rev. Biophys.* **3**, 91–136.

McConnell, H. M., and McFarland, B. G. (1972). The flexibility gradient in biological membranes, *Ann. N.Y. Acad. Sci.* **195**, 207–217.

McConnell, H. M., Devaux, P., and Scandella, C. (1972). Lateral diffusion and phase separations in biological membranes, *In* "Membrane Fusion" (C. F. Fox, ed.), pp. 27–37. Academic Press, New York.

McFarland, B. G., and McConnell, H. M. (1971). Bent fatty acid chains in lecithin bilayers, *Proc. Nat. Acad. Sci. U.S.* **68**, 1274–1278.

McNamee, M. G., and McConnell, H. M. (1973). Transmembrane potentials and phospholipid flip-flop in excitable membrane vesicles, *Biochemistry* **12**, 2951–2958.

Poo, M., and Cone, R. A. (1974). Lateral diffusion of rhodopsin in the photoreceptor membrane, *Nature (London) New Biol.* **247**, 438–441.

Prestegard, J. H., and Fellmeth, B. (1974). Fusion of dimyristoyllecithin vesicles as studied by proton magnetic resonance spectroscopy, *Biochemistry* **13**, 1122–1126.

Sackmann, E., and Träuble, H. (1972a). Studies of the crystalline–liquid crystalline phase transition of lipid model membranes. I. Use of spin labels and optical probes as indicators of the phase transition, *J. Amer. Chem. Soc.* **94**, 4482–4491.

Sackmann, E., and Träuble, H. (1972b). Studies of the crystalline–liquid crystalline phase transition of lipid model membranes. II. Analysis of electron spin resonance spectra of steroid labels incorporated into lipid membranes, *J. Amer. Chem. Soc.* **94**, 4492–4498.

Saupe, A. (1968). Recent results in the field of liquid crystals, *Angew. Chem. Int. Ed.* **7**, 97–112.

Scandella, C. J., Devaux, P., and McConnell, H. M. (1972). Rapid lateral diffusion of phospholipids in rabbit sarcoplasmic reticulum, *Proc. Nat. Acad. Sci. U.S.* **69**, 2056–2060.

Schindler, H., and Seelig, J. (1973). EPR spectra of spin labels in lipid bilayers, *J. Chem. Phys.* **59**, 1841–1850.

Seelig, J. (1970). Spin label studies of oriented smectic liquid crystals (a model system for bilayer membranes), *J. Amer. Chem. Soc.* **92**, 3881–3887.

Seelig, J., and Seelig, A. (1974a). Deuterium magnetic resonance studies of phospholipid bilayers. *Biochem. Biophys. Res. Commun.* **57**, 406–411.

Seelig, A., and Seelig, J. (1974b). The dynamic structure of fatty acyl chains in a phospholipid bilayer measured by deuterium magnetic resonance, *Biochemistry* **13**, 4839–4845.

Shimshick, E. J., and McConnell, H. M. (1973a). Lateral phase separations in phospholipid membranes, *Biochemistry* **12**, 2351–2360.

Shimshick, E. J., and McConnell, H. M. (1973b). Lateral phase separations in binary mixtures of cholesterol and phospholipids. *Biochem. Biophys. Res. Commun.* **53**, 446–451.

Shimshick, E. J., Kleemann, W., Hubbell, W. L., and McConnell, H. M. (1973). Lateral phase separations in membranes. *J. Supra. Struct.* **1**, 285–294.

Smigel, M. D., Dalton, L. R., Hyde, J. S., and Dalton, L. A. (1974). Investigation of very slowly tumbling spin labels by nonlinear spin response techniques: Theory and experiment for stationary electron electron double resonance, *Proc. Nat. Acad. Sci. U.S.* **71**, 1925–1929.

Stier, A., and Sackmann, E. (1973). Spin labels as enzyme substrates: heterogeneous lipid distribution in liver microsomal membranes, *Biochim. Biophys. Acta* **311**, 400–408.

Taupin, C., and McConnell, H. M. (1972). Membrane fusion. "Mitochondria Biomembranes," pp. 219–229. North-Holland Publ., Amsterdam.

Thomas, D. D., and McConnell, H. M. (1974). Calculation of paramagnetic resonance spectra sensitive to very slow rotational motion, *Chem. Phys. Lett.* **25**, 470–475.

Träuble, H., and Sackmann, E. (1972). Studies of the crystalline–liquid crystalline phase transition of lipid model membranes. III. Structure of a steroid–lecithin system below and above the lipid–phase transition, *J. Amer. Chem. Soc.* **94**, 4499–4510.

Wilkens, M. H. F., Blaurock, A. E., and Engelman, D. M. (1971). Bilayer structure in membranes, *Nature (London) New Biol.* **230**, 72–76.

Wu, S. H. W., and McConnell, H. M. (1974). Phase separations in phospholipid membranes, *Biochemistry* **14**, 847–854.

Appendixes

I. Simulated X-Band Spectra—Isotropic Tumbling
II. Principal Values of the **g** and Hyperfine Tensors for Several Nitroxides Reported to Date
III. Commercial Sources of Spin Labeling Equipment and Supplies
IV. Practical Considerations for the Calculation of Order Parameters for Fatty Acid or Phospholipid Spin Labels in Membranes
V. Symbols and Abbreviations

Appendix I

Simulated X-Band Spectra—Isotropic Tumbling

The figure shows simulated X-band spectra for a range of six rotational correlation times τ_2. The nitroxide simulated here has nonaxial **g** and **T** tensors:

$$g_{xx} = 2.00901 \qquad T_{xx} = 6.8 \text{ G}$$

$$g_{yy} = 2.00601 \qquad T_{yy} = 6.2 \text{ G}$$

$$g_{zz} = 2.00241 \qquad T_{zz} = 34.3 \text{ G}$$

The intrinsic electron T_2 was 2.4×10^{-8} sec.

We gratefully acknowledge Dr. Edward Shimshick, Department of Chemistry, Stanford University, for providing the simulation data. This was calculated from diffusion-coupled Bloch equations. (See Chapter 3 for related methods and theory.)

References

McCalley, R. C., Shimshick, E. J., and McConnell, H. M. (1972). The effect of slow rotational motion on paramagnetic resonance spectra, *Chem. Phys. Lett.* **13**, 115–119.

Shimshick, E. J. (1973). (Part I): Rotational diffusion of spin-labeled proteins, Ph.D. Dissertation, Stanford Univ.

Shimshick, E. J., and McConnell, H. M. (1972). Rotational correlation time of spin-labeled α-chymotrypsin, *Biochem. Biophys. Res. Commun.* **46**, 321–327.

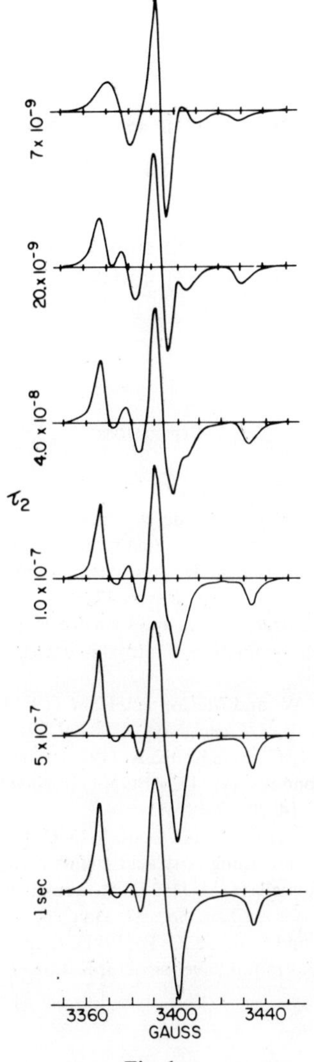
7 x 10⁻⁹
20. x 10⁻⁹
4.0 x 10⁻⁸
τ₂
1.0 x 10⁻⁷
5 x 10⁻⁷
1 sec
3360
3400
3440
GAUSS

Fig. 1.

Appendix II

Principal Values of the g and Hyperfine Tensors for Several Nitroxides Reported to Date

All determinations were made in single crystals unless noted otherwise. See also Chapter 3, Table I (p. 73) and Chapter 6, Table II (p. 247) for additional data.

REFERENCES

1. Libertini, L. J., and Griffith, O. H. (1970). Orientation dependence of the electron spin resonance spectrum of di-*t*-butyl nitroxide, *J. Chem. Phys.* **53**, 1359–1367.
2. Bordeaux, D., Lajzerowicz, J., Briere, R., Lemaire, H., and Rassat, A. (1973). Determination des axes propres et des valeurs principales du tenseur *g* dans deux radicaux libres nitroxydes par etude de monocristaux, *Org. Magn. Res.* **5**, 47–52.
3. Capiomont, A., Chion, B., Lajzerowicz, J., and Lemaire, H. (1974). Interpretation and utilization for crystal structure determination of nitroxide free radicals in single crystals, *J. Chem. Phys.* **60**, 2530–2535.
4. Griffith, O. H., Cornell, D. W., and McConnell, H. M. (1965). Nitrogen hyperfine tensor and *g* tensor of nitroxide radicals, *J. Chem. Phys.* **43**, 2909–2910.
5. Snipes, N., Cupp, J., Cohn, G., and Keith, A. (1974). Analysis of the nitroxide spin label 2,2,6,6-tetramethylpiperidone-*N*-oxyl (TEMPONE) in single crystals of the reduced TEMPONE matrix, *Biophys. J.* **14**, 20–32.
6. Jost, P., Libertini, L. J., Hebert, V. C., and Griffith, O. H. (1971). Lipid spin labels in lecithin multilayers. A study of motion along fatty acid chains, *J. Mol. Biol.* **59**, 77–98.
7. Hubbell, W. L., and McConnell, H. M. (1971). Molecular motion in spin-labeled phospholipids and membranes, *J. Amer. Chem. Soc.* **93**, 314–326.
8. Hsia, J-C., Schneider, H., and Smith, I. C. P. (1971). A spin label study of the influence of cholesterol on phospholipid multibilayer structures, *Can. J. Biochem.* **49**, 614–622.

STRUCTURE	g_{xx}	g_{yy}	g_{zz}	A_{xx}(G)	A_{yy}(G)	A_{zz}(G)	Ref.
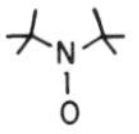	2.0088	2.0062	2.0027	7.6	6.0	31.8	1
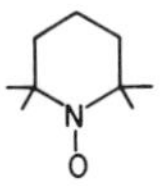	2.0103	2.0069	2.0030	–	–	–	2
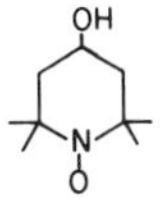	2.0095	2.0064	2.0027	–	–	–	2
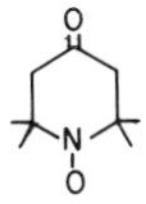	2.0104	2.0074	2.0026	5.2	5.2	31	3,4
	–	–	–	6.5	6.7	33	5
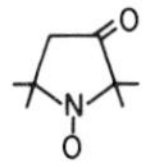	–	–	–	4.7	4.7	31	4
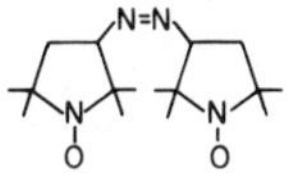	2.0101	2.0068	2.0028	–	–	–	3
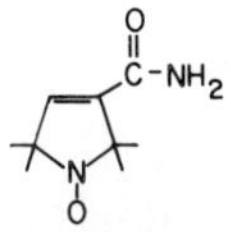	2.0086	2.0066	2.0032	–	–	31	4,3
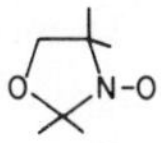	2.0088	2.0058	2.0022	5.9	5.4	32.9	6
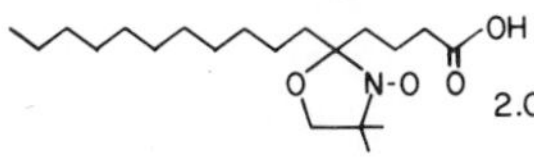	2.0088	2.0061	2.0027	6.3	5.8	33.6	Appendix IV
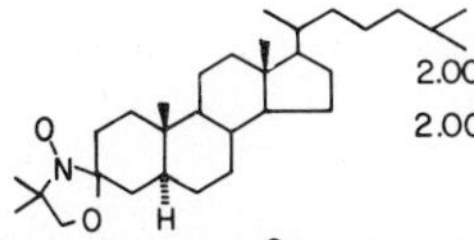	2.0090	2.006	2.002	6.3	5.9	32.	Appendix IV
	2.0089	2.0058	2.0021	5.8	5.8	30.8	7
		($g_{\perp}$) 2.0061	($g_{\parallel}$) 2.0030	6.0	6.0	32	8
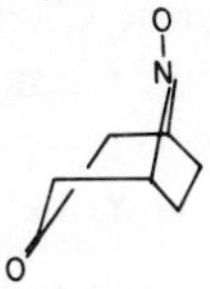	2.0104	2.0066	2.0038	–	–	–	3

Appendix III

Commercial Sources of Spin Labeling Equipment and Supplies

This listing comprises the most common sources and is by no means intended to be complete.

A. ESR Spectrometers

Varian Associates
Analytical Instrument Division
or
Varian Anaspect
611 Hansen Way
Palo Alto, California 94303

Bruker Scientific
Manning Park
Billerica, Massachusetts 01821

JEOL USA, Inc.
477 Riverside Avenue
Medford, Massachusetts 02155

B. Quartz Sample Cells and Glassware

Varian Instruments
611 Hansen Way
Palo Alto, California 94303

Wilmad Glass Company, Inc.
U.S. Route 40 and Oak Road
Buena, New Jersey 08310

James F. Scanlon Company
2428 Baseline Avenue
Solvang, California 93463

C. Spin Labels and Nitroxide Precursors

Aldrich Chemical Company, Inc.
940 W. St. Paul Avenue
Milwaukee, Wisconsin 53233

Frinton Labs
P.O. Box 301
Grant Avenue
So. Vineland, New Jersey 08360

Eastman Organic Chemicals
Eastman Kodak Company
343 State Street
Rochester, New York

SYVA
3221 Porter Drive
Palo Alto, California 94304

Appendix IV

Practical Considerations for the Calculation of Order Parameters for Fatty Acid or Phospholipid Spin Labels in Membranes

BETTY JEAN GAFFNEY

DEPARTMENT OF CHEMISTRY
THE JOHNS HOPKINS UNIVERSITY
BALTIMORE, MARYLAND

For semiquantitative analysis of the paramagnetic resonance spectra arising from fatty-acid-type spin labels in phospholipid membranes, certain spectral parameters are employed to calculate order parameters S. Although computer simulation of spectra provides precise values of the necessary parameters ($T'_{\parallel}$ and $T'_{\perp}$), it is usually possible by using appropriate correction terms to obtain quite accurate values of order parameters directly from experimental spectra, as discussed below.

Figure 1 shows calculated spectra for the anisotropic motion of fatty acid spin labels. The characteristic motions of these labels (Seelig, 1970; Hubbell and McConnell, 1971) include high frequency axial averaging about the long axis of the chain and motion about the long axis which has an amplitude characterized by an order parameter S (see Chapter 10). The order parameter is defined in terms of observed spectral parameters as follows (Hubbell and McConnell, 1971; Gaffney, 1974):

$$S = \frac{T'_{\parallel} - T'_{\perp}}{T_{zz} - \frac{1}{2}(T_{xx} + T_{yy})} \tag{1}$$

The values of T_{zz}, T_{xx}, and T_{yy} are the principal elements of the T tensor and $T'_{\parallel}$ and $T'_{\perp}$, are, to a first approximation, equal to $\frac{1}{2}$ the separation of the outer and inner extrema, respectively, of the spectra shown in Fig. 1. Figure 2 shows a comparison of calculated (input parameters) and "experimental" first approximation values of $T'_{\parallel}$ and $T'_{\perp}$ for the spectra of Fig. 1. It is clear that $T'_{\parallel}$ can be derived from experimental spectra with high accuracy for all

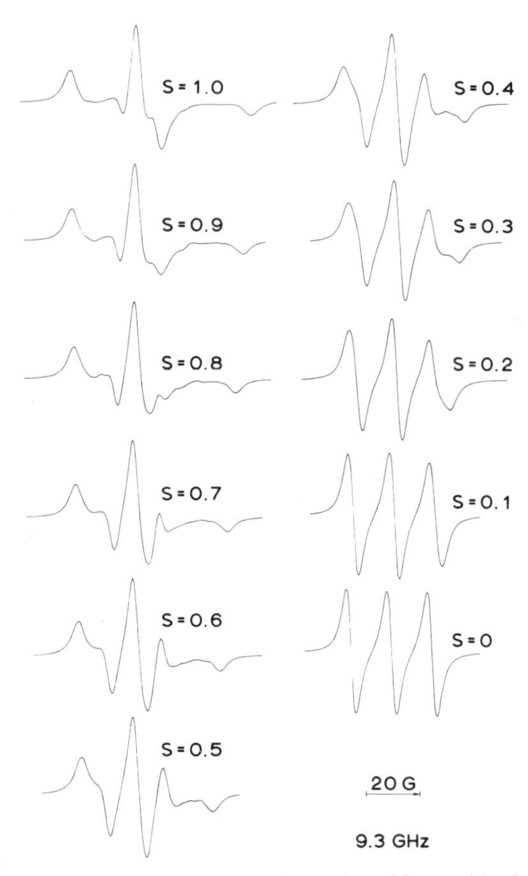

Fig. 1. Spectra calculated for the anisotropic motion of fatty acid spin labels. Axial symmetry is assumed in the calculation. The calculation employed the T-tensor and g-factor elements shown in Table I for label **II**(10, 3). The order parameters S and values of $T'_{\parallel}$ and $T'_{\perp}$ are related as shown in Eq. (1).

order parameters greater than 0.2. However, the true value of $T'_{\perp}$ differs from the observed separation of the inner spectral extrema, $T'_{\perp}$(approx.), the divergence increasing linearly with decreasing order parameter. Thus the calculation of order parameters from data of experimental spectra requires modification of Eq. (1) to include the correction to $T'_{\perp}$(approx.). From the data presented in Fig. 2, it is found that C, the correction term for $T'_{\perp}$(approx.), is given by

$$C = (4.06 \text{ MHz})\left[1 - \frac{T'_{\parallel} - T'_{\perp}(\text{approx.})}{T_{zz} - \frac{1}{2}(T_{xx} + T_{yy})}\right] \tag{2}$$

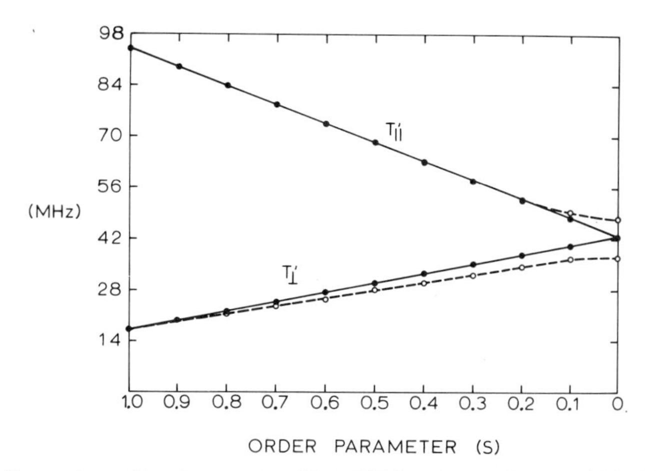

Fig. 2. Comparison of input parameters $T'_{\parallel}$ and $T'_{\perp}$ for computed spectra (●) with the first approximation values (○) of $T'_{\parallel}$ and $T'_{\perp}$ (approx.) derived from the spectra of Fig. 1.

Thus $T'_{\perp} = T'_{\perp}(\text{approx.}) + C$. A second correction to (1) must be made to account for the possibility that the polarity of the environment of the experimental sample may differ from the polarity corresponding to the single-crystal measurement from which T_{xx}, T_{yy}, and T_{zz} are obtained. Thus

$$S = \frac{T'_{\parallel} - T'_{\perp}}{T_{zz} - \frac{1}{2}(T_{xx} + T_{yy})} \cdot \frac{a}{a'} \tag{3}$$

where $a = \frac{1}{3}(T_{zz} + T_{yy} + T_{xx})$ and $a' = \frac{1}{3}(T'_{\parallel} + 2T'_{\perp})$.

Equation (3) may be expressed in convenient form by substitution of appropriate values of T_{xx}, T_{yy}, and T_{zz}. Single-crystal measurements have provided values for the elements of the T tensor (and g factor as well) for cholestane spin label (revised by R. C. McCalley from Hubbell and McConnell, 1971). In addition, computer simulation of the spectra recorded at two magnetic field strengths for a fatty acid spin label immobilized on lyophilized bovine serum albumin (BSA) provided T-tensor elements for the fatty acid labels (Gaffney and McConnell, 1974). These values are summarized in Table I for labels **I** and **II**.

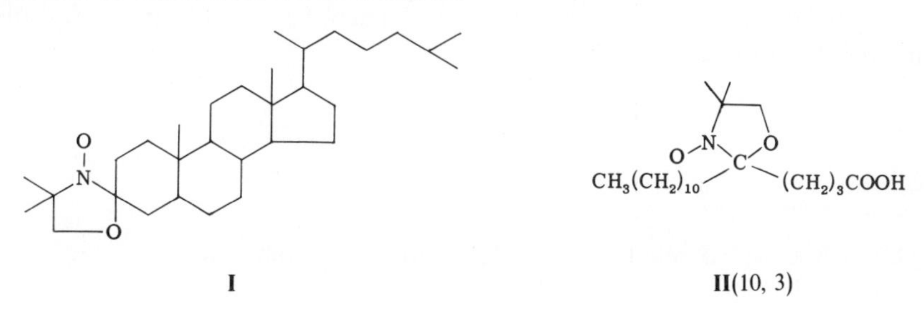

TABLE I

TENSOR ELEMENTS

Label	T_{zz}[a]	T_{xx}[a]	T_{yy}[a]	g_{zz}	g_{xx}	g_{yy}
Cholestane spin label **I**	89.4 ±0.06	17.7 ±0.3	16.4 ±0.3	2.0024 ±0.0001	2.0090 ±0.0001	2.0060 ±0.0001
II(10, 3) + lyophilized BSA	93.8	17.6	16.3	2.0027	2.0088	2.0061

[a] Elements of the T tensor are given in MHz.

Substitution of T-tensor elements for **II**(10, 3) fatty acid spin label and the correction to $T'_{\perp}$(approx.) into Eq. (3) then gives the following simplified expression for the order parameter:

$$S = \frac{T'_{\parallel} - [T'_{\perp}(\text{approx.}) + C]}{T'_{\parallel} + 2[T'_{\perp}(\text{approx.}) + C]} \cdot 1.66 \qquad (4)$$

where

$$C = \{4.06 - 0.053[T'_{\parallel} - T'_{\perp}(\text{approx.})]\} \quad \text{MHz} \qquad (5)$$

Equations (4) and (5) can be used to calculate order parameters for the spectra of fatty acid and phospholipid spin labels in biological membranes and in model membranes. However, in some samples either the inner or outer hyperfine extrema may not be well defined at a particular temperature. In this case, a plot of either $T'_{\parallel}$ or $T'_{\perp}$(approx.) versus temperature is physically significant because both of these parameters have a linear dependence on the order parameter, as shown in Fig. 2. Other spectral parameters that do not vary linearly with order parameter may give erroneous breaks in plots of parameter versus temperature (breaks of this type are often interpreted as phase transitions in membranes).

It should also be noted that the calculations presented here have been made for spectral linewidths in the range observed for fatty acid **II**(m, n) and phospholipid spin labels in phospholipid bilayers. The necessary corrections to $T'_{\perp}$ have not been tested for spectra with narrow linewidths (e.g., alkanol–alkanoate liquid crystals).

REFERENCES

Gaffney, B. J., and McNamee, C. M. (1974). Spin label measurements in membranes, *In* Biomembranes, *Methods Enzymol.* **32**, 161–198.

Gaffney, B. J., and McConnell, H. M. (1974). "The paramagnetic resonance spectra of spin labels in phospholipid membranes, *J. Magn. Res.* **16**, 1–28.
Hubbell, W. L., and McConnell, H. M. (1971). Molecular motion in spin labeled phospholipids and membranes, *J. Amer. Chem. Soc.* **93**, 314–326.
Seelig, J. (1970). Spin label studies of oriented smectic liquid crystals (A model system for bilayer membranes), *J. Amer. Chem. Soc.* **92**, 3881–3882.

Appendix V

Symbols and Abbreviations

Where the same symbol is used with significantly different meaning, or where the symbol is unique to one chapter, the appropriate chapter is noted in boldface type.

Symbol	Meaning
a	Rigid sphere radius of a molecule (**3**)
a, A, A_0	Scalar (isotropic) hyperfine coupling constant
a_i	Scalar (isotropic) hyperfine coupling constant of the ith nucleus (**3**)
$\|a(1)\rangle$	Molecular wave function (**4**)
A	Coefficient of $\tilde{M}^0$ in expansion of δ [**3**, Eq. (79)]
A	Electron–nuclear hyperfine tensor (**10**)
A	Electron–nuclear hyperfine tensor (sometimes refers to the total, isotropic or anisotropic hyperfine tensor)
$\mathbf{A}^{(k)}$	Hyperfine tensor relative to nucleus k (**2**)
$\mathbf{A}'$	Traceless hyperfine dipolar tensor (**2**)
A_0	Tr **A** (**2**)
A_N	Isotropic hyperfine coupling constant for a nitroxide moiety (**11**)
A_{ij}	Hyperfine tensor element
A_{max}	One-half the distance between the outermost lines (first maximum to last minimum in the derivative spectrum) of the spectrum of a randomly oriented sample (**12**)
A_{min}	One-half the distance from the first minimum to the last maximum of the first derivative spectrum of a randomly oriented sample (**12**)
A_{xx}, A_{yy}, A_{zz}, A_i	Principal components of the hyperfine tensor; the superscript c is sometimes used to designate values of these parameters measured from a spin label oriented in a crystalline lattice where there would otherwise be an ambiguity between single-crystal values and principal values
$\langle A\rangle$	$= \frac{1}{3}(A_x''' + A_y''' + A_z''')$ (**3**)

$A_{\parallel}$, $A_{\perp}$, $A'_{\parallel}$, $A'_{\perp}$	Components of the time-averaged electron–nuclear hyperfine tensor (**A**); parallel and perpendicular components of A.
$\bar{A}_{\parallel}$, $\bar{A}_{\perp}$	The motion-averaged hyperfine splittings with the magnetic field parallel or perpendicular to the long molecular axis of the spin label (the axis of rotation R) (**12**)
ΔA_{v}, ΔA_{n}, ΔA_{dl}, ΔA_{ϕ}	The contributions to ΔA_0 from van der Waal's interactions, hydrogen bonding, electric fields of the charged double layer, and the membrane potential, respectively (**12**)
$A_{\mu,i}^{(L,M)}$, $A_{\lambda}^{(l,m)}$	Irreducible tensor component of rank $L(l)$ and component m of a spin operator
$\mathcal{A}$; $\mathcal{A}_{\rm d}$	r-dimensional complex symmetric matrix containing the solution to the slow tumbling problem (**3**); diagonal form of $\mathcal{A}$
b, b'	Components in estimating rotational correlation times (**3**)
B	Magnetic field (**4**) (also denoted as H_0, **H**, and $\bar{H}$)
B	Frictional coefficient for rotational diffusion [**3**, Eq. (79)]
B_L	Model parameter for eigenvalues of $\tilde{\Gamma}_r$ belonging to the Lth set of eigenfunctions (**3**)
$\Delta B_{\rm pp}$	Peak-to-peak linewidth (**4**)
c	Speed of light
c_n	Expansion coefficient of ψ in U_n basis set (**3**)
C	Coefficient of $\tilde{M}^2$ in expansion of δ [**3**, Eq. (79)]
$C(j_1 j_2 j_3; m_1 m_2)$	Clebsch–Gordan coefficient (**2**)
$C_m^{(n)}(\omega)$, $[C_m^{(n)}(\omega)]_\lambda$	Expansion coefficient of $\tilde{Z}^{(n)}(\Omega)$, $\tilde{Z}_\lambda^{(n)}(\Omega)$ in basis set of $G_m(\Omega)$ (**3**)
$C_{KM}^{L}(i)$	Equivalent to $[C_m^{(n)}(\omega)]_\lambda$ where the "general quantum number" m is replaced by the specific L, K, M "quantum numbers" for rotational reorientation and $\lambda \rightarrow i$ (**3**)
C	r-dimensional column vector consisting of the expansion coefficients $C_{K,M}^{L}(i)$ (**3**)
d	Interaction distance for exchange (**3**); spacing between spectral lines (**4**)
$d_{\rm e}$	Transition moment given by $\frac{1}{2}\gamma_{\rm e} H_1$ (**3**)
d_λ	Transition moment given by $\frac{1}{2}\gamma_{\rm e} H_1 \langle \lambda^- \| S_- \| \lambda^+ \rangle$ (**3**)
$d_{l,m}^{j}(\beta)$	Reduced Wigner rotation matrix element
$d_{\max}(\tilde{M})$	Value of the $d_{\rm e}$ at which the $\tilde{M}$th region of the spectrum has maximum intensity (**3**)
doxyl	The 4′,4′-dimethyloxazolidine-N-oxyl group

D	Zero-field splitting parameter; rotational or translational diffusion coefficient
$D, D^{(2)}, D', D^{(2)'}$	Irreducible tensor components of the hyperfine tensor [**3**, Eqs. (76a) and (76b)]
D_λ, D_η	Degeneracy of the λth and ηth allowed ESR transitions (**3**)
$\mathbf{D}$	Dipolar tensor (**2**)
$\mathbf{D}$	Zero-field splitting tensor (**4**); rotation matrix (**10**)
$\tilde{\mathbf{D}}$	Partially averaged zero-field splitting tensor
D_{XX}, D_{YY}, D_{ZZ}	Principal components of **D**
$D^{(2,p)}$	pth component of the irreducible tensor form of **D** (**4**)
$D^{j}_{l,m}(\alpha\beta\gamma)$, $D^{L}_{KM}(\Omega)$	Wigner rotation matrix element (**2**); also referred to as generalized spherical harmonics.
$\mathscr{D}^{(L)}$	Lth-rank Wigner rotation matrix (**4**)
e	Electronic charge $(4.803250 \pm 0.000021) \times 10^{-10}$ esu
E	Zero-field splitting parameter
E	Eigenvalue of the spin Hamiltonian
$\mathbf{E}$	Electric field
$E_i, \Delta E$	Energy of state i; energy
$E_{\lambda\pm}$	Energy of the $\lambda^{\pm}$ spin eigenstates of $\mathscr{H}_0$
f	Tempo spectral parameter (**13**)
f_A, f_M	Fraction of time the nucleus is in the A, M state (**9**)
F_0, F_2	Irreducible tensor components from the g tensor [**3**, Eq. (75a)]
$F'^{(l,m)}_{\mu}$	Irreducible spherical components related to the Cartesian tensor μ (**2**)
$F'^{(Lm)}_{\mu,i}$	Irreducible tensor component of rank L and component m of a spatial function in the molecule-fixed coordinates for the μth type of perturbation and the ith type of nucleus (**3**)
$F^{(l,-m)}_{\lambda}$	Spatial factors (**10**)
$F^{(L,p)}_{\mu}$	pth component of an Lth-rank tensor for interaction μ (**4**)
g, g_0	Isotropic g factor
g_e	Free spin g value $(2.0023192778 \pm 0.0000000062)$
g_{eff}	Effective g value
$\mathbf{g}$	g tensor (**10**)
$\mathbf{g}$	g tensor
g_{xx}, g_{yy}, g_{zz}	Principal components of the g tensor
$g_\perp, g_\parallel$	Perpendicular and parallel components of the (time-averaged) g tensor

$\bar{g}_{\parallel}, \bar{g}_{\perp}$	The motion averaged g values with the magnetic field parallel and perpendicular to the long molecular axis of the spin label (**12**)
g_i	The ith component of **g** in a principal axis system
g_N	Nuclear g factor
$g^{(0)}, g^{(2)}$	Irreducible g-tensor components [**3**, Eqs. (75b) and (75c)]
g_s	Isotropic g factor measured in motionally narrowed region (**3**)
Δg	g-factor difference
$\langle g \rangle$	$= \frac{1}{3}(g_{x'''} + g_{y'''} + g_{z'''})$ (**3**)
G	Gauss
GHz	Gigahertz (10^9 cps)
$G_m(\Omega), G_m^*(\Omega)$	The mth "eigenfunction" of $\tilde{\Gamma}_{\Omega}$ and its Hermitian conjugate (**3**)
h	Planck's constant, $(6.626196 \pm 0.000050) \times 10^{-27}$ erg-sec
h	Transition probability (**4**)
$\hbar$	$h/2\pi = (1.0545919 \pm 0.0000080) \times 10^{-27}$ erg-sec
$\bar{\mathbf{H}}, H_0, \mathbf{H}$	dc magnetic field (also denoted as B)
H_1	Magnitude of rotating microwave field
$H_{1,\max}(\tilde{M})$	Microwave field strength at which the $\tilde{M}$th region of the spectrum has maximum intensity (**3**)
H_{eff}	Effective magnetic field experienced by an electron (**9**)
H_{loc}	Local magnetic field (**9**)
$\mathscr{H}, \mathscr{H}'$	Hamiltonian
$\hat{\mathscr{H}}_1(\Omega)$	Orientation-dependent part of spin Hamiltonian (**3**)
$\mathscr{H}_c$	Fermi contact hyperfine Hamiltonian (**4**)
$\mathscr{H}_d$	Dipolar Hamiltonian (**2**)
$\mathscr{H}_e$	Exchange Hamiltonian (**2**)
$\hat{\mathscr{H}}_{\text{eff}}, \mathscr{H}_{\text{eff}}$	Effective time-independent spin Hamiltonian (**3**); effective spin Hamiltonian (**4**)
$\hat{\mathscr{H}}_{\text{hf}}$	Hyperfine Hamiltonian (**2**)
$\hat{\mathscr{H}}_0, \mathscr{H}^0$	Zeroth-order spin Hamiltonian; orientation-independent part (**3**); scalar spin Hamiltonian (**4**)
$\mathscr{H}_{\text{SS}}$	Heisenberg spin-exchange operator (**3**)
$\mathscr{H}_Z$	Zeeman Hamiltonian (**2**)
$\hat{\mathscr{H}}(t), \mathscr{H}'(t)$	Hamiltonian operator for a single system, usually just the spin Hamiltonian (**3**); time dependent spin Hamiltonian (**4**)
I	Quantum number for nuclear spin angular momentum
I	Moment of inertia of spherical molecule (**3**)

$I^{(i)}$	Nuclear spin operator for *i*th nucleus (**4**)
$\hat{I}_{z_i}$	Operator for *z*th component of nuclear spin of *i*th nucleus (**3**)
$I_{\pm}$	Raising and lowering operators for nuclear spin of *i*th nucleus (**3**)
IF	Intensity factor, a relative measure of integrated intensities of computed spectra (**3**)
$j(\omega_i), j(\omega)$	Spectral density function at frequency ω_i or ω
j_n	The *n*th spectral density
J	Exchange integral or interaction
J'	Exchange integral where $J = 2J'$ (**3**)
$\underline{J}$	Total angular momentum
$J_{\alpha\beta,\alpha'\beta'}(\omega_{\alpha\alpha'})$	Spectral density (**4**)
$\mathscr{J}$	Spectral intensity (**4**)
k	Boltzmann's constant $(1.380622 \pm 0.000059) \times 10^{-16}$ erg °K^{-1}
k	A linear correction factor to adjust the single-crystal parameters for the polarity in the biological sample (**12**)
k	$= (\omega - \omega_e) - iT_{2,a}^{-1}$ (**3**)
$k(\omega_i)$	Imaginary part of $l(\omega_i)$ (**3**)
K	A "quantum number" referring to the projection of $\phi_{KM}^{L}(\Omega)$ along the body-fixed symmetry axis (**3**)
l_{ij}	The direction cosine between the *i*th axis and the *j*th axis (**12**)
L	Rank of spherical harmonic (**3**); rank of rotation matrix or Legendre polynomial (**4**)
$\underline{L}$	Orbital angular momentum
L	A coordinate transformation matrix
m, m_p	Electron rest mass; proton rest mass
$m^{(i)}$	Nuclear quantum number for the *i*th nucleus (**4**)
M	Quantum number for the *z* component of angular momentum
M_I	Total nuclear spin quantum number
M_N	"Quantum number" for the *z* component of the ^{14}N nuclear spin angular momentum
M_0	Equilibrium magnetization aligned along the magnetic field
M_S, M_s	"Quantum number" for the *z* component of the electron spin angular momentum; total electron spin quantum number
M_z	Component of magnetization aligned along the magnetic field (**9**)

$\tilde{M}$	M_N value for a set of equivalent nuclei such that high-field lines have $\tilde{M} > 0$ (**3**)
$\mathcal{M}$	Macroscopic magnetization (**2**)
$\mathcal{M}$	Vector operator which generates an infinitesimal rotation referred to body-fixed coordinates (x', y', z') (**3**)
$\mathcal{M}_{i'}$	Component of $\mathcal{M}$ along ith principal axis (**3**)
$\mathcal{M}_{\pm}$	Raising and lowering operator components of $\mathcal{M}$ (**3**)
n	Number of σ bonds (**4**)
n_L (n_K)	Maximum value for $L(K)$ in expansion in terms of ϕ_{KM}^{L} (**3**)
n_0	Equilibrium population difference of two levels (**9**)
$\mathbf{n}$	Director axis for anisotropic liquid (**3**)
N	Concentration of electron spin (**3**)
N'	Number of spin eigenstates of $\mathcal{H}_0$ (**3**)
$N(L, L')$	A normalization factor [**3**, Eq. (A.11)]
$\hat{N}$	$\equiv \hat{R}_{\parallel}/\hat{R}_{\perp}$ (**3**)
$\mathcal{N}$	Vector operator which generates an infinitesimal rotation referred to lab-fixed coordinates [or to the director frame $(\mathbf{x}'', \mathbf{y}'', \mathbf{z}'')$ for an anisotropic liquid] (**3**)
O, O_{mn}	An arbitrary operator and its mnth matrix element, i.e., $O_{mn} = \langle a \vert O \vert b \rangle$ [**3**, Eqs. (12a)–(12c)]
$\mathbf{O}$	The complex orthogonal matrix that diagonalizes $\mathcal{A}$ (**3**)
p	Ratio of substrate molecules being relaxed by the nitroxide to the total concentration of substrate (**9**)
(p, q, r)	Molecular axis system
$P(\theta_1)$	Distribution function in the Gaussian orientation distribution model, where θ_1 is the angle between the normal to the plane of the sample and the long molecular axis **R** (**12**)
$P_L(\cos \gamma)$	Lth Legendre polynomial
$P_0(\Omega)$	The unique equilibrium distribution in Ω (**3**)
$P(x)$	Probability distribution function (**2**)
$P(x_1 \vert x_2 t)$	Conditional probability (**2**)
$P(\Omega, t)$	Probability of the free radical being in a state Ω at time t (**2**)
$P_\lambda(\Omega)$	Power absorbed by λth ESR hyperfine transition (**3**)
q	Coordination number of the paramagnetic molecule or ion (**9**)
r	Radius; interelectron separation; interradical separation for spin exchange
$\mathbf{r}_i$	Spatial coordinate for ith particle (**4**)
R	Radius; rotational diffusion coefficient (**3**); gas constant

R	The long molecular axis of the spin label (the axis of rotation) (**12**)
R_i	Component of **R** about ith molecular axis (**3**)
$R_{\alpha\alpha'\beta\beta'}$	Element of the Redfield relaxation matrix
$R_{\pm}$	$\equiv \frac{1}{2}(R_x \pm R_y)$ (**3**)
$\overline{R}$	$\equiv (R_{\perp}/R_{\parallel})^{1/2}$ (**3**)
$R_{\perp}, R_{\parallel}$	Components of **R** perpendicular and parallel to the molecular symmetry axis (**3, 4**)
$\mathbf{R}$	Rotational diffusion tensor
s	Type of atomic orbital (**4**)
S	Electron spin quantum number
S	An order parameter
S	$= A_z'/A_z$, where A_z is obtained as one-half the separation of the outer hyperfine extrema and A_z' is the slow tumbling value for the same spectral feature (**3**)
S_e	Ensemble average order parameter (**13**)
$S^{(i)}$	Spin operator for electron i (**4**)
S_{aniso}	Anisotropy parameter defined as the difference between the splittings observed when the magnetic field is parallel and perpendicular to the plane of an oriented sample, divided by the maximum possible anisotropy
S_{ii}	Order parameters
S_{motion}	Motion order parameter, usually designated as S
S_0	A measure of the extent of anisotropic motion leading to $\hat{\mathscr{H}}_{eff}$
S_{space}	Spatial order parameter
$\hat{S}_{\pm}, S_{+}$	Electron-spin raising and lowering operators (**3**)
$\hat{S}_z$	Operator for zth component of electron spin
t	Time
T	Absolute temperature (Kelvin)
Tr	Trace operation over a matrix
T_1	Longitudinal (spin–lattice) relaxation time
T_{1e}	Electron spin–lattice relaxation time
T_{1P}	Spin–lattice relaxation time due to direct interaction between substrate and nitroxide
$T_{1\rho}$	Spin–lattice relaxation time in the rotating coordinate system
$T_2, T_2(m)$	Transverse (spin–spin) relaxation time
T_2^{-1}	Rotationally invariant Lorentzian linewidth (**2**)
$T_2^{-1}(\theta)$	Orientation-dependent (near) rigid-limit Lorentzian linewidth (**3**)
T_{ij}	Cartesian tensor component (**2**)
$T^{(l, m)}$	Spherical tensor component

$T_{\mu}^{(2,p)}$	pth component of an irreducible spin operator for interaction μ (**4**)
Tempo, TEMPO	2,2,6,6-Tetramethylpiperidine-1-oxyl, the piperidine nitroxide moiety
U_i, U_n	An orthonormal time independent basis function
U	r-dimensional column vector [**3**, Eq. (A.15)]
v	Molecular axis about which rapid internal rotation is occurring (**3**)
$V(\Omega)$	Mean field potential of a molecule in an anisotropic liquid (**3**)
W	Energy
W_e	Electron–spin relaxation rate (**3**)
W_i	Defined as Δ_i/Δ_i^r (**3**)
$W(\varepsilon)$	Probability function for elementary diffusive steps by angle ε in a jump diffusion model (**3**)
W_1, W_0, W_{-1}	Widths of the nitroxide hyperfine lines at low-, intermediate-, and high-field regions of the ESR spectrum ($M_I = \pm 1, 0$) (**11**)
(x, y, z)	A Cartesian axis system (molecule, laboratory, etc.) differentiated usually by primes or upper and lower case symbols (see also Chapter 12)
X_{K0}^L, $\hat{X}_{K0}^L$	Coefficients in the expansion of $\tilde{\Gamma}$ for anisotropic liquids [**3**, Eqs. (61)–(65)]
X_L	Lth angular linewidth coefficient (**4**)
$Y_{LM}(\beta\gamma)$	Spherical harmonic of rank L and component M (**3**)
z'	Director axis; axis of motional averaging
$Z^{(n)}(\Omega)$, $Z_{\lambda}^{(n)}(\Omega)$	The nth harmonic of χ or χ_{λ} in a Fourier series expansion of χ or χ_{λ} in the microwave frequency ω; only $n = 1$ is needed for problems without saturation effects (**3**)
$Z_{\lambda}^{(n)\prime}$, $Z_{\lambda}^{(n)\prime\prime}$	Real and imaginary parts of $Z_{\lambda}^{(n)}$, respectively (**3**)
$\tilde{Z}_{\lambda}^{(n)}$	Symmetrized form of $Z_{\lambda}^{(n)}$ [**3**, Eq. (21)]
α	Orientation-independent part of $T_2^{-1}(\theta)$ [**3**, Eq. (29)]
α	The half-angle of motion in the oscillation model for anisotropic motion (**12**)
α_i	Direction cosines of **v** (**3**)
(α, β, γ)	Euler angles
β	Orientation-dependent coefficient for $T_2^{-1}(\theta)$ [**3**, Eq. (29)]
β_e, β	Bohr magneton for the electron $= e\hbar/2mc = (9.274096 \pm 0.000065) \times 10^{-21}$ erg G^{-1}
β_N	Nuclear Bohr magneton $= e\hbar/2m_p c = (5.050951 \pm 0.000051) \times 10^{-24}$ erg G^{-1}
γ	Magnetogyric ratio

γ	The half-amplitude of motion in the wobble model for anisotropic motion (**12**)
γ_e	Magnetogyric ratio of the electron
$\gamma_{n,i}$	Magnetogyric ratio of the nucleus, the *i*th nucleus
γ_2	$\equiv 3/2\varepsilon_0^2$, coefficient for expansion of $V(\Omega)$ for anisotropic liquids [**3**, Eq. (58)]
Γ_Ω	Time-independent Markov operator in variable Ω (**3**)
$\tilde{\Gamma}_\Omega$	Symmetrized form of Γ_Ω, usually just the differential operator for rotational diffusion (**3**)
δ	Angle of tilt of a molecular axis with respect to the sample coordinate system (**12**)
δ, δ_{ij}	Kronecker or Dirac delta
$\delta(x - x_0)$	Dirac delta function
$\delta(\tilde{M})$	Derivative peak-to-peak width motional narrowing (**3**)
Δ_i, Δ_i^r	Measured half-width at half-height for the two outer extrema of a near-rigid-limit spectrum (Δ_i) and rigid-limit spectrum (Δ_i^r); $i = l$ or h, representing low- or high-field extrema, respectively (**3**)
Δ_i^{er}	Effective inhomogeneous width (**3**)
ε	Dielectric constant
ε	Jump angle in a jump diffusion model (**3**)
ε	$\equiv (6)^{1/2}\varepsilon_{2+}^2/2$, a coefficient for expansion of $V(\Omega)$ for anisotropic liquids [**3**, Eq. (58)]
ε, ε'	Empirical correction parameters to $j(\omega_e)$ and $j(\omega_\pm)$, respectively [**3**, Eqs. (94′) and (95)]
ε_{KM}^L	Expansion coefficients of $V(\Omega)$ in the generalized spherical harmonics [**3**, Eq. (54)]
$\hat{\varepsilon}(t)$	Hamiltonian operator for interaction of spin with microwave magnetic field (**3**)
η	Viscosity
θ	Angle between magnetic field axis and molecular z axis
θ, ϕ, ψ	Eulerian angles defined by Fig. 2, Chapter 12; defined in this way, θ and ϕ correspond to the conventional polar and azimuthal angles in spherical coordinates
$\bar{\theta}$, δ, $\bar{\vartheta}$	Tilt angle (**10**)
θ_i	Angle between the molecular i axis and the direction axis (**10**)
ϑ_0	Spread angle (**10**)
κ	Anisotropic interaction parameter (**4**)
λ	$\equiv -\gamma_2/kT$, a dimensionless coefficient for anisotropic liquids (**3**)
λ	Refers to λth ESR hyperfine transition (**3**)

$\lambda^{\pm}$	Refers to the $\pm\frac{1}{2}$ ESR state belonging to the λth allowed ESR hyperfine transition (**3**)
μ	Magnetic moment; macroscopic magnetization (**2**)
μ_e	Magnetic moment of the electron
μ_N	Magnetic moment of a nucleus
ν, ν_0	Frequency
ρ	Density matrix operator; spin density
$\rho_e(\vartheta_i)$	Ensemble distribution function (**13**)
ρ_{nm}	Density matrix element
$\rho(\vartheta)$, $\rho(\Theta)$	Probability or distribution function of the x', y', z' axis system defined in (**13**)
$\rho_0(\Omega)$	Equilibrium density matrix element (**3**)
σ	Type of molecular orbital
σ	The orientation distribution parameter (**12**)
τ, τ_c, τ_i, τ_2, τ_{20}, τ_{22}, τ_r	Correlation times; mean time between successive jumps in a jump diffusion model; mean time of free diffusion in a free diffusion model
τ_A, τ_M	Lifetime in the A, M state (**9**)
τ_1	Lifetime of an interacting radical pair (**3**)
τ_2	Mean time between successive new bimolecular encounters of radicals (**3**)
τ_{1S}	Longitudinal relaxation time of the electron
τ_{2S}	Transverse relaxation time of the electron
τ_R	Rotational correlation time
τ_R	Rotational reorientation time given by $(6\bar{R})^{-1}$ (**3**)
$\tau_m^{-1}(\tau_L^{-1})$	The mth (C^n) "eigenvalue" of Γ_Ω in units of inverse time, i.e., a decay constant (**3**)
$\mathscr{T}(\Omega)$	External torque experienced by a molecule in an anisotropic liquid (**3**)
φ	Angle between the magnetic field and bilayer normal (**10**)
ϕ	Angle defining mixing coefficient (**4**)
$\phi^L_{KM}(\Omega)$	Normalized Wigner rotation matrices or generalized spherical harmonics (**3**)
χ, χ_λ	Deviation of ρ from its equilibrium value and the matrix element corresponding to the λth transition, i.e., $\chi_\lambda = \langle\lambda^-\|\chi\|\lambda^+\rangle$ (**3**)
$\tilde{\chi}_\lambda$	Symmetrized form of χ_λ (**3**)
$\overline{\psi}$	Quantum mechanical wave function (**3**)
$\|\psi\rangle$, $\langle\psi\|$	Dirac ket and bra notation
ω, ω_0	Angular frequency; resonance frequency (in rad/sec)
ω_e	ESR transition frequency neglecting hyperfine contribution

ω_I	Larmor frequency of the nucleus
ω_S	Larmor frequency of the electron
ω_{SS}	Frequency of Heisenberg spin-exchange (**3**)
$\omega_{\pm}$	Transition frequencies for nuclear spin flips for $M_S = \pm\frac{1}{2}$ (**3**)
$\omega_{\alpha\alpha'}$	Frequency separation for states $\lvert\alpha\rangle$ and $\lvert\alpha'\rangle$ (**4**)
ω_λ	Transition frequency of λth ESR transition
Ω	α, β, γ, the Euler angles specifying orientation of molecular principal axis with respect to the lab axes for anisotropic liquid or the director axes for an anisotropic liquid (**3**)

Index

F

G

T

V

W

Y

Z

Molecular Biology

An International Series of Monographs and Textbooks

Editors

HAROLD A. SCHERAGA. Protein Structure. 1961

STUART A. RICE AND MITSURU NAGASAWA. Polyelectrolyte Solutions: A Theoretical Introduction, *with a contribution by Herbert Morawetz.* 1961

SIDNEY UDENFRIEND. Fluorescence Assay in Biology and Medicine. Volume I–1962. Volume II–1969

J. HERBERT TAYLOR (Editor). Molecular Genetics. Part I–1963. Part II–1967

ARTHUR VEIS. The Macromolecular Chemistry of Gelatin. 1964

M. JOLY. A Physico-chemical Approach to the Denaturation of Proteins. 1965

SYDNEY J. LEACH (Editor). Physical Principles and Techniques of Protein Chemistry. Part A–1969. Part B–1970. Part C–1973

KENDRIC C. SMITH AND PHILIP C. HANAWALT. Molecular Photobiology: Inactivation and Recovery. 1969

RONALD BENTLEY. Molecular Asymmetry in Biology. Volume I–1969. Volume II–1970

JACINTO STEINHARDT AND JACQUELINE A. REYNOLDS. Multiple Equilibria in Protein. 1969

DOUGLAS POLAND AND HAROLD A. SCHERAGA. Theory of Helix-Coil Transitions in Biopolymers. 1970

JOHN R. CANN. Interacting Macromolecules: The Theory and Practice of Their Electrophoresis, Ultracentrifugation, and Chromatography. 1970

Walter W. Wainio. The Mammalian Mitochondrial Respiratory Chain. 1970

Lawrence I. Rothfield (Editor). Structure and Function of Biological Membranes. 1971

Alan G. Walton and John Blackwell. Biopolymers. 1973

Walter Lovenberg (Editor). Iron-Sulfur Proteins. Volume I, Biological Properties—1973. Volume II, Molecular Properties—1973

A. J. Hopfinger. Conformational Properties of Macromolecules. 1973

R. D. B. Fraser and T. P. MacRae. Conformation in Fibrous Proteins. 1973

Osamu Hayaishi (Editor). Molecular Mechanisms of Oxygen Activation. 1974

Fumio Oosawa and Sho Asakura. Thermodynamics of the Polymerization of Protein. 1975

Lawrence J. Berliner (Editor). Spin Labeling: Theory and Applications. 1976

A
B 6
C 7
D 8
E 9
F 0
G 1
H 2
I 3
J 4